Comparative Vertebrate Histology

DONALD I. PATT AND
GAÏL R. PATT
BOSTON UNIVERSITY

Comparative Vertebrate Histology

HARPER & ROW, PUBLISHERS
NEW YORK · EVANSTON · LONDON

COVER: Photomicrograph of a Golgi preparation of the cerebral cortex of a rabbit. Magnified 100 diameters.

Comparative Vertebrate Histology

 For information address Harper & Row, Publishers, Incorporated, 49 East 33rd Street, New York, N.Y. 10016.
Library of Congress Catalog Card Number: 69-12864

CONTENTS

PREFACE

THIS BOOK is intended to breathe new life into an old science whose innumerable fascinations have been passed over or ignored in contemporary, anthropocentric texts. Collegiate courses in histology have become almost professional in nature, selecting only such information as relates directly to that all-important mammal, man. This trend has been due, perhaps, to the fact that the only fine books in histology in the English language are medical texts.

We feel that biology, as taught in college, should be broad in its perspective. To focus on man is to leave in peripheral blur the setting in which man became what he is and in which he functions today. We feel it important that mankind recognize and accept its true place in the world of living things, for the assumption that we as a species are central to all the others has led us to commit enormities the consequences of which threaten all life on this planet, including our own. When one studies the microscopic anatomy of the diversified forms of vertebrate life, one cannot avoid being impressed with the ties that bind us to our relations, be they fish or philosopher. Of course a more comprehensive text that would include our invertebrate cousins as well would be even more desirable, but we felt that this would represent too ambitious an undertaking for a single book.

In writing this book, we have tried to select from an overwhelming wealth of information only that which has evolutionary significance or illustrates adaptation in a relevant way. We have omitted much that qualifies for our central purpose, and we have doubtless included some whose relevance is obscure. The omissions were necessitated by the need to keep the book within reasonable limits of length. Unfortunately, many omissions had to be made quite arbitrarily. The inclusions whose relevance is obscure are justified by the principal of including every major organ, even if our knowledge of some of them is imperfect. In spite of our attempts to achieve balanced representation among the vertebrate classes, mammalian examples often predominate. This usually reflects a dearth of information on nonmammalian species. Hopefully this book will stimulate histologists to extend their investigations into nonmammalian groups, where a treasury of information on biological principles is waiting to be discovered.

Although this book will, no doubt, be indexed under histology, that subject, as are all in biology, is merely a subdivision of the larger subject of evolution. Because evolution is the most comprehensive generalization in biology, and because the fundamental unity of all living things stands as an evolutionary beacon, histology, as a discipline, cannot stand alone. Structure in biological systems is always allied with function, and at the submicroscopic level it becomes a moot point as to which precedes which. Hence we have included a liberal measure of physiology and, where applicable, biochemistry. We do this not to confuse the issue of histological structure but to support the relevance of biological unity.

We wish to thank our friends, some of whom are our colleagues, and our children for their patient tolerance during the writing of this book. We wish to thank our parents, both living and dead, for inspiring us to try to transcend our corporeal selves. We acknowledge the enthusiastic interest and help shown and extended by our students, for whom this book is written. We wish to thank Dr. John Albright, Dr. Richard Estes, Dr. Lois Lowenthal, and Dr. Alan Peters for their criticisms, comments, and suggestions. Most especially we wish to thank our many generous colleagues from all over the world who contributed most of the excellent illustrative material for this book. Finally, we acknowledge our gratitude to our technical coworkers: Mrs. Ruth Spencer for the artwork, Mr. Floyd Craft for electron micrography, Mr. Kenneth Linberg and Mr. Frederick Maynard for photography, Mrs. Barbara Miranda for typing, and Mr. Andrew Sinauer and the staff of Harper & Row for patience, guidance, and much-needed explanation of how books finally emerge into the world.

D. I. PATT
G. R. PATT

Comparative Vertebrate Histology

chapter 1
INTRODUCTION AND METHODS

THE PRIMARY CONCERN of histology is the description of microscopic structure and organization of tissues and organs. It is also concerned with the functions of tissues and organs, for it is primarily from function that structure derives meaning. Complete understanding of tissue and organ structure depends also upon knowledge of the manner in which they have differentiated, not only during the development of the individual, but in the evolution of the species as well. Finally, some knowledge of biochemistry is essential to histology, for the chemical reactions characterizing life take place on the myriad surfaces and interfaces that cellular and tissue structure provide. Indeed, the surfaces and interfaces are selectively advantageous because they facilitate these chemical reactions. Histology, therefore, is not an insulated compartment of knowledge, but an intensification of one aspect of the totality.

The material considered in this text is limited to the vertebrates because they demonstrate so well the fundamental kinds of adaptive evolution. Furthermore, the vertebrates present a relatively closely related group of classes, and exhibit evolutionary changes from aquatic, ectothermic forms to terrestrial and aerial endothermic types. And because we are members of the subphylum Vertebrata, we stand to gain a better understanding of ourselves through more complete knowledge of our relatives, both close and distant.

PROTOPLASM

The essential living substance which all forms of life have in common is called protoplasm. Although protoplasm defies precise chemical characterization and the term is somewhat abused, it has enjoyed accepted usage for many years, and still has some operational usefulness. Protoplasm is a heterogeneous mixture, but in final analysis, although as varied as the individuals that populate the earth, it is generally similar enough in all species to enable one to discuss it in a general way. It may be defined as a polyphase colloidal complex organized as a system of membranes, filaments, and particles, involving primarily water, proteins, lipids, carbohydrates, nucleic acids, and soluble salts—all of which bear the kind of dynamic relation to each other that we characterize as life. Life, in turn, has many definitions, depending upon which frame of reference is being used. In the context of histology, a biochemical definition is the most useful one, which, briefly, is the sum total of enzymatically regulated, integrated, physicochemical reactions which manifest themselves in energy transformations, reproduction, and the maintenance of a steady state.

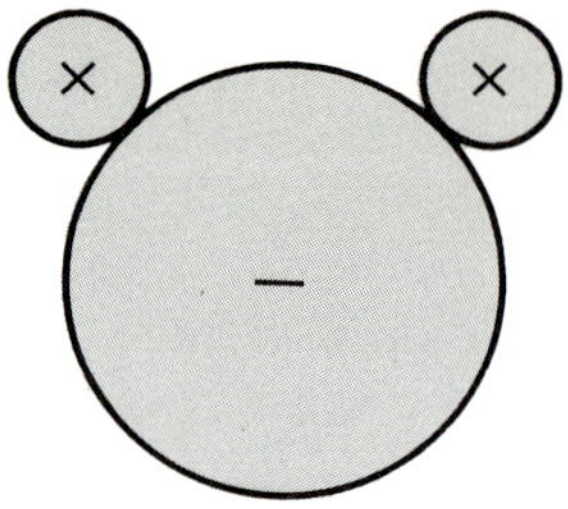

Fig. 1.1 Schematic representation of a molecule of water. Because of the asymmetric position of the hydrogen atoms in a water molecule, the molecule behaves as a dipole.

Chemical Composition

WATER

One cannot reasonably say that any one of the primary components of protoplasm is the most important, because if any of them is lacking, the processes of life fail to occur. Water, however, is the unique solvent on which all metabolic processes depend—not simply because it is the one solvent that happened to be around when protoplasm was being formed, but because it alone possesses the peculiar combination of properties not possessed by any other substance and which are indispensable to all known biological processes (without any one of which life probably would not be possible). Although it is not appropriate to discuss them all at length, a brief résumé of the most striking of these properties is in order.

First, the *solvent power* of water is remarkable. Water is the most general of all known solvents for a broad variety of chemical reactants, such as organic acids, alcohols, esters, and inorganic salts and gases. Secondly, water has the highest dielectric constant of any known liquid within the normal range of temperature encountered on earth, by virtue of which it can permit a high degree of dissociation or ionization of substances dissolved within it. It is for this reason that most chemical reactions occur most rapidly and most completely in aqueous solution. Moreover, water is itself an electrolyte, which accounts for its solvation properties as well as its chemical reactivity, and in the undissociated state it is a dipole, which is of enormous significance in the formation and maintenance of the colloidal state (Fig. 1.1).

The thermal properties of water are important. It has the highest latent heat of evaporation of any known substance, by virtue of which it tends to maintain its liquid state through a wide range of heat exchange. It has the highest thermal conductivity of any nonmetallic substance, permitting rapid and uniform distribution of heat throughout the protoplasmic mass. Save for liquid ammonia, water has the highest specific heat of any known substance, which is significant in maintaining the temperature of protoplasm within the rather narrow limits required by most metabolic processes.

Finally, water has a surface tension which—for liquids in the temperature range prevailing on earth—is exceeded only by that of mercury; further, it has a very low viscosity. The former property is significant in the colloidal chemistry of protoplasm. Its low viscosity enables water to flow freely and with a low energy requirement through pores and channels of submicroscopic dimensions.

PROTEINS

Proteins are compounds of carbon, hydrogen, nitrogen, oxygen, and often sulfur, formed by the polymerization of alpha-amino acids by peptide linkages. They constitute the most complex compounds known, having molecular weights in the tens and hundreds of thousands; but despite this, they have a specificity of structure which is one of their most important characteristics. Protein exists in either fibrillar form, as in collagenous and other fibers of structural significance, or globular form, as in most enzymes and antibodies. Proteins have a low surface tension and they are hydrophilic—properties which cause them as large molecules to form colloidal systems in an aqueous medium. By virtue of their constituent amino acids they are amphoteric, being able to dissociate as either acids or bases or as dipoles (the *zwitterions*) according to pH. The critical pH which is at the dividing line between determining acidity or alkalinity is called the *isoelectric point* and is a constant characteristic of any given protein.

Because of their amphoteric nature, proteins can combine chemically to form salts and complexes with a large variety of other substances and, in view of their low dissociation constant, serve as highly effective buffers. The large molecular sizes of proteins and their derived compounds account for their pronounced osmotic effect, which is significant in the movement of water across cellular and intracellular membranes.

Proteins are rather fragile molecules capable of a phenomenon known as *denaturation*. The property of denaturation is of utmost significance in histology, for it is the basis of fixation and preservation of protoplasm, as well as an essential condition for most kinds of staining. Denaturation is the process whereby the structure of protein is altered in such a way as to affect fundamental physicochemical characteristics, such as solubility, specific activity (as with hormones or enzymes), or crystalizability. Often coagulation follows or is an essential aspect of denaturation, or, as in the case of many proteins, free sulfhydryl groups appear, which, before denaturation, existed as disulfide bonds. Denaturation may also lead to increases in viscosity and surface tension or to a change in optical rotation. Commonly used denaturing agents are heat, many types of organic solvents (mostly alcohol or acetone), detergents, salts of heavy metals (such as mercuric chloride, silver nitrate, or osmium tetroxide) and strong acids and bases.

LIPIDS

The lipids are a heterogeneous group of compounds having the operational characteristic of being generally soluble in nonpolar solvents, the so-called "fat solvents" such as chloroform, ether, benzene, acetone, etc. They can be classified into a number of structural categories, of which the major ones are neutral fats, waxes, phosphatides, cerebrosides, steroids, and the derived lipids (hydrolytic products of the others). Most lipids exist in protoplasm either as component parts of cellular and intracellular membranes, where they are conjugated with protein, or as droplets. The phosphatides are uniquely significant in the former instance because they are polar compounds, consisting of one or more long fatty-acid chains (the *nonpolar* portion) and an ionizable, hydrophilic component (the *polar* portion). Such compounds tend to aggregate within an aqueous medium to form bimolecular leaflets with the nonpolar ends of the molecules of one leaflet apposed to the nonpolar ends of the molecule of the other leaflet, thus leaving two outer surfaces of polar ends (Fig. 1.2). The polar ends, being hydrophilic, adsorb water and/or protein, which is also hydrophilic.

CARBOHYDRATES

The carbohydrates, as contrasted with the lipids, do form a natural group of compounds. They are the aldehyde or ketone derivatives of polyhydric alcohols; that is, they are polyhydroxyaldehydes or polyhydroxyketones. Except for some polysaccharides of high molecular weight, the carbohydrates are soluble in water and poorly soluble in organic solvents.

The basic carbohydrate molecules, the *monosaccharides*, are classified on the basis of the number of carbon atoms in the skeleton. The pentoses (five-carbon skeleton) and the hexoses (six-carbon skeleton) are the most significant ones in protoplasm, the former because of their presence in the nucleic acids, and the latter because of their presence (usually in polymerized or derived form) in membranes, intercellular ground substance, or dispersed in the cellular cytoplasm.

Two or more monosaccharides may be linked together by glycosidic bonds to form polymers. Polymers involving a relatively small number of sugar units, such as two (*disaccharides*), are common in biological sys-

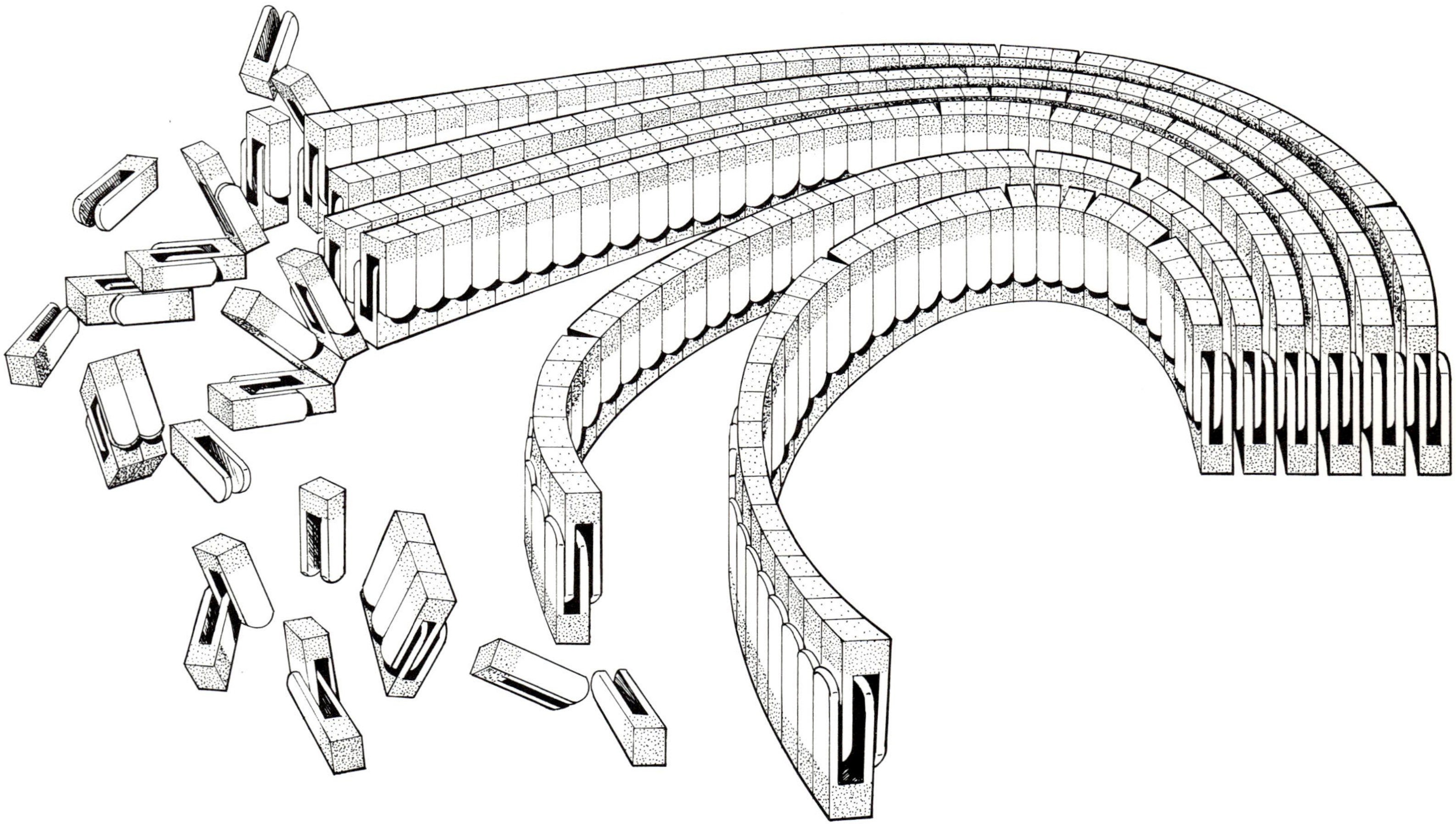

Fig. 1.2 The formation of a membrane by bimolecular leaflets of phosphatide molecules. Each molecule consists of a hydrophilic component (*shaded*) to which are bonded two long fatty-acid residues, which are hydrophobic. This configuration imparts polarity to the molecule. The hydrophobic ends attract each other and form hydrophobic bonds that have sufficient force to maintain their side-by-side arrangement. Whether the two fatty acids fit into each other, as shown in this diagram, or lie end-to-end, has not been determined.

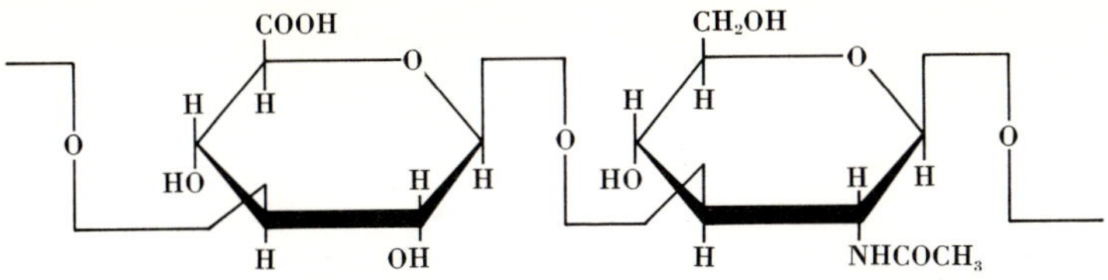

Fig. 1.3 Structural formula of a repeating unit of hyaluronic acid, which is a polymer of these units.

tems, and the component sugars may all be the same kind, as in maltose, or different, as in sucrose. Polymers of high molecular weight, the *polysaccharides,* usually consist of repeating units of the same kind of sugar or its derivatives. For example, glycogen is a high polymer of the monosaccharide glucose, whereas hyaluronic acid hydrolyses to yield approximately equimolar amounts of N-acetyl-glucosamine and glucuronic acid, both of which are monosaccharide derivatives (Fig. 1.3).

Among the important derived sugars from the standpoint of histology are the *mucopolysaccharides.* These are polymers of the derived monosaccharide *hexoseamine,* and usually occur conjugated with protein, or with inorganic acids or bases (Fig. 1.4).

NUCLEIC ACIDS

The nucleic acids are extremely important constituents of protoplasm because they constitute the hereditary material, the material which transfers the genetic "information" to the cytoplasm, and the material which directs the ordering of the amino acids in protein synthesis. (See Chapter 2.)

Deoxyribonucleic acid (DNA) usually exists in the form of a double helix, probably conjugated with basic proteins, in the nucleus of each cell. The component molecules of DNA are the purine bases, *adenine* and *guanine;* the pyrimidine bases, *thymine* and *cytosine;* the pentose sugar, *deoxyribose;* and *phosphoric acid.* A single strand of DNA consists of alternating units of the sugar and

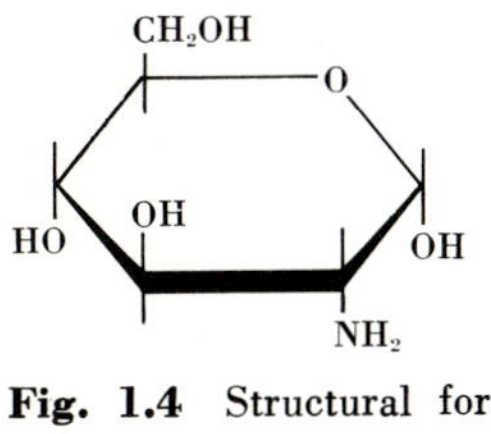

Fig. 1.4 Structural formula of hexosamine.

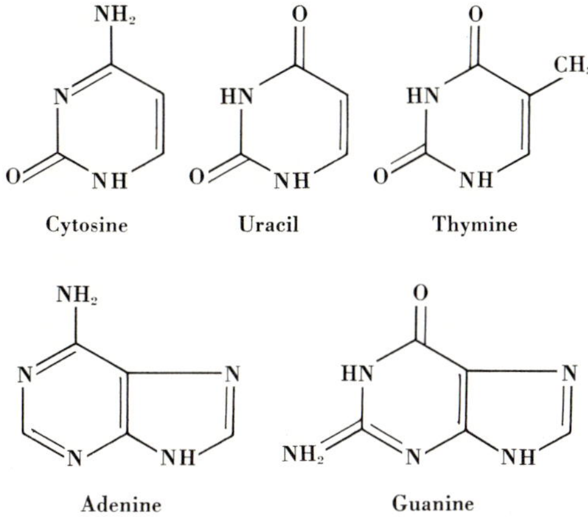

Fig. 1.5 Structural formulae of the pyrimidine and purine bases.

phosphoric acid with side groups of the various bases bonded to the sugars. Two such strands, according to the Watson-Crick model, are bonded collaterally together by hydrogen bonds between corresponding bases of each strand. The base that corresponds with adenine is thymine, and that corresponding with guanine is cytosine (Fig. 1.5). The ordering of the particular bases sequentially is precise and constant, and constitutes the "coding" mechanism of DNA. Both component strands of DNA are wound around each other in such a way as to form an alpha-helix (Fig. 1.6).

When DNA is hydrolyzed, the linkages between the sugars and phosphoric acid are broken, first yielding segments of strands called polynucleotides, and then units consisting of one residue of each principal component, i.e., phosphoric acid, sugar, and base, which is called a *mononucleotide* or simply *nucleotide.* Further hydrolysis attacks the bonds between the phosphoric acid and the sugar, thus yielding phosphoric acid and a nucleoside (sugar plus base), and finally the bond between the sugar and base is broken. Each step in hydrolysis is characterized by increased solubility. This consideration is im-

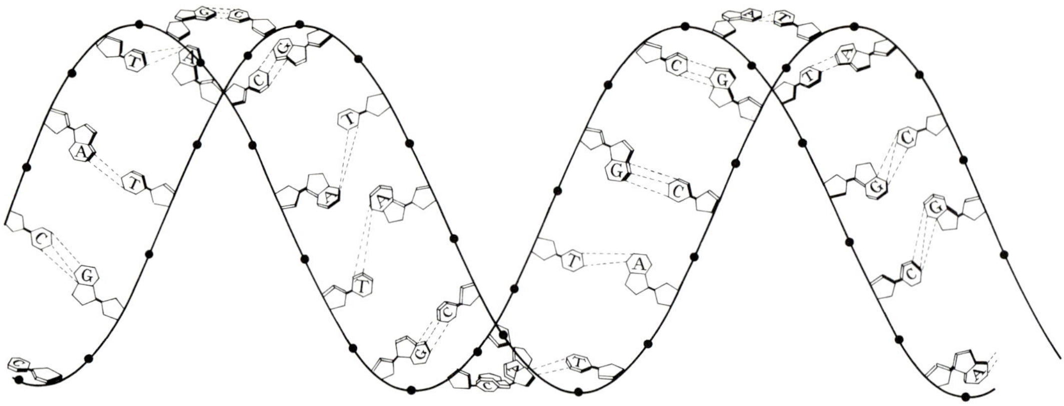

Fig. 1.6 The DNA helix. DNA is a double-stranded helix. The backbone of each strand consists of alternating units of deoxyribose (shown as *unlabelled pentagons* on each strand) and phosphoric acid (*black dots*). The bases are bonded to the sugars of each strand by covalent bonds and to the complementary bases of the contralateral strand by hydrogen bonds (*broken lines*).

portant in fixation and staining procedures, as discussed below.

Ribonucleic acid (RNA) may exist in a variety of forms, but basically all forms are probably modifications of a single-strand form. A strand of RNA consists of alternating components of phosphoric acid and the pentose sugar *ribose,* with one of the purine bases, adenine or guanine, or one of the pyrimidine bases, cytosine or *uracil,* bonded laterally with each sugar (Fig. 1.5). Hence RNA differs from DNA in three fundamental respects: first, RNA is not consistently in helical form; second, RNA has ribose instead of deoxyribose as its constituent sugar; and third, RNA substitutes uracil for thymine. The order of the bases in at least one form of RNA complements their ordering in corresponding units of nuclear DNA. (See Chapter 2.)

INORGANIC CONSTITUENTS

A wide variety of inorganic salts occurs in protoplasm in addition to the substances described above. Many of them are combined chemically with the organic constituents, and many occur in free solution as well as in ionic form. Although most of the elements are represented, certain ones are of more universal and vital occurrence than others. Sodium and potassium chloride, particularly in ionic form, are ubiquitous and constitute the principal substances used in maintaining the ionic balance necessary for the biochemical reactions and transformations that take place in protoplasm. The metals—such as iron, copper, magnesium, and manganese—are well represented, functioning as essential constituents of enzymes or as enzyme cofactors. The role of calcium is vital to the maintenance of the integrity of cell membranes and of the intercellular ground substance, as well as to the formation of carbonate and phosphate salts in bone.

The Colloidal Nature of Protoplasm

Molecules of very large size do not usually form true solutions. The term *colloid* has been used to apply to situations where the "dissolved" particles exceed 1mμ in diameter, or 1500 atoms. The particles may be macromolecular, or an aggregation of molecules. Colloids differ from true solutions in several respects, such as stability and viscosity. In order to distinguish between solutions and colloids, the terms *continuous phase* and *discontinuous phase* have been introduced as equivalents of solvent and solute, respectively.

Eight basically different kinds of colloids can be described on the basis of the states of matter represented by the two phases, whether they are solid, liquid, or gaseous. The principal colloidal systems encountered in protoplasm are *emulsions* (liquid in liquid), *emulsoids* (solid in liquid), *gels* (liquid in solid), and modifications of these.

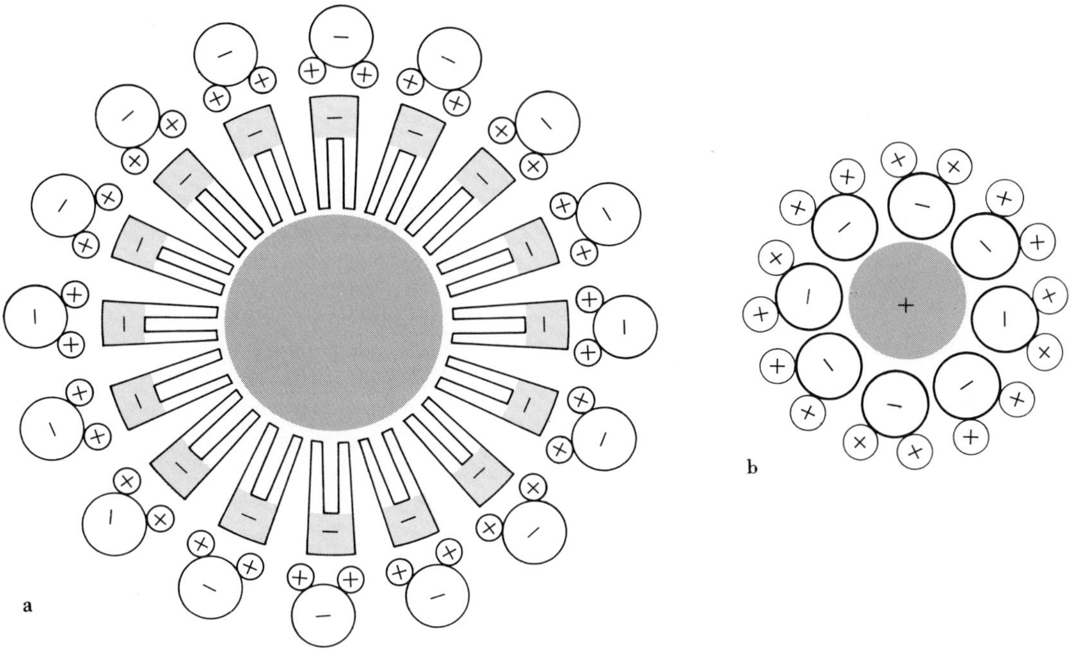

Fig. 1.7 Schematic representation of the particles of *a*, an emulsion; and *b*, an emulsoid. In *a*, polar molecules of a phosphatide are arranged around a lipid droplet (*shaded*), with their hydrophilic ends directed outward. A monolayer of dissociated water molecules is adsorbed on the shell of phosphatides thus formed. In *b*, a hydrophilic particle (*shaded*) with a net positive charge is shown surrounded by an adsorbed layer of dissociated water molecules. The moonlayer of water in both *a* and *b* constitutes the so-called Helmholtz double layer of charges.

EMULSIONS

Emulsions are colloidal states in which one liquid is dispersed in another, immiscible liquid, such as oil in water. Most emulsions require a stabilizing substance in order to prevent coalescence of elements of the discontinuous phase with consequent layering of one component on the other. Soaps and polar compounds, such as some of the phosphatides (e.g., lecithin) are very effective in stabilizing emulsions. Soaps and other such emulsifiers do this by lowering the surface tension of the particles in the discontinuous phase. This in turn causes them to approximate a spherical shape, thus presenting the least amount of surface for coalescence. Polar compounds stabilize emulsions of lipid and water by forming a protective shell around each particle of the discontinuous phase (Fig. 1.7a) in such a way that the polar ends of the molecules project outward. The polar ends, being able to dissociate, acquire a charge, and this charge holds the particles of the discontinuous phase separated in a state of electrostatic equilibrium.

EMULSOIDS

This is the term that applies to a colloid in which the discontinuous phase is *lyophilic*, i.e., has an affinity or "love" for the molecules in the continuous phase. In protoplasm the continuous phase is primarily aqueous, so the term *hydrophilic* is used as being more specific than lyophilic. Invariably the discontinuous phase in protoplasmic emulsoids is protein. The protein molecules, or aggregates thereof, attract and hold water to their surface, a process called adsorption. According to pH, the amino groups or carboxyl groups of the protein dissociate, imparting a weak charge to the particle. The adsorbed water molecules orient themselves according to the nature of this charge as dipoles, thereby in effect forming a tight shell of double charges around the particle, with like charges in each layer (Fig. 1.7b). Particles of like charge, so surrounded, tend to repel each other and thus establish an equilibrated state of suspension. Reagents that remove water so bound—either by absorption of it (as in alcohol or acetone) or by neutralizing the net charge of the particle (which would occur at the isoelectric point)—cause precipitation, agglutination, or denaturation of the protein. Such changes may result in destruction of the colloidal state, or they may produce modified colloidal systems known as *coacervates,* in which swarming of particles occurs to produce large masses, or *tactoids,* in which the particles become oriented with respect to each other to form masses resembling the asters and spindles of mitosis (Fig. 2.12). In fact, it has been suggested that division spindles start out as tactoids which then, through the formation of strong disulfide bonds, become consolidated into well formed and oriented groups of fibers.

GELS

The exact physical structure of gels is not known. They are generally considered to consist of particles that are in contact with each other. The precise physical nature of the contact is not known, but some kind of chemical bonding is believed to be involved, so that the particles constitute a continuous phase. The aqueous medium is presumably trapped in the meshwork or "brush heap" formed by the interlacing particles and, although it may be in continuous contact between the meshes, it behaves somewhat as a discontinuous phase. Many gels are characterized by reversibility, which means they may become fluid: the *sol* state, in which case the colloidal system is an emulsoid. Solation of a gel may result from changes in pressure, temperature, pH, or salt concentration. The process of gelation is accompanied by contraction, which probably reflects the action of the binding forces between particles. Known as *syneresis,* the contraction may exert considerable force and may constitute one of the mechanisms whereby secretion can occur against an opposing osmotic or hydrostatic pressure gradient. The major part of nonmembranous protoplasm exists as either sol or gel, the two states usually being in juxtaposition and reversing from one to the other in response to subtle localized changes in pH and/or salt concentration.

METHODS

Tissues and cells may be studied either alive or dead. Each state imposes certain limitations and thus cells and tissues should be studied under each condition insofar as is possible. In order to permit observation of them in the living state, highly specialized equipment and methods are usually necessary for differentiating between the various structural components, e.g., phase contrast, polarizing, interference, or dark-field microscopy, and tissue culture. Even under the best conditions, the optical differentiation achieved with living tissues leaves considerable to be desired. Moreover, one is faced with the problems of keeping the materials of study under conditions that will sufficiently approximate physiological normality to give one confidence that what one is studying is really normal. For these reasons, among others, histologists ordinarily study tissues that have been killed, preserved, and stained, or otherwise rendered optically suitable. The processes utilized result almost axiomatically in *artifacts,* i.e., the appearance of structures or relationships which do not obtain as such in the living cell. The artifacts produced by carefully executed procedures, however, are constant and for this reason necessarily reflect some condition of the living cell or tissue that *does* obtain in the living material; thus such artifacts can be put to advantage.

Tissues to be prepared in the nonliving state should be killed and all vital processes should be arrested as soon as possible after removal from the organism. For critical work some investigators recommend fixation of tissue *in situ* by perfusing the fixing fluid through the circulatory system. Ordinarily, however, the tissue is removed from the organism, oriented, cut into pieces of dimensions consistent with the rate of penetration characteristic of the particular fixative being used, and immersed as rapidly as possible in the fixative.

The fixative is a fluid material selected for its ability to kill and to preserve the tissue quickly, and either to produce optical differences in structure or to affect the structure in such a way that optical differences may be brought out by subsequent treatment with dyes or metallic salts. Basically most fixatives are protein-denaturing agents, some of which (such as alcohol, mercuric chloride, and picric acid) are precipitants and others (such as formaldehyde and osmium tetroxide) are not. The denaturation of the soluble globular proteins produces a change in cross linkages that are like those of the relatively insoluble fibrous proteins. An incidental effect of such a change is to increase the refractility of the protein. Denaturation frees highly reactive sulfhydryl groups from cysteine-containing proteins, which enhances their affinity for dyes. A widely used fixing reagent is acetic acid, which has the advantage of being an excellent precipitant of nucleoprotein, but which has the disadvantage of dissolving many lipids out of cells or tissues. The particular fixative used depends always upon what part or parts or properties of the tissue one wishes to study, as well as upon the chemical and physical nature of the tissue itself. For example, different fixatives would be indicated for the nuclei, let us say, of muscle and bone.

The following is not intended to serve as a detailed procedure, for which the student is referred to some manual on microtechnique, but rather as an outline of the principles of preparing tissues by commonly used methods.

Infiltration and Embedding

After fixation, the material should be thoroughly washed in order to remove all excess amounts of the fixing fluid. The washing fluid is selected according to the solvent properties of the fixative. The tissue is then prepared for infiltration and embedding in some kind of medium that will hold it rigid and manipulatable for cutting into thin slices. The most generally suitable matrix for this purpose is paraffin. To prepare the tissue for paraffin infiltration all traces of water must be removed. This is done in a series of steps by means of fluids that will produce the minimum amount of distortion of tissue elements. Ethyl alcohol has a strong affinity for water and meets the other requirements fairly well, and so a series of increasing concentrations of ethyl alcohol, up to 100 percent, is generally used.

As alcohol is not miscible with paraffin, another fluid which is miscible with both alcohol and paraffin must be used. Many of the organic solvents meet this requirement, but the most popular ones, for reasons of safety, economy, and effectiveness, are xylene and toluene. After being completely saturated with the organic solvent, the tissue is then placed into melted paraffin and permitted to remain as long as is required to effect complete replacement of the solvent. It is then embedded in a matrix of paraffin, and the entire block permitted to solidify. Care is taken to avoid excessive exposure to heat or to the higher alcohols or organic solvents, for these factors all tend to harden the tissue and introduce technical problems, as well as physical distortion. Ideally the hardness of the tissue is matched by the hardness of the matrix in which it is embedded. Paraffins of various hardnesses and crystalizing characteristics, facilitated by incorporation of various resins, are obtainable for these purposes.

Sectioning and Mounting

The block containing the specimen is trimmed to have parallel faces and is then fixed to a specimen holder (Fig. 1.8) preparatory to sectioning. Sectioning is accomplished by means of a *microtome* (Fig. 1.8) that permits slices of the desired thickness to be made. Such devices advance automatically so that each time the specimen passes across the knife, a slice of the desired thickness is made, and, in the case of most embedding media, each section sticks by its end to the preceding one, so that a ribbon of sections is produced. Segments of ribbons are then floated on a few drops of water on a microscope slide which

Fig. 1.8 A rotary microtome. The specimen to be sectioned, embedded in a block of paraffin, is mounted on a holder which is clamped in a vise that moves up and down with each rotation of the wheel. The specimen is advanced on each upstroke according to the setting of the dial, and is cut on the downstroke. Each slice adheres to the edge of the one preceding, forming a ribbon of consecutive sections on the front surface of the blade, as shown.

has previously been treated with an albumin adhesive. The slides are then warmed sufficiently to cause the sections to spread out flat, and the water is removed by blotting.

After the sections have been fixed to the slide and are dry, the slide is immersed in a solvent to remove the embedding medium. In the case of paraffin, xylene is usually used for this purpose. This solvent is then replaced by whatever serves as the specific vehicle for the stain, as described below. If the stain is in aqueous solution, as many are, the tissue is hydrated by a series of alcohols of decreasing concentration to water. After staining is completed, the slide is then dehydrated again with alcohol; xylene or toluene is substituted for the alcohol; then a drop or two of mounting medium is applied directly to the section. Most mounting media are resins dissolved in either xylene or toluene. A thin cover glass is then applied and the preparation is ready for examination with the microscope.

Staining

Hematoxylin and eosin are the most generally used stains for ordinary histologic work. Hematoxylin is a polycyclical organic compound which slowly oxidizes to the active dye hematein. In acid solution, hematein acquires a positive charge and acts as a basic dye. In order to be effective, it must be used in conjunction with a *mordant* such as ammonium aluminum sulfate, potassium aluminum sulfate, or ferric ammonium sulfate. Mordants are the salts and hydroxides of divalent and trivalent metals which combine both with tissue elements and with certain dyes. The mordants used in routine histological work have an affinity for both nucleoprotein and the so-called basic dyes, such as hematein. Although hematein reacts best as a basic dye in acidic solution, it acquires its most intense coloration when the tissue section is made basic. It is for this reason that after staining, the sections are passed through a solution which is slightly alkaline. In this way the nucleus and the RNA-rich portions of the cytoplasm of many cells are intensely stained, and hence more clearly demarked from the rest of the cell.

Eosin is a nonspecific acid dye which complements the action of a basic dye (such as hematein) and intensifies contrast. Eosin belongs to the chemical family called the xanthenes, which are related to the indicator, phenolphthalein; it is soluble in either water or alcoholic solution. It stains some elements intensely, such as muscle and erythrocytes; others moderately, such as the general cellular cytoplasm; and still others weakly, such as collagenous fibers.

Because hematoxylin and eosin (H & E) are so commonly used in histology, a brief résumé of procedure will be given. After the paraffin ribbons have been secured to the slides, the slides are transferred to xylene in order to remove the paraffin. Then they are transferred to absolute alcohol to remove the xylene, followed by hydration through the descending series of alcohols, such as 95 percent, 80 percent, 70 percent, 30 percent, to distilled water. The slides are then immersed in the hematoxylin stain. Popular hematoxylin preparations are Delafield's, Mayer's, Harris's, and Ehrlich's. The slides are permitted to remain in the dye for about 30 mins at room temperature. Progress of the staining reaction may be checked from time to time by viewing the wet slide through a microscope.

The slides are then washed in distilled water to remove the excess stain, and then, if aqueous eosin is used (approximately 0.1 percent solution), they are stained very briefly in this dye, i.e., for from 5 to 15 sec, depending on concentration, tissue, and temperature. The slides are rinsed again in distilled water and then dehydrated through the ascending series of alcohols. One of the lower alcohols (usually 70 percent) is made slightly alkaline by the addition of a drop or two of concentrated NH_4OH or, better, by use of a 0.1 percent solution of lithium carbonate in 70 percent alcohol. The alkalinity changes the color of the hematein from reddish to blue. Lithium carbonate has the advantage of also being effective in removing excess picric acid from tissues that have been fixed in Bouin's solution or some other picric acid fixative. After absolute alcohol, the slides are transferred to xylene or toluene, depending on the mounting medium to be used, and then mounted with a cover slip.

Although H & E is perhaps the best general histological stain, it has limits of usefulness. The trichrome or double-contrast and the polychrome staining techniques have proved to be very useful for special purposes, and the student is referred to works on microtechnique for details concerning them. For example, Mallory's Triple Stain has proved to be popular. By this technique, various elements are stained in brilliant contrast (nuclei and muscle fibers red, erythrocytes and keratin orange, collagen and cartilage blue, and nervous tissue violet).

Special Staining Techniques

In a separate category are certain staining techniques which have been devised to demonstrate special elements of tissues or cells. Frequently they require special fixation methods as well. A few of the more important ones will be outlined for illustration.

THE FEULGEN REACTION

In the context of histology the Feulgen reaction is specific for the demonstration of DNA. The dye is basic fuchsin which has been reduced through sulfonation to its colorless leucobasic form, in which form it is known as Schiff's reagent. Tissues are fixed most satisfactorily in a 3:1 mixture of absolute ethyl alcohol and glacial acetic acid in order to preserve and to precipitate both cytoplasmic protein and nucleoprotein, and then they may be embedded and sectioned by the usual paraffin method. Deparaffinized slides are hydrated and then incubated at 60°C in 1.0 N HCl, which partially hydrolyses the nucleoprotein, liberating aldehyde groups from deoxyribose. Schiff's reagent is specific for aldehyde groups; the reduced basic fuchsin is oxydized by them, thereby changing its color to a distinctive light purple or magenta. By this technique the distribution of DNA within cells may be demonstrated, and chromosomal material can be positively distinguished from other basophilic substances of similar size and form. The Feulgen reaction has also been used in attempts to determine the quantity of DNA in a given nucleus. Controls for studies of such kinds can be made by use of the enzymes RNAase or DNAase, which are specific for the complete hydrolysis of RNA or DNA, respectively.

METHYL-GREEN PYRONIN

This is a double-stain technique which has been used with some success for distinguishing between DNA and RNA. The DNA stains blue or green and the RNA stains red. It is a capricious technique and so it is not recommended for the novice. Attempts have also been made to quantitate cellular content of DNA and RNA photometrically with this technique.

THE PERIODIC ACID–SCHIFF TECHNIQUES

This technique, commonly abbreviated as PAS, is used for the demonstration of mucopolysaccharides and glycogen. The former occur in basement membranes and in other components of the extracellular ground substance. (See Chapter 3.) Tissues are usually fixed in formalin, embedded and sectioned by the paraffin method, deparaffinized, and then oxidized and hydrolysed briefly in dilute periodic acid solution. This liberates aldehyde groups from some of the sugars which are subsequently stained with Schiff's reagent as described above. If hematoxylin and orange-G are used as counterstains, good preparations will show mucopolysaccharides and glycogen purplish-red, fibrin and collagen pink, and nuclei blue.

METALLIC IMPREGNATION

A number of techniques have been developed for demonstrating the fine fibers of nerves, motor end plates, and reticular fibers. These techniques depend on the deposition of colloidal particles of elemental metal, such as osmium or silver, on the surface of the fibers. The metal is introduced in the form of a soluble salt and then precipitated *in situ*. Similar methods are used for demonstrating the localization of certain enzymes, as the phosphatases and cholinesterase. Details of the techniques are too extensive and involved to be described here.

SUDAN STAINS

The Sudan series of stains or dyes includes a number of related, oil-soluble substances which have a strong affinity for lipid. One of them, Sudan Black B, colors particularly the phosphatides. Others, as Sudan Red and Oil-Red O, color the neutral fats. As lipids are usually lost by solubilization in the various reagents used in the paraffin method, tissues are generally fixed in formalin and then prepared by the frozen-section technique or in the fresh state by the freezing-drying method. A simple partition effect is achieved whereby the lipids are colored by preferential solubilization of the dye. Such preparations are not permanent and must be mounted in some nonfat solvent medium such as glycerin-jelly.

VITAL DYES

A large number of dyes have been described that are relatively nontoxic to living tissues when used in highly dilute form in physiological media. By means of their use living tissues within the intact organism (intravital staining) or removed from the organism (supravital staining) may be more or less differentially stained and studied. Examples of commonly used vital dyes follow.

METHYLENE BLUE. Used in very low concentrations, this dye differentially stains basophilic materials, such as nuclei, ergastoplasm, Nissl substance, and the myelinated sheaths of peripheral nerves. It is one of the least toxic of the vital dyes.

NEUTRAL RED. Neutral red, a basic azine dye, is also quite nontoxic. A chemical indicator, it stains acidic substances such as the nucleus and it also stains the contents of the Golgi apparatus. Stained in this way, the Golgi apparatus appears as clusters of red vacuoles.

TRYPAN BLUE. Trypan blue is an acid diazo dye which is engulfed by phagocytic cells and tenaciously retained by them. For this reason it is one of the very few vital dyes that will remain *in situ* during the course of subsequent fixation and paraffin embedding. Its obvious usefulness is in demonstrating phagocytic cells, such as macrophages and cells of the reticuloendothelial system.

JANUS GREEN. This is another basic azine dye which is relatively specific for staining mitochondria.

Smears

Fluid tissues, such as blood and red marrow, can be spread in a thin film on a microscope slide, dried and stained, and then observed with a microscope. Imprints can also be made of many soft tissues, whereby the cut end of the tissue is merely pressed against the surface of the slide and removed, with a unicellular film of tissue adhering to the slide. Such preparations may be stained with Wright's or Giemsa's stain, which bring out in clear color contrast nuclei, cytoplasm, and inclusion granules.

Microscopy

The useful magnifying power of any optical system is limited by its *resolving power*, i.e., the system's ability to distinguish two adjacent objects. The unaided human eye can resolve two dots or parallel lines from a distance of 10 inches as being two if they are separated from each other by a distance greater than approximately 0.1 mm. If they are closer together than this, they appear as a single dot or line. The resolving power of a microscope is determined by the optical properties of the lens that provides the initial magnification of the object, i.e., the objective lens, and the wave length of the illumination. The former is expressed as the numerical aperture (N.A.) of the lens, which is determined (1) by the focal length and usable diameter of the lens—factors which limit the angle by which a ray of light can diverge from the optical axis and still enter the objective lens, and (2) by the refractive index of the object medium. The object medium includes all of the transparent medium that separates the lowest optical plane of the object from the objective lens, i.e., the mounting medium, cover slip, and the space between the cover slip and the objective lens. The highest numerical aperture that can be attained with glass and materials which have the same refractive index as glass for the object medium (the refractive indices of mounting medium and of immersion oil approximate that of glass) and with minimum chromatic and spherical aberration is about 1.25.

With the N.A. as the first limitation on resolving power, the wave length of the illumination is the second. The mean wave length of visible light is about 5000A (A = 0.0001 μ), that of ultraviolet light (UV) is about 2500A, and that of electrons at 60kv is about 0.05A. The resolving power (R) of an optical system may be expressed by the Abbe formula as $R = \frac{0.61\lambda}{\text{N.A.}}$, where λ equals the wave length and 0.61 is a constant. Accordingly, with a N.A. = 1.25, the limit of resolution of the visible ray microscope is in the order of 0.28μ, and that of the UV microscope is about 0.13μ. The resolving power of the electron microscope is calculated on a different basis, owing to differences in objective medium, among other things; values as low as 10A have been routinely obtained with this instrument.

Fig. 1.9 A compound optical microscope with direct magnifications of X10, X450, and X950.

THE VISIBLE-RAY MICROSCOPE

Three systems of lenses are normally used in the visible-ray, compound microscope (Fig. 1.9). The *condensing lens* serves to focus visible light on the object. The light may come directly from a substage source or be reflected by a substage mirror. If a condensing lens is not used, the light may be focused on the object by a concave mirror. If a condensing lens and a mirror are both used, the flat side of the mirror will provide maximum resolution, which becomes critical at the higher magnifications. The condensing lens is movable vertically on most microscopes so that the focal point can be made to fall directly on the object, thus compensating for differences in thickness and refractive index of the slide, the mounting medium, and the object. Furthermore, the condensing lens is equipped with an iris diaphragm that serves to regulate the cross-sectional area through which the light passes. The smaller the area, the sharper the point of focus on the object. Proper use of the condensing lens implies (1) use of the flat side of the mirror, (2) focusing the light on the object by raising or lowering the condenser to the point of maximum illumination of the object, and (3) stopping down the diaphragm aperture to the extent of maximum resolution or contrast.

The *objective lens* is a compound lens designed to provide high initial magnification with minimum chromatic and spherical aberration. Magnification is achieved at the expense of breadth of field, so that microscopes are equipped with two or more objective lenses in order to provide the range of flexibility that proper study of a given object demands. The 10X objective is useful for demonstrating the relationships of tissues within organs. The 45X, or "high-dry," is useful for revealing the organization of cells within a tissue; the 97X or 100X "oil-immersion" lens is used for discerning the details of cellular structure. Although the distance between the objective lens and the object decreases as the magnifying power increases, the lenses are mounted on a rotating turret in such a way that they are parfocal, or nearly so. That is, once the object is in focus with one objective lens, it will be in relatively close focus with the others when they are rotated into position.

The ocular lens, or eyepiece, magnifies the image formed by the objective lens. Therefore the total magnification of the microscope is the product of the magnifying power of the objective lens and that of the ocular lens. The maximum useful magnification consistent with the resolving power of the oil-immersion objective is in the order of 1250 diameters. As the highest-powered objective lens is about 100X, this implies that the highest-powered ocular that should be used in conjunction with it is 12.5X. Higher magnifications can be used, but they will not increase the resolution and therefore are of questionable value.

THE PHASE-CONTRAST MICROSCOPE

Contrast between the various parts of the viewed object is obtained in visible-ray microscopy by differences in either color or density. Because the differences in density between the various parts of cells and tissues are ordinarily very small, tissues are usually differentially stained (*see above*) in order to obtain color contrast. This generally implies killing and fixation, which proves to be rather inconvenient when one wishes to observe living tissue. Phase-contrast microscopy, however, is a kind of optical system that intensifies slight differences in optical density and thereby provides the necessary contrast between most of the components of cells and tissues.

Transparent objects retard and diffract light according to their optical density. The maximum difference in density between the various components of living tissue is such as to result in a phase difference in the order of ¼ wave length, which is insufficient to be detected by ordinary optical means. The phase contrast microscope introduces a second phase-retarding material in the optical path of the diffracted light which retards it by an additional ¼ wave length. The total retardation achieved in this way is thus ½ wave length. When light which has been thus retarded is brought to the same focal point as that which has not, the phenomenon of extinction or wave interference occurs, which results in the dense material's appearing in *dark contrast.* By a slight modification of the same principle, *bright contrast* can be produced, in which the object appears brighter than the background. Figure 1.10 portrays the essential features of the phase-contrast microscope. Because the phase-contrast microscope provides contrast by virtue of differences in optical density rather than color, it is useful with either live or fixed materials. In all other respects it is like the ordinary compound microscope.

ELECTRON MICROSCOPY

Although students of histology are not likely to be using electron microscopes, so much of value has been contributed to the discipline by electron microscopy that some description of the principles involved should be given.

The electron microscope (EM) is in most ways analogous to the visible-ray microscope, but with three primary differences. The "light" used is a beam of accelerated electrons. The electron beam is spread or focused by electromagnetic forces rather than by lenses. Finally, the image is viewed indirectly rather than directly. Because the EM utilizes accelerated electrons with wave lengths

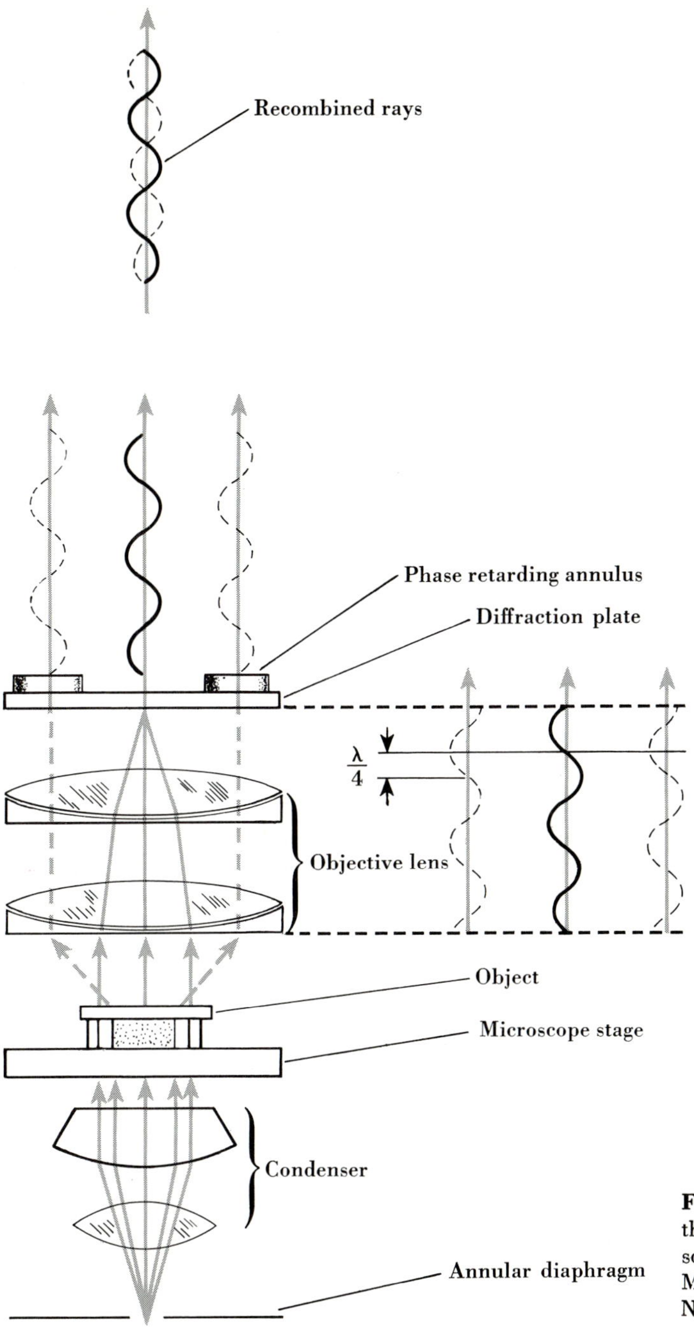

Fig. 1.10 A schematic drawing of the optics of a phase-contrast microscope. (After G. B. Wilson and J. H. Morrison. 1961. Cytology. Reinhold, New York, p. 231.)

around 100,000 times shorter than the wave length of visible light, much higher resolving power (about 10A) can be attained. This permits useful magnifications up to something in the order of 1,000,000 times, with 100,-000X achieving maximum resolution.

Electrons are emitted by a weak electric current from a cathode filament. They are then accelerated by a high voltage potential to form a beam the wave length of which has an inverse relation to the acceleration voltage, as expressed by the formula, $\lambda = \frac{150}{V}$. The beam travels through an evacuated column which incorporates three or four sets of electromagnetic "lenses." The first of these, the condensing lens, focuses the beam on the object, which is supported in the evacuated column by a perforated copper grid. The second lens is the objective lens, which controls the spreading of the beam (thereby producing initial magnification) and focuses it for the rest of the system. The third, which is lacking in some models, is the intermediate lens, which provides additional magnification. Finally, there is the projector lens, which selects the center of the beam, magnifies it, and projects it on a fluorescent viewing screen. Figure 1.11 illustrates the basic features of the EM.

The methods of preparing tissues for electron microscopy are rather different from those for optical microscopy. If cellular detail is magnified, so are artifacts and distortions; thus fixation must be extremely precise and rapid. Pieces of tissue must not exceed 1 mm^3; they must be killed and fixed immediately, preferably *in situ;* all solutions must be isosmotic and appropriately buffered, and alcohol dehydration must be carried out at a very low temperature (4°C has been recommended) to prevent the distorting action of the heat that is liberated when alcohol and water mix. The best general fixative is buffered osmium tetroxide (1 or 2 percent). For special purposes buffered formalin, gluteraldehyde, or potassium permanganate have proved useful.

Because of the short wave length of the electron beam, tissue sections must be extremely thin, i.e., 0.1μ or less. Paraffin is not a suitable medium for obtaining such thin sections; therefore, relatively hard plastic resins are used. A number of products for this purpose are available, each with particular advantages and disadvantages.

The crystalline structure and hardness of steel are not suited for EM work either, but tempered glass is. Glass knives are prepared by breaking plate glass at an angle of about 45° (some electron microscopists prefer a smaller angle), which produces an extremely sharp, smooth, and durable cutting edge. For exceptionally hard materials, such as calcified tissues, diamond knives are used.

Once embedded in polymerized plastic, tissues are mounted in an ultramicrotome, which permits the cutting of sections of the desired thickness. Selected sections are mounted dry on copper grids, which may be coated with a thin film of formvar or collodion for extra support. Although the fixatives used in electron microscopy provide some degree of contrast, additional contrast may be provided by "staining" the sections in solutions of phosphotungstic acid, uranyl acetate, or lead citrate, or by postosmication with osmium tetroxide. All of these serve essentially the same purpose, viz., to deposit a molecular metallic coat on interfacial membranes and surfaces, although some chemical interaction may be involved. With slight modifications of technique, sections obtained in this way have proved to be quite useful for optical study. Photomicrographs of such sections, when the material has been appropriately stained, are much sharper and detailed than those of paraffin-embedded materials (Fig. 6.4 or 11.19).

Objects of macromolecular dimensions, such as virus particles and protein fibers, may more effectively be prepared by "shadowing" them with an electronopaque coating of gold, platinum, or other heavy metal. When prepared in this way, the surface form seems to stand out in three-dimensional relief (Fig. 3.6a).

SPECIAL MICROSCOPY

The foregoing kinds of microscopic instruments have been the most generally useful, but there are others which have been invaluable for more specialized purposes. The *polarizing microscope* is used to study objects which are anisotropic. Anisotropy is the property whereby the wave plane of light is rotated, and it is characteristic of certain kinds of crystals and crystalloid structures. Skeletal muscle and collagen exhibit anisotropy.

THE UV MICROSCOPE. This microscope is useful not only because of its very high resolution (*see above*), but also because of the ability of nucleoproteins and certain other compounds in tissue to absorb UV light. All the system's lenses are made of fused quartz instead of

Fig. 1.11 An electron microscope, *top*. A diagram of the components of an electron microscope, *bottom*. (Both parts courtesy RCA Scientific Instruments.)

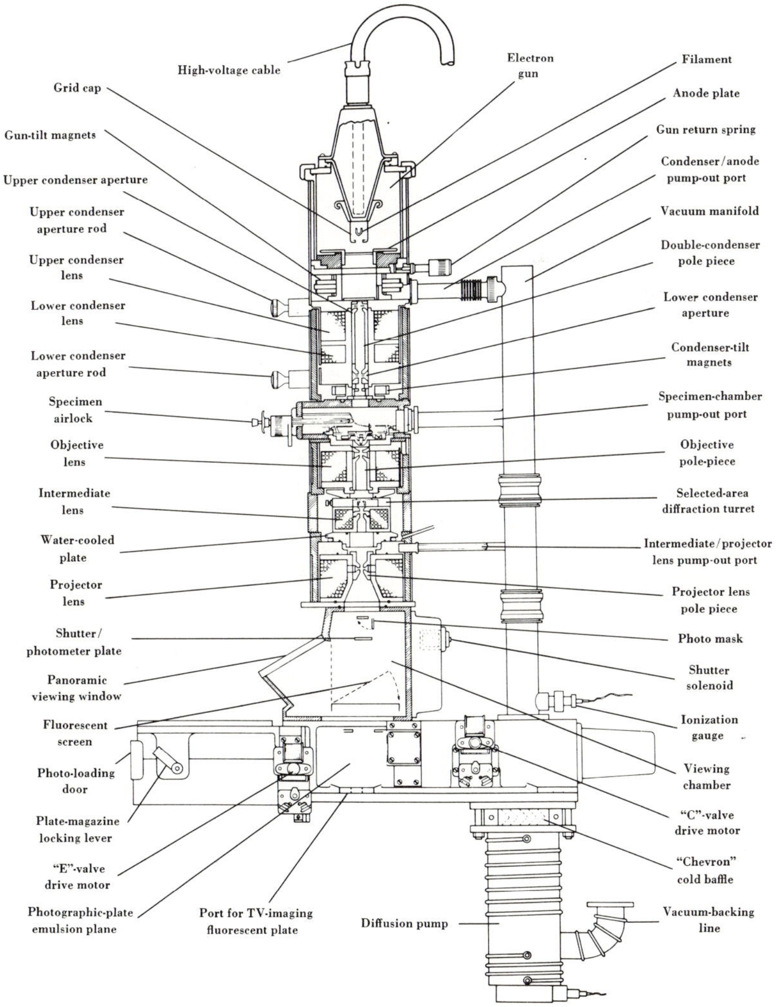

glass, because glass filters out much of the UV. This microscope is most useful in quantitative analysis.

FLUORESCENT MICROSCOPY. Fluorescent microscopy makes use of the fact that some substances (such as riboflavin and the porphyrins) alter the wave length of invisible UV to visible light, and the fact that certain other substances (principally protein) can be rendered fluorescent by labeling them with fluorescent dyes. This method has been especially useful in tracing the course, deposition, and site of interaction of antibodies and antigens.

INTERFERENCE MICROSCOPY. The interference microscope is similar in principle to the phase-contrast microscope, except that part of the light from the source is diverted around the object and recombined with the undiverted light above the object. Both destructive and constructive interference occur, as with phase contrast, but inasmuch as the diverted beam of light has not been significantly altered with respect to phase, it serves as a reference by which to measure the optical density of objects or materials which have affected the undiverted light.

X-RAY MICROSCOPY. X-rays may be used in three basically different ways in the analysis of biological materials. One is by contact photography, by which X-rays pass through a specimen and activate a photographic emulsion directly beneath it according to the scatter and/or absorption of X-rays by parts of the specimen. The emulsion is then developed and examined by visible-ray microscopy. A second method represents a slight modification of this, whereby diverging X-rays pass through the specimen, which is separated by some predetermined distance from the photographic emulsion, which is then developed and examined. By this method, some magnification is gained by virtue of the diverging beam. A third method is X-ray diffraction, whereby the specimen—because of its molecular structure—diffracts the X-rays to produce a diffraction pattern on a photographic emulsion. The diffraction pattern can be analyzed to reveal the structure of crystals and crystalloids, such as collagenous fibers, hemoglobin, and myosin.

SELECTED REFERENCES

De Robertis, E., W. W. Nowinski, and F. A. Saez. 1965. Cell biology. 4th ed. W. B. Saunders Co., Philadelphia.

Downes, H. R. 1962. The chemistry of living cells. 2nd ed. Harper & Row, New York.

Gurr, E. 1962. Staining, practical and theoretical. Williams & Wilkins, Baltimore.

Ham, A. W. 1965. Histology. 5th ed. J. B. Lippincott, Philadelphia.

Mahler, H. R., and E. H. Cordes. 1966. Biological chemistry. Harper & Row, New York.

Pearse, A. G. E. 1960. Histochemistry, theoretical and applied. 2nd ed. Little, Brown and Company, Boston.

Sjöstrand, F. S. 1967. Electron microscopy of cells and tissues. Vol. 1. Instrumentation and techniques. Academic Press, New York.

Wilson, G. B., and J. H. Morrison. 1966. Cytology. 2nd ed. Reinhold, New York.

chapter 2
THE CELL

THE MOST ENERGETICALLY stable form in nature for cells, soap bubbles, or any fluid body for that matter, is one which presents the least surface, i.e., a sphere. Deviations from the spherical form occur, however, when mechanical forces or spatial limitations are imposed on a cell. One criterion of cellular differentiation is the extent of morphological change that a given cell has undergone. Thus, very highly differentiated cells, such as neurones and muscle fibers, are often characterized by their unique spatial configuration. Under optimum conditions, as those which obtain in tissue culture, such cells will often revert to the basic spherical form, indicating that unusual cell shape is a condition imposed on the cell from without as well as from within.

Often related to the shape of the whole cell is the shape of the nucleus. Cells which are spherical, cuboidal, or polyhedral usually have spherical nuclei. Flattened or squamous cells generally contain flattened, ovoid nuclei whereas cylindrical or fusiform cells exhibit ellipsoid nuclei. In both of the latter cases, the longitudinal axes of nucleus and cell are parallel. Nuclei can also be of irregular shape, as in leukocytes.

The degree of morphological or functional cellular differentiation is of major importance in determining the location of the nucleus. Centrally located nuclei are found in undifferentiated or unspecialized cells. As a cell acquires functional specificity, the cytoplasm often is either filled with fibrillar or other structures, as the skeletal muscle fiber, or contains deposits, secretion granules, or inclusions, as fat cells or secretory cells. In such cases the nucleus, being pushed aside by the cytoplasmic contents, is often eccentrically disposed.

Excluding many of the cells of the circulatory and lymphatic systems, most cells of the body neither occur nor function singly. Cells are usually surrounded by other similar cells, each imposing mechanical forces on the other so that potential spheres assume a polyhedral, columnar, cuboidal, squamous, or irregular form, and each cell participates in a function greater than that which it could perform singly, such as forming protective epithelial membranes, contractile muscle sheets, and so forth. Such groups of similar cells working together to perform a group function are termed *tissues*. Four categories of tissues are recognized: (1) *epithelial tissues;* (2) *con-*

necting and supporting tissues, including bone, cartilage, blood and lymph, and all gradations of fibrous and reticular connective tissue; (3) *muscular tissue;* and (4) *nervous tissue.*

Tissues usually combine to form masses of definitive shape and more or less unified function called *organs.* Organs may be composed mainly of a single type of tissue, such as in the brain, or of several tissues, such as the liver or ovary. However, because blood, lymph, and nervous tissue represent the primary transport and integrating systems of the body, and hence provide the necessary functions of delivery of nutrients, removal of waste products, and coordination of activity, each organ, whether composed of primarily one or of more kinds of tissue, maintains a close spatial and functional relation with these three fundamental tissues. Thus, even though the main component of a gross skeletal muscle is muscle fibers, connective tissue forms the enclosing sheaths, carries blood vessels, and nerves, and takes part in the highly structured muscle-nerve junction. The role of the transport tissues is more easily seen in organs such as the liver, which contains many blood sinuses interspersed with cords of epithelial tissue. Organs, then, are centers of tissue interrelationships that make possible the integration of cellular functions and the emergence of higher order group functions.

The acquisition of a three-layered embryonic organization among all but the most primitive metazoans introduced many complications. Foremost was the development of mechanisms to meet the problem of nourishment of internal cells. As these mechanisms evolved, however, cells acquired specialized functions. No longer did one cell have to be a bread winner as well as a bread digester. Each kind of cell, grouped with its fellows in tissues and organs, could perform a specific function more efficiently in addition to maintaining the generalized metabolic pathways necessary for cellular, but not organismal, life. With progressive cellular specialization, increased animal size, and more complex and numerous functional pathways, generalized organs gave way in the organizational hierarchy to organ systems which provide a greater amount of surface area and subdivision of labor than would be available in one generalized organ. Because most biochemical reactions occur at surfaces or along interfaces, one finds that systems responsible for complex metabolic reactions, such as the digestive system, extend throughout much of the length of the organism. Each living cell, then, is capable of vegetative activity; but when organized first as tissues, then as organs, and finally as organ systems, each can demonstrate a unique and necessary property (such as specific molecular synthesis or integrated electrical conduction) necessary in the best interests of the total organism.

Meaningful histology includes function, and in this context the relationships of a specific tissue up the organizational ladder to organs and organ systems are considered. For greater meaning, however, one should be aware of the properties and generalized features of the cells which compose the tissues.

The protoplasm of a cell has long been known to be organized in two structural and functional areas. The basophilic nucleus is the director of synthetic properties and the center of the reproductive and hereditary mechanisms of the cell. The acidophilic cytoplasm contains the substances that respond to the functional orders of the nucleus. Surrounding the nucleus and the cell as a structural and functional entity are limiting membranes, and within the cytoplasm are numerous organelles and inclusions.

THE CELL MEMBRANE

For many years the cell membrane was inferred from physiological studies. Although it was too thin to be visible with the light microscope, the postulation of some sort of structure was necessary to account for such observable phenomena as selective permeability, active transport, and electrical conductivity. Hence, the cell membrane was dutifully included in texts and students' drawings as a thin line delimiting the boundary of the cell.

X-ray-diffraction studies and polarizing and electron microscopy of the red blood cell membrane and myelin sheaths have revealed some aspects of the structure of these unit membranes. About 70A wide, they consist of two concentric layers of protein molecules, sandwiching a radially arranged double lipid layer. Because one "end" of a lipid molecule is polar and hydrophilic, and the rest (the carbon-chain "tail") is nonpolar and hydrophobic, it has been assumed that the tails of the lipid molecules come together side-by-side and end-to-end, forming a bimolecular leaflet, with the polar ends being presented as the outside surfaces. Proteins then become associated with these surfaces by virtue of their common solubility and combining characteristics (Fig. 2.1).

Chemical analyses of red blood cell membranes have indicated that the lipid portion consists primarily of lecithin, cephalin, and cholesterol. The protein layers are acidic and of a fibrous nature. It has been suggested that the inner protein layer consists of extended, simple polypeptide chains, whereas the outer sheet contains mucoproteins, lipoproteins, and possibly mucopolysaccharides as well.

Recent evidence has led to some theoretical

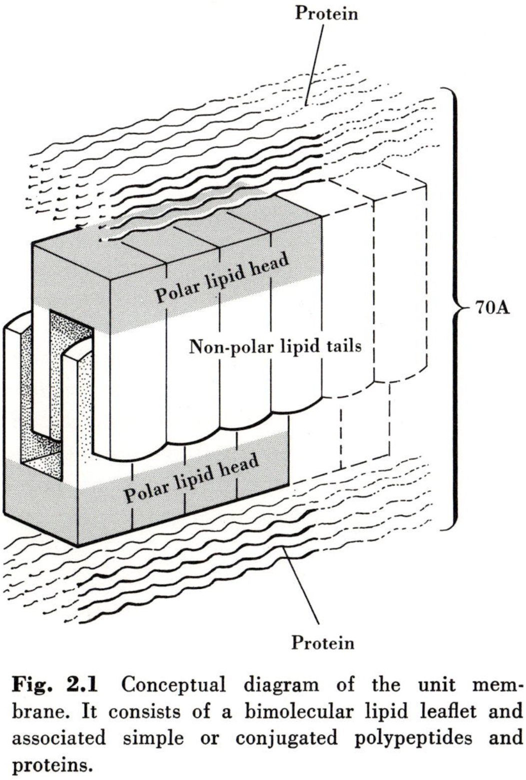

Fig. 2.1 Conceptual diagram of the unit membrane. It consists of a bimolecular lipid leaflet and associated simple or conjugated polypeptides and proteins.

modifications of the basic unit membrane plan. It has been suggested that the lipid and protein layers are not separate, but rather that these molecules are combined to form complex macromolecules which are basically lipoprotein in nature. If one accepts this suggestion, the membrane would be considered to consist of these repeating complex units and would be only one molecule in thickness. Furthermore, although the physical form of the macromolecular complex would be similar in all membranes, the exact protein complement could vary. This could account for some rather specific membrane properties in certain cells and organelles.

The movement of molecules through the membrane is thought to be affected by one or more of the following. (1) Molecules may move into or out of the cell by virtue of their solubility in the protein or lipid layers. (2) The protein and/or lipid layers of the membrane may not be continuous; hence, gaps or pores would exist which perhaps, with an additional electrical charge, would allow passage of molecules or ions of limited size into or out of the cell. (3) Enzyme systems may be present within the cell membrane, and function as molecular carriers. (4) The molecular form of the membrane may be changeable enough to permit smaller molecules to be engulfed by larger ones.

In recent years electron micrographs have revealed the unit membrane character of most of the cellular organelles, and on the basis of this and other evidence, a triphasic cellular system has been postulated. The three phases are (1) the hyaloplasm; (2) the unit membrane of the cell surface, with its ramifications as the nuclear membrane, Golgi complex, endoplasmic reticulum, and lysosomes; and (3) the internal compartment embraced by the vacuoles and spaces created by the infolding and outfolding of the unit membrane. The last is directly or indirectly continuous with the outside environment of the cell (Fig. 2.2). A correlary consideration of this hypothesis is the dynamic nature of the cell membrane. Changes in topography can continually take place during such processes as vacuole formation, pinocytosis, or increase in the number of organelles, the walls of which are ramifica-

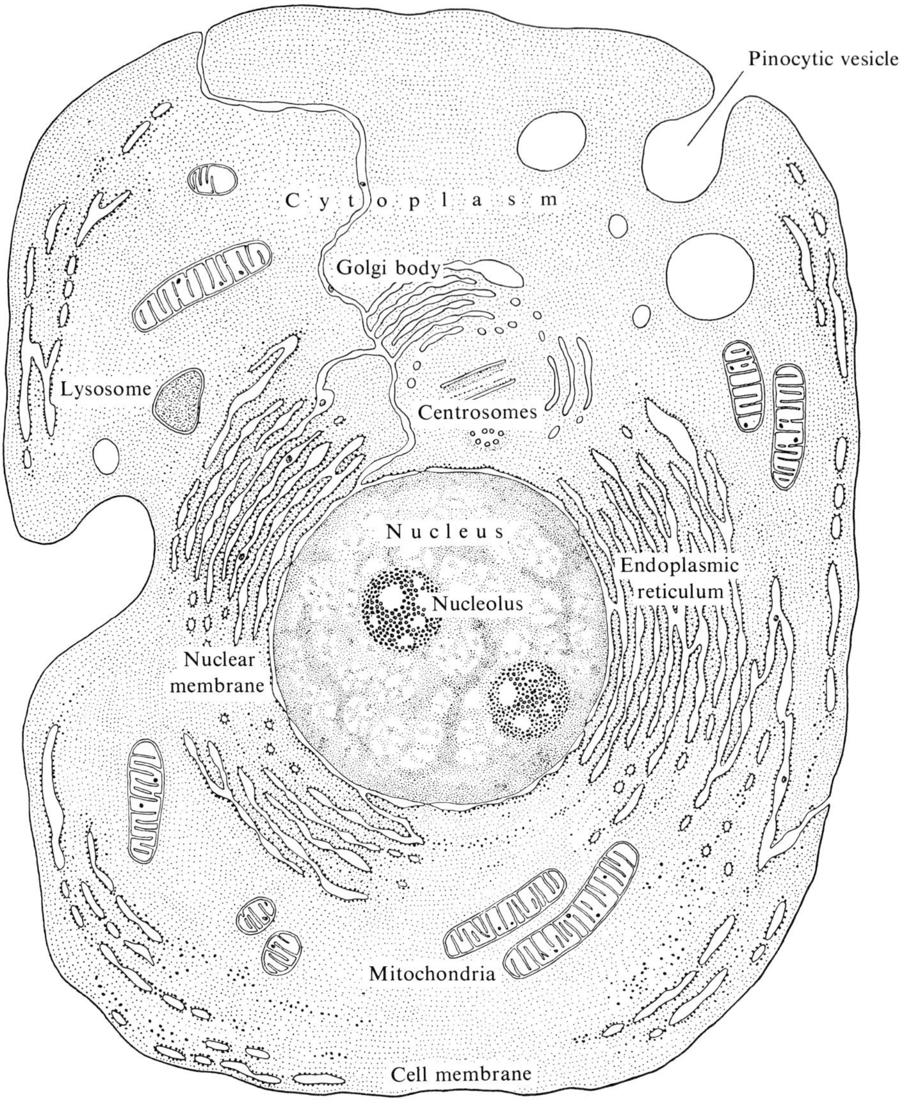

Fig. 2.2 The cell as a triphasic system. This diagram represents a composite of those parts common to all animal cells as visualized with the electron microscope. (After J. Brachet. 1961. Sci. Am. 205:3.)

tions of the basic unit membrane. Such changes can result in the conversion of the simplified, so-called "typical" cell of Fig. 2.2 into a more complex cell as diagrammed in Fig. 2.3.

THE NUCLEUS

The Nuclear Membrane

The interphasic nuclear envelope is a double membrane, each portion of which is similar to the cell membrane; the space enclosed within the membrane is termed the *perinuclear cisterna* (Fig. 2.4). At intervals along the envelope, the inner and outer membranes are continuous around the periphery of pores through which interaction of the nuclear and cytoplasmic protoplasm can occur. According to the triphasic theory, these openings are not so much pores as they are gaps between adjacent infoldings of the limiting membrane.

During mitotic division, the nuclear membrane can not be demonstrated with light microscopy. However, in electron micro-

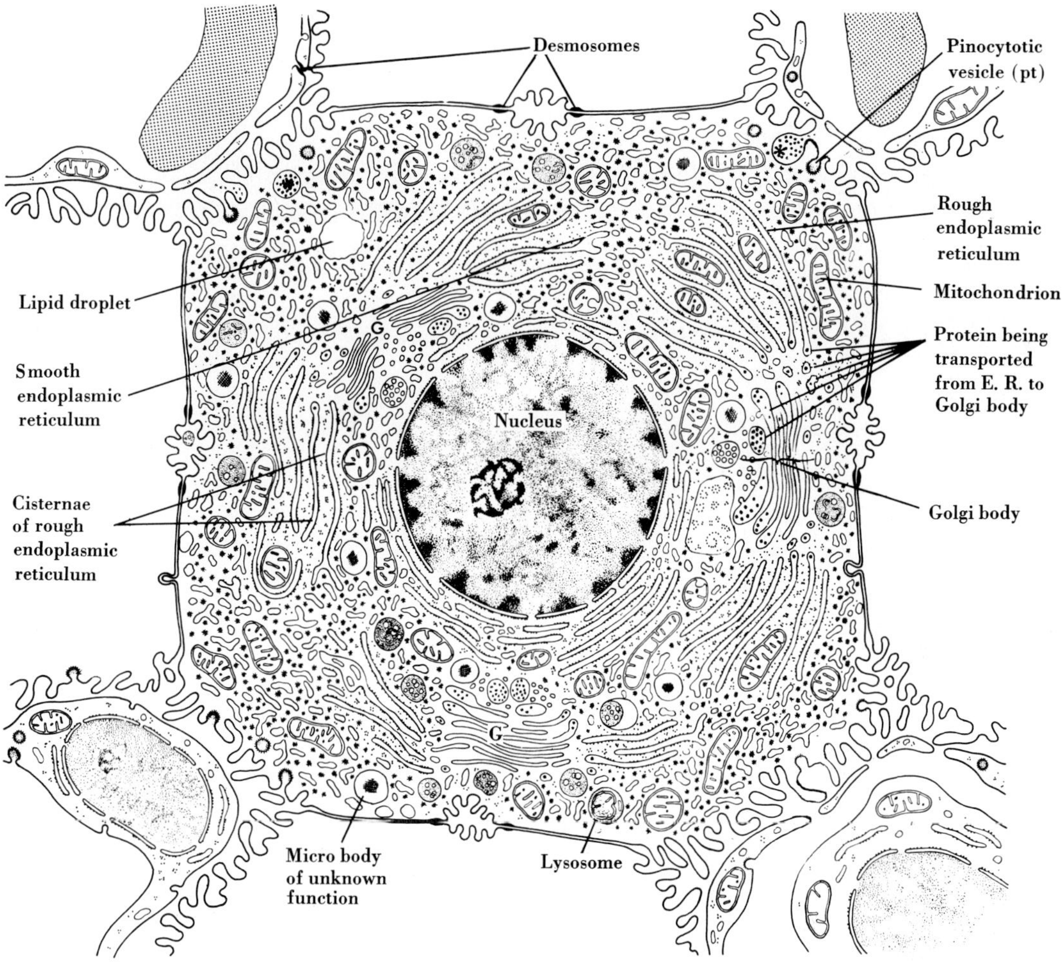

Fig. 2.3 A typical mammalian liver cell diagrammed from observations with the election microscope. X15,500. (After K. R. Porter and M. A. Bonneville. 1964. An introduction to the fine structure of cells and tissues. Lea and Febiger, Philadelphia.)

graphs, the nuclear membrane appears to dissociate, forming vacuoles which retreat into the cytoplasm. At the end of telophase, the membrane reforms, apparently by condensation of these vacuoles around the chromosomal mass.

Nuclear Sap

Enclosed within the nuclear membrane is the protoplasm of the nucleus, the *nuclear sap* or *nucleoplasm.* Optically, the nuclear sap appears to be homogeneous, with little or no structural complexity (Fig. 2.4). Supporting evidence for the substance's homogeneity has also come from microdissection experiments. If a microneedle is inserted into the nucleus, no resistance is encountered and the nuclear organelles may be moved easily from one side to the other. The sap is thixotropic, however, with reversible gelation occurring upon agitation. Under the action of certain fixatives and dyes, a more or less granular appearance can sometimes be demonstrated. However, in the interphase nucleus the chromosomes are in a dispersed state and the chromatin granules which are seen may represent an artifact of fixation. The nuclear sap is composed mainly of proteins, including hydrolytic enzymes and enzymes associated with nucleic acid and

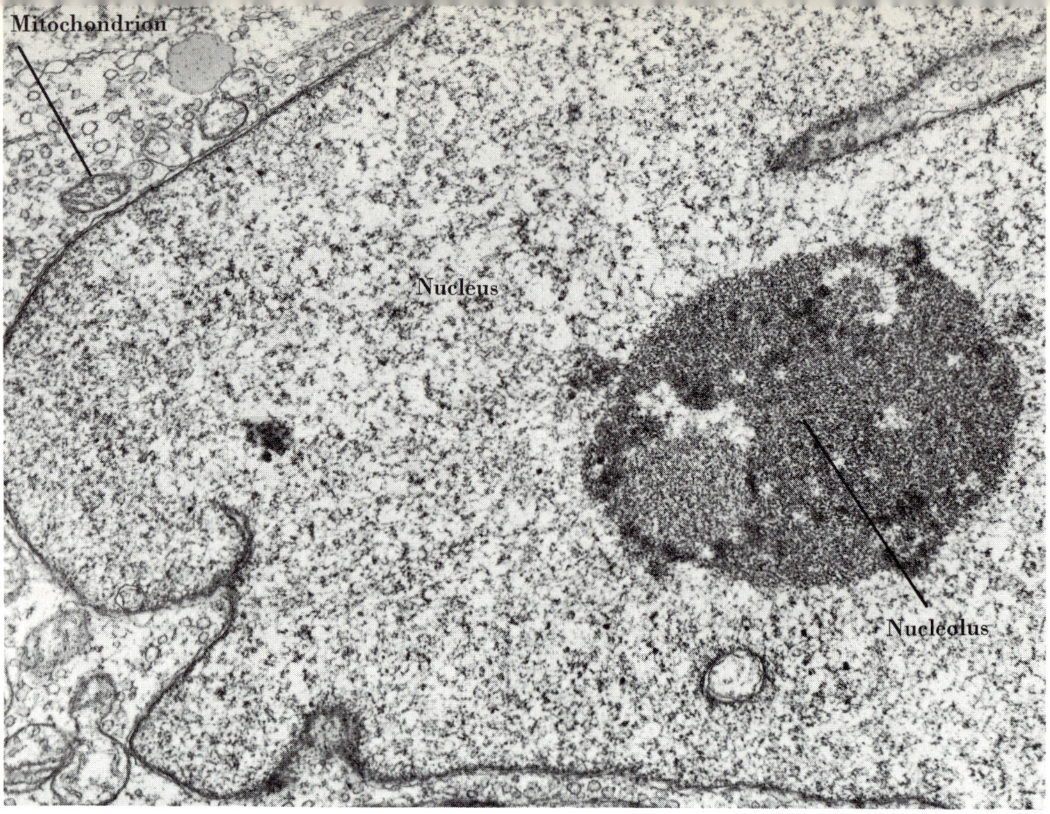

Fig. 2.4 This electron micrograph includes part of the nucleus and adjacent cytoplasm in a cell from a human amnion grown in tissue culture. Note the structure visible within the nucleolus and the double nature of the nuclear membrane. X18,000. (Courtesy of F. B. Merk.)

nucleoside metabolism, such as nucleoside phosphorylase, guanase, and adenosine deaminase. Many other enzymes, such as esterases, catalase, and arginase, can be found in some nuclei some of the time. Soluble ribonucleic acid (RNA) and glycoproteins are also found in the nuclear sap, and recent evidence suggests that ribosomes can also be transitory or permanent occupants of the nucleus.

The Nucleolus

Within the interphasic nucleus there may be one or more smaller, strongly basophilic bodies, the *nucleoli* (Fig. 2.4). Nucleoli disappear during cell division and are reconstituted at the end of the division process—presumably by specific chromosomal regions known as nucleolar organizers. The fine structure of the nucleolus consists of masses of RNA granules, each with a diameter of about 150A, arranged in a network of narrow or occasionally broad fibrils. This structural organization has recently been found to have functional significance because the nucleolus has been shown to be the seat of intense nuclear RNA activity. If one labels the nucleolar RNA with radioactive tracers, such as tritiated uridine, and follows the movement of the label, the active RNA is seen to move out of the nucleolus and the nucleus and into the cytoplasm—where, in association with cytoplasmic particulate RNA, protein is synthesized. The migratory nucleolar RNA, which is characterized by a low molecular weight has been termed "*soluble RNA*" (sRNA), which has been shown to be responsible for recognizing and transporting amino acids within the cytoplasm. Another species of RNA, "*messenger RNA*" (mRNA), of higher molecular weight, is also found in the nucleolus. Coded by the chromosomal DNA in a complementary fashion, mRNA also moves into the cytoplasm, where it becomes associated with particulate or ribosomal RNA. *Ribosomal RNA*, also thought to be synthesized or assembled in the nucleolus, is complexed with protein and appears as bipartite cytoplasmic bodies, the ribosomes.

Other essential components of the nucleolus are large quantities of acidic and basic proteins, some possibly existing in combination with RNA as nucleoprotein, and phosphatides. In the interphasic nucleus, DNA can often be detected condensed around the boundary of the nucleolus.

Chromosomes

The key constituents of the nucleus are the chromosomes; they represent the microscopic visualization of the submicroscopic basis of heredity and metabolic direction. Unfortunately our cytological picture of the chromosomes has not yet been fully reconciled with our molecular picture of the DNA molecule, a situation which results in confusion to students and authorities alike.

As described in Chapter 1, molecular DNA is a double helical polymer of four nucleotides (Fig. 1.6). The order of nucleotides represents the so-called genetic code; i.e., read in a linear direction, each group of three nucleotides is the code for one amino acid. There are more than enough combinations of four nucleotides arranged in three-item groups to specify each of the native amino acids. In addition, as each chain in the helix is a complement of the other, each can act as a template in the synthesis of new DNA. Hence the molecular picture of DNA replication includes the unwinding of the helix, separation of each chain from the other, and the building of one new chain by each old chain. This is presumed to occur during late telophase or interphase, because biochemical investigations of the cell cycle have indicated a doubling in the amount of DNA at these stages. DNA, being acidic, exists in combination with basic proteins as nucleoprotein. Included in the list of basic proteins combined with DNA are histone, which is found as the predominant basic protein in most nuclei; protamine, found in the sperm of many fishes; and a protein of higher molecular weight, known by the negative name of nonhistone protein. The role of these proteins has not been definitely established, but they are thought to have some function in chromosomal metabolism, gene suppression, and/or maintenance of structural shape of the chromosome.

When one begins to enlarge the foregoing picture to the cytological scale, analogous molecular assignments become difficult. The chromosome is indeed composed of DNA, basic protein, and, in addition, some RNA; but the arrangement of each chemical species within the chromosome is not yet clear. Moreover, visible chromosomal structure is cyclic, differing in interphase and during division.

In most interphasic nuclei the chromosomes are in a dispersed state within the nuclear sap. They have not lost their unit character, but they are stretched out sufficiently as to be below the concentration needed for cytological detection. However, regions of the chromosomes exist which differ in such a manner as to appear more dense than other areas after fixation. The dense portions appear as *chromatin granules* or flakes (Fig. 2.4). If tissue specimens of the adult body are taken, most of the chromosomal material will have this appearance because adult, differentiated cells do not possess a mitotic rate statistically high enough to be easily demonstrable. If, however, one chooses embryonic or regenerating tissues or mitotically active sites in the adult (such as those which occur in the basal layers of the skin), cells in the process of division can easily be seen. Naturally, there are exceptions to this generalization. In the salivary gland and other tissues of the dipterous insects, polytene chromosomes exist. These are multistranded structures which are visible during interphase as a series of light and dark bands. Far from being an unwanted exception, the polytene chromosomes have been one of the most useful tools for studying problems of chromosomal structure and abnormality. The second major exception is the lampbrush chromosomes found in oocytes of many classes. These greatly enlarged chromosomes contain much RNA in the lateral loops and are thought to be active synthetic sites in the developing gametes.

Chromosomal structure is most easily seen in fixed and stained sections of cells undergoing division because during division, contraction of the chromosomal material occurs.

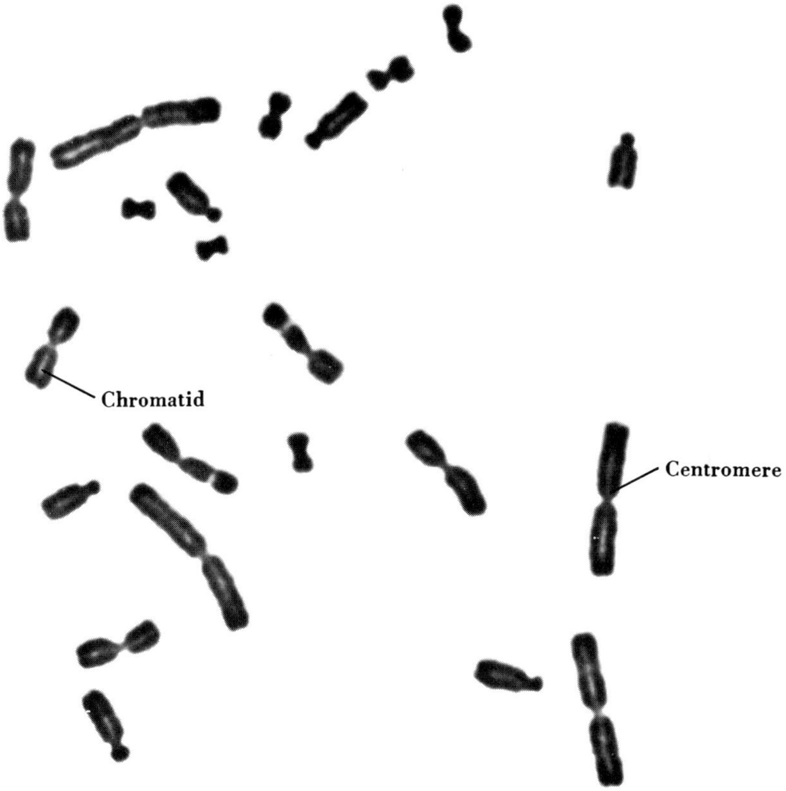

Fig. 2.5 The chromosomes of a female Chinese hamster during colchicine-arrested mitotic metaphase. X2000. (Courtesy of G. Yerganian.)

At such time, the cylindrical chromosomes appear to contain two strands which are seen to be embedded in a matrix and surrounded by a boundary membrane. However, both matrix and membrane may be artifacts of fixation. Each chromosome contains one or more constricted areas, the *centromere,* the number and location of which are specific for the species and particular chromosomal complement studied (Fig. 2.5).

THE CYTOPLASM

Functionally and structurally the cytoplasm is most complex. It is composed of an amorphous matrix in which are located many membranes and granular organelles which provide sufficient surface area for an orderly assembly line of enzymatic reactions (Fig. 2.6).

Amorphous Hyaloplasm

The hyaloplasm is the ground substance or matrix of the cytoplasm. It has been an ever-shrinking "pigeon hole." As new electron-microscopic and biochemical techniques are developed, parts of the hyaloplasm which were thought to be formless are seen to contain structure, whereupon they are named and subtracted from the total hyaloplasm. Although the hyaloplasm is designated as amorphous, under conditions which cause reversion of a sol to a gel a fibrous ultrastructure can be observed. Presumably because of changes in the degree of folding and cross linkages of the protein chains, the resulting micellar and tactoid structure may be important in the economy of the cell.

Contained within the hyaloplasm are many enzyme systems. The most important are those concerned with glycolysis (the breakdown of

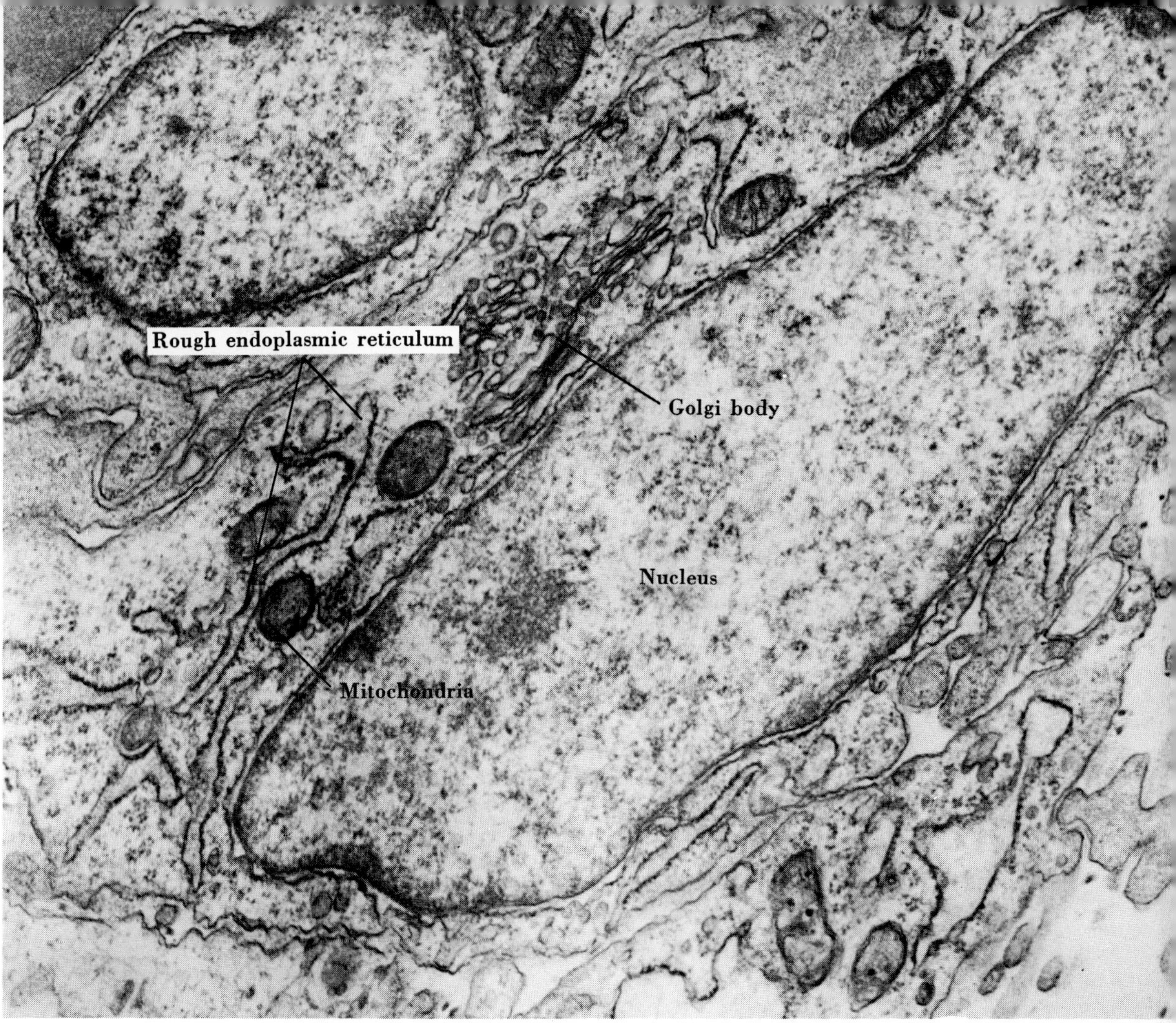

Fig. 2.6 Electron micrograph of a cell from the palate of an alligator. X29,000. Compare this electron micrograph with the diagram in Fig. 2.3. (Courtesy of Floyd O. Craft.)

sugar to pyruvic acid), synthesis of the pentose sugars found in nucleic acids, and synthesis of fatty acids. In addition to the importance of these enzyme systems in the total economy of the cell, the presence of these pathways within the hyaloplasm serves to illustrate the interdependent nature of all parts of the cell. End products of one area become substrates in another. For example, the pyruvic acid produced by the hyaloplasm is further incorporated into nuclear syntheses of nucleic acids or mitochondrial oxidation processes.

Mitochondria

Mitochondria are granular or elongated cytoplasmic organelles composed of a compartmentalized double membrane system, in which the energy-producing, oxidative reactions of cellular respiration occur. Mitochondria vary in size from a few micra to dimensions below the limit of resolution of the light microscope, and because of the nature of the metabolic reactions which occur in mitochondria, they have long been recoognized as the only structures which can initially maintain the vital dye Janus Green B in its oxidized, colored state before reducing the dye to its colorless leucobase.

The general fine structural plan of mitochondria is uniform. The external envelope consists of an osmophilic double membrane. Each half of the double membrane is about 40A thick with a 70-A space between the inner and outer sheets. The outer membrane is cov-

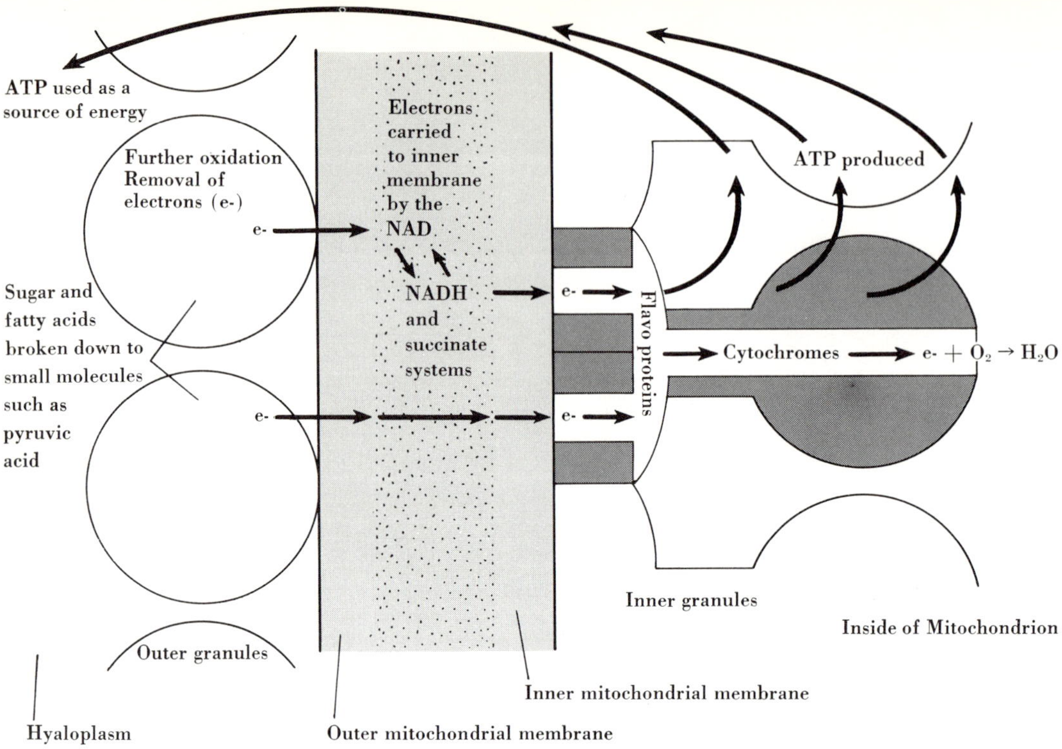

Fig. 2.7 Diagram of the relationship between the hyaloplasm and the mitochondrion, and the proposed activity and shape of the mitochondrial membranes and granules. (After D. Green, 1964. Sci. Am. 210:63–74.)

ered with granules on its outer side and the inner membrane is similarly coated on its inner side. The differences in shape of these two sets of particles reflects a fundamental difference in biochemical function (Fig. 2.7). The mitochondrial interior consists of an amorphous matrix that may or may not contain granules, but which is believed to contain strands of DNA and possibly ribosomes as well. The matrix, composed of lipid and protein, is subdivided into compartments by a series of double membranes or *cristae mitochondriales* which are thought to be extensions of the inner mitochondrial membrane (Fig. 2.6).

Although the general structure of mitochondria is uniform, the degree and shape of compartmentalization can vary. In undifferentiated cells or in primarily vegetative cells, the mitochondrial population is small, and those which do occur contain few cristae. In highly active cells, such as spermatozoa or muscle cells, many mitochondria are present, each possessing many cristae.

Variations in the number and complexity, as well as in the form and density of the cristae and the mitochondria as a whole, can be correlated with biochemical properties. These structures function by oxidizing the pyruvic acid formed in the hyaloplasm to carbon dioxide and water. The stepwise oxidation involves the acceptance of substrate electrons by compounds such as flavoprotein and nicotinamide adenine dinucleotide (NAD). The electrons are then passed along a chain of electron-carrier compounds known as the cytochromes until finally molecular oxygen is reduced to water. Rather than wasting the energy which is liberated every time an oxidation reaction has occurred, the released energy is stored by phosphorylating adenosine diphosphate (ADP) to adenosine triphosphate (ATP; Fig. 2.7). During cellular processes requiring energy, the stored energy can be released by splitting off the terminal phosphate of ATP. The enzymes, substrates, and cofactors needed for the above reactions, known collectively as oxidative phosphorylation, are present in mitochondria, and about 90 percent of all the ATP generated by the cell is produced by the mitochondria. In addition, these organelles are the seat of the energy producing reactions of fatty-acid oxidation. Hence, as biochemical processes occur along surfaces or interphases, it would seem logical that active cells needing much energy contain larger numbers and more complex mitochondria than resting cells.

Theories on the origin of mitochondria have been numerous. The two hypotheses which are most widely held suggest that they are either derivatives of the cell membrane or are self-replicating, metabolically independent organelles. The first theory gains support from the similarity between the membrane systems of the mitochondrion and the cell as a whole. The second theory, one that is rapidly gaining favor, is based on the presence of unique DNA, RNA, and ribosomes within the mitochondrion. Additionally, other organelles have been found to contain nucleic acids heretofore believed to be strictly nuclear inhabitants. Chloroplasts, basal bodies, and centrioles, in addition to mitochondria, all seem to be self-duplicating. Perhaps it is no coincidence that these organelles are all complex in their structure, and collectively form the bulk of the discontinuous cell organelle system. On this basis, the cell can be considered as having a *cytoskeletal system* consisting of all membranes derived from a common origin as the cell and nuclear membranes, Golgi body, endoplasmic reticulum, pinocytotic and phagocytotic vacuoles, lysosomes, and other membrane-bound vesicles, and a *self-replicating* fraction including mitochondria, centrioles, chromosomes, and attendant RNA structures, basal bodies, and, in plants, the plastids.

Endoplasmic Reticulum and Ribosomes

The cytoplasmic membrane-vesicle system known as the endoplasmic reticulum was revealed by the use of the electron microscope (Fig. 2.6). Contained within the cytoplasm and appearing as a three-dimensional membranous network enclosing cavities, the endoplasmic reticulum may be a derivative of the cell membrane and/or the nuclear membrane.

In many cells, especially those synthesizing quantities of protein, there are granules associated with a well-developed reticulum. These bipartite granules, or ribosomes, are about 100A in diameter, and occur on the matrix side, as opposed to the cavity side, of the membrane. According to their presence or absence, the membranes are designated as rough or granular and smooth or agranular, respectively. The granules are highly basophilic and have been found to be composed of RNA and protein. By using many different avenues of approach, we now know that the ribosomes are the seat of protein synthesis. Both mRNA and sRNA are brought together at the ribosomes, the sRNA bringing with it activated amino acids, each attached to specific soluble RNA molecules. There the amino acids are coded in proper sequence and linked together by peptide bonds to form a protein (Fig. 2.8).

Apparently more than one ribosome is needed for the production of a complete protein molecule. *In vitro* synthesis of protein can be obtained by removal of the ribosomes from cells with the addition of amino acids and the proper enzymes to the ribosome mixture. However, care must be taken during the procedure in order that associated ribosomes, or polysomes, do not become disrupted; if disruption occurs, only small polypeptides will be formed.

Often seen in texts is the term "microsome." The microsomal portion of the cell is that part in which, after isolation by differential centrifugation, reactions of protein synthesis can be found. We now know that microsomes represent broken endoplasmic reticulum together with ribosomes.

The cavities or vascuoles enclosed by the endoplasmic reticular membranes may also evidence granules. Because this phenomenon is usually encountered in secretory cells, and because the granules resemble those of zymogen, it has been postulated that the endoplasmic reticulum may also function in collection of secretory granules prior to discharge. This concept is reinforced by the relative rates of synthetic activity exhibited by the ribosomes. Those ribosomes which line or are very close to the membranes of the reticulum are more active synthesizers than those which

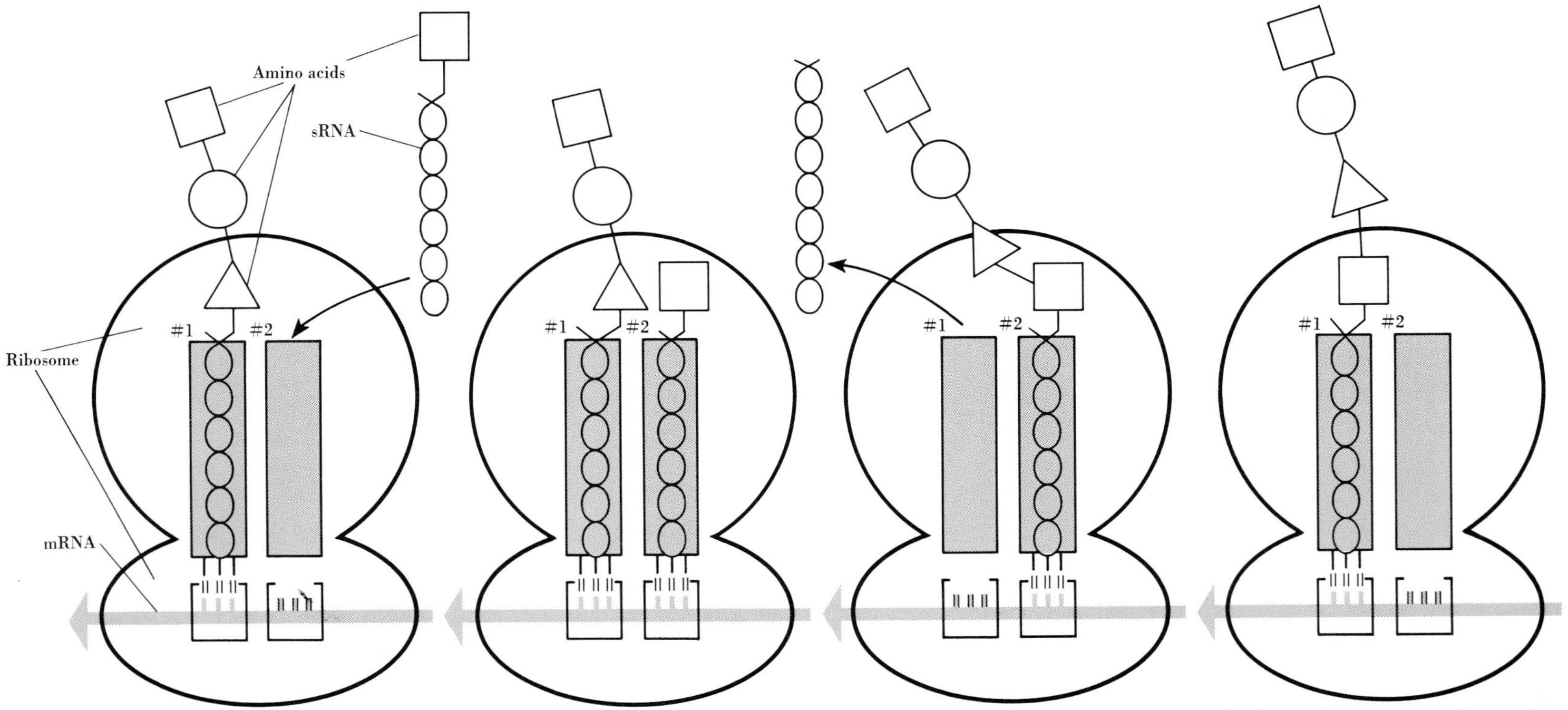

A. As mRNA moves through the ribosome, a molecule of sRNA with the partly formed polypeptide is positioned into binding site #1. A second tRNA with amino acid attached moves into site #2.

B. Site #2 now contains an s-RNA amino acid complex. The particular sRNA which can fill the site depends on the complementary base pairing between the 2 kinds of RNA.

C. All of the amino acids are now shifted to the sRNA in site #2. Note that the chain of amino acids has grown by placing the newest addition at the base.

D. The complex moves to site #1 to await another molecule of s-RNA plus its attached amino acid. The new mRNA triplet exposed at site #2 will determine which sRNA shall be positioned. Each sRNA is specific for one particular amino acid.

Fig. 2.8 Schematic representation of RNA linkage with amino acids to form protein. See text. (After Fig. 11.9, J. Watson. 1965. Molecular biology of the gene. W. A. Benjamin, Inc., New York.)

are located at a greater distance. The protein produced by the membrane-associated granules would enter the vacuoles prior to use within the cell or secretion by the cell. The idea of such an interplay between the granular and membranous components of the cytoplasm is attractive because it attempts to explain the close physical relationship.

In addition to protein synthesis, the endoplasmic reticulum has been shown to have a high lipid content and to function in synthesis of such lipids as cholesterol and glycerides.

Golgi Substance

The uncertainty which for many years has surrounded this cytoplasmic organelle is reflected in its variety of names attached to "Golgi": substance, region, complex, body, apparatus, or network, to name a few. First described by Golgi near the turn of the century, the Golgi substance appears by light microscopy as a system of canals or vacuoles which actively reduces osmium tetroxide, and is therefore known as osmiophilic, and which surrounds an osmiophobic region. Electron microscopy has revealed the Golgi substance as a complex of sacs enclosed by smooth surfaced membranes. Some of the sacs are small, and hence are termed *microvesicles*, while others appear to be distended and are called *vacuoles* (Fig. 2.6). Although the membranes are not associated with granules and are thicker than those of the endoplasmic reticulum, many electron microscopists believe the Golgi substance to be a derivation of the endoplasmic reticulum.

As opposed to the other cytoplasmic organelles, such as ribosomes and mitochondria, it is difficult to remove the Golgi substance by differential centrifugation or by other methods. Hence the function of this organelle has been inferred largely on morphological grounds rather than analyzed directly by chemical means. However, differential centrifugation studies of the Golgi substance have indicated that it is a well-defined mass of lower specific gravity than the cytoplasm.

The location of the Golgi substance can vary with type of cell, but its location is usually characteristic in any given cell type. In secretory cells, the Golgi substance is well developed and is generally polarized between the nucleus and the secreting end of the cell. In addition, the Golgi substance exhibits variations in size corresponding to a state of hypo- or hyperfunctioning of a cell, and secretory granules within the vacuoles of the substance can often be observed with the electron microscope. This evidence has led to the concept that the Golgi substance acts as a surface membrane and system of sacs for collection or condensation of the cellular secretory products (Fig. 2.9). Histochemical investigations of the Golgi substance have revealed that lipid and protein are present, probably as a lipoprotein complex.

Lysosomes

The dense, subcellular bodies known as lysosomes have been found to be the site of many hydrolytic enzymes. Lysosomes range in size from 0.25μ to 0.50μ and are composed of an internal cavity, an outer limiting membrane, and many electron-dense granules. Because the lysosomes contain enzymes which, if free, could hydrolyze the cytoplasm, it is believed that they are essentially closed endoplasmic reticular sacs in which many of the cellular digestive processes occur. In addition, lysosomes are also thought to function as digestive organs during phagocytosis and pinocytosis. Particles or organisms to be phagocytized are done so by the formation of an inpocketing of the cell membrane. After the inpocketing is complete and has broken off from the membrane, it fuses with the sac-like lysosome. Wide acceptance is given to the evidence that the lysosomal membranes are derived directly from the endoplasmic reticulum and the granules from ribosomally synthesized enzymes. Intralysosomal digestion then occurs, resulting in the formation of a *residual* body containing undigestible material that is then egested from the cell (Fig. 2.10).

Centrioles

In the interphasic cell centrioles, self-replicating organelles, usually lie near the nucleus

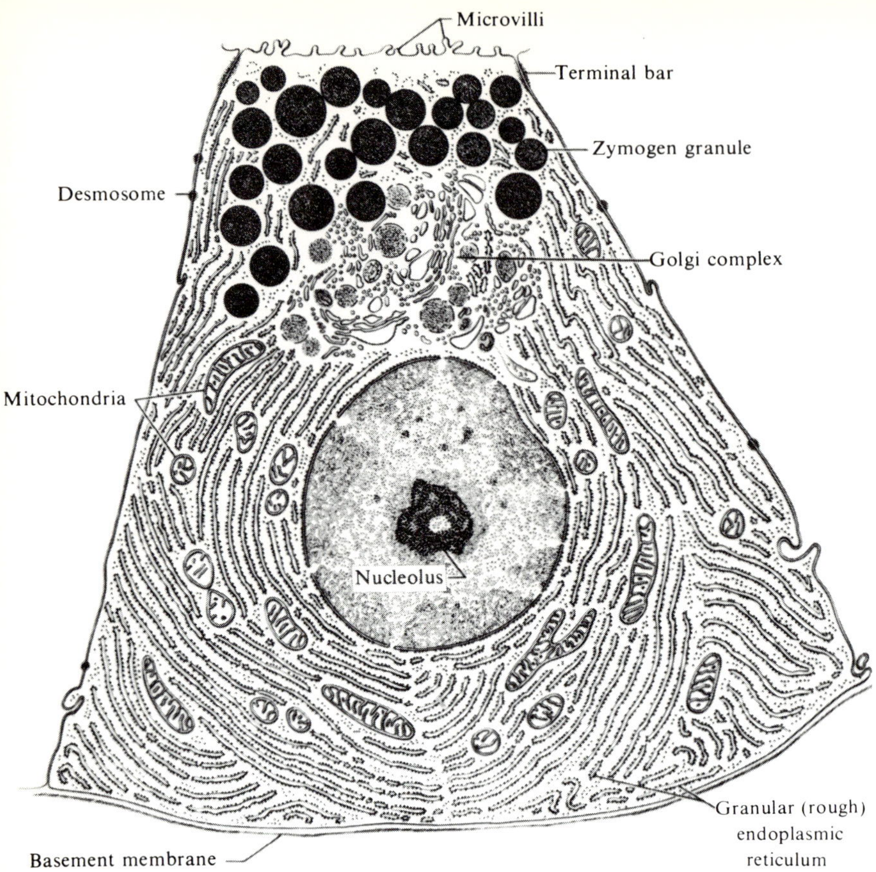

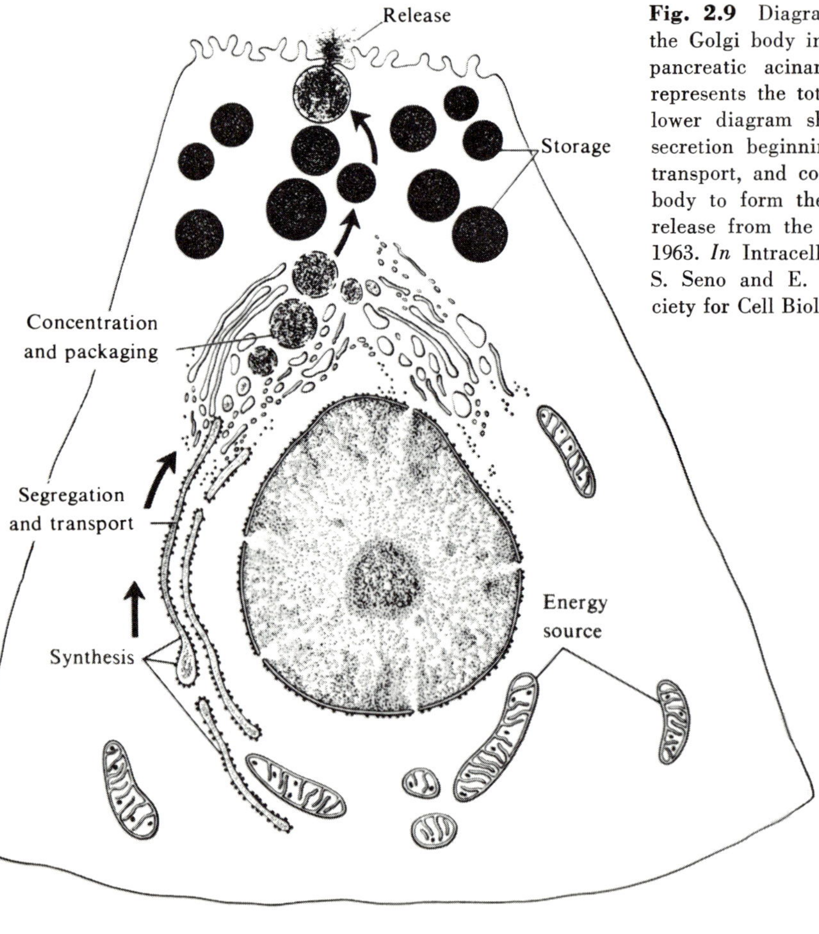

Fig. 2.9 Diagram of the proposed rate of the Golgi body in the secretory pathway of a pancreatic acinar cell. The upper diagram represents the total structure of the cell. The lower diagram shows the current picture of secretion beginning with ribosomal synthesis, transport, and concentration within the Golgi body to form the zymogen granules prior to release from the cell. (After D. W. Fawcett. 1963. *In* Intracellular membranous structures, S. Seno and E. V. Cowdry, eds., Japan Society for Cell Biology, Okayama.)

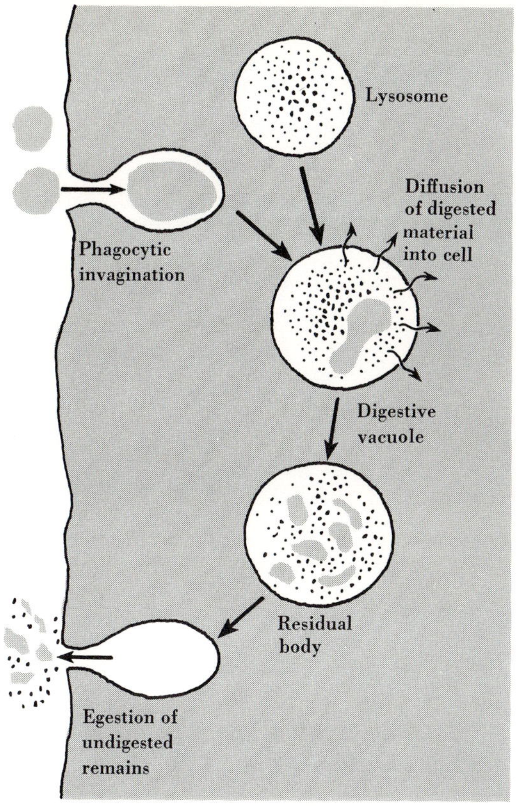

Fig. 2.10 The life cycle of a lysosome.

and as close as possible to the center of the cell. They are surrounded by a zone of dense, globular cytoplasm which, together with the centrioles, is known as the *centrosome.* With the light microscope, the centrioles appear to be granular, there being either one or two in each centrosome. In electron micrographs each centriole is seen to be a hollow cylinder, about 15mμ in diameter by 300 to 500mμ in length. Embedded in the dense wall of each cylinder are nine longitudinal tubules (Fig. 2.11).

At present, our rather meager, functional knowledge of centrioles assigns them a role in determining the polarity of various fibrous elements of the cell. Radiating out from the centrioles are the fibrils of the aster, and during division, the centrioles appear as the poles of the spindle. In addition, centrioles—or structures of similar morphology, the basal bodies—are usually present at the bases of other fibrillar structures, as cilia and flagella.

Microtubules

Small tubules (200–270A in diameter) are often found in interphasic cells. They usually occur in aggregates either near the cell surface or in the area of the centrioles (Fig. 9.2). The microtubules are thought to be cytoskeletal elements influencing cell shape by imparting stiffness to the regions they inhabit. The microtubules which make up the internal structure of cilia and the mitotic spindle are associated with contractility, and all microtubules may function as intracellular transport channels.

Cytoplasmic Inclusions

Classified as inclusions (rather than living, relatively permanent parts of the cell) are many diverse granules and crystals. They occur in cells (usually enclosed within a membranous vacuole or sac, as a result of ingestion, absorption, or accumulation of nutritional elements) and are integral parts of cellular synthesis and secretion. While granules of nonmiscible droplets of lipid, glycogen, pigments, and derived or conjugated protein can be demonstrated histologically by means of specific fixatives and stains, and can often be removed from the cell by differential centrifugation, simple protein or polypeptide reserves are usually soluble in and incorporated as components of the amorphous hyaloplasm.

Histologically defined, secretion granules actually encompass three different functional states. Some granules are to be used within the cell, and, hence, do not leave its borders. Granules or inclusions that do leave the cell may be excretory or secretory in nature, and can be distinguished on physiological and chemical grounds. Products of cellular secretion are usually molecules which have been synthesized under energy-requiring conditions and which must often leave the cell against a concentration gradient. Excretory granules are molecules unchanged by the cell, or products of pathways that produce rather than consume energy. These substances usually move with a favorable concentration gradient.

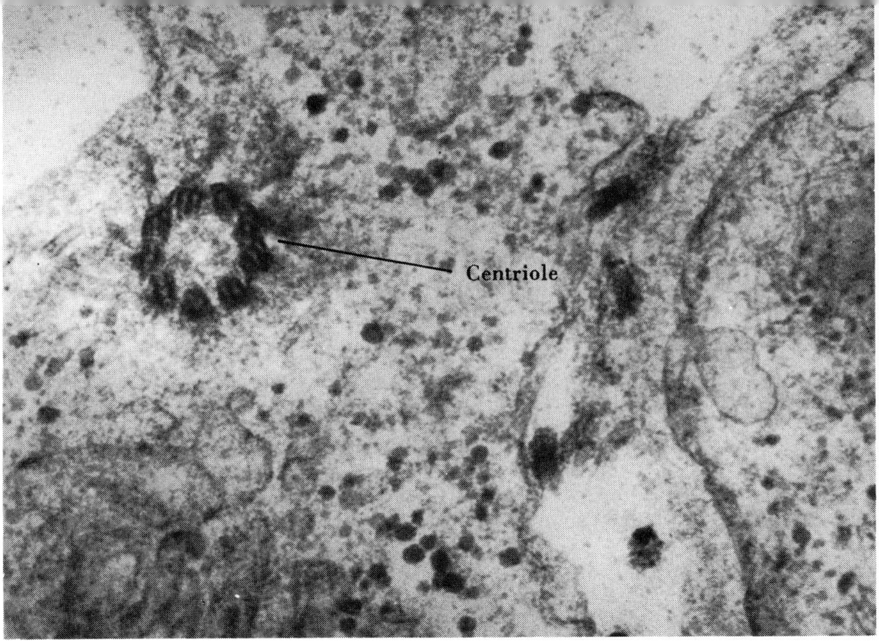

Fig. 2.11 Cross section of a centriole in a cell of the turtle aorta. Note the tubular arrangement of the centriolar wall. Only one centriole appears in this electron micrograph. X35,000. (Courtesy of Floyd O. Craft.)

Although secretion granules have a long history of recognition by light microscopists, and have been the object of extensive electron microscopic study, their life cycles are still not completely known. It is assumed that protein secretions are produced by the ribosomes and that most other compounds are synthesized by or on the membranes of the endoplasmic reticulum. After production, the granules move to and are concentrated in the vacuoles of the Golgi substance. There is some contrasting opinion that some secretory products are manufactured by the Golgi substance, or, alternatively, that the Golgi substance merely represents a specialized part of the endoplasmic reticulum and, hence, that the secretion is made, accumulated, and concentrated within the endoplasmic reticulum itself. Indeed, some secretory cells lack a well-developed Golgi body. In cells which do exhibit Golgi substance, accumulation and concentration of secretion is effected by removal of water and the granules are seen to become enclosed by rough or smooth membranous sacs. As these sacs move toward the secreting end of the cell, imbibition of water may occur with the resultant vacuolozation of the small sacs. The secretion products leave the cell by (1) transfer through the intact cell membrane, (2) decapitation of small microvilli or larger projections from the cell surface, or (3) release of the product through an opening or pore in the cell membrane (Fig. 2.9).

CELL DIVISION

Cell division is one of the fundamental life processes. One kind of cell division, mitosis or somatic division, is the basis of the multicellular nature of organisms, and for each cell of the organism possessing the same genetic complement as that of the one-cell zygote. Another kind of division, meiosis, occurs during formation of the gametes. The extreme importance of meiotic division lies in the separation of homologous chromosomes and the formation of gametes with only one chromosome of each pair. In this way, the diploid state is maintained from one generation to the next.

In mitotic and meiotic division, both nucleus and cytoplasm divide to form either two cells in mitosis or four cells in meiosis. Nuclear mitosis without cytoplasmic division can also occur, yielding cells which are polyploid, i.e., containing more than one homologous pair of each chromosome. Liver cells are commonly polyploid, with two, four, or eight times the standard or diploid chromosomal number.

Mitosis

During the mitotic cycle of a cell, many changes occur in the nucleus and cytoplasm. As an aid to remembering the order of such changes, mitosis has been divided into four stages. However, it should be noted that these stages are categories of the human mind only; the dividing cell does not stop and go to fit these stages, but rather is continuous throughout the process. Figure 2.12 illustrates the major features of each stage of mitosis.

PROPHASE

During this first mitotic stage the most obvious phenomenon is the thickening and concentration of the chromosomes that had previously been dispersed in the nuclear sap. Visible within each chromosome are two intertwined helical bundles, the chromatids, composed of strands, the chromonemata. As the chromosomes thicken, other changes begin to occur. The nuclear membrane and the nucleoli become disrupted and finally both disappear. The centrioles separate and move to opposite poles of the cell. A new structure, the spindle, makes its appearance at this time. Composed of fibrous proteins that are organized as microtubules, the spindle fibers are seen to converge at each side of the cell in the vicinity of the centriole. Another fibrous structure, the aster, is then seen to radiate out from each centriole. During spindle formation, the chromosomes tend to group near the center of the cell, and after spindle formation

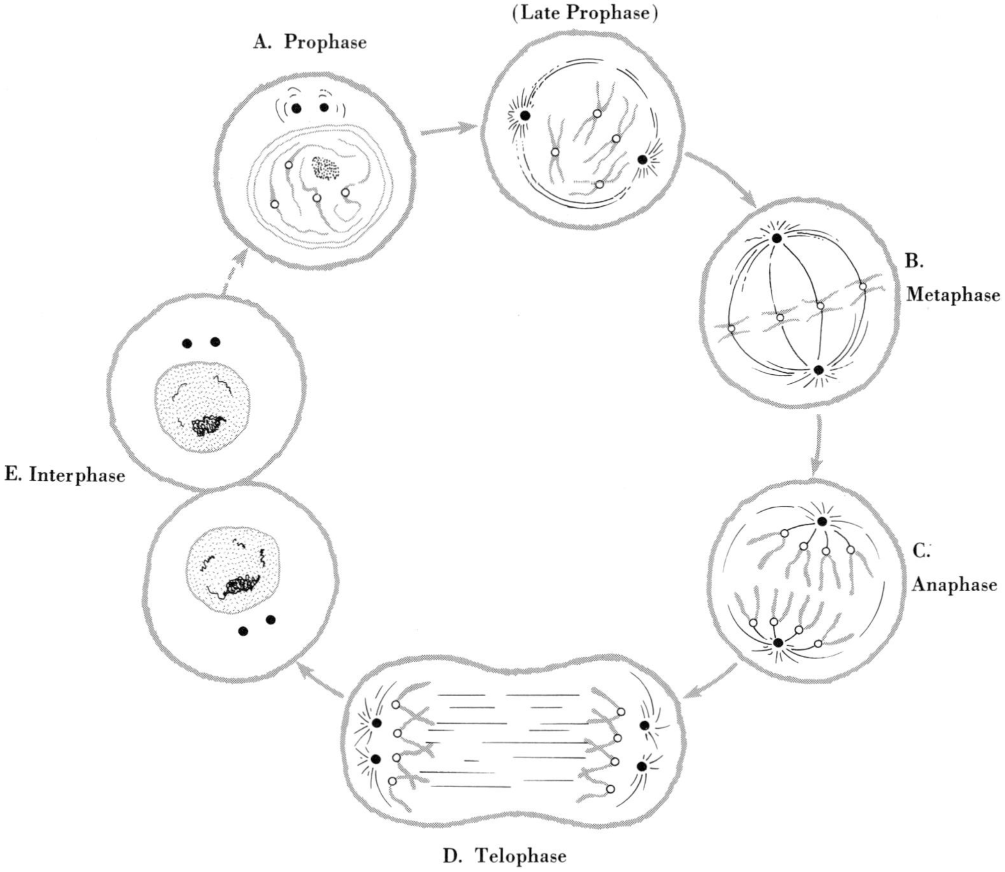

Fig. 2.12 The mitotic cycle. See text for explanation.

is complete, the chromosomes become intimately associated with it.

METAPHASE

Metaphase encompasses the alignment of the chromosomes around the periphery of the midportion of the spindle. The most uniformly aligned parts of the chromosomes are the centromeres or primary constrictions, which almost appear to be attached to the spindle fibers. The arms of each chromosome lie in random arrangement on the spindle, with no relationship between homologous chromosomes.

ANAPHASE

It was commonly thought that during anaphase each chromosome divided longitudinally to form the two daughter chromosomes. We now know that during the previous interphase each chromatid acts as a template for the building of a complementary strand, so that by the time division begins, the chromatids, seen as a single bundle, are really double structures. Hence, before anaphase, a chromosome is composed of two chromatids, each of which is double. During anaphase, the chromosome is seen to split, so that the daughter chromosomes are each composed of two single chromatids. The spindle then elongates and the daughter chromosomes are separated, each moving to an opposite pole. Anaphase also encompasses the beginning of cytoplasmic constriction, or *cytokinesis,* and the arrangement of each chromosomal complement in a cluster at either end of the cell.

TELOPHASE

During telophase, the cytoplasmic constriction is completed with the formation of two daughter cells. Nuclear membranes form around each chromosomal cluster, and nucleoli begin to appear in each nucleus. Toward the end of telophase the chromosomes begin to disperse until each nucleus has a typical interphasic appearance.

INTERPHASE

Interphase is often termed the resting stage, but this is highly inaccurate because during this time the cell is in its metabolically active state. Moreover, because most adult cells spend most if not all of their time in interphase, one would have to say that adult organisms are resting bodies, rather than the dynamic creatures we like to think of ourselves as being.

Meiosis

Meiotic division occurs only within the germinal epithelium of gonads, and can be considered as a double division accompanied by a single chromosomal replication. Although the processes which lead to the production of the ovum and spermatozoan differ in some ways, the same meiotic cycle applies to both. Meiosis, like mitosis, can be divided into several stages, the principal differences between the two taking place during the first meiotic prophase. The main features of meiosis are shown in Fig. 2.13.

PROPHASE I

Meiotic prophase encompasses may important chromosomal changes, and for convenience Prophase I is further subdivided into five sequential phases.

LEPTOTENE. As prophase begins, the chromosomes appear as slender strands, partially condensed.

ZYGOTENE. During this period pairing of homologous chromosomes takes place. A close association, known as *synapsis,* occurs between alleles in each chromosomal pair. It should be noted that pairing of homologues is a feature characteristic only of meiosis, and represents an important difference between meiosis and mitosis.

PACHYETENE. The chromosomes begin to evidence great contraction at this time, and it can be observed that each chromosome is a double structure, the replication having taken place before the onset of meiosis, but not being visible until this time. As homologous chromosomes are paired, each chromosomal set contains four strands or bundles, known as a *tetrad.* Exchanges of material between any

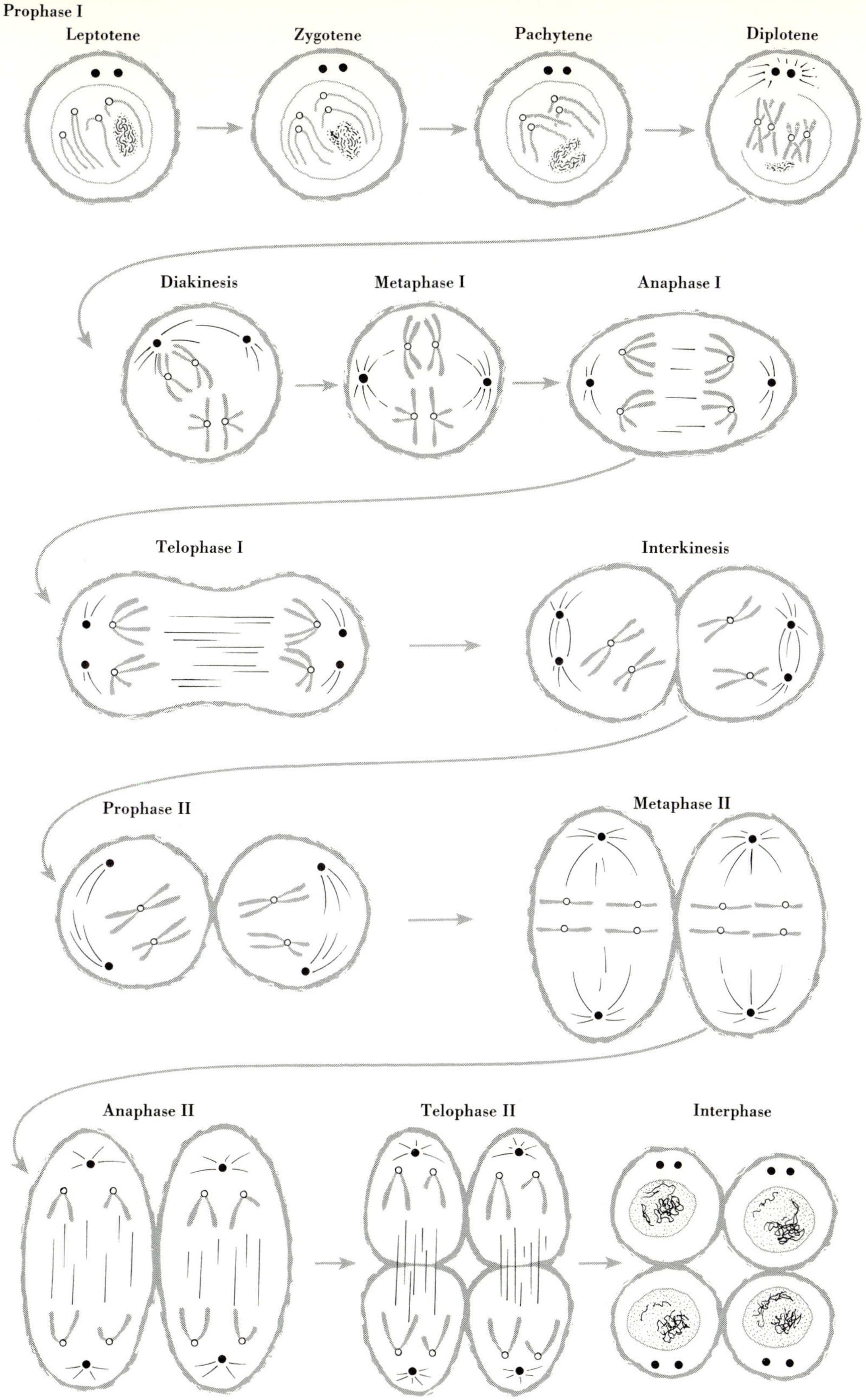

Fig. 2.13 Meiotic division. See text for explanation.

two strands of homologous chromosomes can occur, this exchange being termed *crossing-over*. The physical region where crossing-over takes place is called the *chiasma* (plural: chiasmata). Crossing-over is one of the important features of meiosis because the recombinant chromosomes, i.e., those affected by crossing-over and which contain part of one homologue and part of the other, are different from those of either parent. Hence, crossing-over leads to an increase in genetic variability.

DIPLOTENE. During diplotene, the chromosomes contract even further and the homologues in each pair begin to separate.

DIAKINESIS. As Prophase I nears its end, contraction is at its greatest. The nuclear membrane and nucleoli have disappeared and the chromosomal pairs are scattered throughout the cell. Homologues are still paired, but the association is usually maintained only at the chiasmata. As in mitotic prophase, spindle formation proceeds normally and the tetrads migrate toward it.

METAPHASE I

During this stage, each tetrad becomes aligned around the periphery of the spindle equator with centromeres of homologous chromosomes facing opposite poles.

ANAPHASE I

Separation of homologous chromosomes occurs. As noted before, each chromosome consists of two chromatids, but because crossing-over may have occurred, the sister chromatids may not be identical throughout their length.

TELOPHASE I

As in mitotic telophase, each polar group of chromosomes tends to form a distinct cluster, although this is more evident in some species of organisms than in others.

INTERKINESIS

The duration of this essentially interphasic stage again is species dependent, as in the degree of constriction which divides the original cell in two. Some chromosomal dispersion may occur.

PROPHASE II

Essentially identical to mitotic prophase, the chromosomes contract, become highly coiled, and spindle formation takes place in each daughter cell. In many species this stage is bypassed, the chromosomes entering directly into Metaphase II.

METAPHASE II

Alignment along the spindle equator again is evident, and division of the centromeres holding together the two chromatids in each chromosome occurs.

ANAPHASE II

The sister centromeres begin migrating to opposite sides, dragging their chromosomal tails behind them.

TELOPHASE II

This stage marks the end of meiosis. The nuclei assume the interphasic appearance, although the cytoplasm of the end products of meiosis, ovum and sperm, have a rather specialized morphology.

From the foregoing description, it should be noted that each of the cells produced contains only one of every chromosomal pair, a phenomenon often called *reduction*. The normal diploid state in which cells have both members of a homologous pair is regained after fertilization of the ovum by the sperm.

SELECTED REFERENCES

De Robertis, E., W. W. Nowinski, and F. A. Saez. 1965. Cell biology. 4th ed. W. B. Saunders Co., Philadelphia.

Fawcett, D. 1966. An atlas of fine structure—the cell, its organelles and inclusions. W. B. Saunders Co., Philadelphia.

Green, D. 1964. The mitochondrion. Sci. Am. 210:63–74.
Lehninger, A. L. 1964. The mitochondrion; molecular basis of structure and function. W. A. Benjamin Co., New York.
Mahler, H. R., and E. H. Cordes. 1966. Biological chemistry. Harper & Row, New York.
Malhotra, S. K. 1966. A study of the mitochondrial membrane system. J. Ultrastruc. Res. 15:14–37.
Porter, K. R., and M. A. Bonneville. 1964. An introduction to the fine structure of cells and tissues. Lea & Febiger, Philadelphia.
Robertson, J. D. 1964. Unit membranes: a review with recent new studies of experimental alterations and a new subunit structure in synaptic membranes. *In* Cellular membranes in development. M. Locke, ed. Academic Press, New York. Pp. 1–81.
Schuster, F. L. 1965. A DNA component in mitochondria of *Didymium nigripes*, a slime mold. Exp. Cell Res. 39:329–345.
Scientific American, vol. 205, no. 3, September 1961.

chapter 3
CONNECTIVE AND SUPPORTING TISSUES

THE FUNCTIONS of binding tissues and organs together, of holding them in position, and of providing avenues for the transport of fluids between them are served by a heterogeneous complex of cells and cell products termed *connective* and *supporting tissues*. From the moment that the three primary embryonal layers (ectoderm, endoderm, and mesoderm) are differentiated, cells proliferate freely from the mesodermal component as *mesenchymal cells*. These cells are irregular in shape (suggesting amoeboid activity) and are generally disassociated from each other, hence being free to wander and to insinuate themselves into the spaces that appear between the more compact tissues and developing organs. As they do this, some come to rest and, under the influence of their immediate environment, undergo various kinds of modification. Others, remaining undifferentiated, continue to wander, to proliferate, and to elaborate an intercellular ground substance which is amorphous and has the physical characteristics of a soft gel and the chemical characteristics of globular protein and soluble mucopolysaccharides. Such a physical and chemical nature is well suited to the easy movement of substances by diffusion.

Mesenchymal cells that come to rest, so to speak, become modified in various ways and to varying extents. Cellular modification resulting in some degree of specialization of function is called differentiation (see Chapter 2) and is accompanied by a progressively commensurate loss of ability to reproduce or to change function. Cells showing a wide spectrum of specialization from undifferentiated mesenchymal cells and reticular cells to highly differentiated chondrocytes, osteocytes, muscle fibers, and some of the more highly differentiated elements of the circulatory system are derived from mesenchyme. Between the extremes one finds macrophages, fibroblasts, and the relatively undifferentiated components of the circulatory system, which presumably can undergo certain modifications that result in the appearance of other cell types. Connective tissue retains much of the primitiveness and potential of the mesenchyme from which it is derived, and so it is important with respect to tissue repair and replacement.

Adult connective and supporting tissues

may be considered to consist of two major components, cellular and intercellular. The latter is usually the more conspicuous.

CELLULAR COMPONENTS

Mesenchymal Cells

Although they are the predominant type in embryonic and fetal connective tissues, mesenchymal cells are rarely found in the adult—and then they appear to be confined largely to the connective tissue which covers blood vessels. Mesenchymal cells of vertebrates generally are characterized by a round or oval nucleus with a fine dusting of chromatin and one or two prominent nucleoli. The cytoplasm is agranular and so lightly acidiphilic it is difficult to discern its extent. The cell outline is irregular or stellate, with fine protoplasmic processes which may branch (Fig. 3.1a). Mesenchymal cells are smaller than the immature fibroblasts, which they resemble closely. Electron micrographs of mesenchymal cells show a largely amorphous cytoplasm with very little membranous structure, i.e., "clear" cells (Fig. 3.1b).

Fibroblasts

Fibroblasts are perhaps the most important cellular component of connective and supporting tissues, for they form the several kinds of fibers and presumably the amorphous ground substance as well (*see below*). The term fibroblast is reserved for the cells that are actively producing fibers; *fibrocyte* is used for those that have completed this function. Inasmuch as there is very little difference in the histological appearance of the two types, the term fibroblast is usually used loosely to apply to either. The cell form is flat. with branching processes, and fusiform in side view (Fig. 3.2). The nucleus is oval, with fine chromatin granules and with one or two conspicuous nucleoli. The cytoplasm is rather clear and stains lightly with acidic dyes in fibrocytes, but it is quite basophilic in fibroblasts that are actively secreting fibers. Electron micrographs of active fibroblasts (Fig. 3.1d) show dense areas of rough endoplasmic reticulum, where fiber synthesis is taking place. The fibers are evidently formed in the cytoplasm immediately adjacent to the cell membrane and extruded therefrom, although there is some controversy about this.

Macrophages

Also known as histiocytes and occasionally called clasmatocytes, macrophages (Fig. 3.1c) represent the second most numerous cells in many kinds of connective tissue, specifically the loose and richly vascularized types. Although they are usually immobile, they can respond to inflammatory stimuli and become actively amoeboid. The name, macrophage, derives from the large size and the cells' pronounced ability to ingest particulate matter. As scavengers they constitute an important part of the organism's defense against tissue debris, microorganisms, and other foreign bodies. They occur not only in connective tissue *per se*, but also in association with the sinusoids of the liver and spleen as well as, to a more limited extent, in other organs. They represent one of the elements of the so-called reticuloendothelial system, which will be discussed below.

Macrophages have an irregular cell outline with short, blunt processes. Their nuclei are oval and often indented, with coarse clumps of chromatin obscuring the small nucleoli. Because of their phagocytic activity, the cytoplasm frequently contains lipid inclusions and granules of iron-rich degradation products of phagocytosed erythrocytes. Such phagocytosis has been seen to occur by pseudopodial activity. Electron micrographs of free macrophages of the rat show a normal complement of mitochondria and prominent Golgi material. Free ribronucleoprotein granules form the bulk of the cytoplasmic ground substance. The endoplasmic reticulum is poorly developed in rats, but well developed in mice, especially

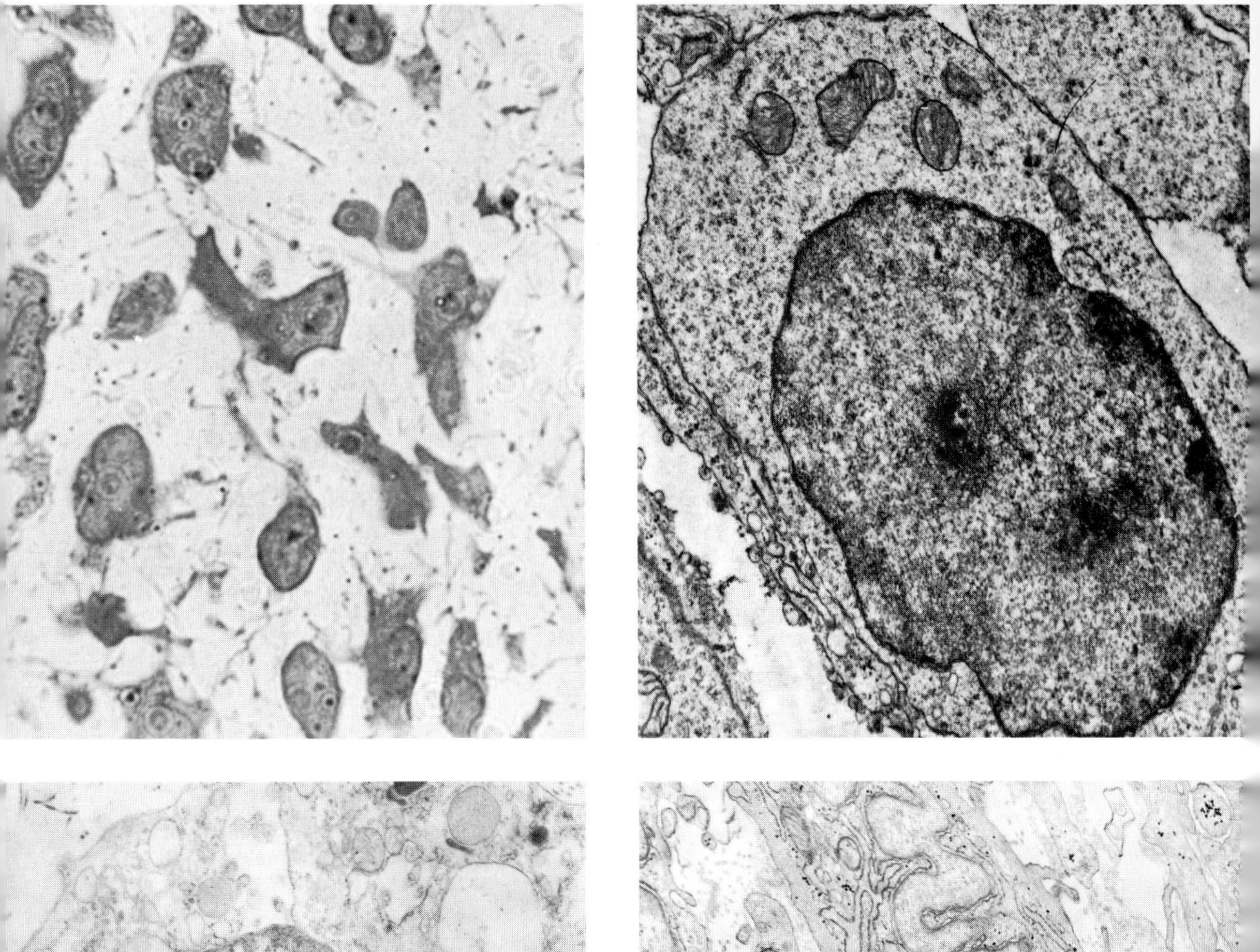

Fig. 3.1 Cells of connective tissue: *a*, mesenchymal cells from a young turtle (X1200); *b*, mesenchymal cell from the thymus of a fetal hamster (X6000); *c*, macrophage from the tonsil of a newborn rabbit (X6000); and *d*, fibroblast from an embryonic chick (X4000). The nucleus of the fibroblast is single, but so irregular in shape that it appears in several parts in thin sections. The distended cisternae of rough-surfaced ER shown here is typical of active fibroblasts.

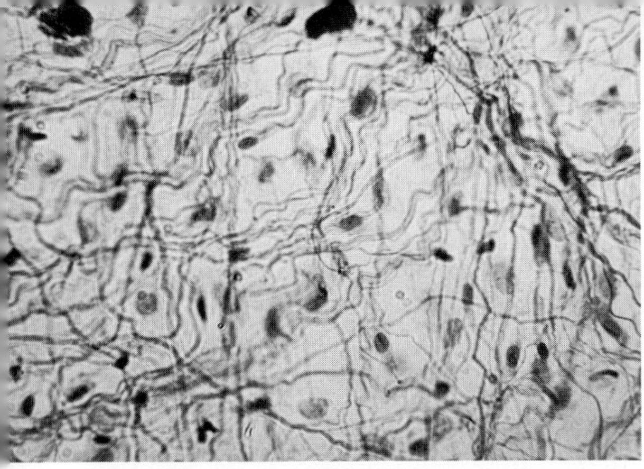

Fig. 3.2 Photomicrograph of a spread of areolar connective tissue, showing collagenous and elastic (*wavy*) fibers, lightly stained fibroblasts, and heavily stained mast cells. X450.

during active phagocytosis. Lamellated inclusion bodies of unknown function have been described in fixed macrophages of the lung.

Mast Cells

Mast cells received their name from Paul Ehrlich in 1879 and since that time they have been a more or less constant subject of interest for histologists and physiologists alike. Mast cells occur in certain invertebrate animals, as well as in all of the vertebrate classes. They vary considerably in form, being profusely branched and irregular in some species, as amphibians, but most generally they are spheroidal. In all species the cytoplasm of mast cells is densely packed with basophilic, metachromatic, water soluble granules that obscure the nucleus. The granules are "dust-like" in some amphibians, somewhat larger in birds, and coarse in most other vertebrates. Electron micrographs of mast cells show a normal development of cytoplasmic organelles dispersed among the dense population of granules (Fig. 3.3). The granules are electron dense, but an internal fine structure has been observed in the granules of mammalian mast cells which has been described as reticular or vacuolar and composed of strands consisting of fine granules embedded in a moderately dense matrix. In human beings at least three kinds of granules have been described on the basis of differences in internal fine structure. In all cases the granules appear to have a definite limiting membrane.

Mast cells arise from mesenchyme and possibly from resting macrophages. Under experimental conditions they may differentiate from fibroblasts. Their antecedent in fishes has been identified as a particular cell called a "blood mast cell." Otherwise little is known about their derivation. They are found in nearly all tissues and organs, but their densest population is found to be along blood vessels in loose connective tissue. In some urodeles and some mammals mast cells are numerous in the serous fluid of the body cavity.

Mast cells are sometimes classed as *trephocytes,* along with lymphocytes and plasma cells. Trephocytes are nonendocrine cells which synthesize and store substances essential for physiological processes. Mast cells have been shown to produce and store heparin and histamine and possibly serotonin, and they participate in the synthesis of hyaluronic acid, collagen, and mucin. Whether they may be amoeboid is controversial; some investigators report that the cells can intrude pseudopodia into the lumina of small blood vessels and others disclaim such activity. Their importance in maintaining the physical and functional integrity of blood vessels by means of their specific secretions is generally recognized.

Fat Cells

Common, although not glamorous, constituents of connective and supporting tissues are fat cells. In amphibia they are concentrated for the most part in discrete masses called fat bodies, which are tucked in spaces where they will not be in the way, usually in the vicinity of the gonads and kidneys. In reptiles, birds and mammals, however, fat cells are more widespread, being dispersed singly or in clumps throughout the loose connective tissues, primarily of the viscera and the skin, and in more or less concentrated depots in the mesenteries and the marrow cavities of bones. When concentrated to form sheets or masses, they constitute *adipose tissue,* as described below.

Fat cells are derived directly from mesenchymal cells that accumulate fatty droplets in their cytoplasm. At one point in their

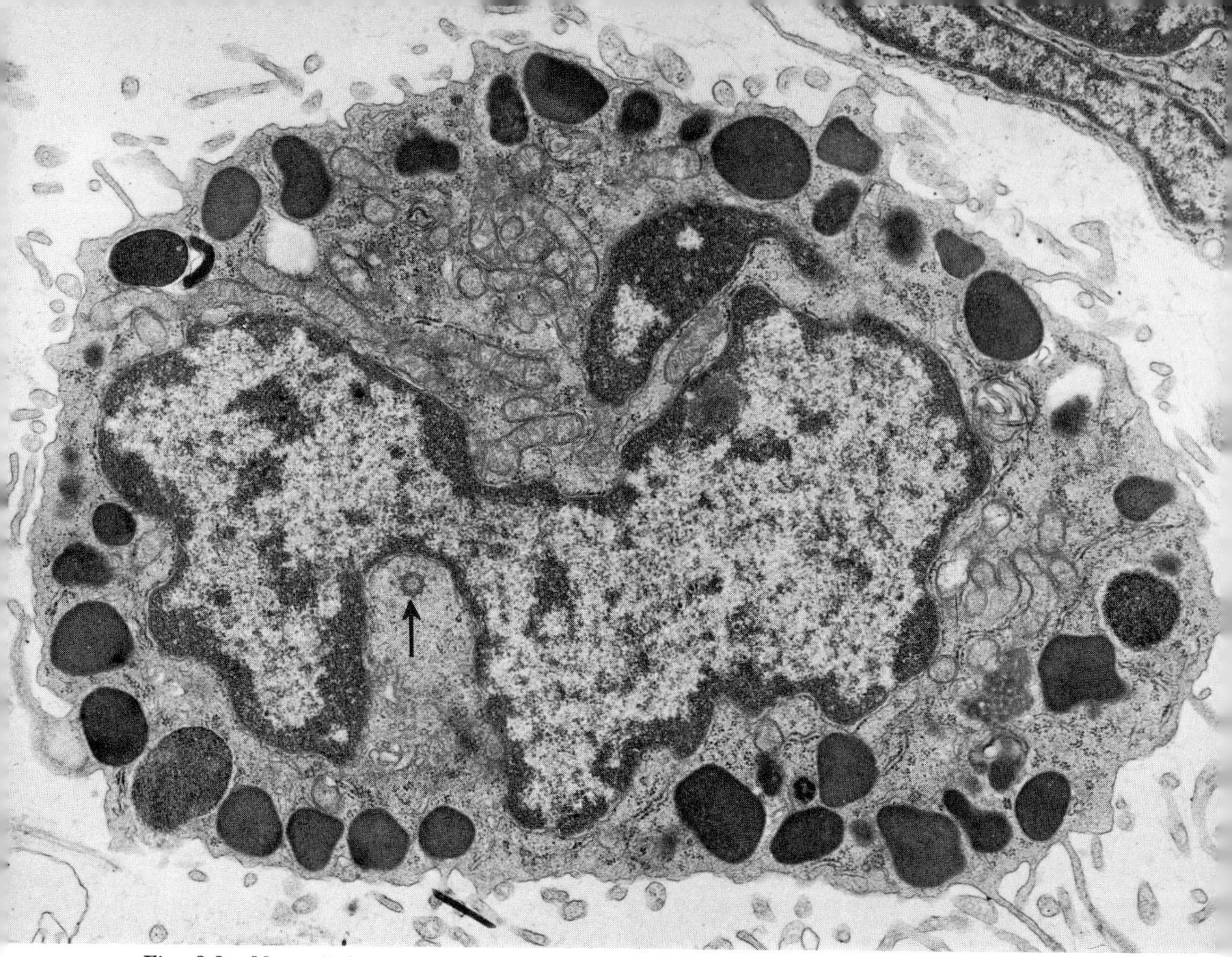

Fig. 3.3 Mast cell from a 10-day-old mouse (X10,000), showing cytoplasmic granules. Note the centriole (*arrow*) and the pseudopodial processes. (Courtesy of Philias Garant, Harvard School of Dental Medicine.)

differentiation the cytoplasm becomes almost completely filled with droplets, at which stage the fat is known as *multilocular fat* (Fig. 3.4). Adult fat, or *unilocular fat,* results from the coalescence of fat droplets in the cytoplasm to form a single vacuole which distends the cell, forcing the nucleus and the remaining cytoplasm into an eccentric position. Hence the adult fat cell has a signet-ring appearance (Fig. 3.4). Fat cells are not inert, as one might infer from their storage function. Electron micrographs show numerous mitochondria in the cytoplasm even in adult fat cells, and glycogen is abundant in actively synthesizing fat cells.

A special kind of multilocular fat, yellowish or brownish in color, is found in specific sites in the adults of some species of vertebrates. Called *brown fat,* it was first discovered in hibernating rodents and for this reason and because of its glandular appearance it was thought to function in hibernation. It is also found in nonhibernating species, however, as in rats and primates. Brown fat does not respond to changes in nutritional state as ordinary fat does, but it does respond to the adrenocorticotropic hormone. Some investigators believe it contains stored cortical steroids, which may account for its color, although the color is generally assumed to be due to a pigment.

Miscellany

There are several other kinds of cells which are normal constituents of connective and supporting tissues, but their description will be deferred to other contexts. These include *leukocytes, plasmacytes, chondroblasts, osteo-*

blasts, osteocytes, osteoclasts, and chromatophores.*

INTERCELLULAR COMPONENTS

The intercellular components may be grouped conveniently into two major classes, fibrous and amorphous.

Fibrous Elements

Four different kinds of fibrous elements contribute to the intercellular substance of connective and supporting tissues. With certain exceptions, they are synthesized by fibroblasts. There is some evidence, for example, that the collagenous fibers of the endoneurium are synthesized by the Schwann cells.

COLLAGEN

The most abundant of all animal fibers, and possibly of all animal proteins, is *collagen.* Collagenous fibers are characterized by a tensile strength which has been estimated to be as high as that of steel wire. The stability of the collagen molecule is attributable to the amino acids, proline and hydroxyproline, which constitute upwards of one-quarter of the links in the collagen molecule. Hydroxylysine and glycine are also consistently occurring amino-acid components of collagen. The constancy of the molecular composition of collagen is illustrated by the fact that every fourth position in the collagen chain is evidently occupied by glycine, with either proline or hydroxyproline in the adjacent position. The collagenous fiber has been postulated to be formed by the parallel alignment of repeating molecular units called *tropocollagen,* which are staggered so as to overlap by one-quarter of their length (Fig. 3.5a). The characteristic banding of collagenous fibers seen by electron microscopy (Fig. 3.5b) reflects the periodicity of the overlap. The tropocollagen molecule is about 2800A long, and the observed periodicity of collagen banding is in the order of 650–700A.

Collagenous fibers are glistening white in the living state, and so they are often referred to as "white fibers." They may be loosely arranged, such as in loose connective tissues, or

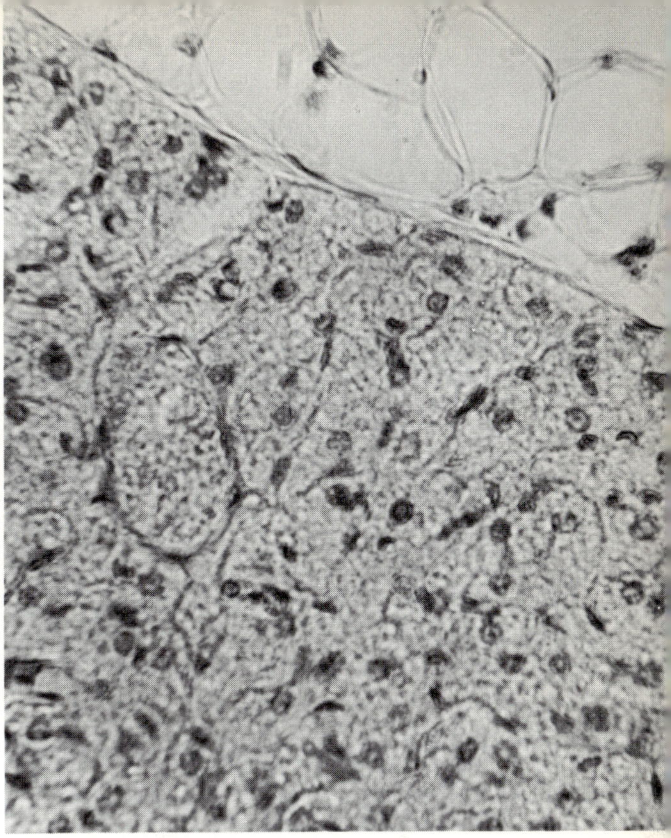

Fig. 3.4 Multilocular fat from a young hamster, with a small area of unilocular fat. X450.

compactly arranged, such as in ligaments and tendons. Their arrangement in bone is extremely precise, following the lines of the mechanical stresses to which the bone is normally subjected and forming the scaffolding for the similarly precise alignment of the bony spicules. Collagenous fibers range from 100mμ to 500mμ (1.0mμ = 0.001μ) in diameter, and they are often aggregated in bundles up to 100μ or more in diameter. The fibers stain readily with eosin and other acid dyes (see Chapter 1) and are readily identifiable by their birefringence in polarized light. Although collagen is not soluble in cold water, it is readily hydrated in boiling water, yielding *gelatin.*

ELASTIN

Elastic fibers (Fig. 3.2) are composed of the protein *elastin,* and possibly also of an associated mucopolysaccharide, for which the name *elastomucin* has been suggested. Elastic fibers are wiespread throughout connective tissues and are especially concentrated in certain areas, as described below. Macroscopically they appear yellow, and they can be demonstrated for light microscopy by several staining methods, of which the most specific are Unna's acid orcein method and the resorcinfuchsin method. Triple staining procedures, as Mallory's and Masson's, differentiate clearly the distribution of collagenous and elastic fibers. Nevertheless, variations in staining reaction are encountered with elastic fibers from different locations in the same organism. Variations are especially pronounced with PAS.

Elastic fibers differ from collagenous fibers in several ways. Elastic fibers branch and anastomose freely with each other, whereas collagenous fibers do not. Elastic fibers exhibit very faint birefringence, or none at all, except when they are stretched. Whereas collagenous fibers have practically no modulus of stretch, elastic fibers can be stretched as much as two-and-one-half times their normal length and show nearly complete recovery. The relative lack of birefringence reflects absence of any highly oriented internal structure, which is confirmed by failure of the electron microscope to reveal any periodicity. X-ray-diffraction studies by Astbury, Bear, and others have failed to show the presence of any pattern that could not be attributed to collagen contamination of the sample.

Although elastin is chemically similar to collagen in that it includes glycine, proline, and hydroxyproline in its composition, it has only about one-seventh as much of the last compound as collagen, and, again, this can be attributed to collagen contamination. Moreover, the majority of the amino-acid residues of elastin are monoaminomonocarboxylic acids, thus precluding the possibility of elastin's being as highly polar as collagen. Two unusual amino acids, desmosine and isodesmosine, have been demonstrated as unique components of elastin, and a third compound, lysinonorleucine has also been reported. Many authorities believe that elastic fibers are structually heterogeneous, consisting of an outer coat and matrix of α-protein and inner fibers of a highly polymerized material, with possibly some mucopolysaccharide and elastomucin, as well.

RETICULIN

The exact status of reticular fibers is not clear. Apparently a physicochemical complex of pro-

Fig. 3.5 The collagen fibril. The staggered alignment of molecules of tropocollagen is shown in *a*. The interval from *x* to *y* is the length of one molecule (about 2800A), and the overlap interval from *y* to *z* (about 650–700A) represents about one-quarter of the length of a molecule. This overlap is responsible for the periodic banding shown in *b*. A collagen fibril from the skin of the shark, *Mustelus canis*, is negatively stained in *b* with 2 percent phosphotungstic acid. X110,000. (Courtesy of F. B. Merk.)

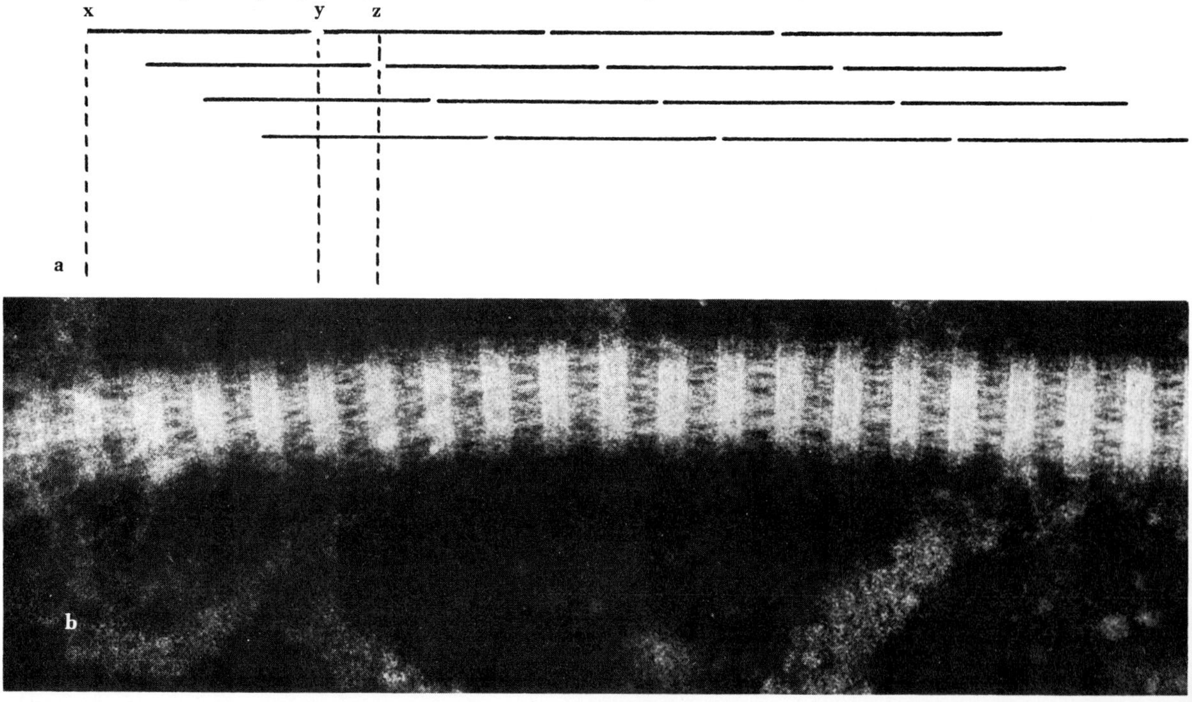

tein and mucopolysaccharide, as demonstrated by histochemical reactions, the fibers are very similar to collagenous ones in all ways except for their dimensions and the amount of mucopolysaccharide present. The term reticulin is used by some authorities to designate the fibers themselves, and by others as the name of the material of which the fibers are composed. Electron miscroscopy and X-ray-diffraction studies indicate that reticular fibers consist of very fine, relatively unoriented fibrils having a diameter of about 100A and a periodicity identical with that of collagen. The fibrils appear to be embedded in a membrane that gives a strong positive reaction to PAS and to silver impregnation and so is presumed to consist of mucopolysaccharide. Reticular fibers differ from collagenous fibers in this respect, therefore, and in the fact that they do not stain with eosin or the other commonly used acid dyes; also, they branch, whereas collagenous fibers do not.

Reticular fibers are found to be associated most commonly with rather primitive cells, such as mesenchymal and fat cells and cells of the reticuloendothelial system, which some investigators interpret as indicative of their being immature collagenous fibers. They are also found associated with delicate nerves, capillaries, and the basement membranes of epithelia.

AMYLOID

When first described, *amyloid* was believed to be a starchlike substance, hence its name; it is now known to consist of protein and glycoprotein. Little is known about its distribution, although it has been observed in all mammalian species in which it has been sought and in several species of anseriform birds. It was thought to be a polysaccharide because it reacts with iodine after brief hydrolysis with sulfuric acid, but hyaluronic acid does not affect it and it stains very lightly at best by the PAS method. Amyloid is strongly birefringent, indicating some organized structure, and its characteristic green color in polarized light is presumed by some investigators to indicate that its molecules are only very slightly hydrated (Fig. 3.6a). In contrast with collagen, fibers of amyloid are totally lacking in hydroxyproline. Amyloid is currently the object of intense investigation because it is most prevalent under pathological conditions, such as rheumatoid arthritis and chronic infectious diseases, as well as in aging. Under such conditions it frequently accumulates between the vascular endothelium and the basement membrane (Fig. 3.6b). Its status as a normal constituent of connective tissue has not been established as yet, although there have been recent indications that small amounts of it do occur in some normal tissues.

Amorphous Elements

The relative amount of amorphous ground substance varies considerably in different kinds of connective and supporting tissues. In the umbilical cord of mammals and in the matrix of hyaline cartilage it constitutes the major part, whereas in ligaments and tendons it is almost lacking. Little is known about the role of the amorphous ground substance in bone; most of the extracellular mass consists of crystalline bony spicules and collagen. Moreover, the composition and the nature of the amorphous ground substance varies from organ to organ. In the vitreous body of the vertebrate eye and in the umbilical cord of mammals it consists almost solely of hyaluronic acid, and in other sites it includes protein and the other mucopolysaccharides in varying amounts and combinations. In loose connective tissue the amorphous ground substance is a plastic fluid rich in mucopolysaccharides, while in dense connective tissue the mucopolysaccharides contribute very little; they are represented mainly by chondroitin sulfate.

Little work has been done with the proteins of the intercellular compartment of connective tissues, aside from the fibrous elements, and so little is known about them; however, a great deal has been learned about the mucopolysaccharides. The latter can be extracted from fresh or dried samples of connective tissue with dilute alkaline or saline solutions or by digestion with proteolytic enzymes. These methods yield in addition to the muco-

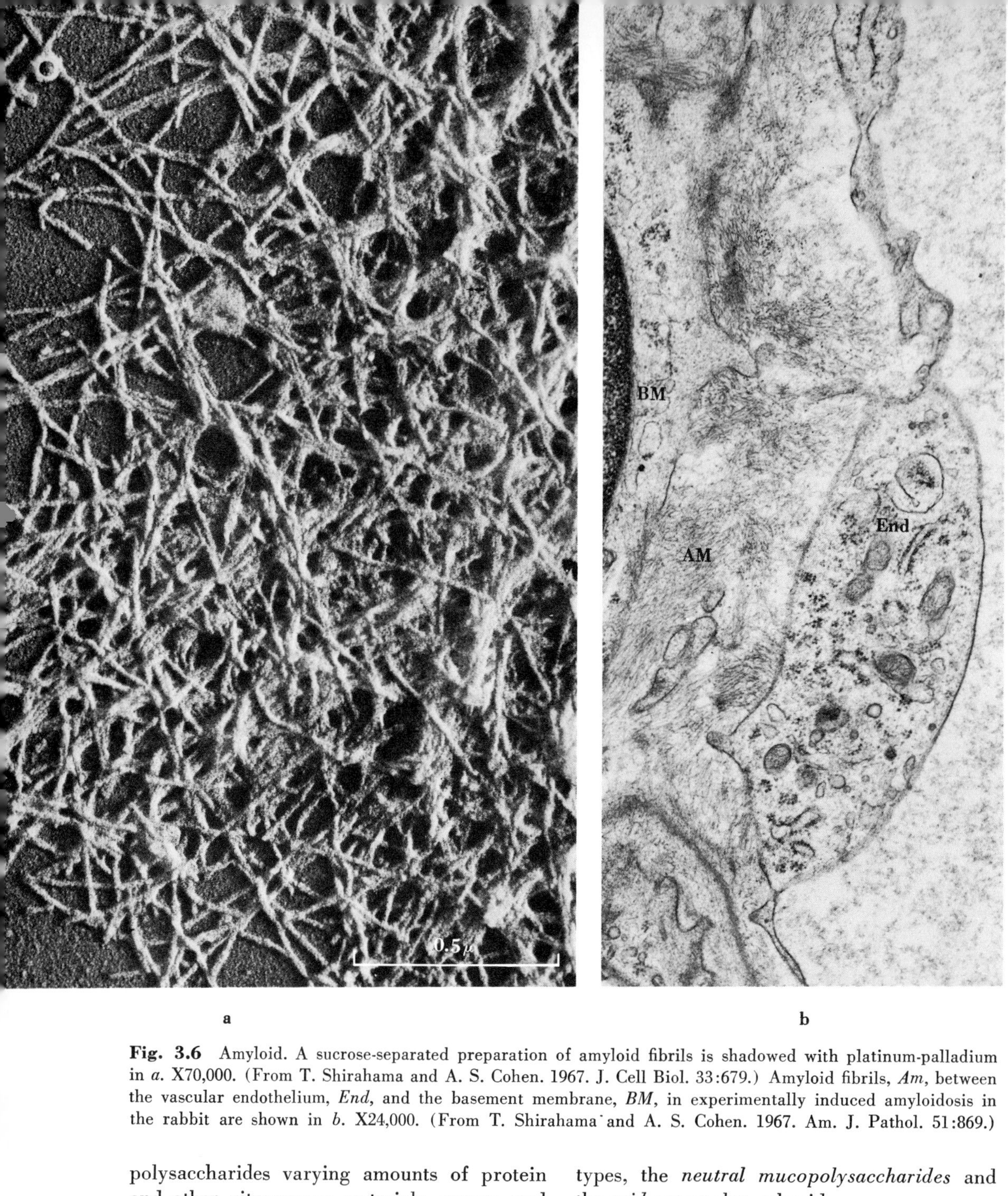

a b

Fig. 3.6 Amyloid. A sucrose-separated preparation of amyloid fibrils is shadowed with platinum-palladium in *a*. X70,000. (From T. Shirahama and A. S. Cohen. 1967. J. Cell Biol. 33:679.) Amyloid fibrils, *Am*, between the vascular endothelium, *End*, and the basement membrane, *BM*, in experimentally induced amyloidosis in the rabbit are shown in *b*. X24,000. (From T. Shirahama and A. S. Cohen. 1967. Am. J. Pathol. 51:869.)

polysaccharides varying amounts of protein and other nitrogenous materials, sugars, and complexes of protein and mucopolysaccharide. The nature of the bonds in these complexes, although poorly understood, is of unquestionable importance in determining the variety and other physical characteristics of the ground substance.

The mucopolysaccharides are nitrogen-containing heteropolysaccharides of two major types, the *neutral mucopolysaccharides* and the *acid* mucopolysaccharides.

NEUTRAL MUCOPOLYSACCHARIDES

Neutral mucopolysaccharides consist of N-acetylhexosamine and hexose (see Chapter 1). They include the mucopolysaccharides of bacteria and the so-called *mucoids* (the immunologically specific blood group substances and the amorphous ground substance of reticular connective tissue). Mucoids are also found in the cornea of the vertebrate eye.

ACID MUCOPOLYSACCHARIDES

Acid mucopolysaccharides may be simple or complex. The simple acid mucopolysaccharides consist of N-acetylhexosamine and hexuronic acid. Examples are hyaluronic acid and chondroitin. Hyaluronic acid consists chemically of repeating units of N-acetyl-glucosamine and glucuronic acid in various degrees of polymerization by glycosidic bonds to form an unbranched polysaccharide (Fig. 1.3). Chondroitin is similar, except that it substitutes N-acetyl-galactosamine for N-acetyl-glucosamine. Hyaluronic acid occurs in significant amounts in loose connective tissue, skin, synovial fluid, the vitreous body, the umbilical cord, and in small amounts in tendons. An enzyme, hyaluronidase, has been isolated from testicular extracts, certain snake venoms, and certain bacteria; it attacks hyaluronic acid vigorously, opening the glucosidic bonds. This enzyme also has a hydrolyzing effect on some of the other mucopolysaccharides, principally chondroitin.

Chondroitin occurs in significant amounts in the cornea and is suspected of occurring in small amounts in other kinds of ground substance. The possibility that chondroitin is a precursor to the chondroitin sulfates has been suggested.

The complex acid mucopolysaccharides contain sulfate in addition to N-acetylhexosamine and hexuronic acid, although some may contain phosphate and one, keratosulfate, substitutes galactose for the hexuronic acid. Examples are *chondroitin sulfate* of cartilage (N-acetylgalactosamine plus glucuronic acid plus sulfate), chondroitin sulfate of skin (N-acetylgalactosamine plus L-iduronic acid plus sulfate) and *keratosulfate* (N-acetylglucosamine plus galactose plus sulfate). Three kinds of chondroitin sulfate, *A*, *B*, and *C*, have further been differentiated on the bases of optical rotation, solubility, and response to hyaluronidase digestion. Keratosulfate has been isolated only from bovine corneas.

The functions of the mucopolysaccharides in the amorphous ground substance appear to be two. Hyaluronic acid, having a strong affinity for water, becomes richly hydrated, forming spheres of about 2100A in diameter that are easily deformed by flow, and thus act effectively as a lubricant and shock absorber. The chondroitin sulfates, on the other hand, do not hydrate, but provide hardness and resiliency to the ground substance; further, they may actually contribute to the physicochemical makeup of the fibrous elements of dense connective tissue and cartilage.

The foregoing components of connective and supporting tissues are organized in ways which effect the specific function or functions to be served. On this basis four rather distinctly different classes of tissues may be distinguished; *loose connective tissue, dense connective tissue, cartilage,* and *bone.* With considerable justification, a fifth class, fluid connective tissue, which includes blood and lymph, may be added to the list; however, because the functions of blood and lymph do not include either connection or support in the mechanical or structural sense, they are dealt with separately in this book.

CLASSIFICATIONS OF CONNECTIVE TISSUE

Loose Connective Tissue

The distinction between loose and dense connective tissues is rather arbitrary in some instances, although generally it is quite clear. Loose connective tissues are characterized by a semifluid, somewhat gelatinous matrix of protein, mucopolysaccharide (primarily hyaluronic acid), and water, which is supported by a loose meshwork of collagenous fibers and, to a considerably less extent, reticular and elastic fibers. When they are removed from the body, membranes of loose connective tissue often entrap air, assuming a frothy or bubbly appearance, for which reason they are called *areolar tissue.* Mesothelial membranes of the peritoneum and pleura and their foldings—the mesenteries, visceral ligaments, and omentum—consist largely of areolar connective tissue supporting the mesothelium (Fig. 3.2). Loose connective tissues also cover skeletal muscles in association with the fascia; appear between the secreting units of exocrine glands; invest blood vessels, nerves, and ducts; and form the deeper, subcutaneous layers of the skin in many vertebrates. They appear where freedom of movement must be provided without a concomitant need for high tensile strength.

The functions of loose connective tissue are very basic ones, and so there is little difference in the tissue's appearance in the various vertebrate classes. The greatest variability is in its thickness in subcutaneous areas.

Included within loose connective tissue membranes are the cell types described above: fibroblasts, macrophages, mast cells, and clusters of fat cells (Fig. 3.2). In addition to these, most of the leukocytic series of blood cells may be found as lymphocytes, monocytes, eosinophiles, neutrophiles; in the loose connective tissues of the digestive tract and omentum, plasma cells are not infrequently found.

Modifications of ordinary loose connective tissue are recognized and described according to their special characteristics. Among these are included the following categories.

ADIPOSE TISSUE

Fat cells are normal constituents of many loose connective tissues, particularly areolar connective tissue, occurring in more or less small clusters or thin sheets. They are concentrated in certain areas to the exclusion of most of the intercellular elements, however, in which case they constitute adipose tissue. The cells of adipose tissue are usually arranged in lobules which are separated from each other by areolar tissue. Each lobule is a functional and morphological unit, receiving its own separate blood supply from a lobular artery and being drained by one, or more commonly, by two lobular veins. Adipose tissue tends to be closely associated with the kidneys, the gonads, and the mesenteries in all classes of vertebrates. In most mammalian species adipose tissue forms an extensive, subcutaneous sheet, the *panniculus adiposus,* which is especially well developed in the cetaceans and other marine mammals.

RETICULAR CONNECTIVE TISSUE

Reticular connective tissue is loose connective tissue the principal fibrous element of which is reticulin and in which the principal cellular elements are macrophages and *reticular cells.* The latter are rather primitive cells which closely resemble mesenchymal cells. The nucleus is large, ovoid, and it stains lightly; the cytoplasm is clear and faintly acidophilic. The cell outline is usually irregular, having many protoplasmic processes that branch out in all directions and anastomose with similar processes from other reticular cells, in this way forming a protoplasmic meshwork (Fig. 9.9). Other reticular cells are intermediate in morphological character between the primitive type just described and the fixed macrophage. Free macrophages also occur in reticular connective tissue.

The precise relation of the reticular fibers to the reticular cells is not well understood, except that it is an intimate one, for protoplasmic processes of the cells may be seen accompanying the fibers and often apparently investing them. The fibers vary somewhat in coarseness and in density, but in all cases they branch and anastomose freely. Reticular connective tissue is characteristic of all lymphoid tissue, such as the spleen, thymus, and lymph nodes and nodules, and it also occurs in red bone marrow and in the sinusoids of the liver. In certain situations the fixed macrophages of reticular connective tissue form an endothelial pavement which lines sinusoids, whence the term "reticuloendothelial system" (RES) has been derived. The RES is discussed in more detail in Chapter 9.

MUCOUS CONNECTIVE TISSUE

Mucous connective tissue is transitional between embryonic and adult loose connective tissue. The cellular elements include primarily mesenchymal cells, immature fibroblasts, and fibroblasts. The protoplasmic processes of these cells form a loose meshwork, which is given some support by a very loose meshwork

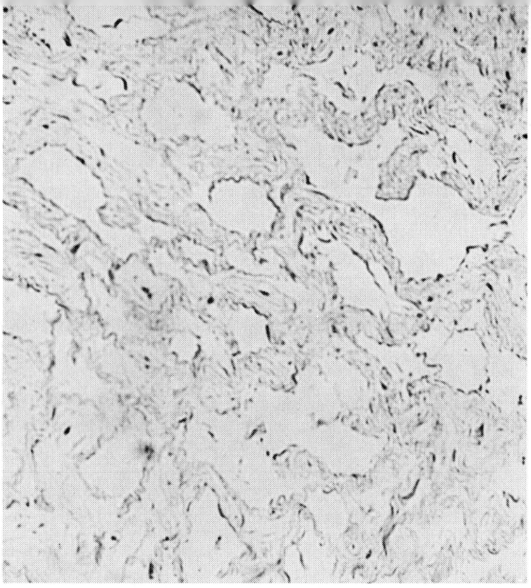

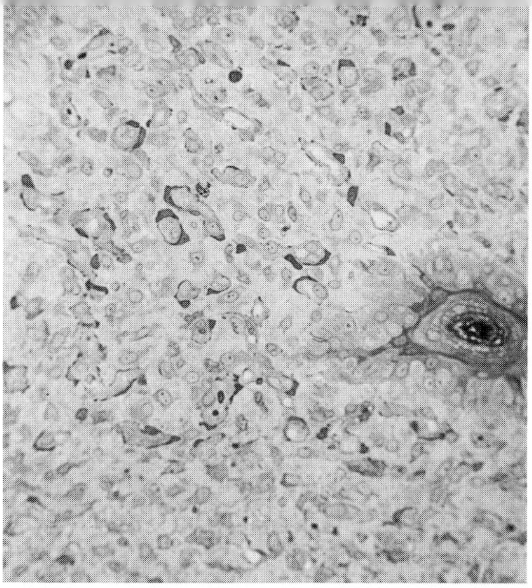

Fig. 3.7 Mucous connective tissue: *a,* transverse section of human umbilical cord (X100) ; and *b,* section from the comb of a hen (X450).

of fine, collagenous fibers. The amorphous ground substance is quite considerable, and consists primarily of water, proteins, and hyaluronic acid. Mucous connective tissue, besides constituting the main connective tissue of the embryo and fetus, comprises the bulk of the mammalian umbilical cord (where it is known as *Wharton's* jelly; Fig. 3.7a) and of the combs and wattles of gallinaceous birds (Fig. 3.7b).

Dense Connective Tissue

The distinguishing characteristic of dense connective tissue is the abundance of the fibrous component, which is principally collagenous. These fibers may be arranged in sheets of bands which crisscross and interlace (*irregular* dense connective tissue) such as in the compact layers of the dermis (Fig. 3.8),

Fig. 3.8 Crisscross laminations of collagen fibers in the dermis of the lizard *Anolis carolinensis.* X15,000. (Courtesy of F. O. Craft.)

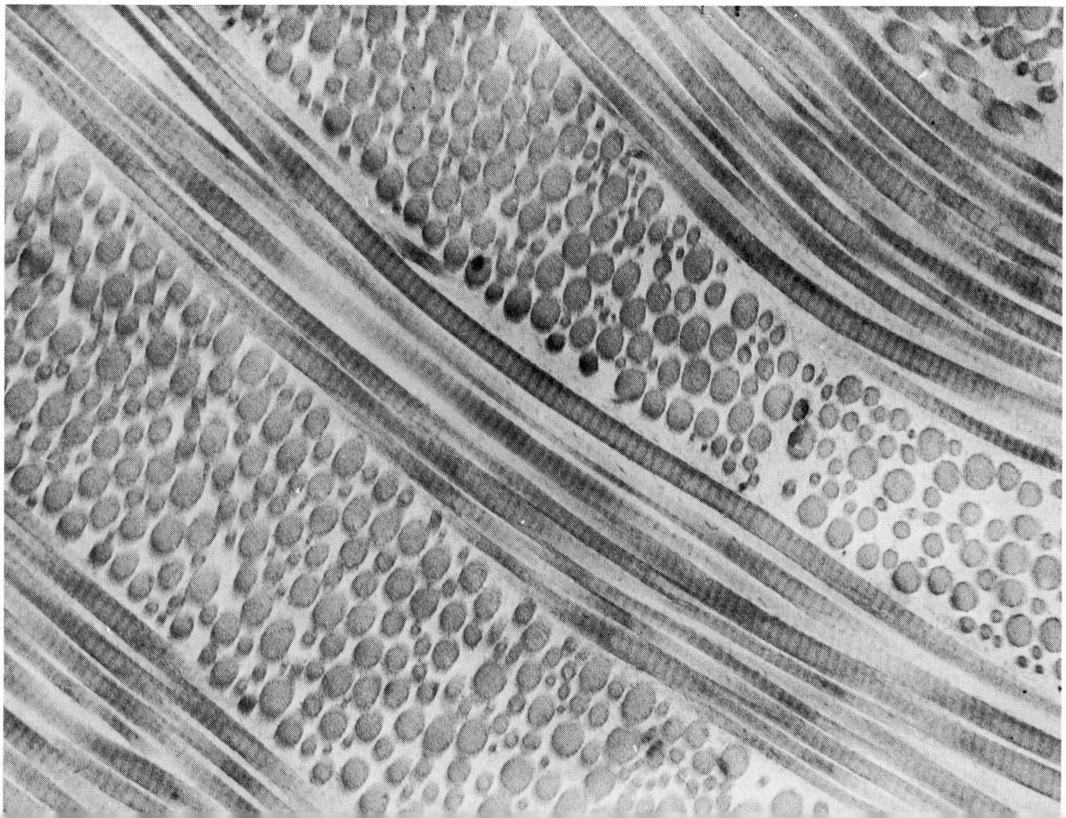

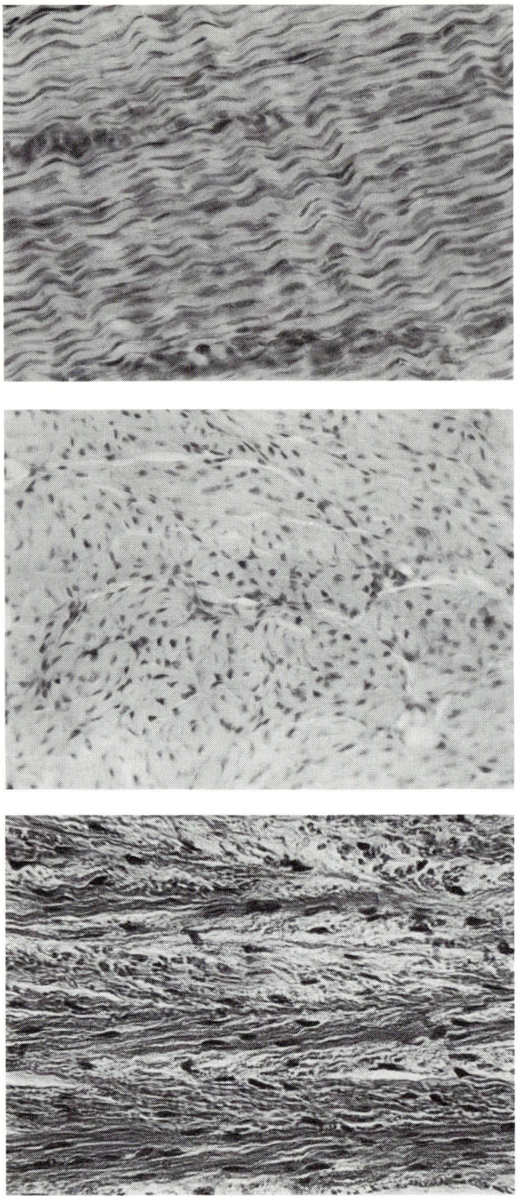

Fig. 3.9 Dense connective tissue: *a*, longitudinal section of kitten's tendon (X430); *b*, transverse section of kitten's tendon (X430); and *c*, ligamentum flava of a mammal (X450).

the periosteum (Fig. 3.20), and the peritoneum, or they form the capsules of many visceral organs. In such cases fibroblasts and macrophages are the most conspicuous cellular elements within the fibrous bundles.

The fibers may be arranged, on the other hand, to form more or less broad cords (*regular* dense connective tissue), such as in *tendons, ligaments,* and *aponeuroses.* In tendons of all vertebrates the fibers are almost exclusively collagenous and they are arranged parallel to each other in compact bundles with fibroblasts as virtually the only cellular components. The latter occur in rows between the fibrous bundles, appearing elongate in longitudinal sections (Fig. 3.9a) and stellate in transverse sections (Fig. 3.9b). Some loose connective tissue is associated with each bundle and surrounds the tendon as a whole.

Aponeuroses are broad, flat sheets of connective tissue bundles, often disposed in two or more layers in each of which the fibers are densely arranged and run parallel to each other. The fibers of adjacent layers run at an angle to each other, however, not unlike the arrangement of the laminae of plywood, except that interweaving of layers may occur. In other respects aponeuroses are like tendons.

Ligaments are also similar to tendons histologically, except that elastic fibers frequently occur and in some instances may constitute the predominant fibrous type, as in the *ligamentum flava* extending between the neural arches of tetrapods, and the *ligamentum nuchae* of ungulates and elephants (Fig. 3.9c).

Tendons and ligaments are most conspicuous in the tetrapods. There has been little displacement of skeletal elements in relation to the muscles which move them in fishes; that is, the muscular elements lie immediately adjacent to their origins and insertions, so that tendinous connections are so short as hardly to qualify as tendons. The evolution of terrestrial vertebrates involved the extensive development of the girdles and paired appendages with attendant growth in length away from the main bodily mass. The muscles functionally associated with the limbs maintained their connection by extending their fibrous connective tissue in the form of prominent tendons. In this manner the muscles were able to maintain a maximum cross-sectional diameter and relatively short length, whereby each muscle cell could be most effective. To understand this, one might imagine two men hanging from a trapeze attempting to lift a weight which is beyond their reach. If one man is suspended from the other, they can reach the weight, which, ignoring the weight of the lower man, either of them could lift alone. But if the two men are side-by-side, reaching the weight by means of a rope or "tendon," they can lift twice as much (Fig. 3.10a and b). It is because of the relative lack of tendinous and dense connective tissue that the flesh of fish is tender, whereas the appendicular muscles of terrestrial vertebrates, which constitute by far the bulk of their muscular flesh, is considerably tougher and requires longer cooking to make it tender. The trunk muscles of tetrapods, as with the filet mignon and tenderloin cuts, are the tenderest, for the same reason that fish muscle is tender.

Similarly, extensive ligamentous development represents a necessary adjunct of terrestrial adaptation. As pointed out in Chapter 6, gravitational forces are highly significant in terrestrial animals, and the elements of the vertebral column are constantly subjected to them, as are the articulations between the elements of the appendicular skeleton. The ligaments are correspondingly better developed and tougher, and where the forces are extreme—such as in the necks of large, grazing animals—elastic fibers are incorporated to provide a range of activity that does not rely on muscle tonus throughout that range. This is illustrated by Fig. 3.11.

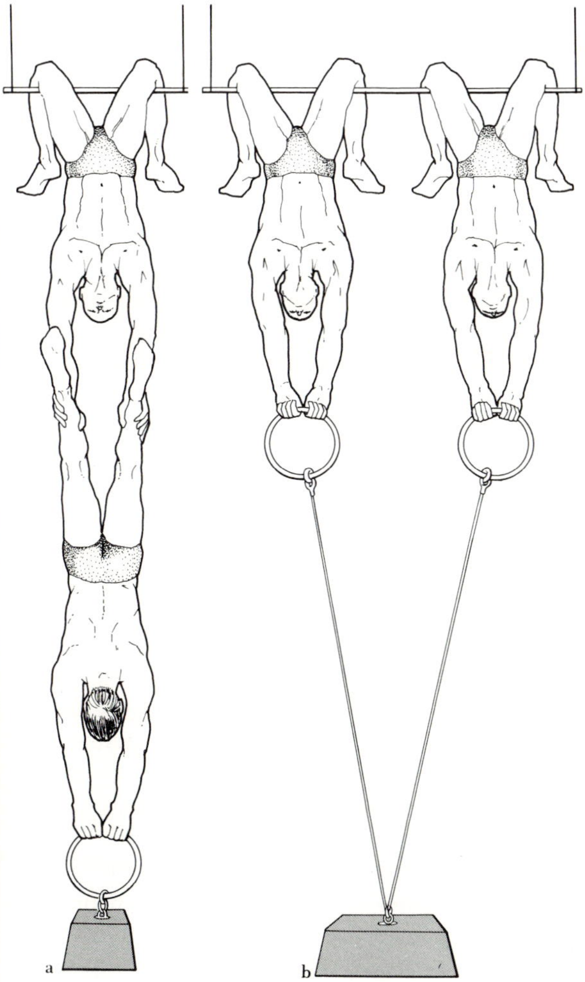

Fig. 3.10 Muscle–tendon relations. See text for description.

Cartilage

The two major kinds of supporting tissues occurring in the vertebrates are cartilage and bone. A third is the notochord, which, because of its limitations and because of its resemblance in consistency to chondroid, is discussed in connection with cartilage. Which is the more ancient, cartilage or bone, is an academic question, for whereas cartilage provides the skeleton of the most primitive contemporary vertebrates, bone served this purpose in their very remote ancestors; the oldest known vertebrate fossils had well developed dermal skeletons of bone. It is now generally accepted that bone appeared early in vertebrate evolution as the dermal skeleton, and that cartilage came later as an internal skeleton, which, as it assumed more functional importance, later became ossified.

Wherever it occurs, cartilage is basically the same. Its principal cellular component is the *chondrocyte,* and its basic extracellular component is a firm gel of chondroitin sulfate supported by a meshwork of collagenous fibers. In this basic form, cartilage is clear and glassy, and is known as *hyaline cartilage.* It may occur in modified form owing to replacement of some of the amorphous, intercellular matrix by elastic fibers (elastic cartilage) by a high concentration of collagenous fibers (fibrocartilage), by deposits of calcium

Fig. 3.11 Function of the nuchal ligament in cattle. The nuchal ligament acts as a spring that holds the head in a neutral position, *a*, against the force of gravity. With relatively little effort, the head can be raised, *b*, by contraction of extensor muscles, or lowered, *c*, by contraction of flexor muscles.

carbonate and other salts (calcified cartilage), or by more or less significant amounts of hyaluronic acid (chondroid, or "mucous cartilage").

HYALINE CARTILAGE

Hyaline cartilage occurs as the embryonic endoskeleton of all vertebrates above the agnatha (*see below*) as well as the adult endoskeleton of all contemporary agnatha and chondrichthyes, and much of that of the chondrostei. It persists as part of the endoskeleton of amphibia and, in decreasing amounts, that of reptiles, mammals, and birds. Hyaline cartilage comprises the articular surfaces of all diarthroses (see page 69) and it is used to maintain passages of the respiratory system and the auditory system patent. In the former it occurs as laryngeal, tracheal, and bronchial rings in all the higher tetrapods and it supports the external auditory meatus, cochlea, and associated structures in most tetrapods.

Hyaline cartilage differentiates from mesenchymal cells which have become transformed into cartilage secreting cells, called *chondroblasts*. Chondroblasts are large, spherical cells with a prominent nucleus which is spherical or ovoid and contains a light dusting of chromatin (Fig. 3.12). The cytoplasm includes numerous vacuoles, some of which may be quite large, and it is rich in glycogen and lipid. In routine histological preparations the vacuoles lose their fluid, with resultant shrinkage of the cell away from the matrix which it has secreted. Electron micrographs of chondroblasts reveal a well-formed Golgi net adjacent to the nucleus, elongate mitochondria, and a well-defined endoplasmic reticulum.

The histogenesis of hyaline cartilage is evidenced by just such a transformation of mesenchymal cells into chondroblasts within the area where cartilage has been induced to differentiate. Simultaneously fibroblasts lay down a meshwork of loosely arranged collagenous fibers within this area, and, surrounding it, a connective tissue membrane differentiates from mesenchyme and fibroblasts. This membrane, the *perichondrium*, soon establishes two more or less distinct layers, an outer *fibrous layer* and an inner *chondrogenic layer*. The former appears as irregular dense connective tissue and the latter is comprised of fibroblasts and prechondro-

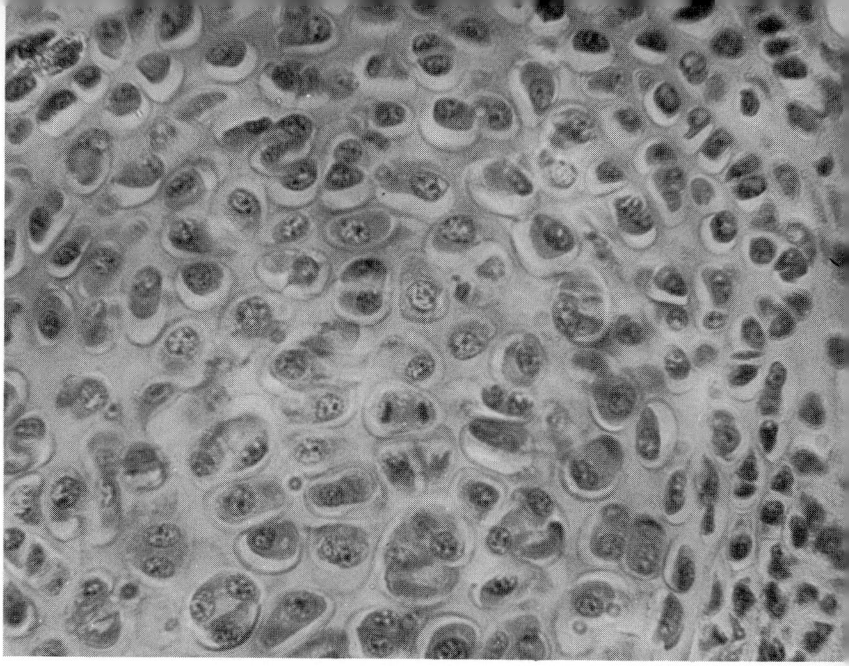

Fig. 3.12 Photomicrograph of growing hyaline cartilage in an embryonic mouse. X450. Note the mitotic telophase near the center.

blasts. The growth of cartilage from this point proceeds in two ways: interstitially, by the laying down of a cartilaginous matrix in the central area, and appositionally, as described below.

INTERSTITIAL GROWTH. Interstitial growth is characterized by mitotic activity of the central chondroblasts and by the laying down of chondroitin sulfate and other mucopolysaccharides and possibly collagenous fibers as well. Soon the cells become entrapped within their own matrical substance, occupying spaces known as lacunae (Fig. 3.12). Usually entrapment occurs before the last mitosis has been completed, so that each lacuna is occupied either by a pair of cells or, occasionally, by a binucleate cell. The chondroblasts continue to secrete for a time after their entrapment, however, so that expansion of the central area of chondrogenesis takes place (Fig. 3.12). Meanwhile, a matrix that is virtually devoid of collagenous fibers and is composed of highly refractile substance forms a *capsule*, immediately surrounding each lacuna. Further expansion may continue for a time, depending on the organ, and then the chondroblasts become inactive, in which state they are known as *chondrocytes*. Chondrocytes are characterized by reduced basophilia and depressed metabolic rate. Their metabolic needs are met by what limited diffusion of nutrients and metabolites can take place through the surrounding matrix, for cartilage, unlike bone, is an avascular tissue.

APPOSITIONAL GROWTH. The events which characterize appositional growth duplicate those described above, except that they occur around the periphery of the cartilaginous mass. Prechondroblasts divide mitotically, their daughter cells synthesize cartilaginous matrix, and fibroblasts give rise to collagenous fibers among them. The perichondrium retreats in measure with the differentiation of cartilage beneath it, so that it always exists intact, with its two layers unaltered, around the cartilaginous mass. Once the chondroblasts thus derived are entrapped, they too may undergo further division and interstitial growth may then follow along on the heels of appositional growth.

ELASTIC CARTILAGE

Cartilages in certain areas acquire a relatively large amount of elastic fibers during their histogenesis, yet in other respects they develop in essentially the same way as does hyaline cartilage. Such areas are the external ear and epiglottis of mammals and the eustachian tube of reptiles, birds, and mammals. In some species elastic cartilage may also be included in the laryngeal cartilages. The density of elastic fibers within elastic cartilage varies enormously, not only from site to site but even within the same organ, where one usually

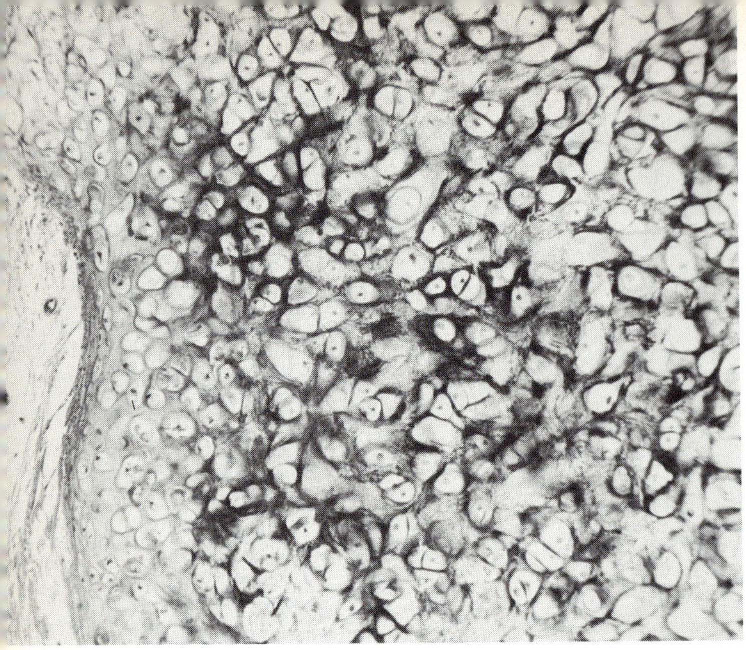

Fig. 3.13 Elastic cartilage in the external ear. X100. The density of elastic fibers increases from the outer edge of the cartilage shown at the left to deeper areas shown toward the right.

finds the fibers thin and sparse at the periphery, and coarse and densely packed within the deeper parts (Fig. 3.13).

FIBROCARTILAGE

Fibrocartilage often resembles dense connective tissue more than it does cartilage, save that the cellular elements are clearly chondrocytes embedded within cartilaginous capsules. In the fibrocartilage which is found at the junction of tendons with cartilage or bone, almost all of the intercellular matrix consists of parallel bundles of collagenous fibers, with the chondrocytes arranged in rows between them. The fibrocartilage of the intervertebral disks of vertebrates resembles more closely irregular dense connective tissue and may show greater amounts of hyaline matrix, at least in the immediate area surrounding each chondrocyte, than is shown by the fibrocartilage associated with tendons. Not infrequently fibrocartilage may have chondroid associated with it—for example, in the center of the intervertebral disk, the *nucleus pulposus*, which is a remnant of the embryonic notochord. Within the fibrocartilage of certain tendons, such as the Achilles tendon in amphibians and the patellar ligaments of mammals, some chondroid is present. Fibrocartilage is also found bridging the space between the articular cartilages (hyaline) of synarthral joints, as in symphyses, where a gradual transition between hyaline cartilage and fibrocartilage is well exemplified. Many diarthral joints, particularly in the larger terrestrial mammals, are characterized by menisci, which are wedge-shaped masses of fibrocartilage intruding into the joint capsule.

CALCIFIED CARTILAGE

When the histogenesis of cartilage bone is discussed below, it will be seen that calcification of cartilage presages its degeneration and destruction. This is true in the situation in which cartilage has a provisional function and is normally replaced by bone. In some species of the chondrichthyes, however, calcified cartilage represents an end tissue in itself, and it may also be found as a normal constituent of small parts of the skeleton of amphibia and reptiles, especially in parts of the hyoid skeleton and occipital condyle. These observations suggest that factors other than calcification alone are responsible for the degenerative processes that characterize hyaline cartilge during osteogenesis, as histologists are generally wont to assume.

Calcified cartilage is always preceded by hyaline cartilage during its histogenesis, but little is known of the details of that histogenesis except in the context of ossification. In this case, the chondrocytes secrete and accumulate an enzyme, alkaline phosphatase, which effects the splitting off of phosphate groups from the phosphorylated sugars. The phosphate then combines with calcium, forming crystals of calcium phosphate, which, in turn, in the case of those vertebrates in which the calcified cartilage is the terminal tissue, catalyzes the precipitation of calcium carbonate from available calcium and carbonate compounds, both of which exist in abundance in tissue fluids. The source of the phosphorylated sugars is also in the chondrocytes, for by histochemical methods glycogen and the enzyme phosphorylase can be demonstrated to exist in very significant amounts within them.

CHONDROID

Possibly the most primitive kind of cartilage, chondroid is fibrous connective tissue which includes a matrix of hyaluronic acid and chondroitin sulfate, with the former in preponderance, and a rich population of chondrocytes. It has a rather limited distribution, occurring as the endoskeleton of cyclosomes, the external ear of some rodents, such as rats, and in bats. As mentioned above, the nucleus pulposus of the intervertebral disc is a kind of chondroid, although its histogenesis is considerably different, and its resemblance may be only a superficial one. The nucleus pulposus represents vestiges of the embryonic notochord which has been replaced entirely in the region of the vertebral centra, but persists to a very limited extent between them—that is, within the intervertebral disks. These vestiges are thereupon transformed into a kind of tissue which has the general characteristics of chondroid.

Notochord

The notochord is in no way to be construed as a kind of cartilage, primarily because its embryonic origin is quite different and unrelated to that of cartilage. The presumptive notochord is one of the very first organs to become differentiated in the embryo, arising initially as a mid-dorsal modification of the archenteron in most vertebrates, and thereafter proliferating from the indifferent tissues of the dorsal lip of the blastopore or its homologue, the primitive streak. Because the function of the notochord is the same as that of cartilage generally (that is, skeletal support), it is perhaps not surprising that it should resemble cartilage in certain ways, as indeed it does. It differs in one very significant way, however. Whereas the important functional component of cartilage, the cartilaginous matrix, is *extra*cellular, that of the notochord is *intra*cellular, with little or no extracellular component being present except in its investing sheaths. The embryonic notochord consists of cells which contain a single, large vacuole and a small amount of cytoplasm which has been forced to a peripheral position, similar to the adult fat cell. Surrounding the notochord *per se* is a series of enveloping sheaths that vary in number and structure according to species but most usually include an inner *elastica interna* and one or two outer, fibrous layers.

The adult notochord of amphioxus has been investigated by electron microscopy and reveals a rather different and more highly organized structure than that of the vertebrate embryo. Two types of notochordal cells are distinguishable. The most numerous ones are laterally compressed cells that give the appearance of plates which are stacked vertically and occupy all of the notochordal substance save along the midventral lines. Within the plate cells are found bundles of microfibrils transversely disposed (Fig. 3.14) and occupying most of the cytoplasm. The nucleus is flattened laterally and lies near the center of the plate cell. Mitochondria, rows of vesicles, and occasional vacuoles also constitute consistent features of these cells.

The second type of cells, constituting *Muller's tissue,* occur dorsally and ventrally. They lack fibers, but have a well-developed endoplasmic reticulum and a system of intracellular canaliculi. Surrounding the notochord *per se* is a distinct basement membrane, which corresponds with the misnamed *elastica interna* described by earlier microscopists and which receives digitations from adjacent plate cells and Muller's tissue. Surrounding this, in turn, are an inner connective tissue membrane consisting of collagenous fibers circularly dis-

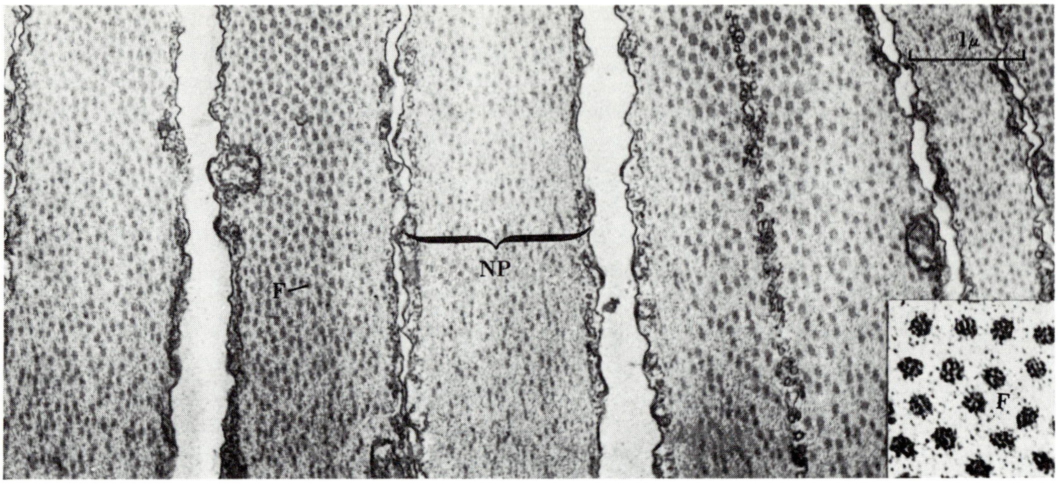

Fig. 3.14 Notochord of Amphioxus in oblique longitudinal section. X15,000. The central parts of six plates (*NP*) separated by spaces (artifacts?) are shown. The insert (X46,000) shows fibrils (F) in cross section. (From R. M. Eakin and J. A. Westfall. 1962. J. Cell Biol. 12:646.)

posed and an outer one with its fibers longitudinally disposed (Fig. 3.15).

Bone

Bone is a kind of calcified connective tissue whose microscopic structure shows it to be a highly and precisely organized tissue which, by the incorporation of hormonal and other feedback mechanisms, is capable of modifying and adjusting its microscopic structure to adapt to changing mechanical stresses and dietary conditions. In contrast with cartilage, which is avascular, bone is a well-vascularized tissue, being richly supplied by blood vessels which penetrate through *nutrient foramina* and are distributed via Haversian canals and interconnecting Volkmann's canals throughout the entire bony substance. The bone cells, *osteocytes* (Fig. 3.16) although entrapped within a rigid, bony matrix, maintain avenues of fluid passage either directly or indirectly with blood vessels by means of systems of anastomosing *canaliculi* (Fig. 3.17).

Two kinds of bone can be distinguished structurally. *Cancellous bone,* also called spongy bone, is characterized by columns or *trabeculae* of bony tissue which interlace and anastomose to form a rigid yet open kind of structure. The spaces between trabeculae are occupied by red bone marrow in most instances, which will be described below. The trabeculae are precisely oriented in such a manner as to provide maximum architectural strength in opposition to the mechanical stresses and strains to which it is ordinarily subjected. Where it exists, cancellous bone is always encased in compact bone. Cancellous bone occurs within all bones during their formative stages, and in adult bones it occurs in the *epiphyses* and *metaphyses* of the long bones and is sandwiched between plates of compact bone in the so-called *flat bones,* such as the bones of the skull, the bones of the girdles, and the ribs.

Compact or *lamellar* bone consists of layers or *lamellae* of bony tissue closely applied to each other. At the immediate surface of a given bone, the lamellae lie parallel with the outer and, in some cases, inner surfaces, thus forming encasing cylinders called the *outer circumferential lamellae* on the one hand or *inner circumferential lamellae* on the other. The remaining lamellae may be arranged in bundles of concentric cylinders, *osteons* or *Haversian systems,* and, tying these together, *interstitial lamellae* (Fig. 3.17). This manner of construction imparts maximum rigidity and shearing strength to the entire member. Moreover, the osteons, by virtue of their centrally contained blood vessels and nerves, are living units capable of metabolism and repair.

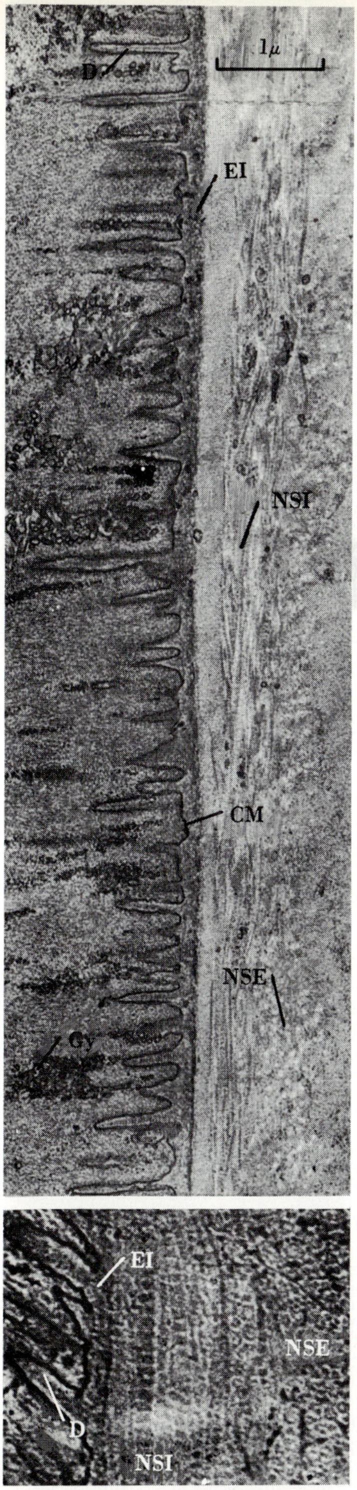

Fig. 3.15 Cross section of the notochordal sheaths in Amphioxus (upper figure, X12,000; lower figure, X25,000), showing the basement lamina, *EI;* inner notochordal sheath, *NSI;* outer notochordal sheath, *NSE;* diverticula of the chordal plate, *D;* cell membrane *CM;* and cytoplasmic strands, *Cy.* (From R. M. Eakin and J. A. Westfall. 1962. J. Cell Biol. 12:646.)

THE HISTOGENESIS OF BONE

The microscopic structure of bone can best be understood by a description of its histogenesis. One can distinguish three different conditions under which bone differentiates, however. It may differentiate *directly* within an area of mesenchyme; *indirectly,* by replacing an existing skeletal member composed of hyaline cartilage; or *heteroplastically,* by displacing connective tissue.

DIRECT BONE FORMATION. The most primitive bone, phylogenetically speaking, is probably formed by the direct method, for it is thus that the bony dermal scales of fishes and their homologues, the dermal bones of the skulls of all higher vertebrates, develop. Direct bone formation is initiated by the differentiation of osteoblasts and fibroblasts in a region consisting otherwise only of mesenchyme and of a loose vascular network. Collagenous and reticular fibers are laid down, changing the character of the mesenchyme to that of a loosely woven felt work or membrane. (It is for this reason that the somewhat misleading terms "membranous bone" and "intramembranous ossification" are often used in connection with this kind of bone development.) Clusters of osteoblasts immediately initiate the process of ossification, laying down a matrix of immature bone along the bundles of collagenous fibers. Immature bone has a relatively low mineral content, as compared with mature bone; it is unevenly basophilic; the spaces (*lacunae*) in which the osteoblasts become entrapped are somewhat larger and less flat than those of mature bone; and the collagenous component is more conspicuous (Fig. 3.18).

Once entrapped, the osteoblast ceases its synthesizing activities and is known as an osteocyte. Osteocytes maintain avenues of contact with each other, however, through the minute canaliculi mentioned earlier, which represent channels that were occupied by the anastomizing protoplasmic processes of osteoblasts during bone formation. The mineral components of bone appear as growing apatite crystals of calcium phosphate, primar-

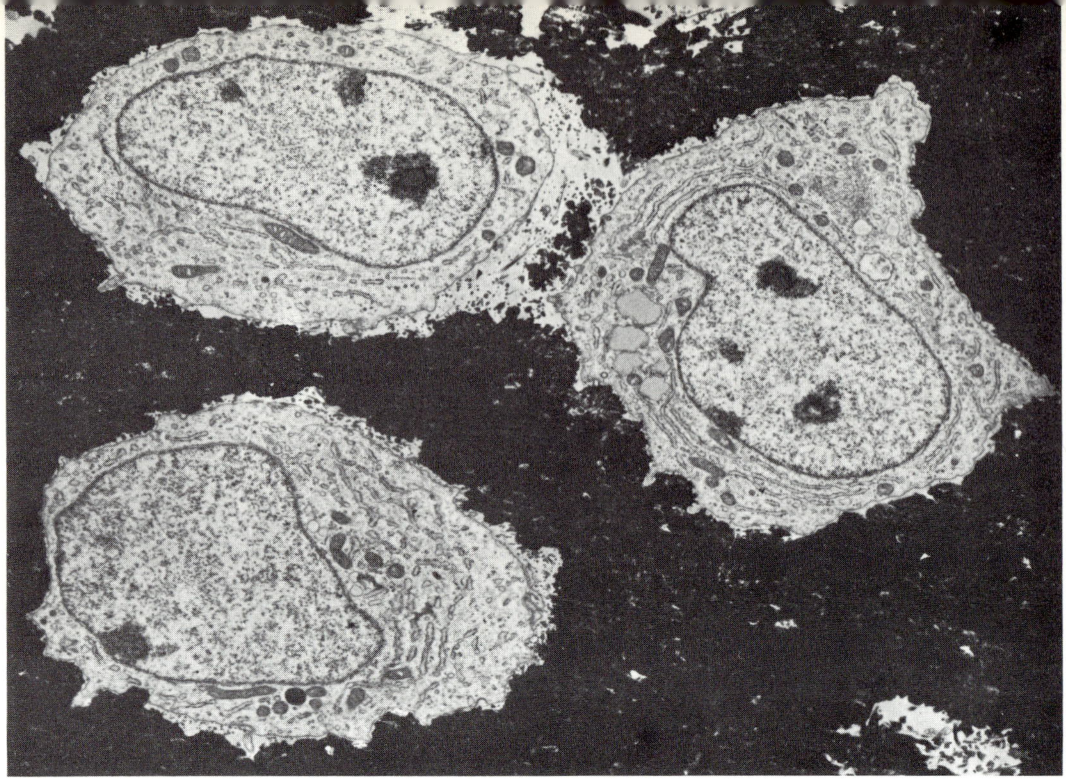

Fig. 3.16 An electron micrograph of three osteocytes in the developing frontal bone of a chicken. The well-developed ER and the presence of fat droplets (to the left of nucleus in the right hand cell) suggests that these cells are still actively secreting bone. (Courtesy of B. Boothroyd.)

ily, in the form of *spicules* which are joined by a cementing substance into branching and anastomizing columns, the *trabeculae* (Fig. 3.19). The unossified areas between trabeculae are occupied by blood vessels, connective tissue, loose mesenchymal cells, and prolif-

Fig. 3.17 Section of compact bone. X150. The canaliculi appear as fine lines running both radially and circumferentially in the osteons and longitudinally and transversely in the interstitial lamellae.

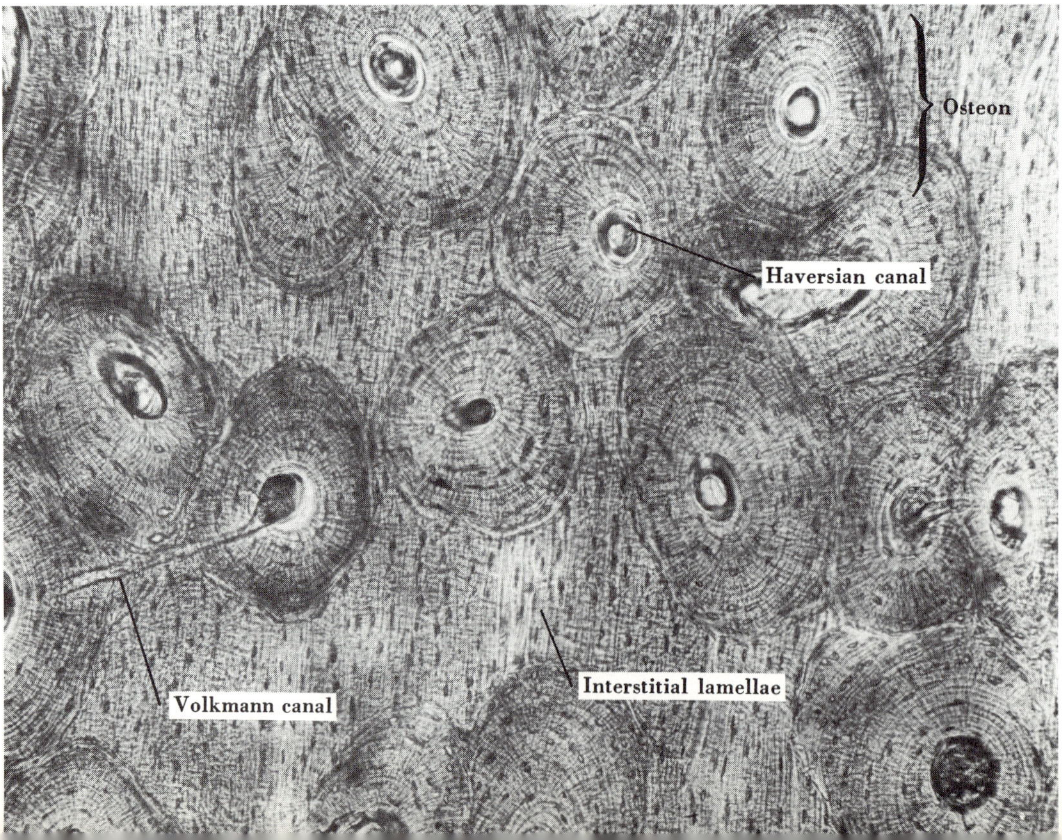

erating osteoblasts. Arising from the mesenchymal component is still another type of cell, the *osteoclasts,* whose activities soon will come into play.

The connective tissue sandwiching the areas of ossification in the meantime has differentiated considerably, establishing itself as a more compact membrane, the *periosteum.* Considerable proliferation of osteoblasts has occurred, and they may be seen adjacent to the ossified interior, constituting an *osteogenic layer* (Fig. 3.20). These osteoblasts lay down bony spicules in essentially the same manner as described above, except that the periosteum with its osteogenic layer retreats in measure as bone is laid down, and the bone thus formed is of the mature type (Fig. 3.20).

All osteoblasts are not involved immediately in bone synthesis. Many may remain within the intertrabecular spaces, as mentioned above. For the most part they tend to be disposed along the surfaces of the trabeculae where subsequently they lay down compact lamellae of mature bone, thus increasing trabecular girth and encroaching on the intertrabecular spaces, causing them to become even smaller, until they are only large enough to include the blood vessels. The maximum diameter of any trabecula is determined by the efficiency of diffusion of nutrients and metabolites between the blood vessels and the center of the trabecula. Accordingly, the extent to which the intertrabecular spaces are obliterated by continued osteogenesis depends on the initial proximity of trabeculae to each other. The trabeculae of the central portions of the bone are widely spaced, so that when it is fully ossified, this portion has a spongy appearance; hence it is known as *spongy* or *cancellous bone.* The trabecular spacing in the bone produced by this method comes to consist, therefore, of two plates of compact bone "sandwiching" a layer of cancellous bone. The cancellous bone and the red marrow within its spaces constitute the so-called *diploe.* Important examples of bones which form by the direct method are the bony plates of fishes' scales, as already mentioned, the bones of the cranial vault, the facial bones, and the bones of the mandible.

Although bone is rigid and inflexible, both

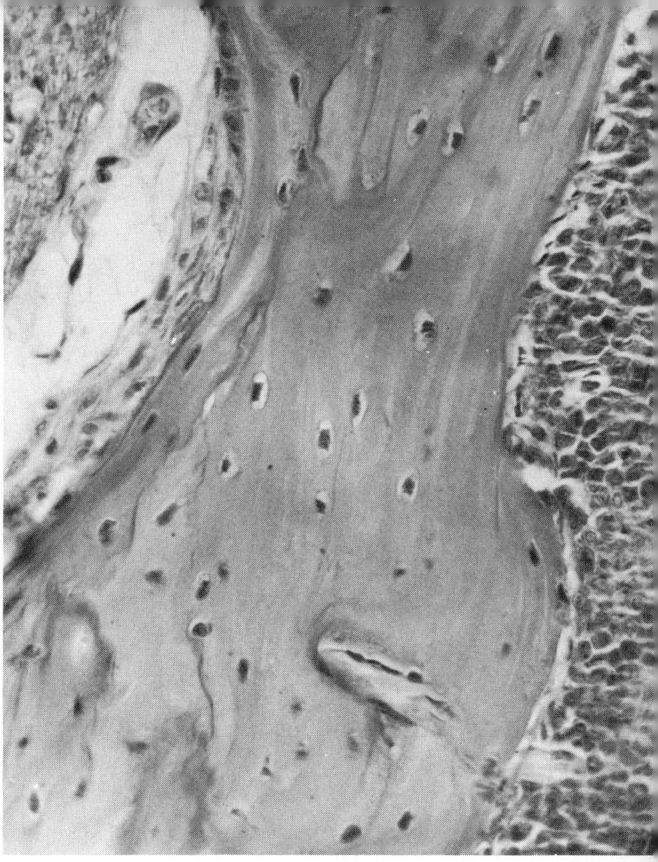

Fig. 3.18 Bone formed by the direct method (X430); from the mandible of a young rat.

Fig. 3.19 Cancellous bone. X100.

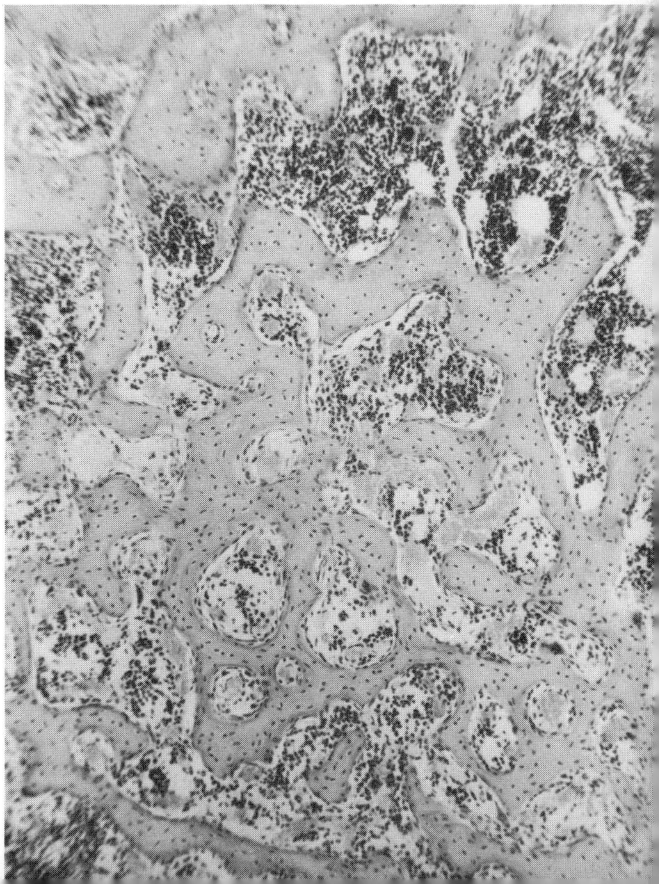

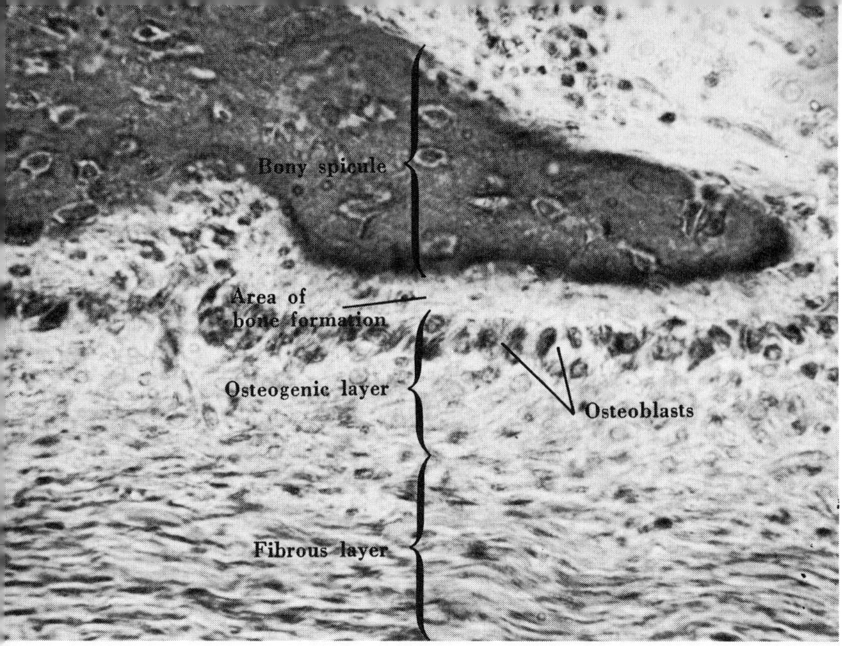

Fig. 3.20 Compact bone, showing both osteogenic and fibrous layers of the periosteum. X450.

its shape and its internal architecture can be changed or modified. Remodelling is effected by the coordinated action of osteoblasts and osteoclasts. The latter (Fig. 3.21) are giant, multinucleate cells, usually oval, with a frothy or vacuolated cytoplasm. The evidence that they are phagocytic is inconclusive. They usually occur singly, closely applied to bony tissue. The cell membrane is smooth except for that portion which is in direct contact with bone, and there the cell has a ruffled border (Fig. 3.21). The exact role played by osteoclasts in bone erosion is not known, but it is clear that they are directly implicated in the process. Whereas some authorities believe that osteoclasts dissolve the inorganic components of bone, others hold that they act only on the organic compartment, and still others postulate that their functions are incidental to bone absorption and that they serve mainly to phagocytose the debris resulting from the process. The last hypothesis is currently in disrepute, for no evidence of phagocytosis has been demonstrated *in vivo*. Active osteoclasts often occur in cavitations in the bone, known as *Howship's lacunae,* where their ruffled border is juxtaposed to a ruffled border of equivalent dimensions on the bone. The ruffled border of the latter is believed to consist of collagenous fibers devoid of the inorganic matrix.

The remodelling of bone involves bone erosion followed closely by or concomitant with the formation of new bone by osteoblasts. The net effect of the two opposed activities is represented graphically in Fig. 3.22.

INDIRECT BONE FORMATION. Indirect bone formation involves the replacement of hyaline cartilage by bone. Hence the portion of the bony skeleton which develops by the indirect method is first represented in almost every detail by cartilaginous skeletal elements. In many species the replacement is incomplete, with only portions of the cartilaginous skeleton being replaced by bone. Striking examples of this are the girdles of urodelous amphibians. The observation that certain genera of primitive fishes have a well-developed bony exoskeleton and a cartilaginous endoskeleton suggests that indirect bone formation is an expedient that was devised later in the evolution of the vertebrates. All the events of indirect bone formation are clearly illustrated in the tetrapod's femur (Fig. 3.23). These events occur simultaneously within and without the cartilaginous model, as *endochondral ossification* and *perichondral* (*periosteal*) *ossification*, respectively.

Endochondral Ossification. In a skeletal member, as the femur, the chondrocytes can be seen to undergo expansion or hypertrophy in three major foci, viz., the middle of the future shaft and, later in growth, the middles of the two future epiphyses. As indicated above, chondrocytes possess stores of glyco-

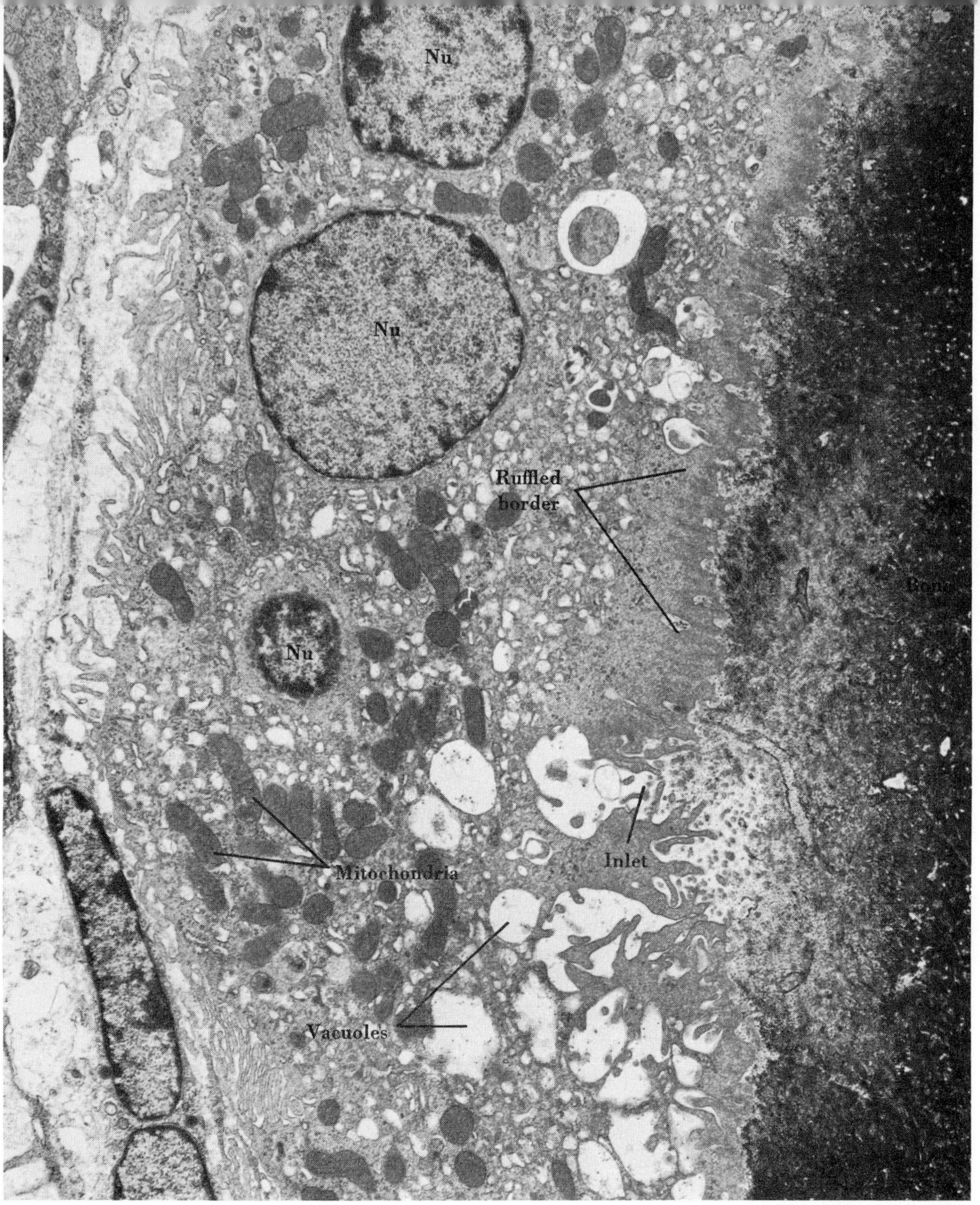

Fig. 3.21 Electron micrograph of an osteoclast adjacent to resorbing bone in the chicken. The ruffled border consists of numerous cytoplasmic projections between which expanded inlets form. X7000. (Courtesy of J. Waldron and J. Nalbandian, Harvard School of Dental Medicine.)

gen in their cytoplasm. They also secrete phosphorylase, an enzyme which effects the phosphorylation of carbohydrate to form sugar phosphates. As the chondrocytes hypertrophy, they secrete the enzyme, alkaline phosphatase, which diffuses out into the surrounding cartilage matrix and splits the phosphates off the phosphorylated sugars. Free calcium ions then combine with the phosphate radicals to form $Ca_3(PO_4)_2$, which precipitates out as apatite crystals in the hyaline matrix. The presence of these insoluble salts interferes with the diffusion of nutrients, oxygen, and metabolites that heretofore occurred between the chondrocytes and the tissue fluids. This interference results in the death of the chondrocytes.

The processes of hypertrophy and calcification spread from the initial foci centrifugally to the periosteum, which in the meantime has

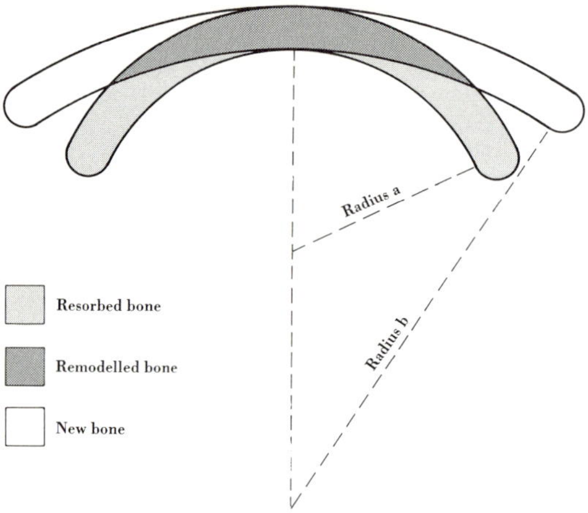

Fig. 3.22 Remodelling of dermal bone. During growth the curvature of a bone, such as in the cranial vault, may change. In remodelling from curvature *a* to curvature *b*, all but the central part of *a* is resorbed.

differentiated by slight modification of the perichondrium. As soon as contact is made between calcified, degenerating cartilage and periosteum, the periosteum sends out shoots (the *periosteal buds*) into the cartilage model. The periosteal buds consist initially of blood vessels, fibroblasts, osteoblasts, and chondroclasts (Fig. 3.24) that invade the degenerating calcified cartilage, dissolving it and replacing it with bony trabeculae in a manner similar to that described for direct bone formation. At the same time, chondrocytic hypertrophy and calcification proceeds lengthwise, delimiting the shaft from the epiphyses at either end (Fig. 3.23). These plates, the *epiphyseal disks,* exhibit chondroblasts and chondrocytes in several activity zones. Starting with the cartilage adjacent to the bone of the epiphysis and progressing toward the shaft, these zones are: (1) a zone of resting cartilage, which is characterized by inactive chondrocytes and extensive matrix; (2) a zone of proliferating cartilage, in which one finds dividing chondroblasts arranged in columns lying parallel to the long axis of the developing bone and having scanty matrix; (3) a zone of expanding cartilage, in which the lacunae are enlarged and the matrix is considerable; (4) a zone of calcification, characterized by degenerating chondrocytes and calcification of the matrix; and (5) a zone of ossification. It is in the last zone that one sees the replacement of cartilage occurring by the development of bony trabeculae (Fig. 3.25) which results in the so-called *metaphysis* of the bone.

The zone of cartilage proliferation sets the pace for growth of the skeletal member by contributing to the successive zones at a rate commensurate with the rates of calcification and ossification. There is frequently a burst of activity in all but the resting zone around the onset of puberty, which results in very rapid growth; but the rates of calcification and ossification gradually exceed that of proliferation, so that ultimately even the erstwhile zone of proliferation undergoes expansion, calcification, and ossification. At this time growth in length ceases, and even the zone of resting cartilage becomes replaced by bone, with resultant fusion of the epiphyses with the diaphysis at the *epiphyseal sutures.*

Periosteal Ossification. Coincident with the first signs of endochondral ossification is the transformation of the perichondrium into the periosteum. This is accomplished by increased vascularization of the perichondrium and by the differentiation of two kinds of cells, osteoblasts and osteoclasts, from mesenchyme. Whereas some of these elements are introduced into the cartilage model by the formation of periosteal buds, others become arranged on the surface of the model as the *osteogenic layer* of the periosteum. This layer proliferates osteoblasts, which deposit provisional bone around the middle section of the cartilage model as a periosteal *collar.* The extension of the collar distally and proximally is synchronized with the degeneration of cartilage internally, so that the skeletal member at no time fails in its supportive function. Surrounding the collar, the periosteum assumes the form of longitudinal furrows which present a convoluted appearance in transverse sections, and its osteoblasts lay down bone in compact lamellae. As they do this, the periosteum retreats and the lips of the furrows close over, entrapping a portion of the periosteum. Further ossification completely encircles the

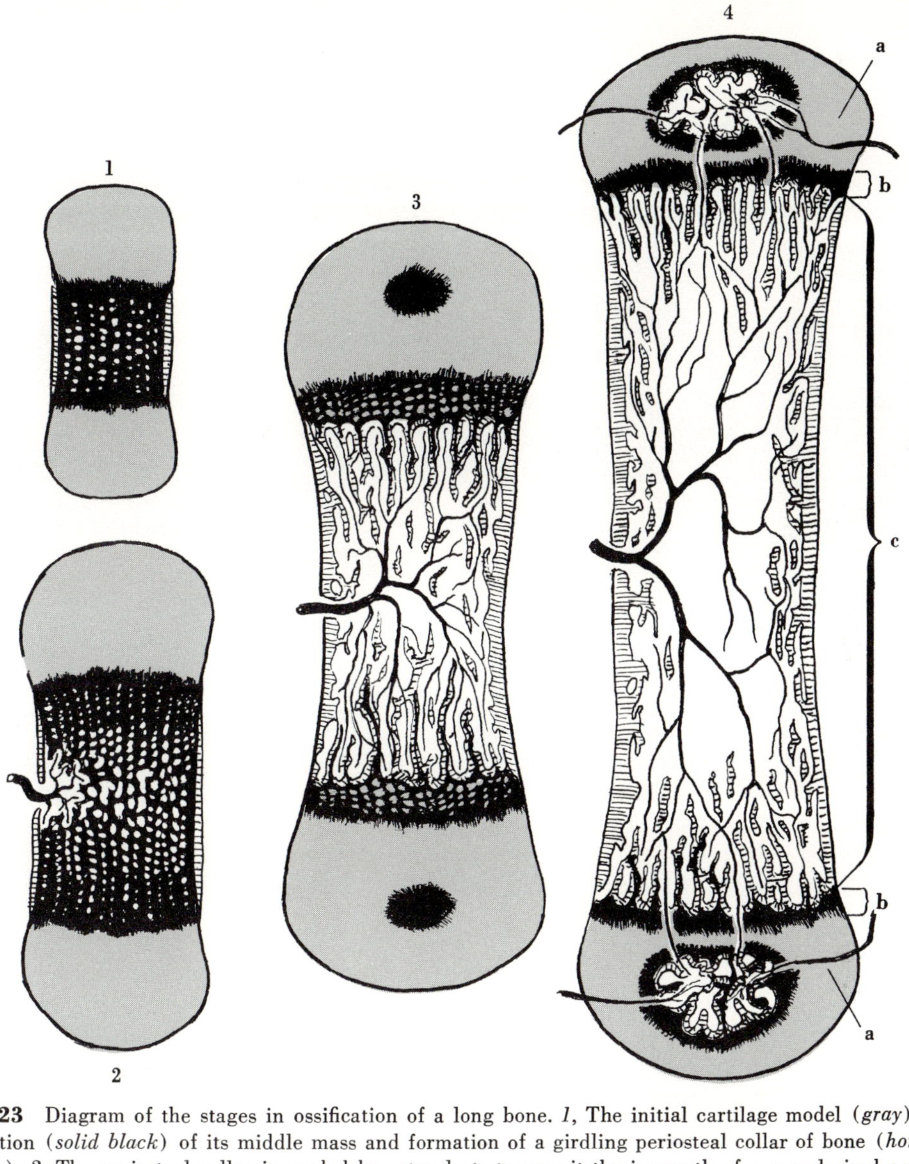

Fig. 3.23 Diagram of the stages in ossification of a long bone. *1,* The initial cartilage model (*gray*) shows calcification (*solid black*) of its middle mass and formation of a girdling periosteal collar of bone (*horizontal hatching*). *2,* The periosteal collar is eroded by osteoclasts to permit the ingrowth of a vascularized periosteal bud; meanwhile, the model has grown and calcification has kept pace. *3,* Chondroclasts, osteoblasts, and fibroblasts, introduced by the periosteal bud, have dissolved the middle mass of calcified cartilage and replaced it with bony trabeculae and vascularized bone marrow; the periosteal bone has increased in length and thickness; and central calcification of the terminal portions of the cartilage model has commenced. *4,* Zonation of the long bone is established, with *a,* ossification and vascularization of the *epiphyses; b,* delineation of the *metaphyseal plates,* which will be the only remaining sites of growth, calcification, and ossification until the adult; and *b,* differentiation of the *diaphysis* or shaft, with its well-vascularized marrow cavity. (After W. F. Windle. 1960. Textbook of Histology, 3rd ed. McGraw-Hill, New York.)

tube thus formed, in this way giving rise to a bony canal, a *Haversian canal.* The enclosed periosteum, as a consequence, is reversed in its position relative to the bone; thus it is termed *endosteum.* Osteoblasts of the osteogenic layer of the endosteum secrete concentric cylinders of bone within each Haversian canal until only a minute central space remains, and this consists primarily of blood vessels and nerves. The unit which is com-

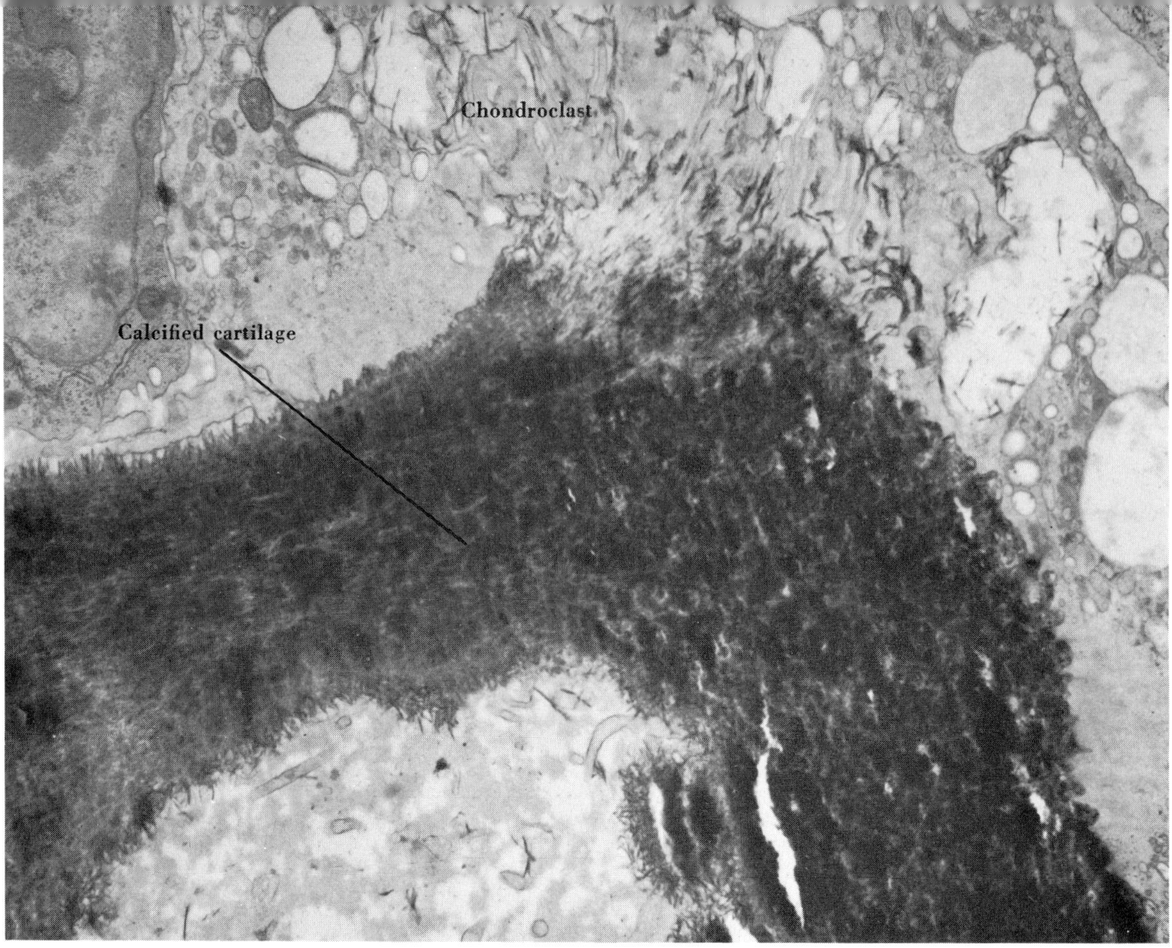

Fig. 3.24 Electron micrograph of a chondroclast at the point of resorption of calcified cartilage. X13,000. Compare with Fig. 3.21. Part of a monocyte appears in the upper left. (From R. K. Schenk, D. Spiro, and J. Wiener. 1967. J. Cell Biol. 34:275–291.)

prised of one Haversian canal and the concentric cylinders of bone surrounding it is known as an *osteon*. By virtue of the manner in which osteons are formed from longitudinal convolutions of the periosteum, they are arranged in parallel bundles which are fused together as fascicles. The Haversian canals of adjacent osteons are connected by the transverse *canals of Volkmann*, which develop at more or less regular intervals at points where the periosteum fails to pinch off the endosteum of a Haversian canal (Fig. 3.26).

Perichondral deposition of compact bone proceeds until the entire cartilage model, except that part adjacent to the epiphyseal disks, is encased. Remodelling also proceeds as a continuous process, as described above, although the results are even more dramatic (Fig. 3.27).

In the diaphysis of long bones the calcified cartilaginous interior is completely removed and is replaced by the cancellous bone of the metaphyses at either end adjacent to the epiphyseal disks and by yellow marrow between metaphyses (Fig. 3.22). In most vertebrates the marrow of the diaphyseal marrow cavity consists chiefly of fat and a sparse network of blood vessels supported by areolar connective tissue. In reptiles and birds the marrow may be rich in both hemopoietic and lymphoid tissue and, in some birds, extensions of the air sacs intrude into the marrow cavities (see Chapter 8). Hemopoietic tissue in mammals is located chiefly in the marrow. The remaining principal cellular elements of the marrow cavity comprise the endosteum, which was derived from the periosteal buds, and this consists of fibroblasts, osteogenic cells, osteoclasts, and collagenous and reticular fibers. Both the periosteum and the endos-

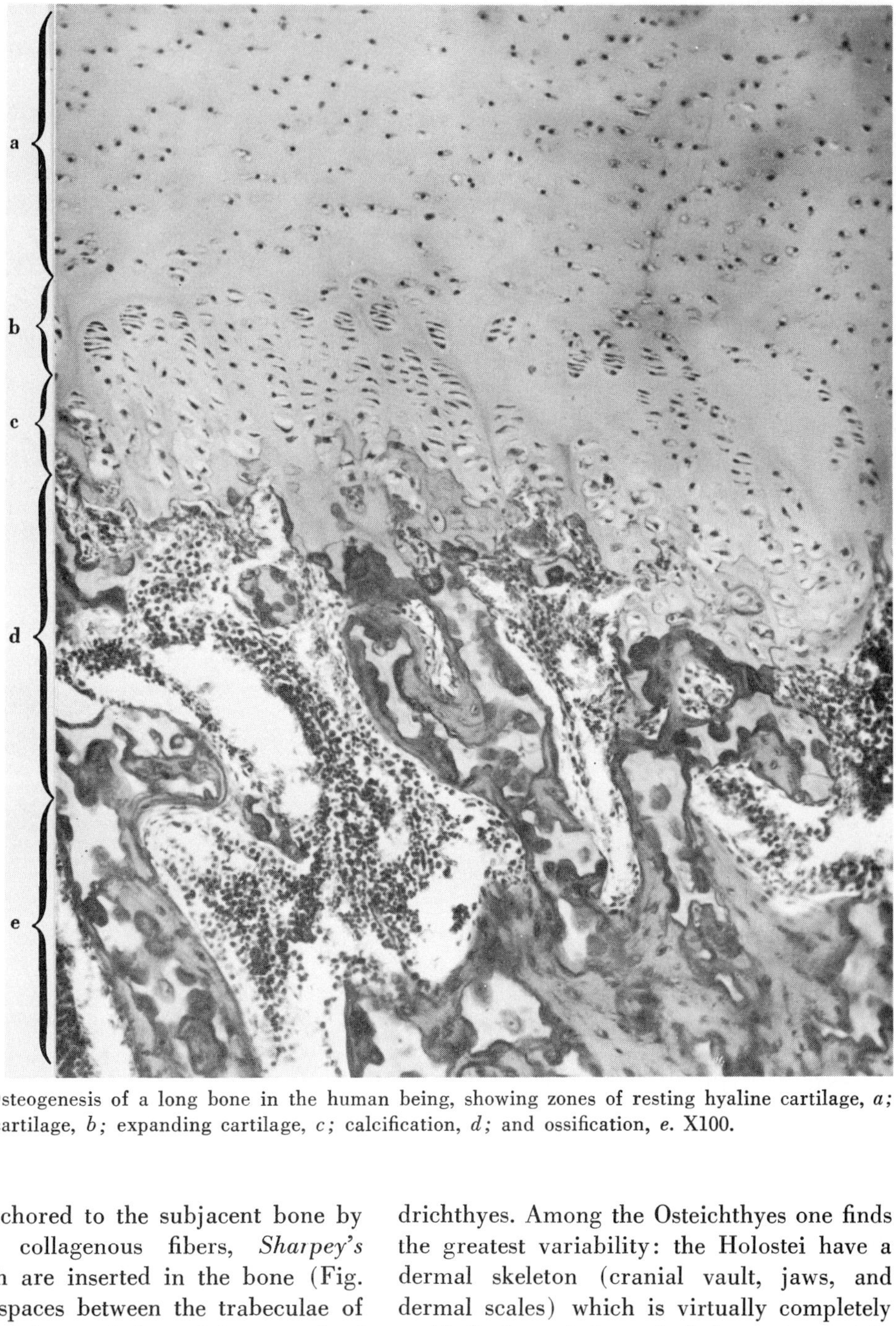

Fig. 3.25 Osteogenesis of a long bone in the human being, showing zones of resting hyaline cartilage, *a;* proliferating cartilage, *b;* expanding cartilage, *c;* calcification, *d;* and ossification, *e.* X100.

teum are anchored to the subjacent bone by bundles of collagenous fibers, *Sharpey's fibers,* which are inserted in the bone (Fig. 3.26). The spaces between the trabeculae of the cancellous bone of the epiphyses and of flat and irregular bones (e.g., skull, ribs, vertebrae, girdles) contain *red marrow,* which is described in Chapter 9.

The extent of ossification varies in the several classes of vertebrates. No ossification occurs in either the Agnatha or the Chondrichthyes. Among the Osteichthyes one finds the greatest variability: the Holostei have a dermal skeleton (cranial vault, jaws, and dermal scales) which is virtually completely ossified, but their endoskeleton is mainly cartilaginous, and the notochord persists to some degree throughout life; the Teleostei generally show extensive ossification of nearly all skeletal components. Ossification is incomplete in the amphibians. The skull, mandible, and vertebral column are nearly completely

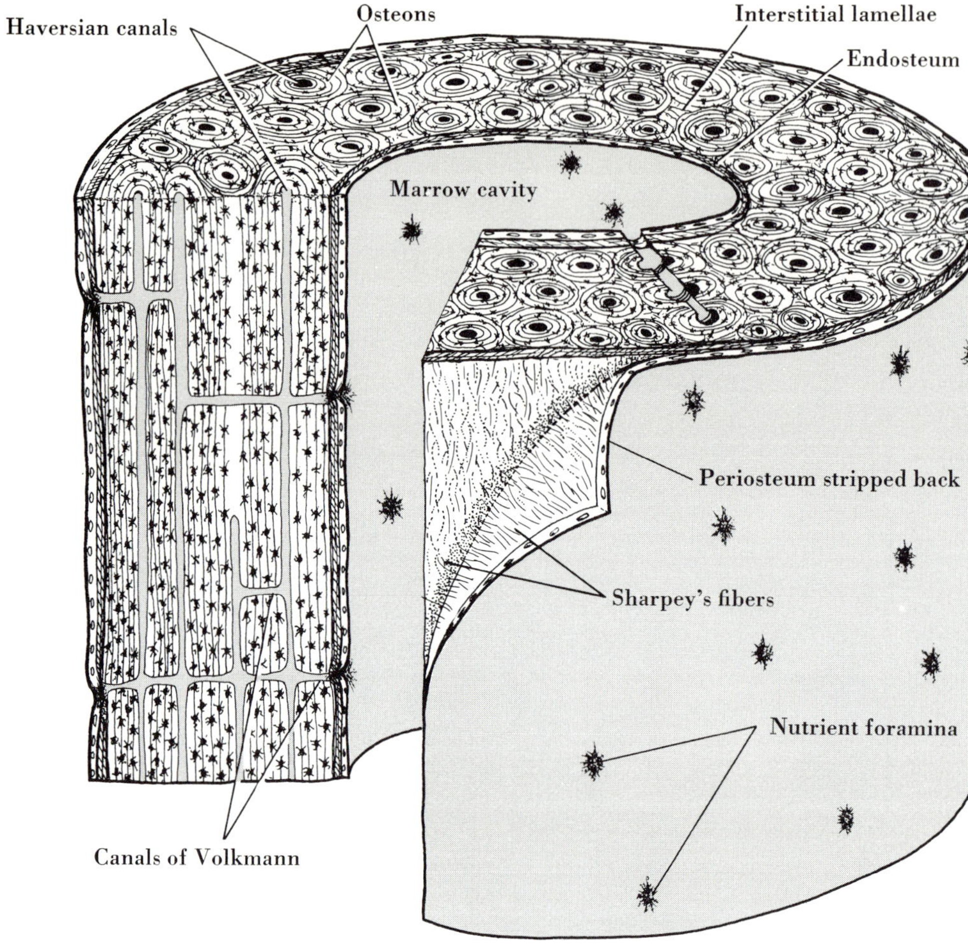

Fig. 3.26 Schema of a section of a long bone.

ossified, but, except in some anurans, the girdles are extensively cartilaginous, and the epiphyses of the long bones consist primarily of calcified cartilage. The diaphyses of the long bones are very thin, compact bone, with little development of osteons.

Reptiles, birds, and mammals show the most extensive ossification. Significant exceptions are the costal cartilages and, in most species, the xiphisternum, both of which consist of hyaline cartilage.

HETEROPLASTIC BONE FORMATION. Although heteroplastic bone formation is not usually considered to be a normal event, it is common enough in tetrapod vertebrates to warrant some mention. It serves also to indicate that osteoblasts are not necessarily restricted to periosteal and bony tissue, but may occur in other kinds of connective tissue. Whether they migrate to these sites from osseous tissue, or differentiate *in situ* from some primitive connective tissue is not known. Common sites of occurrence of heteroplastic ossification are tendons, pharyngeal tonsils, kidneys, and scar tissue. Less commonly it may occur in the heart, in the region of the coronary sulcus. The so-called "splint bones" that have been known to plague athletes are formed heteroplastically. The histogenesis of heteroplastic bone is identical with that of bone which forms by the direct method.

The *sesamoid bones*, which are to be found in certain tendons and ligaments of some

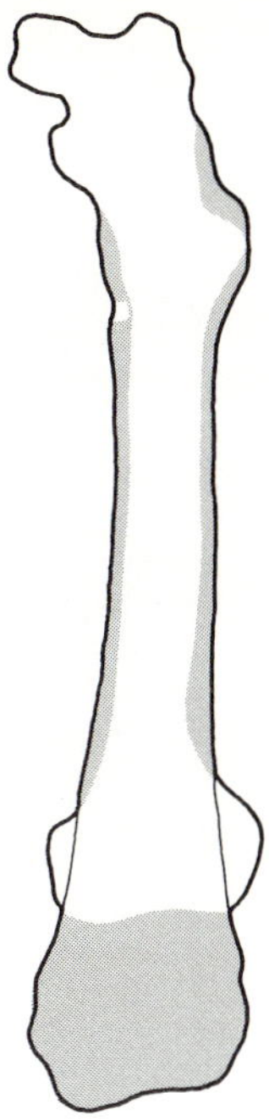

Fig. 3.27 Remodelling of a femur in the shrew, *Sorex cinereus,* as shown by camera lucida drawings of typical immature (*not shaded*) and mature (*shaded*) areas. (After J. L. Felts and F. A. Spurrell. 1967. Am. J. Anat. 120:89–112.

reptiles and most mammals, are similar to heteroplastic bone by virtue of their location, but different in that their occurrence is normal. Examples of sesamoid bones are the *patella* of the knee joint and sesamoids between the proximal phalanx and the metacarpal of the thumb of human beings. The opossum has a pair of sesamoids at the metacarpal-phalanx joint of each digit.

Joints

The juxtaposition of two skeletal elements constitutes a joint. A joint may be rigid (synarthrosis), slightly flexible (symphysis), or freely movable (diarthrosis). In all cases the joint is bridged by fibrous connective tissue which is continuous with the periosteum or perichondrium, as the case may be.

SYNARTHROSES

Synarthral joints are primarily characteristic of the skull. At first consisting of collagenous connective tissue, they then become calcified and finally ossified while the juxtaposed bones are growing appositionally. When fully ossified, such joints are completely rigid and are detectable superficially only as a line or ridge, and as such are known as *sutures.* In some cases there is no external evidence of the suture.

SYMPHYSES

Symphyses, such as the *symphysis mentis* and the *symphysis pubis,* consist most commonly of a plate of fibrocartilage or dense fibrous connective tissue joining a thin cap of hyaline cartilage on the scarified surfaces of each of the abutting bones. The degree of flexibility of symphyses varies considerably. Whereas the symphysis mentis is fairly rigid in most vertebrate species, the symphysis pubis of many female vertebrates is capable of increased flexibility or actually of separation under the influence of the hormone, *relaxin,* during parturition. In some species of mammals a space may be found in the fibrocartilage of the symphysis pubis.

The intervertebral joints are symphyses of a somewhat specialized kind. The hyaline cartilage caps of adjacent vertebrae are joined by fibrocartilage and, surrounding the joint, is a ring of dense fibrous connective tissue. The central part of the joint includes the *nucleus pulposus,* described above.

DIARTHROSES

Also known as synovial joints, diarthroses consist basically of a *joint capsule* which invests the cartilage-capped, juxtaposed skeletal elements; they contain a viscous fluid, the *synovial fluid,* which is made up chiefly of hyaluronic acid and tissue fluid. The capsule is composed of an outer, densely fibrous layer, which is continuous with the periostea of the skeletal elements concerned, and of an inner cellular layer, the *synovial membrane.* The latter is also a mesenchymal derivative, and varies somewhat in composition, depending upon such considerations as pressure, tension, and freedom of movement. At best, it is an ill-defined membrane, consisting of synovial cells, fibroblasts, mast cells, fat cells, collag-

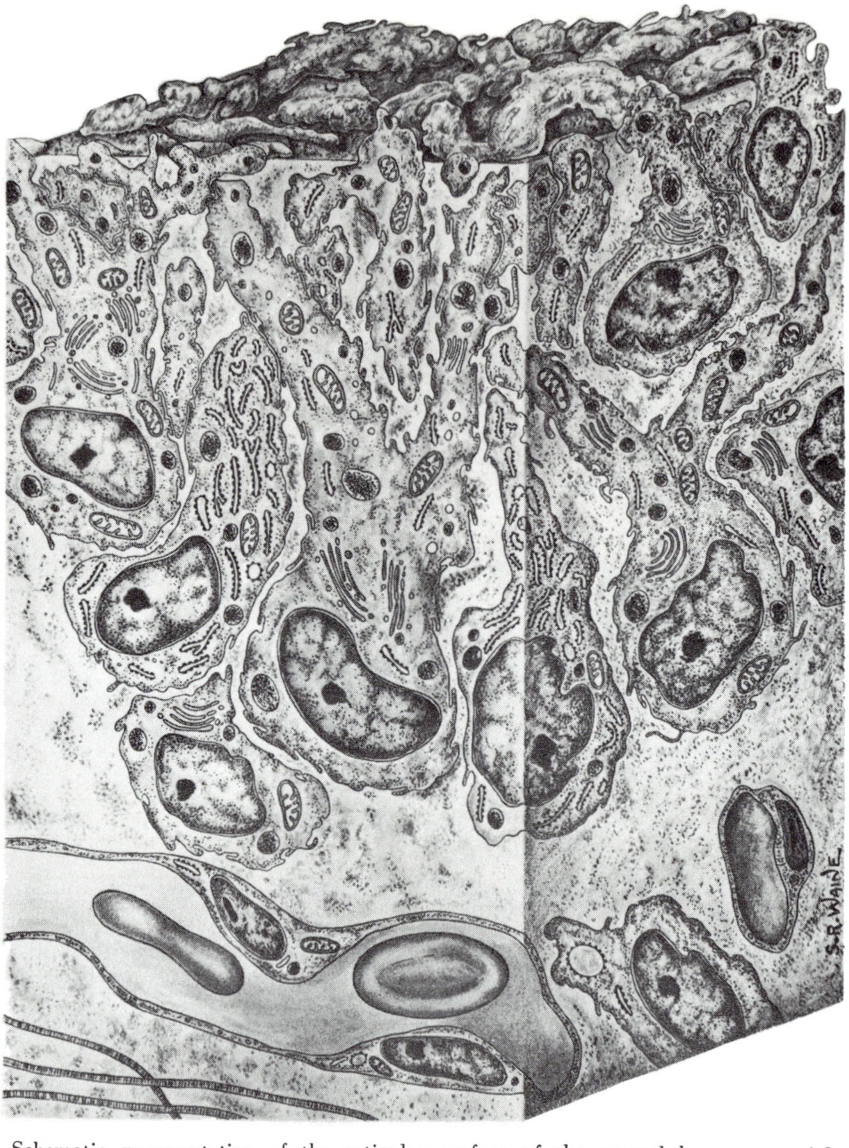

Fig. 3.28 Schematic representation of the articular surface of the normal human synovial membrane. (From P. Barland, A. B. Novikoff, and D. Hamerman. 1962. J. Cell Biol. 14:207–220.)

enous fibers, and elastic fibers. Synovial cells are relatively undifferentiated cells resembling mesenchyme. They are arranged in a kind of pavement lining the joint capsule (Fig. 3.28). The mast cells are believed to be implicated in the secretion of the synovial fluid.

SELECTED REFERENCES

Asboe-Hansen, G., ed. 1954. Connective tissue in health and disease. E. Munsgaard, Copenhagen.

Barland, P., A. B. Novikoff, and D. Hamerman. 1962. Electron microscopy of the human synovial membrane. J. Cell Biol. 14:207–220.

Bohartichuk, F. 1963. A study of bone resorption and the "osteoclast" problem by stain historadiography. Am. J. Anat. 113:117–137.

Dawkins, M. J. R., and D. Hull. 1965. The production of heat by fat. Sci. Am. 213(2): 62–67.

Eakin, R. M., and J. Westfall. 1962. Fine structure of the notochord of Amphioxus. J. Cell Biol. 12:646–651.

Fawcett, D. W., and I. Chester-Jones. 1949. The effects of hypophysectomy, adrenalectomy and of thiouracil feeding on the cytology of brown adipose tissues. Endocrinol. 45:609–621.

Godman, G. C., and K. R. Porter. 1960. Chondrogenesis, studied with the electron microscope. J. Cell Biol. 8:719–760.

Godman, G. C., and N. Lane. 1964. On the site of sulfation in the chondrocyte. J. Cell Biol. 21:353–366.

Ham, A. W., and T. S. Leeson. Histology. 4th ed. J. B. Lippincott Co., Philadelphia, 1961.

Lacroix, P. 1961. Bone and cartilage. *In* The cell, Vol. 2. J. Brachet and A. E. Mirsky, eds. Academic Press, New York.

Murray, P. D. F. 1936. Bones: a study of the structure and development of the vertebrate skeleton. Cambridge University Press, Cambridge.

Nathaniel, E. J. H., and D. C. Pease. 1963. Collagen and basement membrane formation by Schwann cells during nerve regeneration. J. Ultrastr. Res. 9:550–560.

Smith, D. E. 1963. The tissue mast cell. International Review of Cytology. G. H. Bourne and J. F. Danielli, eds. 14:327–386.

chapter 4
EPITHELIAL TISSUES

TISSUE MEMBRANES that cover free surfaces or form the lining surfaces of tubes, sacs, or ducts constitute *epithelial tissues*. The epithelial nature of such tissues as the skin, the lining of the mouth or gut, and the lining of blood vessels is obvious from this definition. Certain other epithelial tissues are not quite so obviously epithelial. For example, most of the liver and pancreas are epithelial, although little free surface is immediately evident in either organ, aside from their ducts. A study of the embryonic derivation of glands such as these, however, reveals their epithelial origins, as does the very form of their cells.

Epithelial membranes and derivatives therefrom develop from each of the three germ layers of the embryo. The ectoderm gives rise to the covering epithelium, and continues also as part of the oral and rectal mucosae. The epithelium lining the central canal and ventricles of the spinal cord and brain has an ectodermal origin, as does the rest of the nervous system. Epithelial tissues derived from endoderm include the lining and glands of the digestive and respiratory systems. Mesoderm or mesenchyme furnishes the lining of the circulatory vessels and coelomic cavities.

Epithelial cells are usually quite regular geometrically. A single nucleus is characteristic, although exceptions can be found, as in liver cells. The shape of the nucleus corresponds to the shape of the cell. Flattened cells contain discoidal nuclei, whereas cuboidal or columnar cells generally exhibit spherical nuclei (Fig. 4.1). Adjacent to the nucleus is a well-developed Golgi network, usually located toward the apical or luminal end of the cell. Its activity and prominence are especially evident in those epithelial cells which are specialized for secretion. Mitochondria are abundant in most epithelial cells, and in those cells which function in the exchange of fluids and salts, the mitochondria are often packed in rows perpendicular to the base of the cell (Fig. 4.2). In such cases the basal plasma membrane may be highly convoluted, with prominent folds lying between the rows of mitochondria. Presumably this represents a mechanism for increasing active surface area, as well as locating the energy source near the site of activity.

Epithelial cells adhere to each other along flat surfaces in the form of continuous sheets, tubes, or cords. The space between cells is

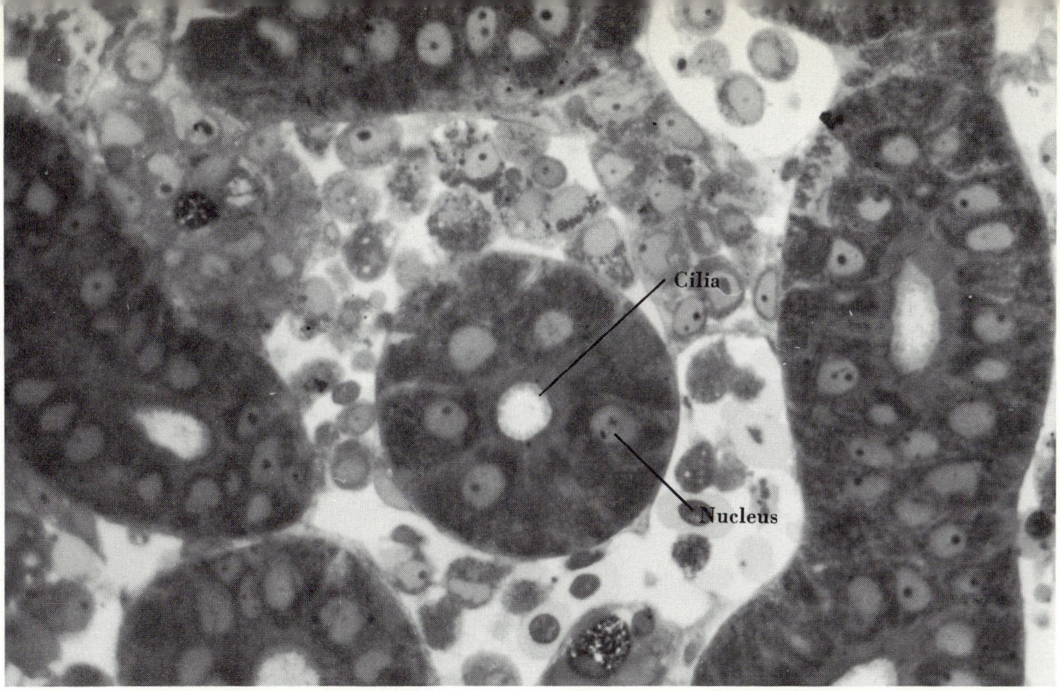

Fig. 4.1 Photomicrograph of epithelial cells in the renal tubules of the toadfish. The tubular cells appear cuboidal to columnar, and have a typically shaped spherical nucleus. Note the presence of cilia on the luminal surface of the cells. X700. (From Fig. 1, R. E. Bulger. 1955. Am. J. Anat. 117:181.)

minimal, rarely exceeding 300A. The results of chemical and microscopic studies suggest the presence of a cohesive force between parallel cell surfaces in epithelial membranes. Furthermore, studies with the electron microscope reveal that epithelial cells are often joined to each other by firm attachment mechanisms of several types, and not infrequently by interdigitating projections from the cell membranes of adjoining cells (Fig. 4.3).

FIRM ATTACHMENTS

The nomenclature used in reference to the various kinds of firm attachments has multiplied to the point of unwieldiness. The student should keep in mind that firm junctions are basically of two types. If apposing cell membranes are fused with obliteration of the intercellular space, the term *occludens* is used. If apposing membranes are separated by a cleft, but differentiated by intercellular thickenings, the term *adherens* applies. Either of these two mechanisms can occur in three different ways. Membranes can be thickened or fused in a band around an entire cell (*zonula adherens* or *zonula occludens*). The firm attachment can be limited to a disklike area (*macula adherens* or *macula occludens*). Finally, the attachment may be more extensive than a disk, but not so extensive as an entire circumferential band (*fascia adherens* or *fascia occludens*). The various kinds of firm attachment mechanisms are perhaps best illustrated by the epithelium of the intestine, which portrays three of the types in what is known as a junctional complex (Fig. 4.4). Running in a band around the cell edges nearest the intestinal lumen is the first of the junctional elements, the *zonula occludens* (Fig. 4.4), which results from a fusion of outer portions of apposing cell membranes. The line of fusion is just barely visible at this magnification. Because of consequent obliteration of the intercellular cleft, the zonula occludens is believed to preclude seepage of materials between the lumen and the intercellular spaces.

Further outward from the lumen, the second of the junctional elements is encountered, the *zonula adherens* or *intermediate junction,* which also occurs as a belt encircling each cell. This type of junction is characterized by a 200A intercellular cleft and the condensa-

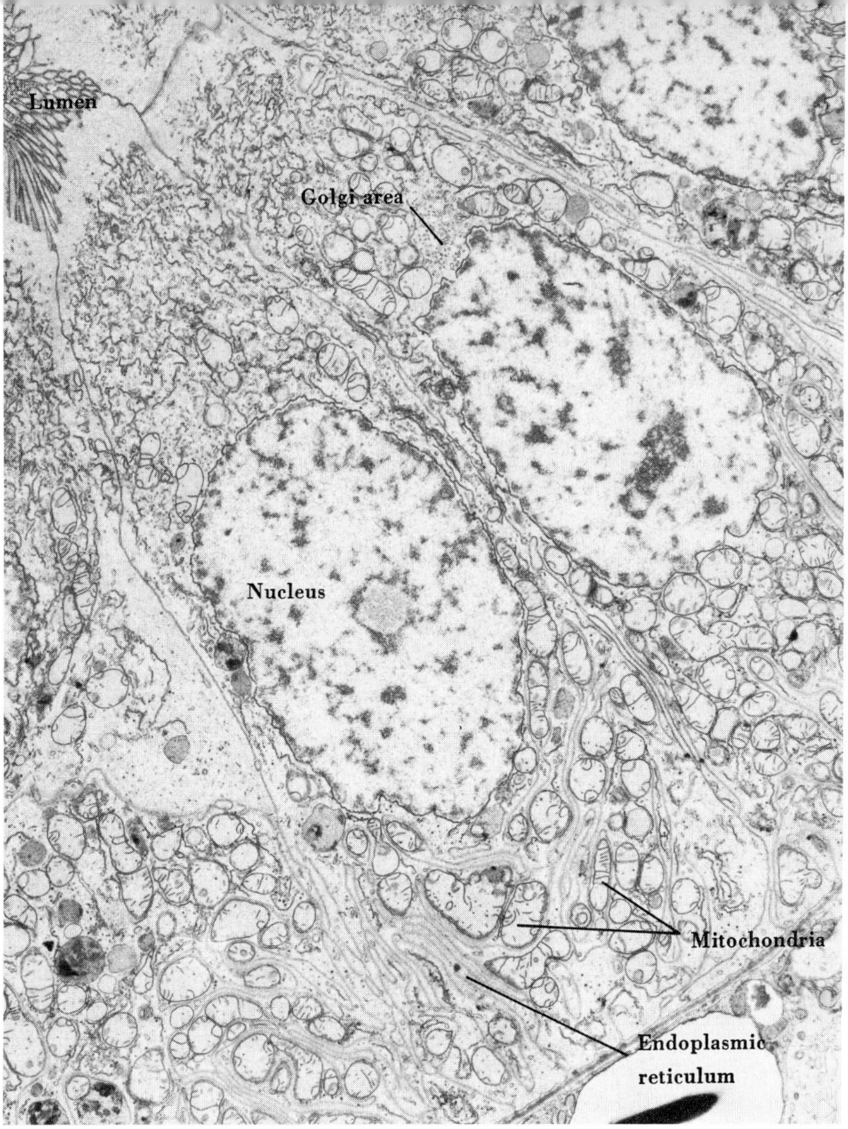

Fig. 4.2 Electron micrograph of the same kidney tubule cells seen in Fig. 4.1. Note the typical arrangement of organelles in secretory epithelial cells. X5900. (From Fig. 3, R. E. Bulger. 1965. Am. J. Anat. 117:181.)

tion of dense filamentous material in the cytoplasm adjacent to the cell membane (Fig. 4.4).

The third junctional element is the *macula adherens,* more commonly known as the *desmosome* (Fig. 4.4). This element does not occur as a belt around the cell, but rather as a pair of platelike densities on apposing cell surfaces. The halves of the desmosome are separated by a 200–250A cleft, and they are coated on their cytoplasmic sides by a dense material that terminates a fanlike array of fine filaments known as *tonofibrils.* The tonofibrils are believed to be keratin. Frequently a thin disk of electron-dense material is found between the desmosomal plaques, as shown in electron micrographs.

Desmosomes do not necessarily appear only as a part of a junctional complex, for they may be present as the only firm intercellular union. Occasionally and under certain situations hemidesmosomes are found. It is believed that a hemidesmosome on one surface induces the formation of a partner on the apposing surface, and hence is not completely differentiated alone. However, hemidesmosomes have been described on the basal surface of cells next to the basement lamina, and in this kind of situation the hemidesmosomes probably are mature.

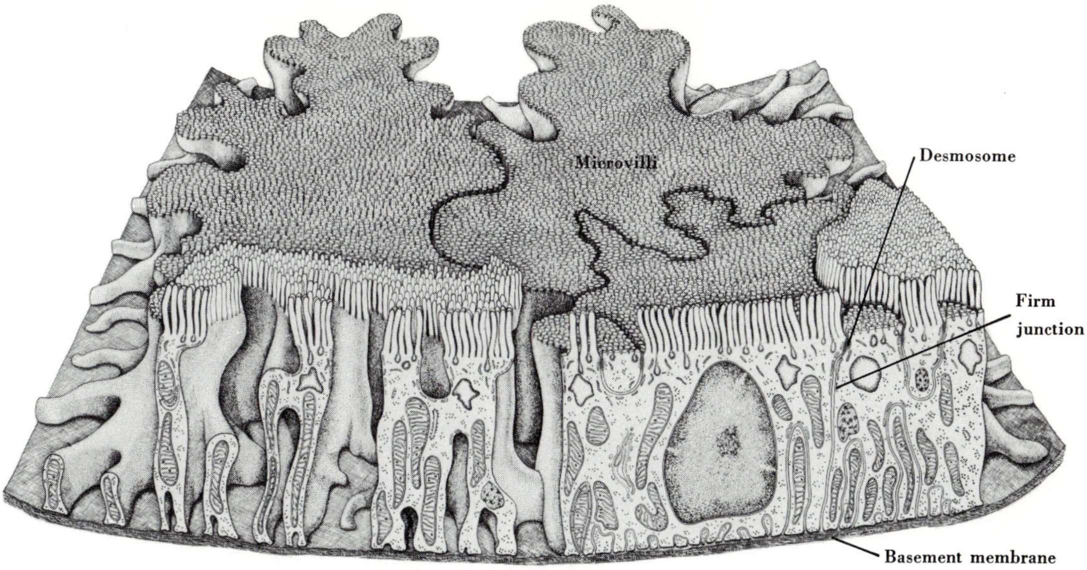

Fig. 4.3 In this diagram of epithelial cells in the rat kidney tubule, note the irregular course of the cell membranes. These interdigitations help to hold the cells together and act in conjunction with desmosomes and other firm junctional mechanisms. (After Fig. 5. R. E. Bulger. 1965. Am. J. Anat. 116:252.)

The electron microscope has failed to support any evidence of protoplasmic continuity between adult cells, i.e., intercellular bridges. Nonetheless, histochemical studies have revealed the presence of PAS-positive material concentrated at the cell interfaces, and this material is believed to represent intercellular cement, presumably a mucopoly saccharide.

BASEMENT MEMBRANES

Basement membranes are thin mats of connective tissue on which many epithelia appear to rest. They are clearly defined in portions of the digestive tract. Not to be confused with the basement membrane is another special structure of rather different dimensions and origin; this is the *basement lamina.* The basement lamina appears as a fine feltwork of fibrils embedded within an amorphous gel. The fibrils are unbranched and are thought to be collagenous in nature, probably tropocollagen (see Chapter 3). The gel consists of mucopolysaccharides with small amounts of glucose and galactose. The basement lamina is antigenically more similar to the overlying epithelial cells than to the underlying connective tissue, hence it is thought to be of epithelial- rather than connective-tissue origin. The basement lamina appears as a dense line about 600A in thickness, which is below the resolution of the compound microscope. A light zone measuring about 400A intervenes between this dense line and the basal cells of the epithelium (Fig. 4.2). Generally the basement lamina reaches its maximum development in very actively secreting epithelia, or in epithelia characterized by a rapid rate of transfer of materials through its cells, such as in the renal tubules, the lung, and the capillaries. It is the only continuous separation between the lumen of the blood capillaries and the lumen of Bowman's capsule, which reflects its fundamental physiological importance. The older term, *basement membrane,* is now used to indicate a composite of basement lamina plus adjacent connective tissue fibers.

FREE SURFACES

The free surface of epithelial cells is commonly modified in some way significant of its function. An epithelium which is primarily protective and subject to abrasion may be cornified (see Chapter 6). Absorptive and secretory surfaces are characterized by a cell membrane which is thrown into a multitude of submicroscopic processes, or *microvilli* (Figs. 4.3 and 4.5). Typically, a microvillus is from 1 to 1.5μ in length and from 0.08 to 0.1μ in diameter. Passing from a surface zone of clear cytoplasm into each microvillus are straight

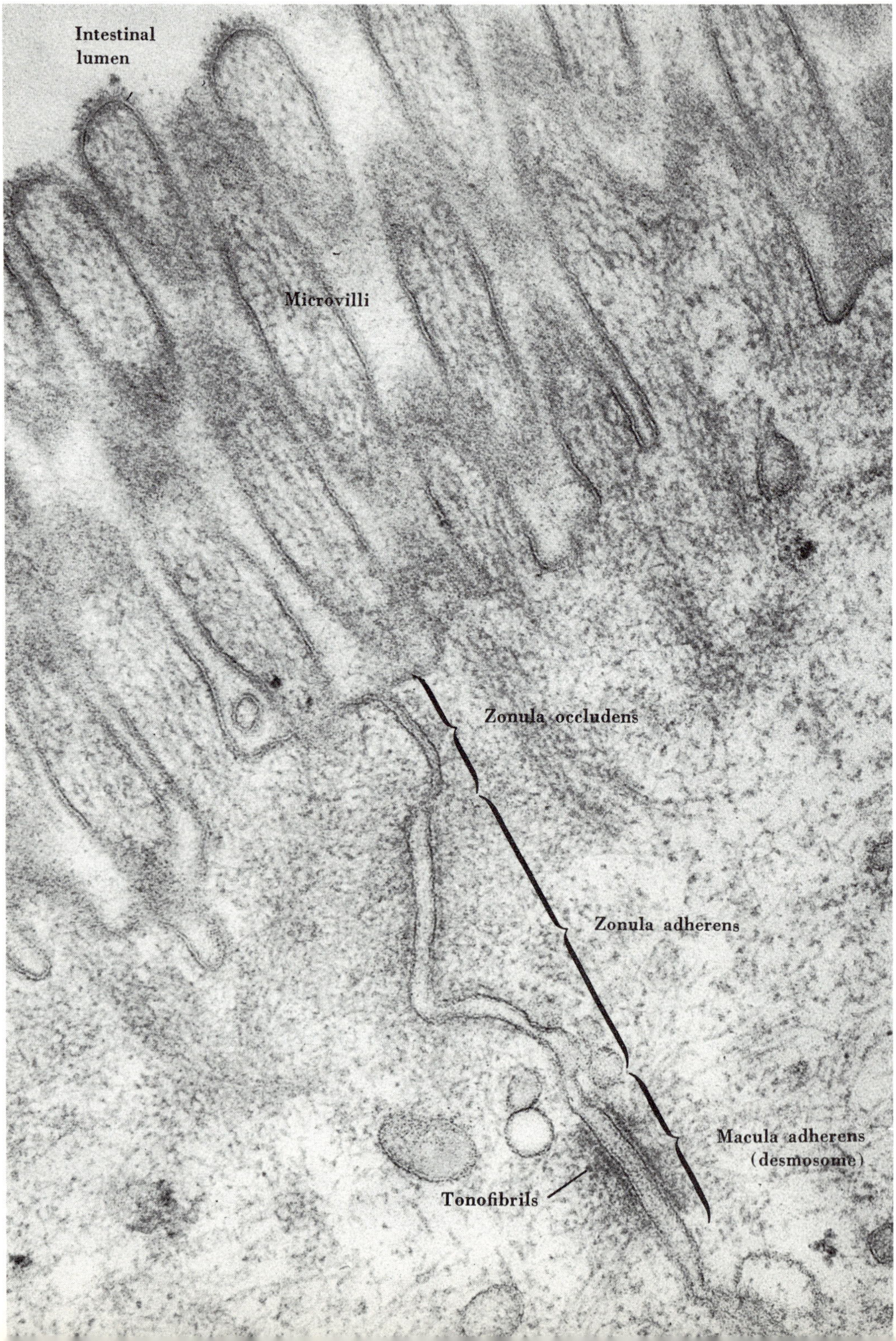

Fig. 4.4 Electron micrograph of the epithelial lining of the rat intestine. All three areas of the junctional complex can be seen quite clearly at this magnification. X96,000. (From M. G. Farquhar and G. E. Palade. 1963. J. Cell Biol. 17:379.)

filaments that suggest a rigid internal skeleton. In intestinal epithelia each cell bears approximately 2000 microvilli, which impart a striated or brushlike appearance to the surface when it is viewed by light microscopy. Such surfaces are designated *striated borders* or *brush borders* accordingly (Fig. 4.1). In some locations, as the epididymal epithelium, the microvilli are of variable length and uneven array. Previously thought to be nonmotile cilia, they are now interpreted as being long, flexible microvilli.

Many epithelial membranes, especially those containing microvilli, have a further modification of surface. Coating the tips and extending between the microvilli is a layer of fine, branched filaments matted together to form a loose lattice. Appearing to be mucopolysaccharide in nature, this surface coat seems to function as a block to passage of large particles that are neither in solution nor in suspension (Fig. 4.5). Magnifications of about 200,000 diameters make it possible for one to detect continuity between the plasma membrane and the filaments coating it. Presumably then, the lattice coat is a product of the surface of the cell.

Mucous epithelia that must be kept free of detritus or for other reasons must maintain a flow of material over their surfaces are usually ciliated. Occasionally cilia may be observed in cells wherein the only explanation so far for their presence is that such cells were derived embryologically from a ciliated epithelium, such as in the gland cells of the interrenal bodies or in the follicles of the thyroid gland.

The fine structure of cilia is remarkably uniform throughout the living world; in all species in which they occur, they consist of a core of nine double tubules, encircling a pair of single, central tubules (Fig. 4.6). The plane in which the two central tubules lie is perpendicular to the plane in which the cilia beat. Tapering toward the tip, the tubule complex is embedded in a matrix and invested in

Fig. 4.5 The free surface of adjacent epithelial cells of the guppy (*Lebistes*) is shown in this electron micrograph. The relatively regular microvilli are in a fibrillar matrix or fuzz. Note the zipperlike arrangement of desmosomes on the lateral cell margins. X32,000. (Courtesy of R. C. Henrikson.)

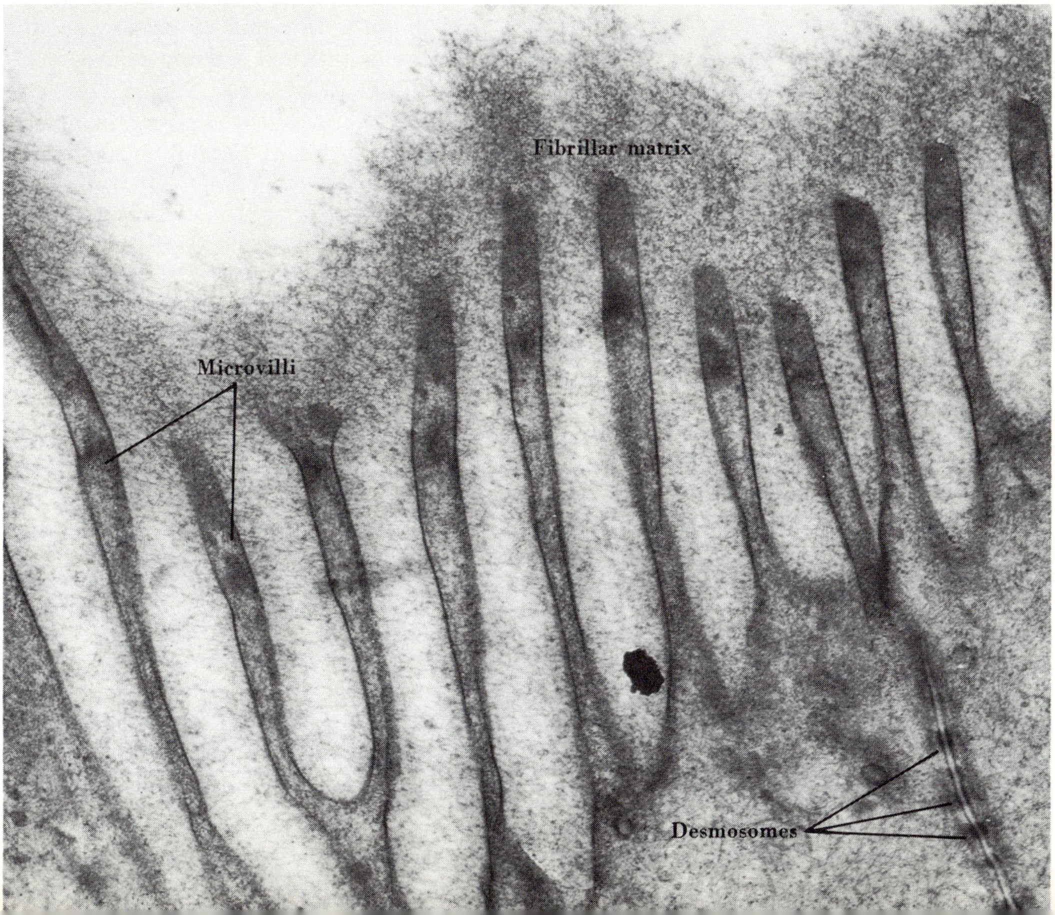

an evaginated portion of the plasma membrane. Basally the tubules terminate in a centriolelike structure called the basal body or basal corpuscle. With few exceptions, the structural plan of flagella is identical to that of cilia. The differences between the two are matters of length and aggregation. Generally cilia are short (up to 10μ) and are found in a forest of other cilia. Flagella, on the other hand, are longer and are usually present singly or in the company of only one or two others.

CLASSIFICATION OF EPITHELIA

Epithelial tissues may be classified conveniently according to the shape of the cells and according to whether one or more layers of cells constitute the membrane. Each of the classes may be further qualified according to whether the cells are ciliated or cornified. On this basis the following classes can be recognized.

Simple Epithelia

Simple epithelia are only one cell in thickness. They may be *squamous*, being composed of rather flat cells with central, disk-shaped nuclei; or *cuboidal;* or *columnar* (Fig. 4.7). Cuboidal or columnar cells, appearing square or rectangular, respectively, if cut perpendicular to the surface, are, in fact, prismatic in shape. The difference between the two types lies in the height of the prism. Cilia or microvilli often occur as adjuncts of these epithelia. In the category of simple epithelia should also be included *pseudostratified* columnar epithelium, so-called because the nuclei occur at two or more levels, giving the appearance of stratification, whereas actually each cell rests on the basement lamina.

Simple epithelial membranes will be encountered frequently in the chapters which follow. However, a few examples of the locations in which each kind of membrane can be found would be instructive with reference to function. Simple squamous epithelium appears, among other places, as the lining of blood and lymph vessels, as well as forming the tubules and small ducts of many glands. In these situations the epithelium presents the smallest separation between the lumen and the tissue spaces, thereby facilitating the exchanges of fluids, dissolved or suspended substances, and occasionally larger particles or single cells. Simple cuboidal epithelium is often found in glands where it can appear as a membrane or a cord of cells. In many of these cases, the cuboidal cells function as the glandular secretory elements (Fig. 4.1). Simple columnar epithelium is also often secretory in nature. Lining the intestine of vertebrates, these cells typically have a brush border. Ciliated colum-

Fig. 4.6 This electron micrograph of trout sperm tail pieces shows a length of a flagellum. The insert clearly shows the 9 + 2 tubular structure of flagella or cilia in cross section. X30,000. (Courtesy of F. B. Merk.)

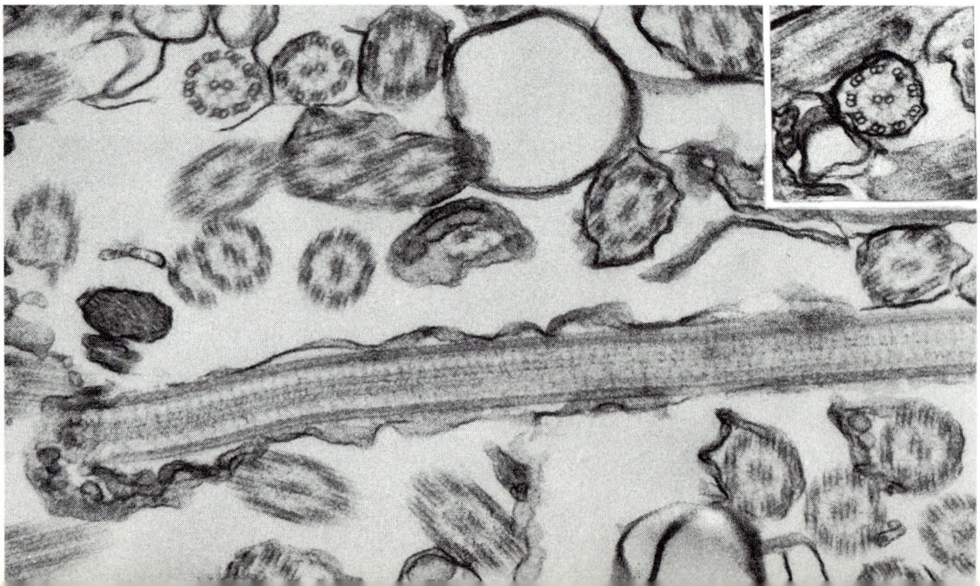

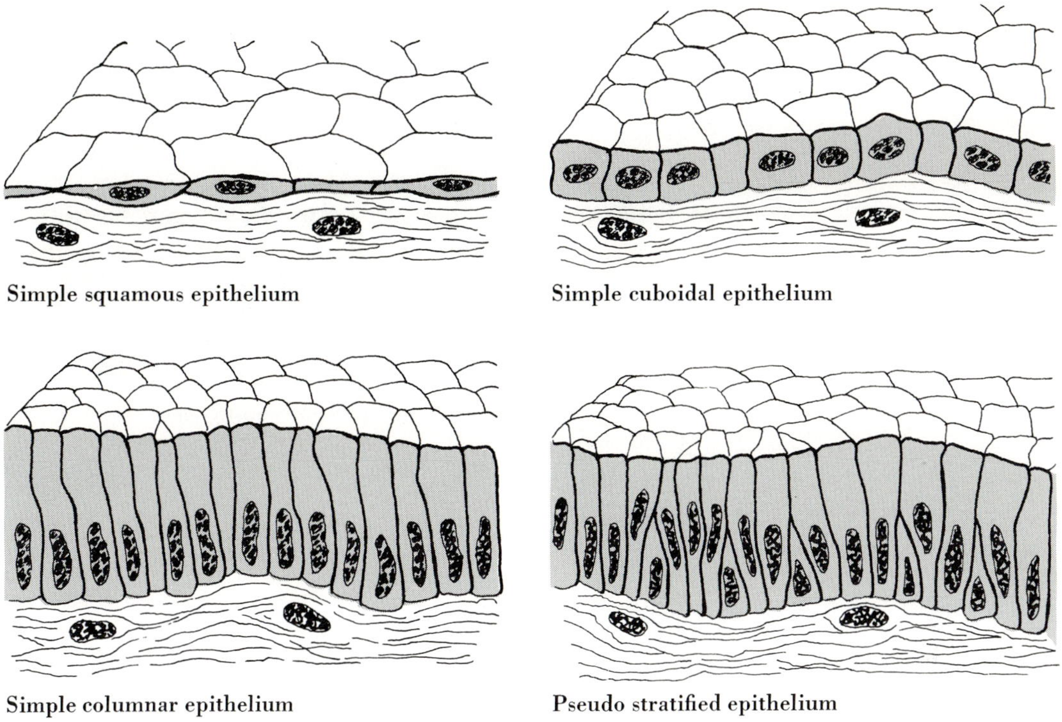

Fig. 4.7 A diagrammatic view of the various kinds of simple epithelial membranes and the connective tissues upon which each rests. Note the appearances of the epithelial cells in section and in surface view.

nar epithelium is common in parts of the respiratory and reproductive tracts.

Stratified Epithelia

When more than one layer of cells is present, the epithelium is called stratified (Fig. 4.8). If the cells of the outermost layer are squamous, the epithelium is called *stratified squamous,* even though the underlying cells may show all stages of transition from columnar at the basal layer, through cuboidal and polygonal near the surface, to squamous at the surface. Stratified squamous epithelium is generally characteristic of surfaces that are primarily protective in function, such as the skin and the lining of the oral cavity and esophagus. *Stratified cuboidal* is uncommon, being generally restricted to certain cutaneous glands (sweat glands and sebaceous glands of mammals) and to the ovarian follicle. *Stratified columnar* is also rare, usually occurring in the transitional zones between stratified squamous and pseudostratified columnar epithelia, such as in the respiratory tract of tetrapods and in the penile urethra of the mammal. Even so, as with stratified squamous, the name is somewhat misleading, for only the surface cells are truly columnar; the underlying ones are usually cuboidal. The last class of stratified epithelia is *transitional epithelium,* which is characterized by a surface pavement of cells with convex apices, much like cobblestones, overlying several layers of polygonal and ovoid cells. The lower sides of the surface cells are concave. Transitional epithelium is restricted to portions of the urinary system, as the ureter and urinary bladder, which may be subjected to varying amounts of distension with fluid. This kind of epithelium accommodates for stretching by the cells' sliding by each other, as it were, so that in the distended state, the number of layers of cells is actually reduced. A distinct basement lamina is lacking in transitional epithelium.

From the foregoing it must be evident that the functions of epithelia include protection, secretion, and absorption. In the skin we shall

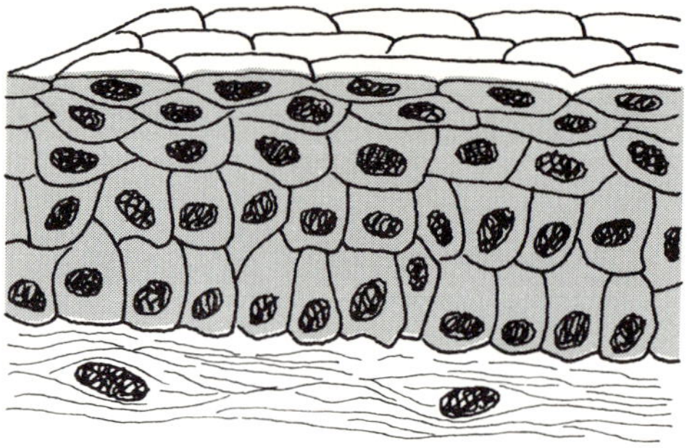

Stratified squamous epithelium

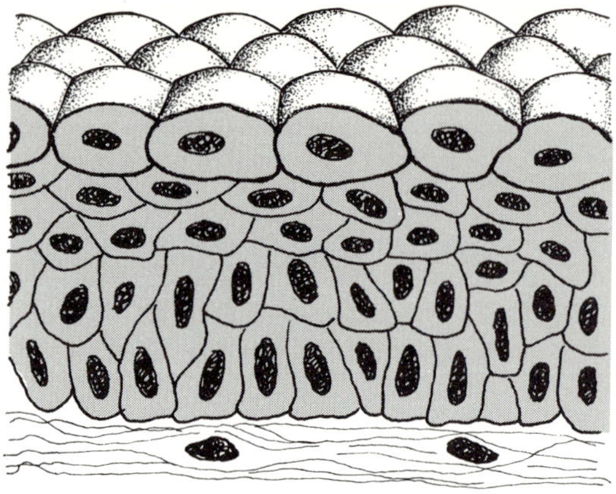

Transitional epithelium unstretched

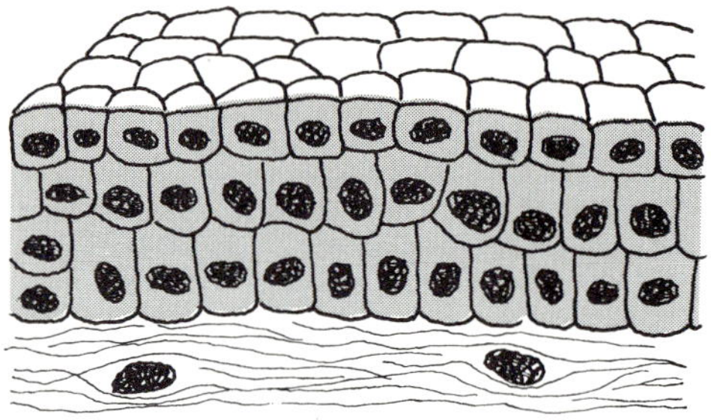

Stratified cuboidal epithelium

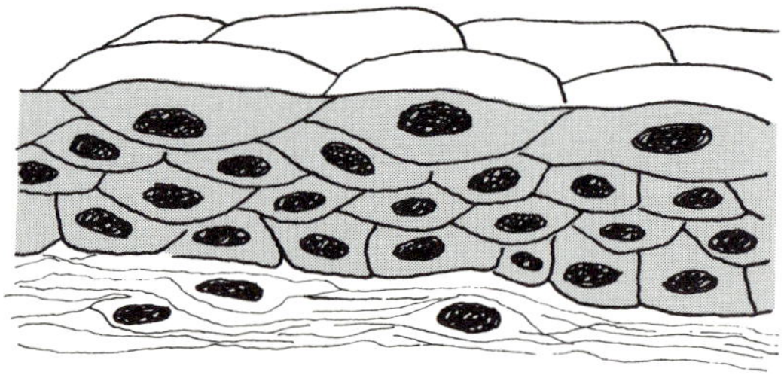

Transitional epithelium stretched

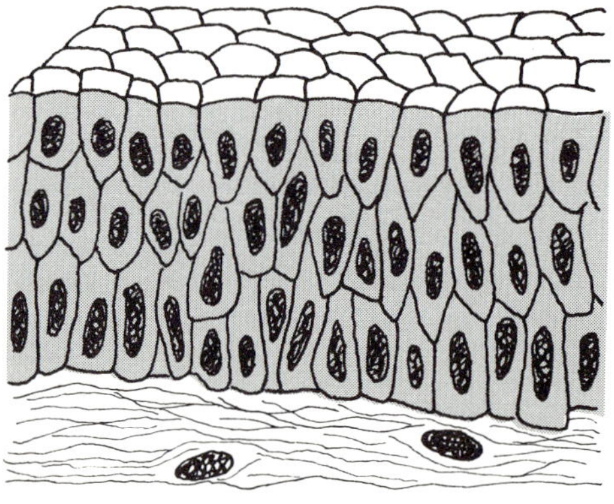

Stratified columnar epithelium

The cells within a transitional epithelial membrane can slide past one another. By doing this, the depth of the membrane decreases as the area increases.

Fig. 4.8 A diagrammatic view of stratified epithelial membranes. The name assigned to each membrane is related to the shape of the surface layer of cells.

see that another important function may be a sensory one. The skin also illustrates most significantly the protective and, in some cases, the secretory functions, whereas absorption and transfusion are best illustrated by the mucous membranes of the digestive tract and the endothelium of the blood vessels and related organs. Each of these and other functions will be elucidated more fully in later sections of this book, but one of them, secretion, will be considered briefly now from the structural standpoint (see Chapter 2 for secretion mechanisms).

EXOCRINE GLANDS

Glands are considered to be *exocrine* if their secretion is released on a free surface, either directly or by way of ducts. Those whose secretion passes directly into the blood or tissue fluid, without ducts, are called *endocrine glands,* such as the hypophysis, the thyroid, or the interrenal glands. With seemingly the perverse intention of thwarting the human tendency to fit everything into neat categories, some glands are both exocrine and endocrine, such as the pancreas and, in a certain sense, the gonads. Moreover, there are some organs which only secondarily may be considered glands, such as the duodenum, which in mammals has exocrine glandular elements and an endocrine function, as we shall see later. Nevertheless, we find that all glands are epithelial in origin, at least, and that exocrine glands may be classified according to their structure and complexity, as follows.

Unicellular Glands

In the skins of lower vertebrates, as well as in the mucous membranes of all vertebrates, single glandular cells occur. The most common of these is the goblet cell (Fig. 4.9), the cytoplasm of which bulges with mucus.

Multicellular Glands

Certain areas of epithelium take on a completely glandular function, so that all of the cells are modified as either gland cells or as ducts, or—as in salivary and sweat glands of mammals—contractile cells that help to move the secretion along. According to their complexity such glands are termed either *simple, branched,* or *compound;* and according to the shape of the secretory portion, they are *tubular, acinar,* or *tubuloacinar.* The several types are shown diagramatically in Fig. 4.10.

Exocrine glands may also be subjected to physiological classification. A *merocrine* gland is one which forms a product and secretes it, free and clear. Goblet cells and the mucous and serous cells of the pancreas and salivary glands are of this kind. *Apocrine* glands extrude a portion of the cell with the secretion. The lactating mammary gland is apocrine. Finally, the *holocrine* gland, of which the sebaceous gland is an example, re-

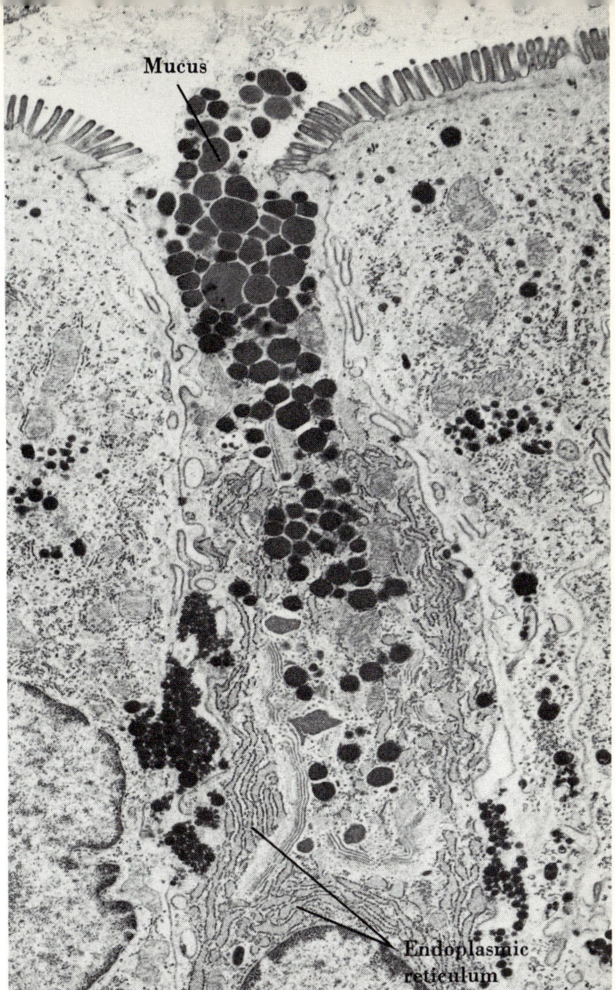

Fig. 4.9 An electron micrograph of a goblet cell in the rat jejunum. Most of the organelles are compressed in the lower portion of the cell. The luminal area contains droplets of mucus prior to secretion. Note the extensive network of endoplasmic reticulum. X9000. (From Fig. 7, J. A. Freeman. 1966. Anat. Rec. 154:141.)

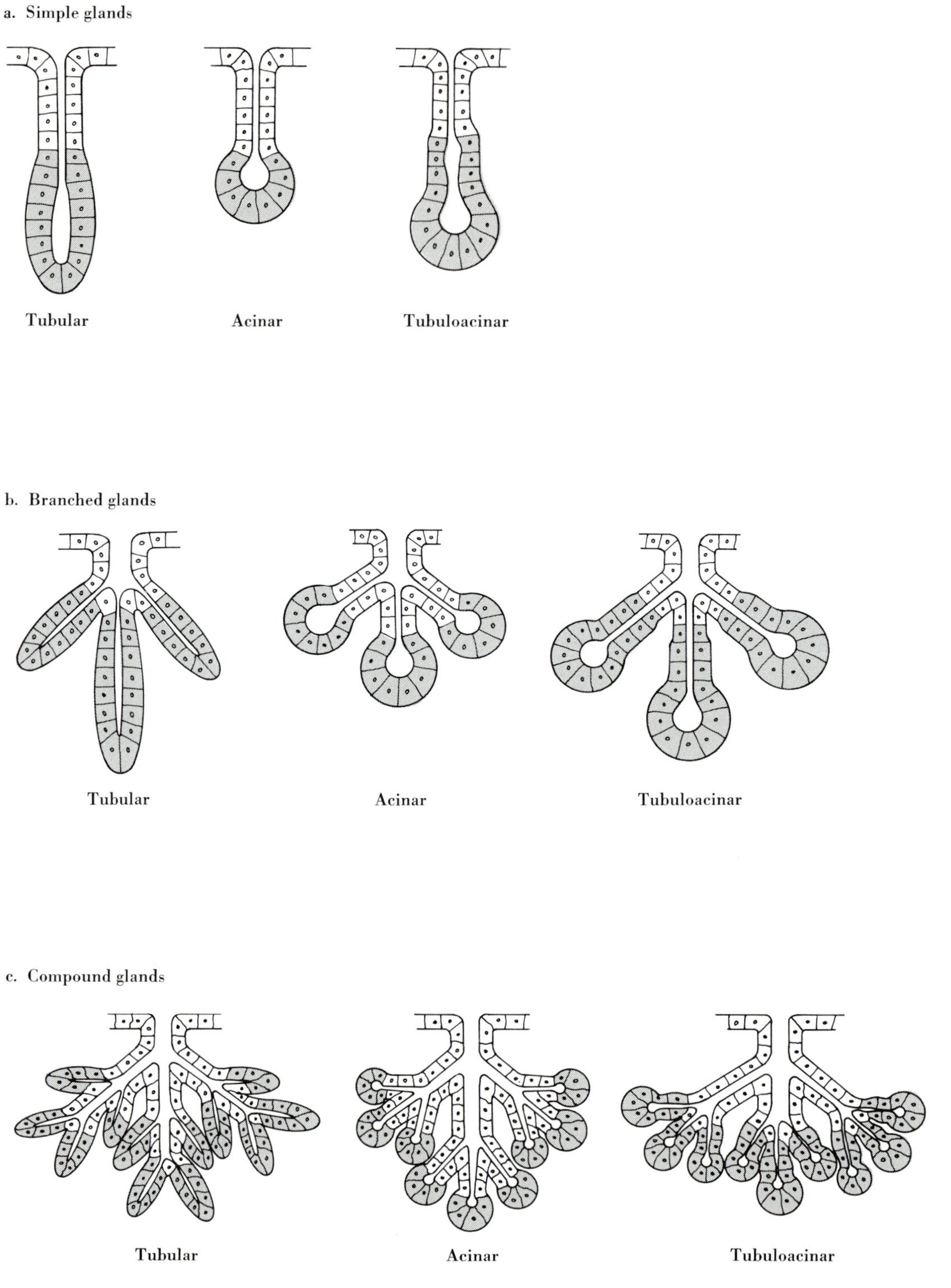

Fig. 4.10 A diagrammatic survey of anatomical types of multicellular exocrine glands. The actual secretory parts of the glands are shaded; it is from the shape of this portion that the gland derives its name. Because they are exocrine glands, the secretory portions are joined to a surface by a duct or ducts. Note that the compound glands contain branches, as do the branched glands, but that they are further subdivided into multiple secretory areas.

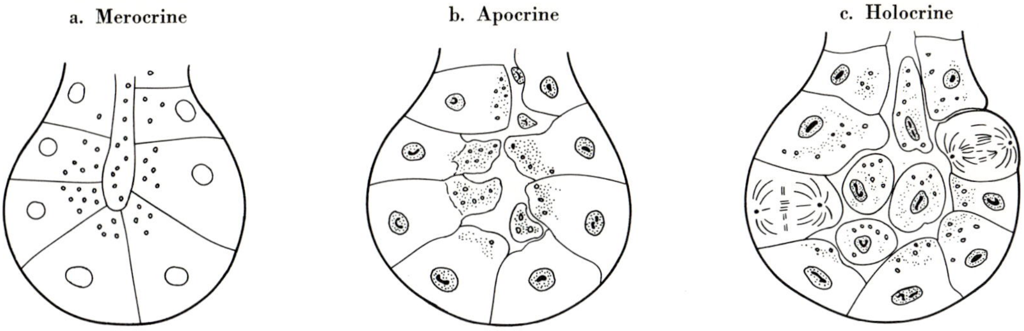

Fig. 4.11 Diagram of the physiological types of glands. Anatomically, the glands depicted here are of the simple acinar variety. Note that in the holocrine gland whole cells are sloughed off; hence, the remaining cells are mitotically active in order to replace the cells that have been lost.

leases the secreting cell in its entirety as a part of the secretion (Fig. 4.11).

Cell structure in each of the foregoing categories of glands can vary. To avoid redundancy, description of cellular architecture for each gland type will be deferred to later chapters in which the individual glands will be considered in their proper context.

SELECTED REFERENCES

Bulger, R. E. 1965. The fine structure of the aglomerular nephron of the toadfish, *Opsanus tau*. Am. J. Anat. 117:171–191.

Copenhaver, W. M. 1964. Bailey's textbook of histology. 15th ed. Williams & Wilkins Co., Baltimore.

Farquhar, M. G. and G. E. Palade. 1965. Cell junctions in amphibian skin. J. Cell Biol. 26:263–291.

Fawcett, D. W. 1966. An atlas of fine structure: the cell—its organelles and inclusions. W. B. Saunders Co., Philadelphia.

Freeman, J. A. 1966. Goblet cell fine structure. Anat. Rec. 154:121–148.

chapter 5
MUSCULAR TISSUES

The muscular tissues are one of the most highly differentiated tissues of the body. Their most obvious and most important characteristic is that of contraction, and to that end muscle cells are elongate structures. Functionally such a shape is important because the degree of shortening possible in one dimension of a long, thin cell is greater than that which could be obtained in a more rounded cell of equivalent volume.

Muscular contraction has two resultants. The first, of course, is movement—not only of skeletal elements but also of the contents of the various body tubes, such as those of the circulatory or digestive systems. Second, contraction functions in maintaining sufficient tension in the body so that forces such as those of gravity do not cause movement. In this respect, muscles act in maintaining posture.

Another characteristic of muscle cells, or fibers as they are commonly called, is their high capacity for conduction of electrical impulses. This specialized development is most easily seen in the cardiac muscle of the heart, which has the innate ability to undergo rhythmic contraction without stimulation by the nervous system. Although most muscles require electrochemical stimulation to function, some muscle fibers have no direct nervous connections. These cells, located especially in the viscera, are stimulated indirectly by conduction of electrical impulses from muscle fibers that do have an intimate nervous association. Hence, muscles are one of the prime energy converters of the body, receiving electrochemical potential energy and transforming it into kinetic or potential mechanical energy, or, in some cases, to heat or even to electrical energy.

Often confusing to the student is the question of tissues, organs, and organ systems within the framework of muscles. When one contracts one's biceps, or dissects the cat trapezius, one is working with an organ, and all of the skeletal muscle masses of the body make up an organ system. The muscle known as the heart and those fibers found in the blood vessels are considered part of the circulatory system. In like manner, individual muscle fibers or groups of fibers embedded in other structures, such as the skin muscles or those of the intestine, are included within the system where they are found. We have defined tissues as groups of like cells performing a group function, and this definition is valid for muscle fibers as well. Bands of fibers are found along the tubes of the digestive, circulatory, urinary, and reproductive systems, and in organs such as the stomach, heart, bladder, and uterus. These bands of muscle fibers can be considered tissues. A skeletal muscle mass contains many muscle fibers in addition to blood vessels, nerves, and connective tissue. The muscle fibers of such an organ constitute the skeletal muscle tissue.

Inherent in the foregoing discussion is an understanding of the morphological and physiological classifications of muscle. Some muscle cells, when seen microscopically, appear transversely striped or striated, whereas others do not show such striation; muscles can be classified on this basis. Striated muscle includes the muscle masses attached to the skeletal elements, and the muscle of the heart. Nonstriated, or smooth, muscles are found in the tubes and layers of other organ systems. Classification can also be on functional grounds. Muscles which are under the control of the somatic-motor nerves are called voluntary muscles, while others, the involuntary muscles, are stimulated by the autonomic nervous system. In the human context only can we say that voluntary muscles are under conscious control and vice versa. Both sets of criteria are used for the ultimate characterization of muscle because in the vertebrates all voluntary muscles are striated, and except for the heart, all involuntary ones are smooth. Because the muscles of the heart are involuntary and striated, they form a third category known as cardiac muscle. The three kinds of muscle cells are differentiated in Fig. 5.1.

With the exception of the ectodermally derived ciliary muscle of the eye and the myoepithelial cells of the skin glands, vertebrate muscle tissue is of mesodermal origin. Smooth muscle is formed from mesenchymal cells that, as parts of the viscera and blood vessel walls, elongate and become contractile. Cardiac muscle develops from anterior splanchnic mesoderm. Skeletal muscles are derived from a number of mesodermal sources. The muscles of the head and branchial regions

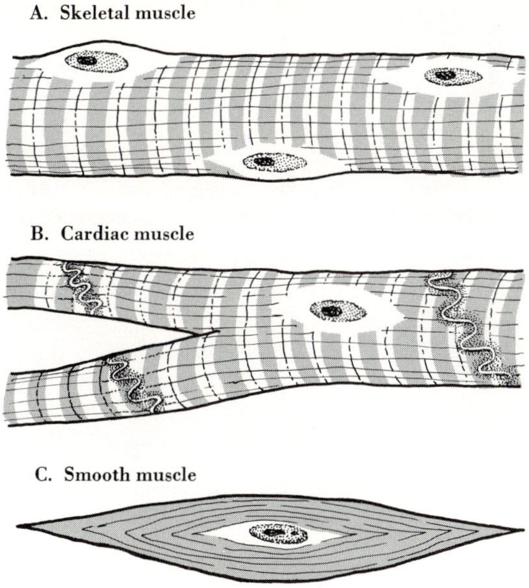

Fig. 5.1 A diagrammatic survey of muscle fibers as they appear with the light microscope. Compare the different types with respect to cell shape, striation, and number and location of nucleus or nuclei.

originate from mesenchyme and anterior myotomes, whereas most of those of the trunk are solely myotomal. The buds of mesenchyme which differentiate into fin or limb muscles may also take their origin, at least in part, from various trunk myotomes.

SMOOTH MUSCLE

Smooth muscle fibers are found in all organs of the body, either intrinsically, such as in the digestive tract and associated glands, ovary, spleen, urinary system, blood and lymphatic vessels, skin, etc., or secondarily as parts of the circulatory vessels draining the organ, such as in the pancreas, heart, adrenal gland, etc. Or they may be found in both circumstances. By virtue of these locations, smooth muscle is primarily concerned with regulation of the internal environment. This is opposed to the situation with skeletal muscles, which serve to adjust the organism to its external environment. Because smooth muscle fibers usually occur surrounding or running parallel to hollow tubes, they lack strict origins and insertions.

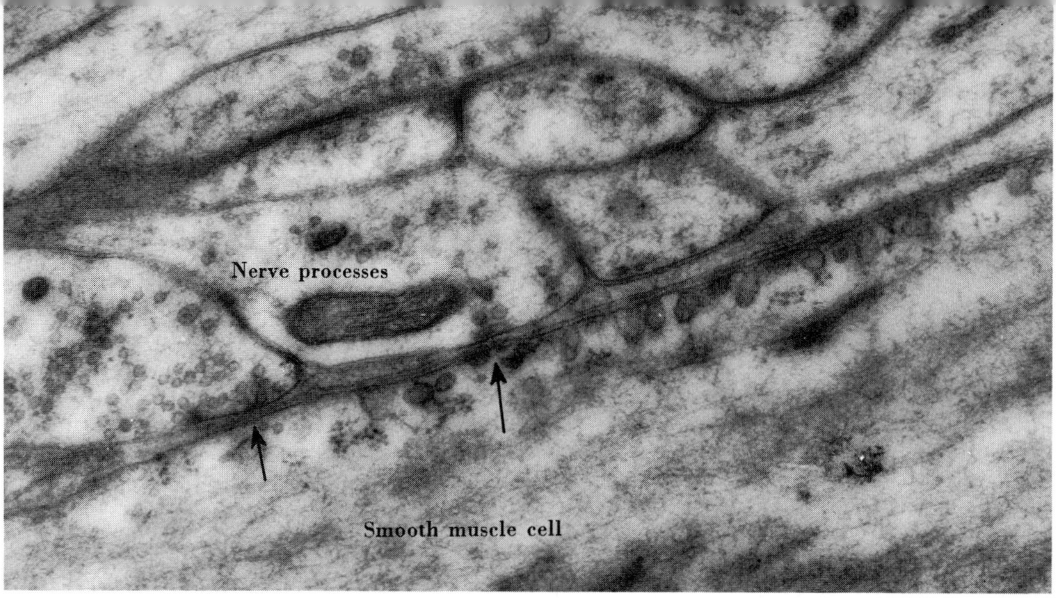

Fig. 5.2 Innervation of smooth muscle. Note the proximity of the cell membranes of nerve and muscle (*arrows*). Synaptic vesicles appear in the nerve processes, and similar ones are found in the apposed muscle cell. Myofilaments and neurofilaments can be seen in the appropriate cells. X40,000. (From J. C. Thaemert. 1963. J. Cell Biol. 16:361–377.)

As would be expected from its nonvoluntary character, smooth muscle is innervated by branches of the autonomic nervous system. Bundles of unmyelinated nerve fibers course between the individual muscle fibers where nerve processes make contact with the surface of the muscle fibers (Fig. 5.2). As in all kinds of myoneural junctions, the nerve twig does not enter the interior of the muscle fiber, but rather, the plasma membranes of both cells occur in close apposition to each other. Not every fiber, however, maintains an association with a nerve twig. The impulses spread to noninnervated muscle fibers by two mechanisms. (1) Surrounding each muscle fiber are interconnecting nets of reticular and elastic connective tissue fibers. When one muscle fiber is stimulated to contract, a pull is exerted on the connective tissue fibers. The force of contraction is thus transmitted to all the integrated muscle fibers. (2) At points along lengths of the surface of each muscle fiber in a bundle or sheet, fusion of apposing cell membranes can be observed connecting each fiber to adjacent fibers, thus permitting the wave of excitation to travel from one muscle fiber to another (Fig. 5.3). This has been termed the *nexus*, but to keep terminology consistent with that used for attachments between epithelial cells, the names *macula occludens* or *fascia occludens* would be preferred (see Chapter 4).

Smooth muscles evidence two states of contractions. Of major importance in maintaining the diameter of blood vessels and other tubes is a condition of partial contraction called tonus. A corollary characteristic of smooth muscle is its ability to maintain such contraction for long periods of time without evidence of fatigue. Smooth muscles also undergo slow, rhythmic waves of contraction and relaxation.

Structure of Smooth Muscle

Microscopically, smooth muscle fibers appear as long strands which are thickened through the middle and have tapering pointed or rounded ends (Fig. 5.4). The length and width of the fiber can vary with animal and organ. Some of the largest, reaching 500μ in length, are seen in the pregnant mammalian uterus; some of the smallest, about 20μ, are located in the walls of small blood vessels. The diameter of the fibers varies throughout the vertebrates, usually being greatest in fish and descending through the amphibia, reptiles, and mammals to birds. Smooth muscle fibers can be organized singly or in small bundles, as seen in the skin, or in sheets consisting of one or more cell layers. In all cases each muscle fiber is surrounded by a network of

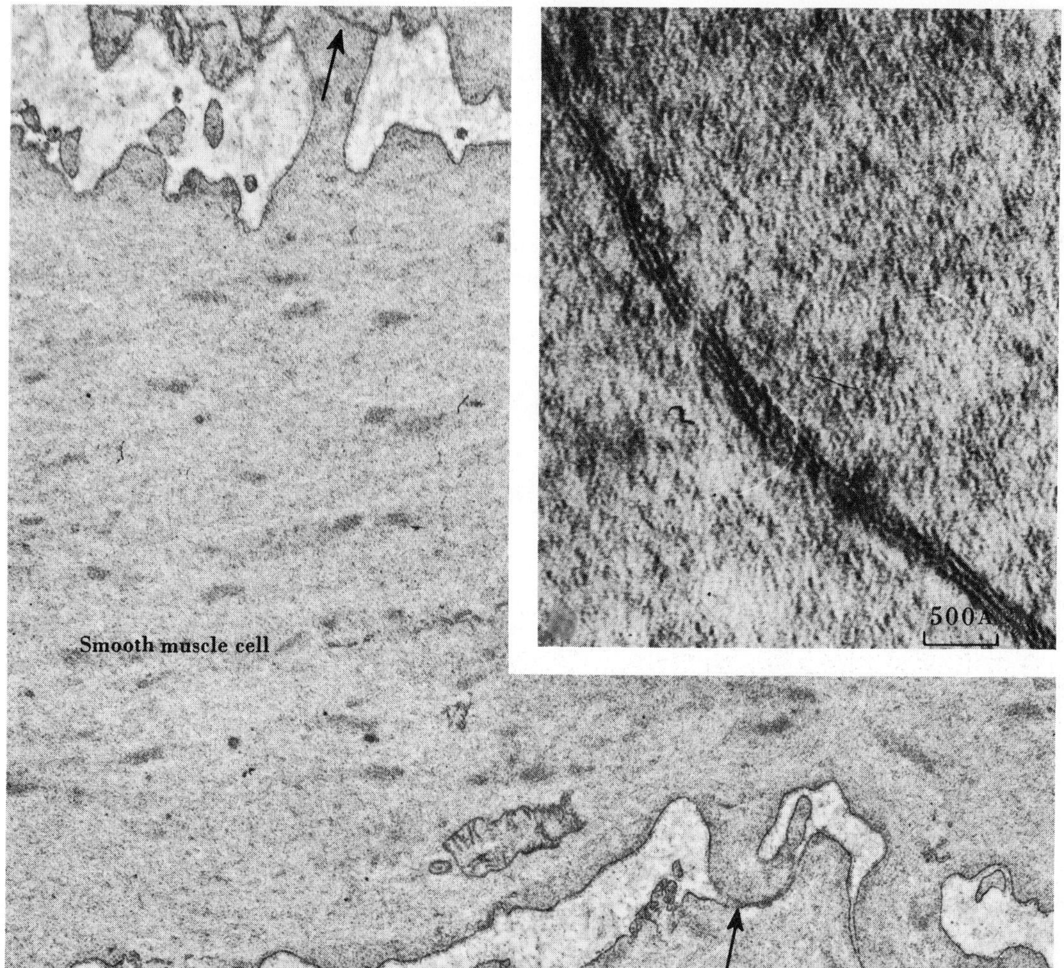

Fig. 5.3 Electron micrograph of muscle cells from the cat small intestine. Cell membrane fusion can be observed at arrows, although the shape of the fusing processes has been altered by fixation. X16,000. The insert illustrates one nexus at higher magnification. X200,000. (From M. M. Dewey and L. Barr. 1964. J. Cell Biol. 23:553–586.)

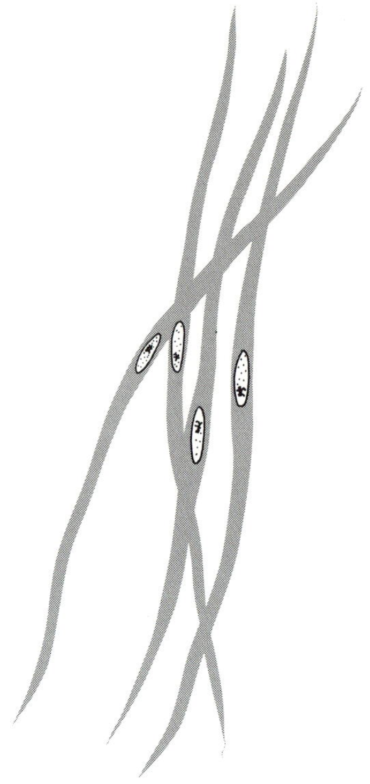

Fig. 5.4 A diagrammatic view of teased smooth muscle fibers. Note the shape of the cell and location of the nucleus.

reticular and/or elastic connective tissue fibers and usually allied with smooth muscle are bundles of collagenous fibers. Interposed between layers of smooth muscle is loose, areolar connective tissue which serves as the carrier for blood vessels and nerves going to and from the muscle cells (Fig. 5.5).

Fig. 5.5 Longitudinal section of intestinal smooth muscle. X450.

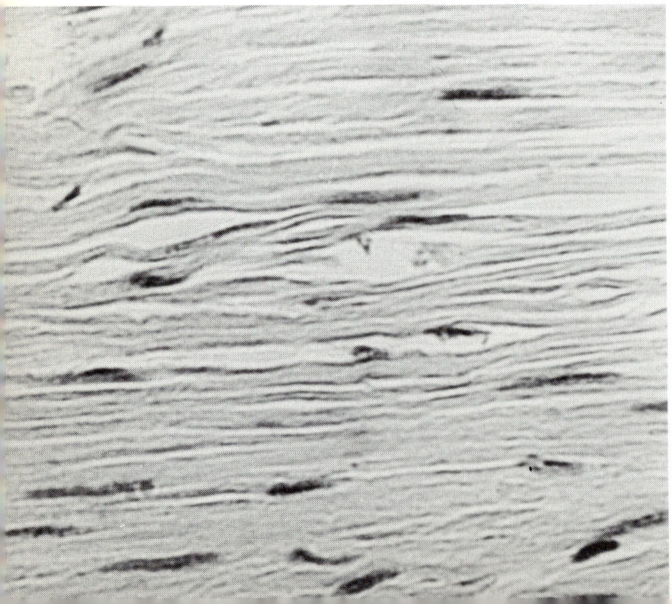

In contrast to skeletal muscle fibers, which are multinucleate, each fiber of smooth muscle has just one oval-shaped nucleus centrally located in the widest part of the cell. The size of the nucleus generally falls between 15 and 20μ in length and 2 to 3μ in width. Several nucleoli are usually visible within each nucleus, and in sections cut in the proper plane a curve or indentation partially surrounding two centrioles can be seen. After maceration in acid, parallel fibrils can be observed running longitudinally through the cell. The fibrils, the *myofibrils,* are composed of collections of smaller *myofilaments* which are evident in electron micrographs. The question of whether the myofilaments actually do aggregate or whether myofibrils are artifacts caused by clumping of myofilaments during preparation has not been resolved at present.

The longitudinal filaments of smooth muscle cells, as opposed to those of skeletal and cardiac muscle, contain no light and dark band periodicity. For the most part, vertebrate smooth muscle myofilaments have been found to be of uniform diameter, about 50–80A, and to resemble physically and chemically the filaments of actin in skeletal muscle fibers. If smooth muscle myofibrils are subjected to protein extraction techniques, residues of actin, myosin, and a third protein, tropomyosin, can be obtained. The way in which these structural proteins are arranged within the myofibril is still a matter of conjecture. Recent evidence supports a model which portrays all longitudinal filaments as those of actin. Extending perpendicularly along the actin filaments are short projections composed of myosin dimers. Contraction would then be envisioned as actin filaments sliding toward each other and powered by the myosin side chains, much as has been accepted as the theory of skeletal muscle contraction. The lack of clear-cut evidence for the localization of myosin, and, hence, the inability to propose a well-documented model of smooth muscle contraction seem to be based on the small polymeric size of smooth muscle myosin (such that it barely reaches the limits of resolution of the electron microscope). Differences in cell shape, however, are evident when different

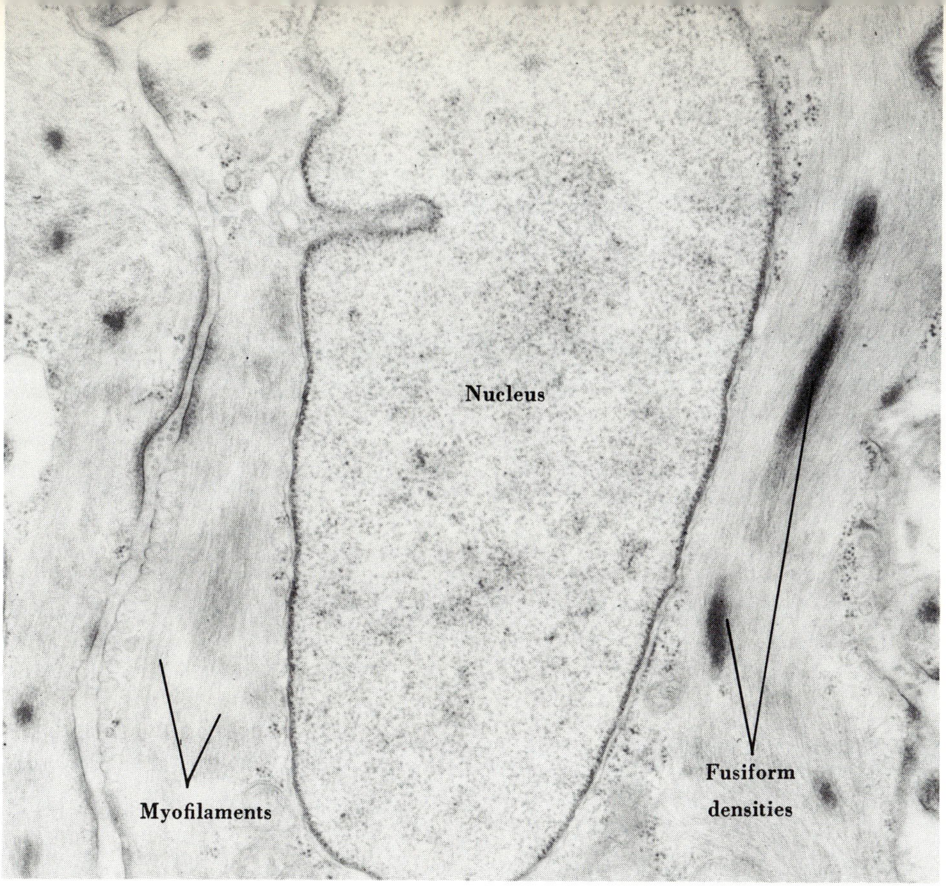

Fig. 5.6 Electron micrograph of a portion of a contracted smooth muscle fiber. X19,500. (From B. J. Panner and C. R. Honig. 1967. J. Cell Biol. 35:303–321.)

contractile states are compared. The relaxed cell shown in Fig. 5.7 is certainly more elongate and contains a smoother nuclear profile than does the partially contracted cell of Fig. 5.6. The myofilaments are anchored to the cell membrane and perhaps to each other at locations which, through the electron microscope, appear to be more dense than the rest of the membrane, and which resemble desmosomes or Z disks of skeletal muscle. Because of the shape of these anchoring points they are called *fusiform densities* or dense bodies, and have been postulated to be functional analogues of the Z disks (Fig. 5.6).

The nonfibrillar cytoplasm of the muscle fiber, known as sarcoplasm, is scanty, the bulk of the cell being taken up with the myofibrils. However, contained within the amorphous matrix of the sarcoplasm, are large, relatively unsegmented mitochondria, usually concentrated at the poles of the nucleus or between myofilaments, a typical Golgi apparatus, and a granular endoplasmic reticulum. Inclusions of various kinds of lipids and glycogen may also be present. The cell membrane may be smooth or thrown into a series of minute inpocketings or folds which perhaps function in pinocytosis or electrical conduction to neighboring fibers (Figs. 5.6 and 5.7).

A limited amount of smooth muscle regeneration is possible by mitotic division of the cells or by differentiation of mesenchymal perivascular cells, although large areas of damage are usually filled in with connective tissue scars. Hypertrophy of smooth muscle, seen especially in the pregnant mammalian uterus, can occur by either of the foregoing mechanisms as well as by increase in cell size (*hyperplasia*).

MYOFIBRILS OF STRIATED MUSCLE

As with smooth muscle, the cells of skeletal and cardiac muscle contain parallel longitudinal strands, myofilaments. These aggregate as myofibrils in skeletal muscle. In cardiac mus-

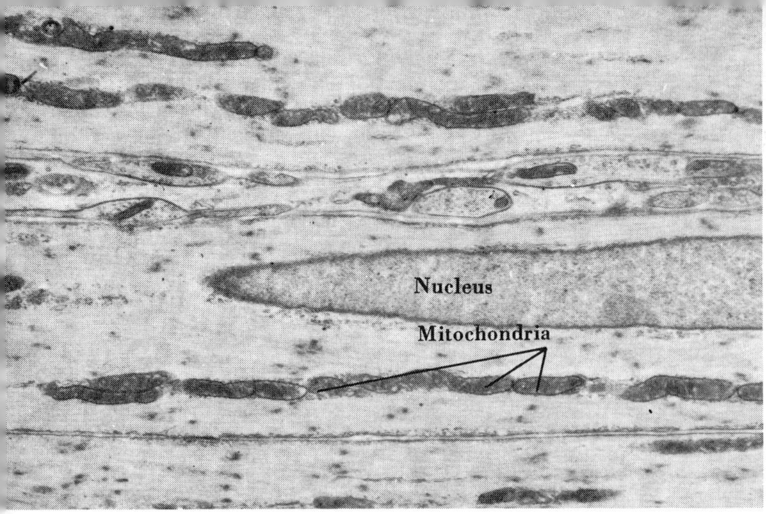

Fig. 5.7 An electron micrograph of a portion of a relaxed smooth muscle cell taken from the mouse jejunum. X8000. (From B. P. Lane. 1965. J. Cell Biol. 27: 199–213.)

cle, groups of filaments are partially segregated from other groups by projections of sarcoplasm; hence, true myofibrils are not found. Unlike the homogeneous fibrils of smooth muscle, those of skeletal and cardiac muscle show a characteristic pattern of cross-banding or striation. While many features of the two types of striated fibers differ, the appearance of the myofilament aggregates in both are essentially the same. Let us first, then, consider the striation pattern of these fibrils, and afterward discuss the differences in these two kinds of muscle tissue.

In a resting skeletal muscle fiber, the fibrils have alternating light and dark regions, the latter usually being slightly longer than the former. Because all the fibrils in one fiber are stacked up in phase or "register," the muscle cell appears to be banded across its width (Fig. 5.8). The colorless or light bands are singly refractile, or nonbirefringent, to polarized light. This condition is referred to as being *isotropic*, and thus the light bands are called the *I bands*. The dark bands, doubly refractile under polarized light, are *anisotropic*, and are known as the *A bands*.

Each of these bands is subdivided by darker or lighter lines or areas. Occurring in the middle of the I band is a dark line, the *Z line*. The distance from one Z line to the next comprises a *sarcomere*, the functional subunit of striation. On either side of the Z line, a faint dark line, the N band, can occasionally be observed. In a contracted or passively

Fig. 5.8 Longitudinal section of skeletal muscle of monkey tongue. Note characteristic banding pattern as described in text. X1000. (Courtesy of F. B. Merk.)

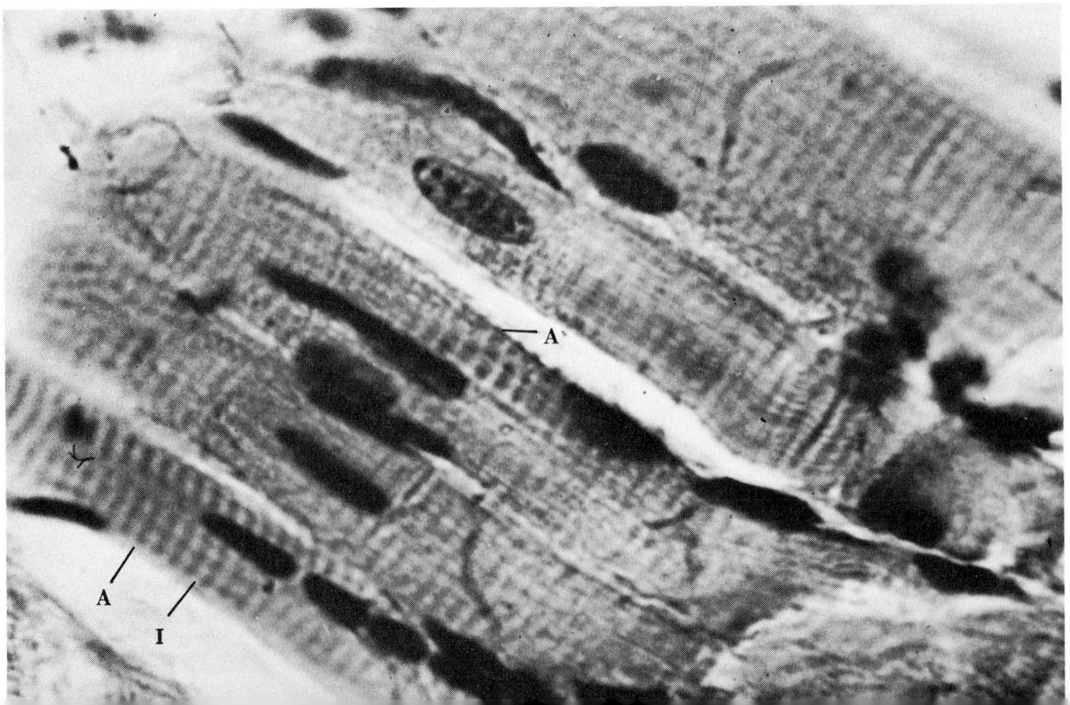

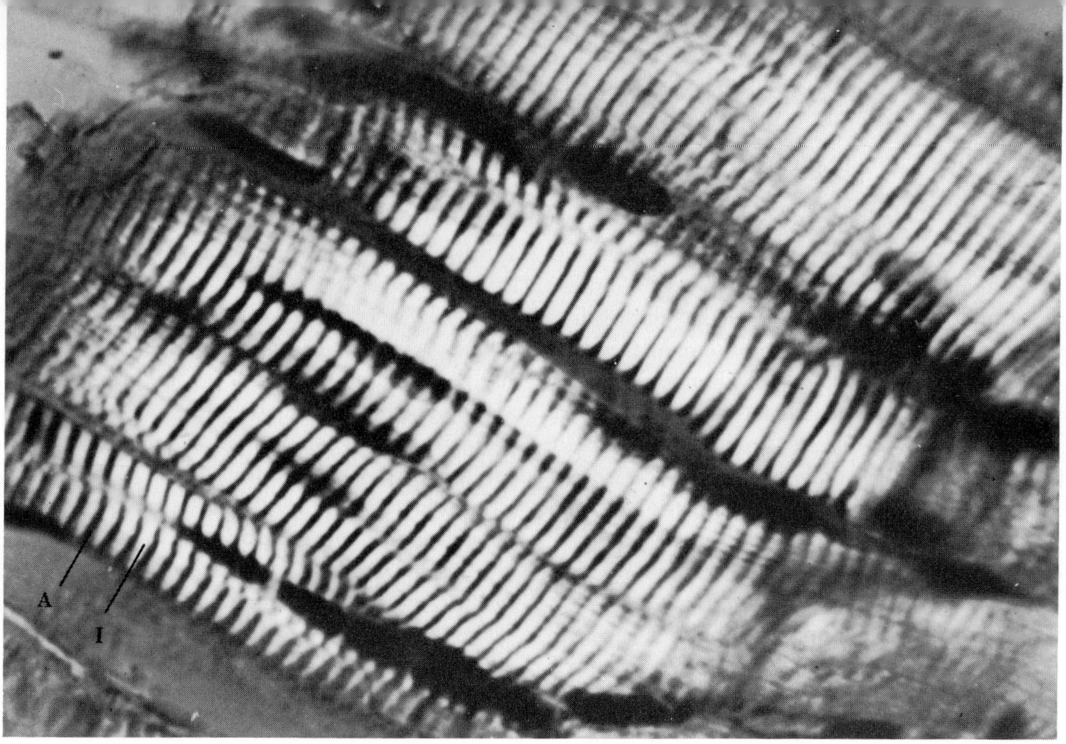

Fig. 5.9 This photograph shows the same cells as seen in Fig. 5.8, photographed with polarized light. X1000. (Courtesy of F. B. Merk.)

stretched fiber the center of the A band may be seen to contain a paler streak, the *H band,* and through the center of this streak runs the *M band* (Fig. 5.9).

This intricate banding pattern is not merely a series of colored and colorless regions. Rather, it is a microscopic reflection of the fine structure of the myofibril or myofilament aggregate. Striated muscle contains two kinds of myofilaments. Those of myosin (80–100A in diameter) are the thick filaments, whereas those of actin, possibly in combination with tropomyosin (about 50A in diameter) are known as the thin components. In regions of the myofibril containing both varieties each thick filament is surrounded by six thin filaments in hexagonal fashion. However, not all areas contain both kinds of fibers, and this fiber-density variation forms the basis of the microscopic banding. The thick filaments of doubly refractive myosin form, and are confined to, the A band; hence, the length of the A band reflects the length of the myosin filaments. The I band contains only thin filaments. These filaments, however, are not confined to the I band, but extend into the A band. The thin filaments of one I band do not extend far enough into the A band to meet those of the next I band, thus creating a lighter, less dense area, the H band. The Z line consists of a dense substance, perhaps tropomyosin, which seems to function in keeping the thin myofilaments in register. Electron micrographs indicate that the I filaments terminate at each Z line with rodlike attachments (Fig. 5.10). There is some recent evidence pointing to the existence of a third myofilament type. Interspersed between the thin filaments, these *very thin filaments* extend for long distances and are not limited to the length of one sarcomere. Their function is thought to be that of maintaining the integrity of the muscle fiber during stretch.

The changes in band structure seen during muscular activity are caused not by folding of the fibrils but rather by the thin filaments sliding past or spiraling around the thick filaments. During contraction, the thin filaments migrate toward the H band, thus interdigitating even further with the thick filaments, and shortening the I and H bands. Under conditions of stretch, the thin filaments move away from the center of the A band and toward the Z line, with concomitant apparent lengthening of the I band (Fig. 5.11).

Strengthening the sliding filament hypothesis of contraction are data gathered about the structure of the thin and thick filaments, as well as from physiological experiments. As we have indicated, the thin filaments, which are smooth-surfaced, consist of fibrous or F-actin. F-actin is a double stranded helix containing polymerized monomers of globular or G-actin. The construction of the thick filaments has not been so well identified. Myosin is a complex, heterogeneous system of small subunits held together by weak bonds. Thus one is stretching a point a bit even to call the myosin polymer a molecule, because whereas molecules are covalently bonded structures, myosin evidences only hydrogen bonds, Van der Waals forces, and dipole movements between the subunits; these are all weak, secondary bonding forces. The subunits of myosin, the protomyosins, fall into two categories. One group, the light molecular weight or L-meromyosins, probably function in the actual shortening of the muscle. The other group, the heavy or H-meromyosins, are responsible for splitting off the terminal phosphate of ATP, thus liberating the energy for contraction. In addition, it is the H-meromyosin portion of myosin which joins actin to form the contractile molecule, actomyosin. Although the exact arrangement of the L- and H-meromyosin units in the myosin dimer is not known, the myosin molecule is a long, thin rod with an H-meromyosin end portion. These are arranged as aggregates representing the thick filaments and, as opposed to actin, are not smooth surfaced. Along the length of the filament, projections or bridges of H-meromyosin appear at intervals of 60 to 70A. They are arranged helically around the filament and each bridge presumably interacts with one actin filament to form the contractile actomyosin unit. If

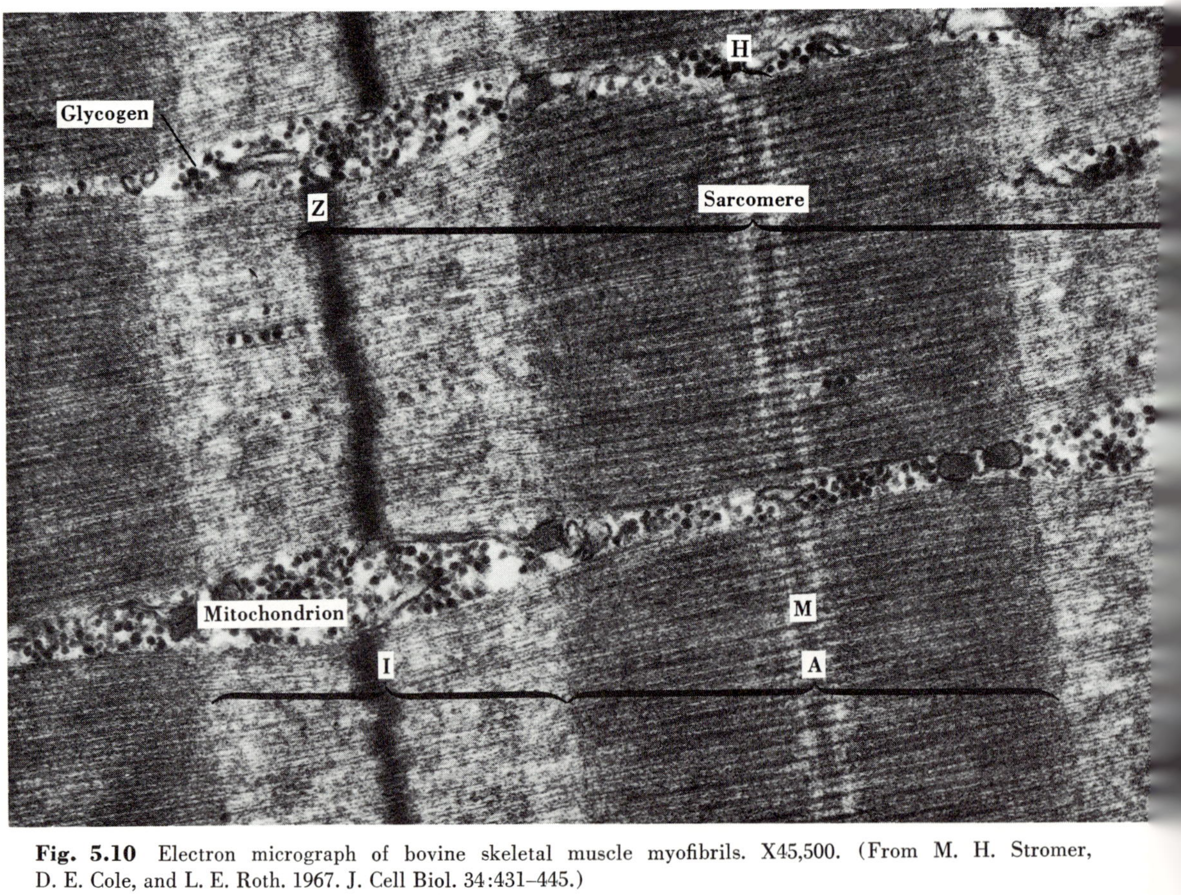

Fig. 5.10 Electron micrograph of bovine skeletal muscle myofibrils. X45,500. (From M. H. Stromer, D. E. Cole, and L. E. Roth. 1967. J. Cell Biol. 34:431–445.)

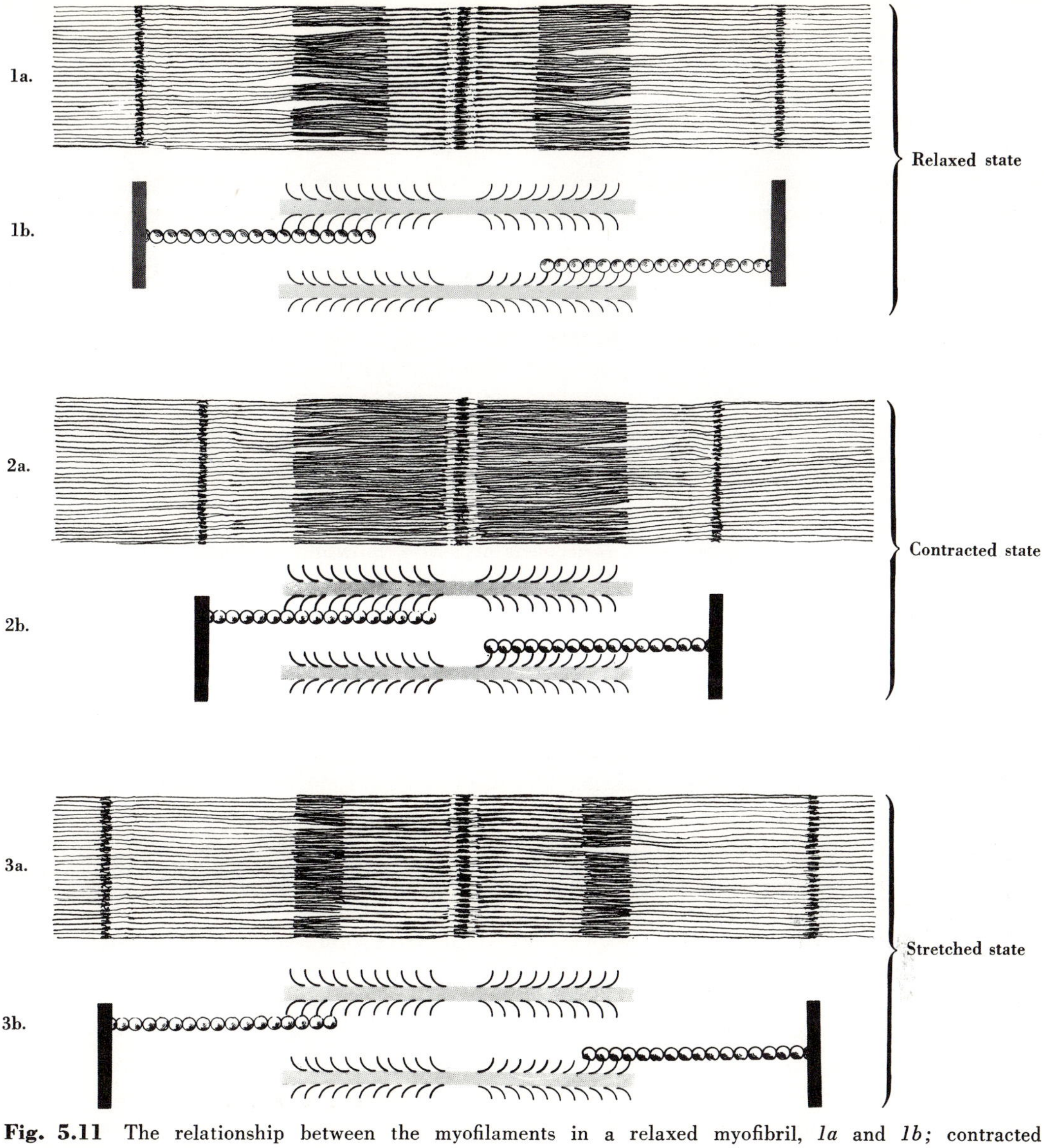

Fig. 5.11 The relationship between the myofilaments in a relaxed myofibril, *1a* and *1b;* contracted myofibril, *2a* and *2b;* and stretched myrofibril, *3a* and *3b.* See text for explanation of the sliding-filament hypothesis.

there is no contact between the bridge of H-meromyosin and actin, no contraction can take place because actin and myosin can not contract individually. If, however, the connection is assumed, and the correct ions as well as ATP are present, contraction will occur.

Although there have been many different hypotheses concerning the mechanism by which the bridges function, it has been generally agreed that the bridge-actin link is not a permanent one, and that as the actin filaments move, the bridges link up at successive locations along the thin filaments, with ATP being split to ADP plus a high-energy phosphate radical at each link. Malfunction of the moving or transitory bridge-actin union has been postulated to be the cause of rigor.

SKELETAL MUSCLE FIBERS

Having discussed the contractile units of muscle, the myofibrils, let us now see how they are

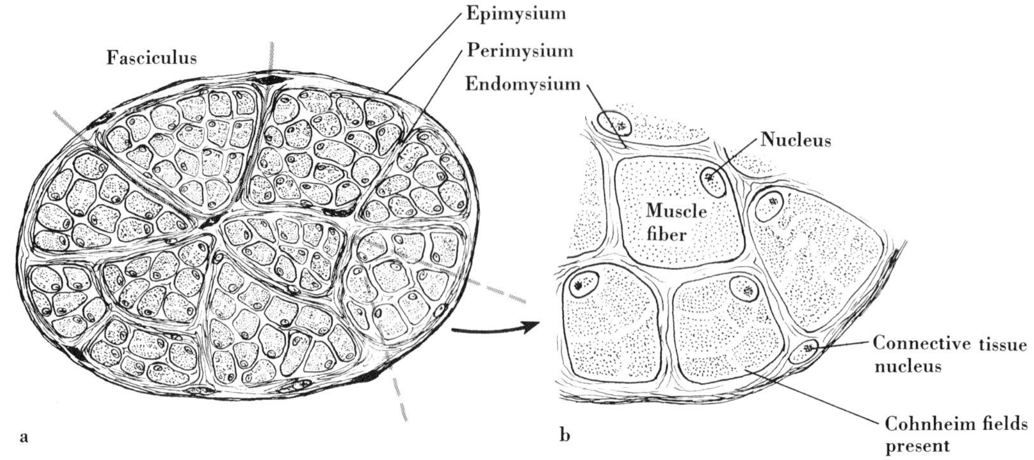

Fig. 5.12 Cross-sectional drawing of a skeletal muscle. Note the arrangement of muscle fibers within the organ, and the distribution of myofibrils within each muscle fiber. The triangular area indicated in *a* is enlarged as shown in *b*. Within a muscle, some fibers may have uniformly distributed myofibrils whereas others exhibit Cohnheim fields. The latter may be fixation artifacts.

integrated into skeletal muscle cells, and how skeletal muscle cells are organized and function in masses.

Classified as skeletal muscle are those striated, voluntary, contractile units which are attached to and move the skeletal elements of the body. Skeletal fibers differ from smooth muscle fibers in that the former contract more powerfully and rapidly, and fatigue more easily than the latter. Skeletal muscles can evidence a short, single contraction, or twitch, in addition to being able to maintain a prolonged state of partial contraction called tonus, useful in postural support.

Most of the volume of a skeletal muscle fiber is occupied by the myofibrils, which may be uniformly distributed through the fiber, or may be partitioned by ramifications of the hyaloplasm so that the fiber contains so-called Cohnheim's fields (Fig. 5.12). Both types of fiber arrangement can be found within an animal and even within one muscle.

Superimposed on the fibril organization within the fiber are the rest of the cell components. In the case of striated muscle fibrils, the prefix *sarco* often appears in reference to components, as sarcolemma, sarcoplasm, etc.

The Sarcolemma

The term *sarcolemma* should be approached with caution, coupled with a glance at the relevant publication date. Within the last decade, the electron microscopists have equated the sarcolemma either with the plasmalemma, or conversely have clearly defined it as a multiple membrane structure consisting of plasma membrane, connective tissue basement membrane and adjacent connective tissue. Previously, the sarcolemma, seen under the light microscope, was considered a thicker, unit structure.

The sarcolemma is notable in two respects. Morphologically it is a membrane which is capable of withstanding stretch and maintaining elasticity. Physiologically, the sarcolemma, more than most cell membranes, is regionally differentiated. Only the sarcolemma underlying a nerve junction can be excited by acetylcholine, whereas this area is impervious to electrical stimulation. However, if an area other than that near a nerve junction is stimulated, a wave of depolarization is propagated over the membrane so that the entire sarcolemma is rapidly and effectively activated. As a corollary consideration, the sarcolemma of a resting fiber is able to maintain a 10-mv potential difference between the inside and the outside of the fiber.

The Sarcoplasm

Qualitatively the hyaloplasm of muscle cells, the sarcoplasm, functions in the same capacity as the amorphous ground substance in other kinds of cells. Quantitatively, muscle cells exhibit little of this "structureless" cytoplasm, the bulk of the cell being packed with myofibrils and other highly structured organelles. Although the sarcoplasm surrounds and permeates the interstices between all of the organelles, it is present most noticeably and in greatest quantity around the nuclei and under the sarcolemma. A small, typical Golgi body can usually be observed in the sarcoplasmic region capping the nucleus. The enzyme systems found in the sarcoplasm indeed are demonstrable in this specialized analogue, and perhaps are present in greater relative concentration because muscle fibers are major energy-requiring cells.

The Mitochondria

These organelles are large, contain many cristae, and are usually numerous in muscle fibers; this, again, can be correlated with the work performed by these cells. Generally muscle mitochondria are ovoid with their long axis parallel to the long axis of the cell. They are located around the nucleus, under the sarcolemma, and wedged between the myofibrils at the levels of the I, Z, or A bands (Fig. 5.13). Occasionally a mitochondrion can be seen partly encircling a myofibril.

The Sarcoplasmic Reticulum

The endoplasmic reticulum of muscle fibers, the sarcoplasmic reticulum, is composed of a highly structured, continuous membrane-vesicle system. In addition to having some role in protein storage and transport, the sarcoplasmic reticulum is believed to perform

Fig. 5.13 An electron micrograph of the bat cricothyroid muscle showing elements of the T-system and sarcoplasmic reticulum (*arrows*). In each triad, the center tubule is the T-system member. The numerous black granules in the sarcoplasm are glycogen. X34,000. (From J. P. Revel. 1962. J. Cell Biol. 12:571–589.)

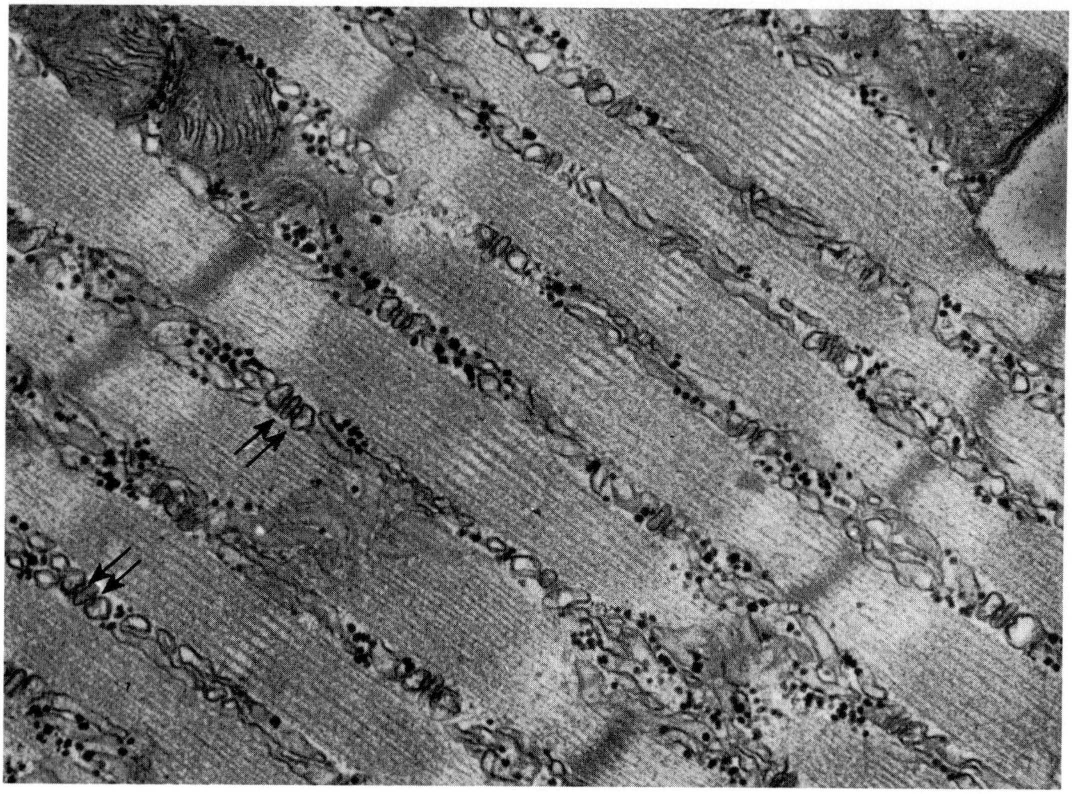

an integrating function so that a nerve impulse which reaches the sarcolemma is transmitted rapidly throughout the cell and most particularly to the myofibrils.

The reticulum is organized as a tubular mesh oriented both longitudinally and transversely around each myofibril and connecting with the mesh of adjacent myofibrils. The longitudinal tubules form a loose vesicle system running the length of the myofibril and enclosing it. At the level of the I band, the I-A junction, or the Z line, the exact location being specific for the species, highly structured transverse canals are seen which are thought to serve as communication channels between the individual myofibrils and the longitudinal systems (Fig. 5.14). The more highly developed the muscle or the more rapid the specific muscle action, the more distinct and organized are these transverse elements. Muscles such as the bat cricothyroid, which functions in the production of very rapid, short bursts of frequency modulated ultrasound (used by the bat for "seeing" by echolocation), or very fast-acting fish muscles, contain a tripartite (triad) arrangement or even a pentad at each H band or Z line. Muscles which act more slowly contain a dyad or perhaps a single channel. However, at least one channel, the middle one in the "triad"-containing muscles, exhibits an ultrastructure identical to that of the sarcolemma, and appears to be an invagination of the outer membrane. It is designated as the component of the *T system*. Hence this channel is directly continuous with the extracellular environment. The other channels of the triad do not seem to connect directly with the T system, but rather serve as links between the longitudinal canals.

Presumably these canals function as shortcut roads for the electrical integration of the muscle fiber, and would, in fact, explain how the entire fiber can be so rapidly activated once an impulse reaches the sarcolemma. This theory is given further support when one compares the typical vertebrate muscle with a skeletal fiber of the cephalochordate, Amphioxus. In Amphioxus, muscle fibers of the body wall are extremely flat and each fiber extends the full distance between the intermuscular septa. Contained within each uninucleate fiber is a single myofibril, which, at many points along its length, comes into contact with the sarcolemma. Because of the simple structure, complete excitation of such a fiber would require less complex mechanisms for impulse spread, and, indeed, in these fibers, little or no sarcoplasmic reticulum can be demonstrated.

Recent evidence suggests an auxiliary function of the sarcoplasmic reticulum. In some fast-acting muscles, the number and location of mitochondria seemingly could not account for the rapid supply of ATP needed. It has been postulated that the sarcoplasmic reticulum can generate ATP via anaerobic glycolysis. Its close proximity as well as its enzymatic properties seem to substantiate the hypothesis. In such a case, the sarcoplasmic reticulum would carry the electrical message, and would act as the intermediate between the nerve impulse and the breakdown of ATP by the myosin within the myofibrils, resulting in sliding filament contraction. Furthermore, the sarcoplasmic reticulum would quickly rephosphorylate the ADP so produced in order that a rapid series of contractions could be maintained. This is not to imply that muscle mitochondria are inactive. They, too, produce ATP, and in great relative quantity. However, mitochondria produce ATP more slowly, and are not as close to the myofilaments, as is the sarcoplasmic reticulum. It may well be that skeletal muscle cells, differentiated as highly active, energy-requiring entities, possess two workable systems for the production of ATP. The sarcoplasmic reticulum would be a short-term, rapid source whereas the mitochondria would represent the long-term, more basic and slower source.

The Nucleus

The nuclei of skeletal muscle are notable for their quantity. In all vertebrates a multinucleate condition is the rule, the number of nuclei varying with fiber size. Some of the larger fibers of amphibians and mammals, measur-

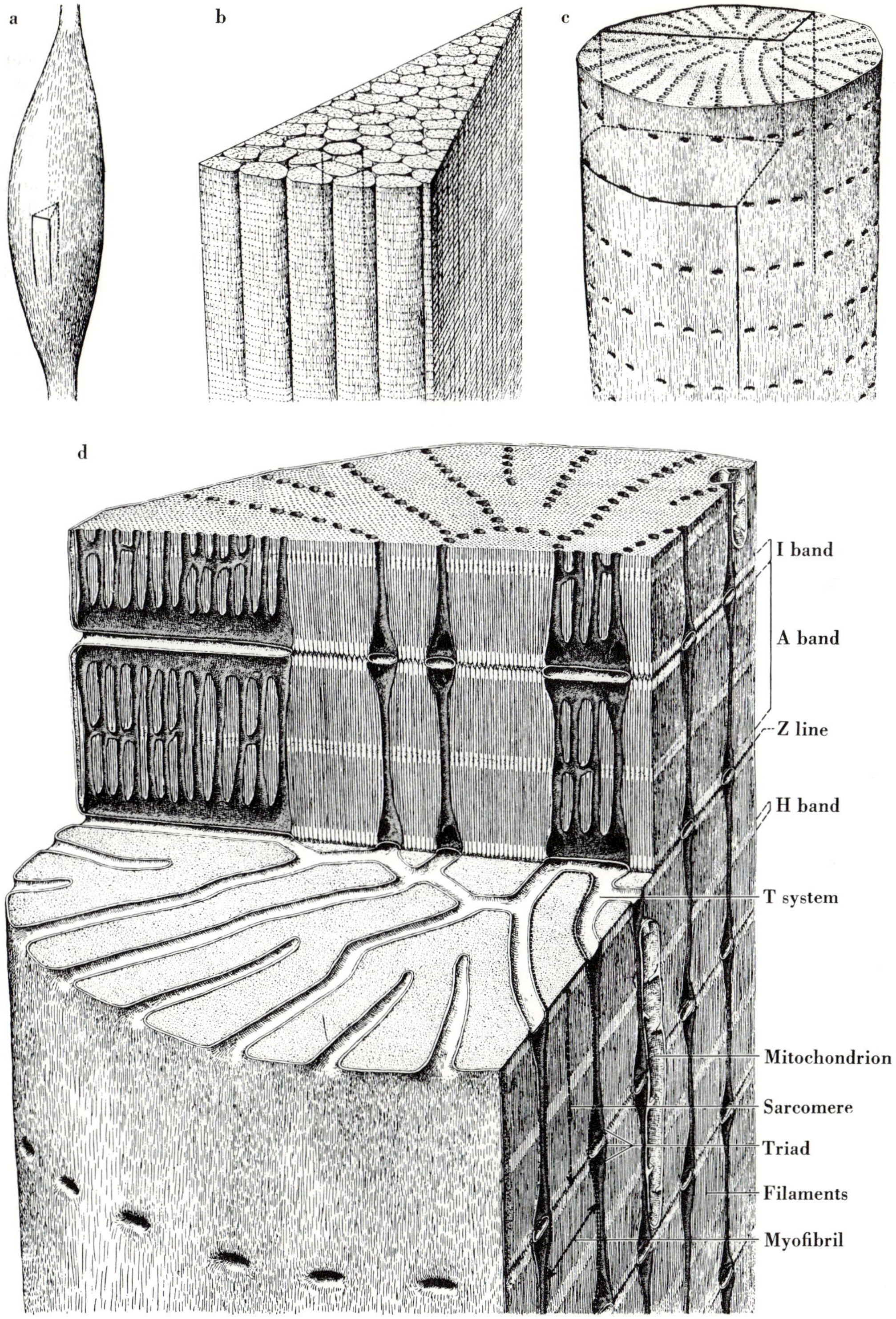

Fig. 5.14 Communication systems and organization within a skeletal muscle fiber. The entire gross muscle, *a*, is composed of many cells, *b*, of which one is shown in greater detail, *c*. The drawing *d* illustrates the amazing complexity of each of these cells. (From K. Porter and C. Franzini-Armstrong. 1965. Sci. Am. 212:74–5.)

ing up to 4 cm in length, contain many hundred nuclei.

Each nucleus is generally elongated, the long axis being parallel to the long axis of the fiber. Visible within the nuclei are loose, granular, networks of chromatin. In adult birds and mammals, the nuclei are peripherally located, lying either immediately under the sarcolemma in the fast-acting fibers, or between outer myofibrils in the slower-acting fibers. In fishes, amphibians, reptiles, and embryonic birds and mammals, the nuclei are generally located toward the center of the fiber.

EVOLUTIONARY TRENDS IN MUSCLE FIBERS

Although skeletal muscle fibers are remarkably consistent throughout the vertebrate classes, variations do exist. Two kinds of fibers are demonstrable in vertebrate muscles: white fibers and red ones. The basic difference between the two is the relative lack of myoglobin, a pigment similar in structure and function to hemoglobin, in the white fibers. Histologically, the white fibers exhibit little sarcoplasm, fewer mitochondria and lipid droplets, and a greater degree of development of the sarcoplasmic reticulum. Chemically, they possess low concentrations of the cytochromes, lipases, and B vitamins, but large amounts of water, glycogen, and niacin. The white fibers are of larger diameter than the reds, and are more densely packed with regularly arranged but shorter myofibrils. The blood supply to the white fibers is less abundant, and the connective tissue wrappings are scanty. In general, the red, myoglobin-containing fibers contain larger amounts of sarcoplasm, lipid droplets, and mitochondria. The fibers are of smaller diameter, with less regularly arranged striated fibrils. The concentrations of lipases and cytochromes are great and the capillary supply abundant.

Whereas white fibers can metabolize their glycogen anaerobically, the energy source available to the red fibers, lipid, requires oxygen—in this case from the oxygen-carrier, myoglobin. Because of these biochemical differences, white fibers are fast-acting, easily fatigued, "twitch" fibers, as opposed to the red fibers (which can maintain a tonic contraction without an early onset of fatigue).

Two evolutionary tendencies are evident when skeletal muscle fibers are studied on this basis. (1) The tetrapods have a greatly increased proportion of red fiber muscles as compared to the fishes, and, in most mammals, the many red and fewer white fibers are intermixed within a muscle mass. In addition, those muscles which may be classified as postural or tonic in function, contain an overwhelming proportion of red fibers as opposed to muscles used for quick twitches. (2) Contained within each vertebrate class are groups of animals which may be classified as predominantly sprinters or stayers. Those animals whose mode of life involves darting or sprinting movements usually contain more white fibers than those forms which live a slower existence or whose migratory or feeding habits require long-lasting muscular action. It should be remembered, however, that this gradient is superimposed on the larger total-vertebrate gradient, so that "staying" fish, such as the salmon or carp, while having more red fibers than their sprinting brothers (pike, perch, etc.), have far fewer red fibers than any mammal.

SKELETAL MUSCLE MASSES

Although isolated skeletal muscle fibers can contract if given proper stimulation, the work they must do to adjust the body in space is a function of the manner in which they are harnessed.

Any given gross muscle may contain three categories of muscle fibers. One category includes fibers with rounded or notched ends that course the entire length of the muscle from origin to insertion. A second type of fiber is shorter than the muscle length, and has neither pole at the origin or insertion. Such fibers usually have tapered ends which terminate on connective tissue septa within the muscle. The last group of fibers, also shorter than the muscle, includes those with one end at the origin or insertion and the other end terminating within the muscle mass. All of these kinds of fibers in any muscle are bound together by connective tissues which, in turn, are continuous with the dense connective tissue structure attaching the muscle to its particular skeletal element.

Surrounding the gross muscle is a layer of dense, fibrous connective tissue, the epimysium. The epimysium is continuous with and, in fact merges with the attaching connective tissue structure such as a tendon, aponeurosis, periosteum, or raphe. Extending from the epimysium is the *perimysium,* which consists of connective tissue extensions containing collagenous, reticular, and elastic fibers as well as connective tissue cells that course within the muscle and divide it into bundles of muscle fibers, or *fasciculi,* of varying sizes. In addition, the bundles of collagenous fibers in the perimysium are continuous with those of the tendon or other connective tissue attaching structure. The number of fibers in the fasciculi of any given muscle often varies with the job of the muscle. Those muscles which demonstrate delicate, quick, finely controlled contractions usually contain fewer fibers per bundle, and vice versa. Finally, extending from the perimysium and surrounding each individual muscle fiber in the fasciculus is the endomysium. This connective tissue sheath is composed mainly of networks of reticular fibers, fibroblasts, and fixed macrophages. It comes in direct contact and, in fact, forms a union with the sarcolemma of each fiber (Fig. 5.12).

Why, one might ask, does a muscle need such an intricate connective tissue meshwork? Three answers may be given for such a question, and each involves a vital part of muscle function.

(1) Muscle fibers, in order to do work, are catabolic centers of activity. They require large and fairly constant supplies of energy rich lipids and carbohydrates, and, in turn, they produce quantities of lactic acid, water, and carbon dioxide. They must, therefore, be adequately vascularized. The connective tissue wrappings provide the needed route for the blood vessels and, by virtue of the endomysium, each muscle fiber can maintain a close capillary connection. (Parenthetically, it should be noted that because blood vessels contain smooth muscle fibers, one component of a gross skeletal muscle is smooth muscle.)

(2) Skeletal muscles require nervous stimulation to contract and must also contain sensory receptors for proprioception. Again, the connective tissue wrappings provide the route for the efferent and afferent nerve pathways.

(3) The interconnection of all the connective tissue elements, from the delicate endomysium to the thick, fibrous tendon, insures that the muscle shall indeed pull as a unified mass against its skeletal or dermal attaching structure, and moreover that this attachment shall not become loose or broken under normal, and sometimes abnormal, wear and tear.

Although each of these reasons for the connective tissue wrappings applies to every muscle, some muscles have different needs than others. In most fish, for instance, the muscles and tendons are short and most fibers extend the entire distance between intermuscular septa. In addition, the white muscle fibers of fish can work anaerobically and therefore require a less extensive vascular bed. Under these conditions a smaller amount of connective tissue will and does suffice. An interesting method of comparing the amount of connective tissue in skeletal muscle tissue is by cooking the tissue to tenderness. Because collagen is a much tougher substance than sarcoplasm, the faster-cooking tissue will demonstrate the smallest proportion of interwoven connective tissue.

EFFERENT INNERVATION AND THE MYONEURAL JUNCTION

Each skeletal muscle fiber is in communication with the axon of a motor neuron or a branch of such an axon (Fig. 5.15). The number of muscle fibers, together with the individual, servicing nerve fiber is termed a *motor unit,* the size of which varies with the function of any given muscle. Those muscles

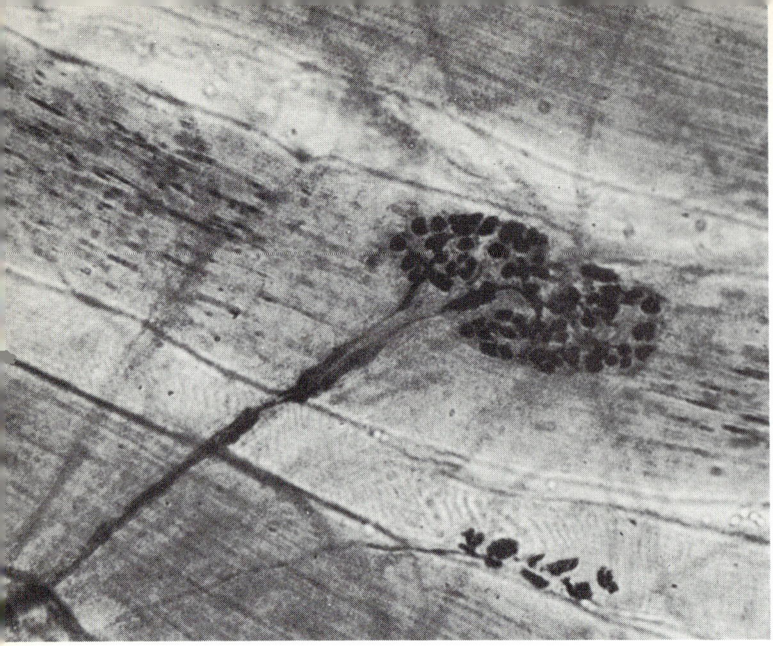

Fig. 5.15 A photomicrograph of a myoneural junction. The axon terminals of motor neurons are expanded as they approach the muscle fiber. X220.

which are responsible for delicate, finely controlled movements are organized in small motor units, whereas the motor units in muscles of more gross functions are large. Two examples will serve to illustrate these extremes. The motor units of the extrinsic eye muscles generally contain one nerve fiber for every muscle fiber. Muscles of the pectoral or pelvic girdle may contain motor units consisting of 100 muscle fibers for each nerve fiber. In such large motor units the muscle fibers innervated by one axon are usually distributed throughout the muscle mass rather than being localized in one small region. This distribution permits graded gross muscle responses even though individual fibers contract in an all-or-none manner. The characteristic graded response is dependent upon the number of stimulated motor units, rather than on partial contraction of all of the fibers.

At point of contact with nerve branches, the sarcolemma of muscle fibers assumes an intricate and specialized morphology, with the underlying sarcoplasm forming a small protuberance. Upon approaching muscle fibers, the nerves first lose their myelin sheaths and then, after dividing into terminal ramifications by which they make contact with muscle, the nerve fibers lose their Schwann sheaths (neurilemmas) as well. Hence, the cell membrane of the axon, the axolemma, is in direct apposition with the sarcolemma. The sarcolemma, in turn, is thickened and profusely grooved with adherence of an amorphous material on the inner surface. The sarcoplasm between and near the grooves contains many mitochondria, nuclei, and characteristic synaptic vesicles.

ELECTRIC ORGANS

One of the most exotic vertebrate adaptations has been the conversion of skeletal masses to electric organs. Forces of parallel evolution have operated in some or all members of seven widely divergent families of fish to produce an electroplaque or electricity-generating cell from a formed or presumptive skeletal muscle fiber. The mere fact of current generation is not unusual; it is, in fact, a commonplace phenomenon occurring in all muscle, gland, and nerve cells, but the potentials generated by these cells are in the millivolt range —barely strong enough for detection by sensitive instruments. On the other hand, the so-called "electric fish" can generate potentials of up to 500 v, quite enough to electrocute other sizeable fishes.

Of the seven families of electric fish, two, the Rajidae and Torpedindae, are elasmobranchs; five, the Gymnotidae, Malapteruridae, Uranoscopidae, Gymnarchidae, and Mormyridae are teleosts. Of these, the torpidines, gymnotides, and malapterurides generate high voltages by means of many electroplaques hooked up in series. These strongly electric fish use their capacity of bioelectrogenesis in offensive and defensive maneuvers, i.e., for gross protection. These fish, therefore, tend to produce current in noncontinuous bursts. The weakly electric fish are presumably the more electrically primitive, and they use their electric organs for quite a different purpose. Many members of this group live in murky waters of South American and African rivers. Under conditions of high external turbidity, vision is obscured and ineffective. In conjunction with their electric organs, these fish have modified their lateral line system and cerebellum to serve as sensory and integrating structures for detection of an electric field. Electricity is generated continuously, hence a constant electric field is maintained. Other living forms and/or metallic objects distort the field, and this distortion serves as a sensory input for purposes of precise navigation as well as for protection.

Although all electric organs are derivatives of skeletal muscle, the fact that the structures evolved independently seven different times suggests differences between the organs in each of the groups of fish. Indeed, gross examination of specimens reveals that the locations of the organs differ, hence different groups of muscles are involved in each case. Whatever the location, however, electroplaques replace rather than supplement muscle groups. Second, fresh water offers higher resistance to the passage of current than does salt water. This is reflected by the fact that fresh-water forms usually have a proportionately greater number of cells hooked up in series, thereby generating a higher voltage than do marine forms. Third, electroplaques develop from differentiated muscle fibers in some fish by selective loss of certain cellular elements, whereas in other families myoblasts give rise to electroplaques directly. In addition, the electroplaques of some fish appear to be derived from a single skeletal muscle fiber, and in others, from more than one fiber. Finally, electroplaques of different fish exhibit variations in ultrastructure and histochemistry.

In general, one can think of an electroplaque as a skeletal muscle fiber which has increased in its width and decreased in its length sufficiently to form a wafer-like cell oriented at right angles to the original fiber. The surfaces of the electroplaque are deeply grooved to provide for increased surface area. Each cell contains many peripherally located nuclei and a scattering of internal ones, each being surrounded by an area of clear cytoplasm. Typical mitochondria and other organelles are usually present. The cell membrane is seen to be regularly invaginated around the surface of the cell, giving a rodlike appearance to the periphery (Fig. 5.16).

A wide spectrum of myofibril modifications may be seen if different species of each family are studied. Some species exhibit almost typical cross-striated myofibrils, whereas others contain far-flung changes. All species appear to contain actin myofilaments, although these structures may be arranged as a meshwork or in parallel lines which are curved rather than being straight (Fig. 5.17). The electroplaques of rays give evidence of containing myosin myofilaments and Z line material in addition to actin.

All electric organs have the same general architecture. The long, thin electroplaques are usually stacked in columns with connective tissue analogues of perimysium surrounding each column. The intercellular spaces between cells in the column are usually wide and are filled with tissue fluid (a good conductor) and connective-tissue fibers. A membrane of dense connective tissue surrounds the parallel groups of columns, merging into tendonlike attachments at each end of the gross structure.

The method of innervation of an electric organ gives rise to its function. Those fish which are electric have large, lateral nerves that run along the electric organ and send ramifications to each of the electroplaques—ending in structures similar to motor end plates. However, in any given species either the anterior or the posterior surface of all of the electroplaques is innervated. Only in very rare cases can two-sided innervation be found. As we will soon see, the side of innervation determines the direction of current flow.

Electroplaques function in a manner similar

to nerves. Normally a high concentration of potassium and a low concentration of sodium is maintained within the cell, a reversed condition obtaining outside the cell membrane. This creates a polarity with the inside of the membrane being electrically negative to the outside. When the surface membrane is stimulated, its permeability changes and ions can flow across it. This results in a reversal of polarity, so that the inside of the membrane becomes positive to the outside. Because only one side of the electroplaque is innervated, and because current flows from positive to negative, the current will flow from the innervated to the uninnervated side of the cell. Because the uninnervated membrane offers little resistance to the flow of current and the intercellular fluid is a good conductor, the next cell in series will be activated. As an example, in the electric eel the innervated sides

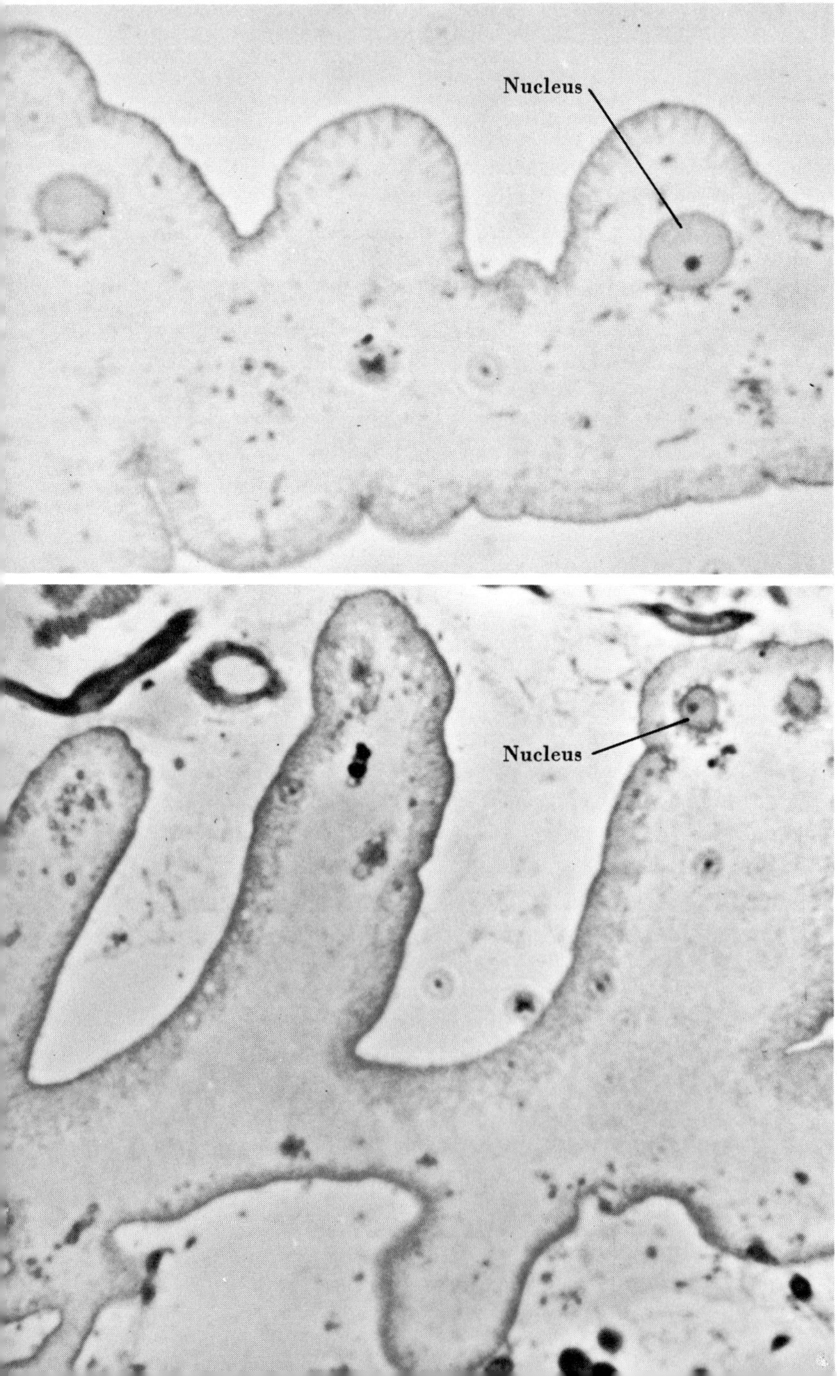

Fig. 5.16 Two electroplaques from the electric eel (*E. electricus*) are shown. Innervation is on the lower surface. X1800. (From Plate 1, A. W. Wachtel. 1964. J. Morphol. 114:335.)

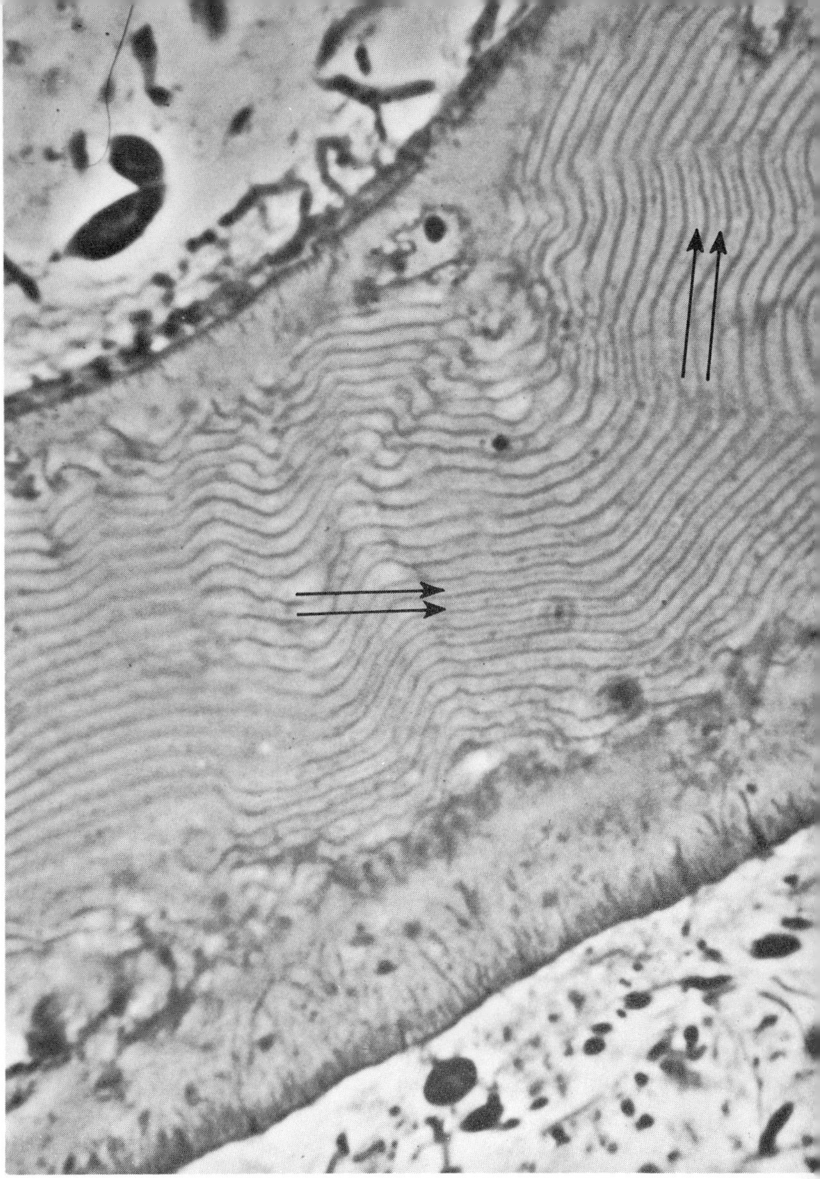

Fig. 5.17 A portion of an electroplaque from *R. erinacea.* Innervation is along top face. Structures similar to myofibrils can be seen at arrows. X2100. (From Fig. 14, A. W. Wachtel. 1964. J. Morphol. 114:351.)

of the electroplaques all face the tail. Hence the current flows from tail to head in the fish, and from head to tail in the surrounding water, thus completing the circuit. Because the voltages of the electroplaques are additive (series circuit), a high voltage can be generated, and because the many columns are hooked up in parallel, the amperage is raised.

CARDIAC MUSCLE

Cardiac muscle cells, the contractile elements of the heart, are in some respects very ordinary, and in others, very extraordinary. Like skeletal muscle fibers, those of cardiac muscle contain characteristic striated elements. Cardiac muscle fibers are under autonomic nervous control. They contract, and usually function in a manner intermediate between fibers of smooth and skeletal muscle. However, the fiber arrangements and gross cell structure of cardiac fibers is so unusual that only within the last few years has the case of individual cells versus one vast syncitium been resolved in favor of the individual cell theory, and, quite predictably, in the attitudes of some histologists, the problem has not yet been settled. In addition, cardiac muscle tissue includes some modified fibers, known collectively as the impulse conduction system of the heart. If all contact with activated nerves and

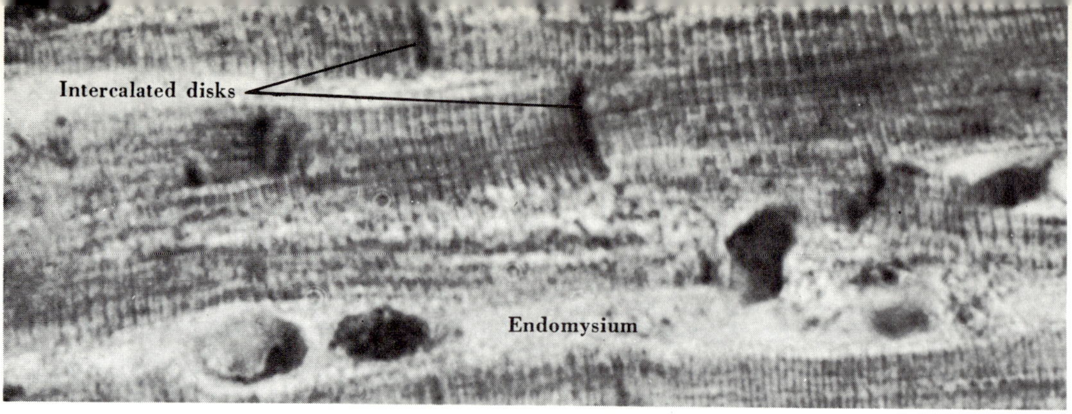

Fig. 5.18 Cardiac muscle tissue. Note striations and branching shape of cells. X400.

circulating physiological agents known to affect the heartbeat is removed, or if techniques of tissue culture are employed, the impulse-conducting tissue continues to stimulate its nonimpulse-conducting neighbors to contract in a coordinated manner. Moreover, even if the conduction tissue is destroyed, cardiac fibers will contract, although coordination of the contraction of the heart as a whole is lost.

Cardiac muscle fibers are arranged in sheets or meshes, the longitudinal dimension of each cell being more or less parallel with that of its neighbors. Some fibers are shaped as regular or irregular cylinders whereas others exhibit branching and anastomosing twigs. The spaces which are created by branching and anastomosing contain loose connective tissue endomysium which serves as a carrier of blood vessels, nerves, and lymphatic vessels (Fig. 5.18). It should be noted in passing that only cardiac muscle endomysium contains lymphatics. Such vessles are found only in the epimysium and perimysium of skeletal muscle.

The nuclei of cardiac muscle fibers are similar to those of muscle fibers in general, but their location is almost always central. The question of number of nuclei per fiber has not been fully resolved. Most current reports indicate a uninucleate or occasionally binucleate condition, leaving the burden of proof with those who espouse the multinucleate theory.

The myofilaments of cardiac fibers show no unique features as compared with skeletal fibers except, perhaps, for their somewhat fainter appearance. In the fish and less specialized amphibians, the filaments appear to be localized around the periphery of the cell. Tetrapod fibers contain more evenly distributed filaments. In either case surrounding each nucleus is a fusiform area devoid of filaments, but usually containing numerous large mitochondria, a small Golgi body, fat droplets, glycogen deposits, and pigment granules. In addition, mitochondria and other inclusions can also be found in the interstices between the myofibrils. The mitochondria of avian cardiac muscle include the most complex cristae ever observed (Fig. 5.19).

With the exception of the end surfaces, the sarcolemma of these fibers is similar to that of skeletal fibers, although somewhat thinner. The end sarcolemmal walls, however, are greatly modified and are known as *intercalated disks*—a name that can be considered an historical remnant stemming from the time when cardiac muscle was considered a syncytium.

Intercalated disks, or end walls, are located at the end of a sarcomere in place of the Z line, and, as in the case with the Z line, they may exhibit concentrations of an amorphous dense material (Fig. 5.20). Three kinds of disks, or rather three kinds of end-wall sarcolemmal modification can be observed; these, however, are similar to, or identical with, the epithelial structures described in Chapter 3. (1) In regions which are free of myofibrils, apposing end membranes may converge and fuse. The line of convergence may be straight or undulating, and oriented diagonally or at right angles to the long axis of the fiber. Each end wall may be irregularly thickened intracytoplasmically with the aforementioned dense, amorphous material. Because this type of function occurs only in limited areas, the term *macula occludens* is applied. (2) The

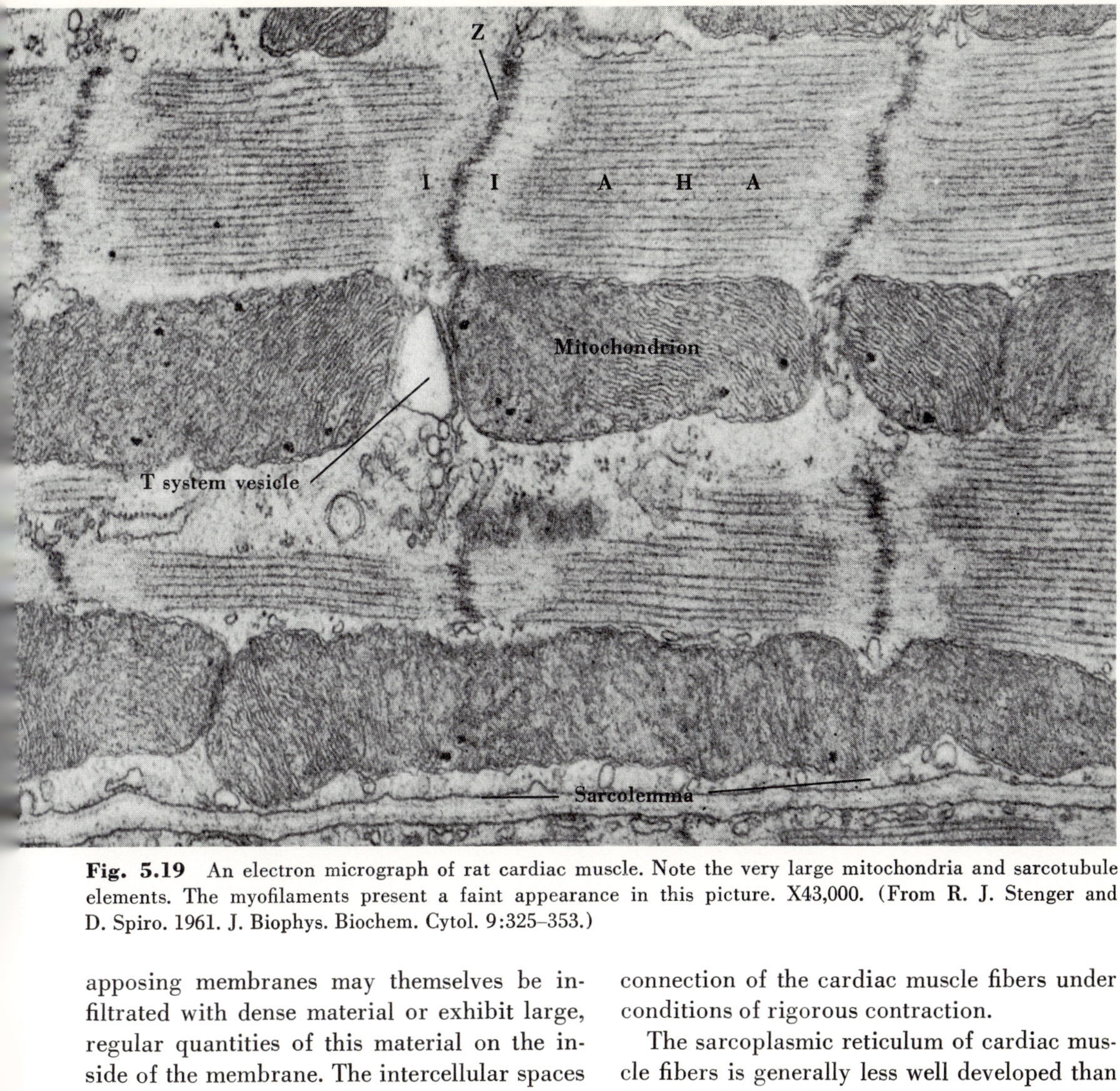

Fig. 5.19 An electron micrograph of rat cardiac muscle. Note the very large mitochondria and sarcotubule elements. The myofilaments present a faint appearance in this picture. X43,000. (From R. J. Stenger and D. Spiro. 1961. J. Biophys. Biochem. Cytol. 9:325–353.)

apposing membranes may themselves be infiltrated with dense material or exhibit large, regular quantities of this material on the inside of the membrane. The intercellular spaces are then narrow and the line of cell union diagonal to the long axis of the myofibril. This type of end wall usually occurs in fiber areas containing myofilaments and extends for considerable distances (*fascia adherens*). (3) In sarcoplasmic expanses of the fiber, structures which appear to be desmosomes (macula adherens) may thicken and strengthen a Type 1 disk (Fig. 5.21).

Regardless of the localized disk structures, an end-on view of the cell boundaries reveals that they are not blunt or tapered, but rather that each end is deeply grooved and interdigitated with the adjoining cell. This hand-in-glove attachment serves to maintain a firm connection of the cardiac muscle fibers under conditions of rigorous contraction.

The sarcoplasmic reticulum of cardiac muscle fibers is generally less well developed than that of skeletal fibers, bearing out the relationship between reticular organization and contraction speed (Fig. 5.19). Often the triad is replaced by a single transverse channel at the Z line, and the longitudinal connections are not so numerous. Here too, however, the membranes of the tubule walls are identical with those of the sarcolemma, which, in turn, appears to be indented at each Z line.

Impulse-Conducting Tissue of the Heart

The two specialized characteristics of muscular tissue, contraction and conduction, are brought to a high degree of development

Fig. 5.20 Electron micrograph of cat heart muscle cell. Note the interconnection (*arrows*) between all intercalated disks in one cell. Because the disks are arranged staircase fashion, there is deep interdigitation between neighboring cells. The numerous black granules are glycogen and the mitochondria are very large. X15,000. (From D. Fawcett. 1966. An atlas of fine structure—the cell, its organelles and inclusions. W. B. Saunders Co., Philadelphia.)

in the heart. Cardiac fibers are thought to contract in the same manner which has been postulated for skeletal fibers. However, the nerve connections of each kind of muscle fiber differ, both in anatomy and physiology. Skeletal muscles are controlled by efferent and afferent nerves of the somatic division of the nervous system, and, as such, require innervation to function in any capacity. Cardiac muscle is governed, but not fully controlled, by impulses from the autonomic division of the nervous system. Although the heartbeat may be speeded up or slowed down by incoming impulses, the origin of the heartbeat is innate. All vertebrates have a region of cardiac tissue in which the ability to respond to changes in electrical potential created by active cells is as well developed, or more highly developed, than is the ability to contract. In cold-blooded vertebrates, this area is the sinus venosus, the heart chamber into which pours the returning venous blood. The heartbeat, or rather the frequency of electrical discharge which initiates the heartbeat, begins in the sinus venosus and the impulses spread over the surface of the muscle fibers resulting in sequential beating of the chambers of the heart. The conduction system of birds and mammals is considerably more complete, consisting of so-called nodal tissue and conduction fibers. In these warm-blooded classes the heartbeat originates in the sinoatrial (SA) node, a bit of modified cardiac muscle and connective tissue located at or near the junction of the right atrium and the base of the anterior vena cava. Because of the demonstrable lack of other specialized cardiac tissue in the atria, current theory states that the impulses, after originating in the SA node, travel across the atria by conduction over the surfaces of ordinary cardiac fibers. In the course of this conduction, some of the impulses reach another bit of specialized tissue, the atrioventricular (AV) node. This area, located in the interatrial septum near the tricuspid valve, exhibits many similarities and some differences when compared with the SA tissue. Arising from the AV node is the AV bundle of fibers; these, composed of small, branching cardiac fibers which have ramifications throughout the ventricular mass and serve to conduct the impulses to the ventricles in a uniform manner.

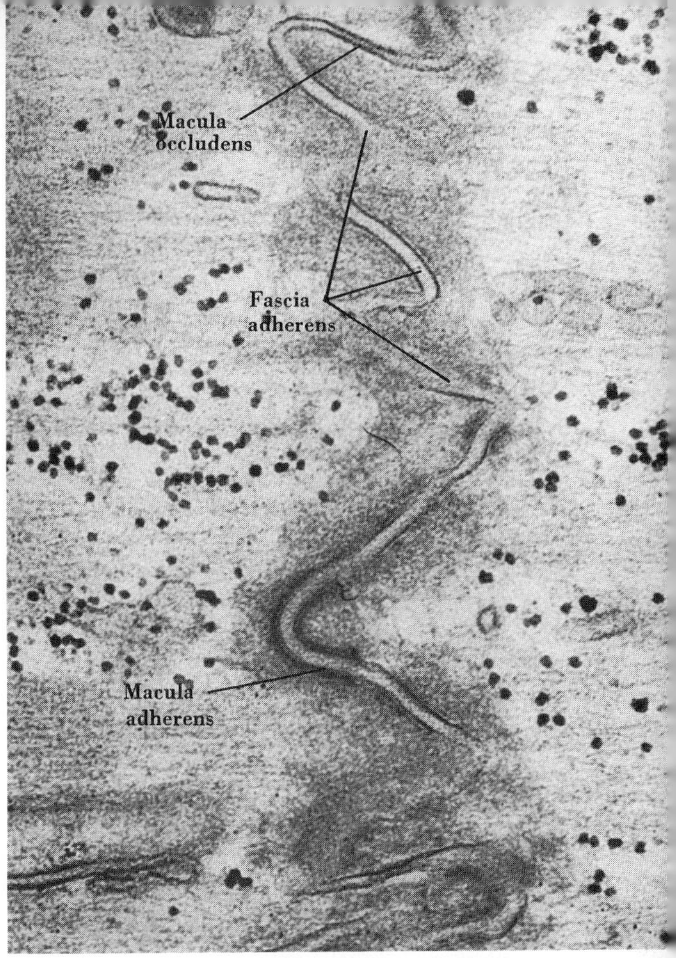

Fig. 5.21 Electron micrograph of part of one intercalated disk. Differentiation along the disk can be seen. X78,000. (From D. Fawcett. 1966. An atlas of fine structure—the cell, its organelles and inclusions. W. B. Saunders Co., Philadelphia.)

Identification of the impulse-conducting tissue is not difficult because, in addition to a somewhat different appearance of the cells as compared to ordinary cardiac fibers, most of the system is segregated from the bulk of the heart by connective tissue membranous partitions which course throughout the system.

The SA node, or pacemaker of the heart, is composed of small, branching cardiac fibers which are more or less fusiform in shape (Fig. 5.22). Surrounding and between the cardiac fibers are quantities of dense connective tissue containing many capillaries and autonomic nerve fibers. In addition, peripheral ganglia of the parasympathetic nervous system are in close proximity to the node and can thus exert a depressant effect upon it. Sympathetic fibers, also present, serve to accelerate the heart rate. There have been fine-

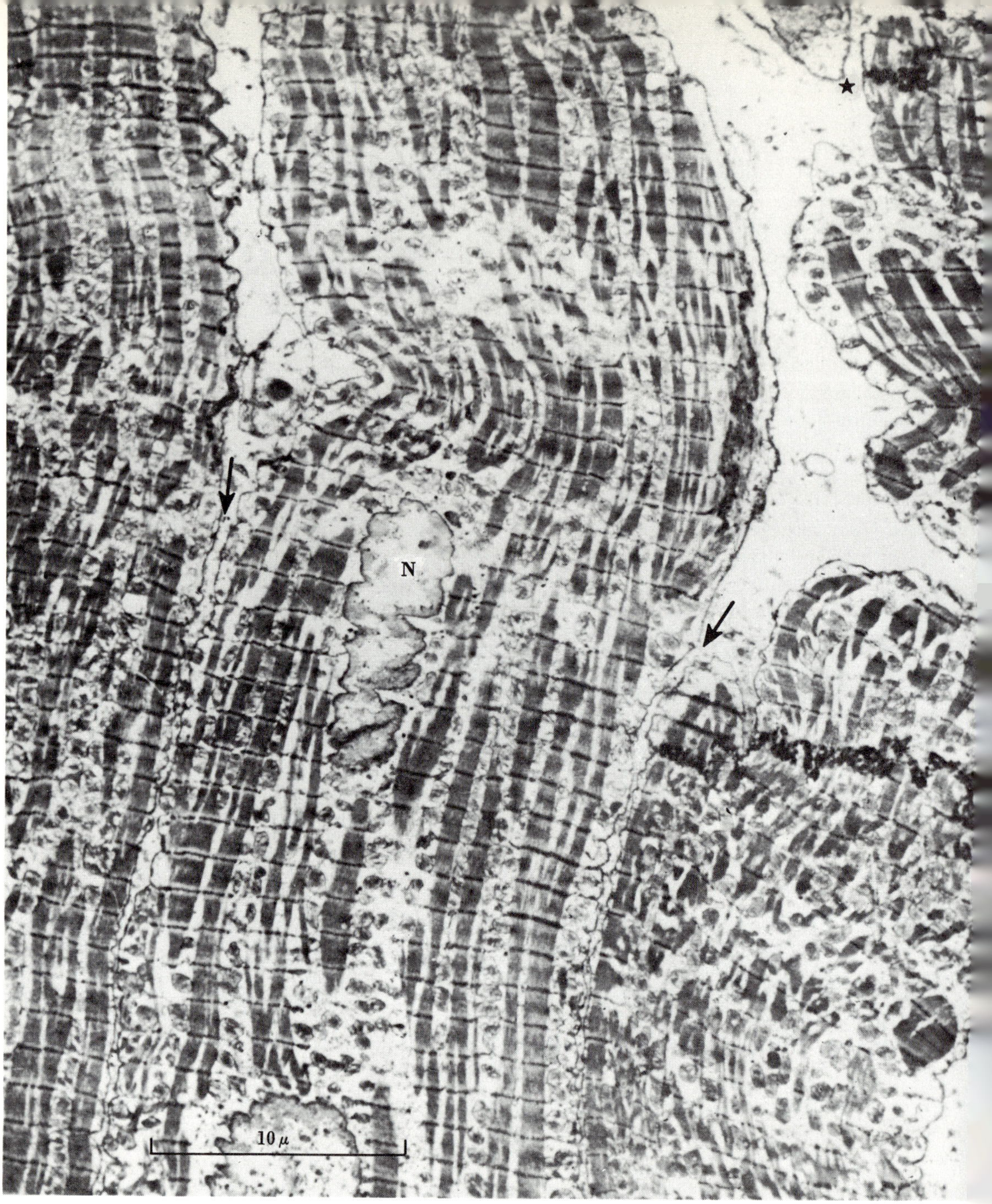

Fig. 5.22 An electron micrograph of longitudinally sectioned fibers of the steer SA node. Note the great quantity of connective tissue between the cardiac fibers. Apposed boundaries are indicated at the arrows and intercalated disks at the stars. X3400. (From J. A. G. Rhodin, P. del Missier, and L. C. Reid. 1961. Circulation 24:362. By permission of the American Heart Association.)

structural reports of two different kinds of cells in the SA node with the possibility of a third type, transitional between the first two, also existing. The first type of cell is similar to an ordinary cardiac muscle fiber. It has myofibrils, extensive sarcoplasmic reticulum, and is located around the periphery of the node. The second cell type is found in a more central nodal location. Occurring in clusters, these small round-to-ovoid cells have few myofibrils and a rather empty-looking cytoplasm. It has been postulated that clusters of these cells ap-

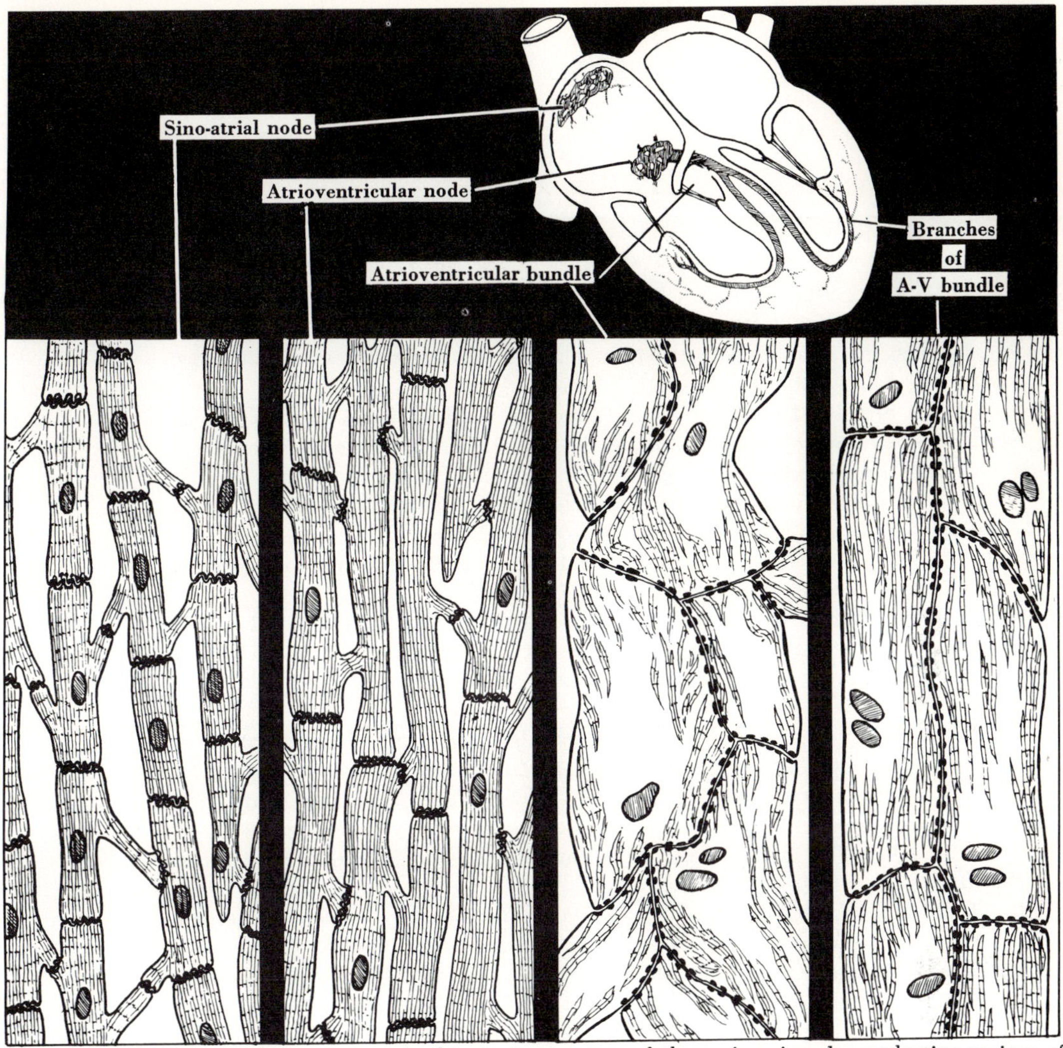

Fig. 5.23 A diagram of the comparative histologic structure of the various impulse-conducting regions of the steer heart. X130. (After J. A. G. Rhodin, P. del Missier, and L. C. Reid. 1961. Circulation 24:350. By permission of the American Heart Association.)

pear as the central, branching fiber seen in the optical microscope. An hypothesis has been advanced that the small cells are derived from the myocardial type and, furthermore, that it is the former which is responsible for the pacemaking capacity of the node.

The cardiac fibers of the AV node form an intricate branching feltwork. The fiber size is larger than that of the SA node, but not as large as ordinary cardiac fibers (Fig. 5.23). Characteristically, the AV fibers appear more closely packed than in the SA node and contain quantities of myofibrils.

The fibers of the AV bundle are irregular in shape and are thought to represent the side-to-side and end-to-end fusion of several cells. Peripheral areas containing striated myofibrils can be seen, although the cells seem otherwise sparsely occupied. More than one nucleus may be seen per cell. About halfway down the interventricular septum, the morphology of these fibers begins to change to that characteristic of true Purkinje fibers. These fibers are large, especially in their length, and contain a central nucleus or nuclei and peripherally disposed myofibrils (Fig. 5.23). The central area, devoid of myofibrils, contains quantities of glycogen-containing inclusions. A true T system is lacking in these fibers.

(A discussion of the arrangement of cardiac muscle fibers and the impulse conduction system within the heart can be found in Chapter 9.)

SELECTED REFERENCES

Comparative aspects of muscle: symposium. 1967. In Am. Zool. Entire issue, vol. 7, no. 3.

Dewey, M. M., and L. Barr. 1964. A study of the structure and distribution of the nexus. J. Cell Biol. 23:553–586.

Fawcett, D. 1966. An atlas of fine structure—the cell, its organelles and inclusions. W. B. Saunders Co., Philadelphia.

Huxley, H. E. Structural evidence concerning the mechanism of contraction in striated muscle. In Muscle. Paul, W. M., E. E. Daniel, C. M. Kay, and G. Monckton, eds. Pergamon Press, Oxford, 1965, pp. 3–29.

James, T. N., L. Sherf, G. Fine, and A. Morales. 1966. Comparative ultrastructure of the sinus node in man and dog. Circulation 34:139–163.

Lane, B. P. 1965. Alterations in cytologic detail of intestinal smooth muscle cells in various stages of contraction. J. Cell Biol. 27:199–213.

Oosaki, T., and S. Ishii. 1964. The ultrastructure of the regions of junction between smooth muscle cells in the rat small intestine. J. Ultrastruc. Res. 10:567–577.

Page, S. 1965. A comparison of the fine structure of frog slow and twitch muscle fibers. J. Cell Biol. 26:477–497.

Panner, B. J., and C. R. Honig. 1967. Filament ultrastructure and organization in vertebrate smooth muscle. J. Cell. Biol. 35:303–321.

Porter, K. R., and C. Franzini-Armstrong. 1965. The sarcoplasmic reticulum. Sci. Am. 212:72–81.

Revel, J. P. 1962. The sarcoplasmic reticulum of the bat cricothyroid muscle. J. Cell Biol. 12:571–589.

Rhodin, J., P. del Missier, and L. C. Reid. 1961. The structure of the specialized impulse-conducting system of the steer heart. Circulation 24:349–368.

Shafiq, S. A., M. Gorycki, L. Goldstone, and A. T. Milhorat. 1966. Fine structure of fiber types in normal human muscle. Anat. Rec. 156:283–302.

Sommer, J. R., and E. A. Johnson. 1968. A comparative study of Purkinje fibers and ventricular fibers. J. Cell Biol. 36:497–526.

Stenger, R. J., and D. Spiro. 1961. The ultrastructure of mammalian cardiac muscle. J. Biophys. Biochem. Cytol. 9:325–353.

Stromer, M. H., D. E. Cole, and L. E. Roth. 1967. Morphology of rigor-shortened bovine muscle and the effect of trypsin on pre- and post rigor myofibrils. J. Cell Biol. 34:431–445.

Thaemert, J. C. 1963. The ultrastructure and disposition of vesiculated nerve processes in smooth muscle. J. Cell Biol. 16:361–377.

Wachtel, A. W. 1964. The ultrastructural relationships of electric organs and muscle: I. Filamentous systems. J. Morphol. 114:325–360.

chapter 6
THE INTEGUMENT

THE INTEGUMENT of vertebrates consists of the skin and all structures that are produced by or are in association with the skin. In a broad sense, the skin is one of the most important organs of the body, for, being in immediate contact with the outside environment, it is the organism's first line of defense against a host of intolerable chemical and physical factors.

More specifically, the integument serves the obvious function of mechanical protection, i.e., protection against mechanical factors, such as abrasion, puncturing, and invasion by other organisms. It also protects against fluid loss or gain, and against the destructive action of chemical substances in the outside environment. Through coloration and pigmentation, it serves either to make the organism more conspicuous, if being so has proved advantageous, or inconspicuous. Through its secretions, it can render the organism attractive or repugnant. The skin may also give rise to structures that are useful in aggression as well as in protection, and only recently have we begun to gain some insight into its importance as an immunologic organ of protection. But the functions of the integument are by no means limited to protection. It may function also in locomotion in certain vertebrates, such as seen in the epidermal scales of snakes, the feathers of birds, and the hoofs of ungulates—as well as the skin of the porpoise, which, by its singular ability to ripple with the flow of water, counteracts the drag effect of turbulence.

The integument serves as an organ of energy storage in many vertebrates that have extensive depots of subcutaneous fat; at the same time these depots may serve as insulation against heat loss, as in aquatic mammals. Coincidental with fat storage occurs the storage of certain oil-soluble vitamins, and the final synthesis of vitamin D has been shown to occur in the skins of some vertebrates. With its many tactile corpuscles and other sensory receptors, it functions as an indispensable adjunct to the nervous system. In certain marine species of reptiles and birds the skin has formed glands for the excretion of salt. Finally, the integument plays a vital role in temperature regulation—primarily in mammals, but to some extent in some species of reptiles and birds as well.

Derivation

The integument is derived from two embryologic sources, and each derivative is sufficiently distinct from the other both morphologically and functionally to be considered as a quite separate part of the skin. The outer portion, the *epidermis,* develops directly from ectoderm of the gastrula, whereas the deeper layer, the *dermis,* is derived from the mesoderm. The epidermis is the more interesting of the two because it gives rise to the receptor organs, the glands, the epidermal scales, feathers, hairs, claws, horns, and other appendages, whereas the dermis simply supports, cushions, and feeds its more important partner. However, inasmuch as they are inseparable partners, they should be considered together.

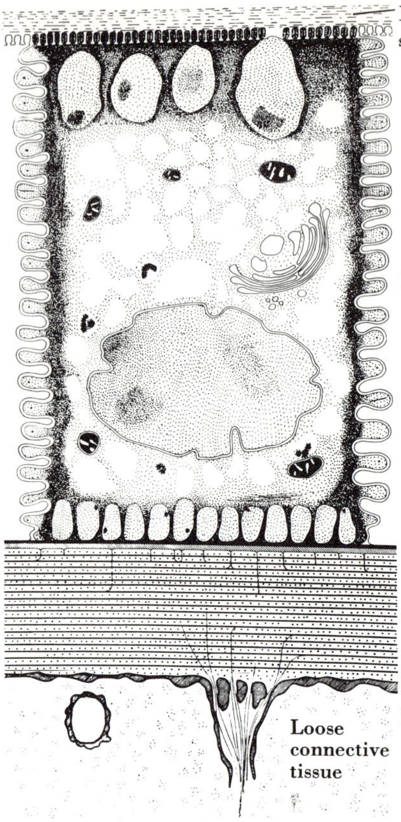

Fig. 6.1 Diagrammatic representation of the skin of amphioxus. X12,000. (From Olsson, R. 1961. The skin of amphioxus. Zeitschrift fur Zellforschung und mikroskopische Anatomie, Bd. 54, S. 90–104. Berlin-Heidelberg-New York: Springer.)

THE BASIC INTEGUMENT

Few organisms exist that expose their bare integumentary cells to the environment. The invertebrates protect their integument by means of various devices, such as cuticles, silicious or calcareous shells, chitin, or by means of mucous secretions. The vertebrates are not very different in this respect. But then, why should they be? They are exposed to the same or similar environmental conditions and if one considers their invertebrate ancestors as part of their lineage, they have been adapting to such conditions for the same extent of time.

In Amphioxus the skin is exceedingly simple, consisting of a thin epidermis which is surmounted by a cuticular membrane, and of a thin dermis that consists largely of intercellular substance, such as fibers and a gelatinous matrix. The cells of the epidermis occur as a single layer resting on a prominent basement membrane. The cuticle, which is secreted by the epidermis, is perforated with minute pores (Fig. 6.1). Whether Amphioxus has the kind of skin which was characteristic of the chordate stock from which the vertebrates evolved is unknown, but certainly a simpler arrangement of epidermis and dermis cannot be imagined. The integuments of vertebrates may be considered to represent modification of such a simple, basic prototype, each modification reflecting the adaptive evolution which the vertebrates have undergone. Most particularly, the nature of the vertebrate integument reflects the nature of the environment to which adaptation has occurred. For example, despite all of the variations from species to species, the integuments of the bony fishes are built on a more or less common plan that reflects the adaptation of an ectothermic vertebrate to an aquatic environment. In a similar way, the integuments of reptiles

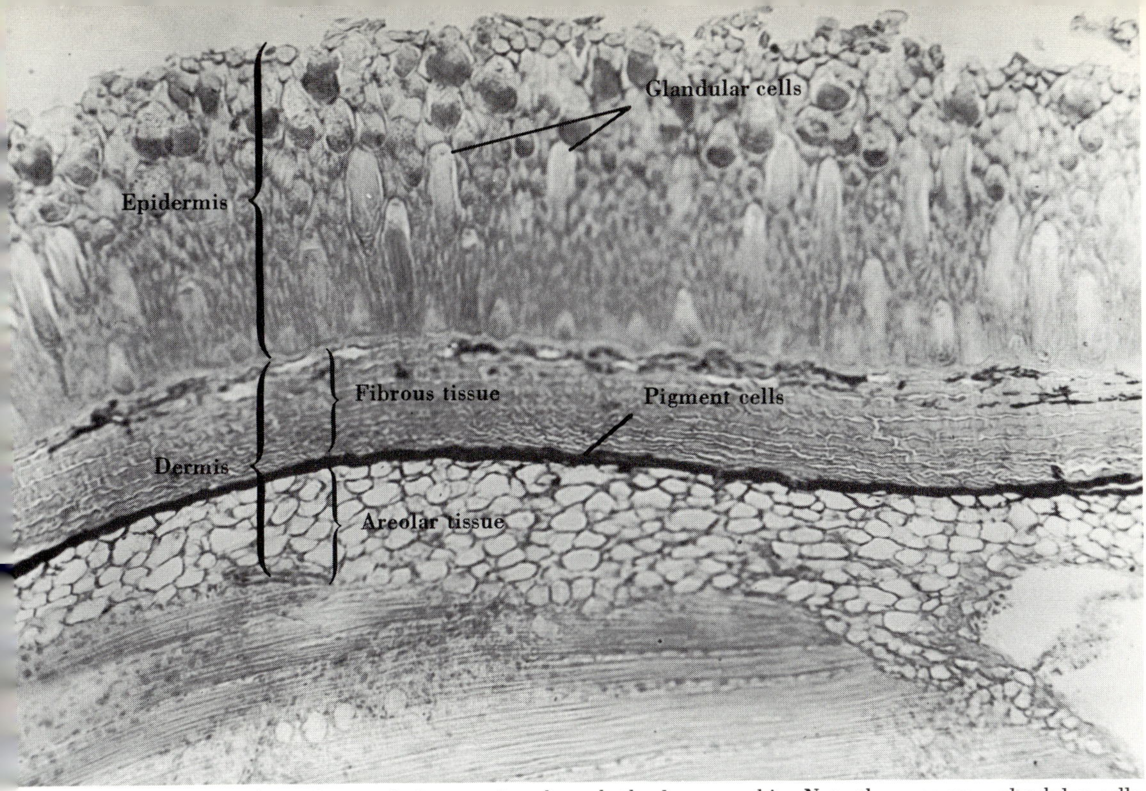

Fig. 6.2 This photomicrograph is a section through the lamprey skin. Note the numerous glandular cells and the highly organized dermis. X100.

demonstrate how ectothermic vertebrates have adapted to a primarily terrestrial environment, and so on. Because the classes of vertebrates represent, by and large, different classes of environmental adaptation, they will be considered in taxonomic order, with comparisons being made between them when such comparisons seem appropriate.

THE INTEGUMENT OF FISHES

The aquatic adaptation of the three classes of living fishes makes it convenient to discuss them under one heading. The cyclostome integument is simple, consisting of (1) a thin, stratified cuboidal epithelium, all of the cells being alive and with no evidence of keratinization, and of (2) an uncomplicated dermis (Fig. 6.2). Among the epithelial cells are numerous glandular cells that secrete rather copious quantities of mucus. They are all unicellular, merocrine glands, although multicellular mucous glands have been described. In the hag fish are found thread cells, which secrete a spirally coiled mucous thread, reminiscent of rather similar devices in some invertebrates. A thin cuticular covering secreted by the superficial epithelial cells has been described, although the authors have failed to find any evidence of such in the lamprey. The dermis includes an outer layer of fairly dense fibrous nature, with pigment cells, and a subcutaneous layer of fatty areolar tissue, blood vessels, and cutaneous nerves. In the vicinity of the mouth and the gills, numerous tactile corpuscles are innervated by conspicuous bundles of such nerves. There is no evidence of scales.

The skins of the gnathostomatous fishes show more extensive and specialized adaptations to the aquatic environment. Such is to be expected in the largest and most diversified assemblage of vertebrates. Despite this variation, there are certain basic features which are characteristic of the integuments of all the major orders of fishes, and these will be considered now together with their more significant modifications. The epidermis in all gnathostomatous fishes consists of a thin, stratified, squamous, nonkeratinized epithelium, with a few glands and minimal glandular specialization. It is most remarkable, how-

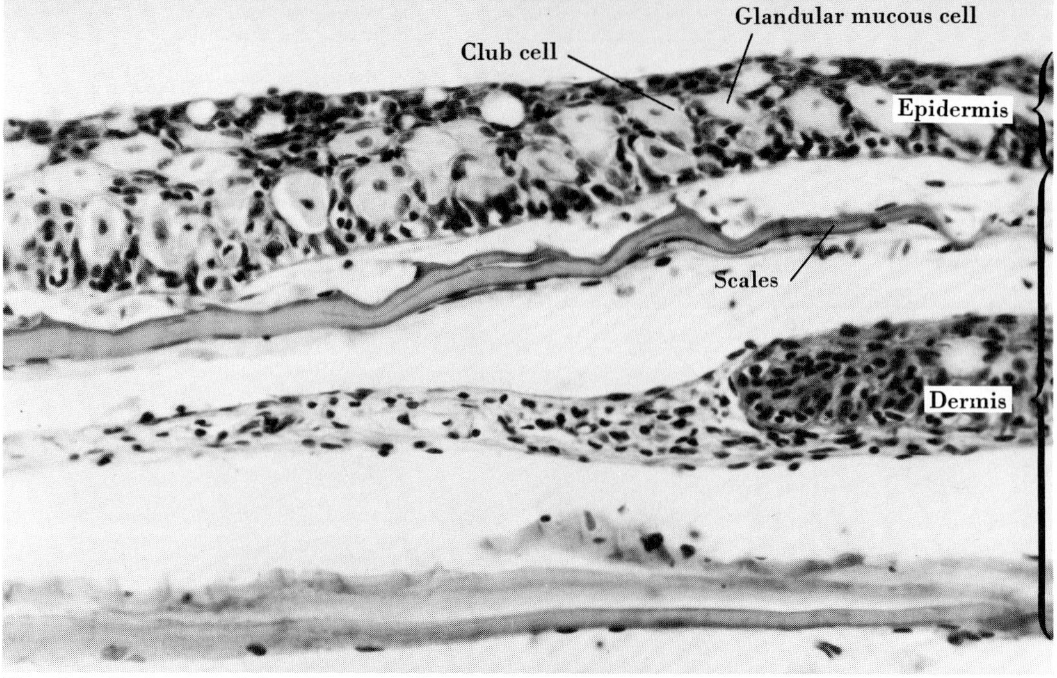

Fig. 6.3 Epidermis and dermal scales in the goldfish. X350. (Courtesy of R. Henrikson.)

ever, for the scales which are associated with it. These are of various types, but they are all produced by the dermis and are intimately associated spatially with the epidermis. In some cases the scales protrude through the epidermis, but in others the epidermis forms a continuous covering over them, even though they may overlap one another (Fig. 6.3). Despite the simplicity of the piscine epidermis, it has served as the basis for the introduction of some very interesting modifications, as will be described.

Glands

Although there is some evidence for the presence of simple, multicellular, tubular glands in fishes, the integumentary glands are generally unicellular. They are mainly mucous and serous cells, some of which are small, and others are rather large (Fig. 6.4). Viewed with the optical microscope, these cells appear to be qualitatively different. However, a study of these cells at higher magnifications has

Fig. 6.4 Section through the skin of *Anguilla*, the eel. Note the extremely thick epidermis and the variety of secretory cell types. The club cells have a characteristic darkly staining nucleus, and a central vacuole which does not stain. X380. (Courtesy of R. Henrikson.)

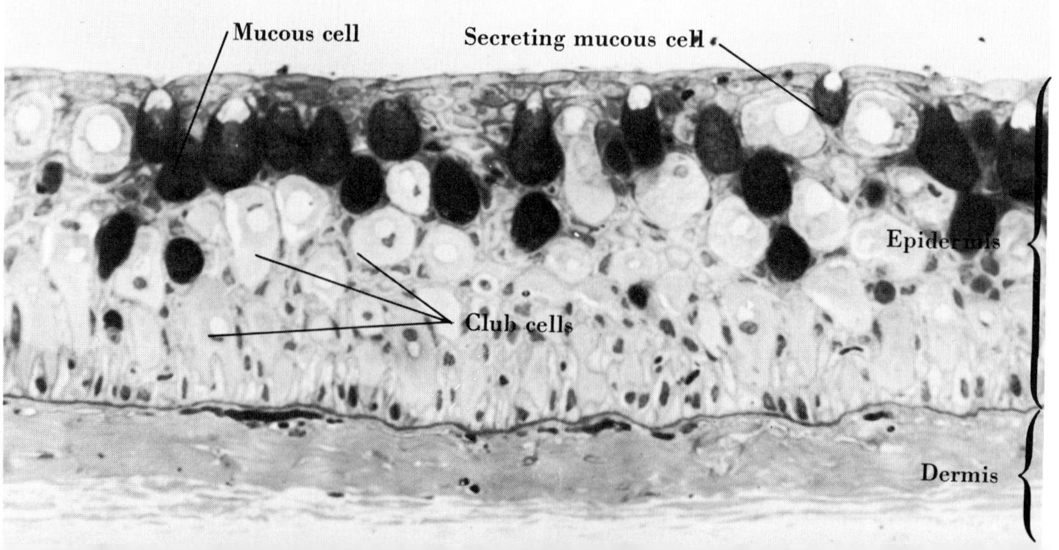

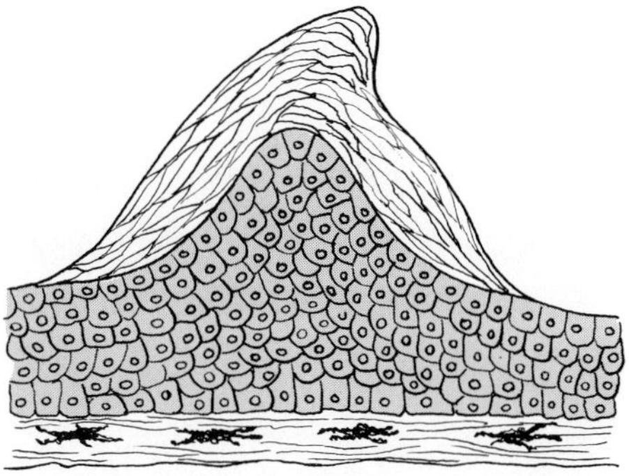

a.

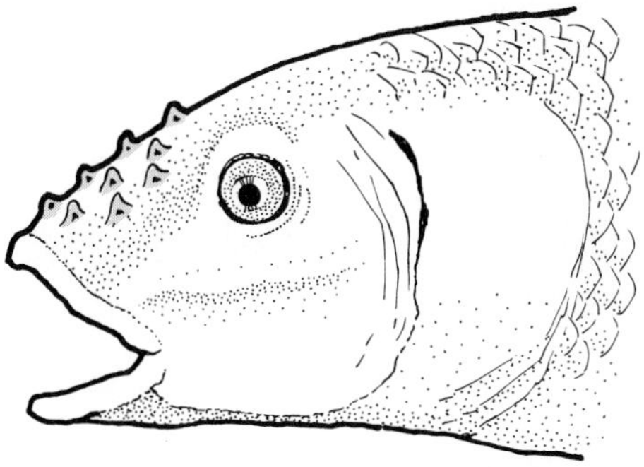

b.

Fig. 6.5 A histologic section through the teleost pearl organ, *a.* (From Maurer, 1895.) The location of these structures on a teleost head, *b.*

shown that the glandular cells differentiated as small cells near the base of the epidermis migrate toward the surface. As they migrate, enlargement occurs—as does the accumulation of secretory granules. *Club cells,* which are presumed to secrete an intercellular cement substance, similarly migrate from the basement membrane to the midepidermal region. *Poison glands,* some of which are multicellular, are associated with the bases of spines in some species. The spines may be hollow tubes, which function as effective hypodermic needles.

Sense Organs

Minute structures called *pearl organs,* so named because of the opalescent horny concentration that sometimes forms a covering over them, occur in some species of fishes (Fig. 6.5). They are presumed to have a tactile function and to be related to the prototrichs of teleosts and to the very primitive tactile organs of elasmobranchs (Fig. 6.6). Typically, they appear in the males of certain species, such as the minnows and suckers, just before the spawning season. The lateral line organs of cyclostomes, fishes, and aquatic amphibians are very prominent integumentary sensory structures, but because of their evolutionary and morphological relation to the internal ear of the tetrapods, they will be considered in some detail in Chapter 13. Also related to the lateral line organs are the *pit organs* of bony fishes, the *ampullae of Lorenzini* of the elasmobranchs, and the *vesicles of Savi,* which occur on the ventral surface of the skin of Torpedo, the electric ray.

Multicellular sense organs are found in the skin of weakly electric fish. Presumably for the detection of changes in the surrounding electric field, they are of two types. The first type, the ampullary organs, essentially consist of a jelly-filled duct which opens to the surface and which expands basally into a saclike enlargement. Beneath the sac are receptor cells which are innervated by unmyelinated nerve fibers. The second type (Fig. 6.7), the tuberous organs, consist of a cluster of receptor cells surrounded by a cellular capsule which is open at the surface. Each cluster is innervated by a single unmyelinated nerve.

Sensory cells which resemble those of the lateral line, by virtue of the cilium at the apical end, are the *olfactory receptor cells.* Although these cells do not occur in the superficial integument of fishes generally, they are found covering much of the surface of the

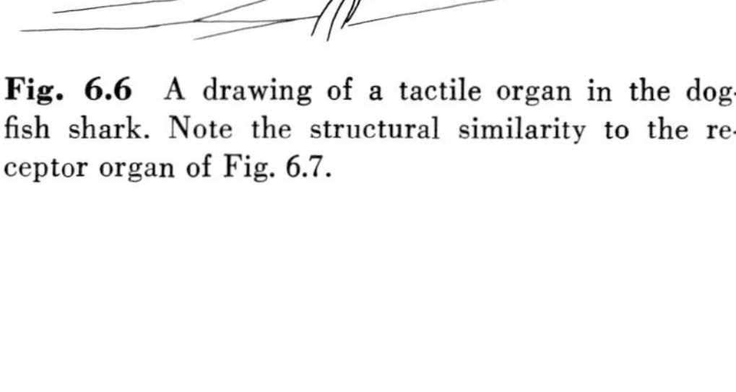

Fig. 6.6 A drawing of a tactile organ in the dogfish shark. Note the structural similarity to the receptor organ of Fig. 6.7.

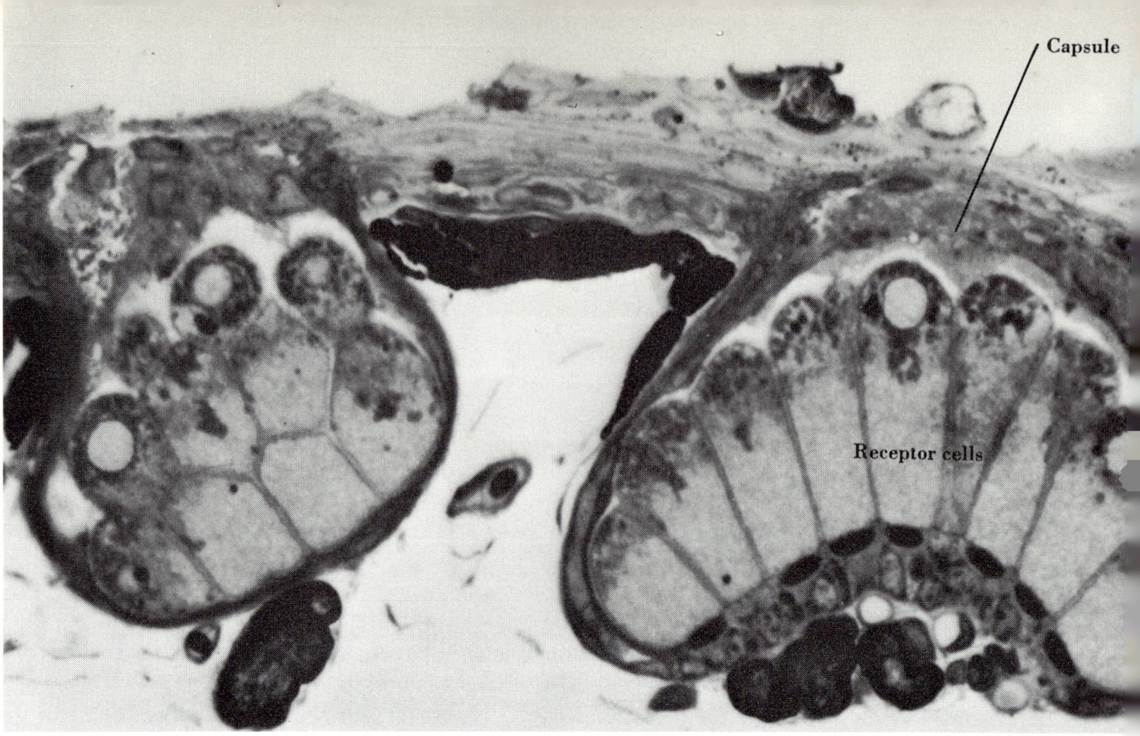

Fig. 6.7 The tuberous organ of a weakly electric fish. (From Fig. 5, A. W. Wachtel. 1966. J. Morphol. 119:67).

body of certain scavenging types, such as catfish and carp. In the former there is a particularly dense population of these cells in the skin of the barbels. Otherwise such cells have become more particularly associated with internal locations, as the mouth, the tongue, and the pharynx.

Scales

The scales of fishes are interesting for many reasons. Their size, shape, and location reflect the nature of the environment in which the fish lives, as well as the way in which the fish functions. They are of evolutionary interest, because there is good evidence that the dermal bones of the skeletons of higher vertebrates have been derived from the bony, dermal scales of primitive bony fishes, and the vertebrate tooth represents a modified kind of scale not unlike the *placoid scale* of modern elasmobranchs. The variety of scales which has evolved is impressive, but the number of basic types is small. In all cases the scales are of dermal origin, as opposed to the epidermally derived scales which are characteristic of the amniote vertebrates. The earliest vertebrate fossil records show that scales, themselves, have evolved from bony plates. Each plate consisted of a layer of spongy bone sandwiched between two layers of compact bone and covered by dentine-containing structures called denticles (Fig. 6.8). The scales are, for

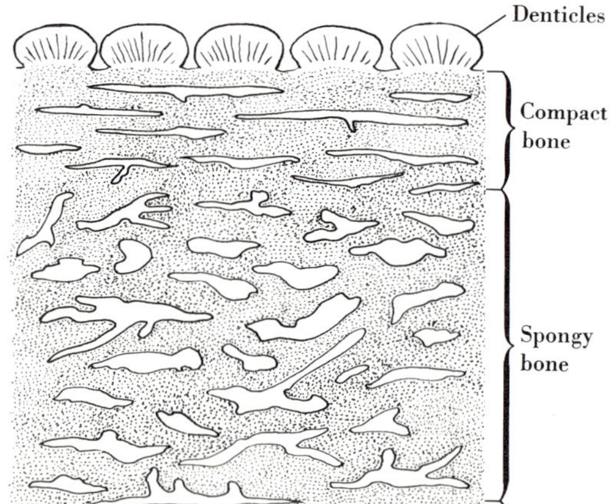

Fig. 6.8 A bony plate of a fossil placoderm. (After Fig. 11.10. T. W. Torrey. 1967. Morphogenesis of the vertebrates. John Wiley & Sons, New York.)

the most part, derivatives of these bony plates, but the placoid scale has been modified by the regression of bone and the prominance of dentine or dentinelike substances (Fig. 6.9). The dentinelike material is a mineral apatite that is hard, lustrous, and has a relatively low content of organic material. As with the dentine of the tooth, it is usually characterized by the presence of dentinal tubules radiating from odontoblasts adjacent to an underlying pulp cavity. Moreover, it may be surmounted, as in the placoid scale, by a capping of a dense layer of dentine that has many characteristics of enamel, being extremely hard and lacking organic constituents almost entirely. It is considered by some authorities to be formed by the outer layers of the dermis, whereas true enamel is formed from the epidermis, and therefore is of ectodermal origin. Others claim that it is formed, at least in part, by an epidermal enamel organ.

The *ganoid scale* (Fig. 6.10), which is characteristic of many fossil holosteans and chondrosteans, compounds the confusion regarding the status of enamel even more. Primarily composed of dermal bone, the ganoid scale has an overlying layer of dentine which is surmounted, in turn, by an enamel substance called *ganoin.* Although structurally similar to enamel, ganoin is clearly a product of the dermis, and so fails to meet the rigid requirements of scientific categorization as enamel. The ganoid scale is a laminated structure, for it grows peripherally from a central focus by the addition of layers of bone beneath, dentine in the middle, and ganoin on top—although in some species alternating layers of dentine and ganoin are produced. Because of this manner of growth, the scale often tends to be somewhat conical or dome-shaped. The dentine is nourished by an underlying vascular network which in some species sends branches by prominent vertical channels, the *canals of Williamson,* through the dentine. These, in turn, subdivide subadjacent to the ganoin layer into a fine network of *dentinal tubules.*

Cycloid scales and *ctenoid scales* are generally characteristic of teleosts. The ctenoid type simply represents a cycloid scale which has been reduced in thickness and has acquired serrations on its caudal margin, as well as short spinous processes on its cranial surface. Both types of scales consist of dermal bone produced by the dermis. The superficial layer of bone is usually acellular and is called hyodentine, which is smooth and hard (Fig. 6.11). Growth is marginal, as evidenced by prominent annular rings, the *circuli.* These scales are covered by a very thin layer of epidermis, except where the spinous processes protrude, and they overlap in most species.

Fin Rays

Some mention should be made of the skeletal rays that support the median and paired fins of fishes. They are lacking in cyclostomes, who have no paired fins and whose median fins are fleshy. The cartilaginous radials of the Chondrichthyes are probably not integumentary derivatives; but in the bony fishes they are conspicuous bony structures of dermal origin. They arise from mesenchyme of the embryonic dermatome immediately beneath the basement membrane of the epidermis. In some species the fin rays are jointed structures, and in most species they are derived from bilateral components. Although their function is primarily support of the fin, some fin rays have a protective function, presenting a formidable array of needle-sharp spines to any predator.

THE INTEGUMENT OF AMPHIBIA

For quasiterrestrial forms, the amphibians appear to have rather ordinary skins with little in the way of dramatic modification. But close scrutiny reveals that the seemingly modest modifications are truly significant. For example, the superficial layer of cells is more or less *keratinized.* This is to say, these cells contain deposits of a fibrous protein, *keratin,* which is highly resistant to desiccation, diffusion of water, and to abrasion. In amphibians which are better adapted to terrestrial life, such as toads, the keratin deposits are considerable—even to the extent of occupying the entire contents of the superficial cells. It is a soft keratin, as opposed to the hard keratin which is characteristic of the

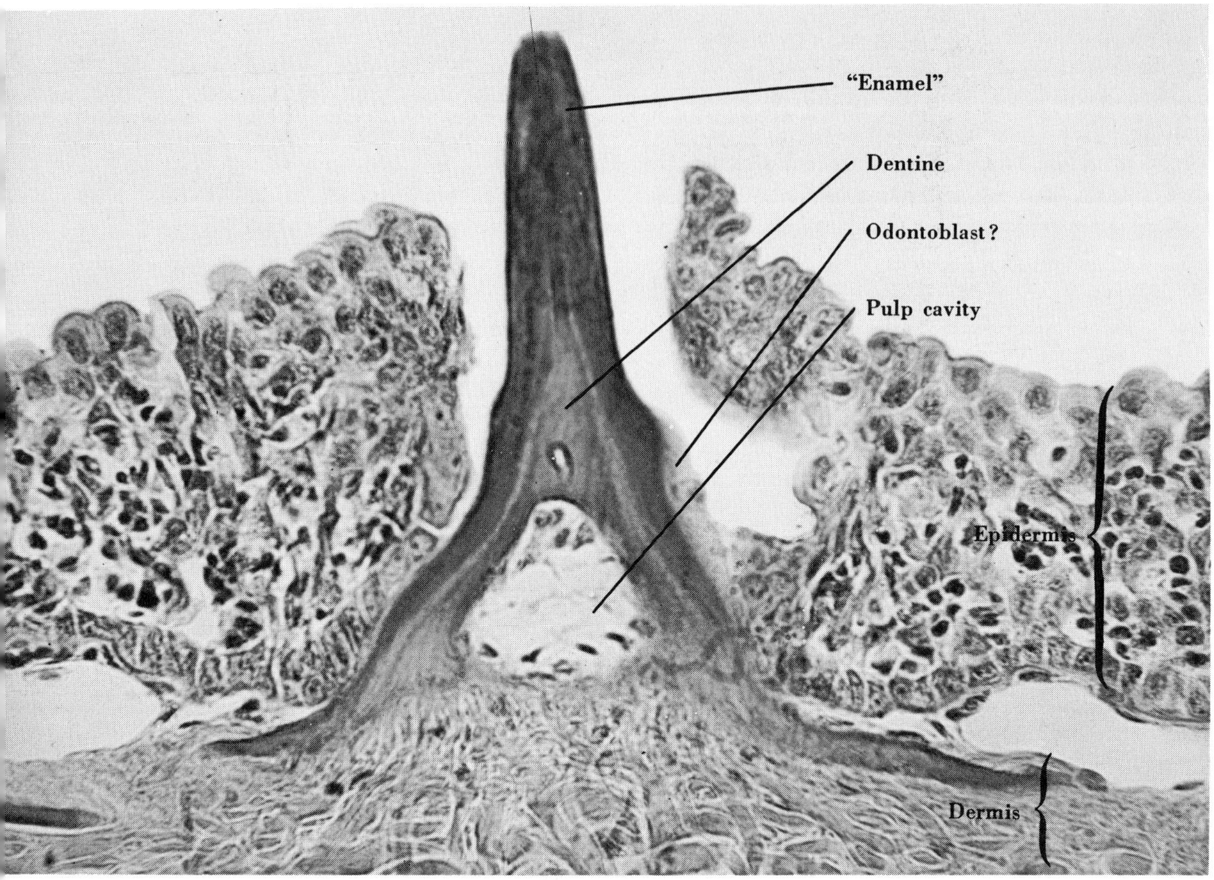

Fig. 6.9 The placoid scale in shark skin. X450.

claws and horns of the amniote classes of vertebrates. Second, one finds that the amphibians have developed the multicellular integumentary gland to a high degree of morphological integrity and functional efficacy. This is important for several reasons that become evident when one considers the kinds of glands which have evolved from them, particularly among the mammals.

The amphibian integument superficially shows conditions ranging from those of the aquatic urodele species to those of tree frogs and toads. The keratinized layer is represented only on the feet in *Necturus,* the rest of the skin being covered by a thin cuticle which may be a primitive form of a keratinized covering, according to some authorities, although this seems an unlikely interpretation, considering that keratin is an intracellular secretion, not an extracellular one. An outer layer of dead, keratinized cells, the *stratum corneum,* is developed to some degree in frogs, and it is well differentiated in toads (Fig. 6.12). The rest of the epidermis is fairly thin, however, except at the claspers, and all of the cells are alive. These cells are arranged in a *transitional layer* of polygonal cells and a *germinative layer* (*stratum germinativum*) consisting of low, columnar cells. The latter rest on a basement membrane which is often considerably obscured by a dense layer of chromatophores in the dermis directly below.

The germinative layer gives rise not only to the two layers above it but to the integumentary glands as well. The latter consist of multicellular, alveolar glands of both *mucous* and *serous* types, which are involuted into the loose layer of the dermis. Occasionally mixed mucous and serous glands occur. Whereas the mucous glands are relatively small and have a regular, low cuboidal epithelium, the serous glands are large and their epithelium is made up of irregularly spaced cells the nuclei of which consequently occur in clusters (Fig. 6.12 and 6.13). Moreover, the serous cells are granular, and they secrete an emetic poison in many species. The connective tissue surrounding them is dense and includes some smooth muscle fibers. The kokoi frog of tropical South America secretes a neurotoxin, *batrachotoxin,* which is the most potent deadly poison known.

The "warts" of frogs and toads are localized thickenings of the epidermis, the cells of which are enlarged and richly supplied with nerve endings. Hence they are, in fact, sensory papillae. The stratum corneum covering them is considerably thickened, giving them an appearance reminiscent of the prototrichs of fishes. Another keratin derivative is the digging "claw" of the spadefoot toad.

In some totally aquatic forms, as *Cryptobranchus* and *Andrias,* in which the skin is the main respiratory organ, the epidermis is richly supplied with blood vessels derived from the dermis.

The amphibian dermis consists of two distinct layers. The outer layer is made up of loosely arranged fibrous tissue containing blood vessels, cutaneous nerves, and chromatophores. Little adipose tissue is found. The chromatophores are of three kinds, depending

Fig. 6.10 Vertical section of a ganoid scale from a cretaceous sturgeon. X100. (Courtesy of M. Mickevich.)

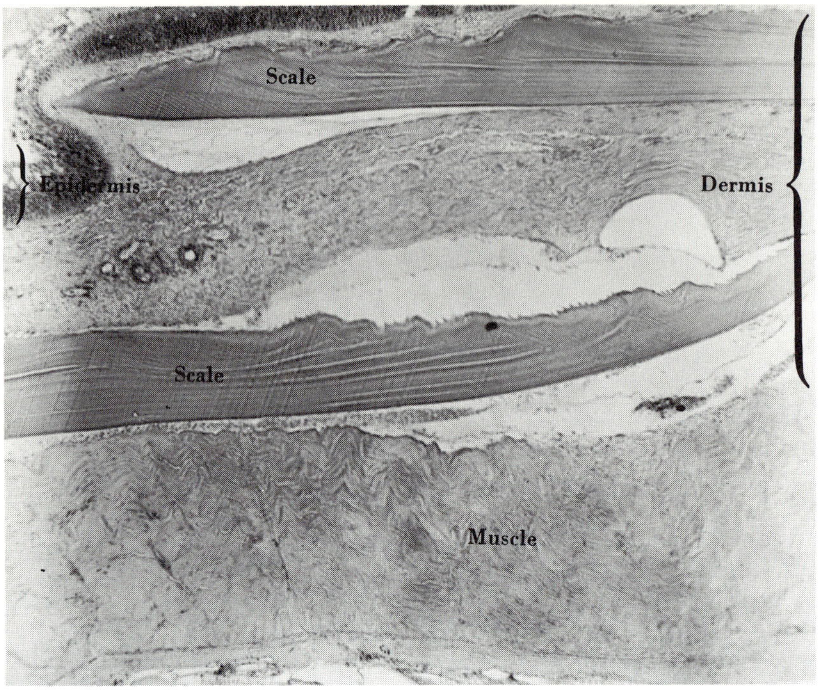

Fig. 6.11 Overlapping teleost ctenoid scale. X90.

on the nature of the pigment which they contain: *melanophores* (Fig. 6.12), containing the brownish pigment, *melanin; xanthophores,* containing a yellowish or golden pigment, *lipochrome;* and *guanophores* (Fig. 6.14), which contain crystals of *guanin,* which impart a silvery or white color. The chromatophores of amphibians like those of many fishes and other forms of life, are branched cells which, by the migration of their contained pigment, may regulate the intensity of coloration of the skin. Their activity is controlled in some cases by the hormone, *intermedin,* which is secreted by the pars intermedia of the pituitary gland.

The deeper layer of the dermis, although usually much thinner than the outer, loose layer, is made up of one or more sheets of dense connective tissue, which is closely applied to the underlying skeletal muscle, except along those areas where the capacious lymph sinuses occur. The fibrous elements are primarily of the collagenous type.

The Apoda constitute the only group of modern amphibians possessing scales. These occur as thin plates of bone diagonally dis-

Fig. 6.12 Cross section through frog skin. Note the very thin surface layer of keratinized cells. X150.

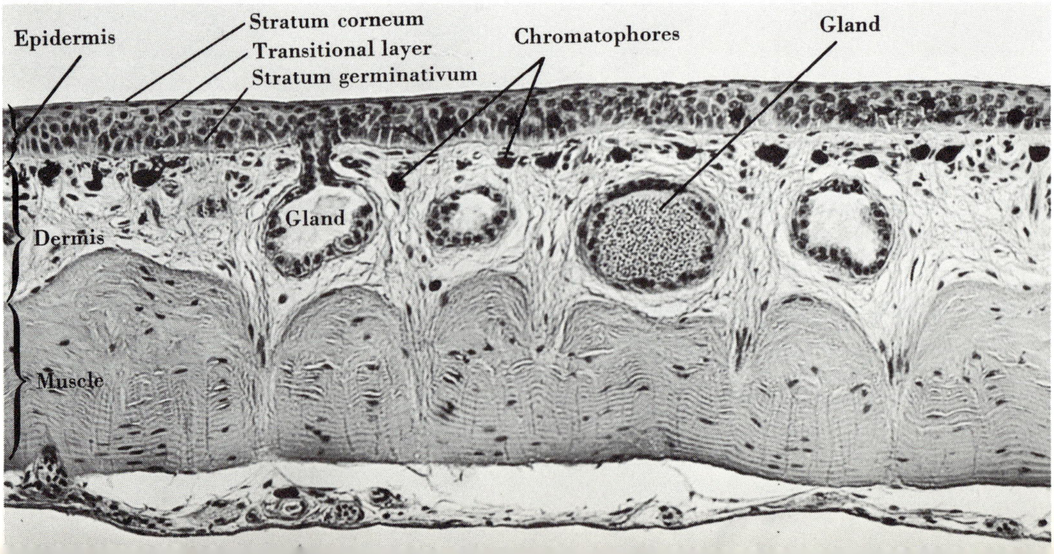

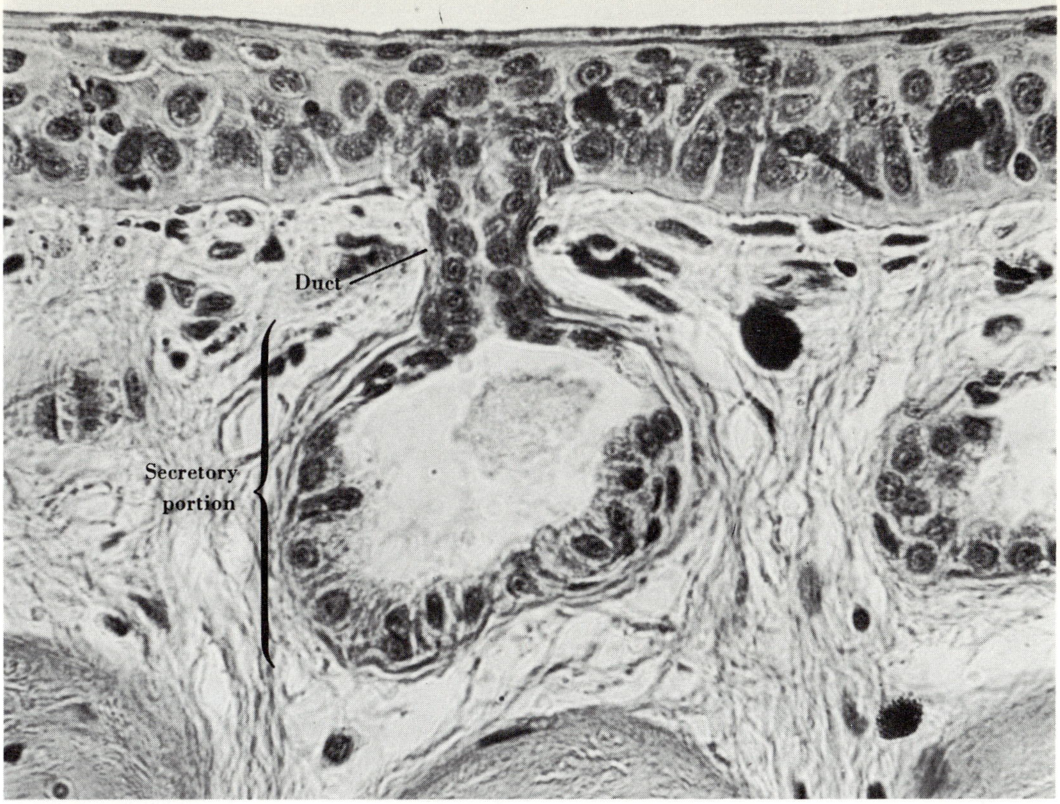

Fig. 6.13 Multicellular glands of frog skin. The secreting part of the gland lies in the dermis and is connected to the surface via a duct. X450.

posed in clusters embedded in the dermis, which is rather thick. The extinct Labyrinthodontia, however, included many species which had prominent, bony scales. In most species these were confined to the ventral part of the body, where they frequently overlapped, but in a few species they also covered the dorsum. The little that is known of them indicates that these scales were dermal bones consisting of a layer of spongy bone sandwiched between two layers of compact bone.

THE INTEGUMENT OF REPTILIA

Reptiles responded to the challenges posed by terrestrial life by developing an integument which was light in weight, flexible, resistant to abrasion and desiccation, and, in most species, capable of replacement at more or less regular intervals. In many respects one might say that they copied the arthropods, except in the choice of material (intracellular keratin rather than extracellular chitin). The turtles even went so far as to use the major part of their integument as part of their dermal skeleton.

It is in the reptiles that the first extensive exploitation of the protein keratin occurs. The introduction of a keratinized integument is important because it enabled terrestrial forms to evolve in the ways in which they have. The first problem of life on land is that of desiccation, and another is that of abrasion. In the aquatic environment, the former is met directly by the formation of bony scales, which, although perhaps heavy and bulky, constitute little or no encumbrance because of the buoyant action of the water. Water loss or gain is also met, although indirectly, by modification of the kidney, which has proved to be feasible. Abrasion does not constitute a serious problem because of the lubricating action of water, for one thing, and because friction is minimized further by the relative lightness in weight of the organism in water. Bony scales do not provide a wholly satisfactory solution to these problems on land, however, where the

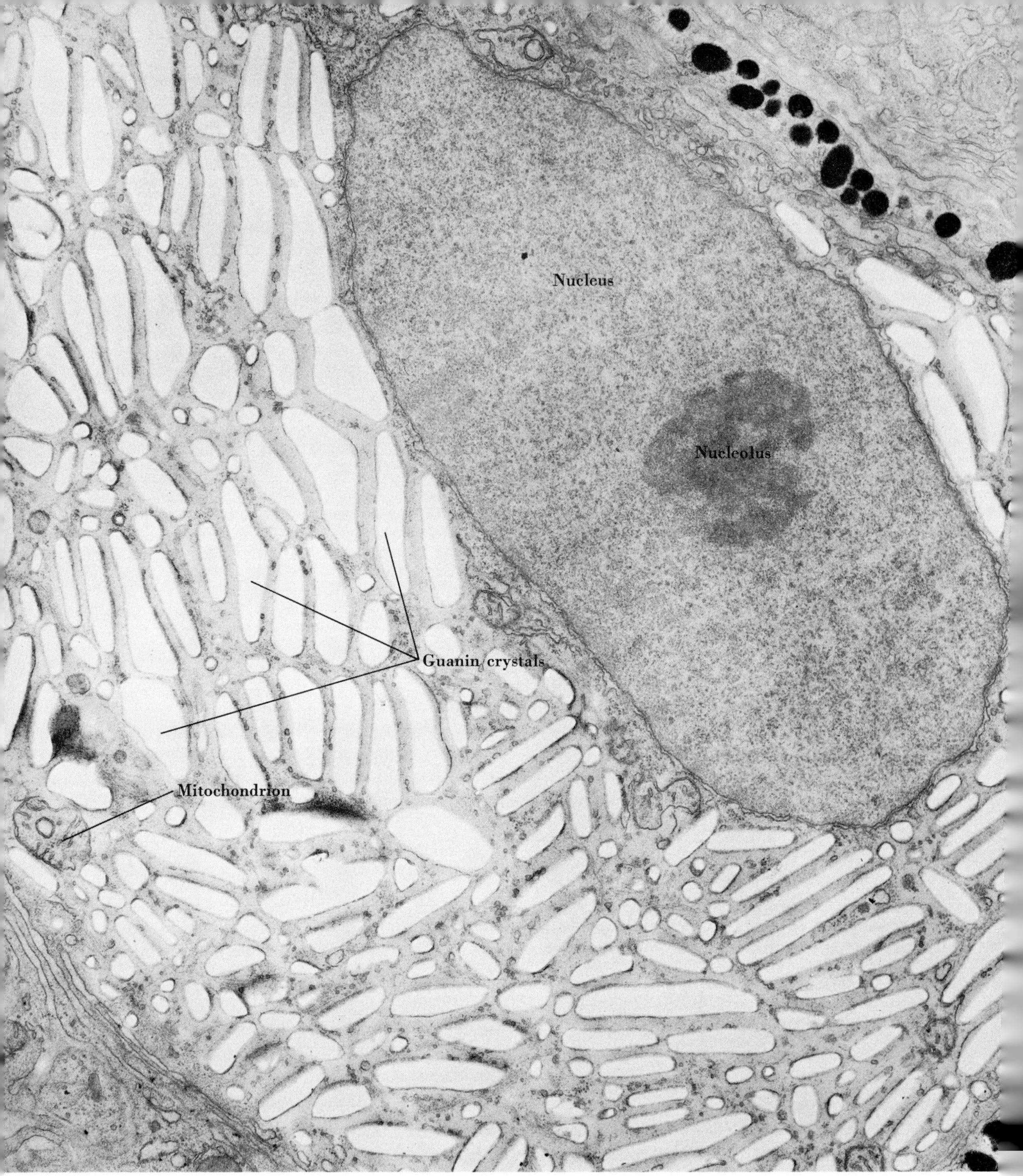

Fig. 6.14 Electron micrograph of a guanophore pigment cell in teleost skin. Note crystals of guanin within cytoplasm. X19,000. (From T. Setoguti. 1967. J. Ultrastruc. Res. 18:324–332.)

tendency for desiccation is more extreme, where sources of water replacement may be limited, and where the differential between body density and air density magnifies friction enormously. Under these circumstances terrestrial animals would be acutely limited in the size which they could attain. With a protective layer of keratin, which is both very

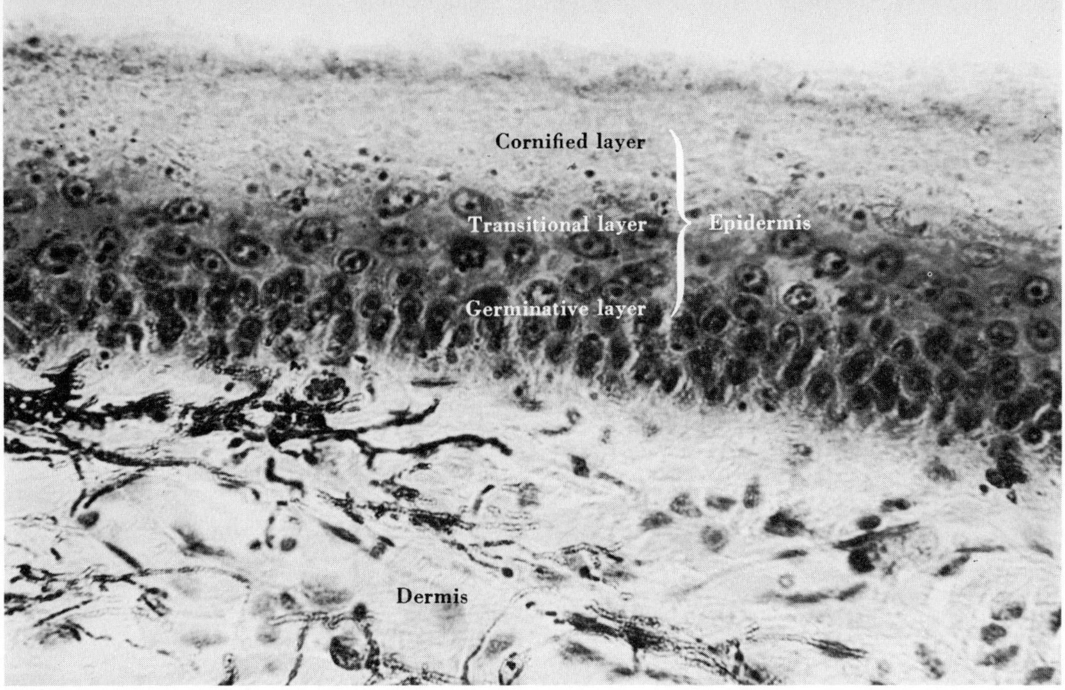

Fig. 6.15 Photomicrograph of a section of turtle skin. X450.

light in weight and flexible, terrestrial animals could cope with both problems very satisfactorily and also attain the large size which many species have in fact done.

The keratin scales of reptiles represent one highly evolved mechanism for dealing with the problems of desiccation and abrasion, but they do not answer well a third one, i.e., the loss or gain of heat. This is a serious problem because air, with its low specific heat and other physical properties which permit it to circulate rapidly, may undergo a wide range of temperature changes in a short time. The mammals owe their phenomenal evolution in large part to the introduction of hair, another keratin derivative, and the birds to feathers, which are modified epidermal scales.

The Epidermis

Basically the reptilian epidermis consists of a cornified layer on the outside, a thin transitional layer, and a thin germinative layer (Fig. 6.15). The first appears as epidermal scales in all species except certain marine turtles, and the last is frequently pigmented (Fig. 6.16). The pigment may diffuse into the cornified layer, as in the well-known "tortoise shell." Glands are notable by their scarcity, being restricted to the margins of the mouth and the cloaca, and in some lizards to the thigh (the *femoral glands*). The alligator has *inframandibular* glands which, like those of the cloacal region, function during the breeding season. Reptilian integumentary glands are all multicellular, alveolar glands, usually holocrine, and they secrete a musky substance which is of interest only to other reptiles.

The Dermis

The dermis varies in thickness in different species. In most it is thin and compact and closely applied to the muscular layers below. It consists largely of collagenous fibers, fibroblasts, blood vessels, nerves, pigment cells subjacent to the epidermis, and scattered fat cells. Strands of muscular tissue, which are presumably homologous with the *erector pili* muscles of mammals and the *erector plumulae* muscles of birds, are encountered in the ventral dermis of snakes. They are attached by connective tissue fibers to the bases of the epidermal scales, functioning as depressor muscles.

Epidermal Scales

The epidermal scales of reptiles exist as either more or less flat plates, such as on the cara-

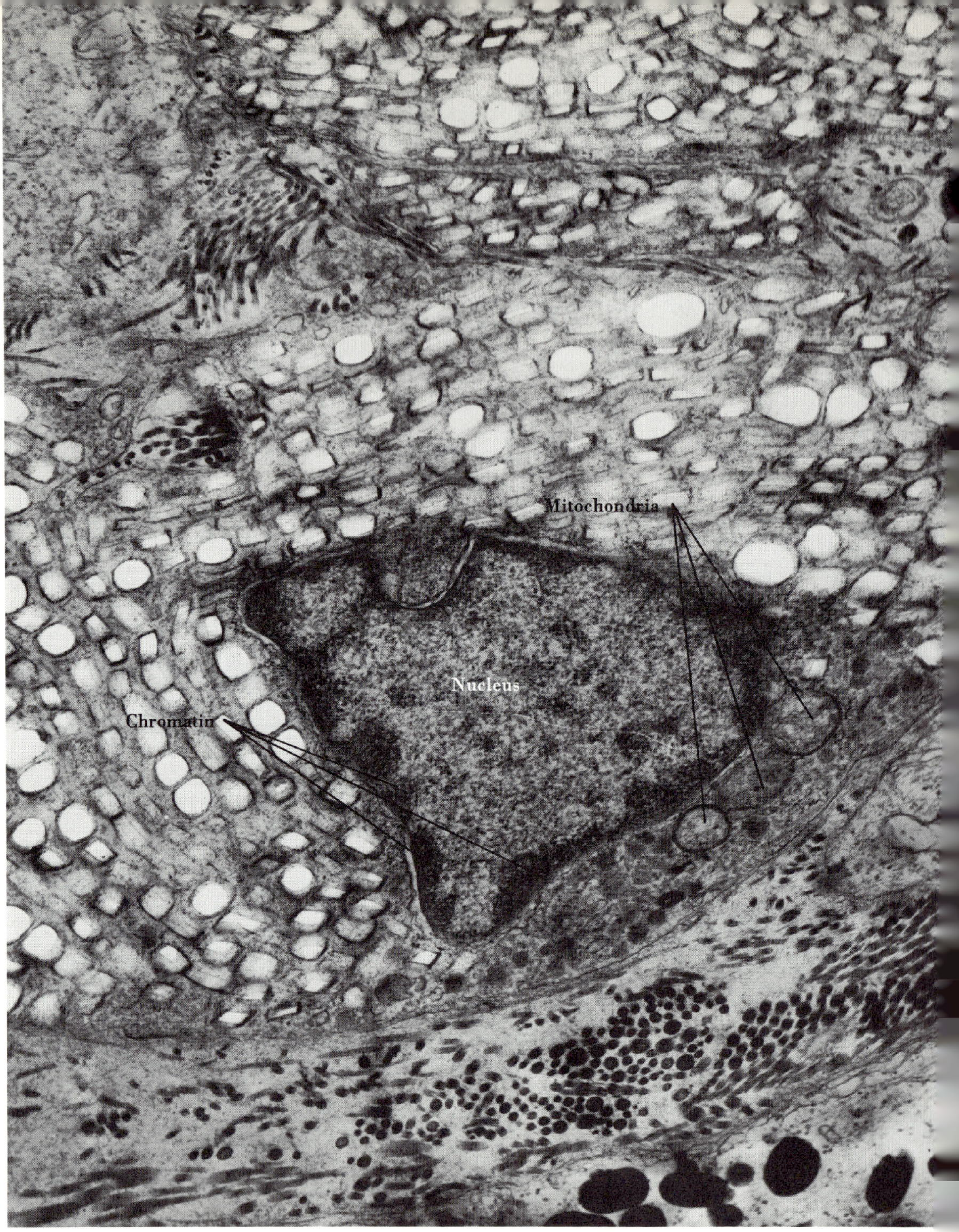

Fig. 6.16 An unknown type of pigment cell from the skin of a chameleon, Anolis carolinensis. X16,000. Note the similarity between this cell and the teleost guanophore in Fig. 6.13. (Courtesy of F. Craft.)

pace and plastron of turtles and tortoises and in most lizards and members of the Crocodilia, or as overlapping structures, such as in some lizards and all snakes. They are produced by cornification of the uppermost cells of the transitional layer, from which they separate as lifeless plates of keratin. They are replaced by the generation of new scales below, in which case they are either sloughed off as a continuous sheet by the process known as *ecdysis,* or individually, as occurs in turtles. Overlapping scales are formed by cornification of either side of a papilla, which appears as an outfolding of both epidermis and dermis, followed by retreat of the living core with consequent apposition of the upper and lower layers of keratin (Fig. 6.17). In all cases the epidermal scales are joined to each other by bridges of cornified epithelium, so that the integument is covered by a continuous layer of keratin which is interrupted only at those points where *prototrichs* occur. Often one finds small scales clustered around the bases or the periphery of the larger scales. The small scales are usually equipped with a fine, cornified process that closely resembles a short hair. This process is richly innervated at its base, indicating a tactile function. The manner in which these "hairs" are associated with the larger scales so closely resembles the pattern of association between the hairs and epidermal scales of the armadillo and of the rat's tail as to suggest a direct homology between them and true hair. Moreover, the feathers of birds bear the same relation—although on the basis of their embryonic differentiation and the presence of a dermal component, feathers are clearly derivatives of scales.

The ventral scales of snakes form wide, transversely disposed bands capable of elevation and depression. Associated with their bases are muscles that effect movement. By coordinated contraction and relaxation, which are in turn coordinated with undulating, sidewise movement of the body, these muscles function in locomotion. In this way the ventral scales serve virtually as legs.

Dermal Scales

Many reptiles have dermal scales in addition to the epidermal scales. These are called *dermal scutes* or *osteoscutes.* They vary from the heavy, butted plates of the carapace and plastron of turtles to the thin plates that underlie the epidermal scales of many lizards. In all cases they are surmounted by epidermal scales, with which they may (in Crocodilia and some lizards) or not (in turtles, tortoises, and other lizards) correspond (Fig. 6.18).

THE INTEGUMENT OF BIRDS

The integument and associated structures of birds are noteworthy in two respects. First, epidermal modifications such as scales, feathers, and claws clearly show the reptilian ancestry of birds. Secondly, the structures of the skin are uniquely adapted to flight.

The skin of birds is thin and quite delicate, having loose attachment to the body musculature and enlcosing numerous air spaces. The skin is connected to the skeletal system at such points as the upper and lower jaws, lower part of the tibiae, and the feet. It extends beyond the skeletal system as webs, combs, and pendant ornaments.

The Epidermis

The avascular epidermis varies in thickness over the body. It is only a few cells deep in feathered areas but reaches many-layered proportions in nonfeathered regions, such as the feet, beak, and comb (Fig. 6.19). The epidermis consists of a nonliving *stratum corneum* and a living *stratum germinativum* (Fig. 6.20). The cells of the stratum corneum are flattened and cornified, being constantly replaced by cells from the mitotically active germinativum. The innermost cells of the germinativum are polyhedral or columnar, and they lie on a basement membrane (Fig. 6.21).

The Dermis

The dermis consists of connective tissue which progresses from the densely packed fibrous type at the epidermal surface to a more loose, reticular organization near its merger with the subcutaneous tissue. In regions covered by feathers the outer layer of the dermis is

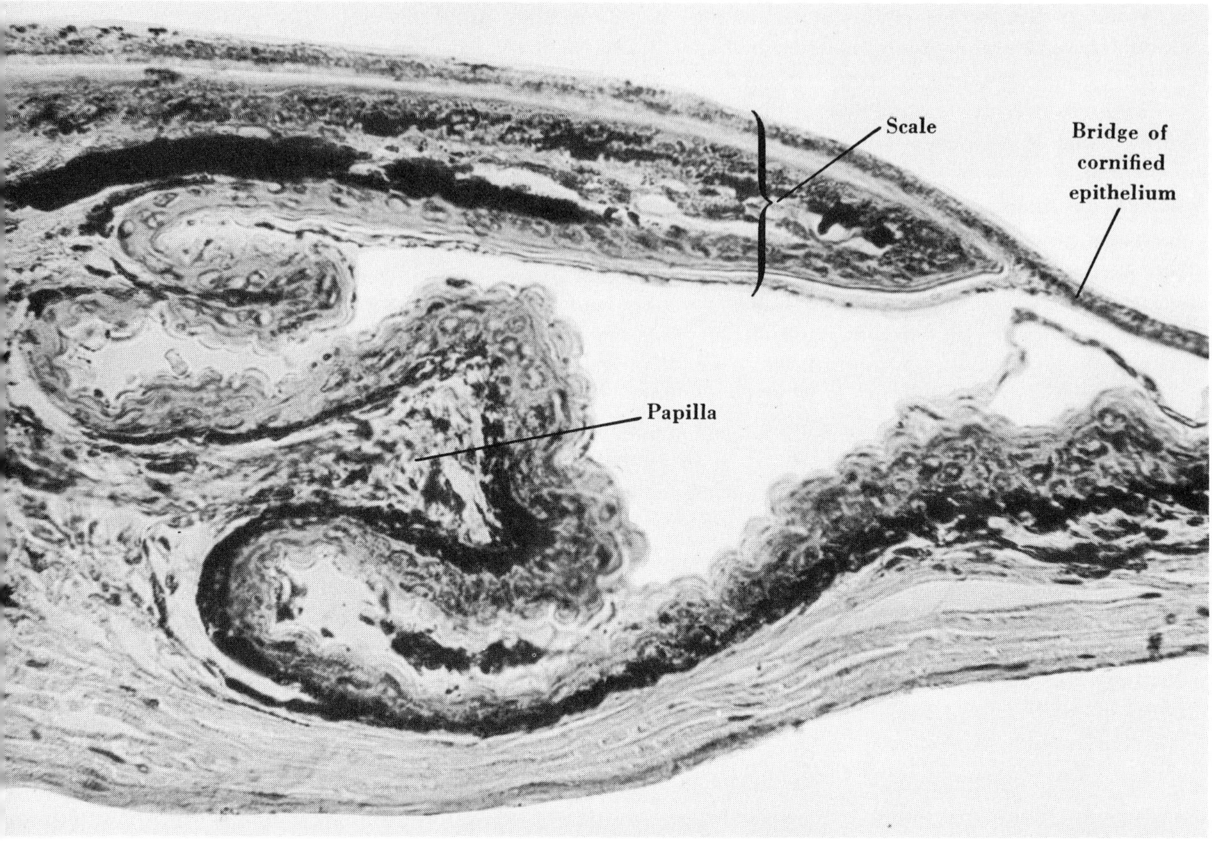

Fig. 6.17 Section through snake skin, showing overlapping epidermal scales. X450.

smooth, but in regions such as the feet and beak, the thickened dermis contains elevated papillae (Fig. 6.19). The dermis is richly supplied with blood vessels, sensory nerve endings, and smooth muscle fibers—especially in association with feather follicles. In addition to variations in dermal thickness in relation to the feather covering, cyclic, hormonally induced changes also occur. During the brooding period, the dermis becomes highly vascularized and thickened in the breast area. This forms the so-called "brood organ" used for incubation of eggs.

Birds possess few integumentary glands. In fact, the uropygial gland has long been considered the only true cutaneous gland. It is present in most birds, and is especially well developed in aquatic forms. Marine birds also contain a well-developed gland for the efficient excretion of the excess salt in sea water (see Chapter 10).

The *uropygial gland* is a large, bilobed gland that is located at the base of the tail. It is a compound, alveolar, holocrine gland closely resembling the mammalian sebaceous gland. Overlying a basement membrane, the secreting epithelium consists of granular squamous cells that give way toward the lumen to cuboidal cells and, lining the alveolus, large polygonal cells which are filled with fatty globules. The dense staining characteristics of the squamous layer reveals the alveolar structure in a striking way. A cavity is present in each lobe for the collection of secretion, which is then discharged through a duct to the surface. The secretory product is composed of lipid material, proteins, and salt and, like the reptilian scent gland of which it may be a homologue, it is most active and odorous during the breeding season. The function of the uropygial gland is controversial, including the obvious function of preening and the less obvious functions of vitamin D secretion and a role in calcium metabolism.

Feathers

Feathers represent the primary modification of the bird integument. Functionally they provide a durable, water-repellent covering which is well adapted to flight. Feathers combine lightness of weight with great tensile strength. As for insulating quality, they are as efficient as anything man has been able to manufacture.

Feathers are richly keratinized epidermal outgrowths. They are arranged in tracts which may show considerable variation with respect to both arrangement and number—variations which, combined with differences in pigmentation, produce the colorful variety of plumages that have helped to make birds one of the most aesthetically fascinating group of animals on earth.

Feathers form from condensations of mesodermal cells, which appear beneath the epidermal ectoderm of the embryo. As in the formation of reptilian scales, these condensations induce a thickening of the ectoderm to form elevations on the surface. As the elevations protrude, ridges form on their surfaces and subsequently form the barb primordia of the *down feathers*. As development proceeds, the papilla sinks downward and filamentous *barbs* bearing *barbules* on each lateral epidermal surface appear. The barbs remain radially arranged, attached to a short, cylindrical *calamus* or quill. Externally the down feathers appear as a coat of soft fluff, which is destined to be lost during the early life of the bird. However, a small mass of mesodermally

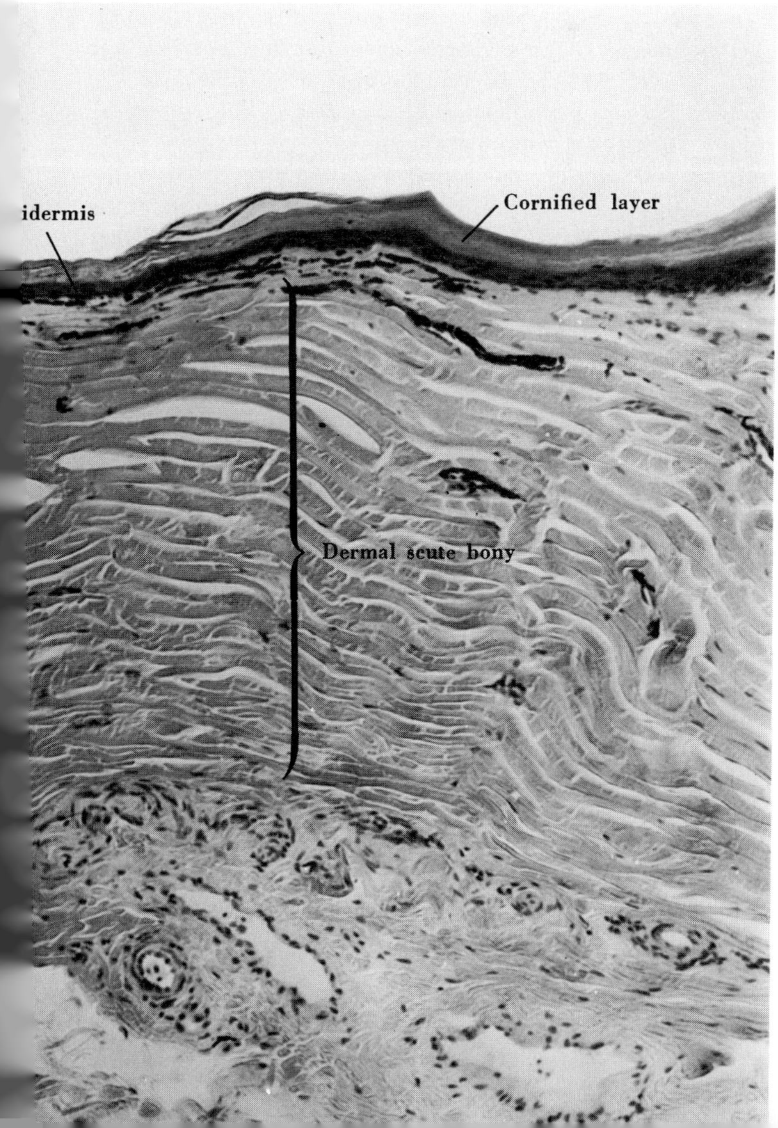

Fig. 6.18 Section through turtle skin. Note bony dermal scale (scute) and cornified epidermal covering. X200.

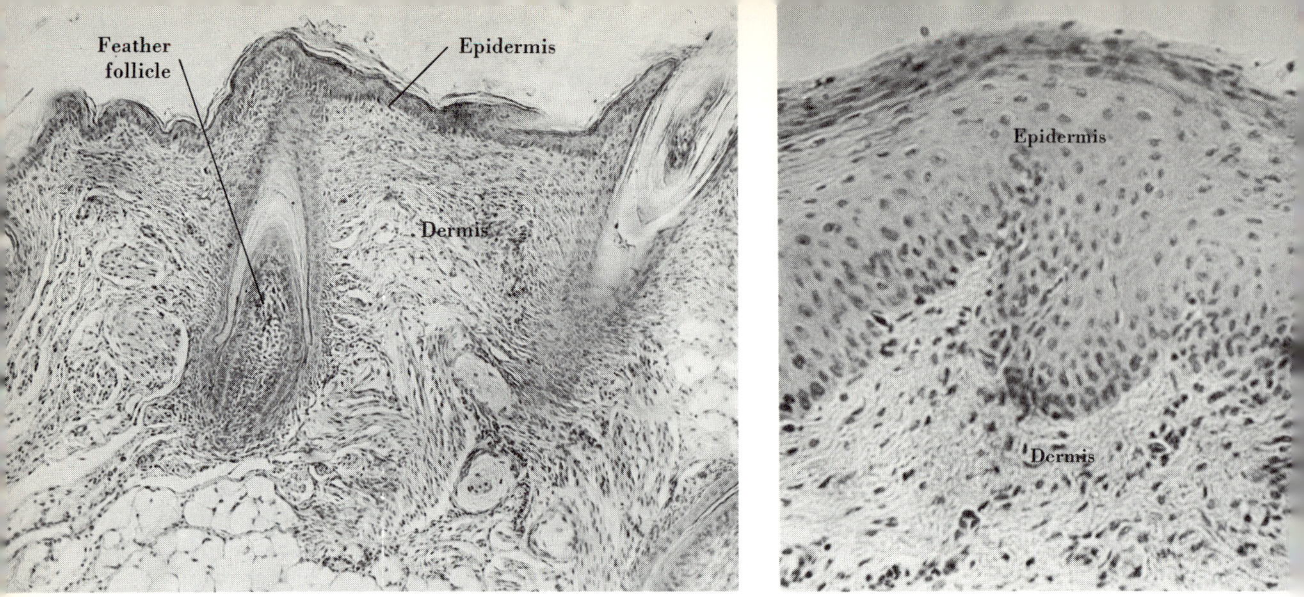

Fig. 6.19 Skin taken from feathered area of a chicken, *a*. Skin from nonfeathered area, *b*. Compare thickness of epidermis. (Slides courtesy of Dr. R. F. Susi.)

derived dermal cells, covered by a layer of epidermal cells, remains at the base of the down feather to form the *follicle* or *feather papilla* from which all succeeding generations of feathers will be formed.

Many generations of juvenile feathers appear after the down feathers are lost. Each generation is less like the simple down feather and more akin to the adult type.

Adult feathers can be classified into three different groups: *contour feathers, plumules,* and *filoplumes* (Fig. 6.22). Contour feathers contribute to the bulk of the adult plumage. They may be moved individually by contraction of the *erector plumulae* muscles attached to the follicular walls. Contour feathers possess a strong, rodlike shaft or *rachis,* the interior of which is packed with pulp made up largely of cornified air spaces. The hollow, proximal part of the rachis is known as the quill or calamus. Located on the proximal and distal surfaces of the rachis are many lateral branches, the barbs, which themselves have subbranches, *barbules* and *barbicels,* which are hooked processes, on one side of the barbule. The barbicels of one barb interlock with the barbules of the adjacent barb in such a way that an almost air-tight membrane is formed, the *vane.*

Plumules are the simplest of the adult feathers, resembling the loose, soft, embryonic down. They are usually located out of sight under contour feathers, and in aquatic birds they form a layer of insulation.

Filoplumes are small, threadlike feathers. They are essentially degenerate, bearing only a few barbs at the apex of a delicate rachis. Like plumules, filoplumes are often hidden by the contour feathers.

The coloration of feathers results from the interaction of two factors. The structure of the barbs and barbules of the contour feather permits the reflection and refraction of light in such a way as to produce the "metallic" or iridescent colors. Such structural color is attributed to a layer of polygonal or prismatic cells overlying the pigmented cells. The second factor is pigmentation. Pigment granules are "fed" into the cells of the developing feather as it grows. Not derived from the papilla at all, the chromatoblasts have differentiated from cells of the neural crest and have migrated to the papilla. The melanin pigments are responsible for the more sombre hues of red, brown, gray, and black, whereas the lipochromes are the basis for the bright reds, yellows, and oranges.

Scales, Claws, and Beaks

The scales, claws, and beaks of birds are indicators of the birds' debt to the reptiles. These epidermal structures arise from cell proliferation of the stratum germinativum. As

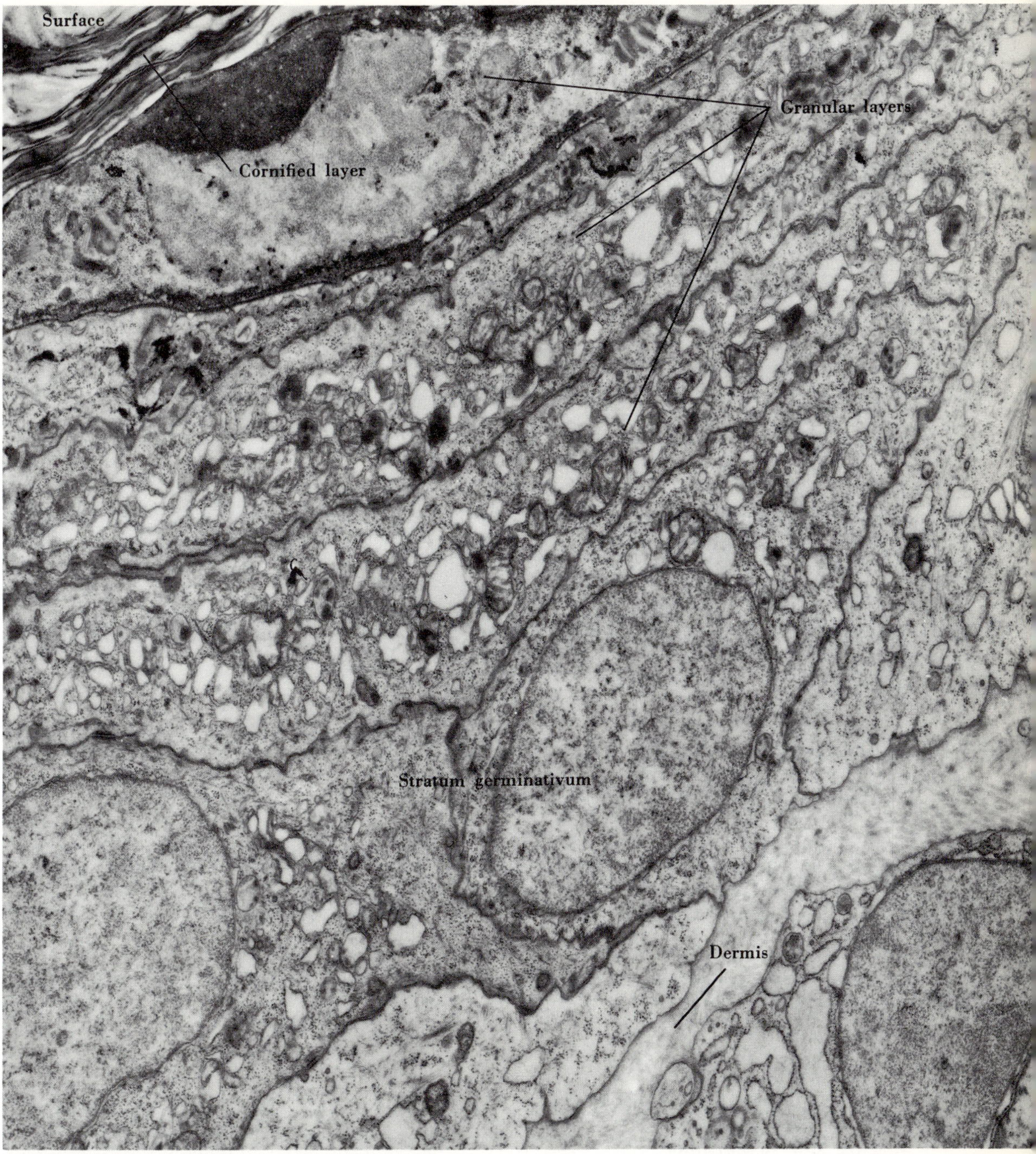

Fig. 6.20 Electron micrograph of a whole thickness of the epidermis and a portion of the dermis of thin skin of the chicken. X20,000. (Courtesy of P. Parakkal.)

the cells are pushed toward the surface, they form densely packed layers of hard keratin and are continuously replaced as the surface is worn away. The cells which form scales are originally bipotential and can also form feathers; in fact, some birds have feathers covering the parts of the feet which are usually scaled. The nail-like claws of grebes, as

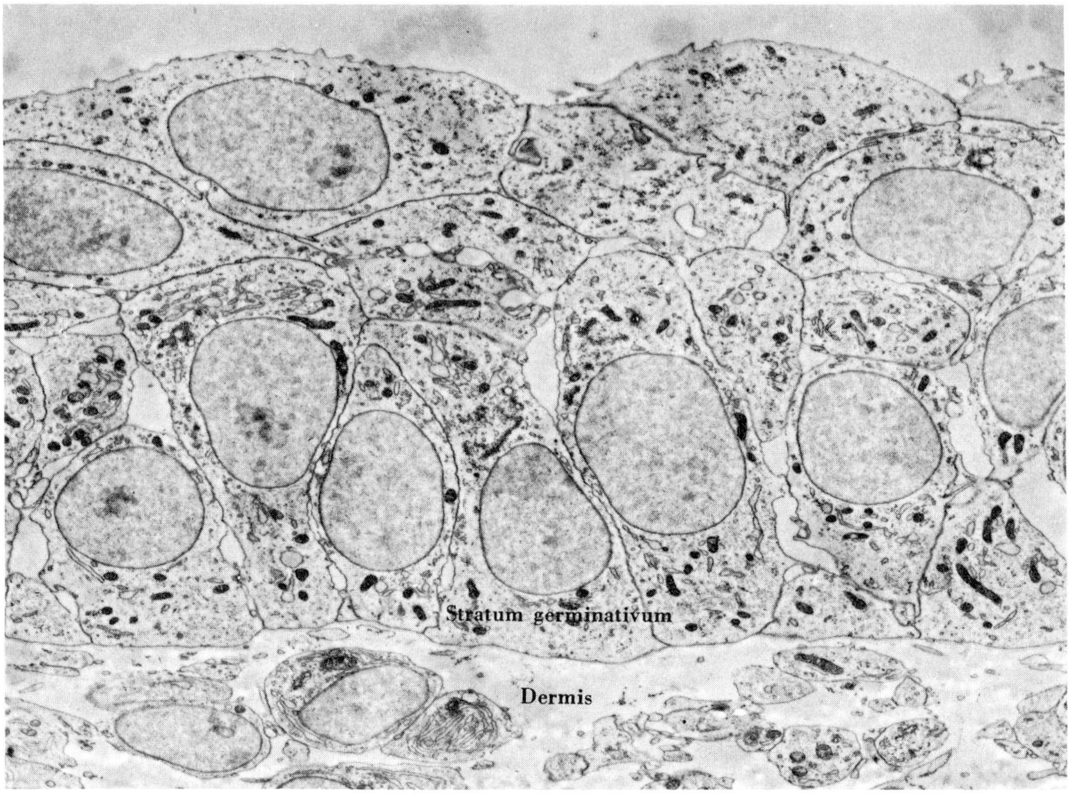

Fig. 6.21 Electron micrograph of the epidermis from the digit of a chick embryo. Stratum germinativum is present, but cornification has not yet occurred. X4500. (Courtesy of R. Henrikson.)

well as other claws are modified scales and vary in shape with the species. The bird's beak is a cornified epidermal sheath covering the upper and lower jaws. Sensory nerve endings may be present here, as in the Anseriformes.

THE MAMMALIAN INTEGUMENT

The integument reaches its maximum cornification and glandular diversity in the mammals (Fig. 6.23). While remnants of reptilian scales are seen on the legs and tails of some rodents and insectivores, and covering the bodies of the African scaly anteater and the American armadillo, new keratinized structures include hair, horns, the ungulate hoof, and the nails of insectivorans, chiropterans, and primates. The depth of cornified cells is greatly increased in such regions as the foot pads and friction ridges, or fingerprint areas of higher primates. Scent glands are found in other classes of vertebrates, but in mammals these glands are joined by (1) *sebaceous glands,* which function in lubricating the skin and hair; (2) *sweat glands,* which aid in temperature regulation and serve as accessory excretory glands; and (3) *mammary glands,* which supply milk for the young.

The Epidermis

Epidermal depth varies in relation to the amount of abrasion and wear to which a given area is exposed. The soles of the feet and the palms of the hands in human beings represent areas of extreme thickness. However, even in areas of moderate depth four distinct epidermal layers can be distinguished (Fig. 6.24). The basal cells of the stratum germinativum are cylindrical and parallel to the surface of the skin. Above the basal layer the cells become flattened and finally polyhedral.

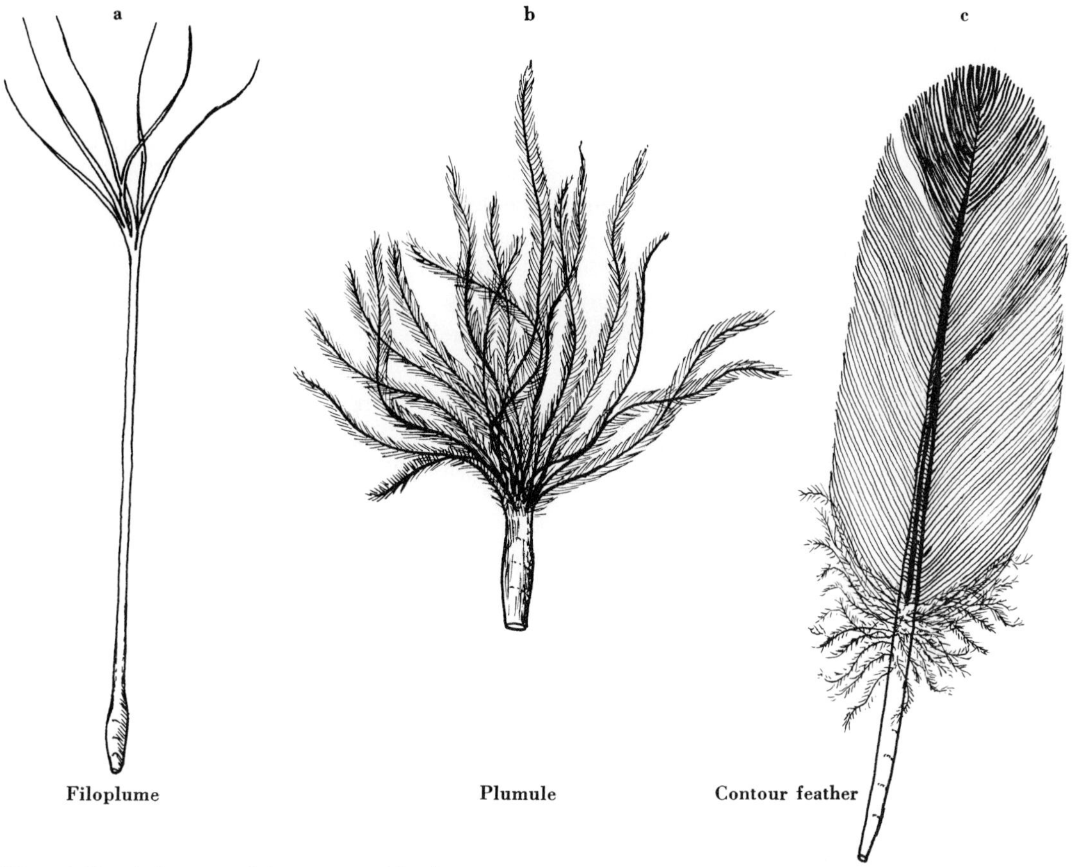

Fig. 6.22 A diagram of three types of feathers.

Desmosomes and tonofibrils occur binding the cells to each other and to the dermis. If neighboring cells are gently pulled away from each other, the greater strength of the intercellular fibers causes the surfaces of the cells to assume a prickly appearance. Hence the upper layers of the stratum germinativum are often seen labeled as the *stratum spinosum*, or prickle-cell layer. Melanin granules are transferred from melanocytes in the dermis to the lower layers of the epidermis in many species. Resting on the mitotically active germinativum are three to five layers of flat, diamond-shaped cells, forming the layer known as the *stratum granulosum*, so called because of the presence of granules of keratohyalin in the cytoplasm of the cells (Fig. 6.25). Keratohyalin, neither keratin nor hyalin, is of unknown composition and function—although it may contribute to the matrix in which keratin is found. Keratin, in turn, is thought to be derived from tonofibrils. Granules of keratohyalin which are of sufficient size to be visible with the light microscope can be seen with the electron microscope to be formed by condensation of smaller granules.

Nuclear distintegration and narrow intercellular spaces are prevalent on this layer. Moving toward the surface, one encounters next the *stratum lucidum*, which consists of several layers of flat cells. Filling most nuclear and cytoplasmic areas are droplets of a refractive substance, *eleidin*, which also is of unknown composition and function, and dense fibrillar bundles of keratin. The surface layer, the *stratum corneum*, contains many layers of flat, elongated, densely cornified cells being constantly desquamated and replaced.

In areas of the body other than the soles and palms, two well-defined layers are always present, the strata germinativum and corneum. The stratum granulosum is reduced to two or three cell layers, and the stratum lucidum may not be distinguishable (Fig. 6.26).

The Dermis

The dermis, or *corium*, also varies in thickness, being quite thin on the eyelids, the prepuce, and the ventral side of the body and paired appendages. The outer, *papillary layer* of the dermis, which is fused to the epidermis, is loosely filled with collagenous fibers that run parallel to the surface (Figs. 6.24 and 6.26). The thicker, subjacent, *reticular layer* is anchored to the subcutaneous tissue by bundles of collagenous fibers and contains dense bundles of intradermal collagenous fibers surrounded by a network of elastic fibers and large quantities of intercellular material. The cellular components of the dermis are fibroblasts, macrophages, and melanoblasts, with fat cells located in the

Fig. 6.23 Section through the cheek of a man. The large complex sebaceous gland is associated with two small hair follicles. Parts of two eccrine sweat glands lie deep in dermis. X160. (Courtesy of R. A. Ellis.)

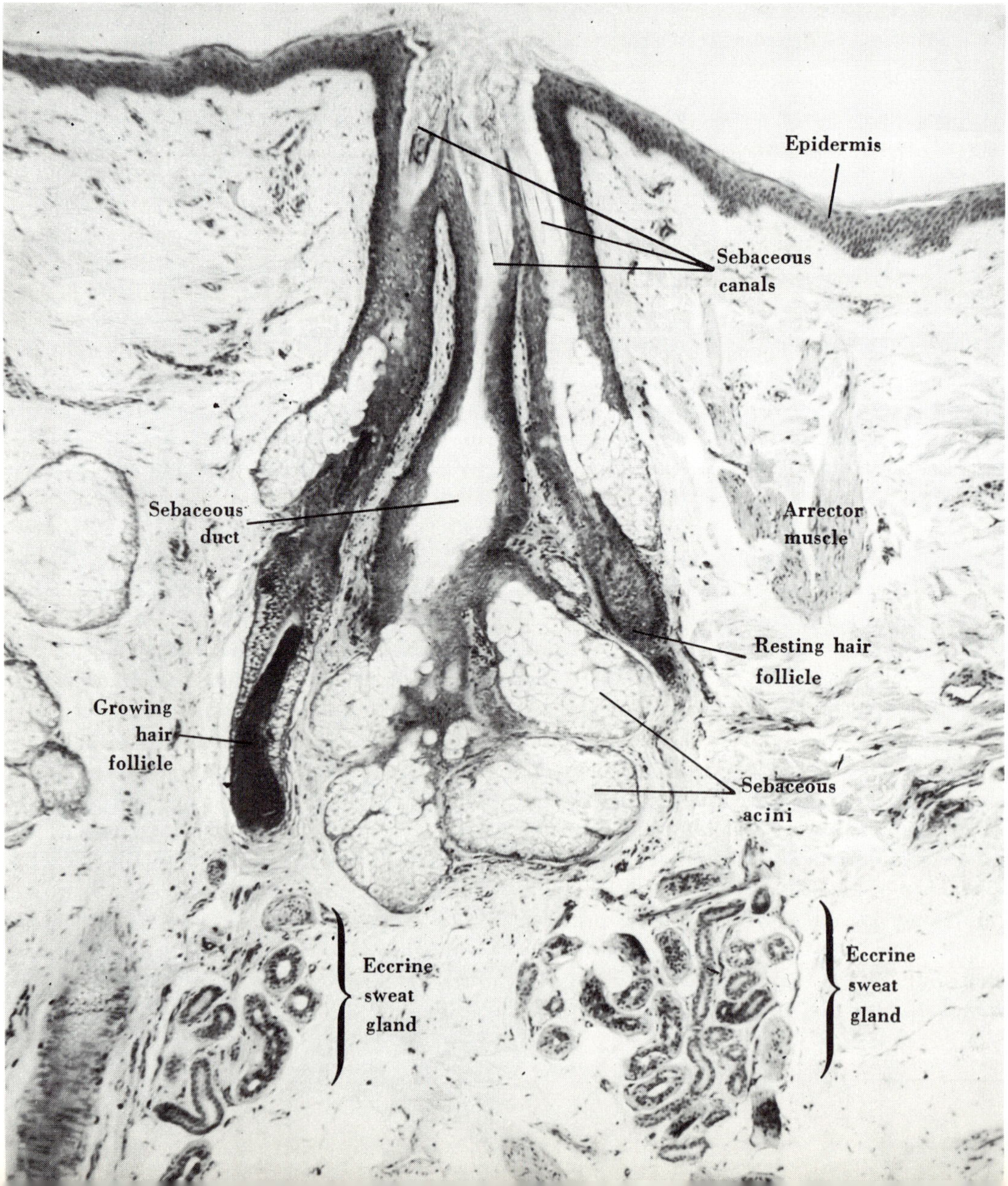

deeper portions. Smooth muscle fibers associated for the most part with hair follicles and blood vessels, and also with nerve endings also occur in the mammalian dermis. Embedded within the dermis are such epidermally derived structures as sweat glands, sebaceous glands, and hair follicles, as indicated above, and in specific areas, mammary glands occur as well.

Sweat Glands

With the exception of some rodents, rabbits, whales, sea cows, and the scaly anteater, in which sweat glands are lacking or limited to the pads of the feet, these simple, tubular, highly coiled glands occur over most of the mammalian body. They penetrate deeply into the dermis through the epidermal interpapillary *pegs*. Two kinds of sweat glands can be distinguished. The more common kind is *eccrine* (Fig. 6.27), which type secretes a watery substance that is expelled from the basal part of the gland by contraction of spindle-shaped *myoepithelial cells* that surround and enclose the tubule. The *apocrine* type of sweat gland is usually larger; it may be branched tubular, and it is located in such limited areas as the axillae, external genitalia, around the nipples of the mammary glands, and around the anus. Its secretion is milky (Fig. 6.28).

The secretory epithelium of sweat glands forms a coiled tube of cuboidal or low columnar cells having spherical or oval mitochondria, a rough endoplasmic reticulum, and well-developed supranuclear Golgi substance. Two kinds of secretory granules, dark ones and a lighter less-dense variety, each enclosed within a membrane, are found within cells, from which the distinctions of clear cell and dark cell derive (Fig. 6.29). The secretory portion of the gland lies deep within the dermis or even in the subcutaneous tissue. The excretory duct of the gland contains a narrow lumen within walls made up of a double layer of cuboidal cells. After the duct reaches and enters the epidermis, the cells of the epidermis merge with and become the cells of the duct walls. Coursing spirally through the epidermis, the duct opens on the free surface of the skin.

Fig. 6.24 Section of mammalian skin. X440. The distinction between the strata granulosum and spinosum is obvious in an H & E slide, but this distinction disappears in a black-and-white photograph.

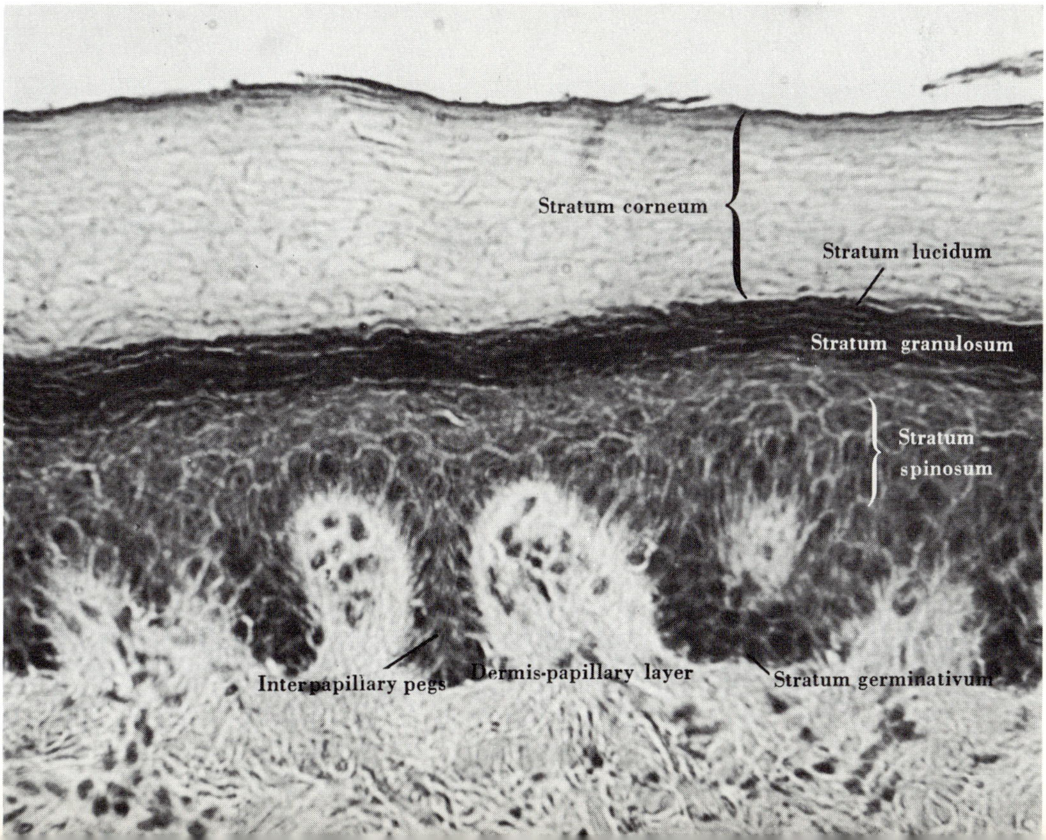

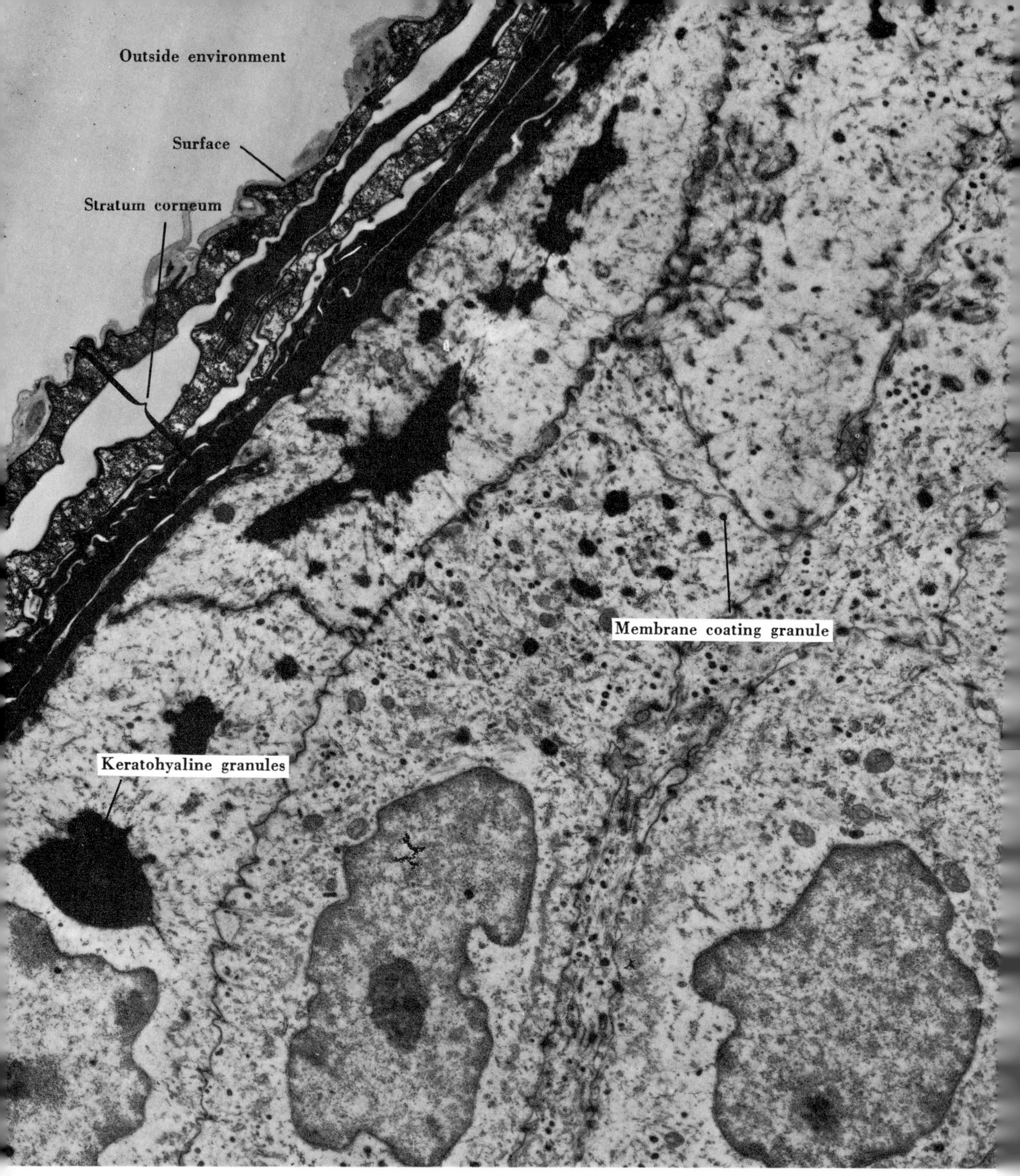

Fig. 6.25 Upper portion of the mouse epidermis, including the strata corneum and granulosum. The small membrane coating granules are believed to contribute to the thickness of the membranes of the desquamating horny layer. X9000. (Courtesy of P. Parakkal.)

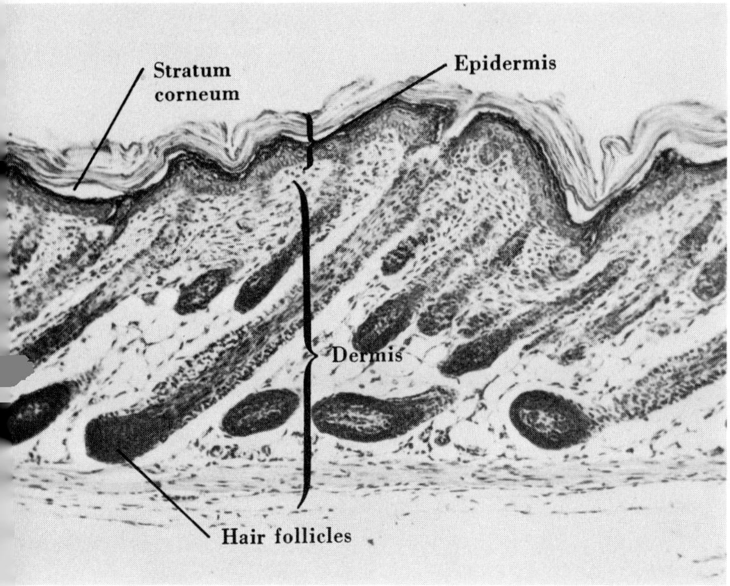

Fig. 6.26 Section through the skin of a newborn mouse. Note hair follicles dipping into dermis, *left*. (X100). Higher-power view of same skin, *below*. (X4440).

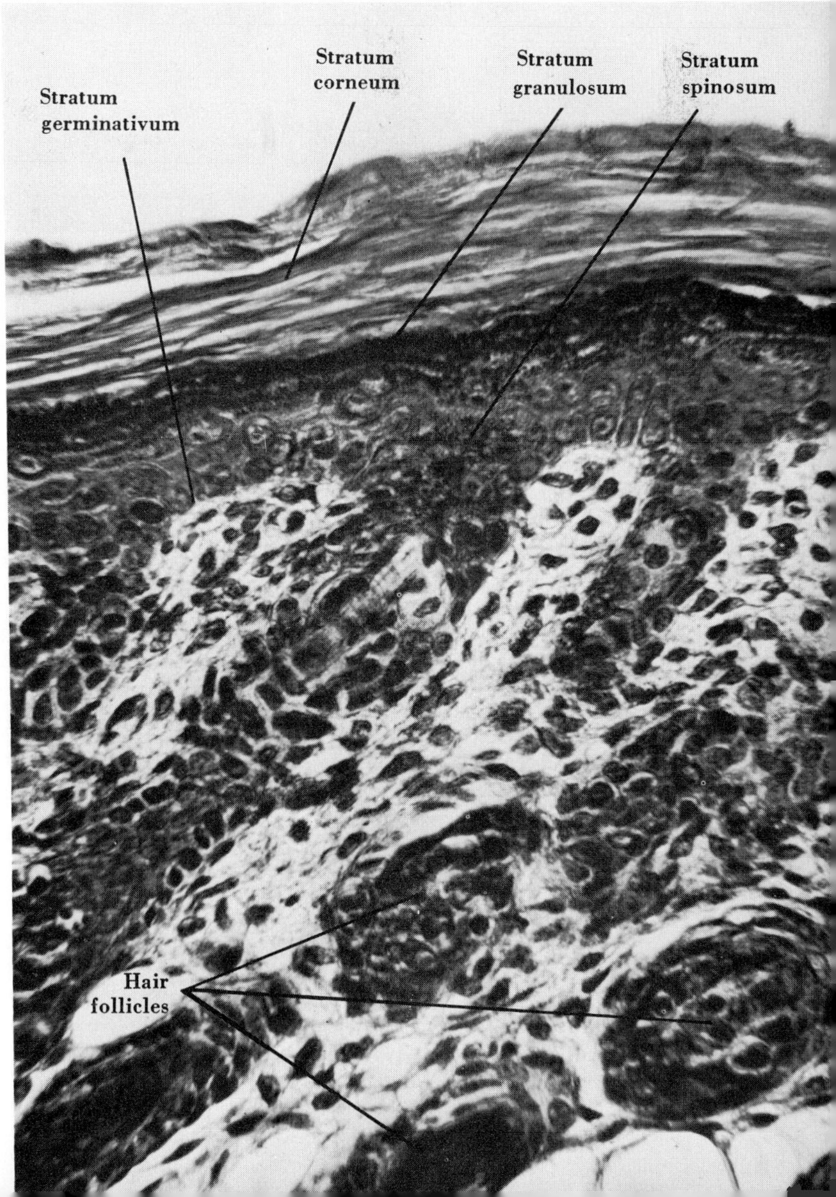

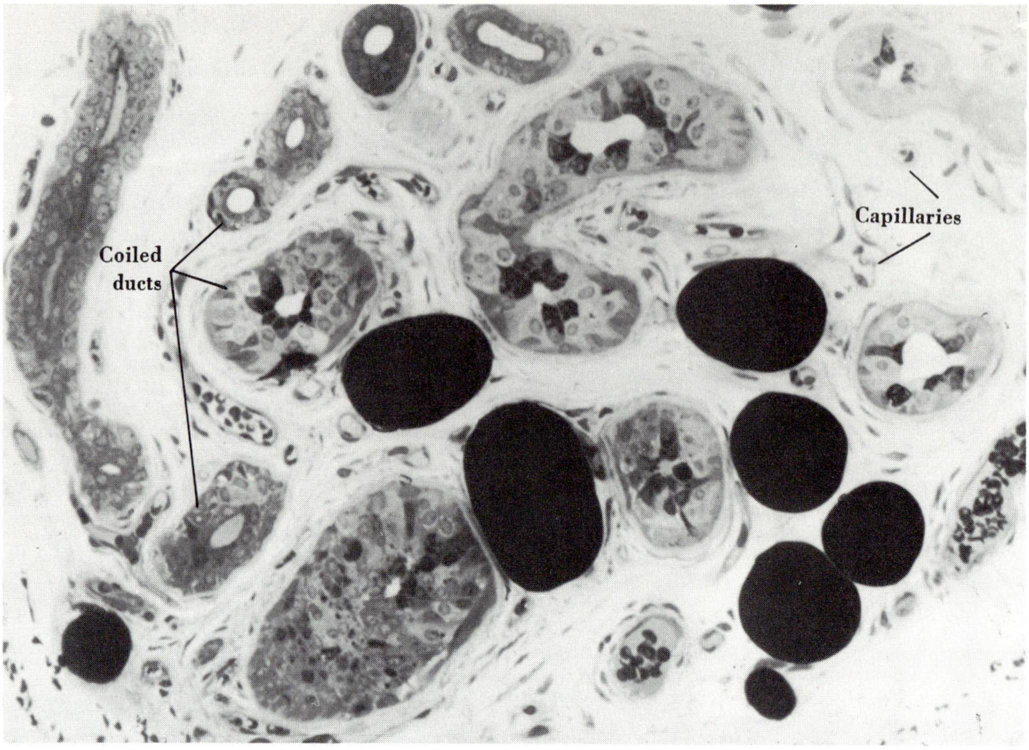

Fig. 6.27 A section through the coiled part of an eccrine sweat gland in human skin, *above*. Several segments of the coiled duct and secretory coil are shown. The contents of several fat cells are blackened by the osmium fixation. Small blood vessels (capillaries) course the connective tissue investing the gland. X480. Longitudinal and cross sections through the secretory coil of an eccrine sweat gland in human skin, *below*, show the characteristic distribution of myoepithelial cells, clear cells, dark cells, and intercellular canaliculi. X1250. (Both microphotographs courtesy of R. A. Ellis.)

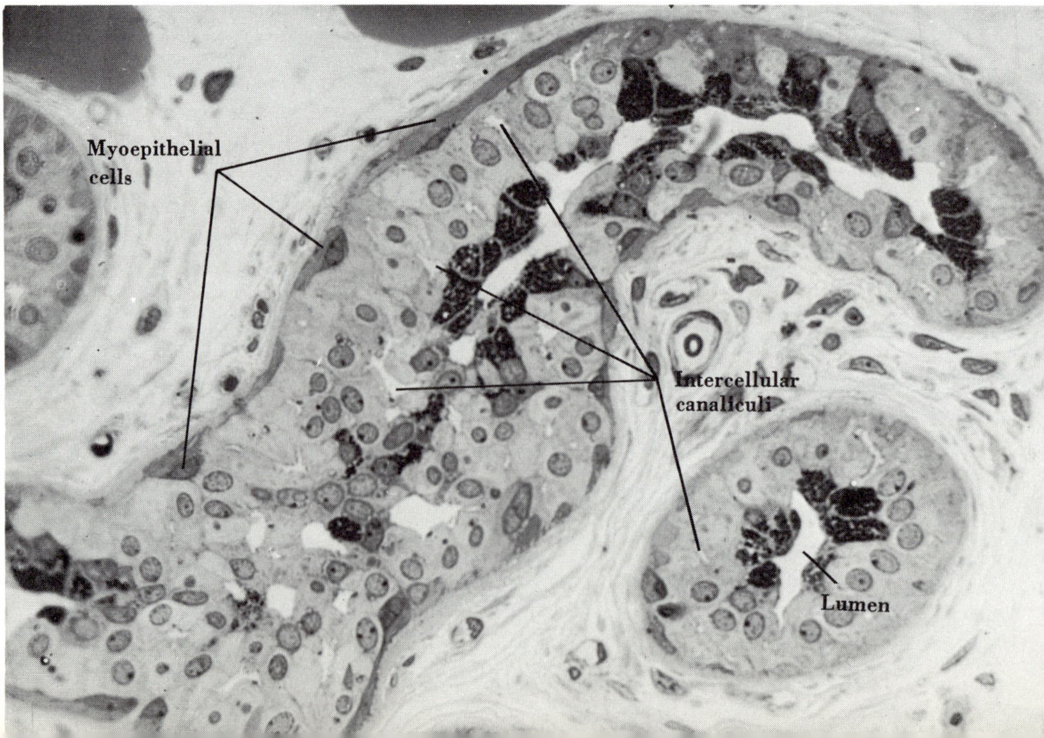

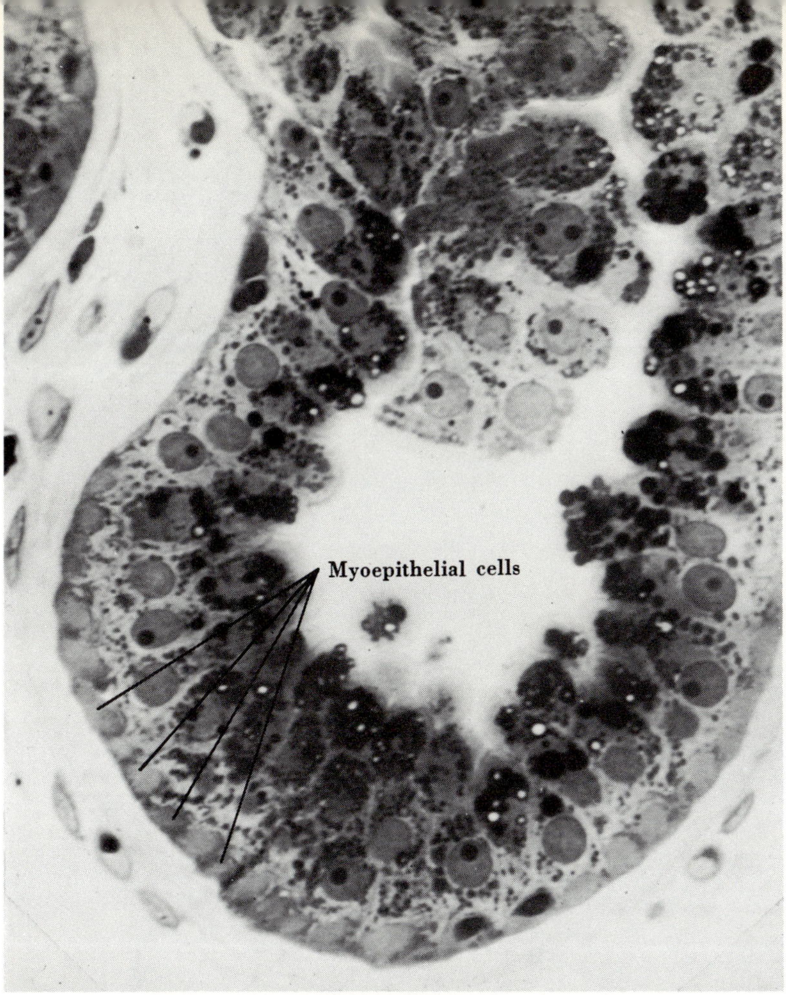

Fig. 6.28 In this section through the secretory coil of an apocrine sweat gland from the axillary skin of man, the columnar epithelial cells contain numerous dense granules. Beneath the nuclei the small granules are normal mitochondria; above the nucleus the dense granules are of several different types: pigment, modified mitochondria, and dense bodies. Although classically these are considered as apocrine glands, evidence from electron microscopy indicates that they have a merocrine mode of secretion. A pavement of myoepithelial cells outlines the perimeter of the secretory coil. X2400. (Courtesy of R. A. Ellis.)

Sebaceous Glands

Sebaceous glands are intimately associated with hair both with respect to origin and function (Fig. 6.23). The epithelial cells which form the gland originate from the neck region of the hair follicle, and the short, wide ducts of one or more glands open directly into the follicle. Hence *sebum,* the fatty secretion of the sebaceous gland, is expelled into the hair follicle, where it functions as lubricant for the hair and skin. In addition, sebum can be rapidly expressed by contraction of the *arrector pili* muscles attached to the hair follicle. In areas not covered with hair, as the soles and palms, sebaceous glands are usually absent. However, two kinds of exceptions can be found. Regions such as the corner of the mouth, labia minora, papillae of the breasts, glans penis, and internal fold of the prepuce, contain no hair follicles but do have sebaceous glands opening directly onto the surface of the skin. In other areas, such as the skin that covers the human nose, the sebaceous glands become the major component, with the hair follicles remaining quite small or degenerating.

Sebaceous glands are branched, alveolar glands usually located on the side of the hair follicle that makes an obtuse angle with the skin. The body or secretory portion of these

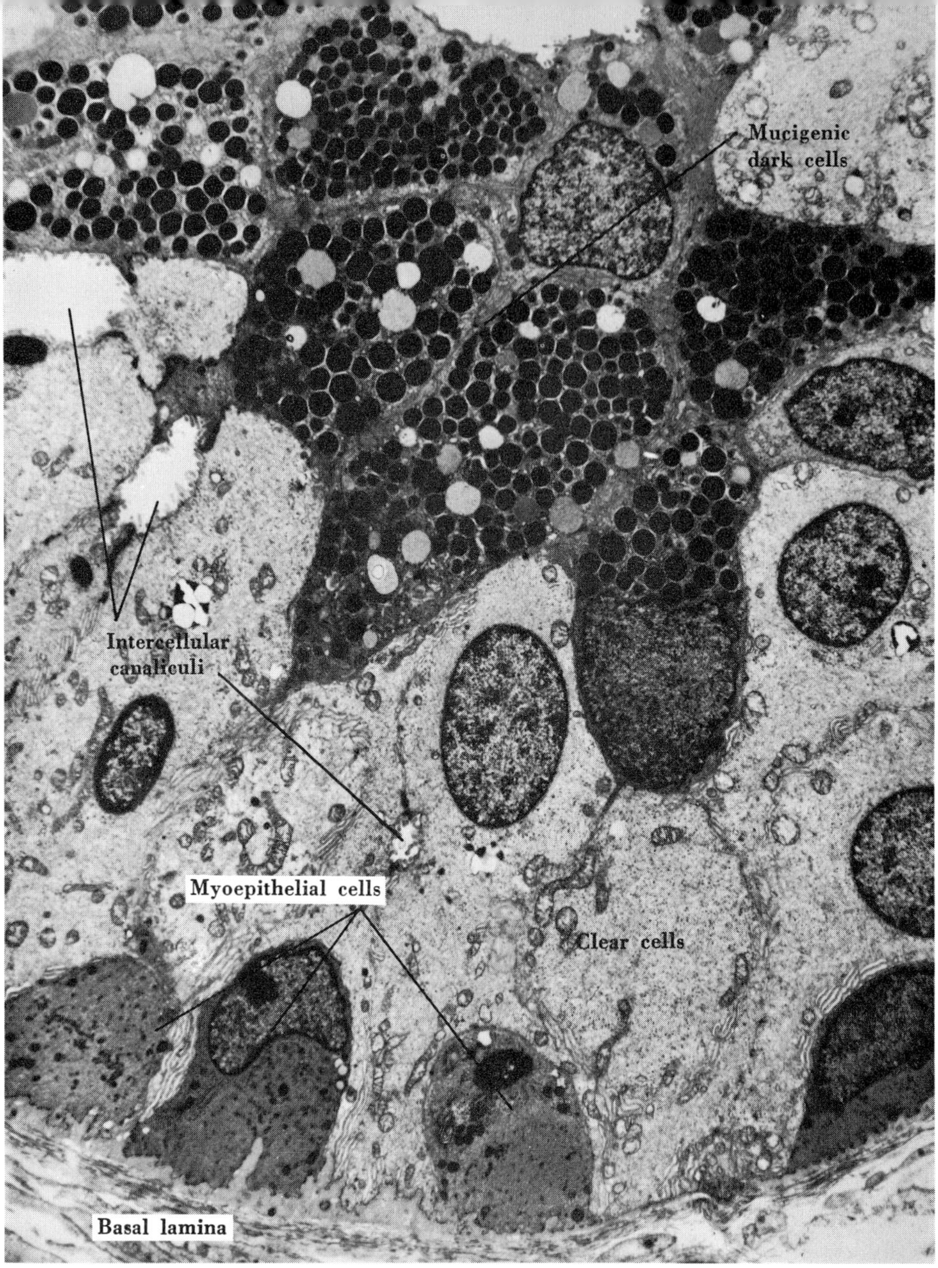

Fig. 6.29 In an electron micrograph of the secretory tubule of a human eccrine sweat gland, all three cell types show distinct characters. The peripheral myoepithelial cells form an incomplete pavement on which the clear cells and the granular dark cells rest. The intercellular canaliculi between adjacent clear cells are continuous with the lumen of the tubule. X6200. (Courtesy of R. A. Ellis.)

holocrine glands consists of many cell layers (Fig. 6.30). Covering the outside of the gland itself is a thin connective tissue sheath. Progressing interiorly, one finds a distinct basement membrane and a germinative, or basal layer of thin, flat, epithelial cells. Typical cells have a poorly developed, agranular endoplasmic reticulum, a small mitochondrial population, and they usually lack a definitive Golgi body. As mitosis occur in the basal layer, the older cells are pushed toward the center of the gland, becoming progressively larger, more rounded, and filled with fat droplets in the process. The more centrally located the cell, the farther it is from the source of nourishment. This is evidenced histologically by pyknotic nuclei and the accumulation of granules of keratohyalin in the cytoplasm. Necrosis of the cell is the final result of the migratory process. The mixture of fat,

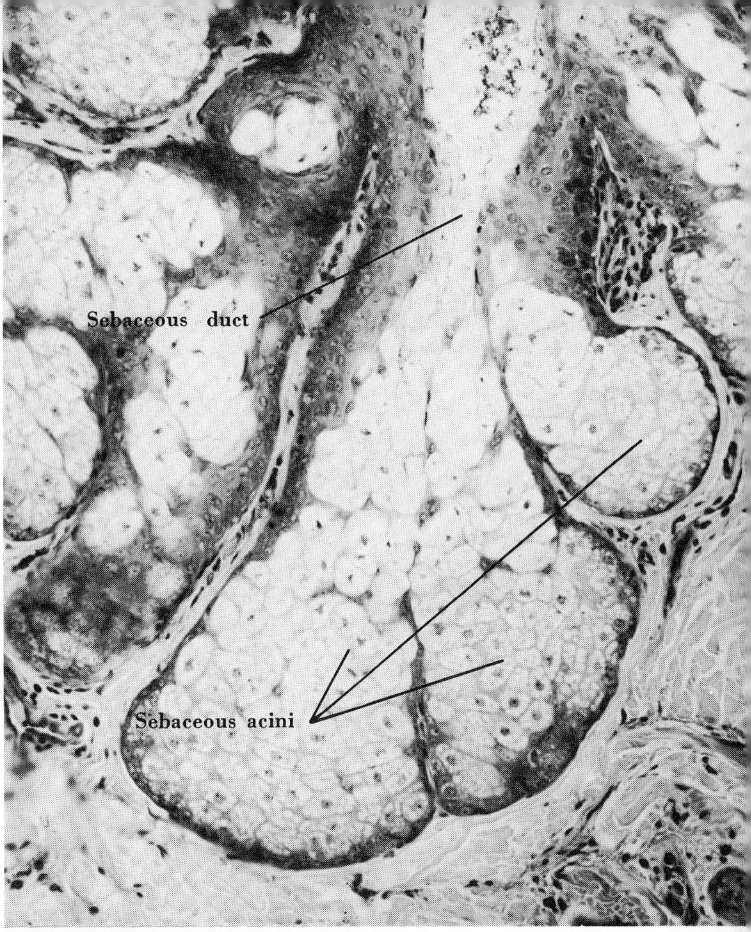

Fig. 6.30 A sebaceous gland in human skin shows the typical features of a holocrine gland. Small deeply stained cells at the perimeter of each acinus differentiate into sebaceous cells filled with lipid vacuoles as they approach the duct. The greatly enlarged sebaceous cells rupture, discharging sebum into the sebaceous duct for transport to the epidermal surface. H & E, 6 section, X360. (Courtesy of R. A. Ellis.)

keratohyaline granules, keratin, and cellular debris forms the sebum, which, in most cases, is expelled into the hair follicle through the neck of the duct of the gland. Glands of the sebaceous type may be slightly modified to serve a purpose other than lubrication, as the *anal scent glands* of Carnivora. The cells of these glands differ from those of ordinary sebaceous glands only in their larger size, greater concentration, and clearer cytoplasm. Surrounding the sebaceous gland proper are saccular groups of simple, watery-secreting epithelia. The combined oily and watery secretions contain odorous materials, presumably mercaptans which, to the human nose, may be extremely repellent, as in the skunk, or attractive enough to be used in perfumes, as in the civet cat or musk deer.

Mammary Glands

The apocrine mammary glands are considered by some authorities to represent modified sweat glands, and by others to be modified sebaceous glands. Like the sebaceous glands and unlike the sweat glands, mammary glands are alveolar structures originally related to mammary hairs. However, the apocrine sweat glands of the human female axilla are similar to the mammary glands in that both exhibit cycles of growth and regression correlated with the menstrual cycle. In addition, the glands around the areola of the nipple have a form which is transitional between the apocrine sweat glands and the mammary glands, and are thought to have originated from lactating hair follicles.

In monotremes the mammary glands are derived from hair follicles along with the ventrolateral milk or mammary ridges. Glandular development is restricted to bilateral areas of the ridges on the lower abdomen. The glands are compound tubular, with no nipple formation. The young grasp tufts of hair issuing from the associated hair follicles and lap the milk from the hairless areas of the skin surrounding the duct opening.

The marsupials and eutherians have compound alveolar glands which originate from thickened epidermal invaginations occurring along the milk ridges. The number and location of these invaginations vary with the species. Hair follicles and associated sebaceous glands develop first in the thickened portions. As cellular proliferation proceeds, branching cords of cells develop which push into the subadjacent dermis to form the rudiments of primary and secondary ducts. The cords then canalize to form a duct system which converges at the surface thickenings or nipple primordia that have formed by this time. With development of terminal acini at the ends of duct branches, the infantile state is reached. During puberty in the female, further growth takes place. Under the influence of estrogen and the lactogenic hormone, large amounts of fat and connective tissue accumulate between the ducts with a slight, concomitant extension of the duct system.

Two types of nipples occur in mammals. In rodents, swine, and other ungulates, for example, the thickened nipple primordium becomes crater-shaped owing to downward movement of the central cells. The ducts open at the innermost end of the cavity that forms, resulting in a true nipple or teat. The edges of the crater and adjacent skin are involved in such formation. In other mammals, such as primates and some carnivores, an eversion nipple is formed by evagination of the thickened epidermis, with ducts opening directly on the surface of the skin.

The nipple parenchyma in all eutherians consists of dense connective tissue and smooth muscle fibers arranged in both circular and parallel fashion to the lactiferous ducts. Encapsulated nerve endings and blood vessels are prominent. The dermal surface is irregular, with many papillae abutting the epidermis. Stratified, keratinized cells form the epidermis, which is quite thin—especially in areas where dermal papillae occur. The main lactiferous ducts ascend through the nipple and open individually onto its summit. Within the nipple the duct lining consists of two layers of columnar epithelium resting on a basement membrane. Close to the opening of the duct, the epithelial lining becomes stratified and keratinized. The brownish-pink skin surrounding the nipple of human females is

Fig. 6.31 A photomicrograph of an active mammary gland. Note alveolar enlargements of the ducts and the dense connective tissue partition around the lobe. X220.

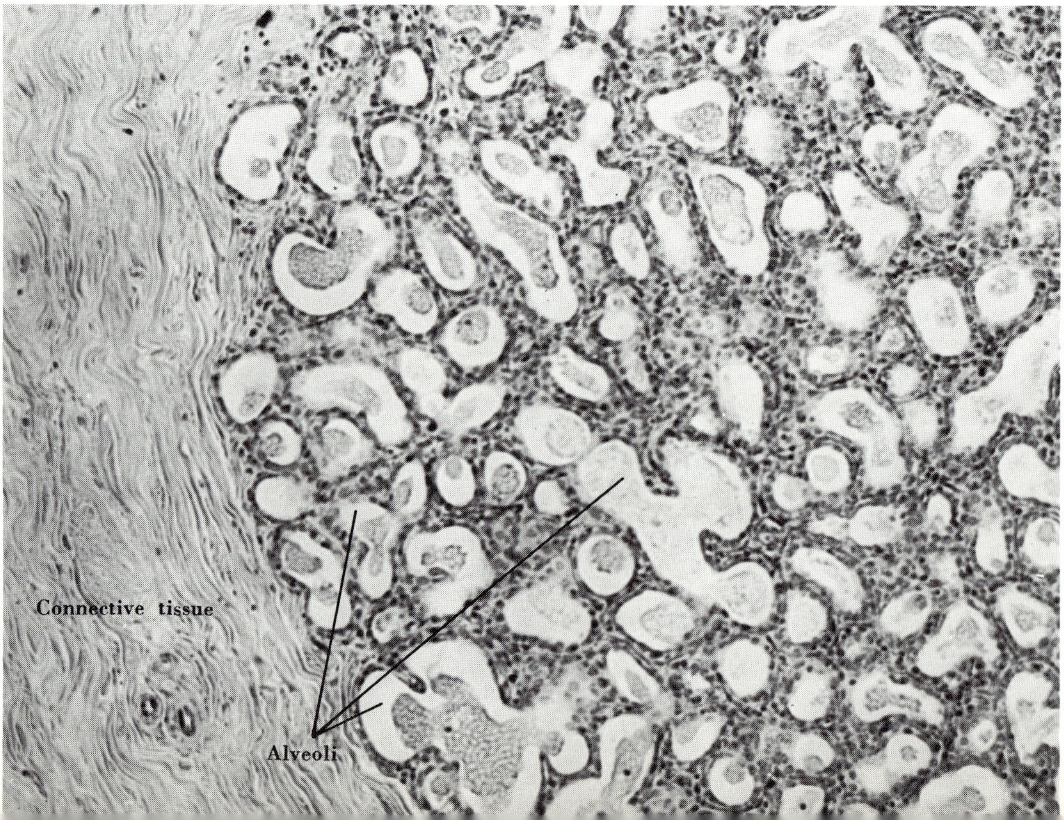

known as the areola. Large sweat and sebaceous glands are found under the areola and open on the surface especially near the periphery. Smooth muscles are present.

When cords of epidermal cells grow down into the dermis, the outer papillary dermal layer is carried along. Each lactiferous duct surrounded by the cellular papillary dermal tissue comprises a lobe. The branches of the main duct running within a lobe are termed *interlobular ducts.* Two layers of epithelial cells form the duct walls. Bundles of noncellular tissue from the reticular dermal layer intervene and act as partitions between the lobes and lobules. Much of the fat which accumulates during puberty is located in these partitions.

During pregnancy the duct system enlarges and alveoli form at the terminations of the primary and secondary ducts (Fig. 6.31). The alveoli are composed of one layer of cuboidal cells and occasional myoepithelial fibers (Fig. 6.32). The secreting alveolar cells contain a well-developed endoplasmic reticulum arranged as parallel lamellae. Alveolar formation and duct enlargement are thought to be under the influence of estrogen and progesterone acting directly and/or indirectly via the anterior pituitary gland.

Milk, the secretion of the alveoli, contains water, fat, protein, lactose, and salts. The alveolar cells contain numerous fat bodies, and the granular appearance of the cytoplasm owes to granules of protein or lipoprotein contained within smooth-walled vacuoles that resemble or form part of the Golgi substance. Milk secretion is under the influence of nervous and hormonal control.

Hair

Hair is as distinctly a mammalian characteristic as feathers are avian. The length, texture, and distribution of these keratinized threads or shafts vary with the species and sex. Modified hairs are present as spines in the hedgehog, porcupine, and echidna. A thick, tightly connected bundle of hairs forms the rhinoceros horn.

Hair formation begins in a manner similar to that of many other epidermal derivatives. Local tubular epidermal thickenings proliferate further and invaginate into the dermis. The papillary dermal layer is carried along during the invagination process to form an enveloping connective tissue sheath. The base of the invaginating epidermis enlarges and becomes the *hair bulb,* while the dermal mesenchyme proliferates and pushes in the distal end of the bulb. Hence the lower end of the developing follicle consists of a knob of nutriment-supplying dermal tissue, the *papilla,* covered by the epithelial cells of the *matrix.* The matrix cells are mitotically active and soon produce a core of cells which move toward the skin surface through the original tubular epidermal downgrowth. The core of matrix cells forms the *hair shaft* and *inner hair sheath,* whereas the original downgrowth becomes the *outer* or *external hair sheath.* The sebaceous gland rudiment appears as an epidermal proliferation of the outer hair sheath about one-third of the way from the skin along the obliquely disposed follicle. Erector pili muscles, arising from the adjacent connec-

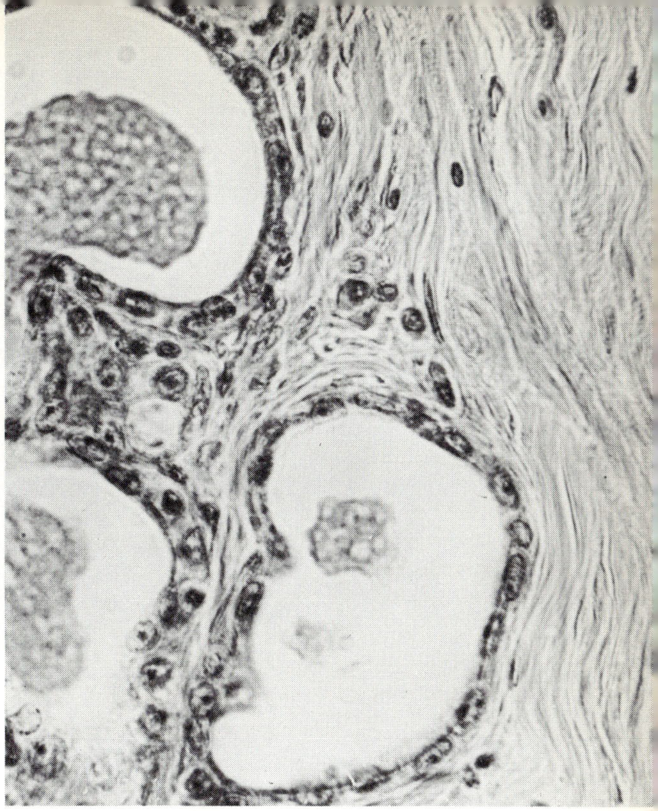

Fig. 6.32 A portion of three alveoli in an active mammary gland. The single layer of cuboidal cells are seen to enclose the milk secretion. X450.

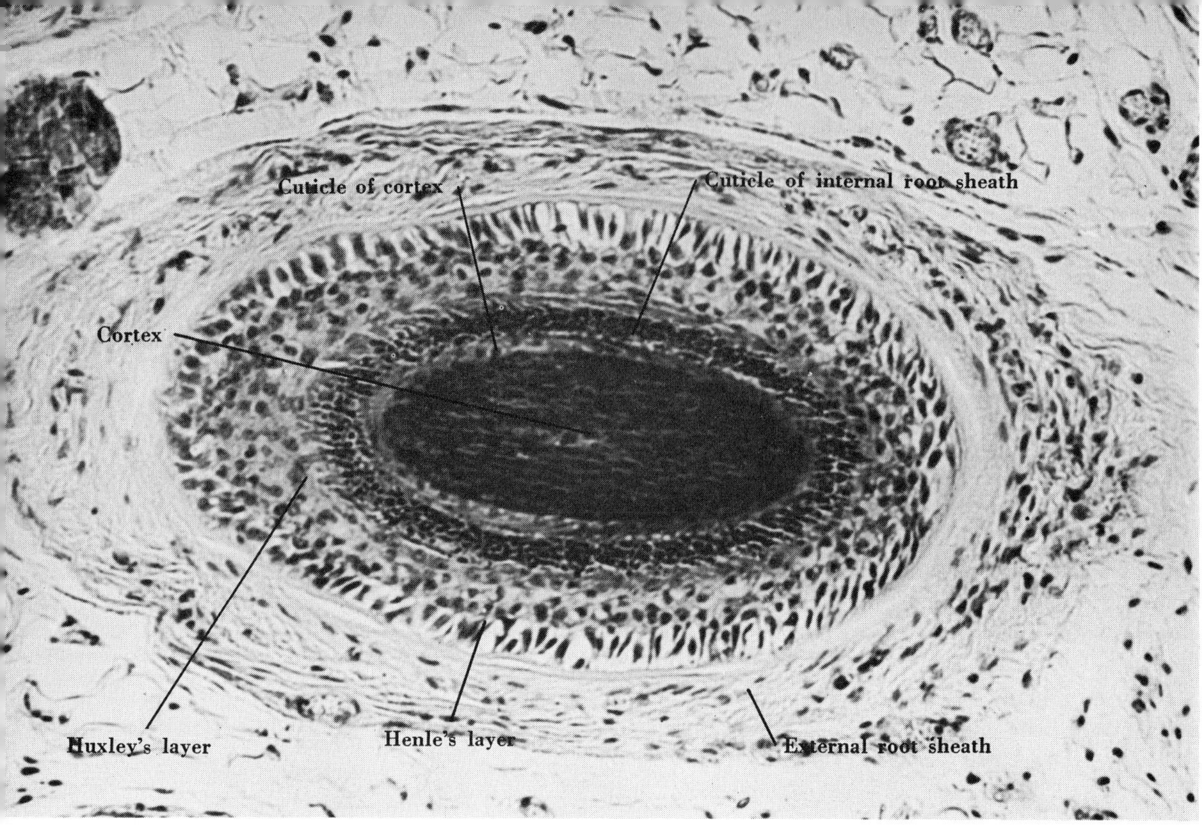

Fig. 6.33 Cross section of a thin hair from human scalp. A distinct medulla is lacking. X220.

tive tissue cells, become anchored to the side of the follicle as well as in the papillary dermal layer (Fig. 6.23).

In cross section (Fig. 6.33) a hair generally appears round or oval, containing many circularly disposed layers. The *medulla* or *core* of the hair shaft contains air spaces surrounded by soft keratinized cells. In thin hairs, the medulla may be absent. Moving outward, the next layer is the *cortex.* It is packed with layers of hard keratin fibers and intervening air spaces. These two layers contain pigment, although cortical pigment usually masks that of the medulla. Surrounding the cortex is the *cuticle of the cortex,* which consists of thin, irregularly shaped, heavily cornified cells. If the section of hair taken is from the upper third of the follicle or beyond, only these three layers will be seen. Sections of hair obtained more basally exhibit an additional compound layer, the *internal root sheath.* Most centrally located in this layer is the *cuticle* of the inner root sheath. This cuticle consists of overlapping keratinized scales. The next inner sheath layer, *Huxley's layer,* contains one to three cell layers which, at the level of the summit of the papilla, exhibit *trichohyalin granules* (Fig. 6.34). In the more distal reaches of Huxley's layer, these granules are replaced by soft keratin. Surrounding Huxley's layer is *Henle's layer.* This is comprised of a single layer of cornified cells which are elongated. The *external root sheath* forms a pocketlike enclosure around the internal root sheath and is continuous with the surface skin at the top and the matrix cells at the bottom. Near the neck region of the papilla, the external sheath consists of one layer of flat germinative cells. This progresses through the two-layer stage at the middle of the papilla and finally assumes the stratified appearance of external skin farther outward. Supporting the entire hair structure is a dermal sheath that contains connective tissue cells and collagenous and elastic fibers.

As noted above, two types of keratin are present in hair. The medulla and internal root sheath contain the soft variety, whereas hard keratin is located in the cortex and hair cuticle. The transition from living cells to solid, hard keratin is gradual but direct, and occurs in the region just above the follicle bulb. Hard

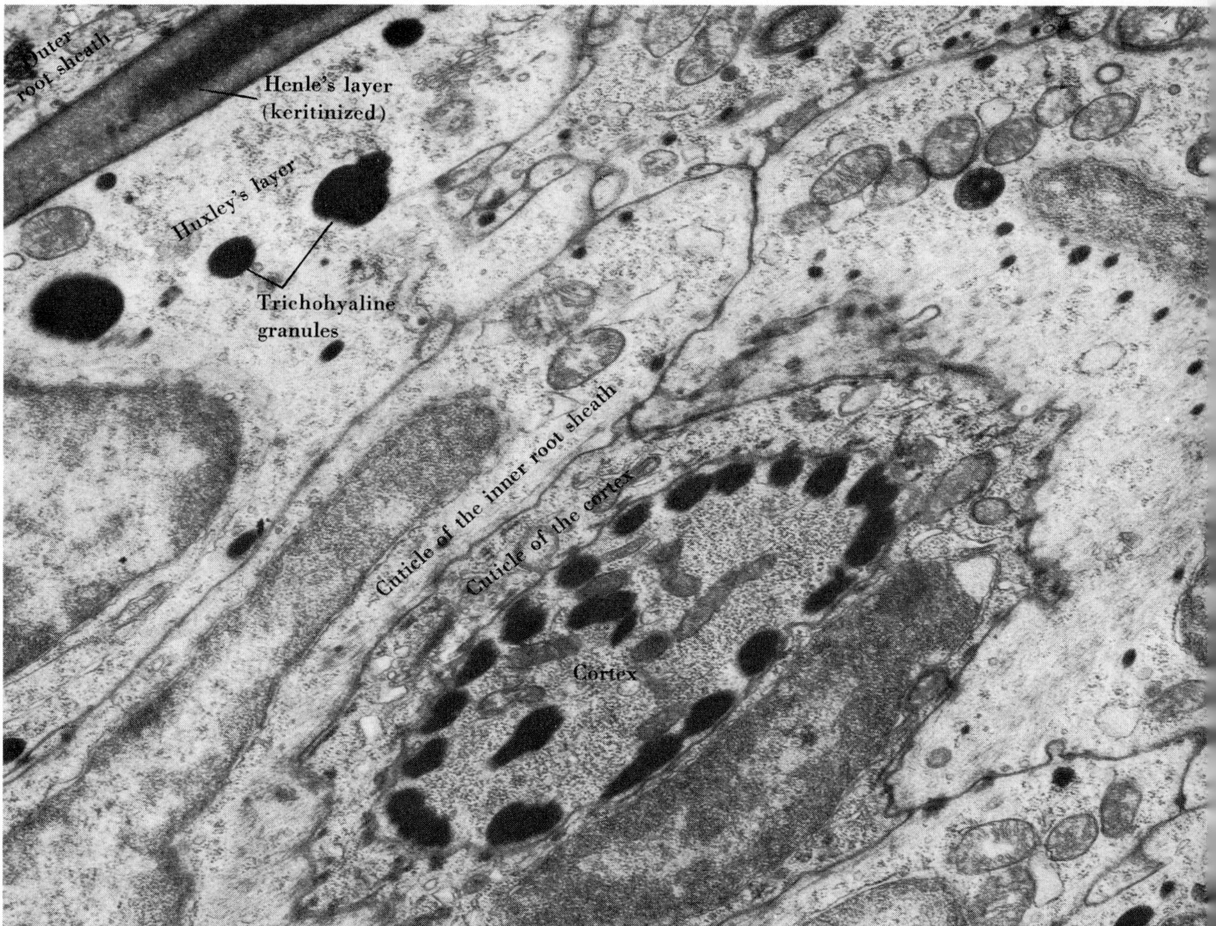

Fig. 6.34 An electron micrograph of a cross section through the lower portion of a hair follicle from a mouse. No medulla is present. X12,000. (Courtesy of P. Parakkal.)

keratin structures do not desquamate. Hair and nails, also hard keratin structures, must be cut to counteract further growth.

Hairs, however, do grow in cycles. During the so-called resting phase, the matrix becomes inactive, the hair root separates from the inactive matrix, and eventually the hair falls out of the follicle (Fig. 6.23). During the growing phase, the matrix and attendant structures are renewed and hair growth occurs.

SCALES AND SCUTES. Mammalian *remnants of dermal scutes* are found as the investing bones of the skull and as bony plates underlying epidermal scales in the armadillo. Epidermal scales are of the reptilian overlapping type. The body scutes of the African anteater are shed singly as opposed to those of the armadillo, which are not shed at all. Rather, armadillo scutes are constantly being worn away by abrasion and renewed by proliferations of the stratum germinativum. The bill of the duckbill platypus is essentially a soft, relatively elastic scale. Friction ridges and tori have been postulated to be scale rudiments.

The scales which occur on the legs and tails of many rodents demonstrate an interesting relationship with associated hairs. In the rat, for instance, three hairs protrude from beneath each tail scale. The middle hair of the three is longer and coarser than the lateral hairs. Over the rest of the rat the hairs grow in the same pattern although scales are absent.

Claws, Nails, and Hoofs

Claws, nails, and hoofs are all hard keratin structures tipping the phalanges. Basically protective, they can also be used in offense, in locomotion, and in food gathering. Claws, characteristic of amniotes in general, and nails, restricted to the insectivorans, bats, and primates, are similar embryologically but they differ in final shape. Nails are broad, flat structures located on the upper surface of the digit. Claws are compressed laterally to form a protective arch for the top, side, and tip of the digit. Some lemurs have claws on certain digits and nails on others.

Claws and nails begin as epidermal thickenings. The proximal part of the thickening invaginates back into the dermis, forming a fold or groove. As the epidermal basal cells on the underside of the fold proliferate, the more superficial ones keratinize, fuse into the nail or claw plate, and are pushed out distally over the nail or claw bed. The proliferative matrix cells maintain their function throughout life. The adult nail or claw, known as the *unquis,* is divided into the visible portion, the *body,* and the part enclosed by the groove, the *root.* The cornified epidermis attached to the distal end of the nail or claw and continuous with the skin covering the underside of the digit is known as the *subungual region.* Its size and degree of cornification are reduced in the primate nail (Fig. 6.35).

As opposed to claws and nails, ungulate hoofs have a definite ventral nail plate, the *subunguis.* Ungulates such as deer, cattle, and sheep have dorsal and ventral matrices which form the unguis and subunguis, respectively. Around the lateral and distal areas these components meet and fuse. Horses, zebras, and donkeys, however, have three matrices. The dorsal matrix forms the unguis, and one of the ventral matrices forms the subunguis. The second ventral matrix produces the *cuneus* or *frog.*

The degree of curvature of claws and hoofs is determined by the shape of the terminal phalanx and the relative rates of proliferation of different parts of the matrix.

HORNS. Mammalian horns consist of a dermal or bony outgrowth covered by a dense epidermal layer of hard keratin. There are two types of true horns and two varieties that are occasionally termed horns.

The *hollow horn* of cattle, sheep, goats, and buffalo is a true horn formed by an outgrowth of the frontal bone that is covered by cornified epidermis. This horn takes its name from a cavity which extends from the frontal sinus into the horn. Neither the bone nor the epidermal horn is shed. The second type of true

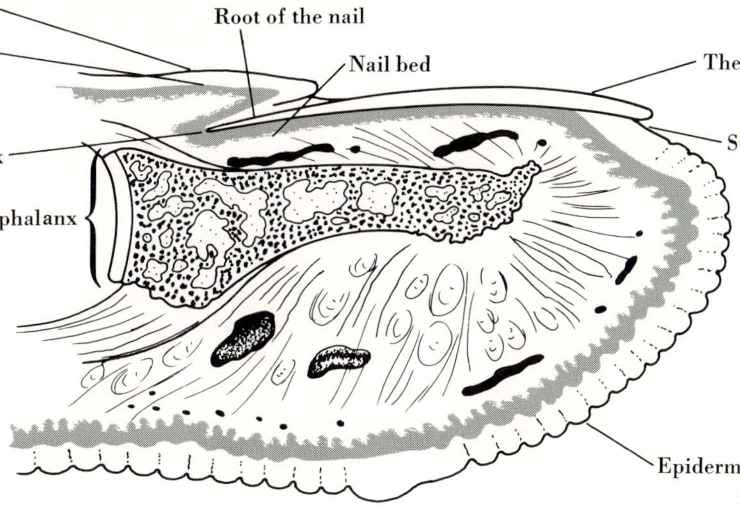

Fig. 6.35 A drawing of the sagittal section of the end of a human finger, showing the structure of the nail.

horn is found in the *pronghorn* antelope. It differs from the horns of cattle in the yearly shedding and replacement of the branching, cornified, epidermal surface. The bony core remains intact.

The rhinoceros horn is morphologically similar to hair. However, the keratinized threads originate in the epidermis and not from follicles located in the dermis. For this reason, some authorities classify the structure as a keratin-fiber horn.

The deer antler is really never a horn at all because the entire antler is an uncovered projection of the frontal bone. However, during the late spring and summer, when the yearly process of growing antlers occurs, the bony outgrowth is covered by the soft, epidermal "velvet." With the approach of the mating season, the velvet becomes isolated from its dermal source of nutriment, atrophies, and is shed. The bony outgrowths are also shed annually with subsequent regrowth in a more branched and complicated fashion. Giraffes possess small "antlers" which are never shed and which remain permanently covered with skin.

SELECTED REFERENCES

Blackstad, T. W., 1963. The skin and the slime glands. *In* The biology of myxine. Brodal, A. and R. Fange, eds. Universitets-forlaget, Oslo.

Ellis, R. A., 1965. Fine structure of the myoepithelium of the eccrine sweat glands of man. J. Cell Biol. 27:551–563.

Matoltsy, A. G., and P. F. Parakkal, 1965. Membrane-coating granules of keratinizing epithelia. J. Cell Biol. 24:297–307.

Montagna, W., 1962. The structure and function of the skin. 2nd ed. Academic Press, New York.

Olsson, R., 1961. The skin of Amphioxus. Z. Zellforsch. 54:90–104.

Parakkal, P. F., and A. G. Matoltsy, 1964. A study of the fine structure of the epidermis of *Rana pipiens*. J. Cell Biol. 20:85–94.

Rawles, M. E., 1960. The integumentary system. *In* Biology and comparative physiology of birds, Vol. 1. Marshall, A. J., ed. Academic Press, New York, p. 190–241.

chapter 7
THE DIGESTIVE SYSTEM

THE DIGESTIVE SYSTEM of all animals is essentially a conveyer of foods from the external environment to the internal environment. In the vertebrates the boundary between these two locales is a double one: the membrane surrounding the lumen of the digestive tube and the membranes that form the walls of the blood capillaries that transport digested food. The interior of the digestive tube itself is considered a part of the external environment for much the same reason that the passage in a drinking straw is considered part of, or contiguous with, the external medium. Only when substances actually move into and through the cellular walls of the tube has the body proper been penetrated. The digestive system, then, functions as the primary supplier of food to the body cells, and these foods must pass through cell membranes to reach the secondary supplier, the blood stream. To meet these needs the digestive system has many specializations along its route.

As a supplier, the digestive system must obtain the food from the environment. The mouth, which serves this function, is located at the anterior end of the animal. It is no coincidence that the anterior end is characterized by both being easily movable and containing a rich supply of sense organs. In many vertebrates the mouth forms the anterior terminus of the body proper. Most carnivores, be they sharks or dogs, have sharp and sturdy teeth for food procurement; herbivores and omnivores have teeth or analogous structures such as beaks, bills, or horny dental pads, especially suited to each individual type of nutrition. Taste buds and olfactory sense organs, useful in searching for and selecting foods are usually located in or near the mouth cavity.

In order for foods to pass through the membrane barriers of the digestive tube, these substances must be physically broken down into particles of small size and chemically broken down into absorbable molecular units in a watery solution or suspension. Specializations such as the teeth and highly muscular organs such as the bird gizzard, supplemented by the contractions of the smooth musculature along the entire tract, masticate the food to produce small particles. Enzymes produced by gland cells and gland masses act in hydrolyzing such food particles so that they may be suspended or dissolved in a watery matrix. By these enzymes, carbohydrates are hydrolyzed to monosaccharides, proteins to amino acids, and most fats to glycerol and fatty acids.

Once the original food has been changed into potentially absorbable molecules, two final conditions are needed before absorption can, in fact, occur. The first condition is sufficient membranous surface to handle all of the absorption that must take place. Various vertebrates possess such structural adaptations as ceca, intestinal villi, elongated intestinal length, and spiral valves, all of which function in the increase of surface area. The second condition that must be met is sufficient time for absorption to occur. Here again, a long, coiled intestine or a spiral valve increases the time spent by food in proximity to absorptive surfaces. Moreover, the digestive tract is supplied with strategically placed circular muscle sphincters which, when contracted, can impede and thereby delay the passage of food. In addition, the tube of the digestive system is muscular; hence it can be distended or contracted around a food mass, thereby moving the food mass along the tract with or without haste.

The digestive system, in its action of breaking down large molecules to small, serves an additional function. Most vertebrates possess an immunological system, which considers large, foreign, biological macromolecules as invaders. Various organs of the body can give rise to cells which, in turn, produce antibodies that destroy the foreign substance by chemical or physical means. This system of immunologic defense can operate against foreign proteins or polysaccharides, but not against the various amino acids or simple sugars of which these macromolecules are polymers. Hence, a vertebrate is immunologically tolerant to the omnipresent monomers but not necessarily to the particular way they are polymerized if that arrangement is not native to the organism. Undigested food, then, if permitted to enter the body proper, would be treated just as any other foreign invader is treated. After chemical digestion, however, the substances produced are immunologically inert.

The digestive systems of the various vertebrates are remarkably similar in gross functional plan as well as in over-all histological organization. Anteriorly, the *mouth* serves as the receptacle for food. There the food may be mechanically broken down by teeth or analogous organs, and, in case of terrestrial animals, lubricated for easier passage through the rest of the tract. Some chemical digestion may also take place in the mouth, although this is mainly a mammalian adaptation. From the mouth the food passes posteriorly into the *pharynx,* and from the pharynx, the food passes via the *esophagus* to the *stomach.* The stomach, which was acquired as jaws developed, originally served as nothing more than a food reservoir. As we progress up the evolutionary scale, and especially as the landmark of endothermy is reached, we find that vastly increased metabolic rates have necessitated the procurement of relatively more food within a given period of time. The simple storage stomach is inadequate to meet these increased metabolic needs, and so we find it has adapted ways to masticate and liquify the food in order to get it moving along faster. The acid secretion of the stomach is thought to have arisen as a means of killing the ingested prey as well as infectious microorganisms. The enzymatic secretions are believed to have evolved from intracellular enzymes.

From the stomach the food passes into the intestine. In the first portion of the intestine, secretions of the liver, pancreas, and, in higher vertebrates, intestinal glands are mixed with the food mass. These secretions serve to carry the hydrolysis of fats, proteins, and carbohydrates to completion. Much of the intestine is concerned with and adapted for absorption of the breakdown products. The residue of unabsorbable material is emptied into the posterior region of the intestine where water is resorbed preparatory to defecation.

Regions of the digestive tube other than the mouth and pharynx are characterized by a basic four-layer, cross-sectional plan (Fig. 7.1).

The innermost layer of the gut wall is termed the *mucosa* or *mucous membrane.* Present throughout the length of the tract, the mucosa consists of an inner epithelial membrane, a middle region of connective tissue,

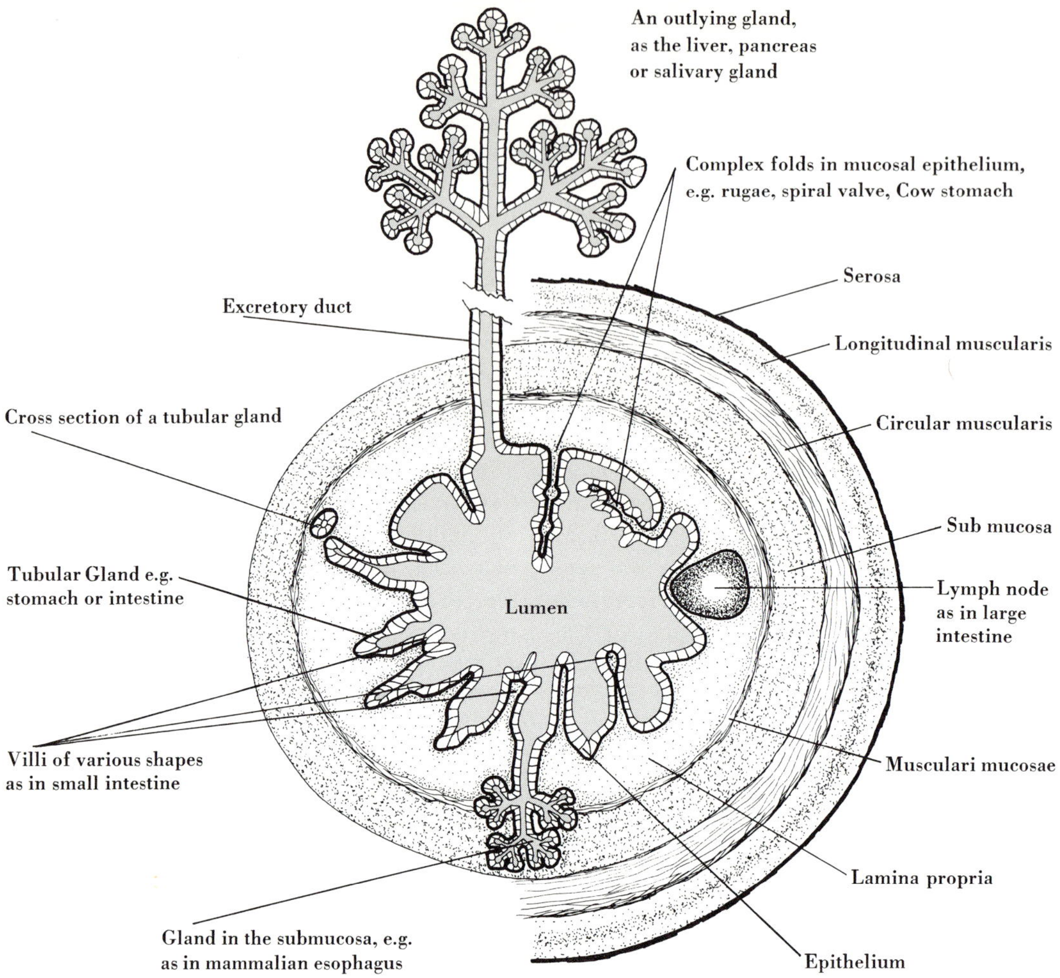

Fig. 7.1 The basic four-layer plan of the digestive tube, including some of the modifications that can be found along its length.

and an outer layer or layers of smooth muscle.

The epithelial membrane, a derivative of the embryonic endoderm, functions primarily as a selective barrier between the lumen of the tube and the inside of the body. In order to maintain the intact nature of the barrier, the epithelium is universally and richly supplied with mucus-producing cells and invaginated mucous glands. Many of the multicellular glands dip into the underlying connective tissue of the mucosa and some reach the second major layer. In the case of the salivary glands, pancreas, and liver, the epithelial outgrowth lies beyond the tube itself, connected to the lumen only by a duct, which represents the location at which the outgrowth began.

The middle region of the mucosa is termed the *lamina* (or *tunica*) *propria*, and it represents, as does the rest of the gut wall, a derivative of the embryonic mesoderm. The lamina propria consists of reticular or fibroelastic connective tissue and often it incorporates lymphatic tissue. In fact, the farther one descends down the digestive tract, the larger is the concentration of lymphatic tissue. This, of course, may be correlated with the increasingly fertile environment for bacteria and other microorganisms in the lower gut. The lamina propria serves to support the overlying epithelium, as well as to provide the pathway by which blood capillaries and lymph vessels nourish and drain the epithelial cells.

The outer region of the mucosa, the *muscularis mucosae,* is variable in structure. Usually beginning in the esophagus, it may consist of scattered smooth muscle fibers surrounded by connective tissue or it may be well developed, consisting of one or two thin smooth muscle layers. Electron micrographs reveal a three-dimensional network of flattened smooth muscle fibers having elongated processes, the processes of one cell contacting those of its neighbors. Where mucosal glands are present, they are usually found to be enmeshed by a net of such cells. The action of the muscularis mucosae is that of allowing localized or limited active movement of the mucosa, although the evidence from electron microscopy seems to suggest an additional or alternate function of expressing and moving glandular secretions.

Surrounding the mucosa is the second major layer, the *submucosa.* Consisting of a loose, meshlike layer of fibroelastic connective tissue, the submucosa carries the larger blood vessels, the branches of which vascularize the mucosa. Located within the submucosa are nerve fibers and ganglion cells which collectively form *Meissner's* (or the *submucous*) *plexus.* The fibers of this plexus are non-myelinated, postganglionic fibers of the sympathetic division, whereas the cells are the ganglionic cells of the parasympathetic division. Fibers from both parasympathetic and sympathetic neurons terminate in the mucosa. The submucosa of fishes often contains smooth muscle fibers, which probably compensate for the lack of a well-developed muscularis mucosae.

The third major layer of the gut wall is the *muscularis externa,* more simply known as the muscularis. This layer determines the tonus of the gut wall and is responsible for peristaltic and mixing contractions of the gut. The muscularis usually consists of an inner circular and outer longitudinal arrangement of smooth muscle fibers, although variations on this pattern can be found. Located between these muscle sublayers are nerve fibers of the *myenteric* (or *Auerbach's*) *plexus.* This nerve plexus contains the terminal portions of preganglionic, parasympathetic fibers which synapse within the plexus with the postganglionic cells that innervate the muscle fibers. In addition, postganglionic sympathetic fibers, also terminating on the muscle fibers, are found throughout the plexus.

The outermost gut layer is the *serosa* or *adventitia.* In regions of the gut that lie within cavities and are suspended by mesenteries, this layer is composed of areolar connective tissue covered with simple squamous or cuboidal mesothelium; it is termed the serosa. In portions of the gut not suspended in cavities, such as the pharynx and part of the esophagus, this layer is spoken of as the adventitia, and it consists of areolar connective tissue which gradually blends with the connective tissue of surrounding structures.

THE DIGESTIVE SYSTEM OF FISHES

Before delving into the histologic characteristics of the fish digestive system, one should keep in mind the great variety of organisms that are classified as fishes. In any attempt to describe a generalized plan of so complex a portion of an animal as the digestive system, and in such a diverse group as the fishes, many exceptions will necessarily occur. The more important ones, from an evolutionary standpoint, will be discussed; the many deviations, presumably adaptive to specific feeding requirements, will not.

One major generalization applicable to the digestive system of almost all fishes is the simplicity of its histology as compared with that of mammals. As mentioned above, foods must be soluble for absorption into the body. Much of the glandular apparatus of tetrapods is concerned with producing a vehicle for the digestive products. Fishes, living in an aqueous environment, have a much simpler problem. One gulp of water can provide the medium for digesting dinner, and lubrication and solubilization of food products easily occurs. Secondly, the low metabolic rate of fishes permits slow digestion of food. A shark may take as long as a week or two to digest a meal completely. Because of the unhurried rate of energy production, fewer and less-

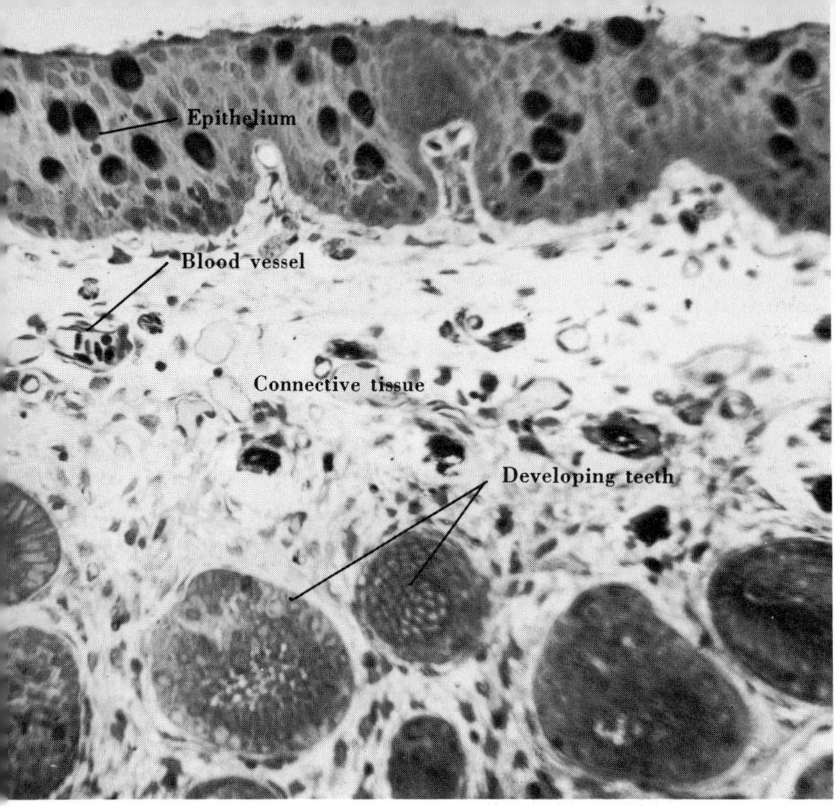

Fig. 7.2 Palatal epithelium of the Gourami. X400.

complex digestive glands and a less highly developed muscular apparatus are needed.

Oral Cavity and Pharynx

The anterior chamber of the digestive tract is subdivided into an oral cavity and a pharynx, the anatomical distinction being based on the presence of gill slits in the latter. Both regions are lined with the same kind of thin, stratified squamous epithelium and supporting connective tissue that is characteristic of the integument (Fig. 7.2). Mucus-secreting cells abound, but multicellular glands are rarely present. A tongue is absent and only a simple covering of epithelium and connective tissue cover the hyoid bones or cartilages.

The mucous membrane of both mouth and pharynx has moundlike protuberances in all classes of vertebrates. These *papillae* are simply a change in the contour of the epithelium and underlying lamina propria, with every papilla having a vascularized connective tissue core and epithelial covering. In fishes that do have papillae, two basic shapes can be noted. Those in which the core and outer epithelial layers are threadlike or conical in shape are termed *filiform* papillae. If the shape more closely resembles a flat-top hillock, the papillae are known as *fungiform*. In most cases, the papillae represent the location of the sensory taste buds (see Chapter 13) and, in addition, may aid the teeth by functioning in a rasping manner.

Most fishes that are not plankton feeders, jawless, or generalized omnivores possess teeth in the mouth and/or pharyngeal regions. Depending on the mode of nutrition, the teeth can vary in shape from pointed spikes and fused, razor-edged cutters in some sharks and teleosts to flat, crushing plates in rays, lungfish, and other teleosts. Some fishes show an *isodont* condition, in which all teeth are alike in shape, whereas others show the *heterodont* condition, in which there is a differentiation in shape and function of those teeth that occur both around the jaw margin and on connective tissue pads in the pharynx. Some fishes have smooth-surfaced teeth, whereas in others the surface is scalloped to a greater or lesser degree. Many sharks continually replace their teeth; other fishes have but one adult set for life. Whatever the shape, true teeth are believed to have originated from the placoid scale, and indeed, in the sharks there is little distinction, other than location, between the two (Fig. 6.9).

The fish tooth consists basically of a simple

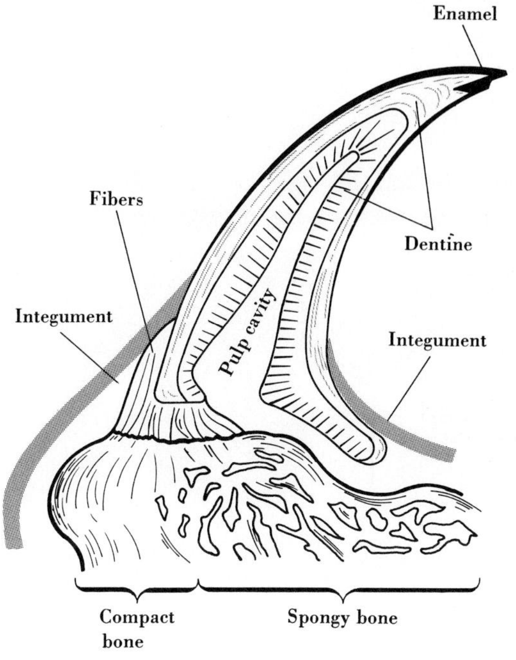

Fig. 7.3 A diagram of a longitudinal section of a fish tooth and its attachments.

covering of enamel or enamel-like material, an inner layer of dentine, and a core or pulp cavity filled with cellular connective tissue, blood vessels, and nerves (Fig. 7.3). In some sharks the pulp cavity becomes filled with osteodentine, a primitive type of bonelike dentine.

Esophagus

The esophagus of most fishes consists of a short, thick-walled, highly distensible tube. Many predacious fishes can and do engulf other fish as large as themselves; yet rarely have fishes been found dead or dying of an overstuffed esophagus!

The mucosa of the esophagus is characterized by an increase in epithelial depth. The stratified cuboidal or columnar epithelium is richly populated with mucus-secreting, *goblet* cells, and most often is ciliated as well. Multicellular glands are usually lacking here, although serous and cardiac glands similar to those of the stomach may be found toward the lower end of the esophagus. Taste buds, usually oral inhabitors, are occasionally seen in the esophageal location as well. The connective tissue element of the mucosa is limited to a strong basement membrane, and the muscularis mucosae is represented by longitudinally oriented smooth muscle fibers. The mucosa is no longer smooth in topography, but follows the outline of the submucosa, which is molded in a series of high, longitudinal primary and smaller secondary folds that partially obliterate the free space of the lumen. Scattered through the areolar tissue of the submucosa, and especially near the junction of the esophagus with the stomach, are granule cells and areas of lymphomyeloid tissue. In the shark these areas are condensed to form a distinct structure, *Leydig's organ,* coursing the entire length of the esophageal submucosa (see Chapter 9). Underlying the submucosa is a highly variable muscularis. Usually composed of two layers of interweaving skeletal muscle fibers, an inner circular and an outer longitudinal, the muscularis may evidence some degree of invasion by smooth muscle fibers. In some fishes skeletal muscle occurs as far posteriorly as the stomach. The trench fish is reputed to contain striated muscle from mouth to anus. The serosa is usually thin, and covered by a layer of simple cuboidal mesothelium.

The posterior end of the esophagus may be delimited by a smooth muscle sphincter, or by the mucosa's being thrown into a series of branched papillae. Usually the longitudinal submucosal folds so characteristic of esophagus disappear—to be replaced by shorter ones, the *rugae,* which are characteristic of the stomach.

Stomach

Of all parts of the fish digestive system, the most highly variable is the stomach. The range in stomachs proceeds from none in such fish as the carp, to straight spindle-shaped enlargements and to curved structures with cardiac and pyloric portions. Even in fishes which do show distinct cardiac and pyloric regions of the stomach, the proportions of each can vary enormously.

The presence of a "typical" stomach is as nonessential for fishes as for man from the enzymatic standpoint. Organisms (of the

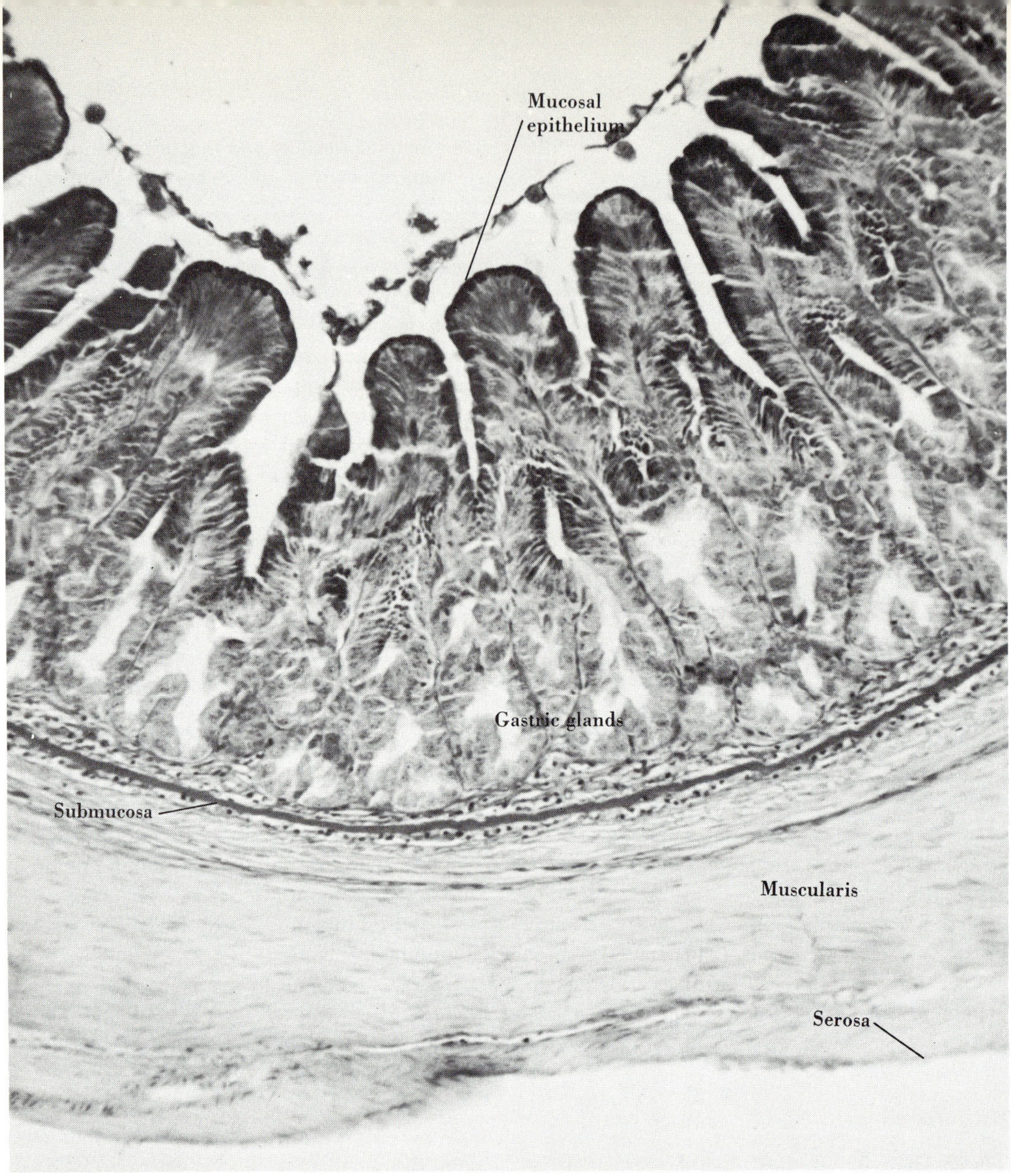

Fig. 7.4 Photomicrograph of a cross section of the cardiac stomach of the rainbow trout. X275. (From E. L. Weinreb and N. M. Bilstad. 1955. Copeia 3.)

human variety) that have undergone surgical procedures for stomach removal need make only minor changes in their mode of living. That this is possible testifies to the role played by the stomach as an enzymatic entity; in most vertebrates its mechanical and lubricating functions can easily be fulfilled by other structures. The stomach is important biochemically as a secreter of certain enzymes, but the enzymes which are characteristic of the pancreas and intestine are easily capable of duplicating the action of the gastric enzymes. Hence, those fishes that lack stomachs show a histological progression from esophagus directly to intestine.

The stomach is the source of a class of enzymes called endopeptidases, which show greatest activity in a highly acidic environment. These enzymes are often lumped together under the singular name of *pepsin.* This terminology is misleading, however, because salmon pepsin is different from perch pepsin, and both, in turn, show marked differences from mammalian pepsins in such characteristics as pH optima, alkali stability, thermolability, and site of action. Each of

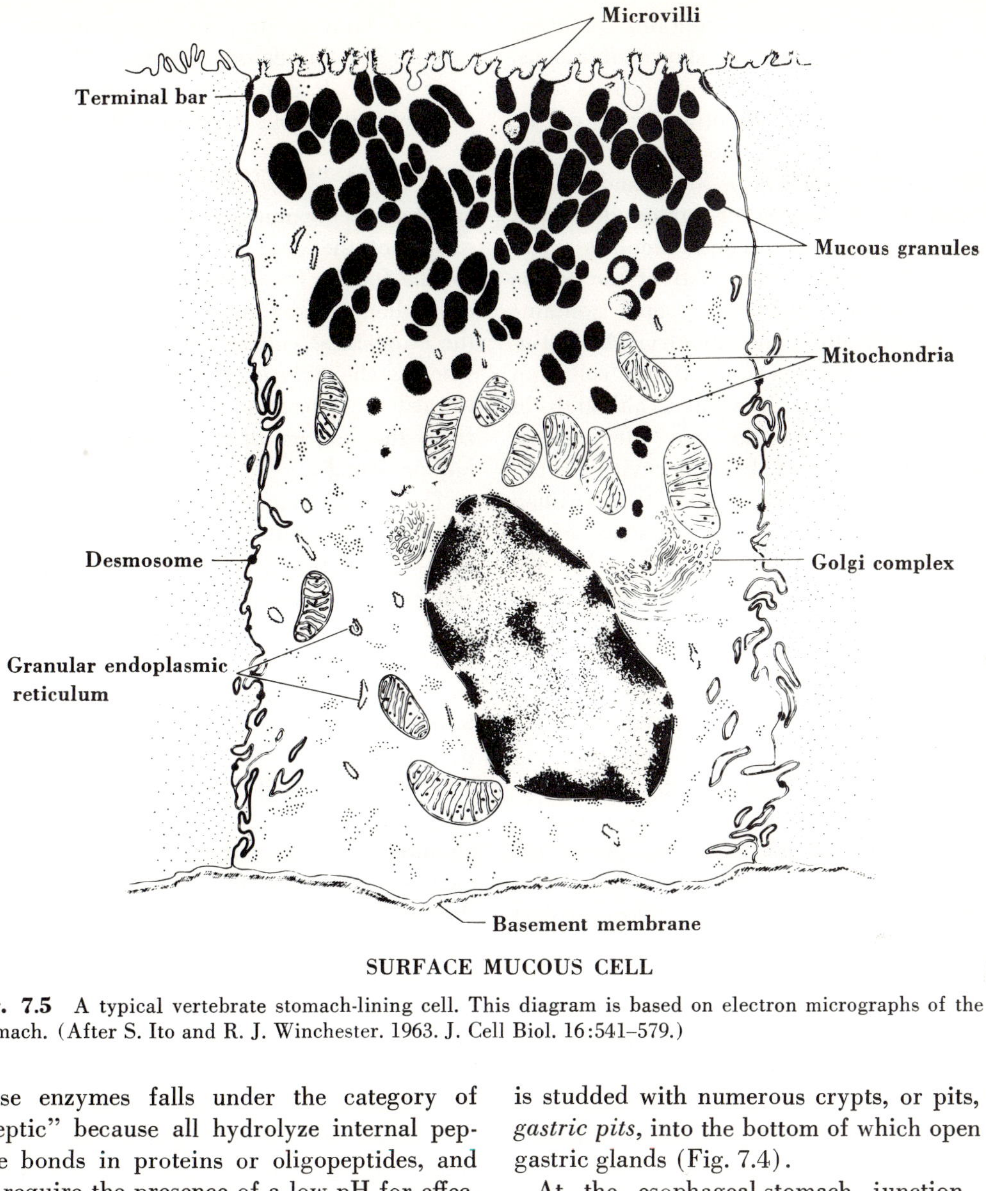

SURFACE MUCOUS CELL

Fig. 7.5 A typical vertebrate stomach-lining cell. This diagram is based on electron micrographs of the bat stomach. (After S. Ito and R. J. Winchester. 1963. J. Cell Biol. 16:541–579.)

these enzymes falls under the category of "peptic" because all hydrolyze internal peptide bonds in proteins or oligopeptides, and all require the presence of a low pH for effective action. The stomach provides the acid pH by accompanying the production of a pepsin with that of HCl. In addition, the stomach, as with all regions of the digestive tract, must provide a lubricant as part of the maintenance of an intact mucosa.

MUCOSA

The stomach mucosa covers longitudinal submucosal folds, the *rugae*. These folds are distinguished from those of the esophagus by their disappearance when the stomach is distended. The mucosa is not smooth, but rather is studded with numerous crypts, or pits, the *gastric pits*, into the bottom of which open the gastric glands (Fig. 7.4).

At the esophageal-stomach junction, the surface epithelium evidences a transition from stratified squamous to a single layer of columnar cells which line the gastric pits as well as the luminal surface. Division of the pit-lining cells is thought to give rise to the surface cells which must be replaced because of physiological wear and tear. With few exceptions these lining cells appear to be typical of the vertebrate stomach in structure and function. A typical lining cell has a high columnar shape with a clear, cup-shaped or banded free border distally (Fig. 7.5). Appropriate staining reactions have been used to identify the clear area as one that contains mucigen granules, the precursor substance of mucus. The secretions of gastric mucous cells is characteristically alkaline. In other areas of the tract, the mucus tends to be acidic. The basally located nucleus is oval and darkly staining, and is capped or surrounded by a sizable Golgi network. The acidophilic cytoplasm in the nonsecretory half of the cell is finely granular or fibrillar, and contains numerous threadlike mitochondria. As distinguished from the intestinal lining cells, no true striated border is visible, although in electron micrographs one may often see a small number of very fine microvilli on the free surface. These cells are also distinguished from true goblet cells by (1) the shape of the mucus or mucigen-containing area; (2) the atypical staining reactions of the mucus; and (3) the fact that these cells discharge their secretion continuously, and not in a single mass as do goblet cells. Functionally, the secretion of these cells is believed to guard against autodigestion by HCl and pepsin.

Exceptions typically exist in the lining epithelium of fishes. Some of the more primitive fishes have ciliated cells scattered among the typical lining cells, and often this condition is accompanied by the presence of special goblet cells which structurally resemble their namesakes, but which differ in having an alkaline secretion. These atypical goblet cells are thought to be homologous with mammalian Paneth cells (see below). The gastric epithelium of the bottom feeder, *Mulloides auriflama,* contains wandering granular cells that closely resemble eosinophils.

The lamina propria of the mucosa is reduced because of the presence of numerous gastric glands in this layer. What is present, however, can often be divided into an outer portion of loose areolar and reticular connective tissue, and a more central zone of lymphoid tissue that contains (in addition to lymphocytes) eosinophilic granulocytes heterophils, and argentophilic fibers. If present, the muscularis mucosae is thin, with longitudinally oriented smooth muscle fibers. If a muscularis mucosae is not present, scattered muscle fibers are found in the outer portions of the lamina propria. Surrounding the mucosa is a thin dense membrane, the *stratum compactum.* It is mainly composed of a feltwork of collagenous fibers and fibroblasts.

GASTRIC GLANDS OF THE MUCOSA

Located in the mucosa are the multicellular glands of the stomach. Two types of glands can be found. The *fundic glands,* also known as cardiac or chief glands, are concentrated in the esophageal end and body of the stomach whereas the *pyloric glands* are located more toward the intestinal end. Both types extend from the bottom of the gastric pits to the lamina propria.

The fundic glands are simple or branched tubular glands with a mouth section that extends from the base of a gastric pit; an elongated, constricted neck region; and an expanded basal body region. In some fishes the mouth and neck regions are lined by modified surface cells which lose their typical clear border and become more prismatic as one progresses down the neck of the gland. In other fishes, the cells lining the neck appear to be similar to large goblet cells. In still others, cuboidal cells containing mucoid secretory granules have been found. The body cells of the fundic glands are those which produce the peptic and hydrochloric acid secretions. In mammals each of these secretions is produced by different cells, but in fishes one kind of cell appears to produce both. The body cells are columnar with lightly staining, spherical nuclei, usually located toward the outer portion of the cell. The cytoplasm is filled with small, strongly acidophilic granules which probably represent the precursor substances for pepsin. Presumably the acid is synthesized from sodium chloride and carbonic acid, both soluble substances, on the outside of the cell—where sufficient alkaline mucus is present to protect the stomach lining from autocatalytic action.

The pyloric glands are fewer in number and shorter in length than are the fundic glands. The neck and body regions are more twisted, however, and both neck and body appear to be lined with cuboidal or low columnar cells which lack a distinctive granularity in their cytoplasm. These cells are not very different from the surface lining cells and are thought to be restricted to mucus production. The number of pyloric or fundic glands opening into each pit varies with the species.

SUBMUCOSA

The submucosa is composed of loose areolar tissue containing lymphatics, arteries, veins, nerve fibers, and the submucous plexus.

MUSCULARIS

In the muscularis one finds two relatively thin coats of smooth muscle. Part or all of this layer may consist of skeletal fibers, depending on the species.

SEROSA

The serosa is characterized by a layer of loose connective tissue, heavily vascularized, and covered externally by simple cuboidal mesothelium.

Intestine

The main functions of the vertebrate intestine are digestion and absorption. In fishes the absorptive role is quite clear-cut, with certain of the lining cells functioning actively as absorbers. In higher vertebrates the digestive functions are just as clear-cut—certain cells produce specific enzymes, and these secretions are added to by the outpouring of secretions through the bile and pancreatic ducts. Many fish, however, have a diffuse pancreas, with nodules of pancreatic tissue often embedded in such sites as the liver, gallbladder, and intestinal wall itself. Under such conditions it is difficult to separate the exact origin of the enzymes which are found in the intestinal lumen. In addition, the histology of the intestinal mucosa is much simpler in fish than in the higher vertebrates. Multicellular glands are usually absent although a very few species (the *Gadidae*) do have such structures. Glandular cells which differ from ordinary mucus-secreting ones are rare or absent. The most one can say is that an enzyme resembling mammalian intestinal peptidase has been found in the lumen of the intestine, and it is presumed that this enzyme originated in the intestinal mucosa. The situation for carbohydrases and lipases is even less clear. Enterokinase, an enzyme responsible for changing pancreatic peptidase from an inactive to an active state, has been definitely traced to the intestinal mucosa, although the specific cellular site of production is still in doubt. Thus, confident that digestion of foodstuffs does occur in spite of our inability to comprehend it fully, let us turn to the problem of absorption.

As stated above, one requisite for adequate absorption is sufficient surface area. Fish have solved this spatial problem in various ways. Many teleosts possess an intestine which is lengthened to a greater or lesser degree. If the intestine is longer than the abdominal section of the organism, coiling or looping of the gut must obviously occur. In many species of bony fish intestinal outpocketings, *ceca,* occur especially near the anterior end of the intestine. Variations from one to more than a thousand ceca per fish have been found. In the sharks, rays, many of the holostean fishes, and lungfishes, the mucosa and submucosa have become elevated in a long longitudinal fold that then forms a spiraling structure, the *spiral valve,* which courses through much of the intestinal length (Fig. 7.6 and 10.15). Still other fish have longitudinal submucosal folds basically similar to the esophageal ones, although the pattern of complexity can vary enormously. A few fish approximate the intestinal villi of mammals, having fingerlike projections of epithelium covering a core of highly vascular connective tissue. Many fish have greatly increased the surface area of the intestine by possessing a combination of all of these features.

Whatever the gross morphology may be, the histology of the intestinal wall, be it studded with projections or thrown into a spiral, varies little from one species to another and from one region to another (Fig. 7.7).

MUCOSA

The epithelial layer of the mucosa usually consists of simple columnar tissue. Two basic cell types are found in this layer: goblet cells and columnar cells. The goblet cells are of

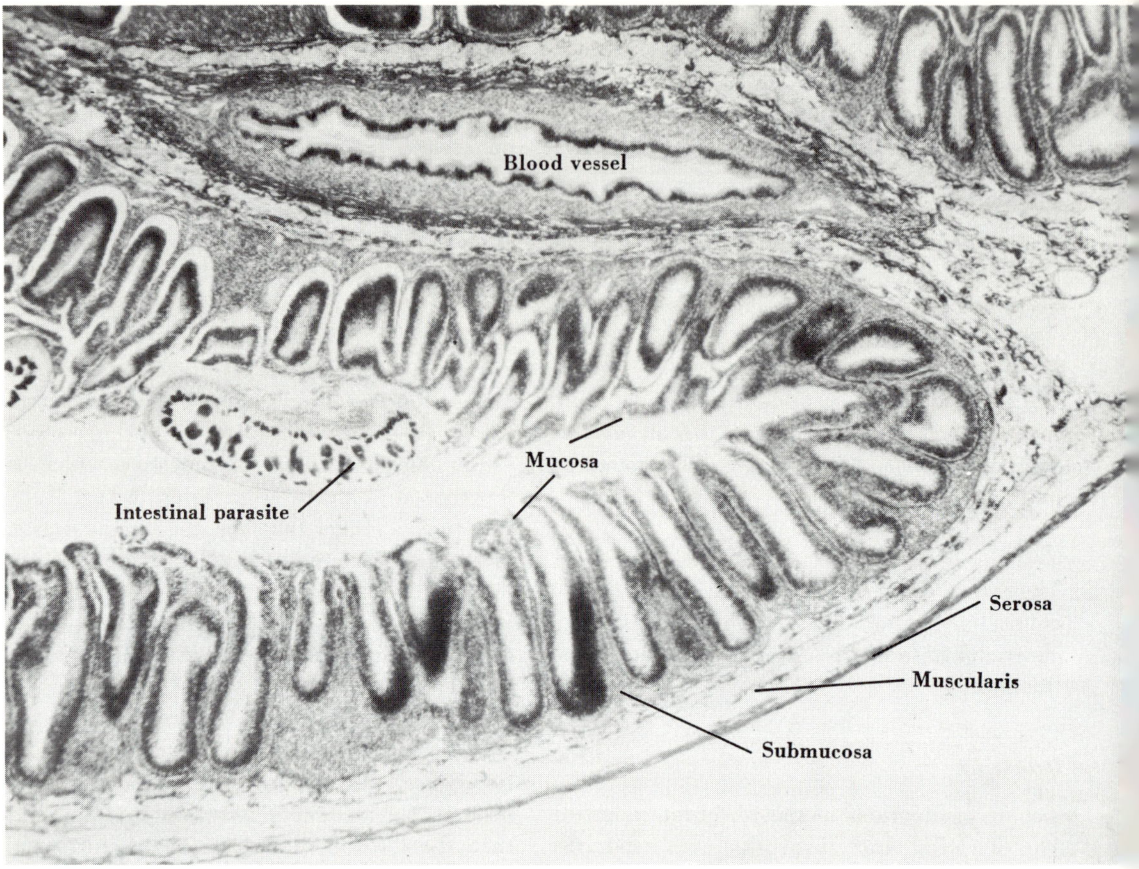

Fig. 7.6 The spiral valve of Squalus. X35.

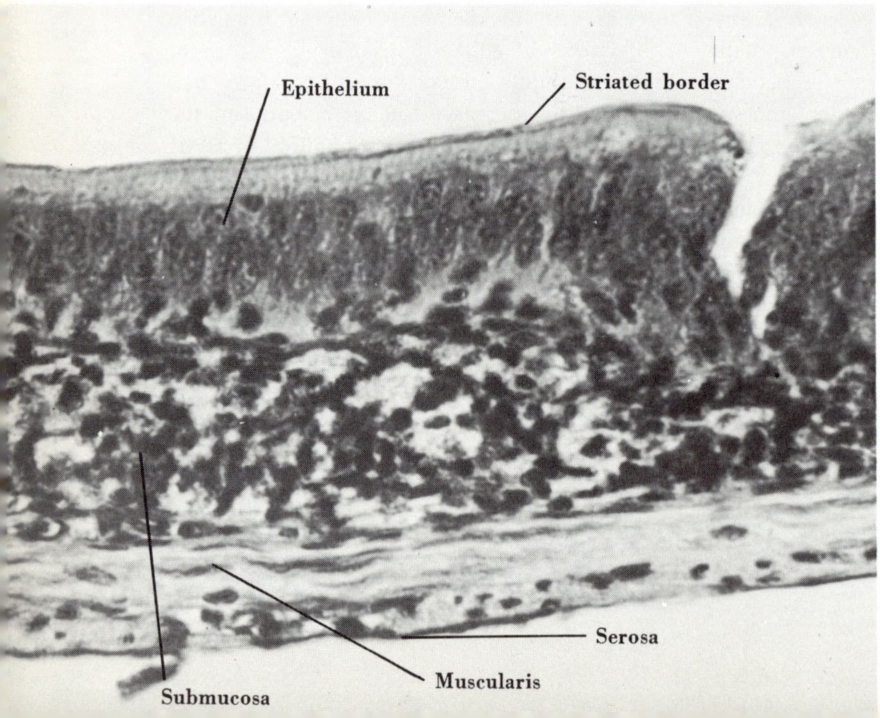

Fig. 7.7 A cross section of a teleost small intestine. X150.

typical chalice shape, with a slender base and a cup-shaped mucigen-containing area permanently open to the lumen. The nuclei of such cells are located in the base of the goblet near the basement lamina, and are surrounded by a granular acidophilic cytoplasm (Fig. 4.9). In electron micrographs of goblet cells, two kinds of cytoplasmic granules have been found. The first is a low-density granule which is presumed to be mucigen, the mucus precursor. The second type is larger (0.4–1.0μ) and opaque, with a typical cytoplasmic membrane surrounding each one. Some relationship must exist between the two types of granules because an inverse numerical relationship is always observed between the two.

The absorptive columnar cells are typical of those in the intestine of vertebrates generally. Under the light microscope these cells are seen to have an apical striped or striated border and a basal nucleus. The cytoplasm between the nucleus and striated border is granular, with the Golgi zone capping the nucleus and with distal and basal zones of mitochondria. When seen under the electron microscope, the striated border appears as a band of closely packed, parallel, thin tubular processes which function to increase the cell surface (Fig. 4.5). These *microvilli* are granular cytoplasmic extensions of the cell and thus are covered by extensions of the cell membrane. It has been estimated that each of the absorptive cells possesses 2000 microvilli. Immediately beneath the microvilli lies a band of nongranular, fibrillar cytoplasm, the *terminal web*, the fibrils of which seem to originate within the microvilli and to condense at the lateral boundary of the cell. This condensation forms a structure which is collectively known as the *terminal bar*. The function of the terminal bar is one of cellular cohesion, and indeed, the bars resemble elongated desmosomes (see Chapter 4). Beneath the terminal web lies a pair of centrioles, numerous mitochondria, and a smooth agranular endoplasmic reticulum. Rough granular endoplasmic reticulum is often found in the basal portion of the cell.

In a few fish ciliated columnar cells can be found. That these cells function in movement of the intestinal contents is attested to by the fact that the presence of cilia can usually be correlated with a poorly developed muscular layer. Eosinophils and wandering leukocytes are often located between the columnar cells, and on occasion cells resembling mammalian chromaffin cells (*see below*) can be observed. The latter are most numerous in elasmobranchs and least abundant in teleosts.

The lamina propria generally consists of a relatively fibrous connective tissue that is rich in granule cells, leukocytes, basophilic mast cells, and eosinophils. Lymphoid nodules can often be found. If a muscularis mucosae is present, it consists of a very thin layer of longitudinally disposed smooth muscle fibers. In many species stratum compactum is present surrounding the intestinal mucosa.

SUBMUCOSA

The connective tissue in the submucosa is of the areolar type, usually containing fewer cells than the lamina propria. Lymphocytes and other wandering cells occur with regularity, as do lymphoid nodules. Some fishes, such as the trout and salmon, lack a true intestinal submucosa.

MUSCULARIS

The usual two layers, an inner circular and outer longitudinal, are present. In many fish both layers are thin; in others the outer longitudinal layer is the more delicate.

SEROSA

The serosa is typical, consisting of areolar connective tissue and a mesothelium which may be either simple squamous or simple cuboidal in type.

The fish intestine is not divisible into regions as it is in higher vertebrates. A vague distinction, however, can sometimes be made between intestine and hindgut or rectum. In sharks and lungfish the end of the spiral valve and the subsequent narrowing of the gut marks the beginning of the rectum. In these forms the epithelium of the mucosa gradually changes from simple columnar to stratified low columnar. The proportion of goblet cells increases as does the thickness of the circular layer of the muscularis. A multicellular rectal gland opens into the posterior portion of the rectum (see Chapter 10). Some teleosts have a rudimentary iliocecal smooth muscle sphincter, but generally the hindgut is marked only by an increase in number of goblet cells and thickness of the muscularis. The absorptive capacities of the hindgut are markedly reduced.

Cloaca

The fish digestive system terminates in the cloaca, a chamber which receives the excretory products and reproductive cells, as well as the fecal residue. The mucosa of the intestine or hindgut, if one can be distinguished, undergoes a transition to the stratified condition or cuboidal condition characteristic of the integument. Connective tissue elements of the gut wall show a transition to the dermal portions of the skin, and the smooth muscle cells of the muscularis are replaced by skeletal fibers.

THE DIGESTIVE SYSTEM OF AMPHIBIANS

The digestive system of amphibians does not contain many morphological advances over what we have seen in the fishes. However, some unique features can be found among amphibians which deserve special mention.

Oral Cavity and Pharynx

The first group of unique characteristics can be found just as soon as an amphibian opens its mouth. As with the histological picture found in most of the other vertebrates, the epithelial mucosa lining the mouth cavity is stratified (Fig. 7.8). The epithelial membrane, however, is usually very thin—with cuboidal rather than squamous surface cells. Furthermore, these cuboidal cells bear cilia in a vast majority of Amphibia, or a distinct striated border in most of the others. Only a few representatives have a cornified and nonciliated mouth lining. Even more unique is the presence of an alveolar capillary network within the epithelium, vascularization stopping either at the basal layer or progressing throughout the membrane to the surface layer of cells. Such a vascular stratified epithelium is highly unusual among the vertebrates. The importance of such an epithelium becomes clear when one remembers that most amphib-

Fig. 7.8 Ciliated epithelium of the frog mouth. Note the mucosal papillae and the mucous cells. X250. (Courtesy of F. O. Craft.)

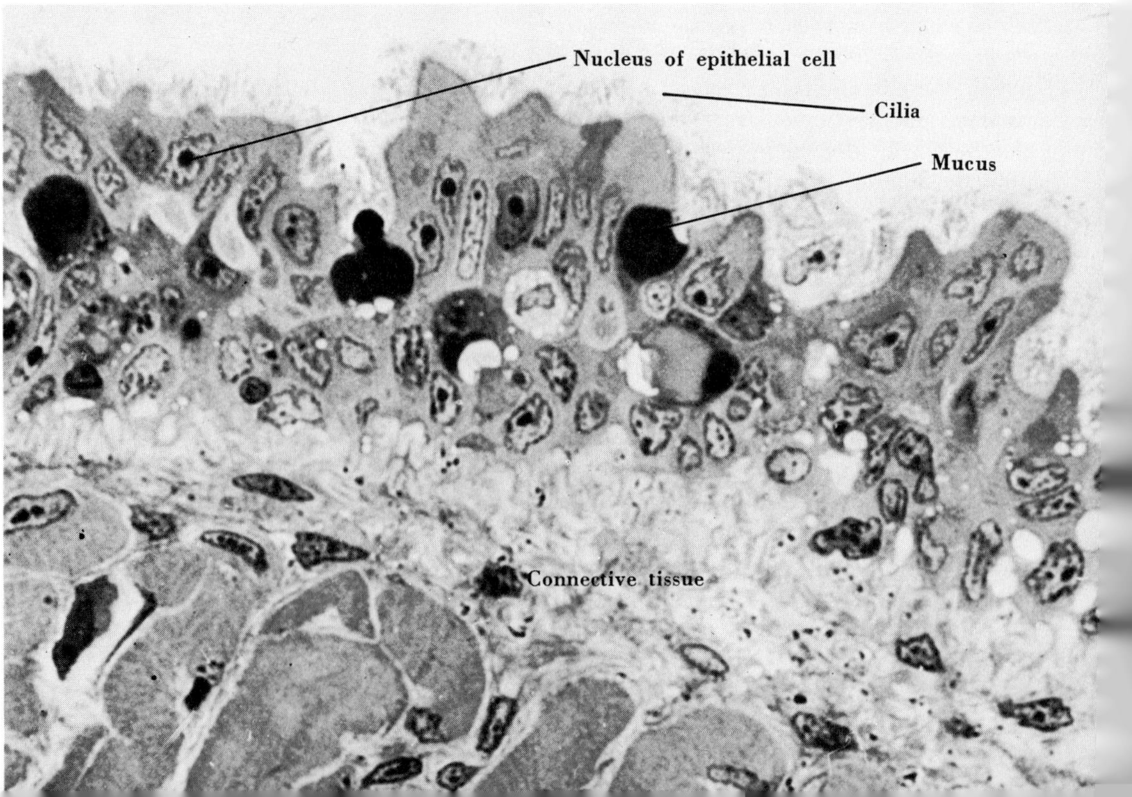

ians have reduced lungs and have adaptively acquired auxiliary methods of environmental gaseous exchange. The one most important method employed by most amphibia is "skin-breathing," and the oral mucosa merely represents a functional continuation of the skin. The relative thinness and vascularity of the membrane allows effective passage of oxygen inward and carbon dioxide outward. By deduction it is assumed that the cilia which are present in most amphibians must play a role in respiration also, because of their lack in the aquatic larval stages and also in those amphibians that retain gills throughout life.

Generally, then, the oral epithelium is thin with noncornified, vascularized, columnar or cuboidal cells. Cilia are usually present except on the summits of the oral papillae (*see below*). Goblet cells and mononuclear or polymorphonuclear leukocytes can be seen interspersed among the epithelial cells. Branched tubular mucous glands can be found dipping into the tunica propria. The secretory cells are cylindrical in shape, with those of the ducts bearing cilia. Usually these glands are considered to be the amphibian's answer to salivary glands, and are named for their location, e.g., *lingual, intermaxillary,* and *pharyngeal.*

The tunica propria, on which the mucosal epithelium rests, contains many nerve bundles and connective tissue fibers. The small blood vessels which vascularize the epithelium appear as a capillary net. The alveolar outpocketings of these capillaries that extend up to the surface and which are so effective in increasing absorptive surface area begin in the tunica propria.

Papillae are found in the amphibian oral cavity just as in that of fish, and, indeed, in all of the vertebrates. Although they are scattered throughout the mouth, they occur most abundantly on the posterior regions of the tongue, and are formed by mounds of connective tissue overlaid with columnar or cylindrical epithelial cells surrounding an epithelial pit. In most amphibians two types of papillae are discernible. Filiform ones are numerous, interspersed with occasional flattened and broader fungiform types. Taste buds are borne only on the fungiform papillae; serous and mucous glands may be found on both papillary types.

The amphibian tongue, though variable in size, can show extensive secondary glandular and muscular growth as compared with that of fishes. The epithelium of the lower surface closely resembles that covering the rest of the mouth. The free surface evidences a transition to a simple cuboidal state, and contains the openings of the numerous branched tubular glands. These multicellular glands of the tongue, containing both mucous and serous cells, often are surrounded by muscle fibers in a forklike fashion. Hence, the muscles, in addition to controlling the movement of the tongue, aid in expressing the glandular secretions. In some amphibians, the striated muscle fibers resemble those of cardiac muscle in that the fiber ends are branched.

The lymphoid tissue of the tongue and mouth cavity is variable in amphibians. In most forms the tongue contains diffuse nodules of lymphoid tissue below the epithelium, and these structures are well drained by lymphatic vessels that usually end in a delicate sac at the base of the tongue. In frogs and toads, homologues of mammalian palatine tonsils can be found, i.e., two large and morphologically distinct masses of lymphoid tissue and invaginated epithelium. Salamanders possess many small epithelial invaginations in the mouth and pharyngeal regions, and these are filled with lymphoid tissue. Often the lymphatic nodules are infiltrated with leukocytes of all types in addition to lymphocytes.

In salamanders and apodans, teeth are always present on both upper and lower jaws, as well as on the palate. In some forms, as the plethodontids, the vomerine teeth may grow backward to form patches on the parasphenoid. Anuran teeth occur only on the upper jaw and vomer, but they may be lost in some forms. Adult amphibian teeth differ from all other tetrapod teeth in being formed of two parts: a basal pedicel and a distal crown. Both are constructed of dentine, and the crown is tipped with enamel or an enamel-like substance. The two halves of the tooth are separated by a zone of weakness composed of fibrous tissue or uncalcified dentine (Fig.

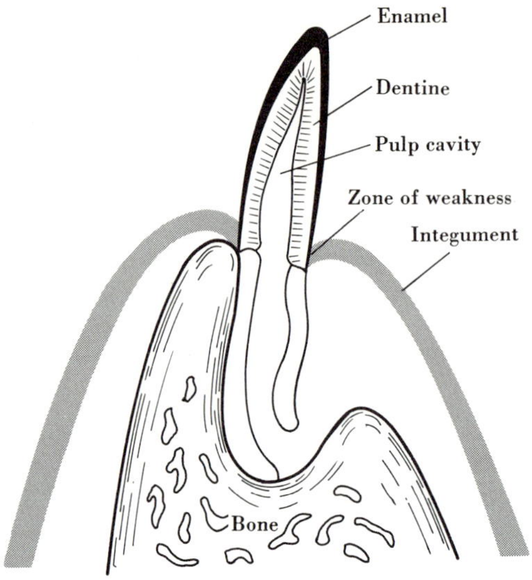

Fig. 7.9 A diagram of a longitudinal section through an amphibian tooth. Note the zone of weakness.

7.9). In apodans, at least, this zone of weakness may be quite complex structurally and form a hinge that is probably capable of movement. The adaptive value of this condition is obscure; it may have been more functional in some long-gone ancestor.

Esophagus and Stomach

The amphibian esophagus is a thin-walled tube leading to the now ever-present stomach. As seen in the fish, the mucosa contains many longitudinal folds which project into the lumen. The epithelium consists of one or two layers of cuboidal or columnar cells which are usually ciliated on the free surface. Goblet cells are abundantly scattered among columnar cells, and in some cases, such as in the salamanders, the epithelium shows a continuation of the vascularization seen in the mouth. Simple or branched tubuloalveolar glandular invaginations consisting of mucous and/or serous cells are found in the esophagus of some amphibians but not all. The lamina propria of many species is well-populated with lymphatic nodules; in toads, the surface epithelium invaginates to form the lining of a cavity that forms the center of the nodule.

The smooth muscle layer of the mucosa is not a constant feature in the esophagus. If present at all, it is usually restricted to the lower third of the tube.

Surrounding an areolar submucosa is the muscularis. Generally the circular fibers are the more strongly represented, the longitudinal fibers forming a distinct layer only at the lower end. The serosa consists of mesothelium covering fibrous connective tissue together with the normal complement of blood vessels and nerves.

The amphibian stomach, though a constant feature, appears as a simple, saclike dilatation of the digestive system. Histologically it closely resembles the one seen in the fishes (Fig. 7.10). The nonciliated surface epithelium consists of clear-bordered cells filled with a mucoid secretory product. Rugae are present, the surfaces of which are dotted with gastric pits extending down through the mucosa for about one-fifth of its thickness. At the junction of the pit with the fundic gland, true mucous neck cells can be found (Fig. 7.11). These cells contain a nucleus which has been flattened at the base of the cell by the acidophilic secretory product which fills the rest of the cell. As in the fishes, the gland proper contains one kind of granular, cuboidal, or pyramidal cell which acts in the dual role of secreting pepsin and HCl. These cells, serous in nature, appear to have basophilic granules. Most of the volume of the upper mucosa consists of fundic glands.

The histologic characteristics of the pyloric glands also mirror the condition seen in the fishes. The cells of these glands are mucoid and resemble the surface epithelial cells. The pits leading into pyloric glands are shorter than those of the fundic glands, and are usually confined to the pyloric region of the stomach.

The tunica propria, found between the bases of the gastric glands contains lymphoreticular tissue and many leukocytes. Surrounding the tunica propria is a well-defined muscularis mucosae with its two characteristic layers. Passing exteriorly from the mucosa is an areolar submucosa, thick circular and thin longitudinal muscularis, and typical serosa.

Small Intestine

True villi are not found in the amphibian intestine. Instead, these animals rely on greatly increased intestinal length and branched mucosal folds for raising the secretory and absorptive capacity of the surface. The epithelium lining the intestinal lumen contains the same major cell types such as are seen in the fishes: goblet cells and columnar cells with acidophilic, granular cytoplasm and a striated border. Cilia, found in some of the fish, are not seen in the amphibian epithelium (Fig. 7.12).

The lamina propria of the mucosa consists primarily of lymphoreticular tissue. A typical double-layered muscularis mucosae may be present, although in *Necturus* and *Rana* it is quite poorly represented.

The submucosa is areolar in nature and overlies a two-layered muscularis. Surrounding the muscularis is a typical serosa.

Large Intestine

The hindgut in the Amphibia consists of a short, broad tube. It can be distinguished from the small intestine by the lack of high slender folds. Short, broad folds may be present, but often they are so low as to appear nonexistent under low magnification. The epithelial lining primarily consists of goblet cells, with columnar cells scattered throughout. The tunica propria and muscularis mucosae are poorly defined. Similarly the submucosa and muscularis are thin except in the posterior regions where each of these two layers may be represented more distinctly.

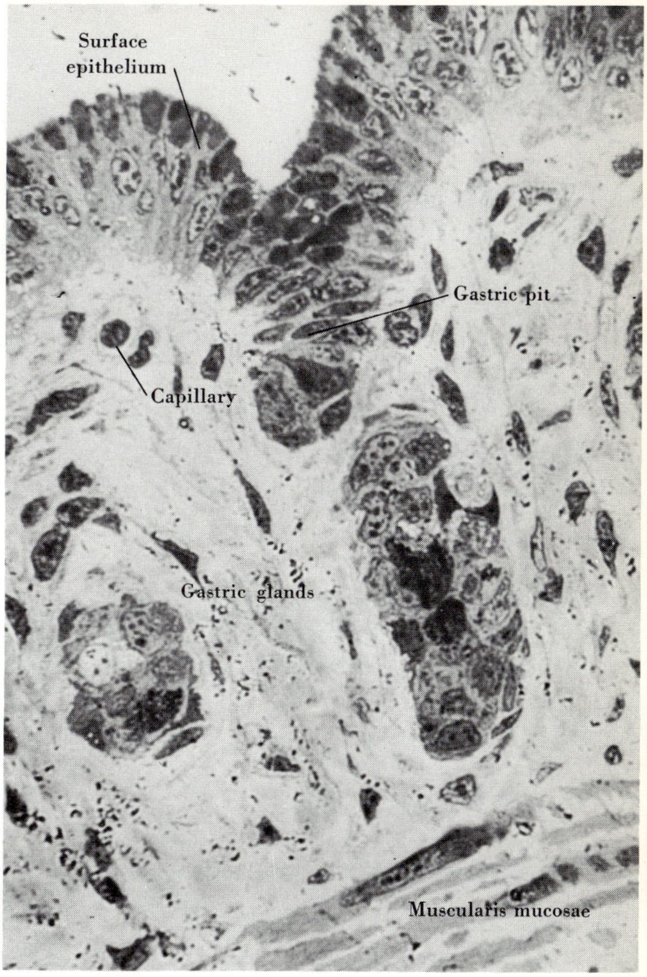

Fig. 7.10 The mucosa of the frog stomach. Note the shape of the mucous accumulations in the surface cells. A portion of a gastric gland is visible. X250. (Courtesy of F. O. Craft.)

Cloaca

The cloacal region contains an epithelium that is composed of two or three cuboidal cell layers and which gradually blends into the stratified covering of the body. The mucosa may be thrown into several low, longitudinal folds. Instead of smooth muscle bands, the fibers appear to be randomly grouped in bundles; at the posterior end of the cloaca, these bundles give way to skeletal muscle fibers.

THE DIGESTIVE SYSTEM OF REPTILES

When we introduced discussion of the fish digestive system, mention was made of variability presumably related to diet. A study of the histological picture in reptiles necessitates a repetition of this statement—perhaps with even stronger emphasis. All of the fish are uniform to the extent that all must fashion a

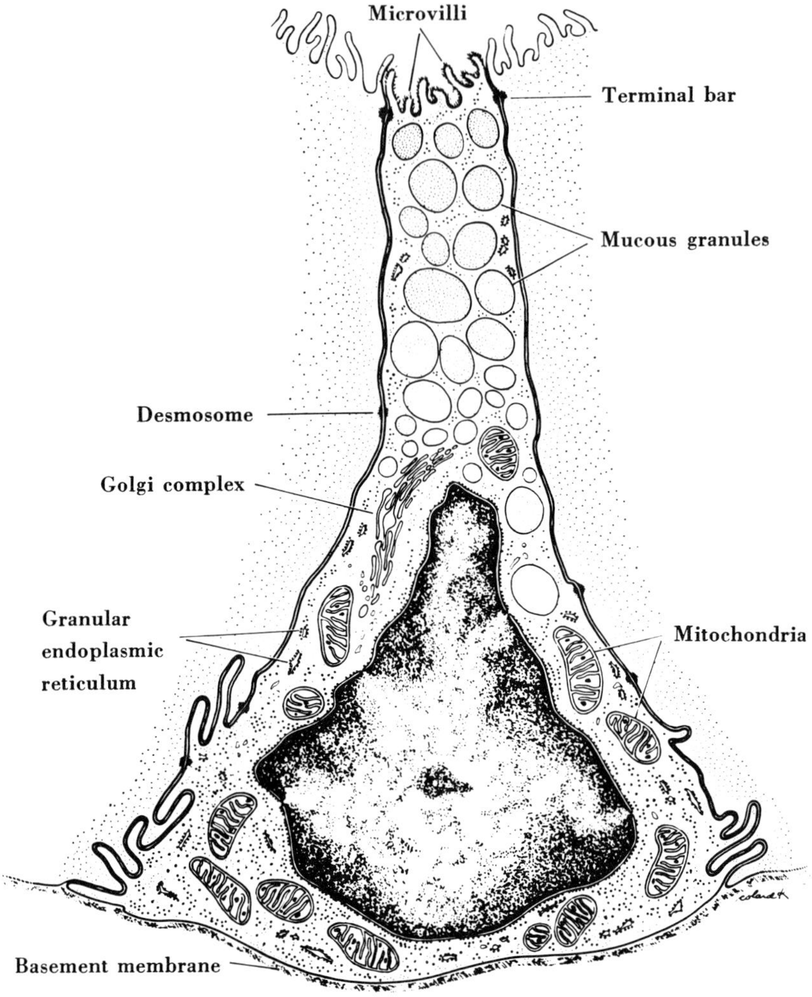

MUCOUS NECK CELL

Fig. 7.11 A typical vertebrate mucous neck cell as diagrammed from electron micrographs of the bat stomach. (After S. Ito and R. J. Winchester. 1963. J. Cell Biol. 16:541–579.)

mode of life which is consistent with a watery environment. The reptiles, however, need not show even that small degree of consistency. A few minutes of thought on the life of the grass snake versus that of the water turtle should serve to make the point. Additionally, we encounter the enormous variety of external form. Obviously the gross picture of the digestive system in a long slender snake will differ from that in a more compact turtle, and this variation, too, can lead to differences in histological structure.

Oral Cavity and Pharynx

The character of epithelium lining the oral cavity differs from species to species and often from region to region within the same individual. Snakes often exhibit a simple, ciliated, columnar epithelium. In the anterior portion of the lizard mouth, one can see a thick, stratified, squamous epithelium with true cornification of the surface layers that blends into a stratified condition in which the surface

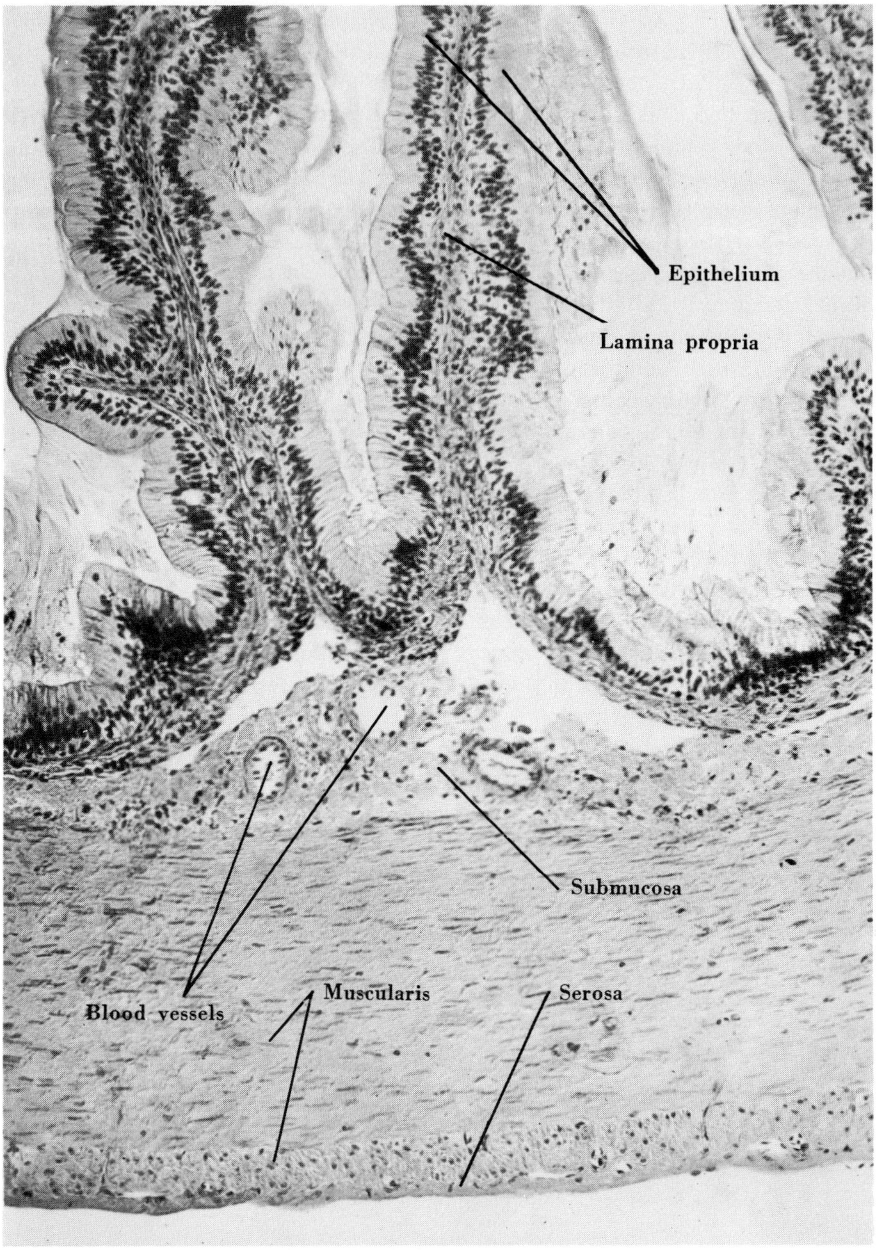

Fig. 7.12 Cross section of the frog small intestine. X125.

layer is ciliated and columnar. Young turtles often possess a simple, columnar lining whereas older ones may show the stratified squamous condition. All reptiles are alike in the possession of great numbers of goblet cells interspersed throughout the membrane, and the presence of typical vertebrate papillae. Many reptiles have the same kind of branched tubular mucous glands (salivary glands) seen in the Amphibia—although, in some reptiles, such as the lizard, these glands may be massed to form definite gross structures. In some cases serous elements may be present as well.

As compared with the Amphibia, the reptilian tongue evidences additional median and lateral tissue, often muscular in nature. The striated muscle fibers appear either in bundles or individually. Some lizards and snakes have a split or forked tongue, other reptiles have a unit structure. The epithelial covering may be of the stratified, squamous type (land turtles and some snakes), pseudostratified (some lizards), stratified columnar (some other lizards), highly cornified (snakes), and simple columnar (*Sphenodon*). One would undoubtedly find all gradations of these conditions simply by examining a large enough sample. Goblet cells are most numerous on the tongue in those reptiles having a simple epithelium. Multicellular alveolar glands can be found in some reptiles, and papillae, some bearing taste buds, are typical vertebrate types.

Variation as a constant feature also applies to reptilian dentition. In some reptiles the teeth are fused to the jaw; others have teeth held in sockets by a cementlike material not unnaturally called cementum. The tooth proper is composed of a porous dentine covered by a thin film of enamel or enamel-like substance. Shapes can vary from the sharp cones seen in the Crocodilia to the long fang characteristic of some snakes. Turtles lack teeth altogether, but do have plates of cornified epithelium around the margin of the mouth.

The fang of poisonous snakes is an interesting adaptation. These teeth are grooved to a greater or lesser degree. In the rattlesnakes the groove is so deep that its margins have grown together, forming a complete casing in which lies a closed canal. The canal is indirectly connected to one of the salivary glands where the poisonous secretion is produced. As with teeth in other living reptiles, the fang is replaceable.

Esophagus

The typical reptilian esophagus is lined with one or two layers of cuboidal or columnar ciliated cells interspersed with goblet cells (Fig. 7.13). There is a general lack of esophageal glands. Lymphoid nodules are abundant in the underlying lamina propria, often

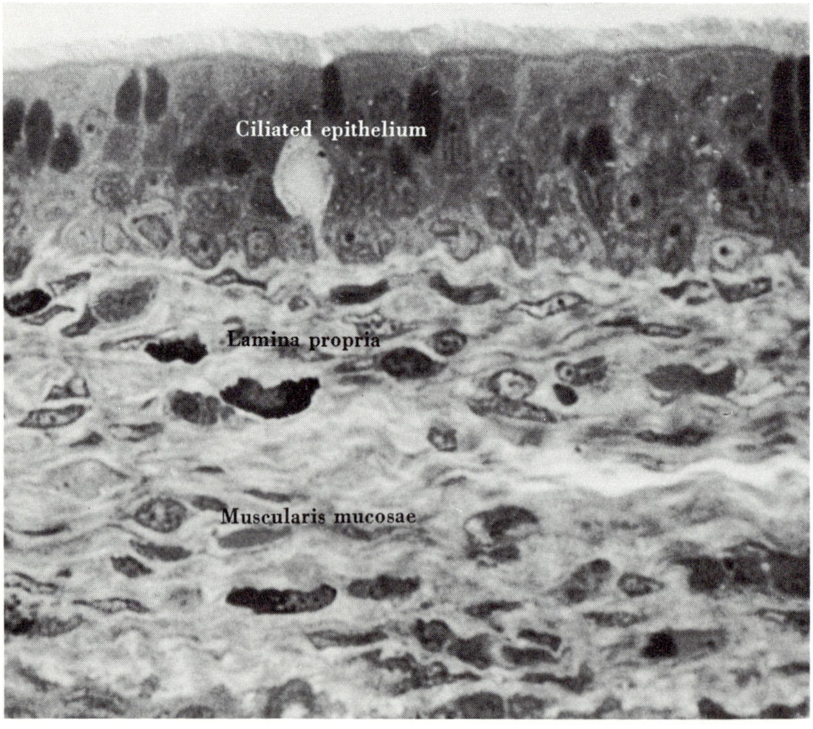

Fig. 7.13 The mucosa of the turtle esophagus. X250. (Courtesy of F. O. Craft.)

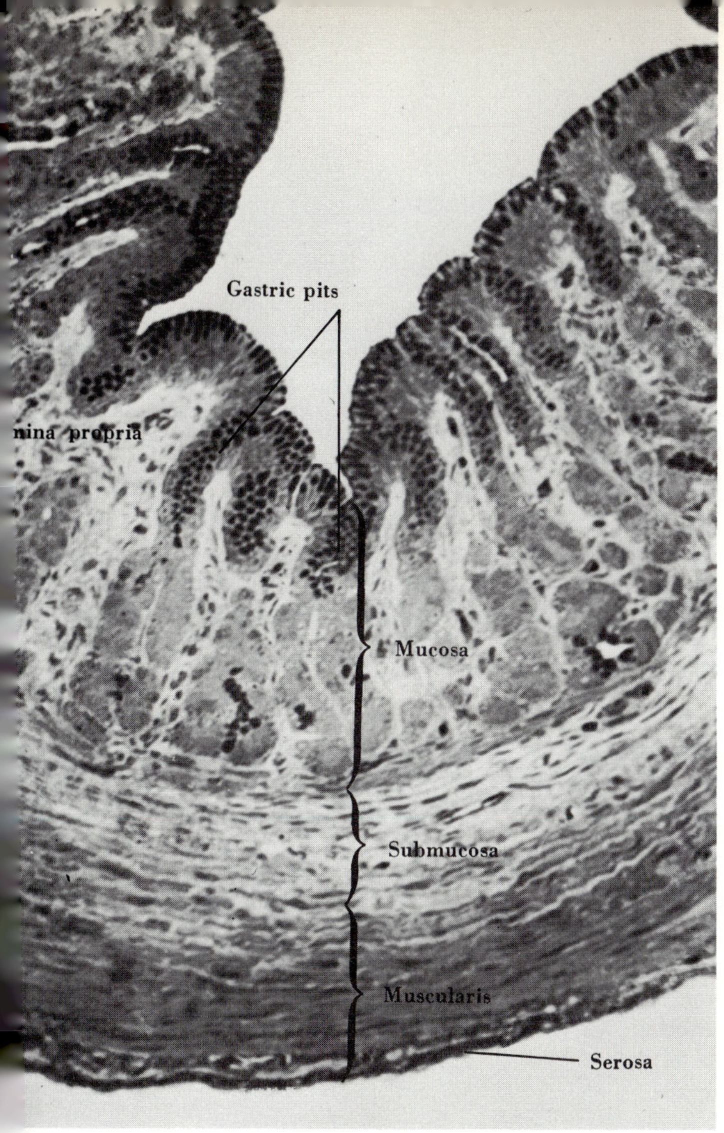

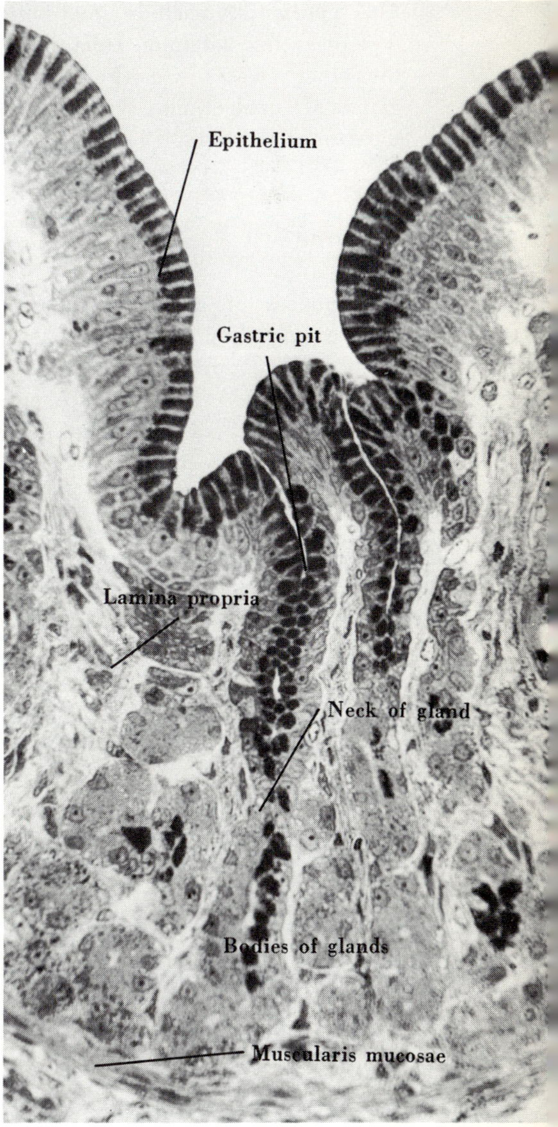

Fig. 7.14 A low-power view, *a*, of the entire stomach wall of the turtle (*P. elegans*). X200. A higher-power view, *b*, of the mucosa and gastric glands. X400. (Both photomicrographs courtesy of F. O. Craft).

obscuring all other detail. The muscularis mucosae is scanty; it may be completely lacking in the anterior portion and represented by only a few smooth muscle fibers posteriorly. The entire mucosa, together with the submucosa, is thrown into high folds from which smaller secondary mucosal projections occur when the esophagus is not distended.

The submucosa, muscularis, and tunica adventitia are similar to that of the amphibian.

Stomach

The reptilian stomach is lined with the same type of gastric columnar cells that are seen in the other classes. Gastric pits and gastric glands are present, closely resembling those of the Amphibia (Fig. 7.14).

Intestine

The mucosa of most reptilian intestines is thrown into deep folds, the bases of which often include portions of the submucosa as well (Fig. 7.15). The surface epithelium is nonciliated and consists of tall columnar cells having striated borders. Goblet cells are represented throughout the epithelium, but they occur most densely toward the posterior regions. The lamina propria contains typical reticular connective tissue in addition to pockets of lymphoreticular tissue and solitary lymph nodules. The muscularis mucosae can vary; in some reptiles the normal two-layered state is seen, whereas in others this layer is represented by scattered groups of smooth muscle fibers, some circularly and some longitudinally arranged. The outer small intestinal layers mirror the picture seen in the Amphibia.

In the turtle the distinction between small and large intestine is indistinct. In other reptiles, as the lizard, the large intestine is marked by a progressive decrease of shorter, broader mucosal folds. The population of goblet cells is often seen to rise abruptly and the muscularis mucosae to become more evident.

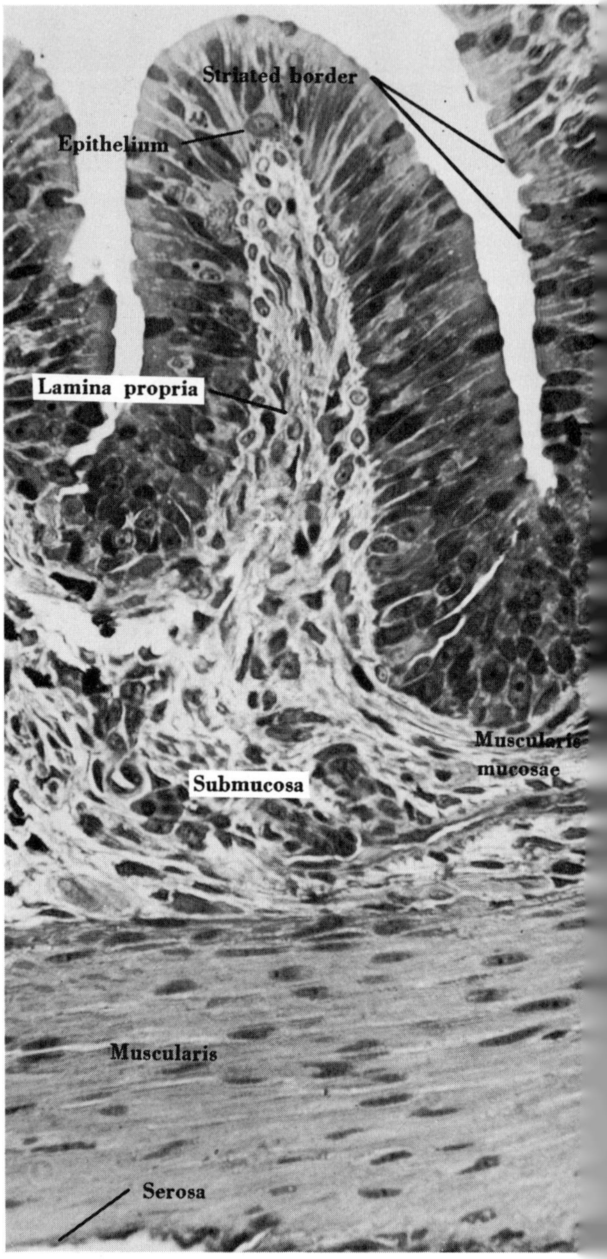

Fig. 7.15 The intestinal wall of the turtle (*P. elegans*). In this preparation the striated border is visible only in a few regions. X400. (Courtesy of F. O. Craft.)

Cloaca

The histology of the cloaca is intermediate between that of the gut proper and that of the skin. The epithelial membrane shows the transition quite nicely, graduating from columnar cells and goblet cells to the stratified condition. Smooth muscle fibers are similarly replaced by skeletal ones in a gradual manner, and the connective tissue elements blend into the dermis.

THE DIGESTIVE SYSTEM OF BIRDS

The digestive system of birds exhibits two major categories of adaptive distinctions. The first category includes specific adaptations related to diet. We have not been primarily concerned with these details in the other classes, but in birds, especially, such adaptations are so extensive that at least some deserve mention. The other major group of histological distinctions bears directly on adaptation to flight; these will be of utmost concern.

Most of the uniquely avian digestive digressions are located toward the anterior end of the system and are associated either with weight reduction (a necessary extension of the capacity to fly) or with assuring balance (also a necessity for the smooth workings of the flight musculature). Hence the relatively heavy tooth complement of the other classes has been eliminated and a light horny beak has been substituted. The chewing apparatus has moved closer to the center of gravity; mastication or its equivalent is taken over by a part of the stomach. Additionally the jaw skeleton has been lightened and the jaw musculature reduced. Without grinding teeth the esophagus must be able to handle larger chunks of food. This structure, usually quite long in birds, is also very wide. Food storage becomes an important problem because of great energy expenditure during flight; a saclike dilatation of the esophagus, *the crop,* has evolved to insure a constant food supply to the stomach. Histological changes accompanying these gross changes will be noted below.

Oral Cavity and Pharynx

The mucosa of the mouth and pharnyx resembles the skin. Formed of stratified squamous epithelium, a thick stratum corneum exists in such areas as the oral roof, the surface of the tongue, and the pharyngeal roof. In such regions the vascular lamina propria mounds upward to form cores for posteriorly directed papillae. The lamina propria is often obscured by diffuse and/or modular lymphatic tissue; no muscularis mucosae can be seen.

The tongue may assume a variety of diet-related shapes, the extreme being the tubular tongue of many insectivores. Generally the tongue contains no intrinsic musculature. The mass of the thick tongue of such birds as the parrot, duck, and goose is composed of fat, tubular mucous glands, and blood vessels. In the two latter birds large rasping barbs are found on the surface. As mentioned above, the dorsal surface is covered with papillae. Filiform and fungiform types are represented with taste buds occurring on the fungiform papillae.

Seven sets of salivary glands can be found, although not all birds possess all seven sets. The degree of development seems to be related to habitat. Aquatic birds eating slippery foods have a reduced complement, seed eaters generally have the most highly developed glands. The glands have been designated according to their location as (1) sublingual, (2) lingual, (3) mandibular, (4) palatine, (5) angle of the jaw, (6) cricoarytenoid, and (7) sphenopterygoid. Usually they are tubular, either simple or compound. Occasionally tubuloacinar or compound acinar structures can be found. Salivary secretions are generally mucigenic; however, in a few birds, such as turkeys and pigeons, a salivary amylase has been detected. In many birds a nest building adhesive is present in the secretions produced during the breeding season.

Esophagus and Crop

The avian esophagus gains a prominence seen nowhere else in the vertebrates; not only used as a passage to the stomach, it often acts as a food-storage organ. In many birds a crop aids in food storage, and acts additionally as a resonating structure for acoustical purposes, as a "milk" producer in doves and pigeons, and as a possible chemical digestive organ in the pigeon and domestic fowl.

The diameter of the esophagus and the degree of mucosal folding vary with the diet. Both of these characteristics increase in birds that eat increasingly larger food masses and in which the esophagus acts as a storage depot. The mucosal epithelium is stratified squamous with a degree of cornification that mirrors diet. The thin, fibrous lamina propria may or may not be formed into papillae, and may, in fact, be obscured by quantities of diffuse and nodular lymphoid tissue. The muscularis mucosae is thick, containing longitudinally oriented smooth muscle fibers (Fig. 7.16).

Surrounding a thin submucosa is a double-layered muscularis. The inner circular muscle fibers are generally well represented; the opposite is the case for the outer, longitudinally oriented layer.

Esophageal glands exist in a wide variety of conditions. Two general parameters, shape and location, exist. The glands may vary from simple tubular to compound tubular to simple or compound tubuloalveolar. They may be confined to the epithelial layer, may extend a short distance into the lamina propria, or may

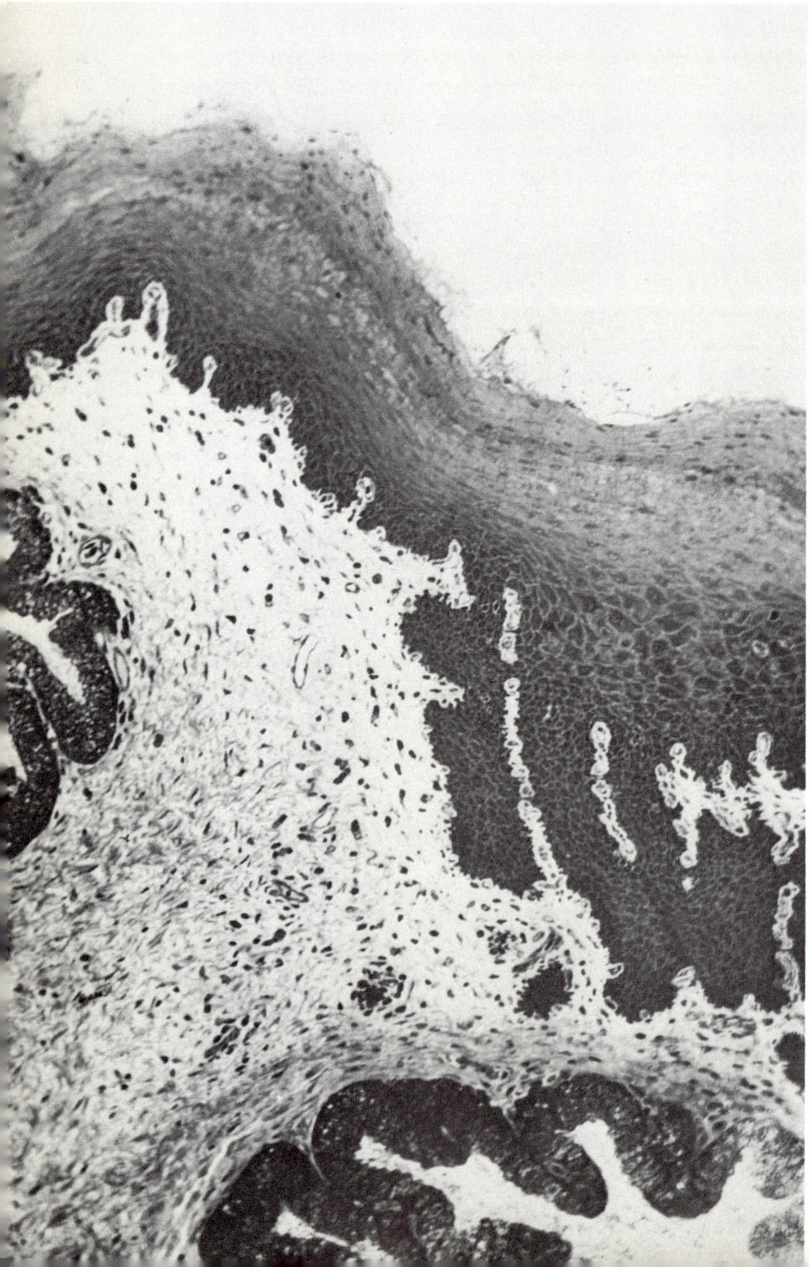

Fig. 7.16 A section through the chicken esophagus. Note the skin-like condition of the cornified epithelium and the bodies of large epithelial glands that empty their secretions into the lumen. X100. (Courtesy of F. O. Craft.)

be primarily in the lamina propria. Practically all combinations of shape and location can be found, but it should be noted that rarely do the glands penetrate to the submucosa. This, we shall see, is a mammalian characteristic.

It would not seem that the owl, woodpecker, penguin, gull, goose, duck, and petrel have very much in common. However, the distinguishing characteristic of this group is the collective lack of a crop. Most other birds have an expandable crop, either as a fusiform expansion of the lower esophagus or as a lobular diverticulum. Generally the crop presents an histological picture similar to the rest of the esophagus except for a reduced number of glands and an increase in concentration of lymph nodules (Fig. 7.17). The glands of the pigeon and domestic fowl are thought to have both mucous and serous functions. There have been conflicting reports of amylase, invertase, and proteases originating in these glands.

In contrast is the "milk secretion" in the lateral lobes of the crop of the pigeon and dove. In reality, this secretion results from desquamation of fat-laden surface cells. The

Fig. 7.17 A section through the chicken crop. Note the desquamating cells and the thick, cornified epithelium. X400. (Courtesy of F. O. Craft.)

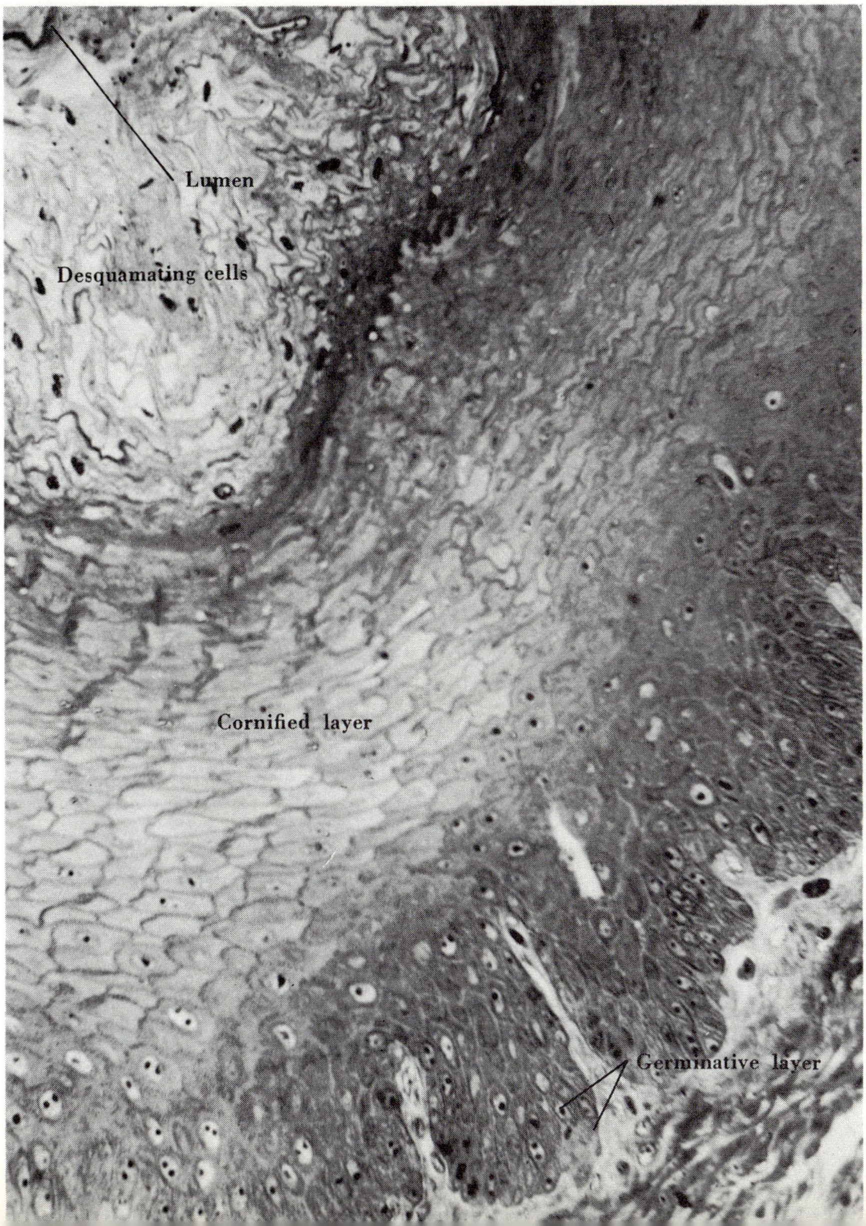

so-called milk, produced by both males and females, contains no sugar or casein, being composed rather of fat, lecithin, and proteins.

Stomach

The bird stomach is a dual structure. The thinner-walled anterior section, the *proventriculus,* is glandular, and functions in food storage, protein digestion, HCl secretion, and possibly in some fat breakdown. The posterior region, the *ventriculus* or *gizzard,* is highly muscular and acts as the tooth replacement. In addition to its food-grinding function, the gizzard has a role in storage, and, to some degree, in protein digestion. In a few birds, such as the heron, stork, and cormorant, a third section, the pyloric stomach, is evident.

PROVENTRICULUS

The mucosa of the proventriculus is thrown into many long folds. The epithelial lining consists of typical, clear-bordered stomach lining cells. Gastric pits are again present at the terminus of short mucus-secreting simple tubular glands. These structures extend into the lamina propria. In addition, as a result of mucosal folding, papillae are present either in distinct bands or are uniformly distributed. They represent the terminus of compound glands which are highly complex in nectivores and herbivores, but less so in carnivores. The glands have a compound tubular or tubuloalveolar or cylindrical or spherical shape. Each possesses a collecting chamber into which secretions from the secretory ducts are poured, and from which leads a duct to the lumen of the proventriculus. The cells of the collecting duct are mucus-secreting, and closely resemble the mucous neck cells of amphibians and reptiles. The remainder of the gland is composed of cuboidal cells with homogeneous acidophilic cytoplasm. They are thought to be analogous to the pepsin- and HCl-secreting cells of the mammalian stomach.

A delicate lamina propria consists of lymphoreticular tissue with a scattering of eosinophilic granulocytes. The muscularis mucosae is double layered. The inner circular layer lies beneath the simple glands; the outer longitudinal fibers are interposed between the compound glands and submucosa.

The proventricular submucosa is scanty. Surrounding it are the thick, inner circular and thin, outer longitudinal layers of the muscularis, and, finally, a typical serosa.

GIZZARD

The ventriculus or gizzard is noteworthy in two respects: (1) the musculature of this organ can generate one of the highest tensions recorded for smooth muscle, and (2) the inner lining of the gizzard is noncellular and tough (Fig. 7.18). These peculiarities are exhibited to varying degrees that are, not surprisingly, related to diet. Herbivores and grain eaters have the most highly developed gizzards, followed by omnivores, nectivores, carnivores, and, finally, fruit eaters.

The epithelium of the mucosa consists of columnar cells, crypts, and a dense array of simple to branched tubular glands. The secretory cells are cuboidal or columnar with a granular, acidophilic cytoplasm and centrally located nucleus (Fig. 7.19). Because some protein digestion begins in the gizzard, and because the acidity is high, it is assumed that acid and proteolytic enzymes are secreted by the glands. In addition, the glands function in producing a material known as koilin. Chemically slightly different from keratin, the koilin lining is diagnostic of an avian gizzard.

When viewed under a microscope the koilin lining appears to have a horizontally laminated structure subdivided by clear, perpendicular striations. The laminations contain desquamated or molted epithelial cells, debris, and regurgitated bile pigments. The perpendicular striations are formed of the secretory products of the epithelial glands.

The remainder of the mucosa consists of a lymphoreticular lamina propria and scanty muscularis mucosae. Surrounding the mucosa is a layer of concentrated elastic and collagenous fibers, the *stratum compactum,* and following that a thin, areolar submucosa.

The muscularis of the gizzard is unusual because the thickness can vary temporally in relation to the hardness of the food, and, additionally, because of the extreme thickness that

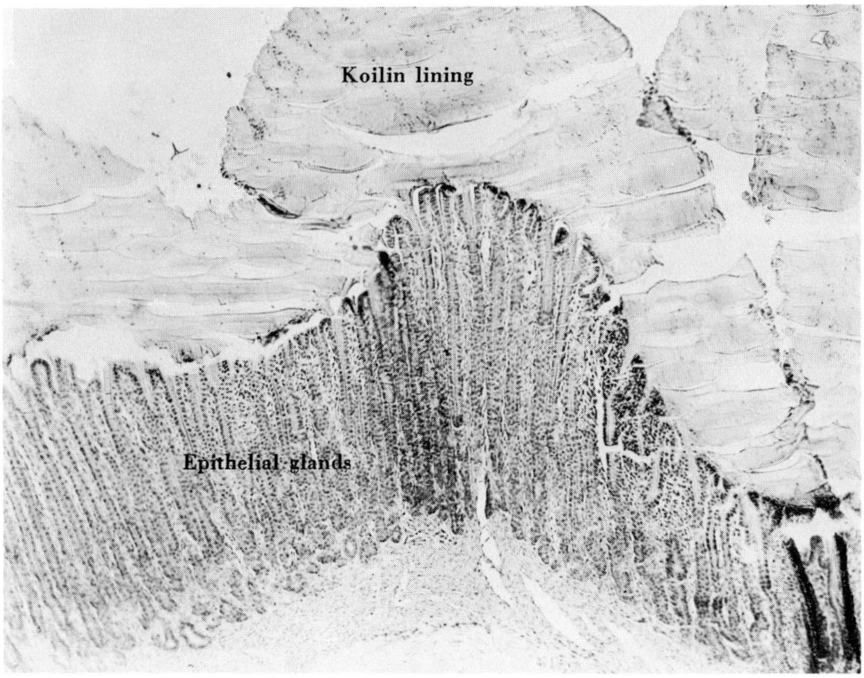

Fig. 7.18 The mucosa and koilin lining of the pigeon gizzard. X100.

can be reached. This ability to hypertrophy and regress is specifically a property of the circular fibers, the longitudinal ones being poorly represented. A typical serosa surrounds the muscularis.

Intestine

The intestinal mucosa of birds contains the same kind of epithelial lining as we have met before: simple columnar cells and goblet cells (Fig. 7.20). However, the bird intestine shows an abrupt change from the preceding vertebrates, and more closely resembles the mammalian intestine in two important respects. The mucosa dips down toward the underlying layers to form *crypts of Lieberkühn,* and extends upwards in fingerlike or leaflike projections as *villi.* A villus contains a core of lamina propria surmounted by an epithelial covering and enclosing a blind, terminally distended lymph sac (a lacteal) and a blood capillary network. Villi, characteristic of birds and mammals, provide a greatly increased absorptive surface, and because a few muscle cells from the muscularis mucosae extend upward into the villus, these structures are individually mobile.

Fig. 7.19 An oblique section through the chicken gizzard. You are looking into the glands that produce the koilin lining. X100. (Courtesy of F. O. Craft.)

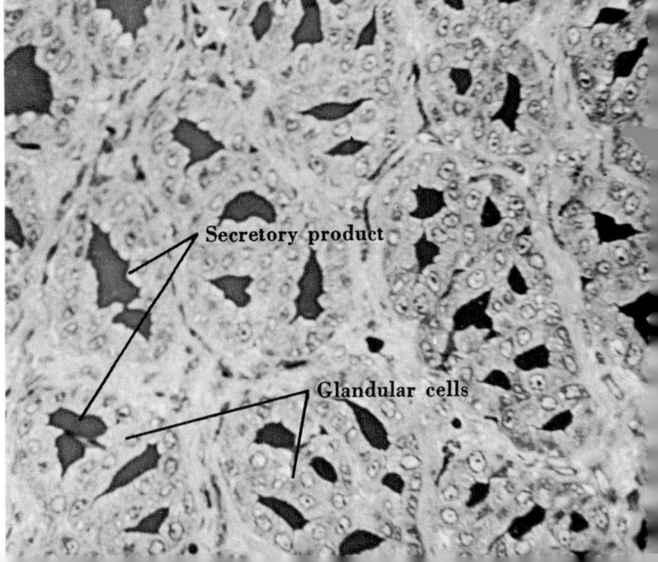

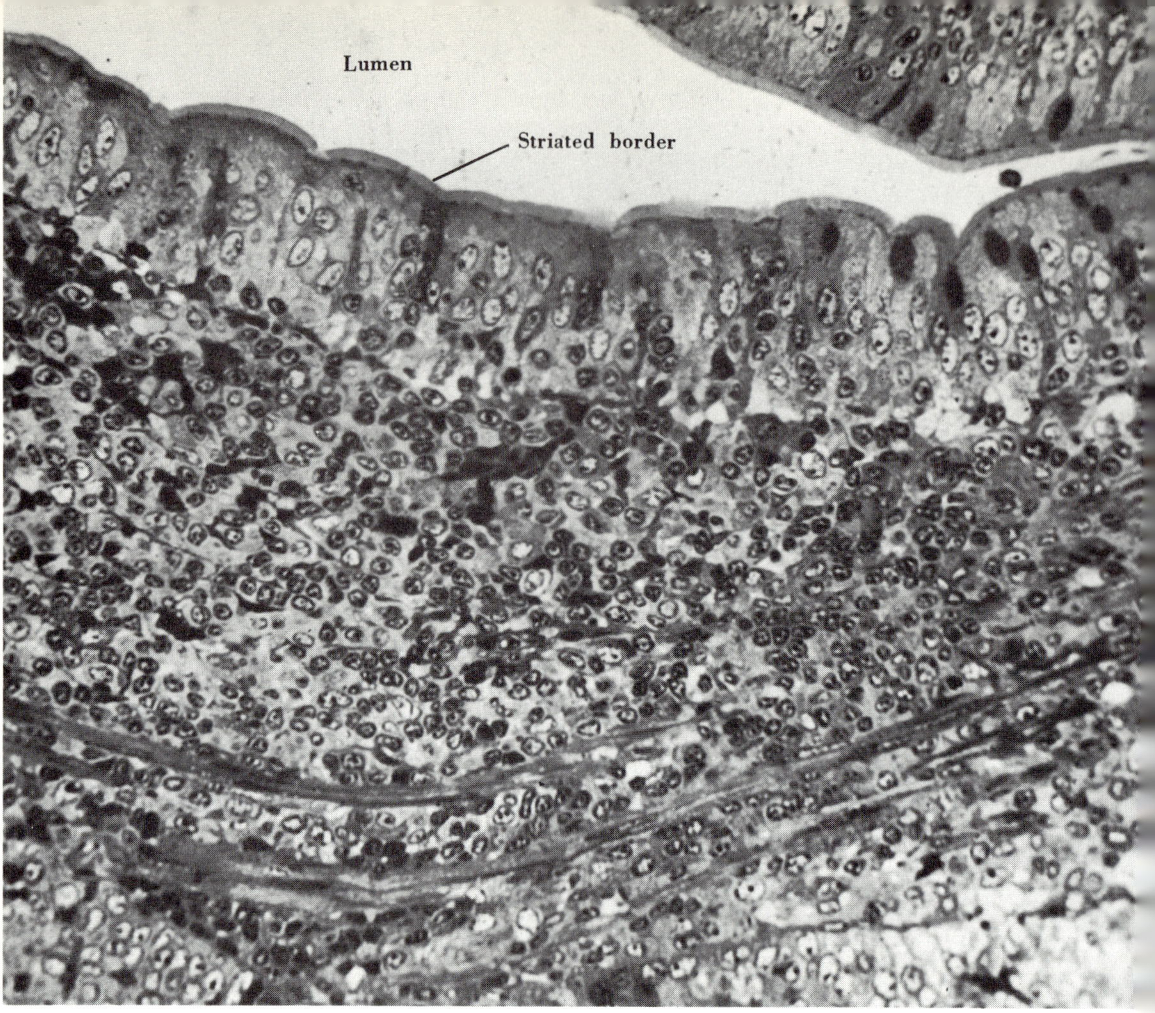

Fig. 7.20 The mucosa of the chicken intestine. Note the striated border. X400. (Courtesy of F. O. Craft.)

The crypts of Lieberkühn occur between the bases of the villi, and act as the opening for short single or branched tubular glands. In birds the secretory cells are columnar with acidophilic, granular cytoplasm, and are presumed to be the source of definitive intestinal enzymes. Heretofore the enzymatic activity in the intestinal lumen was attributed to enzymes produced in the pancreas and merely activated in the intestine. Intestinal enzymes of the birds have been found which hydrolyze starch, disaccharides, most proteins, and fats.

The lamina propria is reticular in nature, with many lymphocytes and lymph nodules. A stratum compactum may be present. The muscularis mucosae is represented by longitudinal smooth muscle fibers. The submucosa tends to be fibrous and is surrounded by a thin, inner circular and a thin, outer longitudinal muscularis. The serosa is typical.

At the junction of the small and large intestine, intestinal ceca may be present. Usually the histology is similar to that in the small intestine, but in some herbivores, the ceca are enlarged with masses of lymphoid tissue.

The large intestine differs from the small only in the decrease in number and height of the villi and the increase in goblet cell concentration.

Cloaca

The junction of the large intestine with the cloaca is notable for an abrupt loss of villi and a sharp increase in number of goblet cells. The epithelium gradually changes to that of the skin as the smooth muscle fibers are replaced by skeletal ones. In young birds a dorsal diverticulum occurs at the intestinal-cloacal boundary. This structure, the *bursa of Fabricius,* is discussed in Chapter 9.

THE DIGESTIVE SYSTEM OF MAMMALS

The mammalian digestive system generally follows the plan seen in the lower vertebrates. A few major modifications can be found, but these may, in some cases, represent the greater interest and accumulated study given the mammals. In contrast to the other vertebrates the mechanisms of enzymatic secretion and food absorption have been fairly well studied so that definitive structures and processes can be identified as characteristic. At some future date similar interest, especially on the part of electron microscopists, may come to focus on the other classes, resulting in structures which we now say are new in the mammals being identified in other classes as well.

Oral Cavity and Pharynx

The gross anatomy of the anterior end of the digestive system is more complex in the mammals than in the other vertebrates. Muscular, mobile cheeks and lips make their appearance —as does a complex, muscular tongue.

The boundary between the integument and oral mucosa is a variable one. In rodents a skinlike condition, replete with hair, is seen on most of the lip surface; in dogs, cats, sheep, goats, and other animals, external encroachment of a mucosal condition obtains.

Generally, the mucosal epithelium is composed of stratified squamous cells (Fig. 7.21). A few, many, or no layers of cornified cells may be found in various parts of the mouth; however, regions subject to abrasive wear have thickly cornified surfaces in all mammals. These regions include the inside of the lips, dental pads of ruminants, dorsum of the tongue, cheeks, and the roof of the pharynx. Pigment granules may be found in the epithelial cells of all of these areas—often paired with cells containing translucent eleiden. In the dog, rabbit, and other mammals, a distinct stratum granulosum may be found, reminiscent of the structure of the skin.

Especially in abraded areas, but often in smooth regions as well, many papillae are found. These have a core of highly vascular lamina propria and a well-cornified stratified

Fig. 7.21 A section of the human buccal mucosa. Note the great numbers of glands lying within the connective tissue. The cells of the epidermis are living and not keratinized. X100.

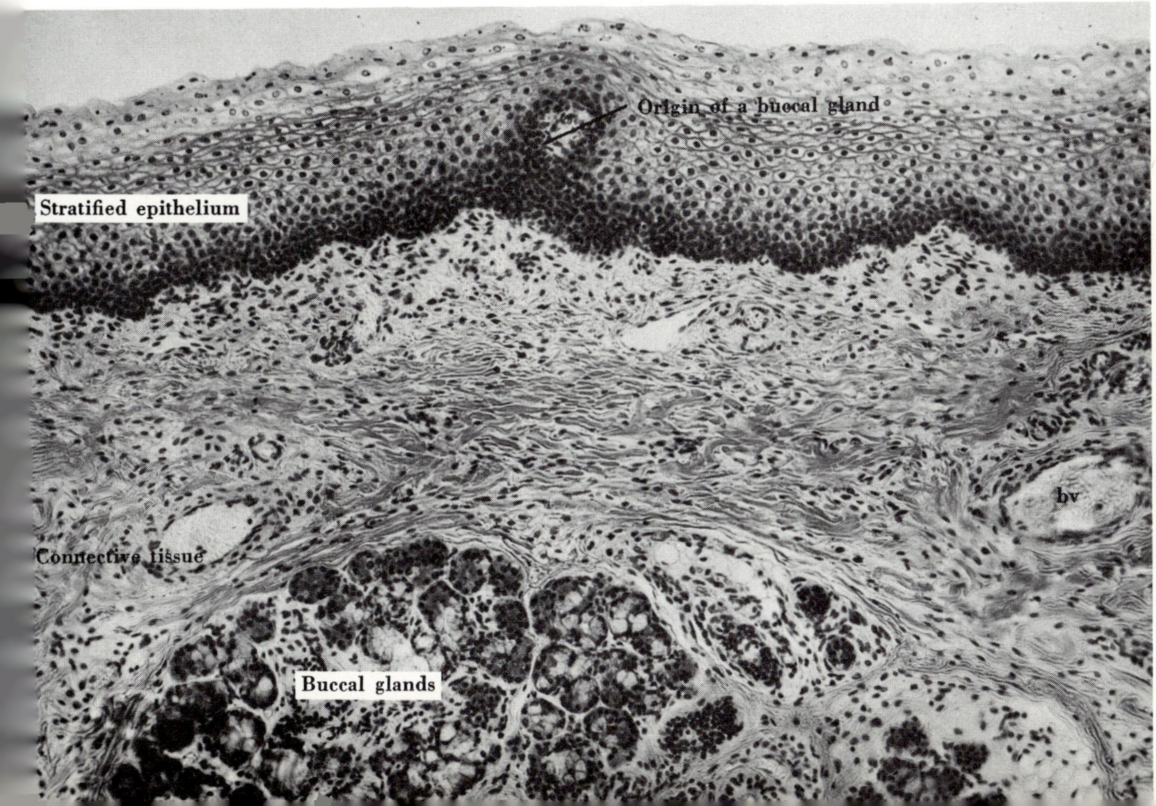

squamous covering. Serous and/or mucous glands may be found dipping into the lamina propria, and emptying on to the surface. No muscularis mucosae is present, but a thick, loose fibroelastic submucosa is in evidence, attaching the mucosa to underlying structures in such a manner as to preclude frequent biting and injuring of the mucous membrane. In man a distinct submucosa is lacking on the hard palate. The submucosa of the swine is noteworthy for the unusual presence of solitary oral lymph nodules. This is perhaps a continuation of the comparatively great amount of lymphoid tissue found in all parts of the digestive tube of the animal. Wild pigs have not been much studied, and the predisposition of the digestive tract of domestic pigs to acquire lymph nodules may be a reflection of their rather unappetizing diet. The oral submucosa of the horse, an animal widely used for preparation of antisera, is unusual in that it includes structures which look very much like the Hassal's corpuscles of the thymus (see Chapter 9).

Mammalian teeth consist of an enamel-covered dentine crown and one to many roots of dentine that fit into depressions or sockets in the jaw. A layer of cementum, thickest in herbivores and thinnest in carnivorans, forms a tooth covering beginning at the gum (a thick mucosal and submucosal fold over the jaw bone) and extending to the basal root canal (Fig. 7.22). It is through the root canal that the pulp enters the interior of the tooth. A thin connective tissue membrane, the *periodontal membrane*, surrounds the cementum and is reflected back as the lining of the jaw socket. The mammalian tooth is thus held in place, but is not fused to the jaw as are the teeth of many of the lower vertebrates.

Most mammals contain teeth differentiated in shape (heterodont) but in some of the Cetacea the homodont condition prevails. In other whales a uniform dental picture is obtained by the lack of teeth in the adult.

Tooth succession is variable among the mammals. The Edentata (such as the sloth and armadillo), so named because they lack teeth on the premaxillary bone, usually have but one set of nonenameled teeth for life. Most mammals, including man, have two succeeding sets of teeth. The elephant, the "Texas" of

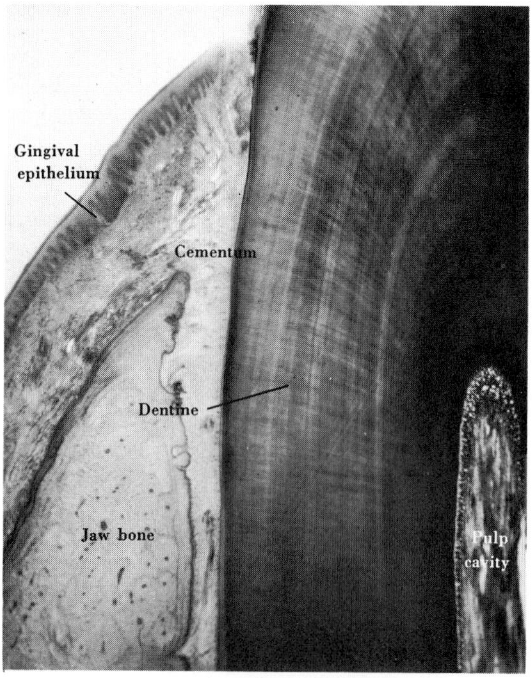

Fig. 7.22 A section through part of a mammalian tooth and gum. Note the dentinal tubules. X35.

the mammalian class, is usually considered to have six sets—but this condition actually results from delayed emplacement of certain teeth.

The tongue is a heavily muscular organ containing intrinsic musculature for flexion and rotation and also extrinsic musculature for retraction and protrusion. A vertical partition, the *medial* or *lingual septum,* runs the entire length of the tongue from base to tip. Usually consisting of dense fibroelastic tissue, it may contain islands of cartilage as well.

The stratified squamous covering is thickest on the dorsal aspect of the tongue and thinnest on the lower surface. The degree of cornification, again thickest on the dorsal side, varies from mammal to mammal. In man the lower surface is noncornified. Below the epithelium is a layer of fibrous connective tissue. This layer extends upward dorsally to form the core of numerous papillae, and inward to form thin septa that separate small bundles of skeletal muscle fibers oriented longitudinally, vertically, and transversely at right angles to each other (Fig. 7.23).

The tongue is roughly divisible into two main areas. The posterior one-third contains large lymphoid masses, the *lingual tonsils,* just below the mucosa. Mucosal glands are found in this region, with ducts opening into the crypts of the tonsils and into depressions between them. Just anterior to the tonsil area is a V-shaped shallow depression, the *foramen cecum,* which marks the posterior boundary of the papillary region of the tongue.

The dorsal surface and sides of the papillary area contain four different kinds of lingual papillae (Fig. 7.24). The most numerous are the filiform structures already described. They occur all through the papillary area, have a core of highly vascular connective tissue, and are usually heavily cornified. The tips of the filiform papillae may branch to form secondary and even tertiary points. Taste buds are not present on filiform papillae, but rather they function as a rasping adjunct to teeth.

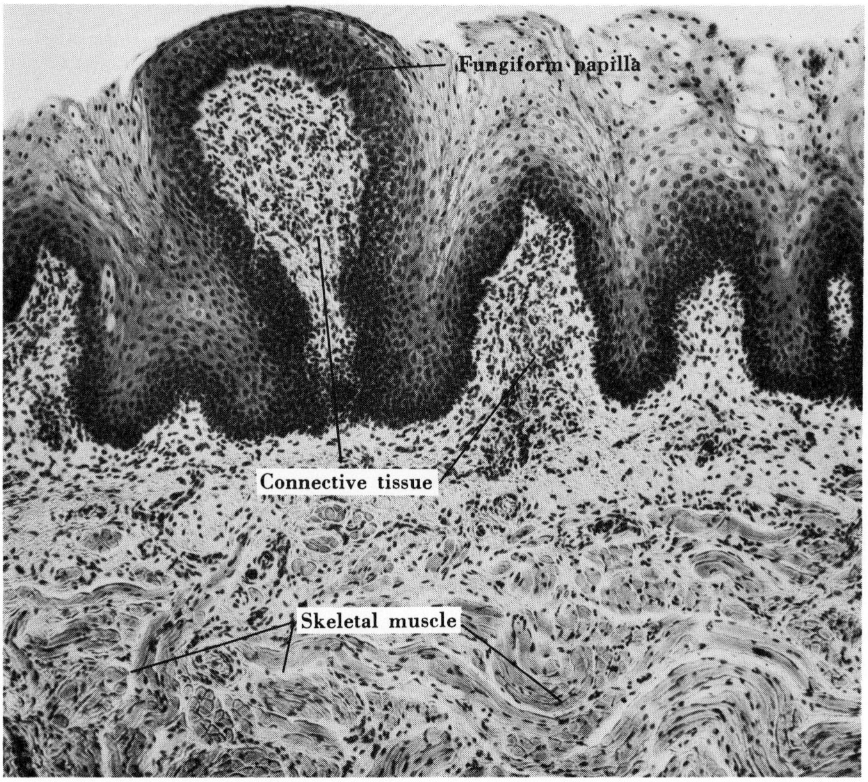

Fig. 7.23 A section through the dorsal surface of the human tongue. X100.

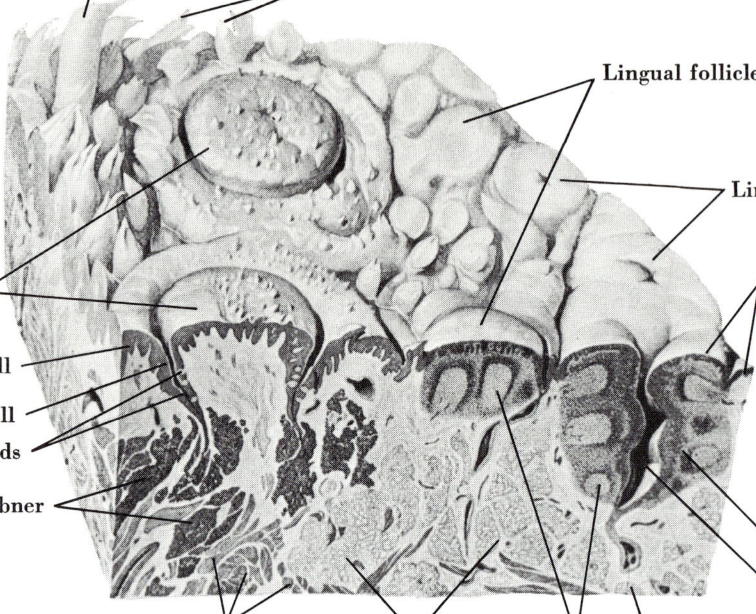

Fig. 7.24 Stereogram of the surface of the tongue. X16. (After Braus. 1924. Anatomie des Menschen.)

Fungiform papillae, also described above, can be found scattered among the filiform. Often the tops of these papillae are not cornified. Taste buds are protectively located on the side walls; only occasionally can any be found on the summits.

Circumvallate or *vallate* papillae are the largest of the four kinds. About nine to twelve are located just in front of the foramen cecum. Peculiar to mammals, these structures are similar in shape to the fungiform papillae, but are surrounded by a deep moat. The summits of the papillae are usually level with the surface of the tongue. In monotremes the papillae lie in depressions so that the summits are below the surrounding surface. The connective tissue core, heavily vascularized and supplied with nerves, is often branched to form secondary papillae supporting the stratified squamous covering. Serous glands, the *glands of von Ebner,* are located below these papillae, with ducts opening into the bottom of the moats. Functionally this is an advantageous relationship for insuring the taste buds of the side and moat walls a steady wash of dissolved foods.

The fourth type, the *foliate papillae,* are also found only in mammals. They are leaflike and occur in parallel rows laterally at the rear of the papillary zone. Separated by furrows, the foliate papillae are thought to represent modified, serially arranged circumvallate papillae. Evidence supporting this argument is found in the monotreme tongue, in which the circumvallates substitute for the foliates. In addition, von Ebner's glands enjoy the same relationship with these papillae as they do with the circumvallates. Man has a poor display of foliate papillae; the rabbit and opossum are noted for well-developed ones. In rabbits, each primary papilla branches to form three secondary structures. Taste buds are found on the lateral walls, the surface being smooth.

Salivary Glands

In the embryo three pairs of epithelial buds arise in the mouth area. These buds, the presumptive salivary glands, invaginate and eventually lie beyond the oral mucosa—connected to the surface only by a duct. There is marked species specificity in these structures, presumably related to habitat and diet. The aquatic mammals, living in a food-dissolving medium, have poorly developed glands. Herbivores, needing great amounts of lubricant saliva, are at the other end of the spectrum. Additionally, there are many small mucosal glands which add to the saliva a secretion composed of water, salts, some proteins, and salivary amylase. The particular quality of the saliva in any given animal depends on the proportion and rate of activity of each of the glandular elements.

Each compound salivary gland is composed of masses of glandular cells, the *parenchyma*, plus a connective tissue supporting framework, the *stroma*. The stroma may be continuous with a covering capsule; however, the presence of a capsule is not universal. The stroma forms thin but dense fibroelastic partitions, or septa, which divide the parenchyma into major *lobes*. The lobes are further subdivided by thinner septa into *lobules*. In addition the septa provide a route for interlobar, interlobular, and intralobular ducts that drain the secretory cells, and which, in certain areas, add to the secretion. These ducts connect with the oral mucosa via larger, branching ducts (Fig. 7.25).

The main duct and its large branches leading from the oral cavity are termed the *excretory* ducts. These divide to form large secretory ducts (*interlobar*), which further subdivide to form smaller secretory ducts (*interlobular*). The smaller secretory ducts give rise to a system of intermediate or intercalated ducts (intralobular), each one of which eventually terminates at a hollow ball of secreting cells known as an end piece or *alveolus*.

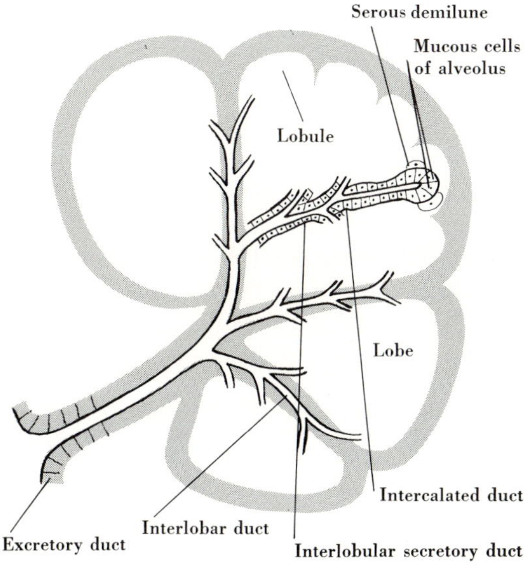

Fig. 7.25 A diagram of the structure of a salivary gland. See text for explanation.

Three distinct kinds of alveoli may be found (Fig. 7.26). Serous or albuminous alveoli are lined with pyramidal cells having a spheroid, basally located nucleus and cytoplasm containing zymogen granules. The cell boundaries are indistinct and the lumina are narrow. These cells secrete a watery, albuminous substance which may or may not have salivary amylase (depending on the animal). The serous cells contain a well-developed, basally located rough endoplasmic reticulum. Its presence, plus the basal aggregation of rodlike mitochondria, lend a striated appearance to the cell. In electron micrographs one can see cisternae of the endoplasmic reticulum filled with a dense material prior to secretion. With specialized techniques, infoldings of the cell membrane, intercellular canaliculi, can be seen. These seem to conncet with a terminal bar at the free surface, and to continue intercellularly. Presumably this represents a mechanism for increasing cellular surface area.

Mucous alveoli have wide lumens and are lined with tall, clear cells. The cytoplasm is deep-staining, with a basally located, compressed nucleus. The boundaries between these cells are sharp, as is the basement lamina upon which the cells lie. The cytoplasm is filled with a delicate, granular, basophilic network often seen to enclose precipitated mucogenic material. The endoplasmic reticulum is far less abundant than in serous cells, but, as with the serous cells, the free surface is seen to contain a terminal bar.

Mixed alveoli contain both serous and mucous cells. Often the serous cells are crescent-shaped and are located around the base of the mucous cells. In this state, the serous cells are known as *serous demilunes* and communicate with the lumen either directly or via intercellular canaliculi.

All salivary alveoli contain myoepithelial

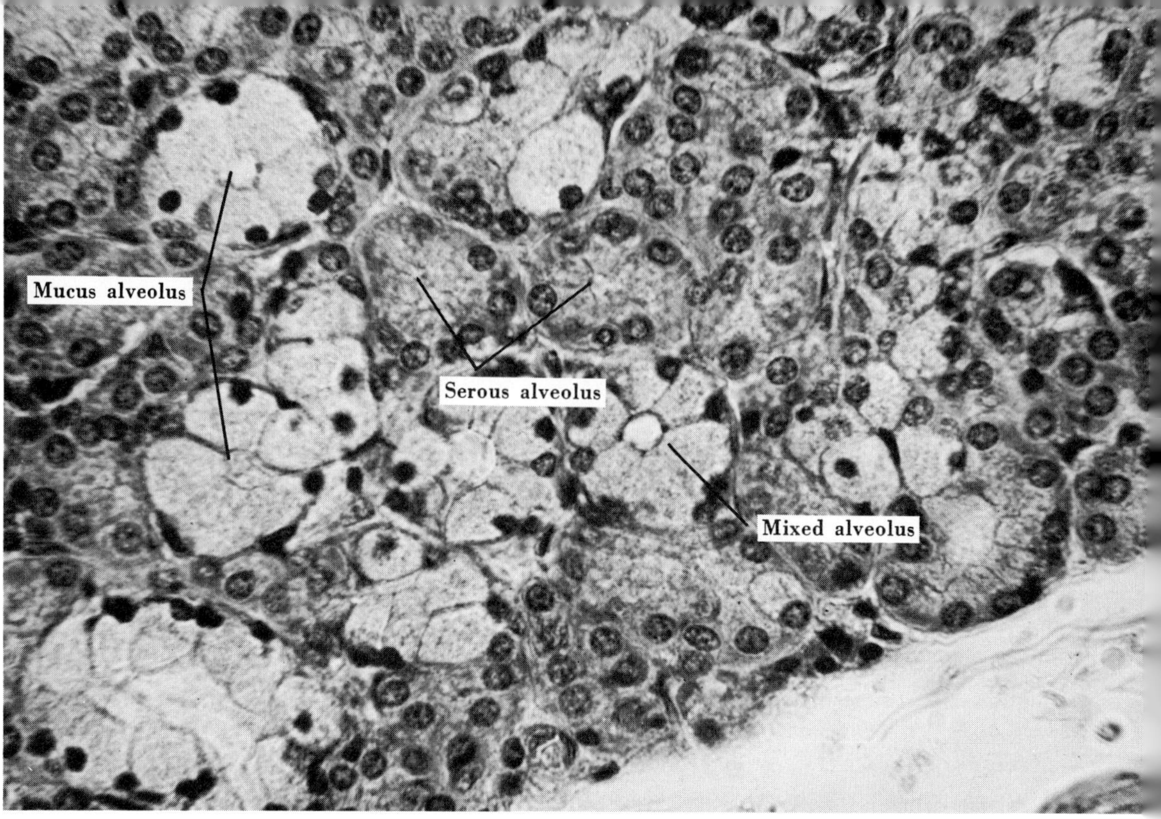

Fig. 7.26 A section through the human sublingual gland. Note the three kinds of alveoli. X450.

cells similar to those of the sweat glands. Often called *basket cells,* they are stellate in shape and seem to aid in expression of cellular secretory products.

Let us now turn to a consideration of the three pairs of glands themselves.

The parotid glands are located near the base of each external ear. Usually the largest of the salivary glands, they are small in the Edentata. In most mammals a thick connective tissue capsule is present which continues interiorly as the stroma. The main excretory duct is lined with pseudostratified columnar epithelium which changes to simple columnar in the larger excretory branches. The interlobar and interlobular secretory ducts are also lined with one layer of columnar cells, but these can be distinguished from those of the excretory ducts by the stratified appearance of the basal portion of the cell caused by rodlike mitochondria being oriented perpendicularly to the base between deep infoldings of the plasma membrane. This condition, a basally striated appearance, seems to be common to many cells in which water and ion exchange is a common and primary process; the secretory product of these cells contains water and calcium salts. The intralobular intercalated ducts are lined with low cuboidal epithelium and terminate in serous alveoli.

The second pair of salivary glands is known as the *submaxillary* glands—except in man, where they lie *submandibularly.* These glands are mixed glands containing either mixed alveoli, often with serous demilunes (as in man), or separated alveoli, such as in the rat and cat. The proportion of mucous to serous cells also shows variation. In man serous cells predominate; the cat has a preponderance of mucous cells. The system of ducts is similar to that in the parotid with a few exceptions: (1) pseudostratified columnar epithelium continues as far as the larger interlobular ducts; (2) the secretory ductwork is more extensive; and (3) the intercalated ducts are shorter. In the Edentata these are the largest of the salivary glands; connected to the main excretory duct is a bladder to hold the copious and sticky secretion.

The third set of glands, the *sublinguals,* are also mixed glands. Located at the base of the tongue, these are the smallest of the three pairs. As in the other glands, pseudostratified columnar epithelium lines the large excretory duct. The branches, however, are lined with simple columnar. Secretory, and especially intercalated ducts, are greatly reduced—with alveoli often opening into large interlobular ducts. The alveoli themselves are composed of mucous cells with serous demilunes.

Esophagus

The mucosa of the esophagus is thrown into long folds which can be partially erased with distension. Additionally, large papillae are present, and are cornified in such mammals as the sheep, goat, and cat. Although the epithelium is generally composed of stratified squamous cells, the thickness and degree of cornification are diet-dependent, being greatest in the herbivores and least in the carnivorans (Fig. 7.27). When nuclear remnants still remain in the upper epithelial layers, such as in rodents, a stratum granulosum can be found. Simple, branched tubular mucous glands dipping into the lamina propria can be found often in the upper, and usually always in the lower, region of the tube. The granular secretory cells are cuboidal or low columnar.

The lamina propria, composed of fibroelastic tissue with many lymphocytes is not smooth. Rather, it contains projections that form the cores of the mucosal papillae. The muscularis mucosae usually begins in the lower esophagus and is represented by longitudinal smooth muscle fibers and elastic fibers.

The highly elastic submucosa contains loose areolar or dense fibroelastic tissue. Diffuse lymphoid tissue occurs in most mammals. It is well vascularized and contains compound tubular or tubuloalveolar glands. Such submucosal glands in the esophagus are unique in mammals. In man the glands contain alveoli which are typically mucus-producing. Other

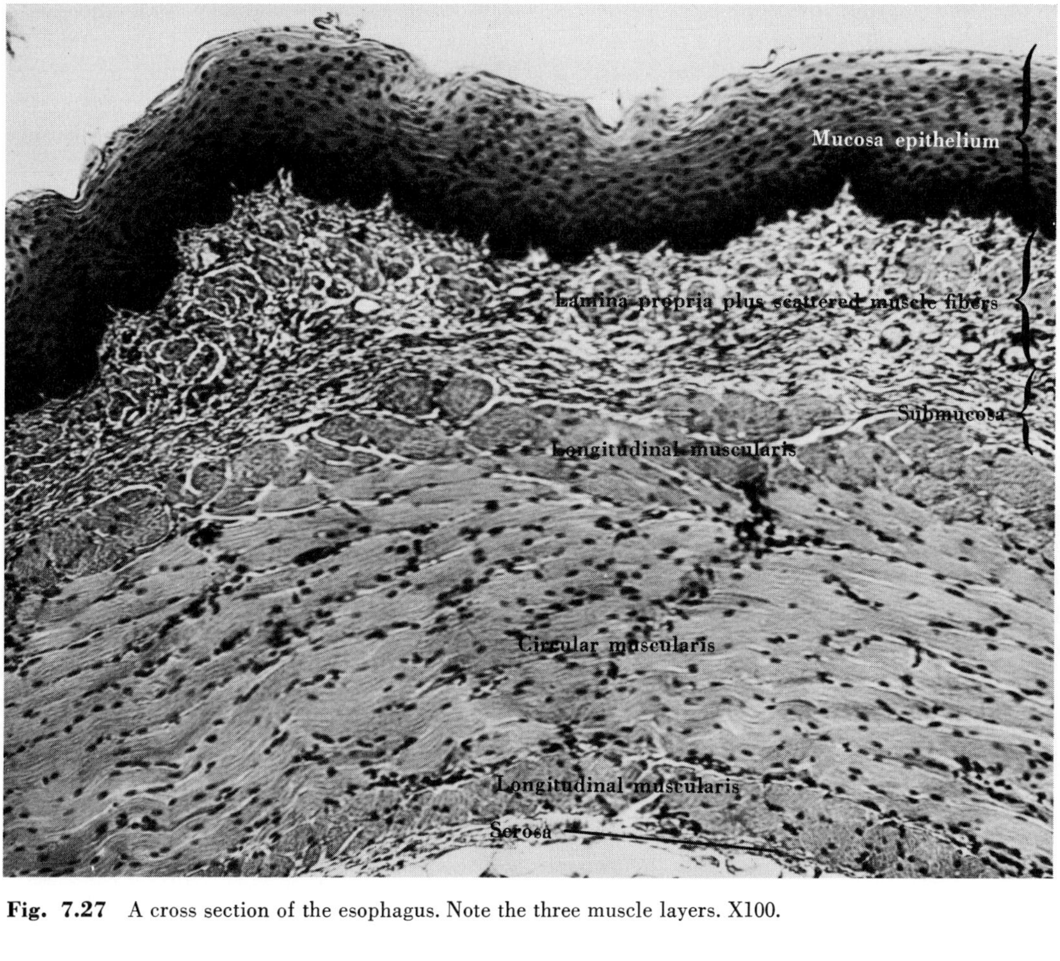

Fig. 7.27 A cross section of the esophagus. Note the three muscle layers. X100.

mammals have mixed mucous and serous glands, and some possess mucous cells with serous demilunes. In all mammals except in some rodents, the horse, and the cat, these glands are plentiful in the esophagus. The bodies of glands which secrete into the lower esophagus can often be seen in the wall of the cardiac region of the stomach.

The muscularis is variable in composition and number of layers. In man and many other mammals, two layers are typically present. Usually the muscle fibers are of the striated variety in the upper region; mixed, with smooth muscle appearing first in the inner circular layer, in the midregion; and smooth in the lower portion. Three roughly spiraling muscle layers occur in many carnivores, ungulates, and ruminants. Four layers can be seen in pigs. In this latter animal the inner and outer layers are thin and contain longitudinal fibers, the middle layers being thicker with circularly disposed fibers.

The upper part of the esophagus, located in the neck region, is covered with a loose adventitia of fibroelastic tissue. In the thorax a typical serosa appcars.

Stomach

The stomach assumes a wide variety of forms in mammals. It is also noteworthy for the division of labor seen in the secreting cells of the glandular apparatus. Let us begin our discussion with the single-chambered "simple" stomach found in insectivores, carnivorans, and man, and then proceed to the interesting complications found especially in the ruminant mammals.

The mucosa of the simple stomach is wide, in man from 0.3 to 1.5 mm. It is folded and packed with numerous branched, tubular glands. The lining epithelium can assume three possible conditions. (1) In man, most carnivorans, some edentates (such as the anteaters), in insect-eating marsupials (such as the opossum), and a few others, the lining consists of those typically vertebrate gastric cells: glandular columnar cells with apical clear borders. (2) In many of the lower primates, the Cetacea, many artiodactyls (such as the pig), perissodactyls (such as the tapir), the rhinoceros and the horse, some edentates, herbivorous marsupials (such as the kangaroo), and a few monotremes, the upper portion of the stomach is lined with a stratified squamous epithelium while the lower portion is columnar and glandular. (3) In many rodents and a few monotremes and marsupials, most of the stomach is lined with a nonglandular, keratinized, stratified squamous epithelium.

In glandular areas, i.e., in areas covered by columnar epithelium, the surface is not smooth, but rather is raised to cover submucosal rugae. Gastric pits are present, leading into three kinds of branched tubular glands.

Cardiac glands are present in those animals possessing a stomach completely lined with columnar epithelium. They are located near the esophageal junction. Few in number, they consist of a short excretory duct lined with columnar cells, and tortuous secretory tubules. The mucus-secreting cells of the tubules are cuboidal or low columnar and are not too dissimilar from the surface cells. Occasionally parietal cells (*see below*) are present.

Fundic glands, from which gastric juice is derived, are the most numerous. Concentrated especially in the midregion of the stomach, they can, however, be found throughout the organ—wherever columnar lining cells are present. These glands have a short excretory duct lined with columnar cells and an opening into a gastric pit (Fig. 7.28). The glandular portion begins with a neck; it is long and ends blindly in a widened base or fundus. Four types of cells may be discerned in the gland. The neck and often the base of the gland contains *mucous neck cells* similar to those found in amphibians, reptiles, and birds. These cells are cuboidal, with a finely granular cytoplasm. The nucleus is basally located and flattened. Small PAS-positive granules are abundant, especially in the apical region of the cell; they are the precursor substance to a mucus which is less viscous than that produced by the stomach lining cells. Some evidence exists which points to the presence of

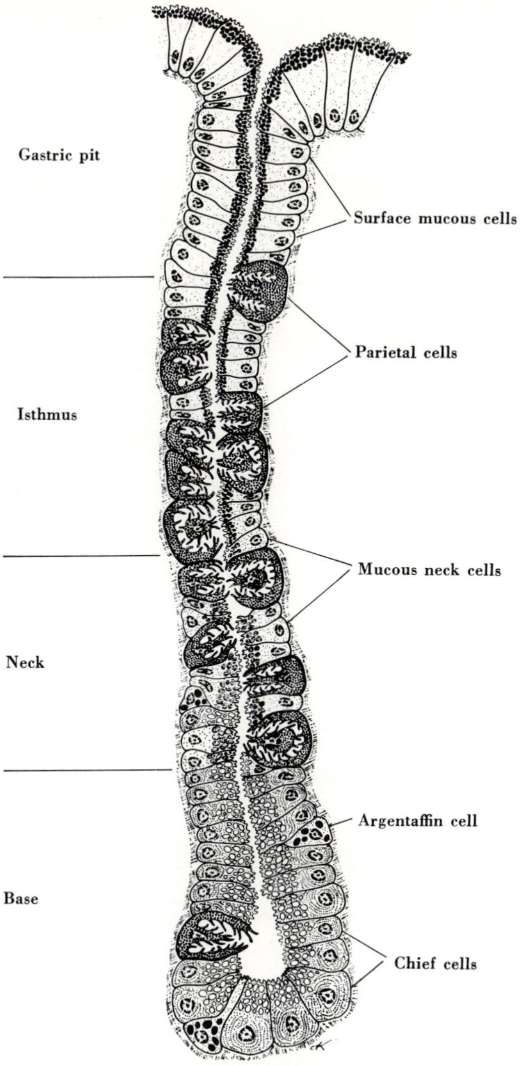

Fig. 7.28 A typical gastric gland. This diagram represents a composite of information gathered from electron micrographs of the bat stomach. (After S. Ito and R. J. Winchester. 1963. J. Cell Biol. 16:541–579.)

the "intrinsic factor" of Castle, a substance which facilitates absorption of Vitamin B_{12}, a substance that combats pernicious anemia. The free surface of these cells bears microvilli and the lateral surfaces interdigitate with neighboring cells (Figs. 7.11 and 7.29).

Located throughout the gland are many *chief cells.* These pyramidal cells have their bases abutting a basement lamina and their apices pointing into the lumen of the gland. A spherical nucleus is located toward the base of the cell. Between the base and the nucleus, the cytoplasm has a basophilically striated appearance. The cause can be seen in electron micrographs. Arrayed in parallel fashion is a lamellar, rough endoplasmic reticulum with mitochrondria lined up between the lamellae. Pepsinogen, the inactive pepsin precursor, is produced in this area of the cell, and is then moved apically to a point at which a delicate cytoplasmic membrane system is seen to enclose membrane-bound precipitated zymogen granules. The Golgi material is concentrated supranuclearly, and presumably has a role in consolidating the secretory product. Short microvilli appear on the free apical surface (Fig. 7.30).

Parietal cells, producers of hydrochloric acid, although concentrated in the upper regions of the gland, can often be seen scattered throughout the base. Larger and fewer in number than the chief cells, they are usually located between them and the basement lamina. The cells are oval or polygonal, with a finely granular eosinophilic cytoplasm. Centrally located is one, or sometimes two, large spherical nucleus or nuclei. The surface of these cells contains many slender microvilli, often giving the cell a striated border appearance. Free acid can be detected in the microvilli, which continue as a system of intracellular canaliculi. These structures, in turn, continue extracellularly as intercellular canaliculi so that each parietal cell maintains an intact connection with the lumen of the gland. In electron micrographs, one can additionally see an extensive smooth endoplasmic reticulum and many large, highly structured mitochondria (Fig. 7.31).

The mechanism of acid secretion by the parietal cells is not clear. The cytoplasm of the cell itself tends toward the alkaline end of the spectrum. As we have noted, free acid is detectable in the microvilli and presumably indicates the extreme importance of the cell membrane in selective secretion and ion exchange. Secretion of HCl into the lumen goes hand in hand with an equivalent release of bicarbonate ion into the bloodstream. It is thought that parietal cells convert sodium

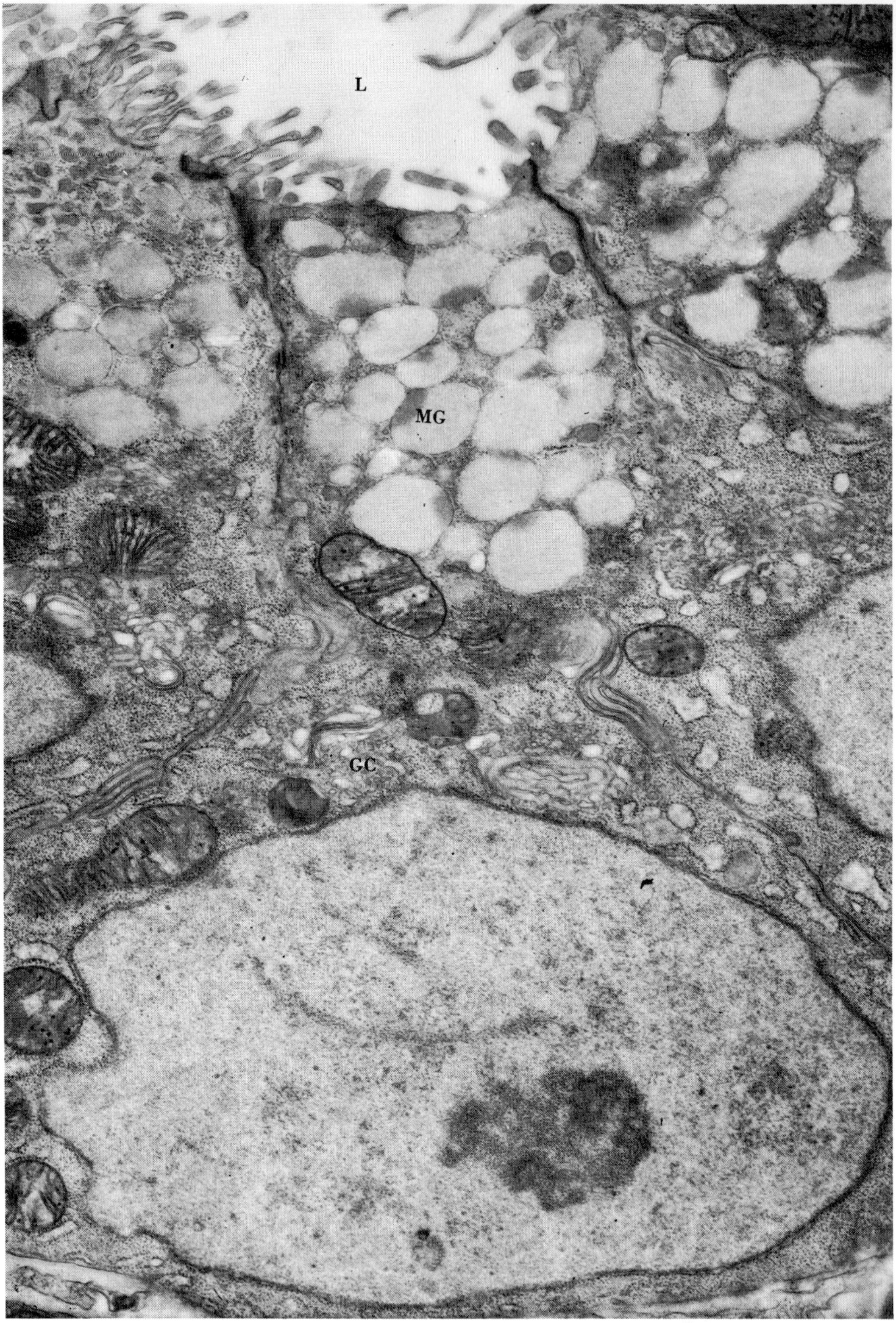

Fig. 7.29 An electron micrograph of one mucous neck cell and portions of two others from the bat stomach. *L*, lumen; *MG*, apical mucous granules; *GC*, Golgi complex, which is well developed in this cell. The nucleus and its contents occupy most of the basal part of the cell. Mitochondria and endoplasmic reticulum can be seen near the nucleus. Compare this figure with Fig. 7.11. (From S. Ito and R. J. Winchester. 1963. J. Cell Biol. 16:541–579.)

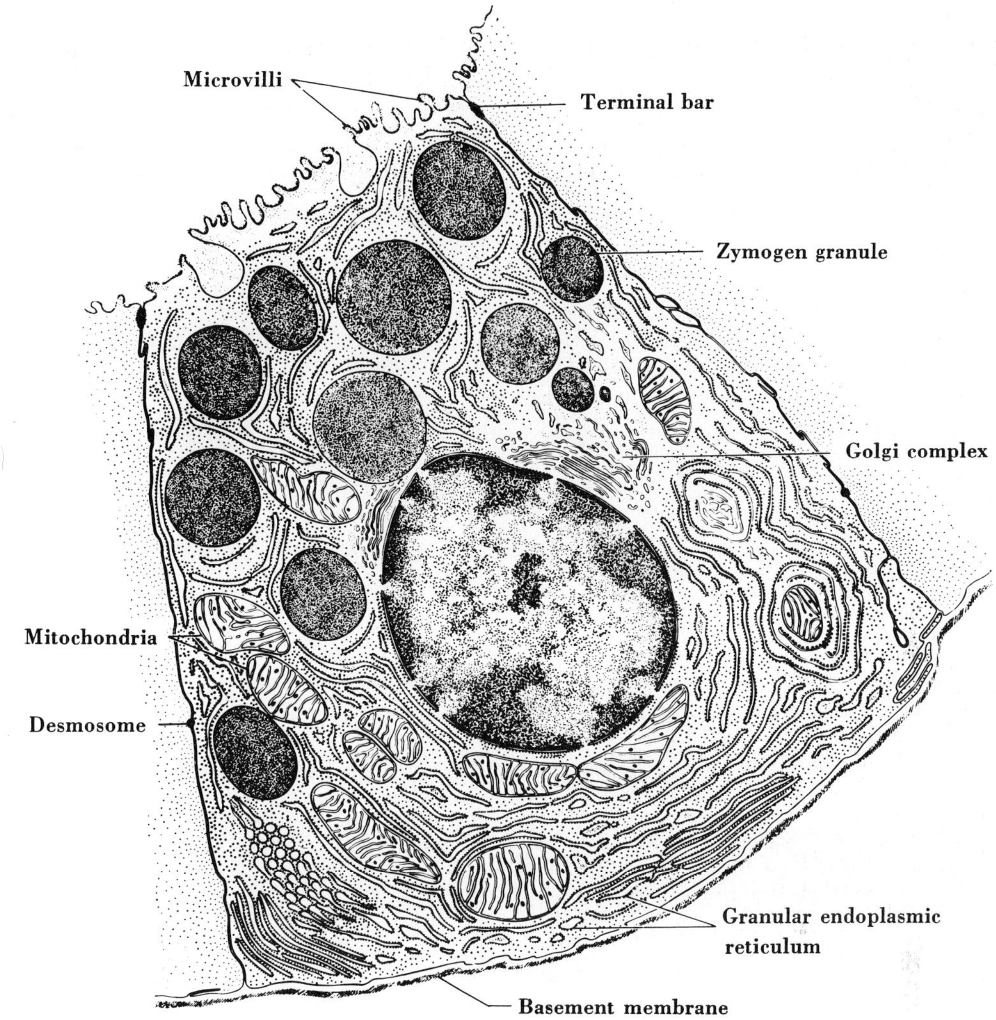

CHIEF CELL

Fig. 7.30 A diagram of a gastric chief cell of the bat stomach. (After S. Ito and R. J. Winchester. 1963. J. Cell Biol. 16:541–579.)

chloride and carbonic acid into HCl and bicarbonate. Carbonic anhydrase, the enzyme necessary for the production of carbonic acid from carbon dioxide and water is always abundant in parietal cells.

Diffusely scattered throughout the gland are *Argentaffin* cells. So named because they stain brownish-yellow with chromate dyes, these cells have a flattened pyramidal form and finely granular appearance. As seen in electron micrographs, these cells contain a large, irregularly shaped nucleus and a dense array of spherical granules, each surrounded by a loose membrane (Fig. 7.32).

In the pyloric region of the stomach, a few pyloric glands can be found. The gastric pits of these glands are deeper than those of the fundus, and the secretory tubules are more twisted and have larger lumina. The secretory cells are primarily mucous cells resembling the mucous neck cells. A few parietal and/or chief cells may be found.

The lamina propria is compressed between and below the glands. A meshwork of connective tissue fibers encloses a few histiocytes, plasma cells, leukocytes, and fibroblasts. Lymphocytes and lymph nodules can be seen especially in the pyloric region. A typical two-layered muscularis mucosae is present in almost all mammals. In a few instances a third, outer circular layer is present. Processes of the muscle cells extend upward to the epithelium and cause slight contraction of the membrane preparatory to secretory expression.

The submucosa is ridged with rugae. It is formed of fibroelastic connective tissue and contains many fat cells, mast cells, lymphoid cells, and leukocytes.

The muscularis contains three well-defined layers in many mammals, including man. The inner layer is obliquely disposed, the middle layer is circularly arranged, and the outer layer contains longitudinally oriented fibers. In the pyloric region the two inner layers, but especially the circular one, thicken to form a sphincter. Covering the muscularis is a typical serosa.

The stomach we have been discussing is termed "simple" because it is not divided by internal partitions. In this stomach one can distinguish a cardiac region, body or fundus, and pyloric region, but the internal lumen is

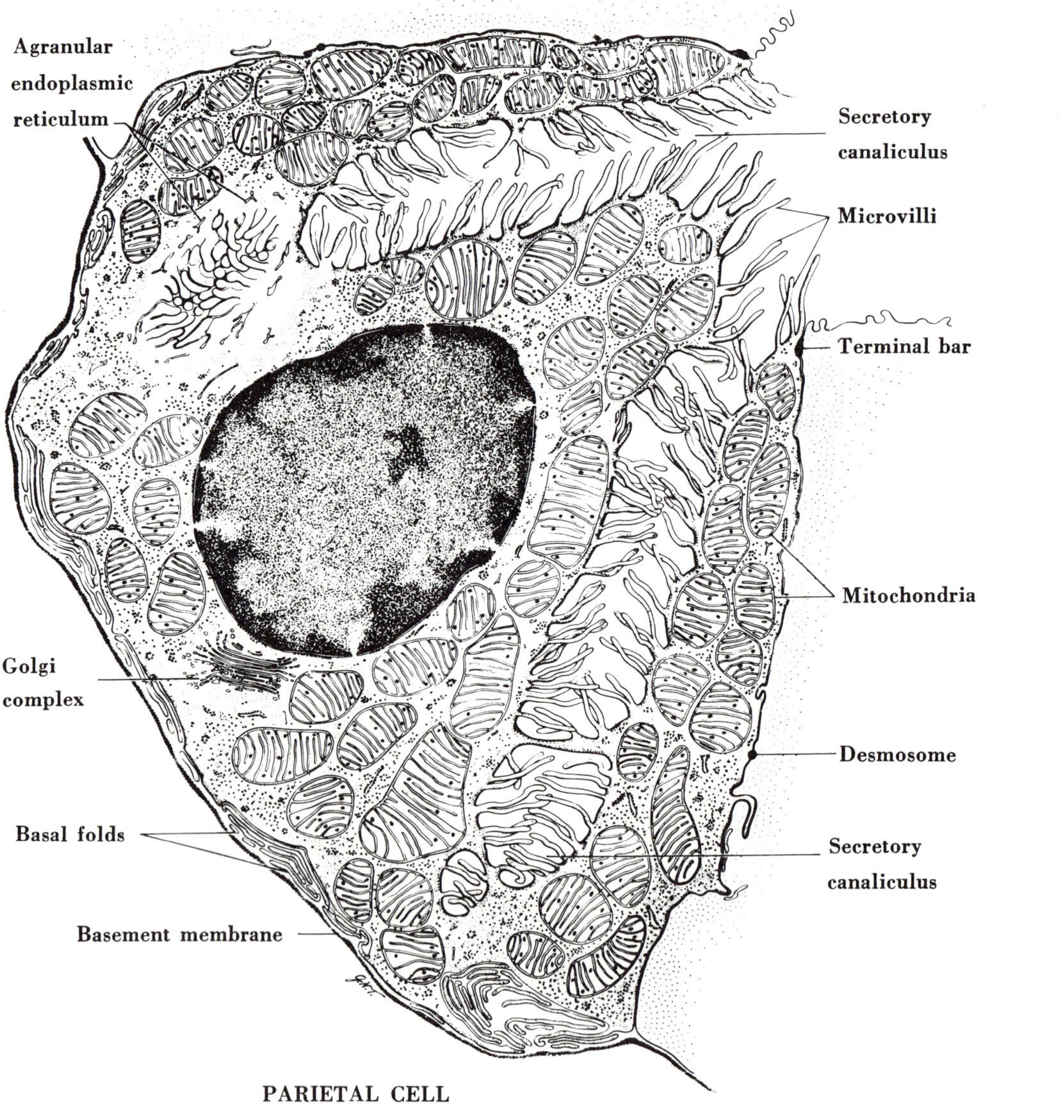

PARIETAL CELL

Fig. 7.31*a* A diagram of parietal cells of the bat stomach.

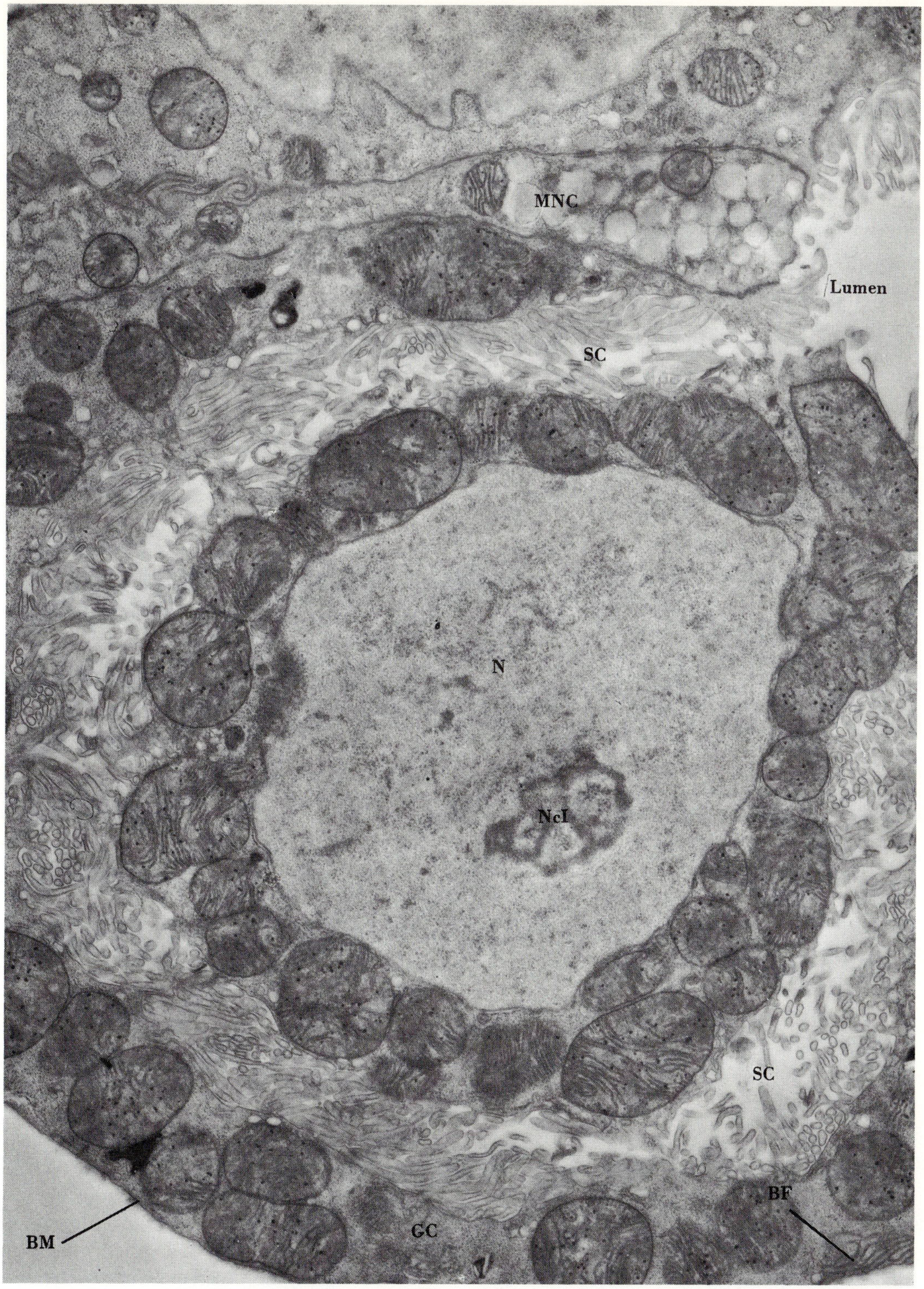

Fig. 7.31b An electron micrograph of a parietal cell of the bat stomach. The nucleus, *N*, and nucleolus, *Ncl*, are large. Surrounding the nucleus are large mitochondria. The central part of the cell is separated from the periphery by the secretory canaliculi, *SC*. A small Golgi complex *GC* and a basal fold *BF* can be seen. *MNC* is a neighboring mucous neck cell. X11,000. (From S. Ito and R. J. Winchester. 1963. J. Cell Biol. 16:541–579.)

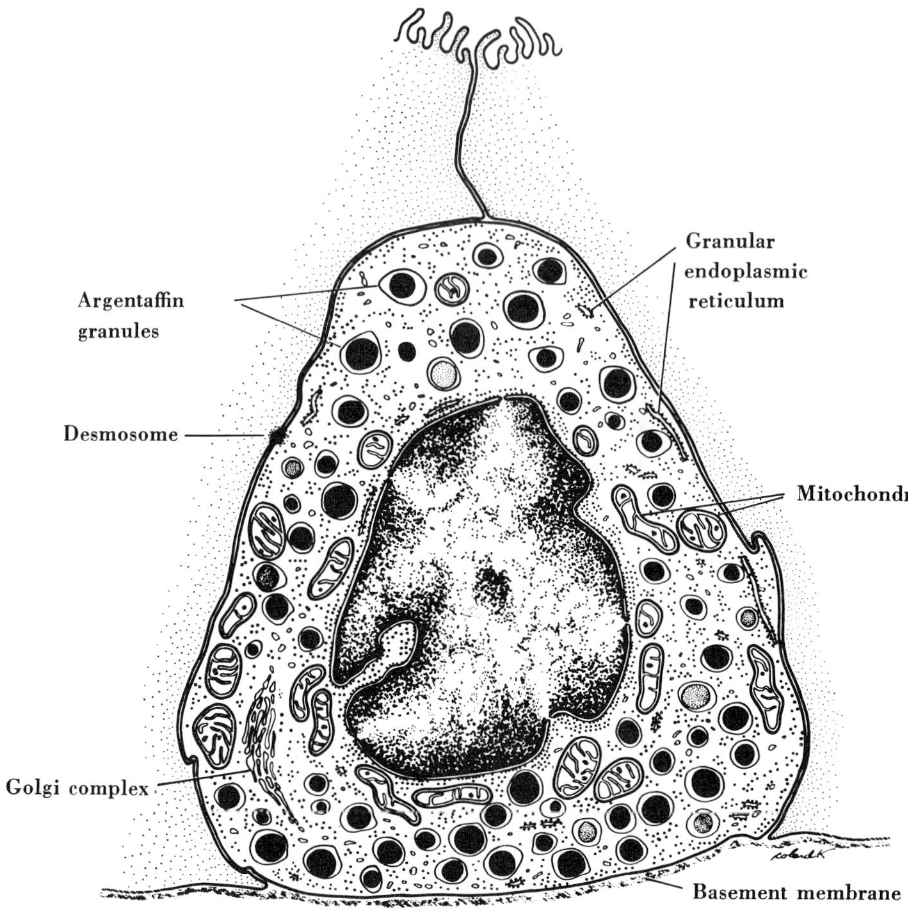

Fig. 7.32 A diagram of an argentaffin cell in the bat stomach. (After S. Ito and R. J. Winchester. 1963. J. Cell Biol. 16:541–579.)

completely contiguous and the epithelial wall shows no drastic regional changes. In pigs, the cardiac region is much more highly developed than are the other areas. In the cetacean stomach the upper portion is saclike and almost forms an appendage to the rest of the organ. Hamsters have a two-chambered, or bilocular, stomach.

It is in the ruminant artiodactyls, such as the cow, sheep, and goat, that the esophageal-stomach complex assumes its full majesty. Food coming from the esophagus on its first trip around enters the *rumen.* It then passes to the *reticulum,* is regurgitated, chewed, swallowed, and then passes to the third chamber, the *omasum.* Finally it enters the *abomasum* and then goes to the intestine (Fig. 7.33).

The rumen functions mainly for storage and further moistening of the food. It is subdivided by ridges and these, in turn, are covered by papillae. The pigmented, stratified squamous epithelium is surmounted by a thin layer of cornified cells, and rests directly on the muscle fibers of the muscularis mucosae. No lamina propria is present, but a few elastic fibers are interposed between the mucosa and submucosa. Submucosal mucous glands are present. The muscularis is double-layered and the serosa is typical.

The lumen of the nonglandular reticulum is partitioned by many complex septa. Highly cornified papillae are present on a stratified

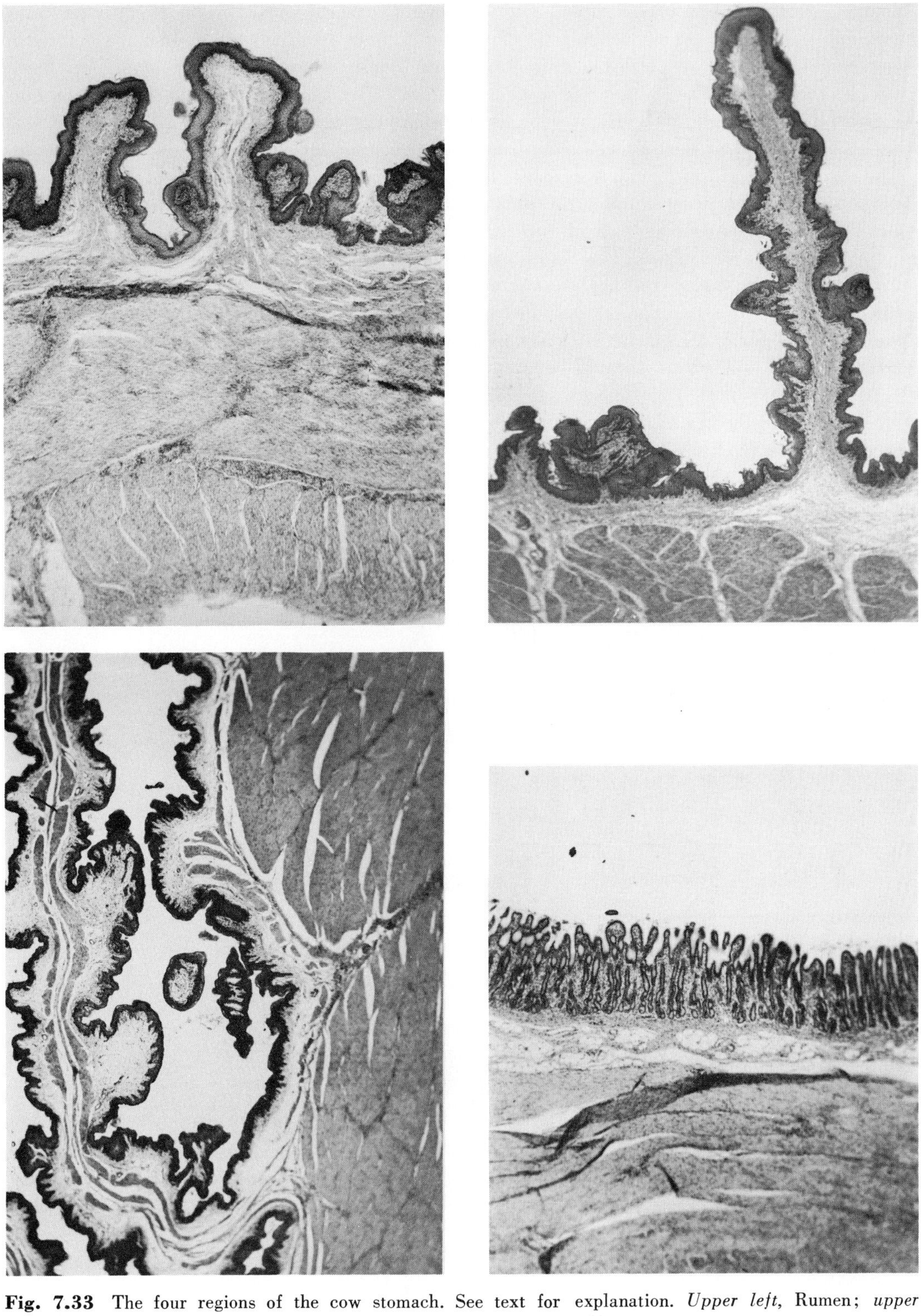

Fig. 7.33 The four regions of the cow stomach. See text for explanation. *Upper left,* Rumen; *upper right,* Reticulum; *lower left,* Omasum; *lower right,* Abomasum. All X35.

squamous epithelium. The lamina propria is sparse, containing a few elastic fibers, and a muscularis mucosae is present only in the larger ridges. In other respects the reticulum resembles the rumen. The legendary ability of the camel to withstand drought owes to the water-storing ability of the animal's reticulum.

The omasum, lacking in some ruminants, is distinguished by its great number of partitions. These leaflike structures extend into and practically occlude the lumen. The leaves are covered with cornified stratified squamous epithelium and large papillae. A bundle of smooth muscle fibers that arise in the muscularis mucosae and elastic connective tissue comprise the leaf center. The omasum can also be recognized by the dense capillary net subadjacent to the epithelium.

The abomasum is the glandular stomach. Its histologic characteristics are similar to those of the simple stomach.

Small Intestine

The small intestine can vary in length. It is four to five times the body length in carnivorans, and upwards of twenty times the body length in herbivores. Usually it is divided into three main sections: the duodenum, jejunum, and ileum; however, as we shall see, these areas differ only in proportionate representation of certain structures.

The mucosal topography is complex (Fig. 7.34). Unclosed circular and oblique folds up to 8 mm in height, the *valvulae conniventes* (also called the *valves of Kerckring* or the *plicae circulares*) increase the absorptive capacity of the surface. They reach maximal height in the lower duodenum and jejunum, and are shorter in the ileum. These folds remain even under conditions of distension. Additionally, projecting from the entire surface, are motile villi. They tend to be leaf-shaped in the duodenum, finger-shaped in the jejunum, and clublike in the ileum. The number of villi is usually greatest in carnivorans.

Resting on a distinct basement lamina, the epithelium contains typical lining cells, columnar with a striated border, a basal nucleus, and a distal mucus area (Fig. 7.35). As we have seen before, the number of goblet cells progressively increases posteriad. Argen-

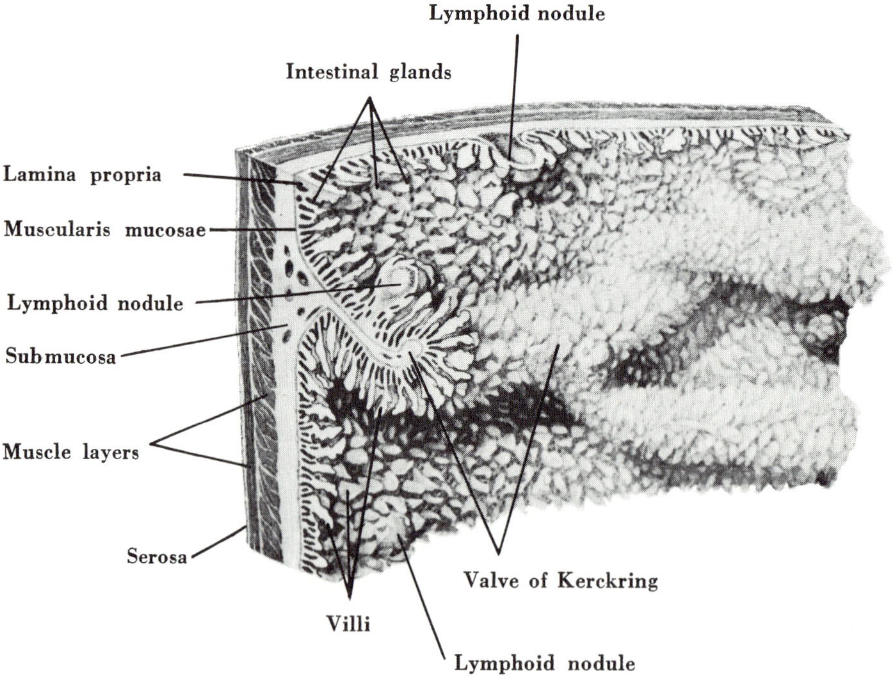

Fig. 7.34 Stereogram of a region of the small intestine. X17. (After Braus. 1924. Anatomie des Menschen.)

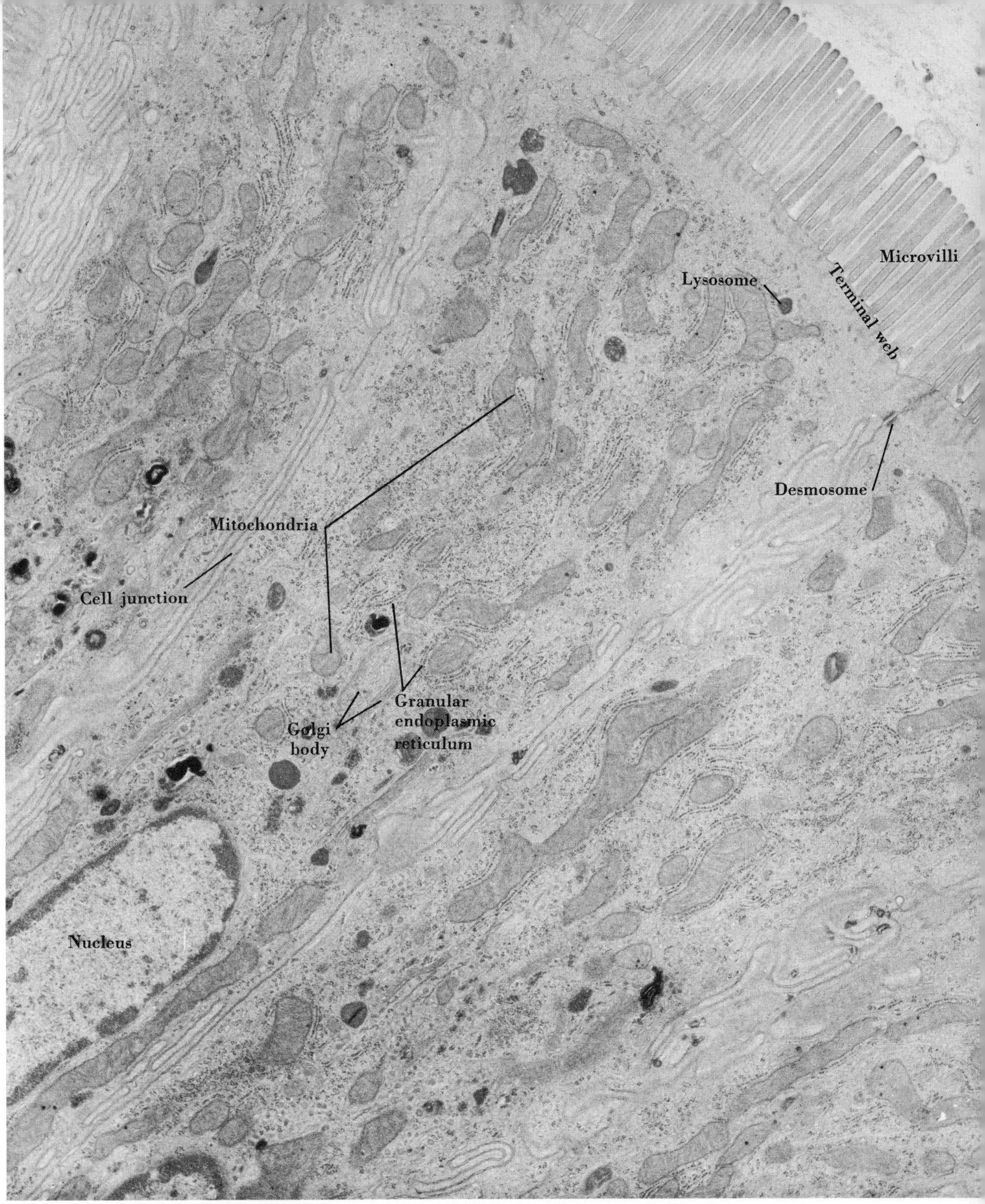

Fig. 7.35 An intestinal lining cell of a fasted rat. Note the complexity of the cell junctions. X12,000. (From R. J. Cardell *et al.* 1967. J. Cell Biol. 34:123–155.)

taffin cells are scattered throughout, decreasing to the ileum, and then increasing to the rectum.

Crypts and simple tubular glands of Lieberkühn are again present within the mocosa. As one progresses into the crypt and down the tubule, the cells shorten and lose their striated border (Fig. 7.36). Goblet cells are often found interposed between the columnar cells of the crypt. The base of the glands contain

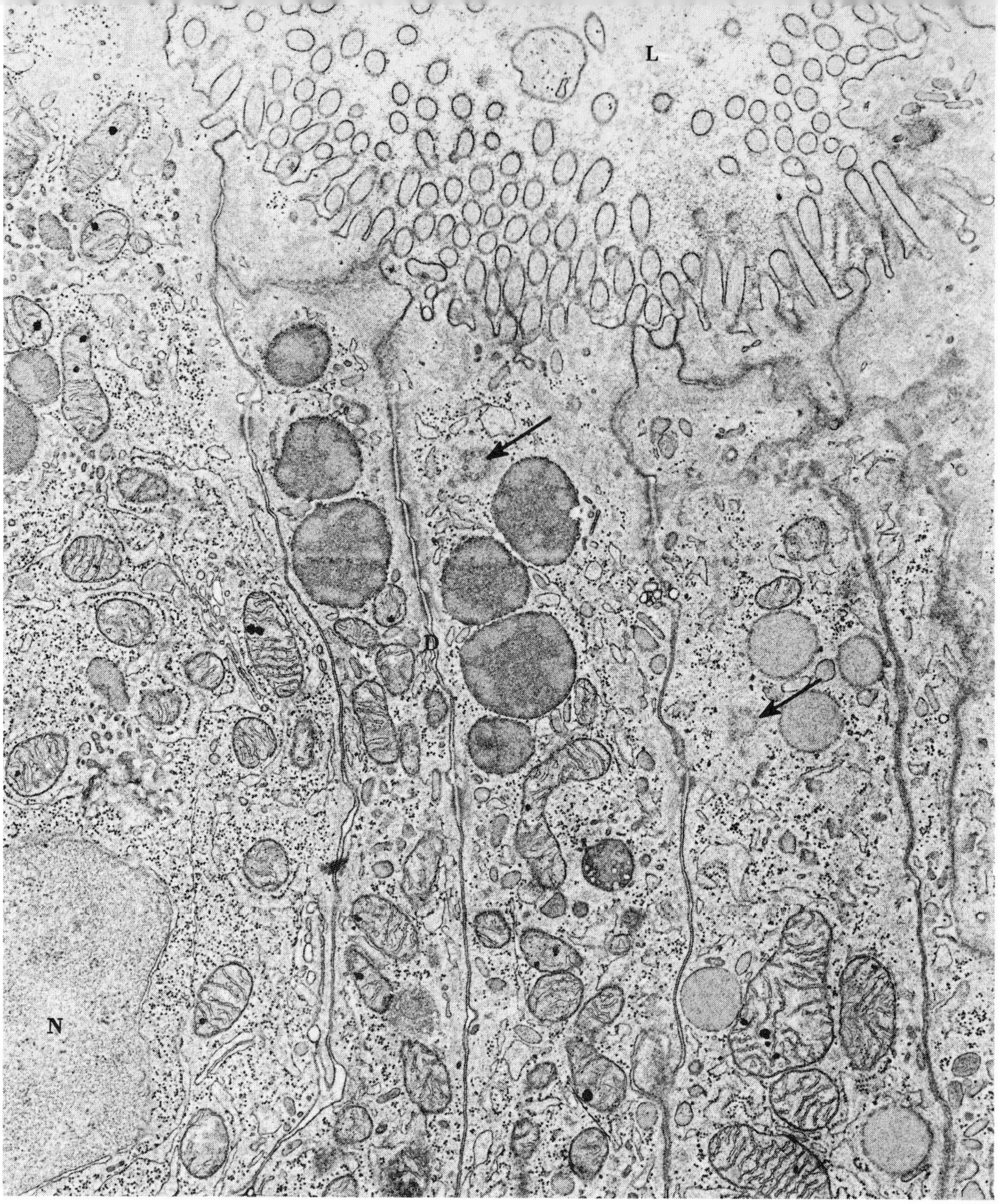

Fig. 7.36 An electron micrograph of portions of five crypt cells from the human small intestine. Membrane-bound secretory material can be seen especially toward the lumen, *L*. Desmosomes, *D*, are found along the cell junctions. Mitochondria, ribosomes, and granular endoplasmic reticulum are present in abundance. A portion of a large nucleus, *N*, appears at the bottom right. Two centrioles (*arrows*) can be seen. X12,500. (From J. S. Trier. 1963. J. Cell Biol. 18:599–620.)

Paneth cells (Fig. 7.37). These cells, numerically decreased in carnivorans, may function in acid regulation. The acidophilic cytoplasm is granular and appears basally striated because of parallel granular endoplasmic reticular lamellae. Typical wedge-shaped serous cells having a granular acidophilic cytoplasm abound near the base of the tubule. The zymogen granules presumably represent the precursors of intestinal lipase, peptidase, and amylase. Disaccharide splitting enzymes are thought to reside in the border of the surface cells.

A delicate, lymphocyte-filled lamina propria is present between the glands and in the villi. It consists of loose reticular and fibroelastic tissue with cellular elements of larger size in the villi than between the glands. Diffuse

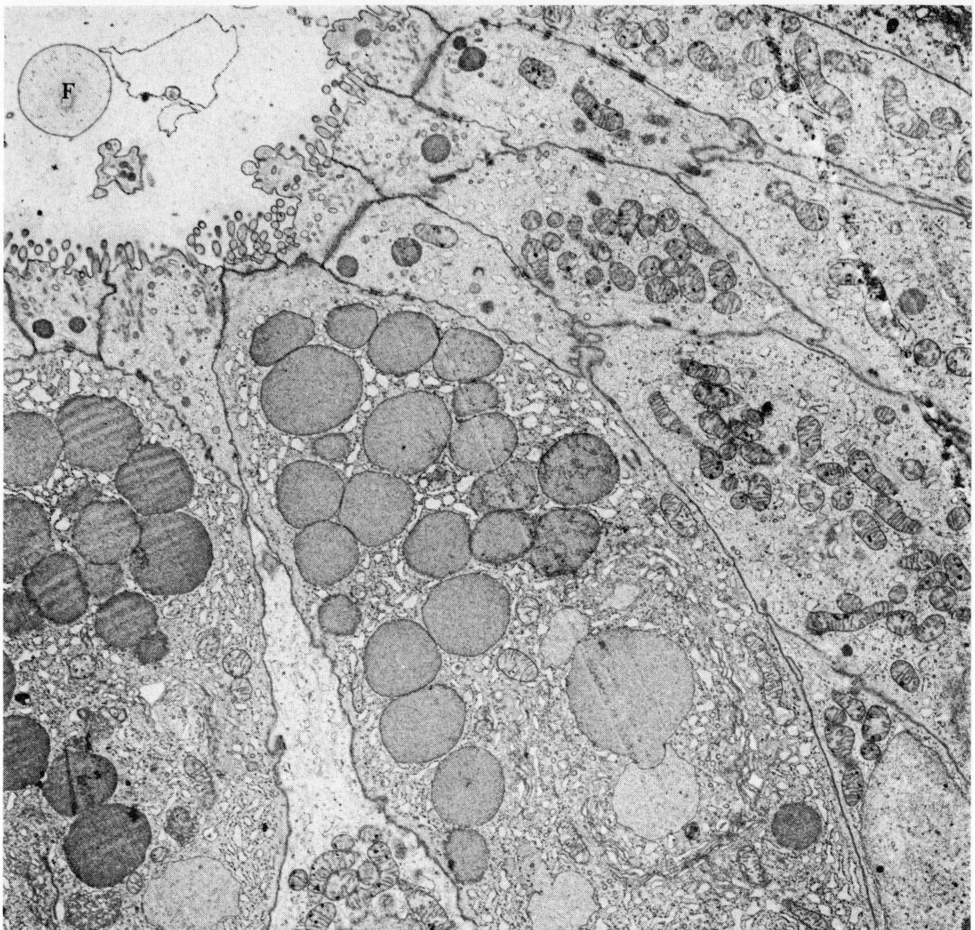

Fig. 7.37 In this electron micrograph two Paneth cells and several tangentially sectioned epithelial cells can be found. Note the quantity of subnuclear mitochondria, *M*, in the epithelial cell. The Paneth cells are noteworthy for the density of granular endoplasmic reticulum. Large accumulations of membrane-bound secretory material can be found in the supranuclear area. Dense bodies that resemble lysosomes are indicated by arrows. In the lumen, one can find fragments of cytoplasmic material, *F*. X5000. The inset shows the fragments at higher magnification. They are seen to include membrane-bound granules, ribosomes, and endoplasmic reticulum. X22,000. (From J. S. Trier. 1963. J. Cell Biol. 18:599–620.)

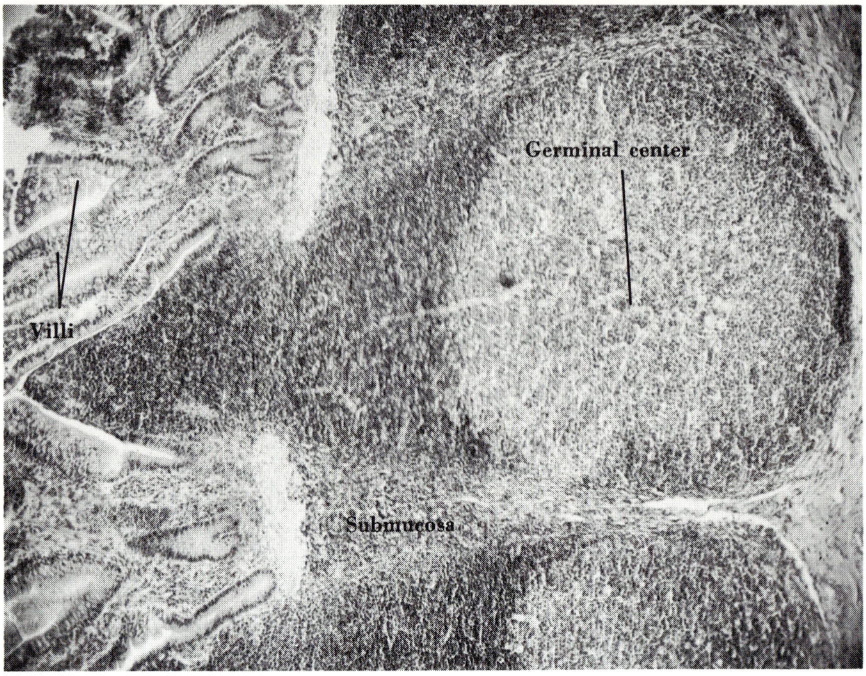

Fig. 7.38 A photomicrograph of a Peyer's Patch. Note the projection into the lamina propria. The germinal center of the lymphocyte-filled node is located in the submucosa. X90.

lymphoid tissue abounds, and solitary lymph nodes can readily be found. In the ileum, nodules are grouped together to form large macroscopically visible *Peyer's patches* (Fig. 7.38). These structures always lie under regions free of villi and goblet cells, and may extend into the submucosa. A typical muscularis mucosae is present in most mammals; however, this layer is reduced or absent beneath a Peyer's patch.

The submucosa is of variable thickness, consisting of loose areolar or fibroelastic connective tissue and large vessels. Except in the duodenum it is nonglandular. In the duodenum, and, surprisingly, in the ileum of the ox, branched tubular or tubuloalveolar glands (Brunner's glands) penetrate the submucosa. The secretory cells are pyramidal or columnar, forming predominantly mucoid alveoli (Fig. 7.39). An occasional serous alveolus can be seen, and, indeed, the secretory product has been found to contain a proteolytic enzyme.

The well-developed muscularis is double-layered, with the longitudinal fibers actually coursing as an open helix and the inner fibers as a closed one.

At the junction of the small and large intestine of many mammals, an outpocketing, the caecum, can be found. Similar in histology to the small intestine, it nevertheless is noteworthy for its dense population of lymph nodules. A definite caecum is lacking in many insectivorans, most bats, and in bears. It can reach lengths in herbivores which approach that of the rest of the intestine combined.

Extending as a blind pouch from the end of the caecum is the vermiform appendix. Lacking villi, it resembles the caecum with its dense array of tonsil-like lymph nodules (Fig. 7.40).

Large Intestine

The large intestinal lumen is much simpler topographically than that of the small intestine is (Fig. 7.41). Few or no villi are present, and usually the plicae circulares have been lost as well. Long tubular mucous glands, homologues of the glands of Lieberkühn are present. The tubal cells are columnar, with a thin striated border. Toward the base of the gland, the cells become shorter, less differentiated, and are accompanied by many goblet

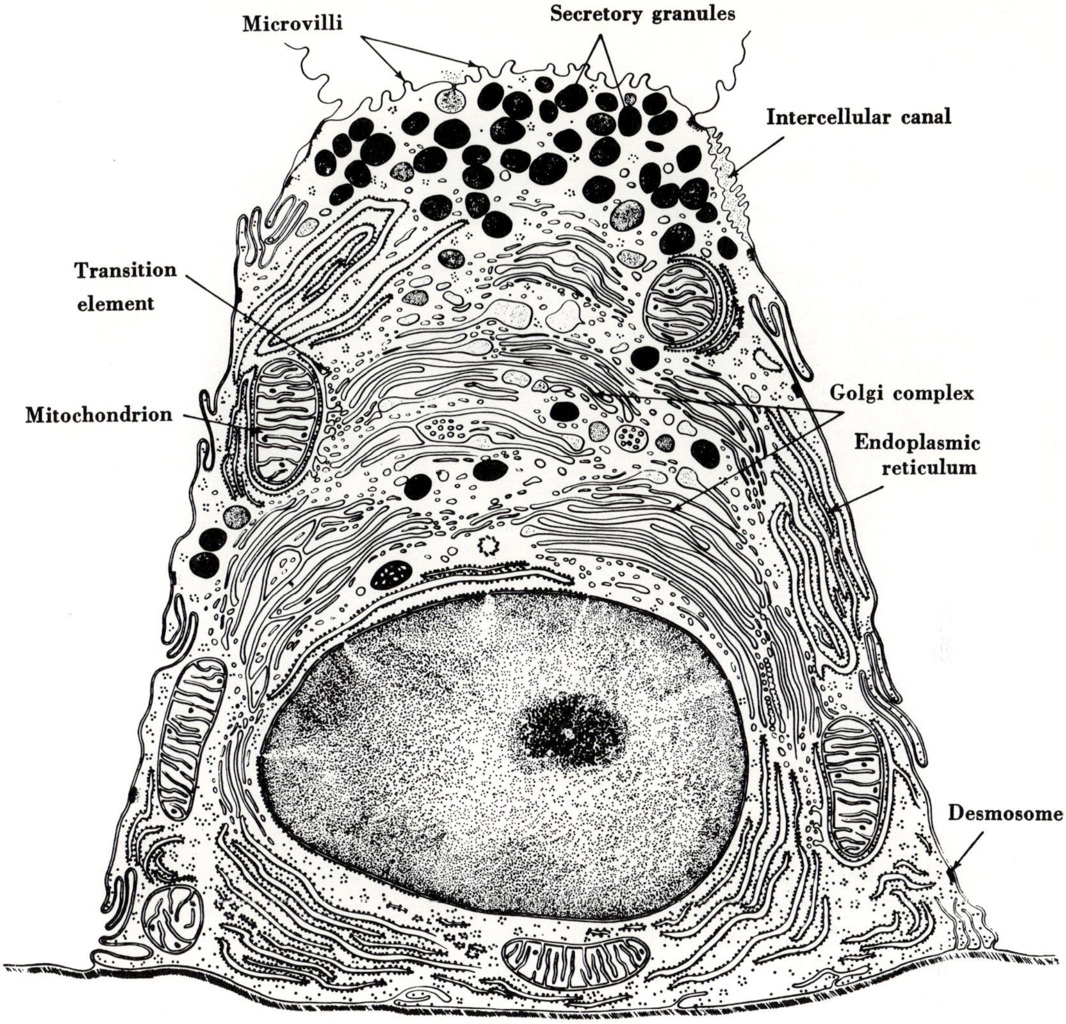

Fig. 7.39 A diagram of a Brunner's gland cell from the mouse duodenum. (After D. Friend. 1965. J. Cell Biol. 25:565–577.)

cells. The lamina propria is highly lymphoid and the muscularis mucosae is double-layered.

The areolar submucosa contains sizable amounts of adipose tissue. A strong muscularis, especially in carnivores, is present. In the kangaroo, sloth, ungulates, ape, and man, the muscularis is localized, forming three longitudinal thickenings, the *taeniae coli,* which gradually reform into a continuous layer in the rectum.

The rectum is notable for its long, large tubular glands, and for great numbers of lymph nodules in the lamina propria and submucosa. The mucosa of the lower rectum and anus is thrown into longitudinal folds and, at that point, the glands of Lieberkühn are lost. In the lower rectum the epithelial layer begins its transition to the skinlike condition as striated muscle fibers replace the smooth and condense to form a sphincter.

PANCREAS

In most vertebrates, outlying glandular masses are associated with the digestive system at two major loci. The salivary glands, at the entrance to the tube, function in a lubricatory and/or hydrolytic manner. The other locus of massive glandular activity lies just beyond the entrance to the intestine, namely the pancreas and the liver.

It is thought that originally all digestive

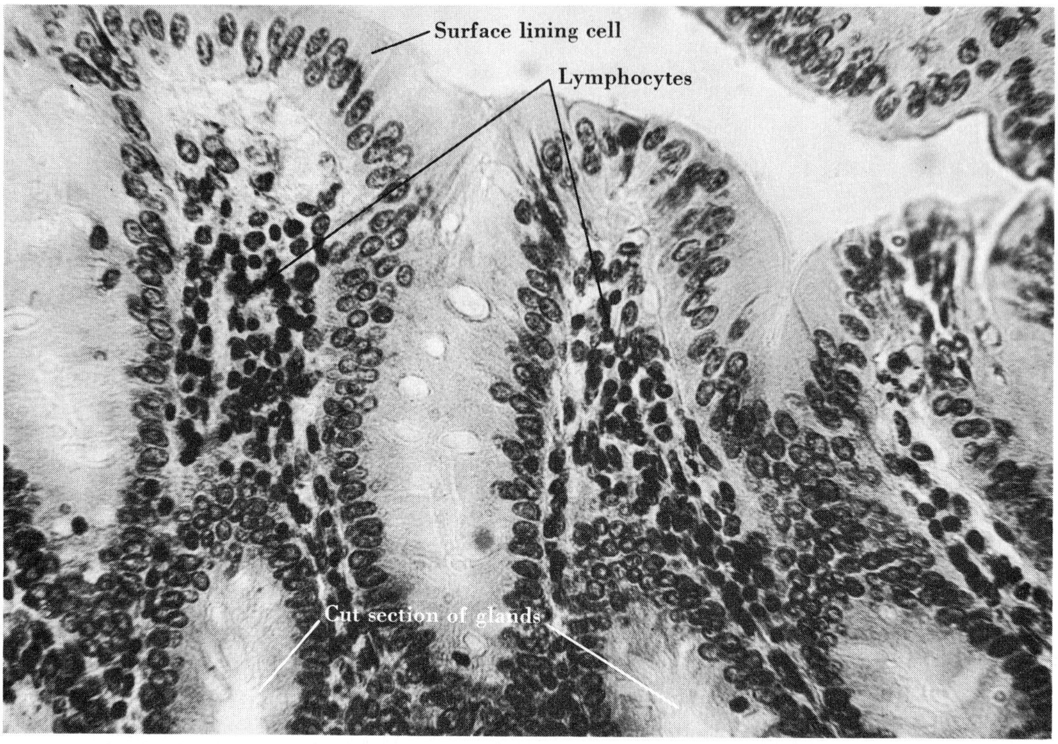

Fig. 7.40 A photomicrograph of a section of the caecum of the squirrel. Note the large number of lymphocytes in the lamina propria. X450.

enzymes were products of small masses of glandular tissue buried in the intestinal wall. Supporting this hypothesis is the lack of a distinct pancreas in Amphioxus. Additionally, the cyclostomes have small separate glands, each opening individually into the bile duct or intestinal lumen. The secretory portions of these glands can be found around or embedded in the intestinal serosa or the wall of the bile duct. In higher fishes the transition from a diffuse pancreas to a compact one can be found, and in all of the higher vertebrates the pancreas exists as a distinct entity that secretes many of the hydrolytic enzymes that are vital to digestion. *In vitro* studies have shown that pancreatic lipases, polysaccharases, and peptidases can break down food into absorbable molecules. Moreover, in all lower vertebrates the job has been left entirely to the pancreas, with little or no known assistance from intestinal lining cells.

The pancreas embryologically arises as two or more diverticula of the intestine near the entrance of the bile duct. One evagination, the dorsal diverticulum, forms the bulk of the pancreas in most vertebrates. One or two ventral diverticular arise from the bile duct primordia; usually these grow dorsally and fuse with the dorsal pancreas. Initially each of these outgrowths is ducted separately. In some vertebrates both ducts are retained; in others, one or the other atrophies.

In addition to the tubuloalveolar pancreas, in many vertebrates an endocrine pancreatic component is present as well. These pockets of endocrine tissue, the islets of Langerhans, contain cells that secrete the hormones insulin and glucagon (see Chapter 14).

The exocrine pancreas is composed of serous alveoli or acini which in many respects resemble the salivary glands (Fig. 7.42). The secretory cells are pyramidal and contain a spherical nucleus with prominent nucleoli. The cytoplasm is highly basophilic, reflecting the concentrated apparatus for protein synthesis. The basal portion of a typical acinar cell exhibits a dense aggregation of granular endoplasmic reticulum, numerous nonmembrane-associated ribosomes, and complex filamentous mitochondria. The supranuclear area

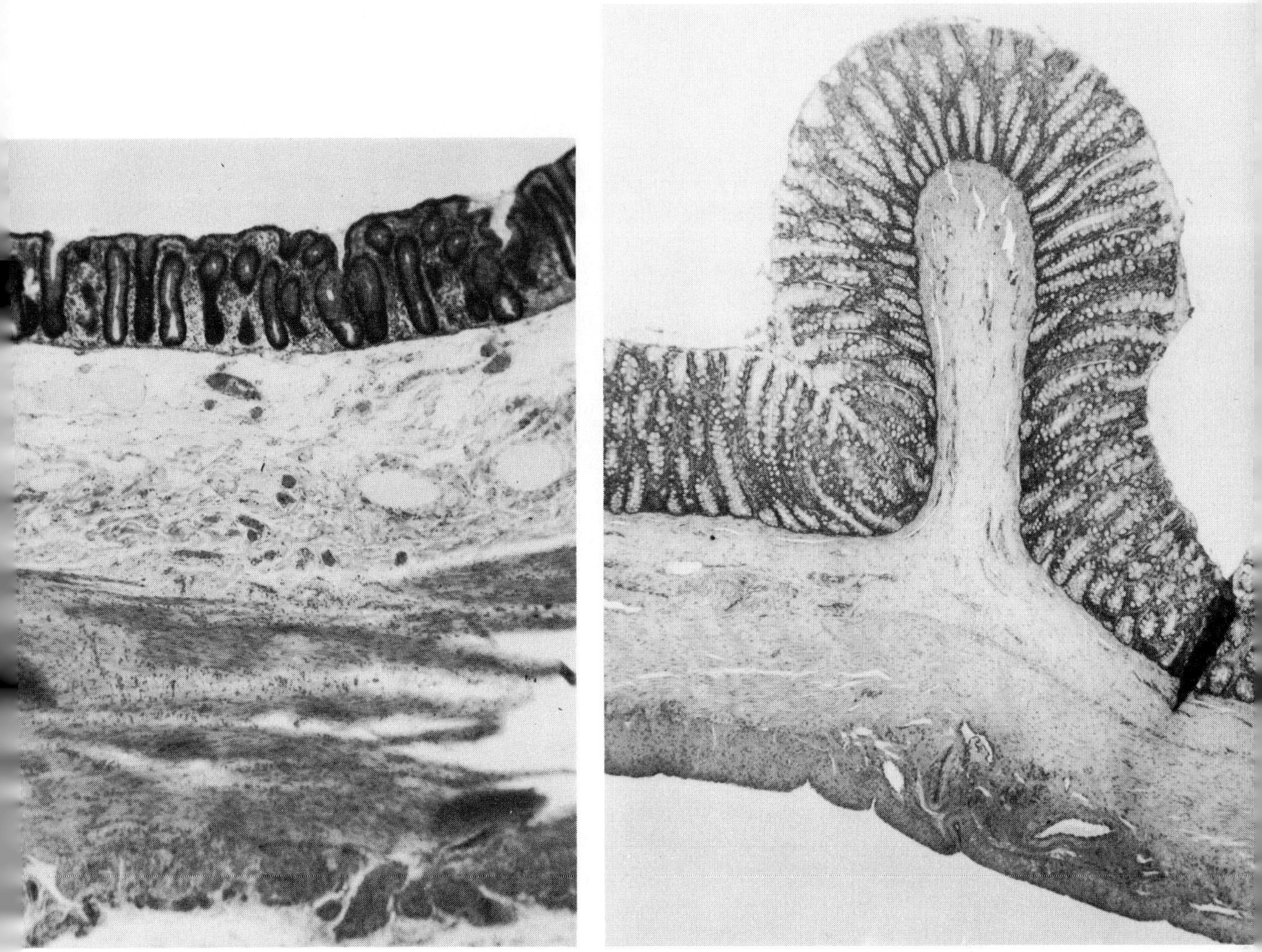

Fig. 7.41 A section through the human large intestine, *a*. Note the absence of villi. X35. Section through the human colon, *b*. Note the density of goblet cells. X50.

is occupied by a prominent Golgi network consisting of vacuoles, vesicles, and membrane-bound sinuses. Golgi vacuoles containing electron-dense material represent membrane-bound insoluble zymogen which was funneled there from the basal area of synthesis via the endoplasmic reticulum. In the course of the journey the secretory product is rendered increasingly insoluble and concentrated. Dramatic changes in relative predominance of the various organelles can be demonstrated before and after feeding. In a resting cell, i.e., before feeding, the Golgi network swells by virtue of an increased number of prozymogen granules. Hence, the supranuclear area is the most visibly dominant. After feeding, during the period of discharge of secretion, the Golgi vacuoles appear shrunken and highly reduced. At this time increasing synthetic activity can be detected basally, with a corresponding increase in basal basophilia (Fig. 2.9).

Arrangement of pancreatic cells among the fishes is highly variable. In some, such as the pike, eel, and some catfish, pancreatic acini are all gathered together to form one distinct macroscopic organ. Such a pancreas, typical of all higher vertebrates, is called *compact.* The *diffuse* pancreas of the cod and some flounders occurs as many distinct but macroscopically visible masses of acinar tissue. The perch, carp, and most cyclostomes have a *disseminated* pancreas. Such a classification refers to the fact that no mascroscopically visi-

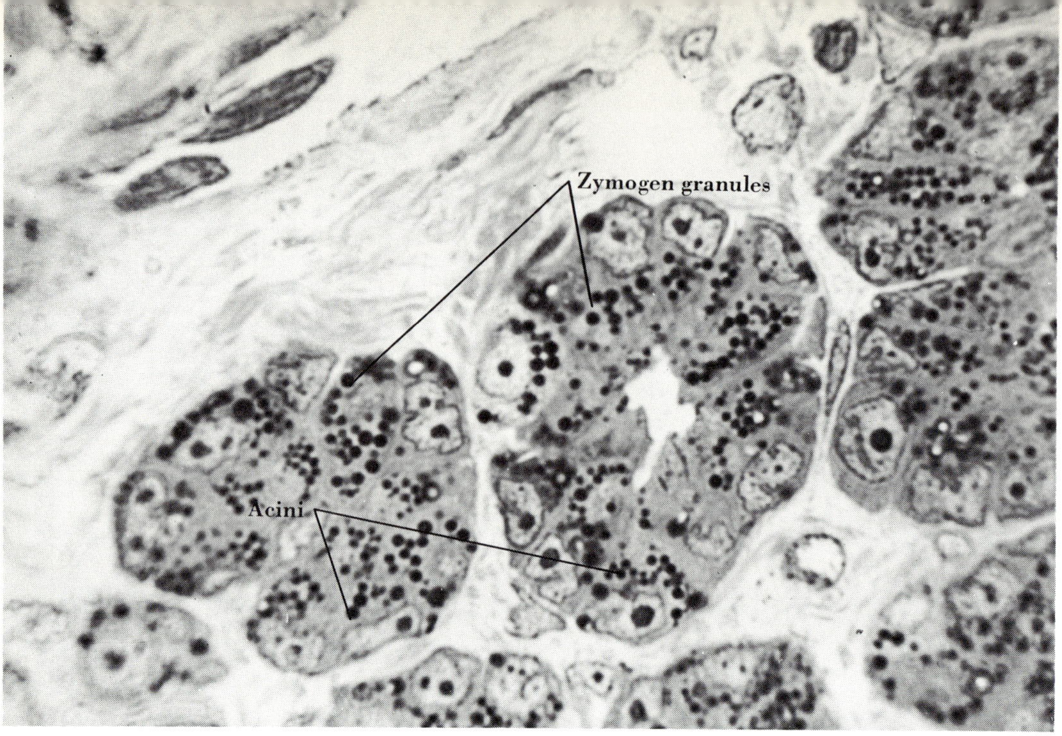

Fig. 7.42 A section through the pancreas of the turtle, *P. elegans*. X1200.

ble organ can be found. Rather one finds only microscopic clusters in and around other organs. The cell clusters of a diffuse or disseminated pancreas are most often found within or on the walls of the mesenteric vessels, hepatic vessels, or the bile duct. In some cases pancreatic tissue can be found within the liver or the small intestinal wall and contains unusual numbers of lymphocytes which are presumed to have invaded the pancreatic tissue from surrounding structures.

In fishes a typical compact pancreas lies in the mesentery between the stomach and intestine. The one or more excretory ducts subdivide to form inter- and intralobular ducts within the pancreatic tissue. The tall columnar cells that are most often found lining these ducts are supported by a connective tissue lamina propria and muscularis in the larger channels, or by a solitary, thin lamina propria in the smaller interlobular ducts. Secretory ducts, the most delicate of the channels, are formed by a lining of low columnar cells backed by a fragmentary lamina propria. The entire pancreas is covered with a thin connective tissue capsule which, in turn, is surmounted by mesothelium, the cells of which may be ciliated in one or more regions. Branches of the capsule continue into the gland mass, providing supporting tissue for the ductwork. The secretory cells of each acinus are typical and have been described above. A condition whereby some of the secretory cells reflect back over the end of a secretory duct is found in all vertebrates. Such an arrangement results in the positioning of the terminal duct cells within the lumen of the acinus. Such cells, the *centroacinar cells*, appear smaller, less basophilic, and more homogeneous than the secretory acinar cells. This, of course, reflects their relatively lower rate of synthetic activity.

The amphibian pancreas is separated into a dorsal portion, located in the dorsal mesentery, and a ventral one between the liver and intestine. Each has a separate opening into the bile duct or intestinal lumen. Moreover, some pancreatic tissue can often be found along the bile duct and/or hepatic portal veins. The acini differ little from those of fish, but there tends to be less connective tissue between and around the ducts and lobules—giving the amphibian pancreas a more packed appearance. Masses of endocrine tissue with its needed vascular support often appear in close proximity to the smaller ducts.

The reptilian pancreas is noteworthy for the great amount of endocrine tissue contained within it. Variation occurs with respect to density of acini and connective tissue, so that some reptiles, such as the snakes, have a pancreas which appears to be packed with glandular tissue, whereas others, such as the lizards, contain an organ that is less concentrated.

The avian pancreas is usually a bilobed organ. However, in some birds, such as the Ardea and Casuarius, only one lobe is present and in others three lobes can be found. One, two, or three main excretory ducts may be present, but the number of ducts does not always correlate with the number of lobes. The acini are typical, although the islets of birds are noteworthy for the segregation of cell types into separate aggregates.

The mammalian pancreas typically lies within the mesentery between the stomach and duodenum. A distinct connective tissue capsule is lacking, although loose fibroelastic partitions separate the lobes and lobules. The pancreatic mass is most compact in the higher mammals. The prototheria and metatheria have a more diffuse and/or branched organ. Although the number of endocrine islets shows species variation, all mammals appear to have islets in which mingle the different islet cell types.

LIVER

The vertebrate liver is a large, compact organ that represents the major recipient of food absorbed from the gastrointestinal tract. Its primary blood supply is from the hepatic portal vein, which drains the intestinal tract and spleen. A secondary supply is from the dorsal aorta via the hepatic arteries. The liver parenchyma arises typically from two ventral diverticula of the foregut, which develop into profusely branching tubules that penetrate the adjacent mesenchyme. Thus the diverticula are endodermal and are homologous with the intestinal mucosal epithelium. The main stems of the initial diverticula persist in most species as the *cystic* and *hepatic ducts,* which carry the exocrine secretion of the liver, the *bile,* to the small intestine. In most species a *gallbladder* develops in association with the cystic duct. In many mammals, such as human beings, the two ducts fuse more or less adjacent to the gallbladder to form a single *common bile duct.*

In the cyclostome, *Myxine,* the liver appears in its most primordial condition and provides insight into liver structure and function in the higher vertebrates. The branching tubules of the liver of *Myxine* undergo no anastomosis, but form a tortuous mass of tubules embedded within mesenchyme. The mesenchyme differentiates into a complex system of blood vessels and irregular vascular channels called *hepatic sinusoids,* and reticular connective tissue (Fig. 7.43). The si-

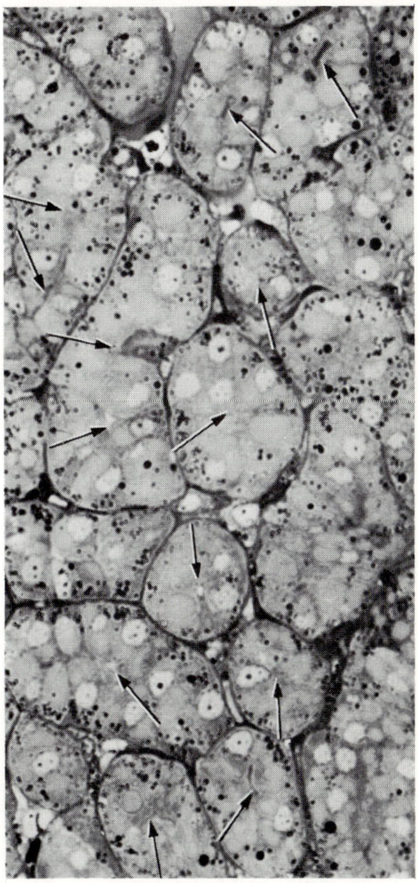

Fig. 7.43 Liver tubules of Myxine. The tubules are cut in different planes and are surrounded by a thin membrane. The arrows indicate the lumina. X405. (From E. Mugnaini and S. B. Harboe. 1967. Zeitschrift für Zellforschung und mikroskopische Anatomie Bd. 78, S. 341–369. Berlin-Heidelberg-New York: Springer.)

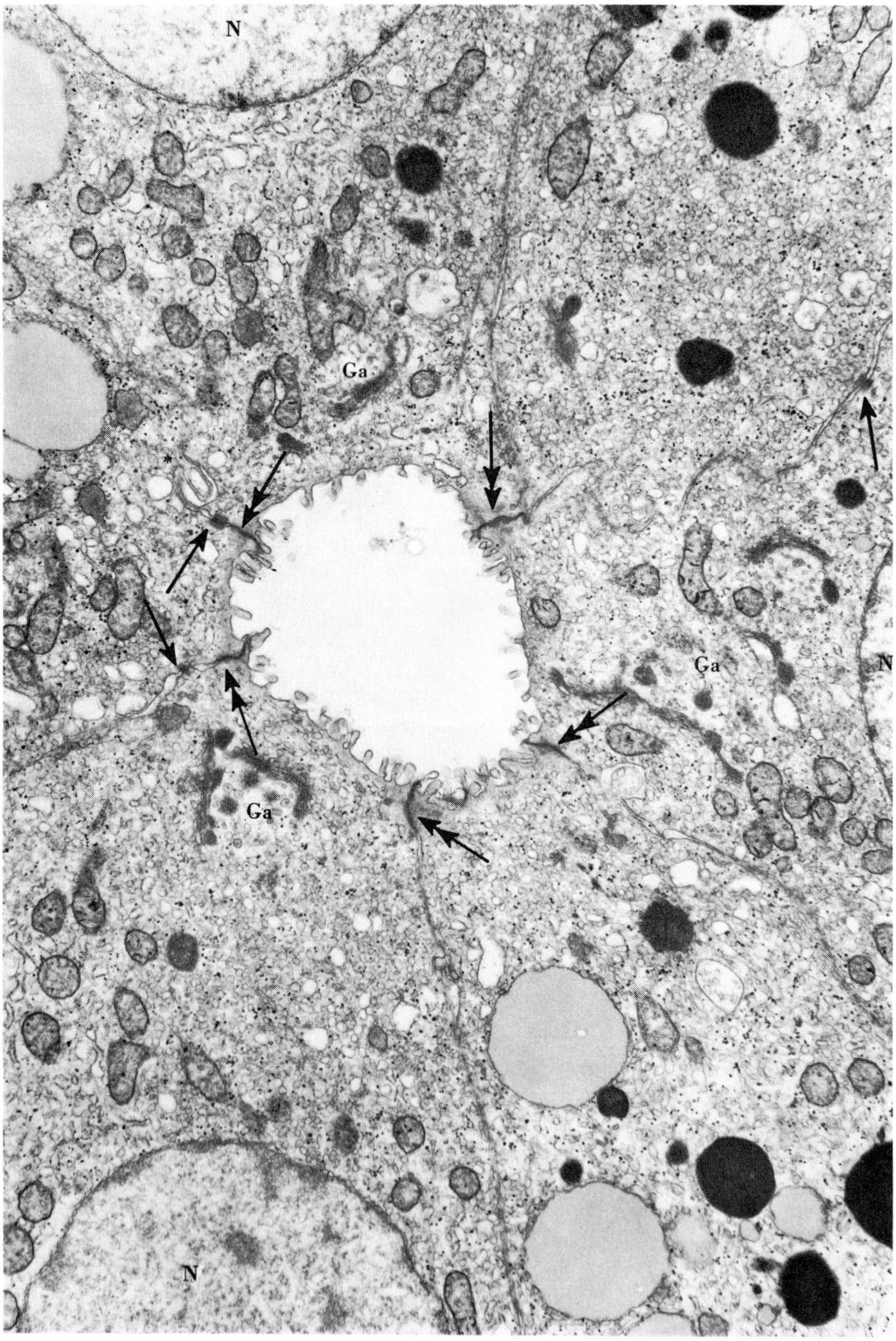

Fig. 7.44 An electron micrograph of the central area of a liver tubule of Myxine. The asterisk indicates an interdigitating cell junction. *N*, nuclei; *Ga*, Golgi bodies. Desmosomes are marked by single arrows; tight junctions by double arrows. Note the microvilli which project into the lumen. X9150. (From E. Mugnaini and S. B. Harboe. 1967. Zeitschrift für Zellforschung und mikroskopische Anatomie Bd. 78, S. 341–369. Berlin-Heidelberg-New York: Springer.)

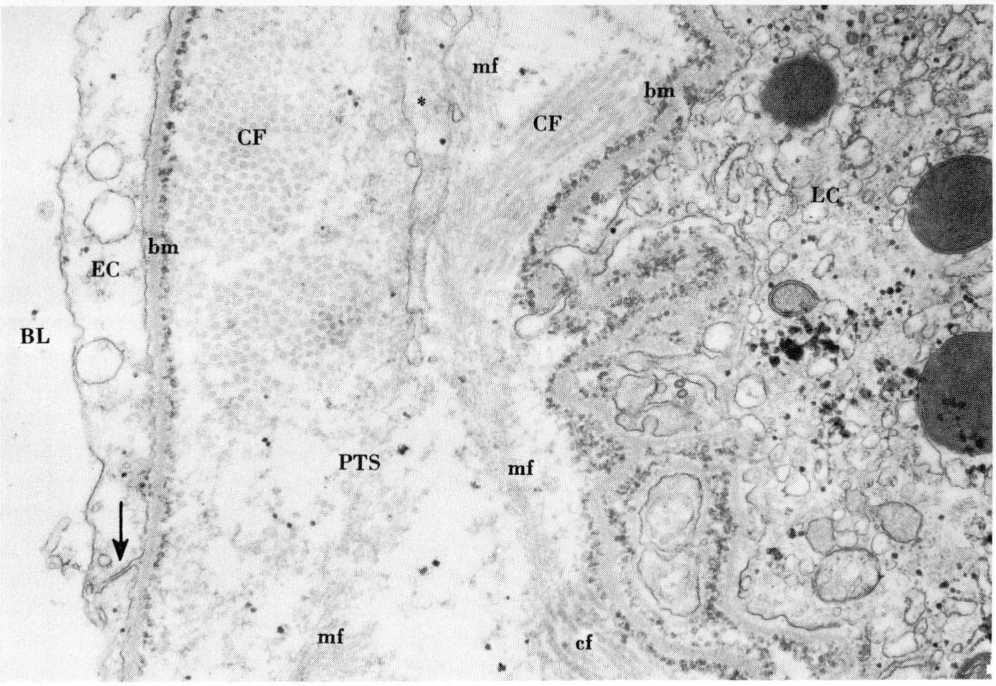

Fig. 7.45 An electron micrograph of the periphery of a liver tubule of Myxine. *BL*, blood lacuna; *EC*, endothelial lumina; *bm*, boundary membrane; *PTS*, peritubular connective-tissue space with collagen fibrils, *CF*. Microfibrils, *mf*; Liver-cell cytoplasm, *LC*. A tight cell junction is indicated by an arrow. The asterisk is located in a paratubular cell. X20,300. (From of E. Mugnaini and S. B. Harboe. 1967. Zeitschrift für Zellforschung und mikroskopische Anatomie Bd. 78, S. 341–369. Berlin-Heidelberg-New York: Springer.)

nusoids receive blood from both the portal vein and the hepatic artery, as in higher vertebrates, and empty into venous sinuses that drain into the common cardinal veins and sinus venosus.

Electron microscopic studies have revealed that the liver tubules in Myxine surround a lumen that is lined with microvilli (Fig. 7.44). The tubules themselves, except where they contact neighboring tubules, are surrounded by a peritubular space that is occupied by collagenous or reticular fibrils, scattered microfibrils, and a double array of dense globules. One array of globules is closely applied to the basement lamina of the tubule epithelium, and the other lies next to the basement lamina of the sinusoidal epithelium (Fig. 7.45). Fenestrations have been observed in the sinusoidal laminae, but it is not clear whether or not they are fixation artifacts. Occasionally, elongated cells have been described in the peritubular space, which may represent reticuloendothelial elements. The lumen of the tubule is sealed at the point of junction of component cells by tight junctions as a fascia occludens (Fig. 7.44).

The organization of the liver of gnathostomatous fishes and lower tetrapods is less straightforward or clearly understood. There is some evidence that the hepatic cells have lost their tubular arrangement distally in some areas and form laminae that branch and anastomose to form a complex meshwork, with blood vessels and sinusoids within the meshes. Double nuclei and polyploid nuclei are common characteristics of the hepatic cells of all higher vertebrates. Whereas in Myxine the bile secreted by the hepatic cells passes at all levels into well-differentiated tubules, in the higher forms it passes into bile canaliculi that in some areas are nothing more than channels between adjacent cells. In other areas a more conventional tubular arrangement is evident. In the goldfish the cell membrane of each parenchyma cell is involuted to form a single intracellular bile canaliculus that penetrates to the nucleus. Its lumen is lined with numerous, irregular microvilli, and the cytoplasm adjacent to the canaliculus contains clusters of Golgi vesicles. The terminal bile duct in the goldfish appears in transverse section as a single cell that is formed into a circle, with

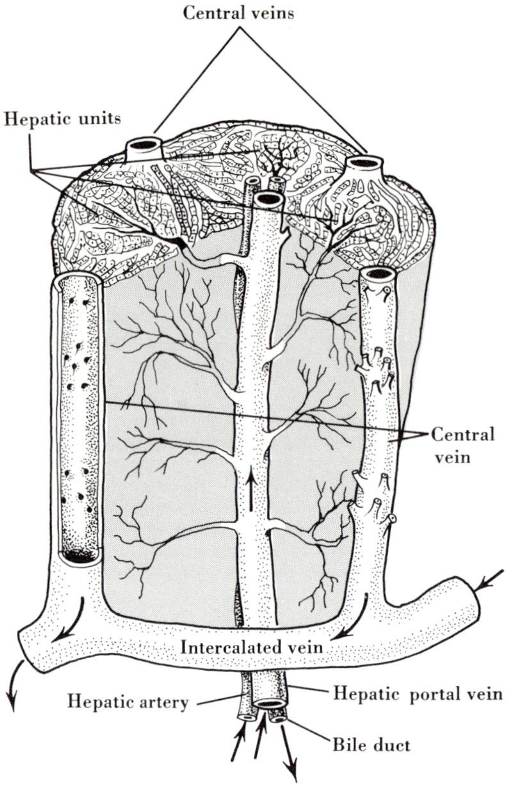

Fig. 7.46 A diagram showing the organization of the hepatic lobule in the mammalian liver. (After A. M. Rappaport *et al.* 1954. Anat. Record 119:11.)

terminal bars sealing the junction where the two ends of the cell come together to close the circle. The larger ducts are similar to those of higher vertebrates.

The organization of the mammalian liver has been the subject of controversy for many years. The parenchyma cells give no evidence of tubular arrangement in most if not all species. The liver is organized into morphological units, the *hepatic lobules,* each of which consists of branching and anastomosing laminae of parenchyma one or, in some areas, two cells in thickness (Fig. 7.46) that appear to radiate out from a central point. This point is occupied by a vein, the *central vein* (Fig. 7.47). The lobule is generally hexagonal in shape, and each angle of the periphery is usually occupied by a complex of vessels and connective tissue called a *portal area.* A portal area includes a portal vein, an hepatic artery, a variable number of lymphatic capillaries and vessels, and a bile duct. The lymphatics are generally collapsed and not easily delineated. The functional unit of the mammalian liver, however, is the so-called liver acinus (Fig. 7.46) which consists of the parenchyma situated around the terminal branches of a

Fig. 7.47 A section of an hepatic lobule of the pig liver. X100.

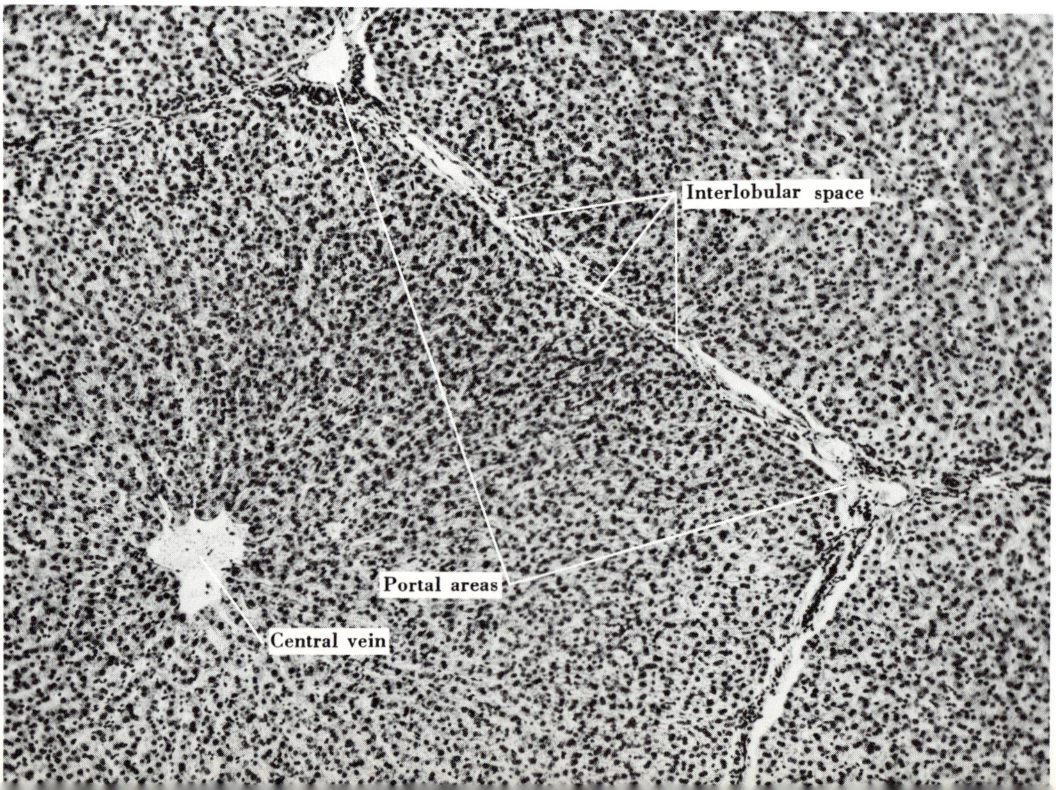

portal vein and hepatic artery. Blood from these vessels percolates through the sinusoids into the central veins of two lobules.

The bile canaliculi of mammalian livers all are formed as contiguous spaces between adjacent parenchyma cells (Fig. 7.48). One might visualize the parenchyma cells as polygonal blocks girdled by a groove and stacked in vertical laminae or sheets in such a way that the grooves approximate each other to form a network of canaliculi that encircles every cell (Fig. 7.49). In this way a system of tubules is formed that is intercellular rather than cellular, which permits the cells to be arranged in laminae only one cell in thickness, with two faces of every cell adjacent to a sinusoid. This adaptation has raised the efficiency of the mammalian liver to a level consistent with the metabolic demands of endothermy.

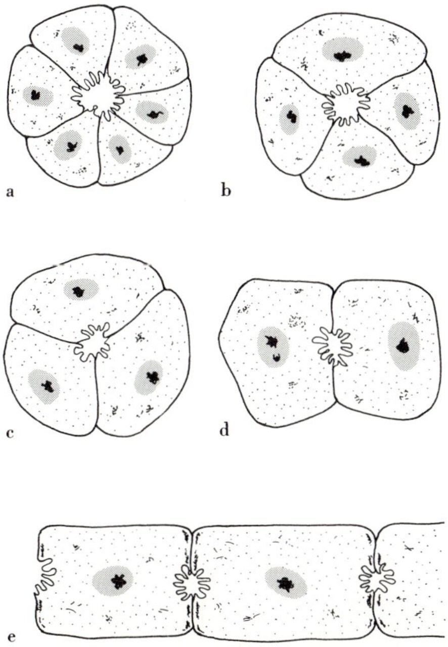

a b c d e

Fig. 7.48 A series of schematic diagrams to show the evolution of the bile canaliculi of the vertebrate liver. The liver parenchyma started out as a typical tubular gland, *a*, which soon became simplified by reduction in the number of cells forming the circle, as in contemporary Agnatha, *b*. The condition seen in most nonmammalian vertebrates, *d*, evolved through a hypothetical transitional arrangement, *c*. The mammalian arrangement, *e*, was attained by stacking parenchyma cells in sheets or laminae, with bile canaliculi formed as shown.

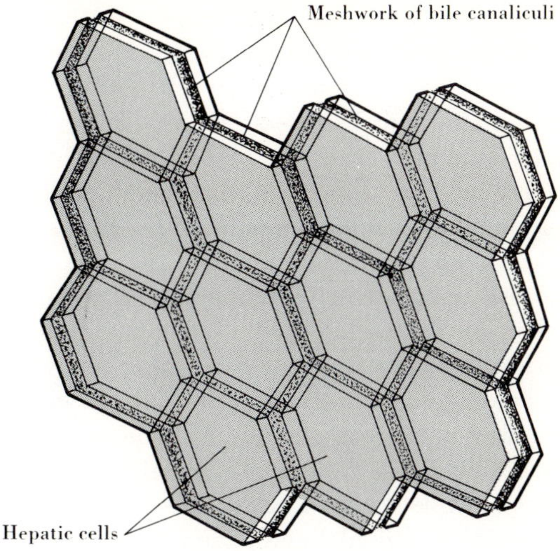

Fig. 7.49 Schematic representation of a parenchymal lamina in the mammalian liver. See text for description.

The liver sinusoids are the vascular channels that lie between the parenchymal laminae. They receive blood from the terminal twigs of both the portal veins and the hepatic arteries, and they drain into the central veins. As vascular channels they are lined with endothelium, but of a sort that is different from that of regular blood vessels. Its cells include more or less typical endothelial cells, reticular cells, and stellate macrophages (*Kupffer's cells*) and constitute part of the reticuloendothelial system. The junctions between these cells are evidently neither so firm nor so continuous as in regular endothelium. The reticular cells, in addition to being somewhat phagocytic, secrete a loose meshwork of collagenous or "reticular" fibers outside the sinusoidal wall. The Kupffer's cells are very active phagocytes, and so they contain numerous vesicles of ingested material, lysosomes, and microbodies called peroxisomes. As described in Chapter 2, lysosomes are rich in phosphatases and proteolytic enzymes, and peroxisomes contain uricase, which functions in nitrogen metabolism. The sinusoidal cells are supported by a basement lamina, which is discontinuous at points, and loose reticular connective tissue. A thin space, the *space of Disse* (Fig. 7.50) separates the sinusoidal epithelium from the parenchymal cells. This space is occupied in part by collagenous bundles of the reticular connective tissue and by irregular microvilli of the parenchymal cell membrane. Because of spaces between the sinusoidal cells and imperfections in their basement lamina, plasma can filter into the spaces of Disse. This doubtless constitutes the source of the lymph that is produced in copious quantities by the liver. The lymph so formed finds its way to the connective tissue of the portal areas, where it is taken up by the blindly ending lymph capillaries.

The bile canaliculi described above come together at the bases of the laminae and join the *terminal bile ducts* in the portal areas. The terminal bile ducts of vertebrates are formed of a simple cuboidal epithelium supported by a basement lamina and the loose connective tissue of the portal area. They join to form larger bile ducts, the epithelium of which becomes simple columnar, and eventually the hepatic ducts.

The circulation of fluids in the liver is therefore complex. It receives blood essentially of three different kinds: blood from the intestinal tract, which contains nutrients and toxins absorbed therefrom; blood from the spleen, containing the products of red blood cell degradation; and arterial blood from the hepatic artery. These bloods enter the sinusoids, where they come in close or even intimate contact with the parenchyma cells and with phagocytic cells. Nutrients and toxins are absorbed by the parenchyma cells, and the red blood cell breakdown products are engulfed by the phagocytes, so that the blood that leaves the liver is adjusted in respect to these substances. The parenchymal cells metabolize the substances absorbed and secrete bile into the biliary canaliculi. The flow of bile in the canaliculi is opposite in direction to that of the blood in the sinusoids (a counter-current effect). Lymphatic drainage is also counter to the direction of blood flow, and it occurs not in lymphatics but probably in the tissue spaces of Disse.

Gallbladder

The gallbladder (Fig. 7.51) is a thin-walled sac that collects and stores bile secreted by the liver. Its mucosal lining is a nonglandular, tall columnar epithelium supported by a thin lamina propria. The luminal surface of the

Fig. 7.50 An electron micrograph of portions of two hepatic cells bordering on sinusoid. X41,000. (From N. L. Trotter. 1967. J. Cell Biol. 34:703.)

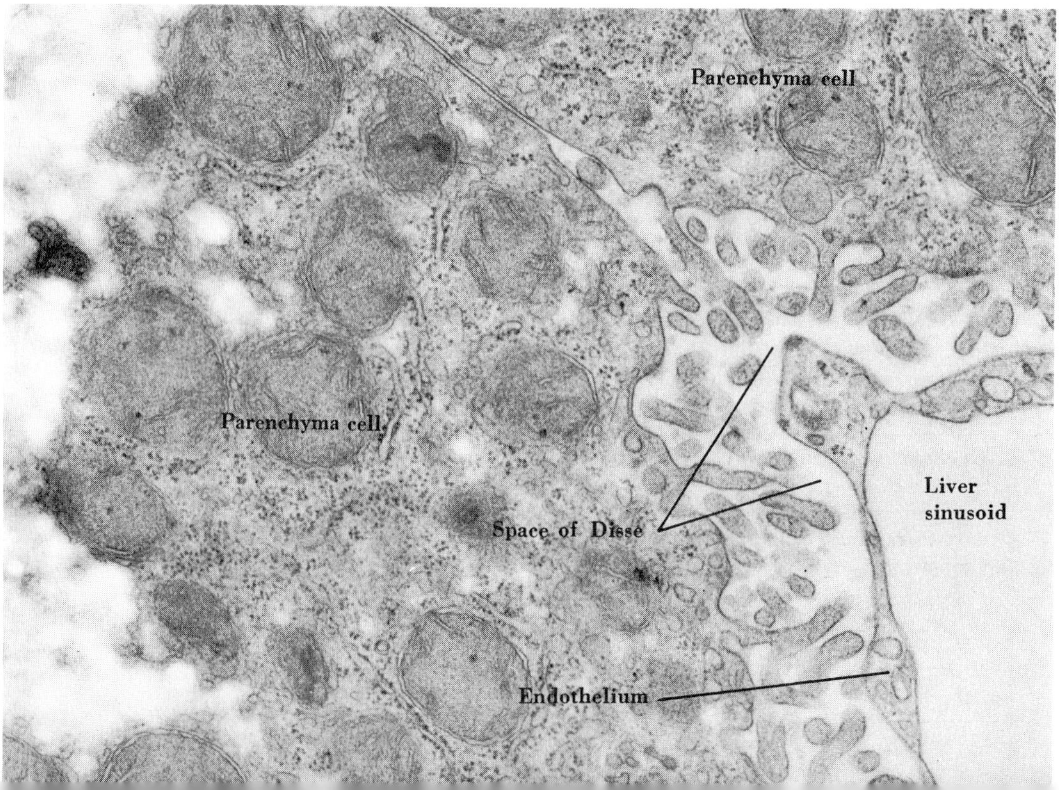

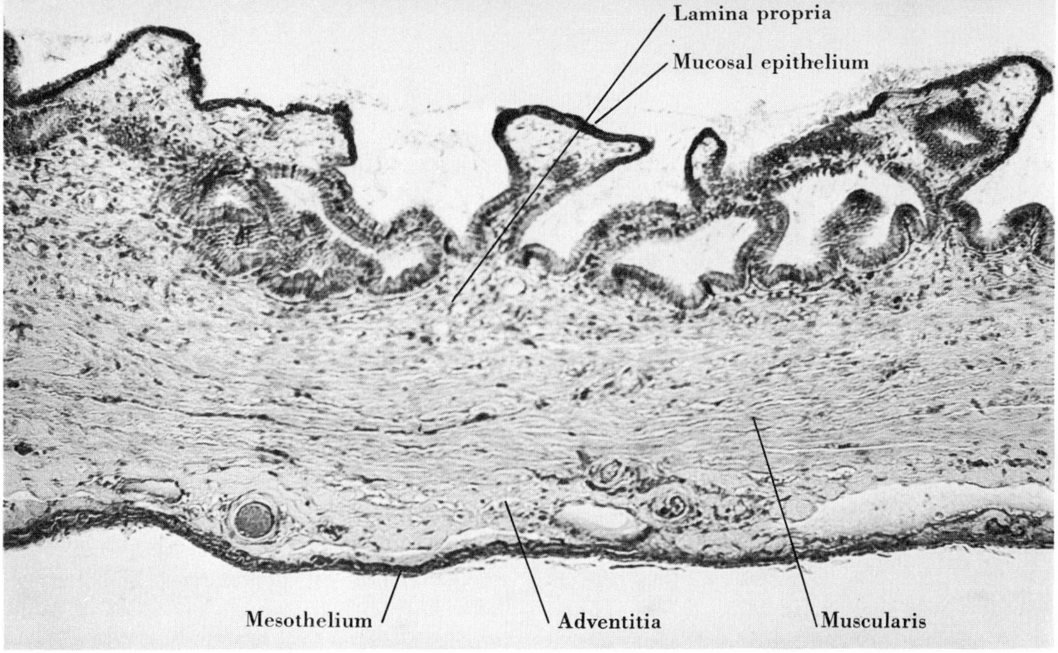

Fig. 7.51 A section through the gall bladder of the cat. X100.

epithelial cells appears smooth by light microscopy but shows short, irregular microvilli by electron microscopy. At the neck of the gallbladder the epithelium is cuboidal, and numerous simple tubuloalveolar glands occur in the lamina propria and muscularis. The mucosa is thrown into folds that flatten out with distention. The muscularis is also thin and includes smooth muscle fibers that run in various directions, collagenous and reticular fibers, and an elastic network. The muscularis in reptilian gallbladders may be exceptionally thin. Surrounding the muscularis is an adventitia of fairly dense connective tissue in which may be found collagenous and elastic fibers, and scatterings of fibroblasts, lymphocytes, and macrophages. The surfaces of the gallbladder not embedded in the liver are covered finally by the mesothelium of the peritoneum.

The gallbladder is present in most vertebrates, but there are noteworthy exceptions. Although myxinoid cyclostomes have a large gallbladder, the petromyzonts, such as *Lampetra,* lose the gallbladder and the major bile ducts after metamorphosis. Penguins and other birds that feed on oil- or fat-rich fishes have a large gallbladder, but many others lack a gallbladder altogether, such as ostriches, pigeons, parakeets, and hummingbirds. The mammalian orders that reputedly lack a gallbladder are the cetaceans and the ungulate orders.

Functions of the Liver

As the main metabolic "factory" of the body, the liver serves several very basic functions as well as certain specialized ones. The broad categories of vertebrate liver function are metabolism, storage, and the secretion of bile. None of these is unified, however, for they are all interrelated and to a large extent concurrent.

METABOLISM

Glucose that has been absorbed into the blood by the gastrointestinal tract is taken up by the liver parenchyma cells and polymerized to glycogen, a process known as glycogenesis. Other sources of glycogen are the products of protein and lipid metabolism, which are transformed to glucose and then to glycogen. The opposite process, glycogenolysis, is the means by which glycogen is degraded by phosphorylation to glucose-6-phosphate, which in turn may be converted to glucose prior to release into the bloodstream. Under hormonal regulation, an appropriate balance between glycogenesis and glycogenolysis achieves the maintenance of a concentration of glucose in the bloodstream that varies but little during the day.

Synthesis and conversion of fat and other lipids takes place in the liver parenchyma from carbohydrate, protein, or other lipid precursors. Fats and lipids that are synthesized in the liver are conjugated with protein to form small lipoprotein particles, which may either be stored in the liver or liberated into the bloodstream for use elsewhere as a lipid source.

The liver parenchyma is the site of synthesis of the plasma proteins, particularly of serum albumin, fibrinogen, and prothrombin. It is also the site of deamination, the process whereby NH_2 groups are split off from amino acids preparatory to lipid or carbohydrate metabolism. The amino groups are immediately coupled enzymatically and converted to urea, which is nontoxic and readily excreted by the kidneys. A wide variety of toxins absorbed by the intestinal tract, either directly ingested or originating in the intestine from bacterial fermentation, are either degraded to harmless derivatives in the liver or are reacted with glucuronic acid or potassium acid sulfate to form innocuous byproducts. The products of detoxification such as these may be excreted with the urine or with the bile.

STORAGE

The liver is important as a storage organ for certain substances: principally glycogen, fats or oils, and a variety of vitamins (chiefly A, D, K, and several of the B vitamins). The liver of some species of sharks has been likened to twin barrels of oil, for they may comprise up to one-tenth the total body weight and consist of 80–90 percent oil. In some fishes the oil content of the liver is reputed to be significant enough to provide considerable buoyancy to the animal, thus compensating for lack of a swim bladder in these species. Fish liver oils are high in Vitamin A and D content, and include unusual substances, such as clupanodonic acid, certain alcohols, and peculiar hydrocarbons of high molecular weight. Tuna oil polymerizes and hardens sufficiently on exposure to air to be as useful in paint as linseed oil is.

SECRETION OF BILE

Bile represents both an excretion and a secretion of the liver parenchyma. The excretory components are primarily the bile pigments, which are derived from the hemoglobin of degenerated red blood cells, and cholesterol and lecithin. The major secretory products are the bile salts and other emulsifying agents (of which lecithin itself may be one) that facilitate the digestion and absorption of fats in the intestine. Other secretory products may include alkaline salts, such as sodium bicarbonate, that aid in adjusting the pH of the intestinal juices.

SELECTED REFERENCES

Andrew, W. 1959. Textbook of comparative histology, Oxford University Press, New York.

Barrington, E. J. W. 1957. The alimentary canal and digestion. *In* Physiology of fishes. Vol. 1. Margaret Brown, ed. Academic Press, New York.

Brauer, R. W. 1963. Liver circulation and function. Physiol. Rev. 43:115–213.

Cardell, R. R., S. Badenhausen, and K. R. Porter. 1967. Intestinal triglyceride absorption in the rat: An electron microscopical study. J. Cell Biol. 34:123–157.

De Duve, C., and P. Bandhuin. 1966. Peroxisomes (microbodies and related particles). Physiol. Rev. 46:323.

Dumont, J. 1965. Prolactin-induced cytologic changes in the mucosa of the pigeon crop during crop-"milk" formation. Z. Zellforsch. 68:755–782.

Ells, H. A. 1954. The gross and microscopic anatomy of the liver and gall bladder of the lizard, *Sceloporus occidentalis biserialus* (Hallowell). Anat. Rec. 119:213–229.

Friend, D. S. 1965. The fine structure of Brunner's glands in the mouse. J. Cell Biol. 25:563–577.

Helander, H. F. 1964. Ultrastructure of the secretory cells in the pyloric gland area of the mouse gastric mucosa. J. Ultrastruct. Res. 10:145–159.

Ichikawa, A. 1965. Fine structural changes in response to hormonal stimulation of the perfused canine pancreas. J. Cell Biol. 27:369–387.

Ito, S. 1965. The enteric surface coat on cat intestinal microvilli. J. Cell Biol. 27:475–493.

Ito, S., and Winchester, R. J. 1963. The fine structure of the gastric mucosa in the bat. J. Cell Biol. 16:541–579.

Mugnaini, E., and S. B. Harboe. 1967. The liver of *Myxine glutinosa:* a true tubular gland. Z. Zellforsch. 78:341–369.

Rappaport, A. M., S. J. Borowy, W. M. Lougheed, and W. N. Lotto. 1954. Subdivision of hexagonal liver lobules into a structural and functional unit: Role in hepatic physiology and pathology. Anat. Rec. 119:11.

Trier, J. S. 1963. Studies on small intestinal crypt epithelium. I. The fine structure of the crypt epithelium of the proximal small intestine of fasting humans. J. Cell Biol. 18:599–621.

Trotter, N. L. 1967. Electron-opaque bodies and fat droplets in mouse liver after fasting or glucose injection. J. Cell Biol. 34:703-712.

Weinreb, E. L., and N. M. Bilstad. 1955. Histology of the digestive tract and adjacent structures of the rainbow trout. Copeia 3:194–204.

Yamamoto, T. 1965. Some observations on the fine structure of the intrahepatic biliary passages in goldfish (*Carassius auratus*). Z. Zellforsch. 65:319–330.

chapter 8
THE RESPIRATORY SYSTEM

The respiratory system consists of the organs that effect the exchange of carbon dioxide and oxygen between the blood of the organism and the outside environment. In vertebrates the respiratory system is associated both structurally and embryologically with the anterior part of the digestive tract—more specifically, the pharynx. In fishes and aquatic amphibians it consists primarily of the gills, and in other vertebrates it consists of the lungs and associated pulmonary passages. Actually, any epithelium in contact with the outside environment has the potentiality of functioning in respiration. Most amphibians and some fishes, for example, use the integument as an accessory respiratory organ, even to the extent of its being the principal organ for this purpose in some species, such as in most frogs and many urodeles. In some fishes other respiratory organs appear as specializations of the rectum or of the cloaca.

Although the organs for aquatic and aerial respiration appear superficially to be so completely different as to be unrelated, a study of their evolution and morphogenesis reveals that both represent the adaptation of the pharyngeal epithelium to two different respiratory media. It is probably not coincidence that the lungs receive their blood supply from the sixth gill arch. On the contrary, this fact, together with the site of derivation, innervation, and function suggests that the pulmonary epithelium is in every way homologous with the gill epithelium and, in the broadest sense, that the lung represents a gill which has invaginated instead of evaginated.

There is little doubt that gills preceded lungs in vertebrate evolution, probably having been inherited with only minor modification from a primitive chordate ancestor. It was not long before lunglike structures made their appearance, however, for indications of pharyngeal and adjacent esophageal outpocketings have been described in all orders of gnathostomatous fishes. Such outpocketings were especially well developed in the Crossopterygii, a primitive order of bony fishes that flourished during the Devonian period. The crossopterygians are important to biologists because they were ancestral to the terrestrial vertebrates and the dipnoans.

Palaeontological evidence suggests that the crossopterygians lived in shallow inland waters which, according to one school of thought, were subjected to seasonal drought. In response to this situation, these fishes are believed to have developed a lunglike outpocketing of the pharynx that permitted sufficient respiratory exchange between the blood and gulped air so as to permit survival during the times when the pools were so concentrated with organic matter and perhaps mud that gill breathing would be seriously impaired. Along with lungs, the crossopterygians developed limb skeletons which enabled them to support themselves and to get about in a minimum amount of water—perhaps even, according to one speculation, to traverse the mud flats to adjacent and more favorable pools. A modified form of this hypothesis holds that the ancestral crossopterygians lived in lush, swampy areas that were so choked with plant life that the fishes had to develop a kind of lung to augment respiratory efficiency. This would follow, because in an environment which would support a rich flora, one would expect the water to be very warm and consequently to have a low oxygen tension. The development of limb skeletons would also have adaptive value in such an environment, for they would facilitate locomotion through the dense vegetation. Acquisition of both lungs and limbs, then, would make the rich insect life along the shore more accessible and set the stage for emergence on to land. Any anatomical change that would open up new environments for exploitation would be of greater significance to progressive evolution than would changes which merely facilitated survival. The latter are more likely to lead to specialization at the expense of potential adaptability to new changes in the environment. Among the descendants of the crossopterygians were the dipnoans, or lungfishes, which did just this and have changed little since. On the other hand, a second line of descent took advantage of the new environment and led to the evolution of the amphibians and, from them, of reptiles, birds, and mammals.

Most fishes have an outpocketing of the foregut which is generally held to be homologous with the pharyngeal outpocketings described above. They have developed it as a hydrostatic organ rather than as a lung, however, and as such it is known as a swimbladder or *gas bladder*. It may be single or paired, and the members of a pair may be equal or of different sizes. Moreover, in many species, the *physostomous* fishes, the gas bladder is connected to the esophagus or pharyngeal chamber by a patent tube, whereas in others the connection is lost and gas is secreted into the bladder from the blood. The latter are known as *physoclistous* fishes. The histological characteristics of gas bladders are described below.

Whether the respiratory epithelium covers a gill or lines a lung, it consists of simple squamous or low cuboidal cells supported by a richly vascularized fibroelastic connective tissue component, as will be described.

GILL RESPIRATION

Early in the embryogenesis of vertebrates, four to seven bilateral outpocketings, the *pharyngeal gill pouches,* appear in the pharynx. The epithelium of the pouches, which is of endodermal origin, comes in contact and fuses with the epithelium of the ectodermally derived integument. The mesenchyme, which heretofore has occupied the space between ectoderm and endoderm, becomes restricted thus to vertical columns, the *gill arches,* between successive pouches. In fishes and amphibians the gill pouches then break through to the outside through *clefts* which appear along the line of fusion of endoderm and ectoderm. Hence water can be drawn in through the mouth and expelled through the gill clefts, in this way effecting a water pumping mechanism.

The mesenchyme within the gill arches, meanwhile, differentiates cartilaginous supporting arches, blood vessels, and skeletal muscle. The blood vessels form a rich vascular network in intimate contact with the epithelium lining the clefts. In most gill breathers the epithelium that is on the endodermal side of the margin of the cleft undergoes the extensive outgrowth which characterizes the differentiation of the gill filaments; but in the larvae of some species (such as lungfishes and most amphibians) the gill filaments develop from the epithelium of the ectodermal side. The gill arches undergo further differentiation to form gill rakers, which prevent food, debris, and other solid matter from passing through the gill clefts and over the delicate gill filaments. The epithelium of the gill filaments undergoes further outpocketing to form the *gill lamellae.* The dissolved oxygen in the water is absorbed by the epithelium of the lamellae and carbon dioxide is released, thus effecting respiratory exchange. In some species of fishes, such as the mackerel, water runs through the mouth and out the gill clefts continuously as long as the animal is moving forward and keeps its mouth open. In fact, some species can breathe only in this way—they cannot close their mouths and will suffocate unless they keep moving. Elasmobranchs can inhale water through spiracles, rather than through the mouth. The spiracles represent the first pair of gill clefts that have become modified for this purpose. The Sarcopterygii, which include the crossopterygians and dipnoans mentioned above, have nostrils through which either water or air can be inspired, as do all amphibians and higher vertebrates. The movement of water in cyclostomes appears to be tidal, surging in and out of the branchial chamber through the branchial pores.

The gills of teleosts are representative of fishes' gills. Here each gill constitutes a demibranch, which in turn consists of gill filaments on either the cranial or caudal border of a gill cleft. The gill filaments radiate out like a lady's fan. They are supported internally at their bases by skeletal elements, the gill septa. Outfoldings of the filaments occur as lateral plates, the gill lamellae, which represent the functional parts of the gill.

The gill lamellae consist of a single layer of squamous or low cuboidal epithelial cells supported by a basement lamina and in direct contact with the basement lamina of a rich capillary network (Fig. 8.1). Blood enters a given gill lamella by a branch of an afferent branchial artery and leaves by a branch of an efferent branchial artery. The direction of blood flow is counter to that of the flow of water over the lamellae, so that the blood leaving the lamellae is in closest contact with water in which the oxygen tension is at a maximum and the carbon dioxide tension is at a minimum (Fig. 8.2). The number of gill lamellae per gill varies considerably, but it can be correlated directly with the customary activity of the species. For example, the filaments of the mackerel, an exceedingly active fish, have a far denser population of lamellae than do the gill filaments of the toadfish, a rather sedentary species.

Gill filaments are supported usually by cartilages or bony elements along one edge or by a single, centrally located element. Branchial

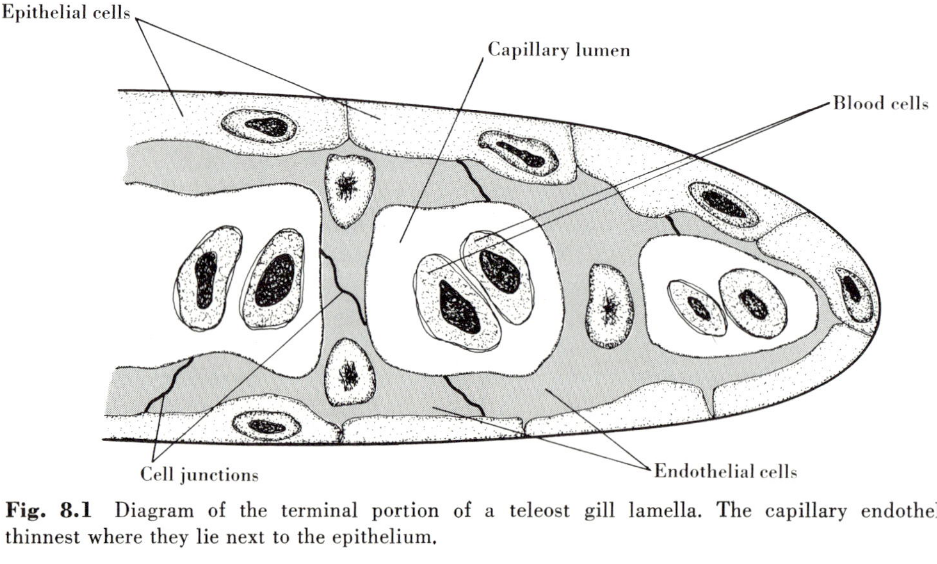

Fig. 8.1 Diagram of the terminal portion of a teleost gill lamella. The capillary endothelial cells are thinnest where they lie next to the epithelium.

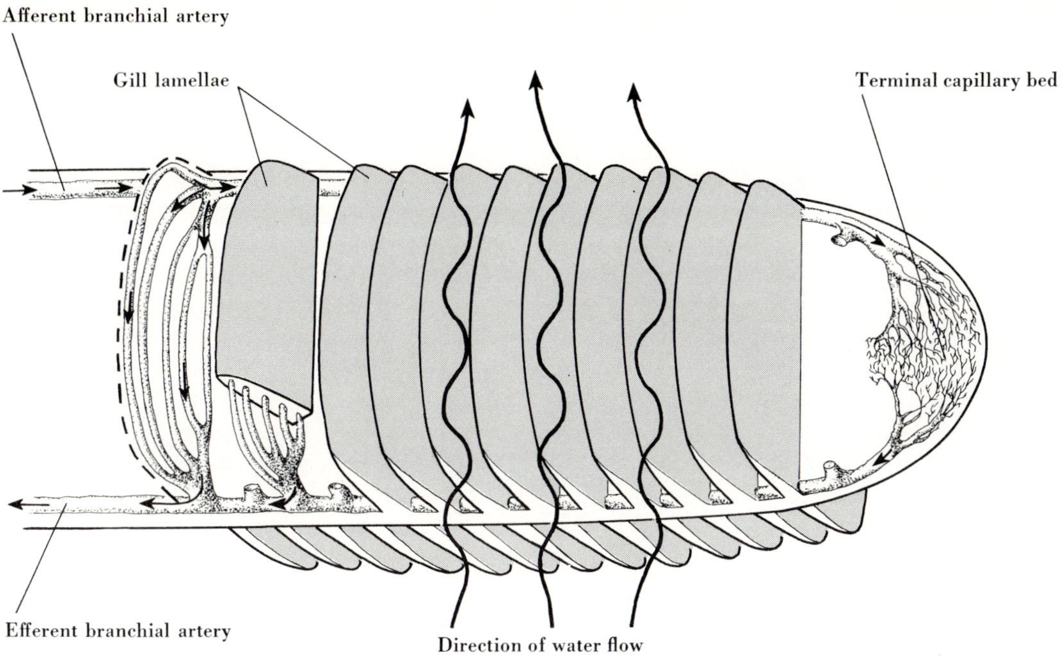

Fig. 8.2 Schematic representation of the circulation in a teleost gill filament. Blood enters the gill filament via an afferent branchial artery, from which branches enter the gill lamellae, where they undergo further branching to form the gill capillaries. The direction of blood flow through the gill lamellae is counter to that of the water that flows between the lamellae.

muscles insert on these for adduction and abduction of the filaments.

In addition to the epithelial cells covering the lamellae are occasional goblet cells and large, *acidophilic cells*. The latter occur most densely in the stratified epithelium between the bases of adjacent lamellae (Fig. 8.3), and they may also be found in clusters in the epithelium which is closest to the afferent branchial artery. Electron micrographs show the acidophilic cells to have numerous mitochondria and a well-developed endoplasmic reticulum of the smooth or agranular type (Fig. 8.4). At the apex of an acidophilic cell in marine fishes may be found a vesicle which is presumed to contain excreted salt ready to

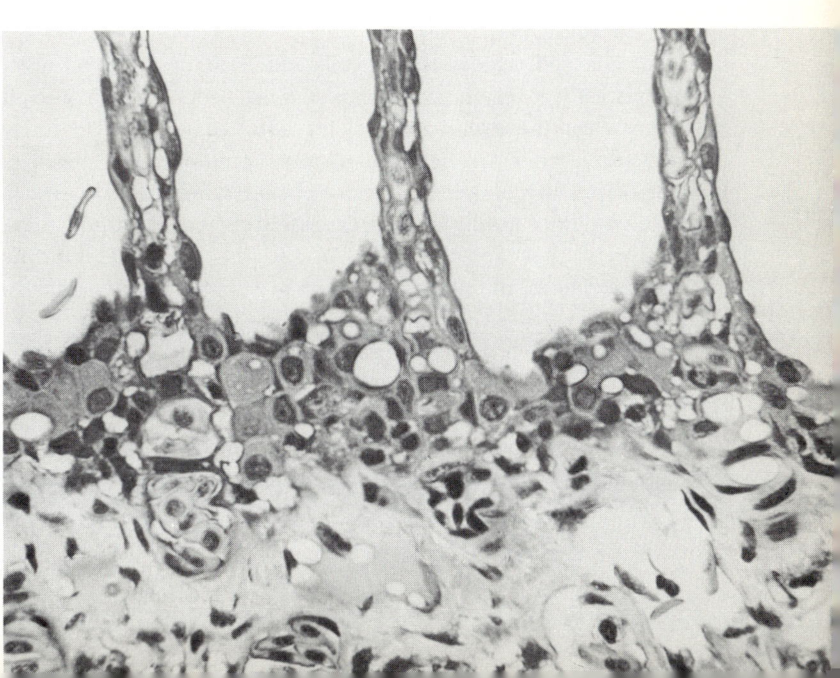

Fig. 8.3 A longitudinal section through a gill filament of the shark *Mustelus canis* showing the origin of three gill lamellae. The epithelium of the lamellae is simple squamous, but at the bases it is partly stratified. Note the plump acidophilic cells in the base epithelium. The "holes" are blood capillaries, some of which contain blood cells. X450.

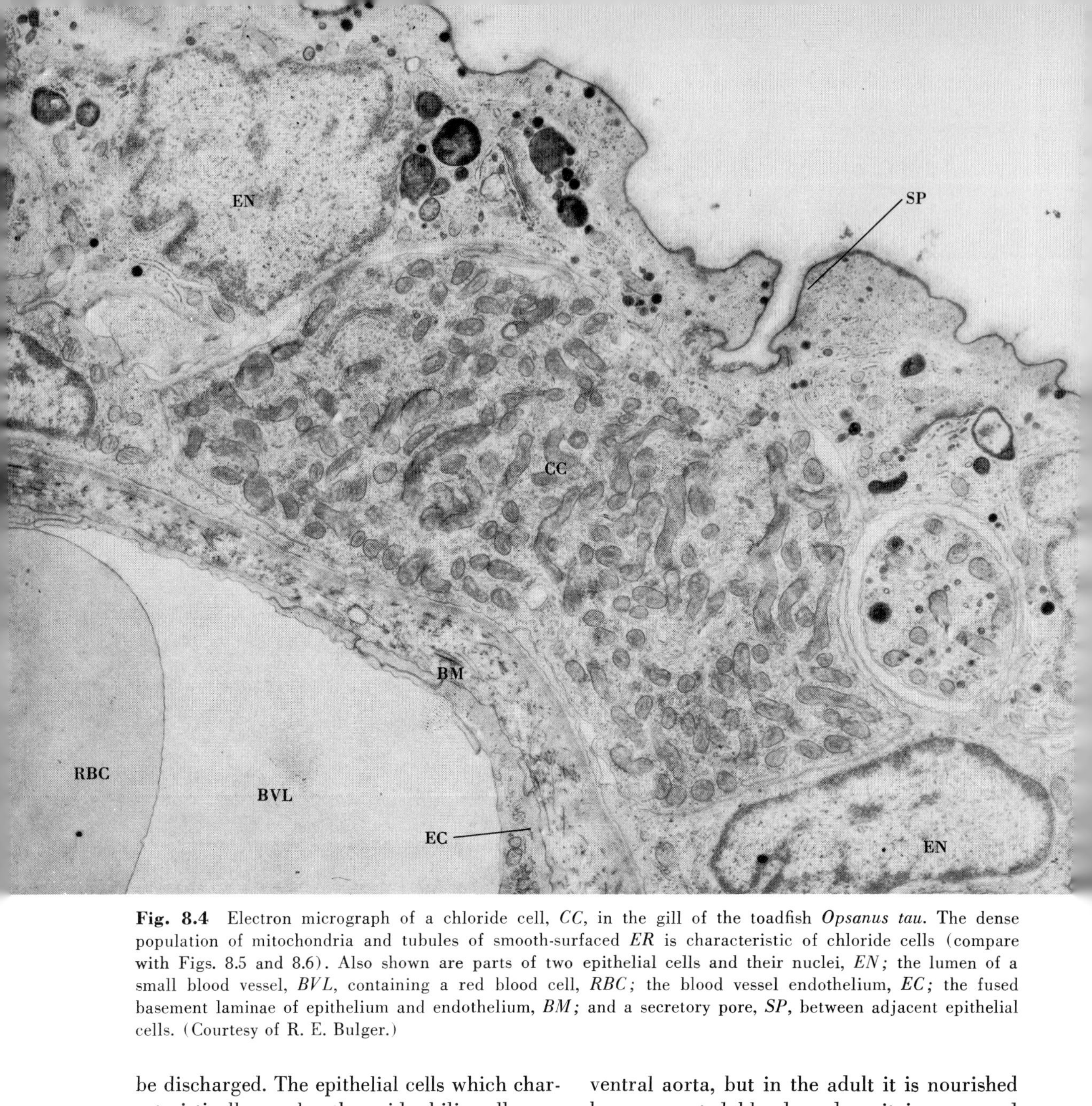

Fig. 8.4 Electron micrograph of a chloride cell, *CC*, in the gill of the toadfish *Opsanus tau*. The dense population of mitochondria and tubules of smooth-surfaced *ER* is characteristic of chloride cells (compare with Figs. 8.5 and 8.6). Also shown are parts of two epithelial cells and their nuclei, *EN;* the lumen of a small blood vessel, *BVL*, containing a red blood cell, *RBC;* the blood vessel endothelium, *EC;* the fused basement laminae of epithelium and endothelium, *BM;* and a secretory pore, *SP*, between adjacent epithelial cells. (Courtesy of R. E. Bulger.)

be discharged. The epithelial cells which characteristically overlap the acidophilic cells separate when the excretory vesicles burst. The acidophilic cells are believed to remove chlorides from the blood against a concentration gradient, thus aiding in osmoregulation. Accordingly, they are not found in the gills of elasmobranchs, the tissues of which are nearly isosmotic with sea water. Some of the acidophilic cells, such as in the pseudobranch, described below, are believed to secrete the enzyme carbonic anhydrase.

The first demibranch is an unpaired gill. In the embryo it is supplied by a branch of the ventral aorta, but in the adult it is nourished by oxygenated blood, and so it is presumed not to have a respiratory function; hence it has been known as a *pseudobranch* (false gill). In elasmobranchs and in some of the ganoid fishes (such as sturgeon and gar) the pseudobranch is exposed and has essentially the same histological structure as any gill, except that acidophilic cells are lacking. In some of the teleost fishes, such as the cod, the pseudobranchs have become covered by the buccopharyngeal epithelium, and in the others it appears as a glandlike cluster of cells embedded in the subepithelial connective tissue, in which case it is known as the pseudobranch gland. Characteristic of the pseudobranch and the pseudobranch gland, except in the elasmobranchs and ganoids referred to above, are large, granular, acidophilic cells. These cells contain a dense population of mitochondria, which are arranged in packets or bands sandwiched by an abundant system of tubules of endoplasmic reticulum (Figs. 8.5 and 8.6). These cells are presumed to secrete carbonic anhydrase, an enzyme which breaks down carbonic acid to carbon dioxide and water. Secretion of carbonic anhydrase by the pseudobranchs has been shown to be significant in oxygen utilization by the choroid gland in the eye, which receives its blood supply via the pseudobranch. A functional relation between the pseudobranch and the swimbladder has also been demonstrated in some species of fishes.

LUNG RESPIRATION

The lungs of vertebrates exhibit extreme variation from the simple sacs found in certain fish and urodelous amphibians to the complex system of air capillaries and accessory air sacs found in birds. The degree of complexity may be most directly correlated with the metabolic rate characteristic of the species. Typically the respiratory organs of air-breathing vertebrates include the larynx, trachea, bronchi, bronchioles, and lungs, which shall be dealt with separately.

Larynx

A larynx is lacking in lung-breathing fishes, and in urodeles it is still a rather insignificant structure, amounting to little more than a short tube connecting the lungs with the pharynx through a slit, the glottis. In urodeles the larynx is lined by a simple columnar epithelium supported by fibrous connective tissue and some smooth muscle fibers. It is also supported adjacent to the glottis by a pair of lateral cartilages which have been derived from the last pair of gill arches. In anurans the larynx is somewhat more complex, due to the introduction of vocal cords. These consist of two lateral ridges of fibroelastic tissue which project into the lumen of the larynx. The mucous epithelium covering the glottis changes abruptly to simple cuboidal as it enters the larynx, and then becomes stratified ciliated, although nonciliated patches are present (Fig. 8.7). The lateral cartilages are replaced in the anurans by two pairs of cartilages, the *arytenoid cartilages,* around the glottis and also by ring-shaped *cricoid* cartilages within the walls of the larynx. Skeletal muscle fibers attach these structures to the hyoid apparatus, and serve to dilate or constrict the glottis and the larynx.

The larynx of reptiles is essentially similar to that of the anurans, except that the laryngeal cartilages are more extensive and fused. In some lizards an *epiglottis* appears as a fold of the pharyngeal wall immediately anterior to the glottis. It consists of a stratified squamous epithelium supported internally by a thin layer of elastic connective tissue and some skeletal muscle fibers. The larynx of birds is essentially reptilian, but the connective tissue layer immediately underlying the epithelium forms a well-defined *lamina propria* that separates the epithelium from the laryngeal cartilages and the laryngeal muscles. Occasional lymph nodules may be observed in the lamina propria. Vocal cords are lacking in birds, as they are in most reptiles as well.

The mammalian larynx is somewhat more complex than the foregoing. Its lining consists of stratified squamous epithelium covering well-defined muscular vocal cords, but otherwise it is of the pseudostratified ciliated variety. Adjacent to the epithelium is a distinct lamina propria containing many elastic fibers and incorporating lymph nodules, isolated lymphocytes, and the bases of scattered laryngeal seromucous glands of the tubuloalveolar type. A poorly defined *submucosa* of denser fibroelastic tissue is present in some mammals, but it is rendered inconspicuous in most areas by the laryngeal cartilages and their muscles. The cartilages, which are interconnected by ligaments, include the arytenoids, the cricoids, the thyroids, and, in most mammals, the cuneiforms and corniculates.

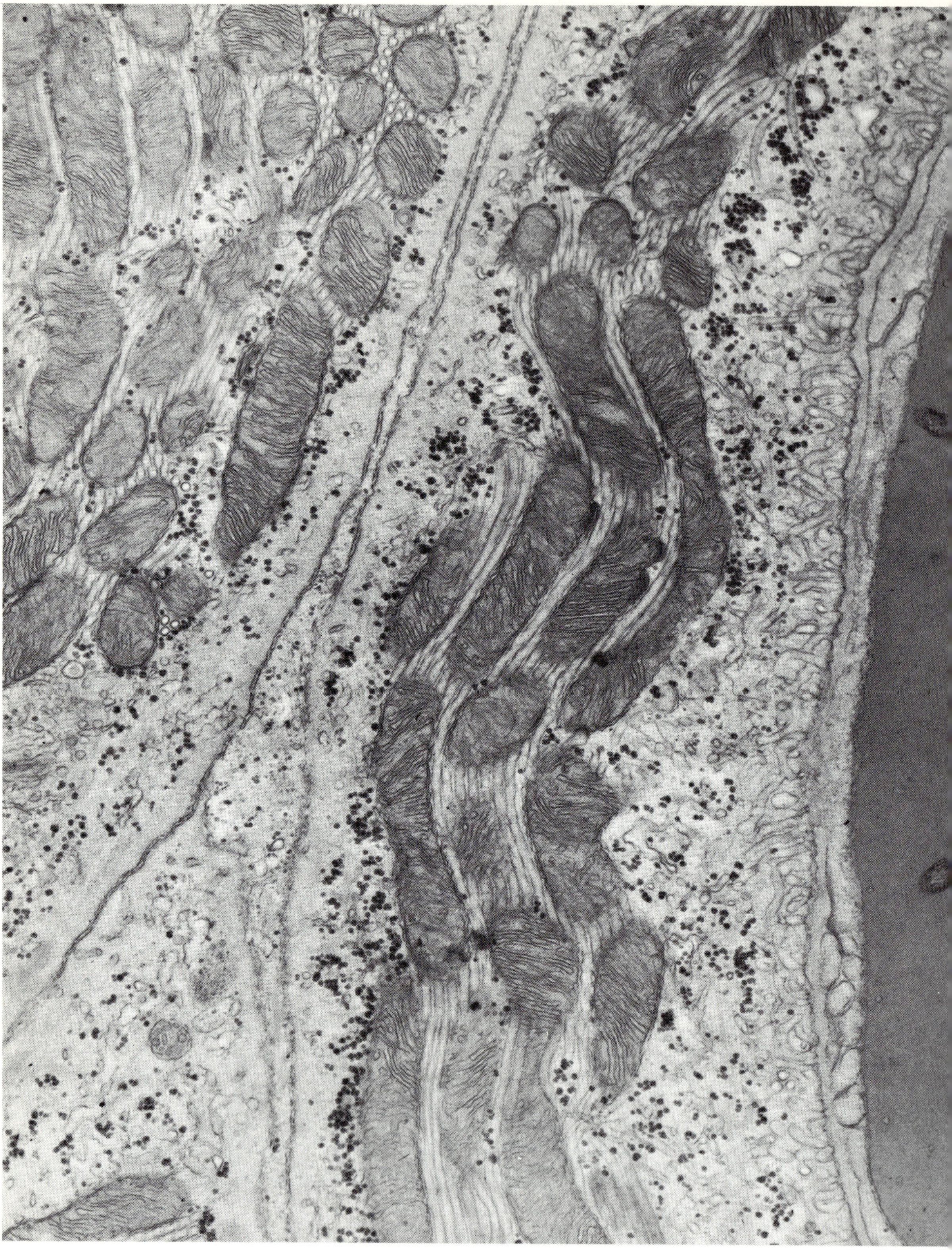

Fig. 8.5 Electron micrograph of parts of two cells of the pseudobranch gland in the teleost *Fundulus heteroclitus*. This section is longitudinal to the mitochondria and shows the tubular connection of smooth-surfaced *ER* to the cell surface. (Courtesy of D. E. Copeland.)

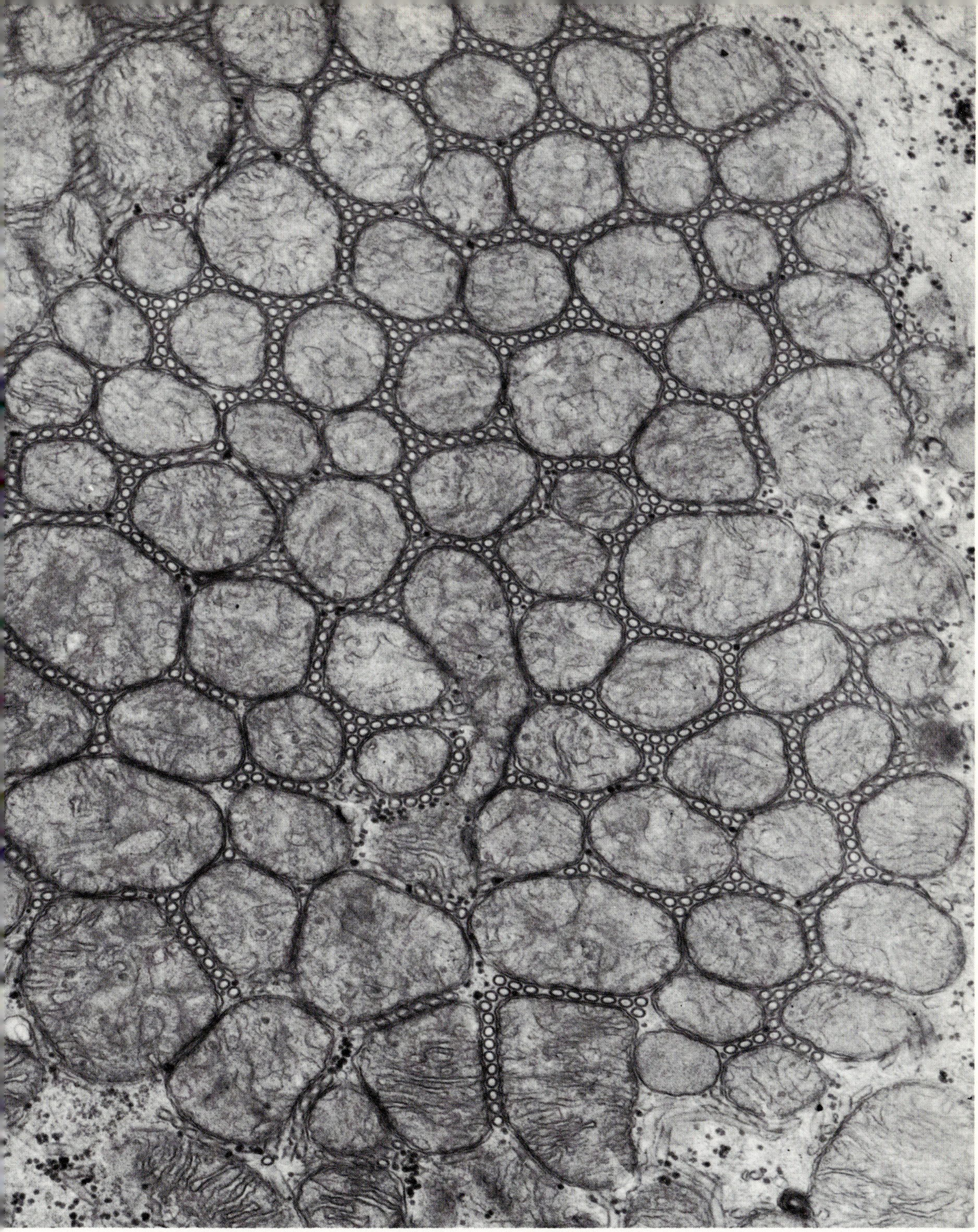

Fig. 8.6 Same as Fig. 8.5 but sectioned across the mitochondria and the tubules of smooth-surfaced *ER*. (Courtesy of D. E. Copeland.)

The first three consist of hyaline cartilage that later becomes calcified or ossified in the larger species. The cuneiform and corniculate consist of elastic cartilage. All mammals have a well-formed epiglottis, which constitutes a part of the larynx, rather than the pharynx. Covered for the most part by stratified squamous epithelium, the epiglottis consists largely of elastic cartilage with occasionally a central zone of adipose tissue (Fig. 8.8). The caudal surface is thrown into a fold, the aryepiglottic fold, which is also covered by stratified squamous epithelium. Taste buds may be found on the epiglottis. The muscles of the larynx may be divided into the *extrinsic muscles,* which, anchoring the larynx to adjacent organs and tissues, move the larynx as a whole, and the *intrinsic muscles,* which alter the shape of the larynx and control the vocal cords.

Trachea

The trachea is relatively insignificant in amphibians, for their lungs branch off more or less directly from the larynx. In reptiles, where for the first time a true neck may be distinguished, the trachea is well differentiated, although in some species it may be very short. In all the classes above the amphibians the trachea is supported by bars or split rings of hyaline cartilage at regular intervals along its length (Fig. 8.9). These *tracheal cartilages* keep the trachea patent, thus enabling it to present minimum resistance to the flow of air through it. Gaps between the ends of the tracheal cartilages permit limited constriction or dilatation, however. Ciliated pseudostratified epithelium containing goblet cells forms the mucosal lining of the trachea (Fig. 8.10). The secretory ends of scattered seromucous glands originating from the epithelium are found both within the lamina propria and in the submucosa. A distinct basement lamina is present. The lamina propria is thin and con-

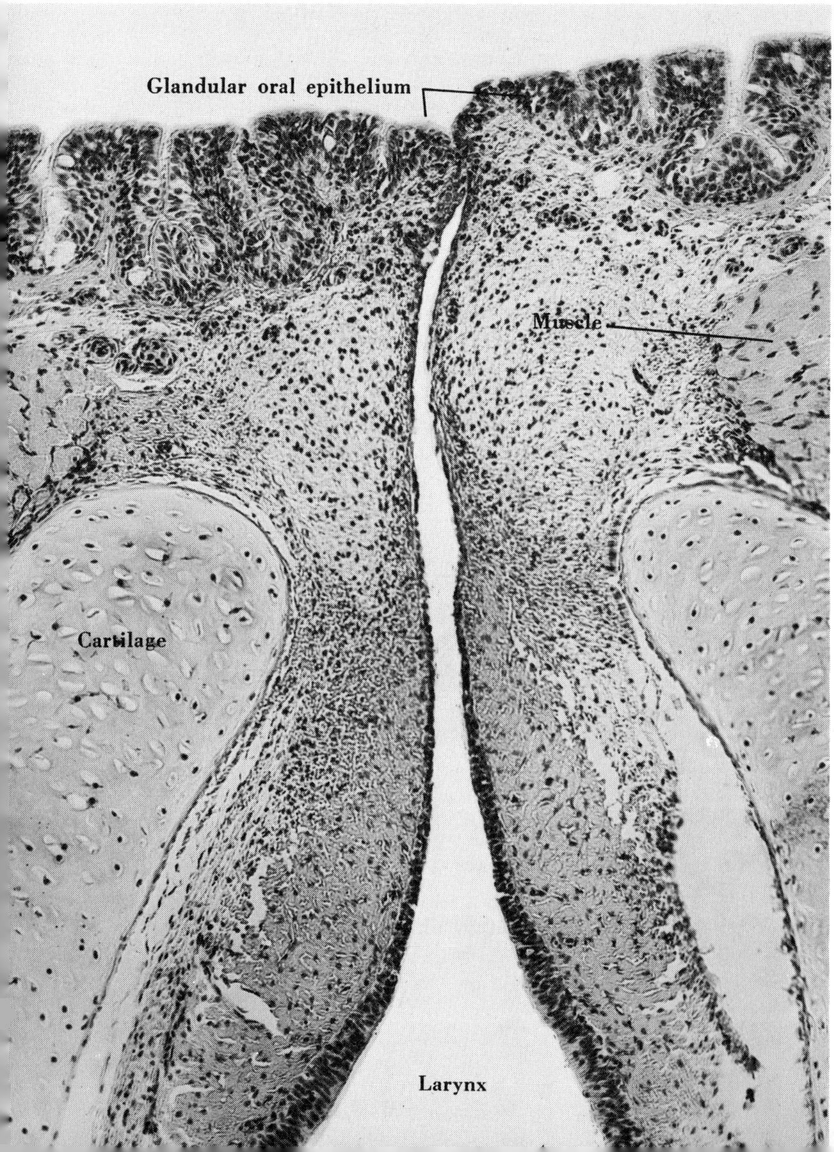

Fig. 8.7 Photomicrograph of the glottis of a frog, *Rana pipiens,* showing the transition from the glandular, ciliated epithelium of the oral mucosa to the nonglandular, simple squamous epithelium lining the glottis to the stratified squamous epithelium of the larynx. X150.

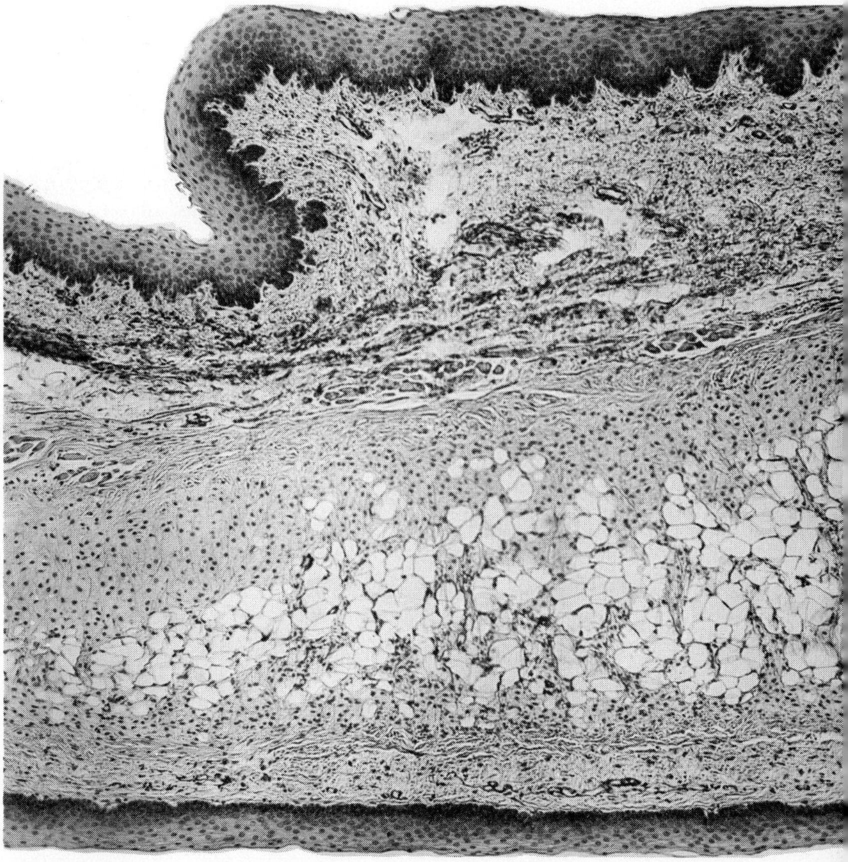

Fig. 8.8 A longitudinal section of the mammalian epiglottis showing the nonkeratinized, stratified squamous epithelium on both dorsal (caudal) and ventral (cranial) surfaces, a thick dermis adjacent to the aryepiglottic fold, and a central band of elastic cartilage. X100.

tains many elastic fibers. The submucosa is richly supplied with lymphatic vessels. The tracheal cartilages are invested in a tough, fibrous membrane that bridges the gap between the open ends of the rings. Smooth muscle is also found in the gap, forming a thick layer that brings the ends of the cartilages together (Fig. 8.11). Near its bifurcation into the two bronchi, the trachea exhibits glands which are larger, and the secreting ends of these glands may extend out into the muscularis.

Bronchi

The bronchi, which are also lacking in the amphibians, are similar in the other three

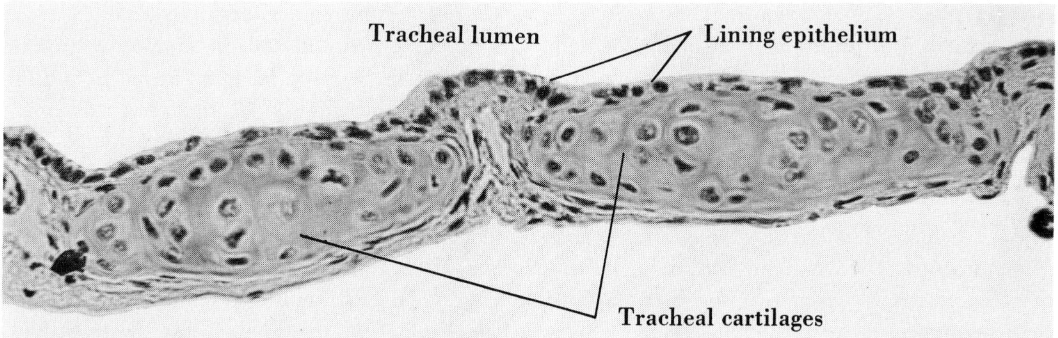

Fig. 8.9 Longitudinal section through the wall of the trachea of a lizard, *Anolis carolinensis*. X300.

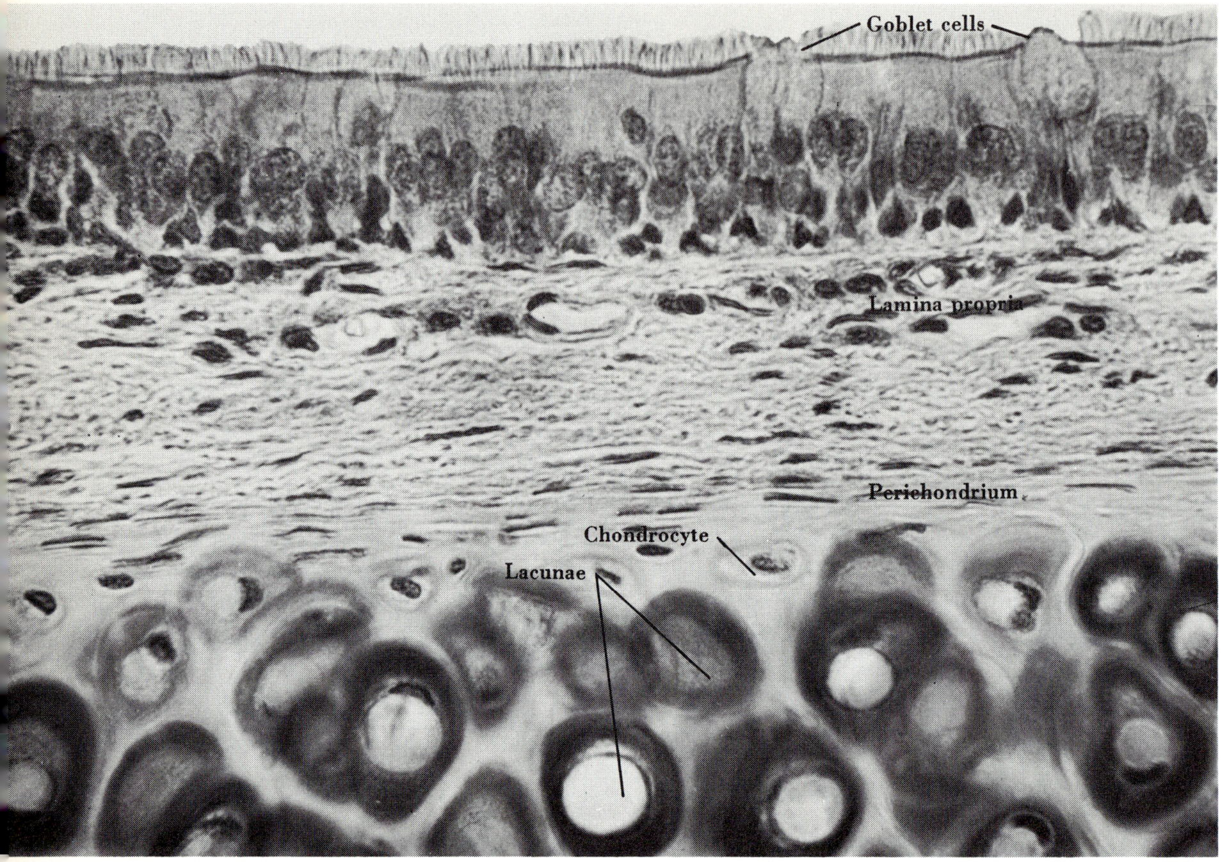

Fig. 8.10 Longitudinal section through a mammalian trachea. Note the pseudostratified ciliated epithelium and the characteristics of the chondrocyte lacunae. X450.

classes and differ only slightly from the trachea. Smooth muscle fibers largely replace the elastic fibers of the lamina propria in the smaller species, however, and the epithelium is characterized by a more extensive glandular component. As the bronchi enter the lungs, they are accompanied by the major blood vessels, lymphatics, and nerves. They immediately branch into secondary bronchi, which have a slightly different microscopic appearance in tissue section. The lining epithelium remains essentially the same, but the lamina propria acquires more smooth muscle, the fibers of which are arranged in two crisscrossing helices. The cartilaginous rings completely encircle the bronchi, but they are so irregular that they appear as numerous isolated segments in histological sections (Fig. 8.12). The secondary bronchi branch into the several lobules of the lung, where they subdivide to form *bronchioles*.

Bronchioles

Bronchioles are distinguished from secondary bronchi primarily by their lack of cartilage. The epithelium undergoes a transition from pseudostratified ciliated to simple columnar ciliated, with decreasing numbers of glandular elements, and finally becomes of the nonglandular, nonciliated, low columnar type. This is supported by a thin, elastic lamina propria, an encircling muscularis, and, outside of this, the connective tissue merges with the elastic connective tissue of the alveolar sacs (Fig. 8.13). The bronchioles undergo further branching in mammals into thin-walled, terminal extensions, the *respiratory bronchi-*

oles. In birds the bronchi give rise to unique tubes called *parabronchi,* which are discussed below. As is the case with bronchi, bronchioles are lacking in amphibians and in some of the reptiles.

Lungs

The evolution of lungs among the terrestrial vertebrates has been remarkable, and probably represents one of the best examples whereby reconstruction of the evolutionary process can be made from contemporary species. In urodelous amphibians, lungs, when present, are represented in their simplest, if somewhat degenerated, form—consisting of little more than richly vascularized, elongate sacculations of the respiratory epithelium. The epithelium is simple, low cuboidal overlying a basement lamina that is relatively indistinct. It is thrown into loose, longitudinal folds and is supported by a thin, richly vascularized, areolar connective tissue layer. A denser fibroelastic membrane, in which smooth muscle elements, lymphatics, and larger blood vessels occur, surrounds this and is covered in turn by mesothelial cells of the pleura with a scattering of goblet cells.

Some of the details of the respiratory epithelium deserve special mention. The epithelial cells occur in groups of three or four, and the outer ones of such a cluster are characterized by cytoplasmic flanges that arch over a blood capillary as a thin layer so that there is minimal protoplasmic separation between the lumen of the lung and the lumina of the blood vessels below (Fig. 8.14). Similarly, the endothelial cells are thin where they lie in contact with the epithelium, a relationship which is reminiscent of fishes' gills. The total surface available for respiratory exchange in urodeles is highly limited, however, and they depend more on either external gills or the skin, or both, to meet their respiratory needs. Some urodeles lack lungs altogether and depend entirely on the skin and the buccopharyngeal mucosa for respiration.

Frogs and toads show a slight advance over the urodeles in their lung structure, reflecting better adaptation to terrestrial life—although here, too, there has probably been some degeneration. The wall of the lung is thrown into folds or septa which protrude into the lumen of the lung. The epithelium, in turn, has evaginated to form *infundibula* between the septa (Fig. 8.15). Within each infundibulum

Fig. 8.11 Transverse section of a mammalian trachea at the point of approximation of the two ends of a ring of tracheal cartilage. The muscle that draws the two cartilage ends together is partially contracted, throwing the mucosa into folds. X100.

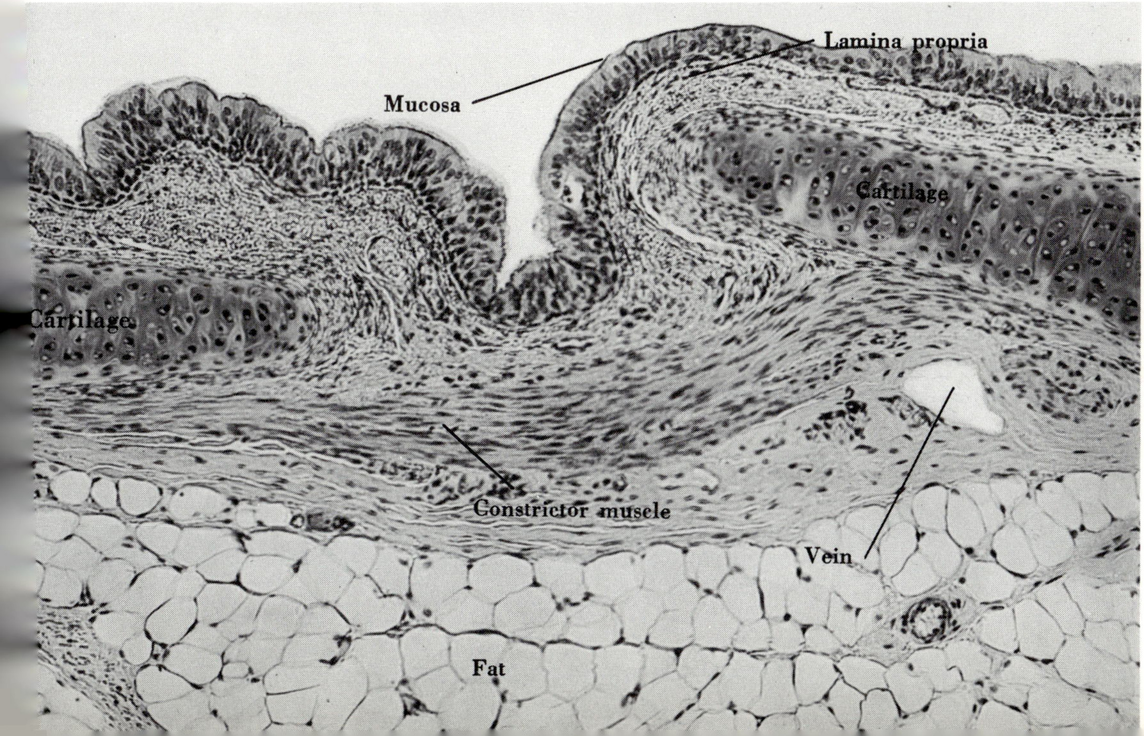

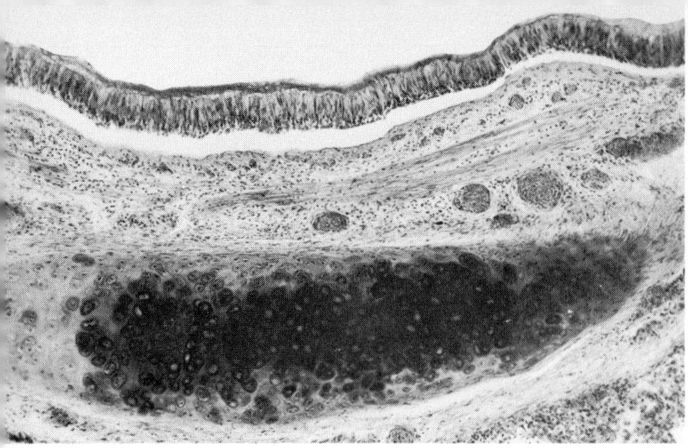

Fig. 8.12 Secondary bronchus of a mammal. The space between epithelium and lamina propria is an artifact. Note the strands of smooth muscle and the cartilage plaque. X100.

the epithelium forms smaller, blindly ending outpocketings, the *pulmonary alveoli.* The alveoli of toads' lungs are smaller and consequently more numerous than those of frogs' lungs, enabling toads to depend less on their skins for respiration than do frogs. The epithelium of the infundibular openings is ciliated columnar or cuboidal, with scattered goblet cells. It becomes simple squamous where it lines the pulmonary alveoli, with neither cilia nor goblet cells (Fig. 8.16). The relation between capillary endothelium and the respiratory epithelium is similar to that described for urodeles, except that the basement laminae of each are more distinct and are fused where they meet.

The widest range of pulmonary complexity is encountered in the reptiles. *Sphenodon,* which has been referred to as a living fossil, exhibits perhaps the most primitive condition, for its lungs are not unlike the frog's. Infundibula are present, but there is little alveolar development. Ophidians have only one well-developed lung, the other (left) one being very small or absent. Infundibula are present in only part of the lung of snakes, and alveolar development is not extensive. Lizards are the first to show a lobulated lung, with infundibula and alveolar development, and structurally they are very like mammalian lungs. Special adaptations in reptiles are illustrated by turtles and African chameleons. The lungs of the former are very extensive, occupying much of the thoracic and abdominal cavities. Their great volume serves more for the storage of air, however, than for respiratory efficiency, because there is virtually no alveolar development, only large infundibula that open into a spacious central cavity. The African chameleon is unique among reptiles in having saccular extensions of the lung which branch out into the abdominal cavity. The extensions have thin walls and no alveoli. They are so like the air sacs of birds as to suggest evolutionary affinities with them.

Mammalian lungs represent the culmination of the evolutionary trends described above. They are divisible into several lobes and many lobules. The latter are the functional units. Each lobule consists of an apex and a base. A bronchiole and a branch of the pulmonary

Fig. 8.13 Cross section of a human bronchiole showing the simple squamous epithelium, circular smooth muscle fibers, and adjacent alveoli. X1000.

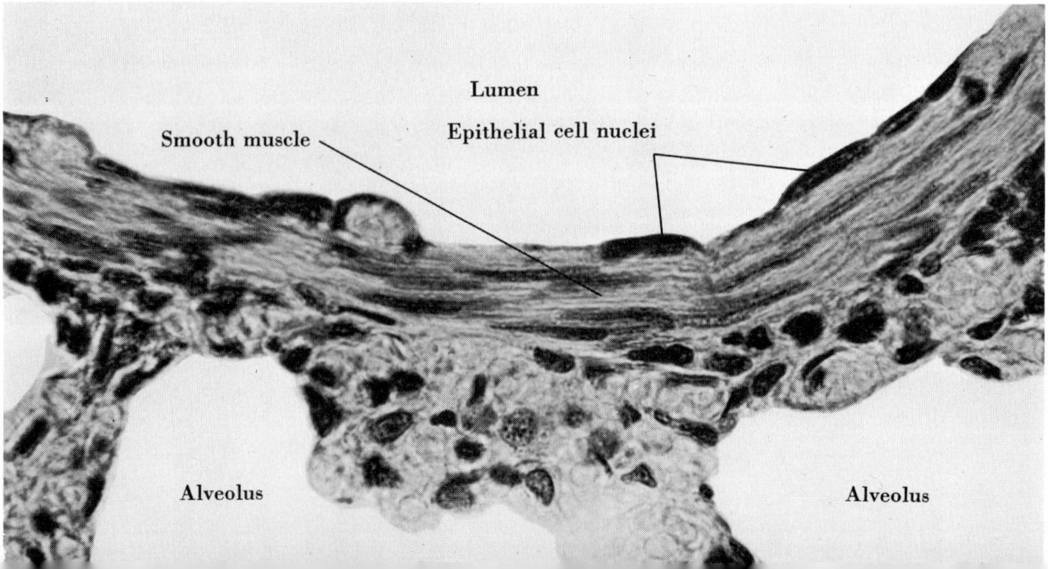

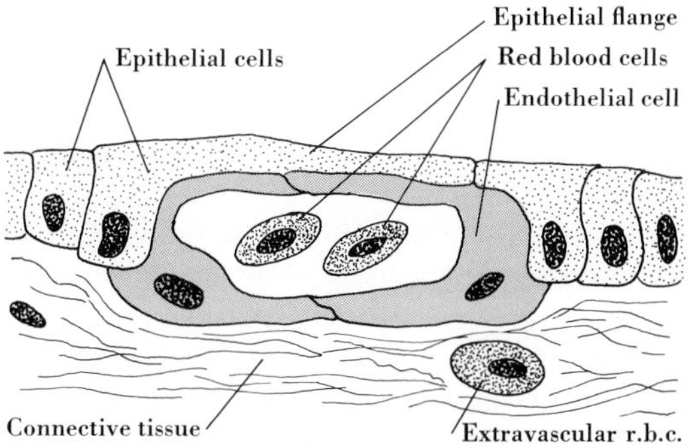

Fig. 8.14 Diagram of the respiratory epithelium of a urodele lung. Compare with Figs. 8.1 and 8.19.

artery enter, and a series of veins and lymphatics leave the apex. The lobule is surrounded by elastic connective tissue which forms the interlobular septa. The bronchiole, artery, and a lymphatic vessel form the core of each lobule, so to speak; the alveoli and their capillary networks form the main mass; and the interlobular septa, in which the veins and lymphatic vessels are carried, form the covering (Fig. 8.17). Within the interlobular septum is a rather profuse network of coarse elastic fibers, which ramify as more delicate branches into the interalveolar septa. By virtue of these fibers, the lungs tend to collapse when intrapulmonic and intrapleural pressures are equalized. Hence the lungs are inflated in mammals by muscular contractions which cause lowered intrapulmonic pressure, and conversely, they are deflated by relaxation of the same muscles, the work of expiration being done by the contraction of the elastic fibers. Expiration can also be aided and accelerated by contraction of the expiratory muscles, however, as in labored breathing (dyspnea), singing, shouting, and the like.

Mammalian lung alveoli make use of the same principles of construction as have already been described. The respiratory epithelium consists of a single layer of squamous cells that is supported by a thin basement lamina. The cytoplasm of these cells is drawn out into such an extremely thin layer where it overlies a blood capillary as to be almost at the limit of resolution of the light microscope (Fig. 8.18). Similarly, the endothelial cells of the capillaries are exceedingly thin where they contact the alveolar epithelium. The only separation between the blood plasma and the alveolar air, therefore, is a membrane about 0.3 to 0.6μ in thickness. This membrane consists of four layers (Fig. 8.19): (1) an endothelial cell (0.1–0.2μ); (2) basement lamina of the endothelial cell (0.05–0.1μ); (3) basement lamina of the alveolar epithelial cell (0.05–0.1μ); and (4) the alveolar epithelial cell (0.1–0.2μ). See Fig. 8.19. In addition to epithelial and endothelial cells are the so-called *alveolar cells,* or *septa cells.* They are distinguished from the others by their lighter nuclei and more or less oval outline. Alveolar cells are of unknown origin, but inasmuch as they function like macrophages, they may be presumed to be derived from mesenchyme. Some are vacuolated, others are not. They may wander into the alveolar spaces as free *alveolar phagocytes,* and be discharged to the outside via the pulmonary passages.

The lungs and associated respiratory passages of birds present a different and highly specialized line of adaptation from the basic reptilian type. The larynx is poorly developed and lacks vocal cords. The number of laryngeal cartilages is reduced in most birds to just the arytenoids, which are often ossified, and the cricoids; a pair of procricoids may also be found in some species. The trachea is long, and in some species it may form a loop just under the skin, or, as in swans and cranes, the loop is housed in a cavity in the sternal keel. In penguins the trachea is double throughout most of its length. The tracheal cartilages of birds usually occur as complete rings, which may be partly calcified or ossified.

The region at which the trachea branches into the primary bronchi is expanded and modified into a sound-producing organ, the *syrinx.* Although the precise structure of the syrinx is somewhat variable, it consists essentially of a resonating box enclosing vibratile membranes whose pitch is regulated by the contractile tension of skeletal muscle fibers. An accessory structure, the *pessulus,* is present in the syrinx of many species. It is a modified tracheal cartilage which projects from the bronchial bifurcation into the syringeal chamber.

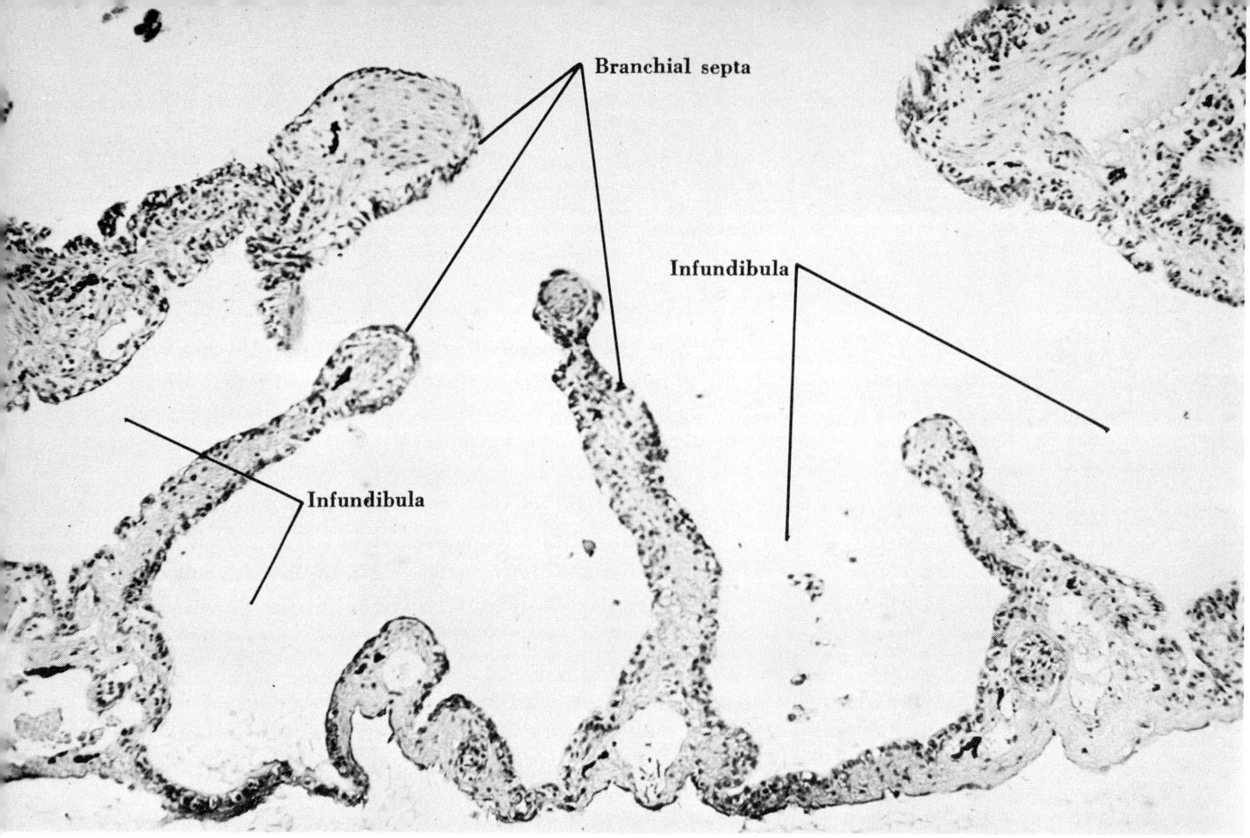

Fig. 8.15 Cross section through the anterior region of a lung of the frog *Rana pipiens.* The ends of the branchial septa are expanded to accommodate smooth muscle bundles. X100.

As each bronchus enters the lung on its side, it expands slightly, forming a *vestibulum,* and progressively loses its cartilages. It narrows again as it leaves the vestibulum, and is then known as a *mesobronchus.* As the mesobronchus, it continues caudally through the lung, emerging therefrom to enter the *abdominal air sac.* The epithelial lining of all tracheal, bronchial, and mesobronchial components is pseudostratified ciliated and in-

Fig. 8.16 Section through the posterior region of a frog's lung showing the abrupt transition from the ciliated infundibular epithelium shown in the upper part of the micrograph to the smooth alveolar epithelium in the lower part of the micrograph. X900.

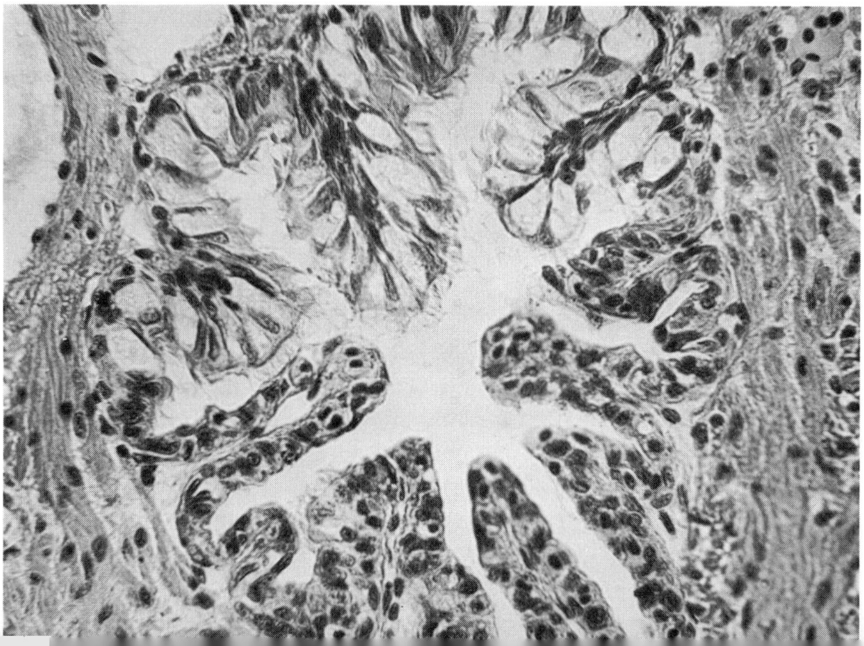

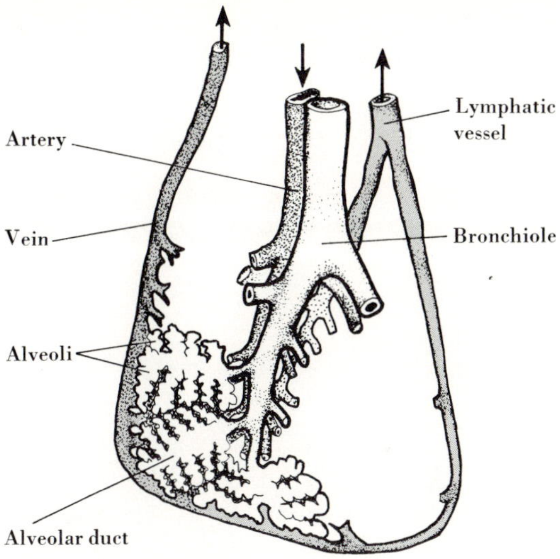

Fig. 8.17 Diagram of a mammalian lung lobule showing the relations between air passages and blood vessels and lymphatics.

cludes simple alveolar mucous glands and goblet cells. The lamina propria contains conspicuous lymph nodules. There is no submucosa, but there is a connective tissue adventitia which includes longitudinally oriented smooth muscle fibers.

Just before expanding into the vestibulum, the bronchus gives rise ventrally to eight secondary bronchi, four of which lead into three other air sacs, the *cervical, interclavicular,* and *anterior thoracic* air sacs. Additional secondary bronchi, six to ten in number, arise dorsolaterally from the caudal part of the vestibulum and from the mesobronchus. One of these enters into the fifth air sac, the *posterior thoracic* (Fig. 8.20). The secondary bronchi of the first group (cranioventral) sweep upward and anastomose with those of the second group (dorsolateral) either directly or by an extensive system of ramifications, the *parabronchi,* of which there are about a thousand in each lung. The secondary bronchi consist of a simple ciliated columnar epithelium with a few alveolar mucous glands and goblet cells, a lamina propria containing rather prominent lymph nodules, and an adventitia which is virtually a muscularis because of its rather dense, circularly disposed, smooth muscle fibers. The parabronchi are smaller in diameter (approximately 0.5mm), their lining epithelium is simple squamous, and the muscular coat is well developed.

Branching off radially from the parabronchi is the respiratory epithelium *per se* in the form of minute branching and anastomosing tubules, the *air capillaries* (Fig. 8.21). This epithelium shows a transition from low cuboidal to squamous where contact with the rich vascular network of the lung is effected. The relation between the respiratory epithelium and the blood vessel endothelium is essentially the same as in the alveolar lungs of other tetrapods.

The air sacs, of which there are typically nine (one interclavicular and paired cervicals, anterior thoracics, posterior thoracics, and abdominals) represent extrapulmonary extensions of the major bronchi. They do not function directly in respiration, however, as is apparent from their smooth, rather avascular walls. The thin walls consist of a simple epithelium which is ciliated cuboidal near the openings from the lungs and squamous elsewhere, and a thin layer of connective tissue. Numerous diverticula extend from all but the

Fig. 8.18 Photomicrograph of mammalian lung. X900.

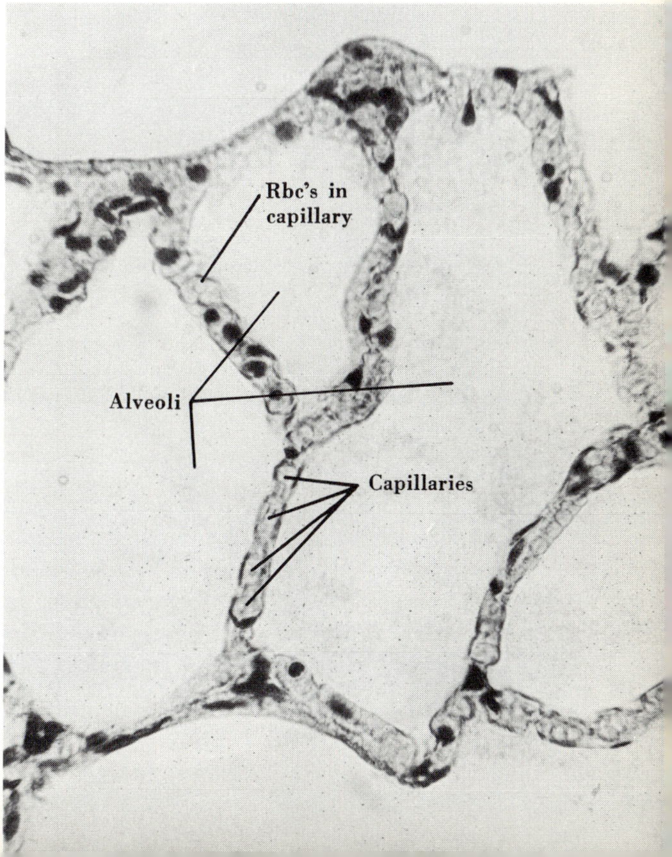

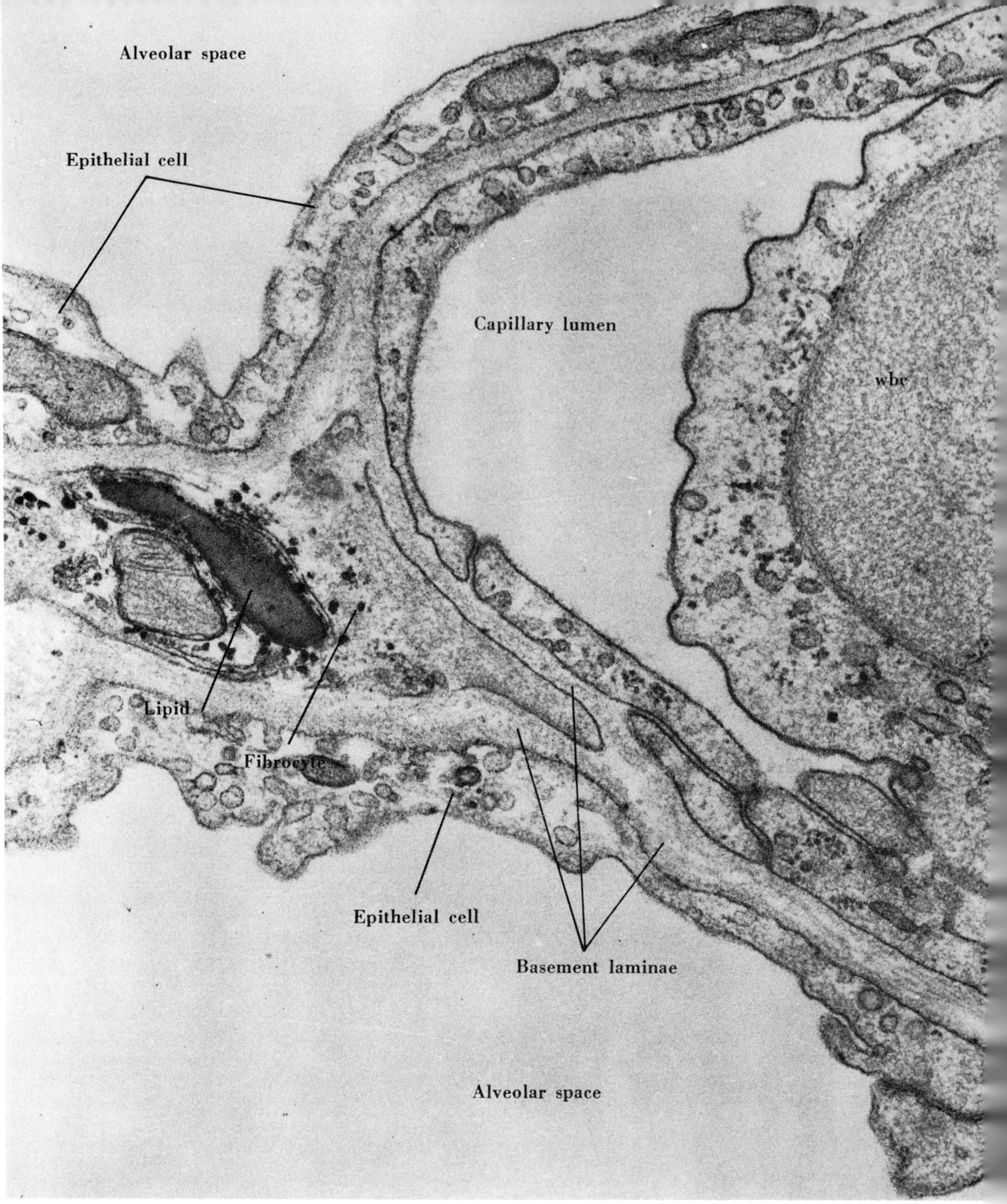

Fig. 8.19 Electron micrograph of a normal rat lung showing the relation of a blood capillary to two alveolar spaces. Note the conjunction of the basement laminae of endothelial and epithelial cells to form a single basement lamina. X51,000. (From G. S. Kistler *et al.* 1967. J. Cell Biol. 32:605–628.)

thoracic air sacs into spaces in the vertebrae and other major bones of the skeleton and into spaces between muscles. These diverticula are usually well vascularized. Present in all birds, except notably the flightless Kiwi, in which only rudiments of them are found, the air sacs appear to serve three functions: (1) they act as a sort of bellows in respiration, so that air can pass completely through the lungs both during inspiration and expiration, thereby greatly increasing respiratory efficiency; (2) they and their numerous diverticula contribute to the animal's buoyancy; and (3) they function in temperature regula-

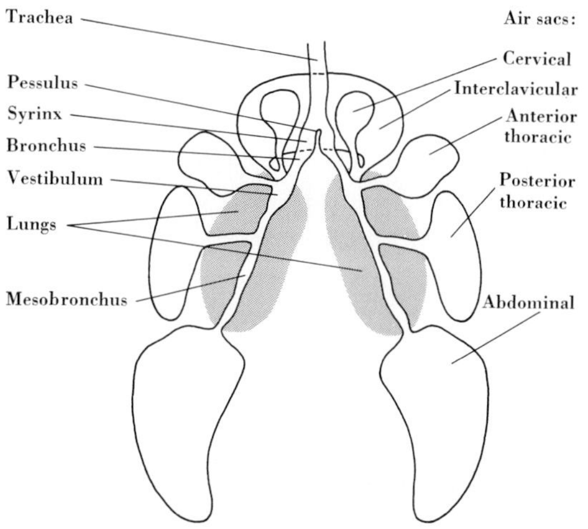

Fig. 8.20 Diagram showing the relation of the air sacs to the lungs and major respirator passages in birds. Dorsal aspect.

tion, inasmuch as the glandless skin is too well insulated by the feathers to serve this function.

Precisely how the avian lung functions is not known. Many bird physiologists favor the aerodynamic theories of lung ventilation, one of which proposes that because of the angle at which the secondary bronchi supplying the air sacs branch off from the bronchi and mesobronchi, more air passes through the parabronchi and, hence, the air capillaries during expiration than during inspiration. If this is so, it follows that the principal respiratory exchange occurs on expiration and under conditions of positive pressure relative to the ambient, which would contribute materially to respiratory efficiency. Very little is known of the mechanisms of respiration during flight.

GAS BLADDERS

The gas bladders, or swimbladders, of fishes are generally considered to represent adaptations of primitive lungs. They are associated with the foregut, however, instead of the pharynx, and their histological organization is like that of the gut, inasmuch as a mucosa, a submucosa, and an external muscular layer are clearly defined. Connection with the digestive tract is by a *pneumatic duct* which is patent in physostomatous fishes and closed or lost in physoclistous forms, as mentioned above.

Toward the anterior end of the gas bladder is a glandular organ, the *red body* or *gas gland,* which owes its appearance to dense networks of capillary loops, the *retia mirabilia.* The gases secreted by the gas gland vary somewhat in their proportions from species to species, but generally they consist of up to 95 percent oxygen, 4 percent carbon dioxide, and 1 percent nitrogen and argon. Removal of the pseudobranchs in teleosts may result in loss of ability of the gas gland to secrete carbon dioxide, presumably because the gas gland derives its carbonic anhydrase from the pseudobranchs via the circulatory system, as mentioned above. There is some evidence that oxygen secretion is also impaired under these conditions. Posterior and dorsal to the gas gland in physoclistous fishes is a gas resorbing gland, the *oval.*

The functions of the gas bladder include respiration, hydrostatis, sensory reception, and sound production. A respiratory function

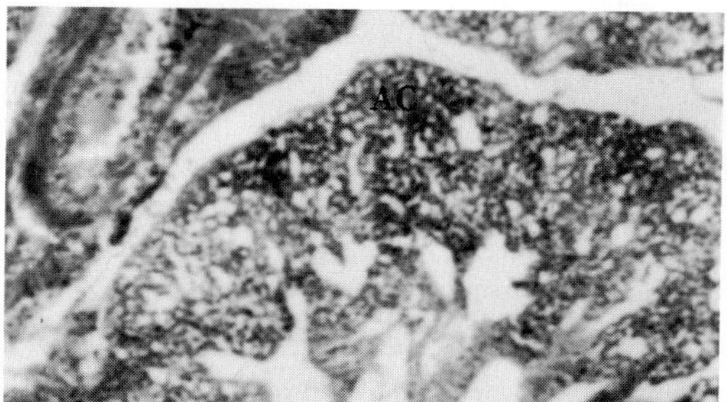

Fig. 8.21 Photomicrograph of a lung lobule of a chicken, showing air capillaries, *AC;* a parabronchus, *TB;* and several atria, *At*. The atria represent the primary branches off the parabronchi from which the air capillaries arise. X700. (From W. S. Tyler and J. Pangborn. 1964. J. Cell Biol. 20:157–164.)

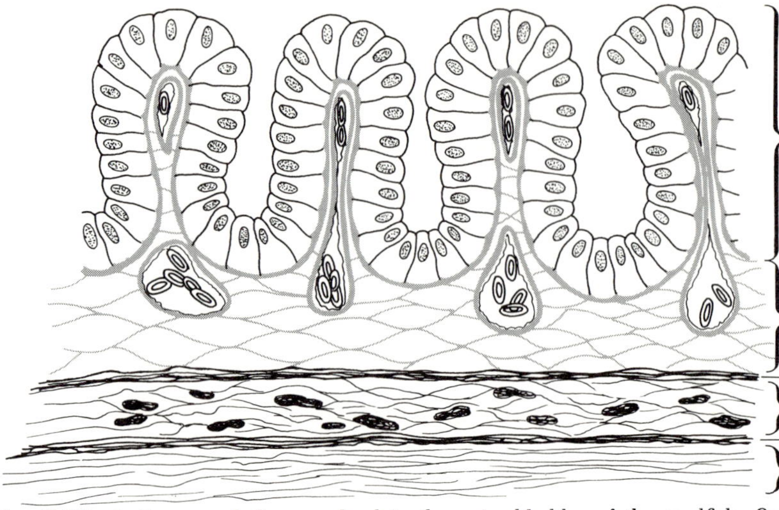

Fig. 8.22 A diagram of the gas gland in the swim bladder of the toadfish, *Opsanus tau*. (After R. Fänge and J. B. Wittenberg. 1958. Biol. Bull. 115:172–179.)

is served in certain physostomatous fishes which gulp air, and also to a limited extent in physoclistous fishes who may, under emergency conditions of oxygen deprivation, utilize the oxygen stored in the gas bladder. The hydrostatic function, which is perhaps the primary function of the gas bladder in teleosts, is served by controlling the volume of gas in the bladder, thereby regulating the buoyancy of the fish. Through its sensory innervation, the gas bladder is sensitive to pressure changes, according to some authorities. In most teleosts the gas bladder acts as a sort of tympanic membrane, transmitting sound vibrations to the inner ear by the Weberian ossicles or by its own diverticula. Finally, the swimbladder may function in sound production either as (1) a resonator for vibrations originating from grating of the teeth or certain bones, (2) by audibly releasing gas through the pneumatic duct, or (3) by the contraction of muscles (such as the "drumming muscles" of certain catfish) attached to the wall of the gas bladder.

The mucosa of the gas bladder consists usually of a low columnar epithelium supported by a basement lamina, a lamina propria, and a muscularis mucosae of several layers of smooth muscle cells. The mucosa of the gas gland, however, is characterized by a convoluted, simple columnar epithelium and, extending into the lamina propria, vascular loops of the retia mirabilia. The submucosa is loose connective tissue. The tunica externa consists of both dense fibrous connective tissue and muscle fibers. The muscle fibers may be smooth or striated, or both types of fibers may be present. The ventral wall of the gas bladder is usually covered by a serosa (Fig. 8.22).

SELECTED REFERENCES

Bertalanffy, F. D. 1964. Respiratory tissue: structure, histophysiology, cyptodynamics. Part I: Review and basic cytomorphology. International Review of Cytology. G. H. Bourne and J. F. Danielli, eds. 16:234–328.

Bertalanffy, F. D. 1965. Respiratory tissue: structure, histophysiology, cytodynamics. Part II: New approaches and interpretations. International Review of Cytology. G. H. Bourne and J. F. Danielli, eds. 17:214–297.

Bevelander, G. 1935. A comparative study of the branchial epithelium in fishes with special reference to extrarenal secretion. Morphol. 57:335.

Brown, M. E., ed. 1957. The physiology of fishes. 2 vols. Academic Press, New York.

Copeland, D. E. Function of the glandular pseudobranch in teleosts. Am. Physiol. 167:775. 1951.

Copeland, D. E., and A. J. Dalton, 1959. An association between mitochondria and the endoplasmic reticulum in cells of the pseudobranch gland of a teleost. Biochem. Biophys. Cytol. 5:393.

Fänge, R., and J. B. Wittenberg, 1958. The swimbladder of the toadfish. (Opsanus tau L.). Biol. Bull. 115:172.

Gans, C., and G. M. Hughes. 1967. The mechanism of lung ventilation in the tortoise *Testuda graeca* Linne. J. Exp. Biol. 47:1–20.

Holliday, F. G. T., and G. Parry. 1962. Electron microscopic studies of the acidophil cells in the gills and pseudobranchs of fish. Nature 193:192.

Keys, A., and E. N. Willmer. 1932. Chloride secreting cells in the gills of fishes, with special reference to the common eel. J. Physiol. 76:368.

Lagler, K. E., J. E. Bardach, and R. R. Miller. 1962. Ichthyology, the study of fishes. John Wiley & Sons, New York.

Marshall, A. J., ed. 1960. Biology and comparative physiology of birds. 2 vols. Academic Press, New York.

Philpott, C. W., and D. E. Copeland. 1963. Fine structure of chloride cells from three species of *Fundulus*. J. Cell Biol. 18:389–404.

chapter 9
THE CIRCULATORY SYSTEM —BLOOD AND LYMPH

THE CIRCULATORY SYSTEM is considered in this work as consisting of those organs which are directly concerned with the propulsion, circulation, and production of blood and lymph, and, by corollary, it is considered to include blood and lymph as well.

THE CIRCULATING MEDIA

Blood

Blood is a fluid tissue which differentiates in intimate association with connective tissue. It consists of plasma, the liquid matrix, and of formed elements, the blood corpuscles. The functions of blood are to convey substances and heat from one part of the body to another. More specifically, the blood transports food materials absorbed by the digestive tract, oxygen absorbed by respiratory epithelia, and a variety of substances produced by the body for its own use, such as hormones, antibodies, and other specific secretions and cells. The blood also conveys waste materials away from their sites of production to sites of elimination. Finally, the blood serves as the major source of the tissue fluid, the chemical constancy of which is essential for the survival and proper function of all cells and tissues. Such constancy is reflected in the narrow ranges of variation, in pH, osmotic pressure, acid-base ratio, and concentration of certain ions characteristic of the blood.

Plasma

The plasma of the blood of vertebrates is a clear liquid consisting primarily of water in which are dissolved or suspended a variety of substances, principally the serum proteins, serum globulin and serum albumin, and the plasma protein, fibrinogen. The globulins comprise primarily the antibodies, but they also include enzymes. Serum albumin contributes materially to the osmotic properties of the blood and tissue fluid. Fibrinogen is a protein which polymerizes under the influence of certain enzymes and other substances to form fibrin, a fibrous protein that forms the clot. Blood plasma also contains glucose, hormones, dissolved gases and inorganic salts as constant components, and a variety of other substances in more or less widely fluctuating quantities, such as the food materials and metabolic wastes mentioned above. Blood plasma whose composition shows rather extreme variability is normally sequestered in portal vessels, such as the hepatic portal vein and the renal portal vein (see Chapters 7 and 10).

Blood Corpuscles

Two major categories of blood corpuscles are characteristic of vertebrates: erythrocytes and leukocytes. In mammals the blood platelets are usually considered to represent a third category.

ERYTHROCYTES

Erythrocytes, or *red blood cells,* are typically vertebrate structures. Whereas in invertebrates the blood pigments, when present, are dissolved or suspended directly in the blood plasma, in vertebrates the blood pigment, hemoglobin, is contained exclusively within specialized corpuscles, the erythrocytes. The advantage of this arrangement is that blood vessel walls may be porous to molecules larger than the hemoglobin molecule, and yet hemoglobin will not be lost. Thus hemoglobin, contained within its specific carrier corpuscles, can pass through countless circulatory cycles, giving off CO_2 and picking up O_2 in the gills or lungs, giving off O_2 and picking up CO_2 in the capillary beds of the tissues, and returning again to the gills or the lungs, over and over again. In abnormal circumstances wherein hemoglobin escapes from the erythrocytes, it also escapes very quickly from the blood plasma at the kidneys and is lost with the urine. Erythrocytes occur in all vertebrate classes, although they are lacking in certain species of fishes. The Chaenichthyidae include a few, small, Antarctic species which, because of their low metabolic rates and the fact that the water they inhabit has a very high oxygen content, have no need of hemoglobin and hence have lost the ability to synthesize it. Certain larval eels (*Anguilla*) and some species of deep sea fishes also have colorless blood.

In all vertebrates except the mammals,

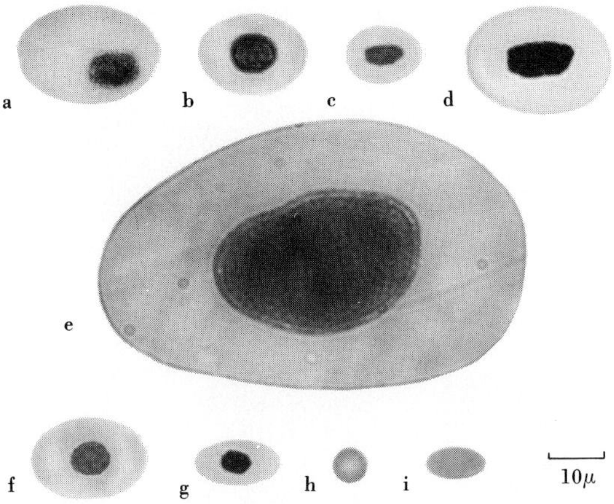

Fig. 9.1 Erythrocytes representative of the various classes of vertebrates, all shown to the same scale. Agnatha (*Petromyzon*), *a;* Chondrichthyes (*Mustelus*), *b;* Osteichthyes (*Carassius*), *c;* Amphibia (*Rana* and *Amphiuma*), *d* and *e*, respectively; Reptilia (*Chrysemys*), *f;* Aves (*Gallus*), *g;* Mammalia (*Homo and Camelus*), *h* and *i*, respectively.

mature erythrocytes are flattened, ovoid or eliptical cells containing a single, oval, basophilic nucleus (Fig. 9.1). The cytoplasm is virtually devoid of organelles, except for an occasional, stray mitochondrion and for a circumferential "hoop" of microtubules (Fig. 9.2) which may function as a sort of internal skeleton. Vestiges of ER and Golgi vacuoles have been described in mature teleost erythrocytes. Otherwise the cytoplasm consists of hemoglobin, which is strongly eosinophilic. The nucleus contains rather dense clumps of chromatin in fishes, but tends to be uniformly dense in amphibians, reptiles, and birds. Mature mammalian erythrocytes, on the other hand, have no nucleus and are biconcave discs (Fig. 9.1), except in the Camelidae (camels, alpacas, llamas), whose erythrocytes, although enucleate, are biconcave ovals. Enucleate erythrocytes account for 90 to 98 percent of the red blood cells in the urodele *Batrachoseps,* however, and they have been observed from time to time in cyclostomes, in fishes, and in other amphibians. The enucleate condition is quite reasonably assumed to be more advanced and more efficient, for it provides for maximum hemoglobin content per unit corpuscle volume and, allowing for the biconcave shape, it enables the corpuscle to present

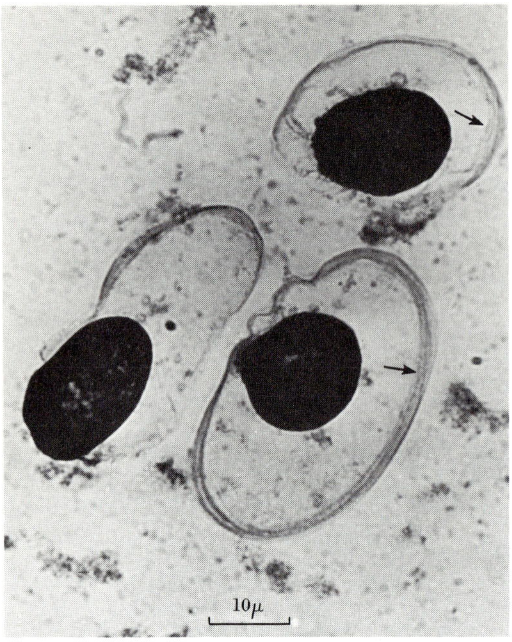

Fig. 9.2 Microtubules forming marginal bands in the erythrocyte of the newt (*Diemictylus viridescens*). (From M. D. Maser and C. W. Philpott. 1964. Anat. Rec. 150:365–382.)

somewhat more surface area. The price paid by mammals for this state is that their erythrocytes are incapable of further proliferation, and so must be replaced by the division and maturation of antecedent cells in erythropoietic centers.

The erythrocytes of all species have smooth, even contours, enabling them to flow freely through capillaries. They are somewhat flexible, as well, and thereby able to bend around the sharp "corners" where capillaries join arterioles or venules. Probably because of the surface forces implicit in their shape, mammalian erythrocytes tend to "stack" like coins (*rouleaux*) under certain conditions, although the relationship is readily reversible.

Erythrocytes vary widely from species to species in respect to both size and number per cubic millimeter of blood. Among the smallest that have been described are the erythrocytes of mouse deer (*Tragulus*), which measure about 1.5μ in diameter, and the largest are probably those of the urodele, *Amphiuma means tridactylum* (Fig. 9.1), which are about 80μ in their long dimension. Mean erythrocyte counts as low as 73,000/mm^3 have been described in the dogfish, *Squalus;* as high as 3,900,000/mm^3 in the highly active mackerel, *Scomber;* and in the order of 16,000,000/mm^3 in goats. Adult human males have a mean value of about 5,400,000 erythrocytes/mm^3. A more meaningful measure of hemoglobin content of blood is its "hematocrit" value. This value approximates the percent volume of erythrocytes in a given sample of blood. For example, although goats have an erythrocyte count that equals $16 \times 10^6/mm^3$, their hematocrit is about 33ml/100ml (33 percent) whole blood; human blood, with an erythrocyte count of only $5.4 \times 10^6/mm^3$, has a hematocrit of 47 percent. The discrepancy owes to the difference in size of goat (4.0μ) and human (7.5μ) erythrocytes.

LEUKOCYTES

The leukocytes, or *white blood cells,* occur in two general classes: granulocytes and agranulocytes. The former are characterized by the presence of distinct, membrane-bound granules in the cytoplasm which, although highly variable with respect to size and number, are classified into three types on the basis of their reactions to standard blood stains: acidophilic, basophilic, and heterophilic. The granules of acidophilic leukocytes stain brilliantly with the acid dye eosin (hence they are also known as *eosinophils*) whereas the granules of basophilic leukocytes stain deeply with basic dyes and those of heterophilic leukocytes tend generally to stain lightly with either acid or basic dyes. Moreover, the granulocytes of many vertebrates have lobed nuclei. In heterophils this reaches an extreme condition, whereby lobes may be connected only by very slender strands; hence they are known by the alternative name, polymorphonuclear leukocytes. Agranulocytes, on the other hand, have a more or less homogeneous cytoplasm, and their nuclei are more likely to have a fairly regular outline, being round or only slightly indented in most cases. The agranulocytes include lymphocytes and monocytes.

GRANULOCYTES. Classifying granulocytes on the basis of the staining reaction of their granular inclusions may be compared with the ancients' system of classifying animals according to where they live, i.e., the land, the sea, or the air. We know now, of course, that the relationship that exists between a clam and a starfish is much more remote than that between a whale and a mouse. We are not quite so secure in our knowledge of granulocyte relationships, however. Many authorities believe that some kinds of eosinophils of lower vertebrates are more comparable with the heterophils of mammals than they are with other kinds of eosinophils. However, until the functions and relations of granulocytes have been clarified, it is probably more practical to classify them on the basis of a clearly definable criterion, such as color, however superficial this criterion may be, than on some obscure or questionable basis. The student simply must bear in mind that the following classification is used with these reservations in mind.

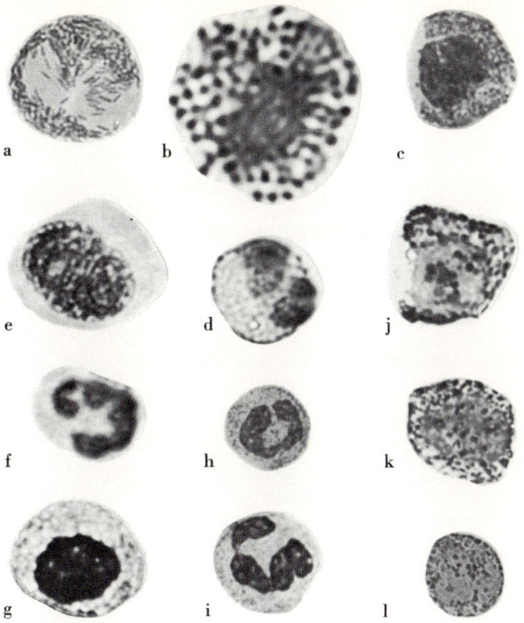

Fig. 9.3 Granulocytes from representative vertebrates. Eosinophils, *a–c;* heterophils, *e–i;* basophils, *j–l.* Granulocytes *a* and *e* are from *Mustelus canis; b, g,* and *j* from *Chrysemys picta; c, i, k,* and *l* from *Homo sapiens; d* and *f* from *Rana clamitans;* and *h* from *Felis domesticus.* The eosinophil of the turtle, *C. picta,* had ruptured, showing the granules spread out. Approximately X1000.

Eosinophils. Eosinophilic leukocytes have been found in most species of vertebrates. According to one report, the adult lamprey *Petromyzon* is an example of a species which lacks eosinophils and yet eosinophils have been described in another lamprey, *Lampetra.* In most fishes eosinophils constitute the most abundant form of granulocyte. They are generally somewhat larger than their companion erythrocytes; usually they are round, and their nucleus varies considerably in shape among the various species. The nucleus may be round, irregular, bilobed, or polymorphic. In the lungfish, for example, both bilobed and polymorphic types predominate. In many rodents doughnut-shaped nuclei occur. The nucleus of eosinophils tends to stain lightly, however, with rather uniformly dispersed chromatin clumps (Fig. 9.3). The size and shape of the granules also vary. In the smooth dogfish (*Mustelus*), for example, both fine-granular and coarse-granular eosinophils and eosinophils with rod-shaped (bacillary) granules occur. Eosinophils with bacillary granules have also been described in dipnoans, in holosteans (*Amia*), in many reptiles, and in birds. One curious phenomenon is that although both horses and asses have eosinophils with granules of modest dimensions, the hybrid, the mule, has eosinophils characterized by extraordinarily large granules. As in most types of mammalian eosinophils, the granules are spherical, however. Electron micrographs of eosinophils reveal that the granules are banded by transverse zones of high electron density (Fig. 9.4).

The function of eosinophils is unknown. It is thought that they may have an antihistaminic action, because eosinophils are most abundant at sites of potential invasion by foreign protein to which the individual may have become sensitized. For example, they are especially numerous in the intestinal submucosa, or in the connective tissue adjacent to or surrounding parasitic infestation. Their numbers in the circulating blood increase dramatically under conditions of allergic sensitivity. The number of circulating eosinophils is also high in individuals suffering adrenal insufficiency, which is suggestive, considering that at least one of the cortical steroids is effective in counteracting the action of histamine. Eosinophils are motile cells, however, and occasionally they have been observed to engulf particles, and so

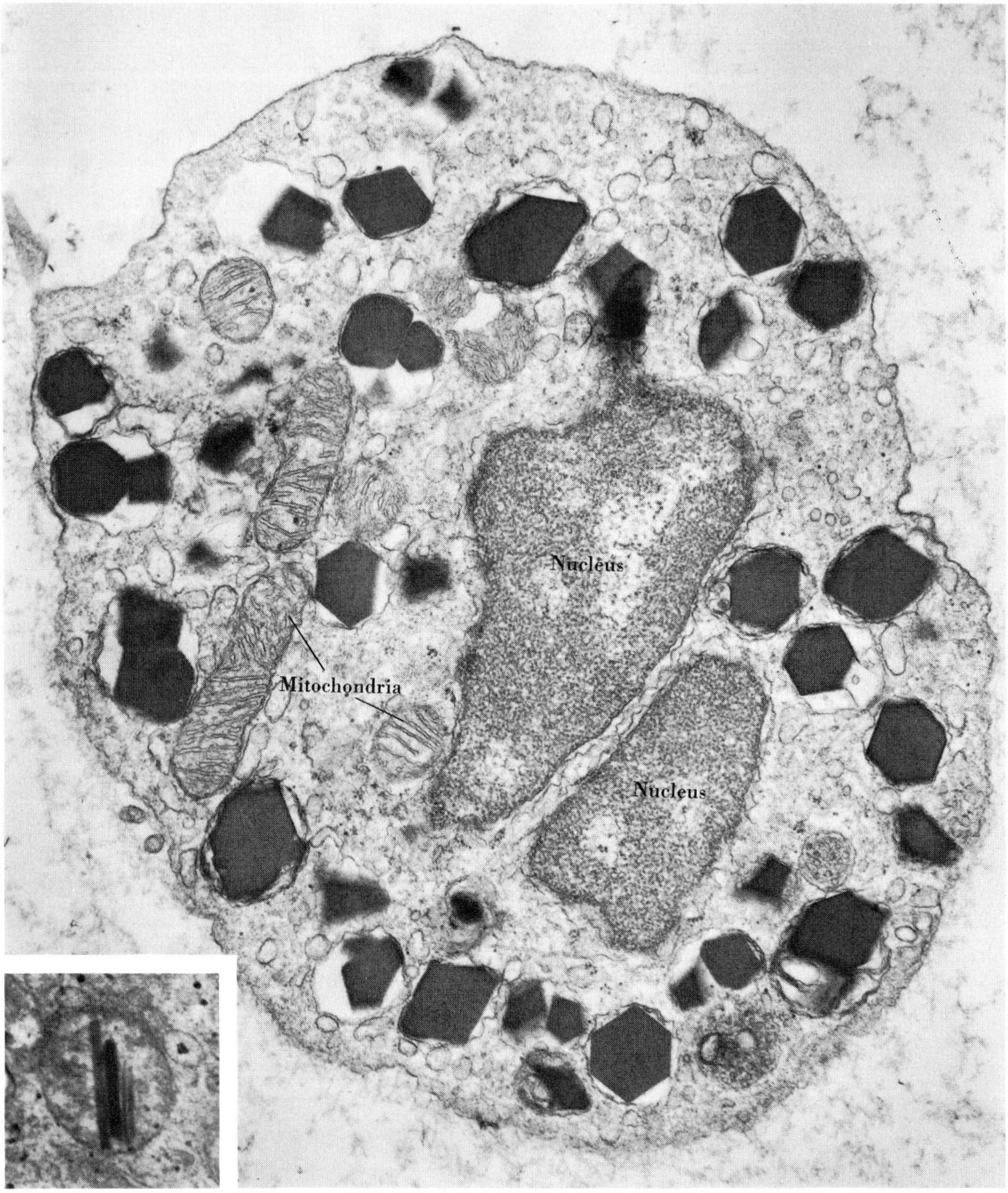

Fig. 9.4 Electron micrograph of an eosinophil from the smooth dogfish shark, *Mustelus canis*, showing the crystalline structure of the eosinophilic granules and various cytoplasmic organelles. X26,250. (Courtesy of F. Merk.) Inset: banded eosinophilic granule from the goldfish *Carassius auratus*. X64,800. (From E. L. Weinreb. 1964. Anat. Record 147:219–238.)

they appear to have some phagocytic capabilities. Furthermore, they are very numerous in tadpole tails and gut during metamorphosis, when these organs are being resorbed, which also suggests a phagocytic function.

Basophils. Basophils (Fig. 9.3) are the least numerous of the granulocytic series of leukocytes in most vertebrate species. They are reported to be lacking in the circulating blood of cyclostomes and elasmobranchs, and

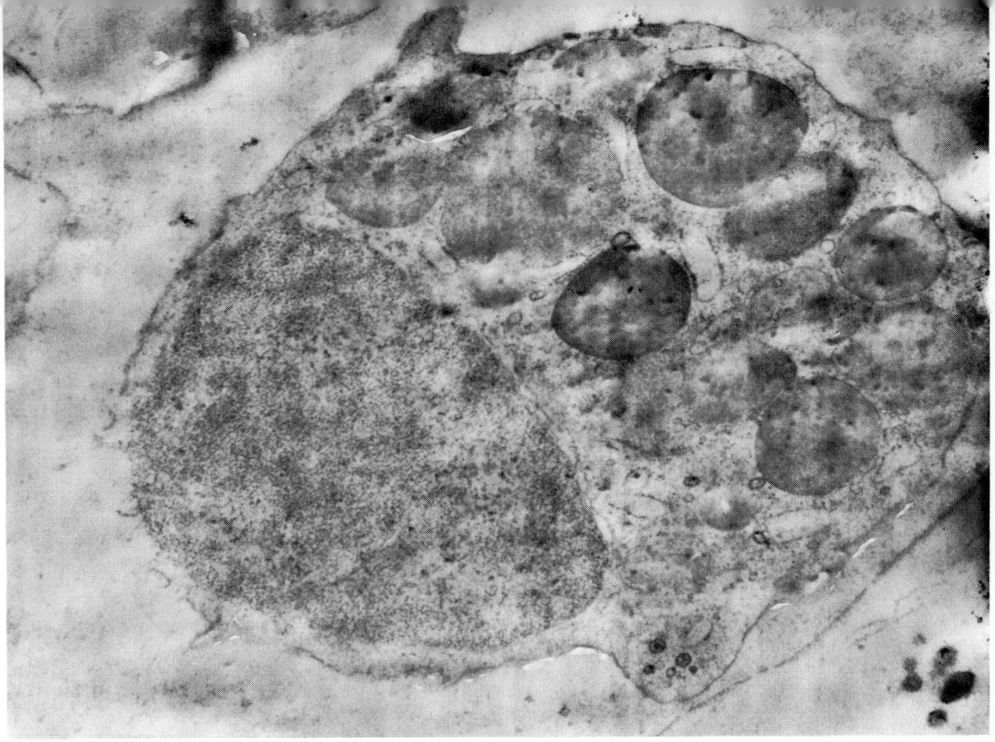

Fig. 9.5 Electron micrograph of a basophil from the goldfish *Carassius auratus*. X42,440. (From E. L. Weinreb. 1964. Anat. Record 147:219–238.)

they are rare in most bony fishes, except for the holostean, *Amia*, in which they are abundant. In dipneusti two types of basophils have been described, differing in granule size. Basophils of the goldfish (*Carassius*), although rare, have been studied by light and electron microscopy and have been found to be similar to those of higher vertebrates. The nucleus is round, eccentric, and more homogeneously basophilic than in other granulocytes of goldfish. The cytoplasmic granules are large and either round or chunky and, like those of eosinophils, membrane-bound (Fig. 9.5).

Basophils are also rare in most amphibians, but in others (such as *Cryptobranchus* and *Necturus*) they are as numerous as eosinophils. In reptiles and birds basophils are not infrequently encountered and their granules are large and strongly basophilic. The basophils of mammals have been the most extensively studied, although their numbers are typically small and they are virtually lacking in the circulating blood of some species (e.g., *Felis*, *Rattus*, and *Mus*). Their granules, while strongly basophilic, do not stain uniformly, and they are quite variable in size. The granules are water-soluble in most mammalian species, and so they are easily lost in some histological procedures.

Basophils are perhaps the least motile of the granulocytic series, and it is doubtful if they have any significant phagocytic function under normal conditions. Their granules have histochemical characteristics similar to those of mast cells, and presumably their function is similar. That is, they are believed to be trephocytes (see Chapter 3) and to carry heparin and possibly histamine; thus they probably play a significant role in endothelial permeability and in control of the blood clotting mechanism.

Heterophils. Heterophilic leukocytes are the scavengers of the blood stream and they are found in all vertebrates. They were first studied in human blood, where their granules were found to stain best with a mixture of acid and basic dyes. Because of this, they were named neutrophils. Perhaps the name is appropriate for human heterophils and for the heterophils of many other species, but it is not generally appropriate. For example, in rabbits and guinea pigs the granules are usually eosinophilic, and yet it is known that these cells are functionally identical with human neutrophils. Leukocytes with eosinophilic inclusions with a fusiform shape are found in many vertebrates, such as lungfish, certain reptiles, and birds. Because of their phagocytic action and nuclear form, they are considered to be heterophils. How distinct these are morphologically and/or functionally from eosinophils with bacillary inclusions is a matter of conjecture at this stage of our knowledge.

In fishes and amphibians typical heterophils have an eccentric nucleus, which may be slightly indented or lobed (Fig. 9.3). Polymorphic heterophils have been described in the hagfish (*Myxine*), however, and they occur occasionally in fishes and are fairly common in amphibians. The chromatin is lightly to moderately basophilic and dispersed in small clumps in most species. The cytoplasmic granules are typically very fine, as in higher forms, and quite numerous. Electron micrographs of heterophils of the goldfish (*Carassius*) show the granules to be enclosed in a unit membrane. Some of the granules of fishes and amphibians are lightly basophilic, but generally they are lightly eosinophilic or neutrophilic, except for those heterophils having the large, fusiform granules discussed above. Mitochondria and elements of the endoplasmic reticulum are present, but in small amounts.

Reptiles have surprisingly few heterophils, compared with other granulocytes, but they are very distinctive. The heterophils of birds are primarily the type having fusiform, eosinophilic inclusions, and the nucleus is usually bilobed. Mammalian heterophils, although varying somewhat in staining reaction, as mentioned above, are quite uniformly polymorphonuclear (Fig. 9.3) and the granules are very fine. Whereas the granules of human eosinophils measure from 0.5 to 1.0μ, those of human neutrophils measure from 0.1 to 0.2μ in diameter. In many rodents the heterophils are characterized by ring nuclei and nuclei in the form of the figure eight.

Heterophils are the most phagocytic of the granulocytic series. Their phagocytic activity can be observed both within blood vessels and in tissue spaces. As is the case with other leukocytes, they are attracted to areas of infection or injury in large numbers, where they immediately set to work engulfing invading microorganisms. It has been proposed that the cytoplasmic granules are vacuoles containing hydrolytic enzymes which digest ingested materials. In heavy infections, the heterophils become gorged with ingested microorganisms and die in great numbers, forming in this way the main ingredient of pus.

In a significant proportion of the neutrophils of human females a small pyknotic body can be observed appended to a nuclear lobe. This is believed to represent a heteropyknotic sex chromosome and is of diagnostic value in determining the genetic sex of pseudohermaphrodites. It may be found in an occasional male neutrophil as well.

AGRANULOCYTES. The term *agranulocyte* is somewhat misleading, for cytoplasmic granules are by no means uncommon in the blood cells included in this category. The granules, when present, are few in number, as a rule, and are azurophilic, i.e., they stain a reddish-purple color with methylene-azure, an ingredient of the common blood stains. The kinds of agranulocytes encountered in circulating blood are numerous in lower vertebrates, where hemopoiesis is to a greater or less extent intravascular, and few in higher vertebrates. The major categories of agranulocytes are two, lymphocytes and monocytes, but in cyclostomes and some fishes the primitive stem cells from which all blood cell types apparently differentiate, as well as their transitional forms, also occur in the circulating blood. This stem cell is the hemocytoblast.

Hemocytoblasts. Hemocytoblasts are about the largest cells in the blood, although in some species the erythrocytes are larger. In the hagfish (*Myxine*) they have been described as abundant cells, characterized by a large, imperfectly round nucleus that stains purple, and a rather homogeneous cytoplasm that stains a deep blue. Occasional vacuoles occur in the cytoplasm, and in some preparations paired centrioles may be observed lying in a small indentation in the nuclear membrane. In the lamprey (*Lampetra*) hemocytoblasts are considerably smaller, as are all the blood cells, and the nucleus is only lightly basophilic. In most other vertebrates the hemocytoblasts are confined to hemopoietic centers and thus are not common inhabitants of circulating blood.

Lymphocytes. Lymphocytes are very numerous in all vertebrates, and they come in a variety of sizes, which, for want of better names, are called small, medium, and large lymphocytes. The differences between them are primarily caused by the amount of cytoplasm surrounding the nucleus, although the nucleus tends to be less dense in the larger forms. There are functional differences, which will be considered in connection with hemopoiesis. It is interesting that lymphocytes are very similar in all classes of vertebrates, both structurally and, as far as we know, functionally. Because of this, a description of mammalian lymphocytes will be given as representative of all vertebrates, except where noted.

Small lymphocytes in mammals are only slightly larger than the erythrocytes. The nucleus nearly fills the cell and it has a spherical shape with a small indentation on one side (Fig. 9.6). The chromatin tends to be concentrated about the periphery, causing the nucleus to stain quite densely; but coarse clumps of chromatin occur centrally, so the staining is not uniform. The stained chromatin obscures an otherwise prominent nucleolus, which shows clearly by phase-contrast microscopy in live lymphocytes. The cytoplasm is lightly basophilic, owing to a rather good population of RNP particles as revealed by electron microscopy; also, a variable number of lipid droplets may be encountered. A small number of azurophilic granules occur in many small lymphocytes. Although the chemical composition of the granules is unknown, they may correspond with the compound vesicles described in electron micrographs of teleost lymphocytes. A pair of centrioles lies in the indentation of the nucleus in company with a group of Golgi vesicles. Mitochondria are present, but they are not very numerous.

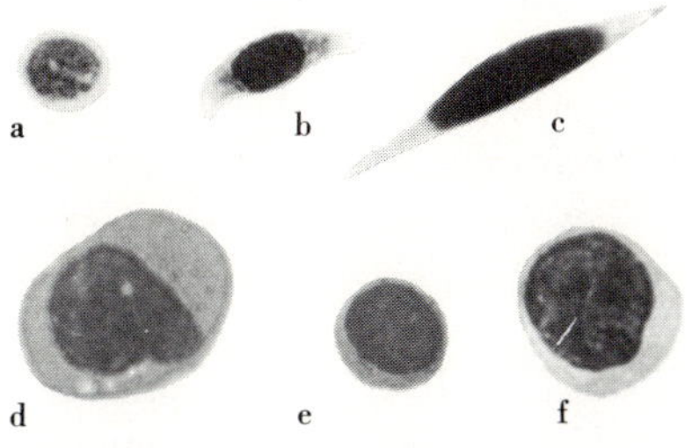

Fig. 9.6 Representative agranulocytes. Agranulocytes *a–c* are thrombocytes: *a*, *Petromyzon*; *b*, *Mustelus*; and *c*, *Rana*. Agranulocyte *d* is a monocyte (*Homo*); *e* is a lymphocyte (*Homo*); and *f* is a large lymphocyte (*Homo*). Approximately. X950.

Despite their large numbers and their ubiquitous presence extravascularly, the functions of lymphocytes are only partly understood. They are known to give rise to erythrocytes and to granulocytes in many species, as well as to plasma cells and thrombocytes, as will be discussed below. Under some conditions they give rise to monocytes as well. They were believed to form antibodies in response to antigen stimulation, until studies with fluorescein-labeled antigens showed that this is more likely done by plasma cells. And yet they occur in great numbers surrounding and infiltrating tissue grafts. They are also abundant immediately under most epithelia, and they often migrate out through the epithelium into the luminal spaces. Lymphocytes have even been thought to replace epithelial cells of the gut, which show a very rapid turnover. At this juncture little more can be said than that (1) lymphocytes may differentiate into other cell types, and (2) they are implicated in some way in the antigen-antibody defense mechanisms of the organism.

Monocytes. To comparative histologists, monocytes are the poor relations of the leukocytes; they have been ignored, overlooked, or confused with large lymphocytes in most nonmammalian species. To be sure, they are similar to large lymphocytes, but the nucleus is usually less basophilic, the chromatin is finer, and one or two nucleoli are observable in good preparations. The shape of the nucleus varies with the species, and may depend upon the age of the cell. Monocytes of hagfish (*Myxine*) have a large, irregular nucleus which stains a rather dark blue. In birds the monocytes have a deeply indented nucleus. In mammals the nucleus of young monocytes is

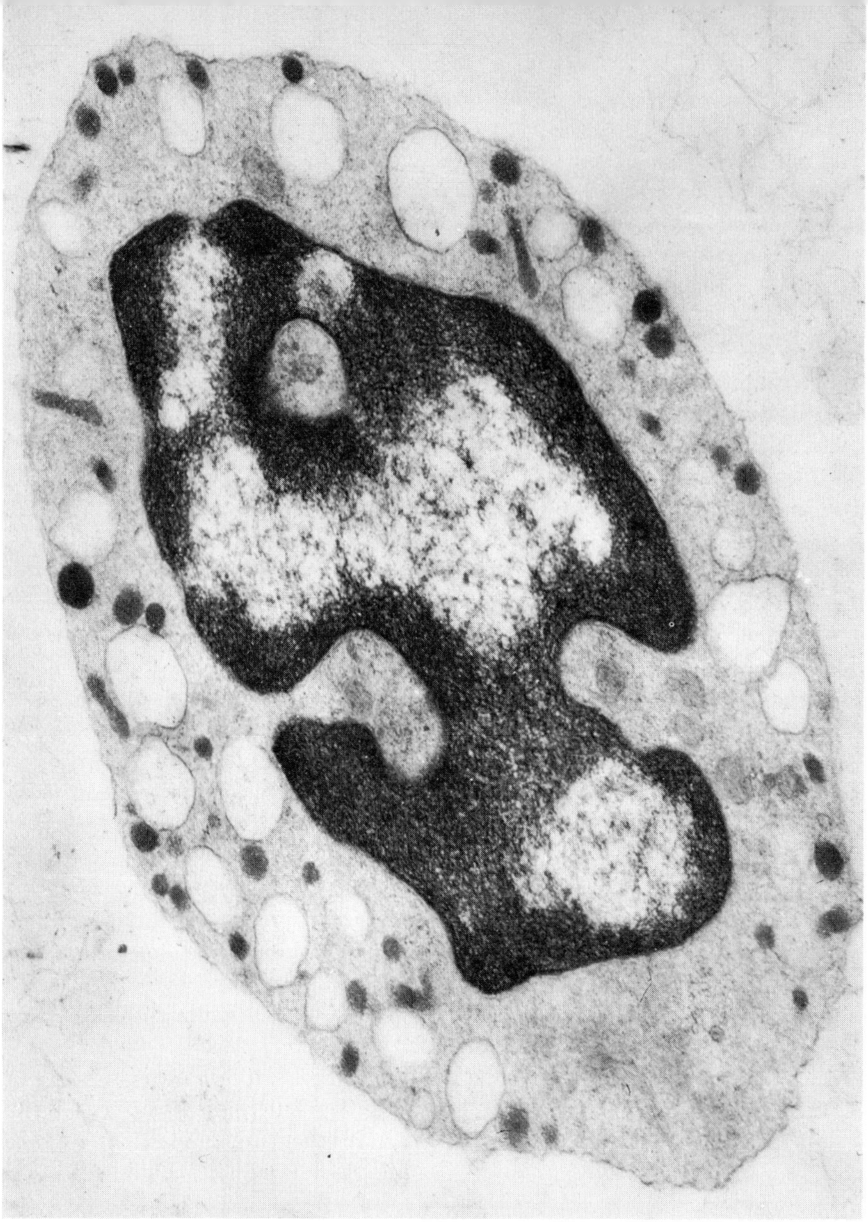

Fig. 9.7 Electron micrograph of a thrombocyte from a frog, *Rana pipiens.* X28,000. (Courtesy of E. M. McKinnon.)

similar to that of a large lymphocyte, bean-shaped in older monocytes and horse-shoe shaped in still older monocytes (Fig. 9.6). The cytoplasm is generally clear, although azurophilic granules occasionally occur. Mitochondria and Golgi vesicles are arranged like the sepals and petals of a flower around the paired centrioles, in a pattern similar to that sometimes found in septal cells of the lung, in plasma cells, and in macrophages. Monocytes are motile cells and they are aggressive phagocytes in extravascular areas, where they may differentiate further into macrophages.

Thrombocytes. Although thrombocytes are functionally distinct from lymphocytes and monocytes, they are generally believed to be derived from the former, and so they are considered here. They are found in all vertebrates except mammals, in which their function has been assumed by unrelated structures, the blood platelets. Thrombocytes are quite numerous, in fish blood being second in number only to the erythrocytes. Young thrombocytes are elliptical or spindle-shaped (Fig. 9.6). The nucleus varies from round to elliptical and it often has longitudinal grooves along its surface. Electron micrographs of amphibian (*Rana*) thrombocytes (Fig. 9.7) show that the nucleus is deeply grooved. The nucleus is strongly basophilic, resembling very closely that of small lymphocytes. The cytoplasm is considerable in young thrombocytes, and it is filled with PAS-positive granules. The granules are lost and the cytoplasm becomes greatly reduced with age of the cell, so that the older thrombocytes are virtually naked nuclei.

Thrombocytes can be harvested in large numbers from fish or amphibian blood by differential centrifugation of whole, citrated, or heparinized blood. The erythrocytes are thrown into a compact layer at the bottom of the centrifuge tube, and, forming a thin "buffy" coat on top of the erythrocytes, are the leukocytes. At the junction of the buffy coat and plasma the thrombocytes may be found. They contain an enzyme similar to thromboplastin, which is released under conditions of tissue damage and catalyzes the transformation of prothrombin into thrombin, which, in turn, influences the conversion of fibrinogen to fibrin. The details of the clotting mechanisms are not fully understood and are beyond the scope of this text. There have been reports of a fortuitous structural similarity between thrombocytes and blood platelets in that bits of cytoplasm have been observed to become separated from the nucleate body of the thrombocyte. It is doubtful if the significance of this is more than a matter of coincidence.

BLOOD PLATELETS. Found only in mammals, blood platelets are enucleate fragments of certain giant cells of the bone marrow, the megakaryocytes. They serve the same function as the thrombocytes, and so they have replaced them. In normal circulating blood the platelets are spherical or oval bodies measuring usually between 1 and 4μ in diameter. They are packed with membrane-bound granules (Fig. 9.8) that in dry blood smears stain purple. Platelets respond very rapidly to the slightest chemical or physical alteration of their environment by sending out streaming protoplasmic processes and by becoming "sticky." In injured vessels they adhere to each other and to the endothelium so as to form plugs over the site of injury. In very small vessels, such as capillaries or small venules, the platelet plug may form an ob-

Fig. 9.8 Electron micrograph of blood platelets from an ox. X30,000. (Courtesy of F. Chao.)

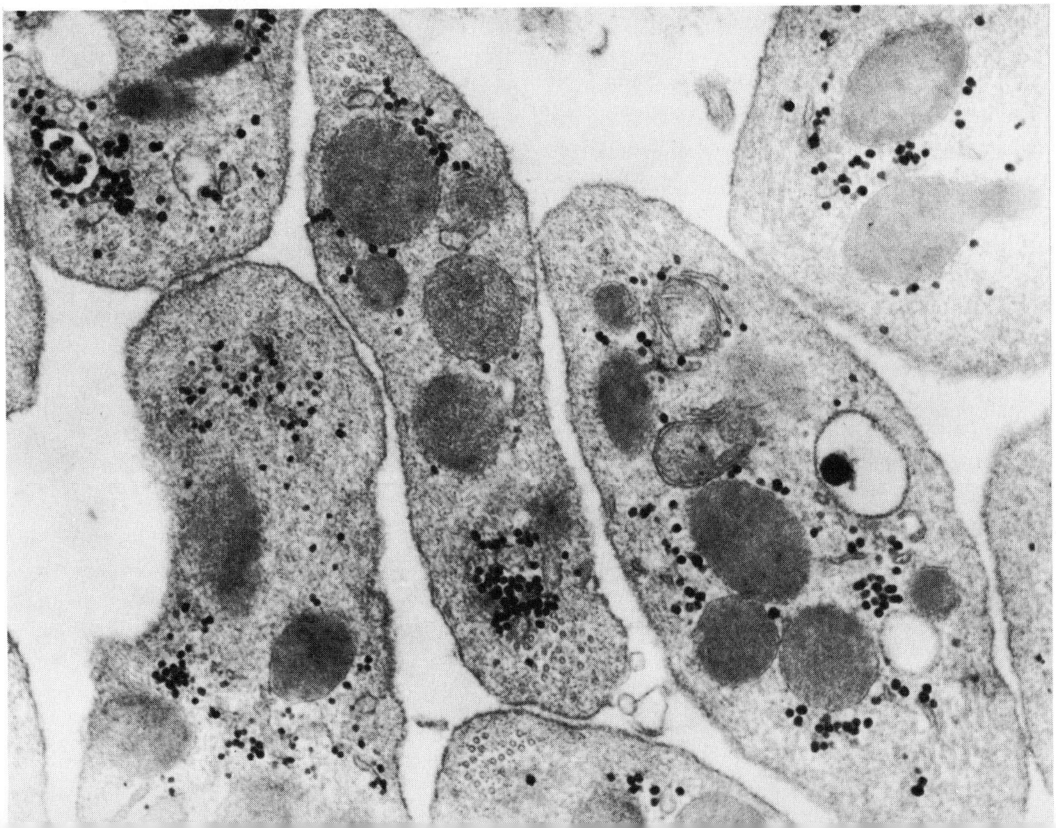

struction or *thrombus* that may stop blood flow. Under pathological conditions such a thrombus may initiate intravascular clotting in the immediate area. Even in a rapidly prepared blood smear the platelets will clump together and present an appearance that is far from that which has been observed in normal circulating blood. Blood platelets are extremely fragile. When a tissue is lacerated or otherwise injured, the platelets disintegrate, releasing their granules, and the clotting reaction is triggered. Platelets have been shown to contain a contractile protein which is presumed to be responsible for the contraction of clots. Serotonin, histamine, and fibrinogen have also been shown to be associated with platelets, although the nature of the association is not clear.

Lymph and Tissue Fluid

The blood of vertebrates is contained within a closed system of vessels, with the qualified exception of the blood in the spleen (*see below*). Therefore, blood does not come into direct contact with tissues other than with the endothelial lining of the heart and blood vessels. Instead, these other tissues are bathed in an extravascular fluid of their own, which is called *tissue fluid.*

Tissue fluid is derived from two sources: the blood plasma and the tissues themselves. The blood contributes essentially all of its kinds of ingredients except its corpuscles and its macromolecules (the serum and plasma proteins), which are simply too large to diffuse in any significant quantity through the walls of capillaries under ordinary conditions. Despite this, however, an occasional vascular defect will permit a few erythrocytes to leak out, and leukocytes are able to sneak out by amoeboid movement, which involves poking a protoplasmic process into an endothelial cell junction and then flowing the rest of the corpuscle through the opening thus made (a process known as diapedesis). Thrombocytes or platelets have never been observed moving out either passively, as with the erythrocyte, or actively, as with the leukocytes. On the contrary, they devote their dedicated lives to plugging up any holes by which transcapillary migration of cells could occur! For the most part, however, the fluid leaving the blood capillaries is cell-free and devoid of macromolecules. The cells of the tissues contribute water as a byproduct of metabolism; metabolic wastes, such as carbonic acid, lactic acid, and ammonium compounds; secretion products; and miscellaneous metabolites. The principal components of tissue fluid, nevertheless, are water and the inorganic salts derived from the blood plasma. The proportions of these are remarkably constant. The tissue cells are everywhere bathed by the tissue fluid; they extract their needs from it and pour their excretions and secretions into it. The relation between the cells and tissue fluid is similar to that between free-living primordial protozoa and the sea. One might consider that we really have not evolved very far, after all, for we are still little more than a collection of unicellular organisms swimming in a skin-full of sea water!

Tissue fluid circulates in the sense that as rapidly as it is produced by the cells and by transfusion from blood capillaries, it is eliminated. There are two main avenues of exit for tissue fluid; one of these is the blood vascular system, whereby the tissue fluid returns to the blood vessels by the simple expedient of flowing back into the blood farther down the capillary stream, and the other is the lymphatic vessels.

When fluid passes out of a capillary, there is a fall in hydrostatic pressure and a rise in osmotic pressure within the capillary, which may be represented as a gradient of pressures extending from the arterial end to the venous end of the capillary. The midpoint of this gradient is in equilibrium with both the hydrostatic and the osmotic pressures of the tissue fluid. Hence at the arterial end of the capillary the forces favor flow from capillary to tissue fluid, and at the venous end the opposite prevails.

The foregoing mechanism alone is not sufficient to maintain an equilibrium, however, for just as macromolecules cannot pass *out* of capillaries, so the macromolecules produced by cell synthesis and by cell breakdown (as a normal consequence of wear-and-tear) cannot pass *into* capillaries. This potential dilemma is solved by an alternate escape route, the lymphatic vessels, described below. The lymph capillaries *are* permeable to macromolecules and, inasmuch as they originate in the tissue spaces as a fine network of blind vessels (Fig. 9.30), the lymph capillaries cannot carry macromolecules to the tissue fluid, only away from it.

The fluid contents of the lymphatic vessels is called *lymph.* It does not differ appreciably from tissue fluid in its basic composition, except for having a slightly higher osmotic pressure, as one might expect from the foregoing remarks, and being high in CO_2 and low in O_2. The lymph which drains the intestinal wall does show wide variation in composition, for it carries most of the fats absorbed by the intestinal mucosa after a meal. The fat is emulsified by the colloidal components of the lymph, thereby imparting a milky appearance to the lymph. The lymphatic vessels draining the small intestine are called *lacteals,* because of this milky appearance. Fatty lymph is called *chyle.* The fatty droplets in chyle are extremely small (about 1μ) and many of them eventually find their way to the blood vascular system, where they are known as *chylomicrons.* There are rarely any blood cells in the lymph capillaries, but as the lymph filters through the lymph nodes (*see below*) small lymphocytes are added to it. The lymph in the large lymphatic vessels that lead to the systemic veins consequently contains a large number of cells, and they are almost exclusively small lymphocytes, with an occasional erythrocyte or eosinophil. Some fibrinogen finds its way into the lymphatic vessels, for lymph is capable of clotting, although the clot is soft and rather loose.

HEMOPOIETIC TISSUES AND ORGANS

The blood of vertebrates arises ontogenetically in small clusters of mesenchymal cells, the *blood islands,* which are abundant in the extraembryonic splanchnic mesoderm of the yolk sac. The outer cells of the blood islands differentiate into endothelial cells which, joining up with adjacent blood islands, form an anastomosing network of extraembryonic blood vessels. These, in turn, are joined by intraembryonic blood vessels which have similarly differentiated from mesenchyme. The inner cells of the blood islands, meanwhile, become *hemocytoblasts,* from which all kinds of embryonic blood cells differentiate. Later in embryonic life other tissues and organs take over the function of blood cell production (*hemopoiesis*)—principally, the liver, the kidney capsule, the intestinal submucosa, and the spleen. These organs continue their hemopoietic function throughout life in the lower classes, but in the course of vertebrate evolution, two kinds of hemopoietic tissue emerge in adult forms, *myeloid* and *lymphoid* tissue. Constant and clear-cut distinction between these tissues is not achieved lower than the level of birds and mammals, however, as described below.

In the teleosts the kidneys and the spleen become the principal adult centers for the formation of erythrocytes, while the granulocytes differentiate mainly in the intestinal submucosa and the spleen, and the agranulocytes develop in all of these sites. In adult urodeles erythrocyte and thrombocyte formation is restricted primarily to the spleen, and granulocyte production to the subcapsular region of the liver, although granulocytes continue to be formed to some extent in the intestinal submucosa. In some urodeles, such as *Plethodon,* and in the anurans the bone marrow has been introduced as an adult site for the production of leukocytes, but not of erythrocytes. Hemopoietic tissue has been described in the cranial cartilages of some sharks and holocephalans and also in the large cranial bones of the gar (*Lepisosteus*) and of *Amia,* anticipating the condition found in higher tetrapods.

The reptiles evidence most clearly the transition from spleen to bone marrow as adult centers of hemopoiesis. In some, such as the horned lizard (*Phrynosoma*), the spleen is the principal site. In others, such as the Chelonia, hemopoiesis is shared about equally by spleen and bone marrow; in still others, such as *Lacerta,* the bone marrow is the primary

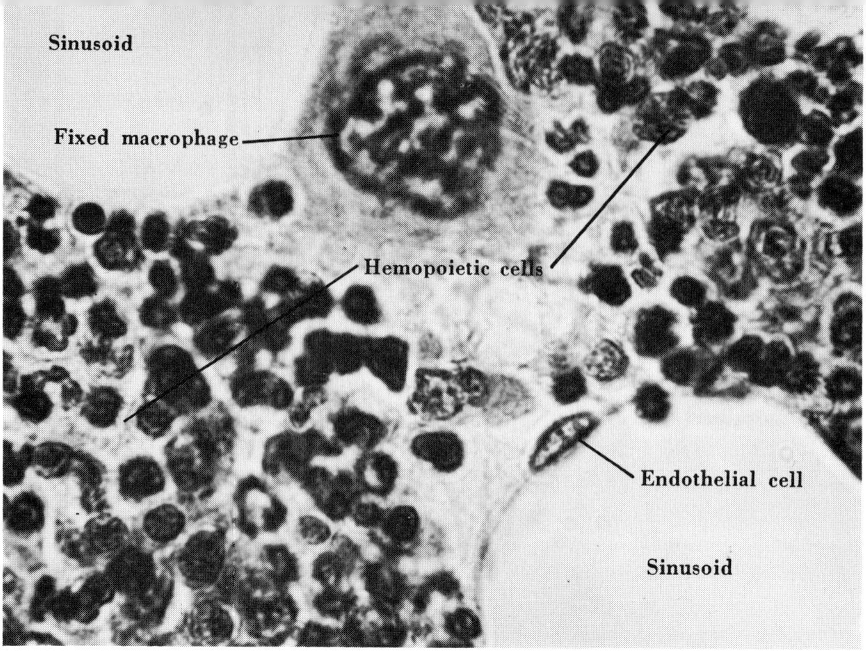

Fig. 9.9 Red bone marrow from a rabbit. X1200.

source of erythrocytes and granulocytes. In birds and mammals the red marrow is typically the sole center for erythrocyte and granulocyte production, whereas agranulocytes are produced in the spleen and in other lymphoid organs and tissues, as described below. One should be aware, nonetheless, that lymph nodules occur in the marrow cavities of birds, and that production of erythrocytes and granulocytes can, under severe conditions of hemorrhage or anemia, take place in birds or mammals in the liver or kidneys.

Myeloid Tissues

Myeloid tissue is connective tissue which is primarily concerned with the production of erythrocytes and granulocytes. With few exceptions it is a distinct tissue only in the birds and the mammals, and so the term is applicable only to these two classes of vertebrates. In the mammals myeloid tissue also serves as the source of the blood platelets. Myeloid tissue of adults is found as red marrow in the marrow cavities of the ribs, the sternum, the vertebrae, the diploe of the cranial vault, and the epiphyses of the long bones. Red marrow is distinguishable from yellow marrow, which consists primarily of depots of fat, although it does retain potential hemopoietic function.

Red marrow consists of a stroma of reticular cells and reticular fibers which form an anastomosing network loosely embracing fat cells, primitive mesenchymal cells, reticuloendothelial cells, and free cells of various types (Fig. 9.9). Because of their close association with the reticular fibers, the mesenchymal cells are referred to as *primitive reticular cells.* They are the most primitive cells in the marrow and, as such, they give rise to two other important types of cells, the *reticuloendothelial cells* and the *myeloblasts.*

RETICULOENDOTHELIAL CELLS

Some of the primitive mesenchymal cells of the red marrow enlarge and become phagocytic cells known as macrophages (see p. 41). Most of these become fixed and arranged to form a kind of endothelium lining sinusoids (Fig. 9.10). In this situation they are known as reticuloendothelial cells, although in other situations they are known either as reticular cells or simply as fixed macrophages. Some of them disattach themselves from the reticular stroma and become known as free macrophages, and they may be found within the sinusoids. Not only does blood cell production occur in red marrow, but also blood cell destruction does; and it is by macrophages, both fixed and free, that disintegrating blood cells are engulfed and removed from the circulation. The reticuloendothelial cells which form the sinusoids have been shown by electron microscopy to be joined to each other by

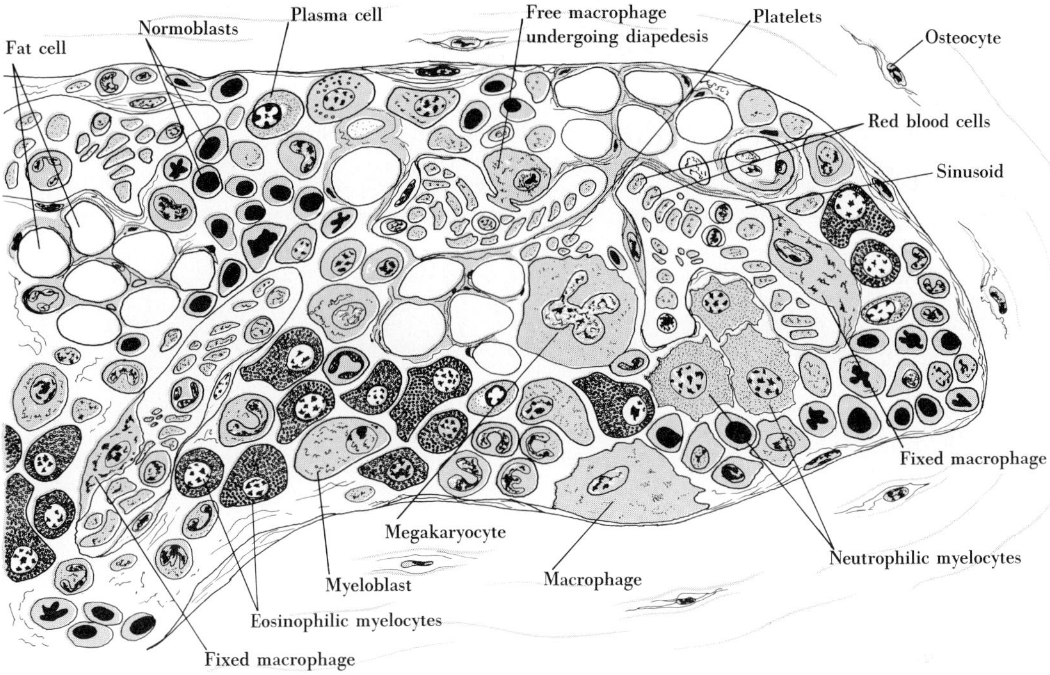

Fig. 9.10 Diagram of mammalian red bone marrow, showing the relation of reticular cells and macrophages to the sinusoids and other elements of the hemopoietic tissue.

the tips of protoplasmic processes, with small openings between them through which blood cells can percolate into and out of the sinusoids.

MYELOBLASTS

Myeloblasts are distinguished from other red marrow cells as being rather large (about 15μ in human marrow) primitive cells characterized by a large, round, pale nucleus with one or more prominent nucleoli and by lightly basophilic cytoplasm (Fig. 9.10). They are indistinguishable, however, from either the hemocytoblasts described above or the lymphoblasts described below. According to the unitarian theory of hemopoiesis, there is a common stem cell for all kinds of blood cells, the hemocytoblast. In myeloid tissue this cell is known as a myeloblast, and in lymphoid tissue it is known as a lymphoblast. According to the dualistic theory of hemopoiesis, myeloblasts and lymphoblasts are different cell types because they have different developmental potentials. Evidence from comparative histology and from experimental studies favors the unitarian concept and attributes the differences in the courses of differentiation taken by hemocytoblasts in myeloid tissue, on the one hand, and in lymphoid tissue, on the other hand, to differences in environmental fields of influence.

The processes of differentiation that myeloblasts undergo in the genesis of erythrocytes and granulocytes in the red marrow of mammals can serve as a model for blood cell differentiation in all vertebrates, although it is difficult to find two hematologists who agree on the details even in a single species. The sequence of cell types is presented in Fig. 9.11. In fishes, all the counterparts of types shown in the erythrocyte cell lineage can be found in the circulating blood, although in nonmammalian classes erythrocyte differentiation proceeds only as far as the equivalent of the normoblast stage. In amphibians, too, some of the intermediate types of erythrocytes may be found in the circulating blood. In amniote vertebrates (reptiles, birds, and mammals) immature forms of erythrocytes are rarely found outside of the hemopoietic

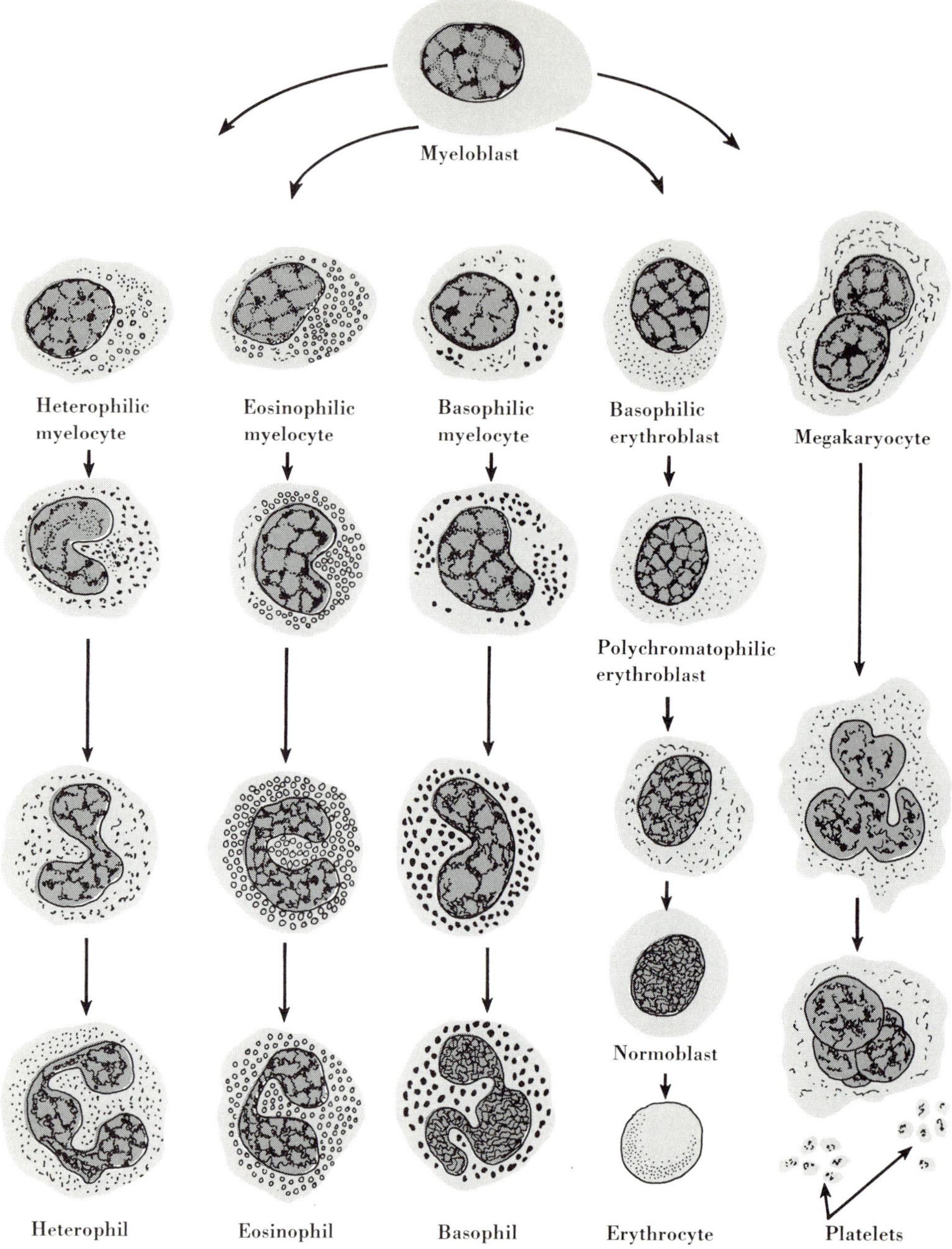

Fig. 9.11 A scheme of the ontogeny of the formed elements in the myeloid series.

centers, and when they are, some abnormal condition is indicated. Granulocyte differentiation, however, is almost entirely extravascular in all classes, with only mature types occurring as a rule in the circulating blood. Nevertheless, differences between young and old mature granulocytes can be described in terms of the form of the nucleus; i.e., the older the cell, the more polymorphic the nucleus, so that in the oldest heterophils, for example, the lobes are connected by only a very fine strand.

It must be understood that between each of the cell types shown in Fig. 9.11 mitotic divisions occur, with a residual pool of undifferentiated cells remaining. The lines of differentiation taken by daughter cells of hemocytoblasts are evident at once. The most primitive progenitor of erythrocytes, the basophilic erythroblast, has a very deeply basophilic cytoplasm, reflecting the high concentration of RNP for hemoglobin synthesis. Older erythroblasts show a progressive increase in hemoglobin content which, superimposed upon a basophilic cytoplasm, causes the color to vary from gray to lilac to purple when stained with Wright's or similar blood stain. Because of this, these cells are called polychromatophilic erythroblasts. Most of their daughter cells continue to differentiate and become normoblasts. Normoblasts have a very dense nucleus and the cytoplasm consists almost exclusively of hemoglobin, with an occasional mitochondrion, Golgi vacuole, and ER cisterna. After an undetermined number of mitotic divisions, they transform to erythrocytes by losing the nucleus. Whether the nucleus is lost by karyolysis or by extrusion is not clear, but recent evidence favors the latter. Immature enucleate erythrocytes can be shown by supravital staining with cresyl blue to contain a delicate network; hence these cells are known as reticulocytes. Normally the reticulum is lost before these cells leave the bone marrow and enter the circulating blood stream.

Cells of the granulocytic series are also identifiable at an early stage as eosinophilic, basophilic, or heterophilic myeloblasts. The granules, although sparse, are distinct. The nucleus is oval or bean-shaped in the myelocytes, and mitoses are commonplace. As the cells are ready to mature, the cytoplasm fills with its specific granules and the nucleus becomes U-shaped. In this form the cells are known as metamyelocytes. The nucleus then undergoes one or more constrictions, thereby becoming polymorphic to varying degrees, and the mature cells result.

Certainly as distinctive as any of the foregoing are the megakaryocytes of mammalian red marrow. These cells become very large (30 to 40μ in human marrow) and the nucleus undergoes several endomitotic divisions. The daughter nuclei usually rejoin each other, resulting in a compound-nucleus condition. The cytoplasm is copious, irregular in outline, and it acquires numerous azurophilic granules. Mitochondria, Golgi vacuoles, and a number of centrioles also are present. Older megakaryocytes show aggregations of the azurophilic granules around the periphery, followed by sloughing or pinching off of bits of cytoplasm containing the granules. These pieces of cytoplasm are the blood platelets. Old megakaryocytes, stripped of most of their cytoplasm, may often be seen undergoing degenerative changes in the myeloid tissue of the marrow.

Other inhabitants of mammalian red marrow are plasma cells, lymphocytes, and monocytes. Plasma cells (Fig. 9.11) have been shown to be derived directly from lymphocytes and, according to some evidence, from particular lymphocytes which have arisen in the thymus gland (see page 252). Plasma cells are slightly larger than lymphocytes, with a smaller nucleus and correspondingly more cytoplasm. Two diagnostic characteristics of mammalian plasma cells are the "cartwheel" nucleus and the "Golgi ghost." The distribution of chromatin in the nucleus is radial, like the spokes of a wheel; the Golgi complex, which is juxtanuclear, is dissolved out by usual techniques so that a clear area marks its location in an otherwise densely basophilic cytoplasm. Electron micrographs of plasma cells (Fig. 9.12) reveal a rich endoplasmic reticulum with associated ribosomes. Plasma cells have been shown to be directly concerned with antibody synthesis. This is consistent with the high degree of development of its rough endoplasmic reticulum.

Lymphoid Tissues and Organs

Lymphoid tissue is most simply defined as reticular connective tissue which is more or less densely populated with lymphocytes. Two classes of lymphoid tissue are recognized on the basis of the degree of organization imparted by the density of the lymphocyte popu-

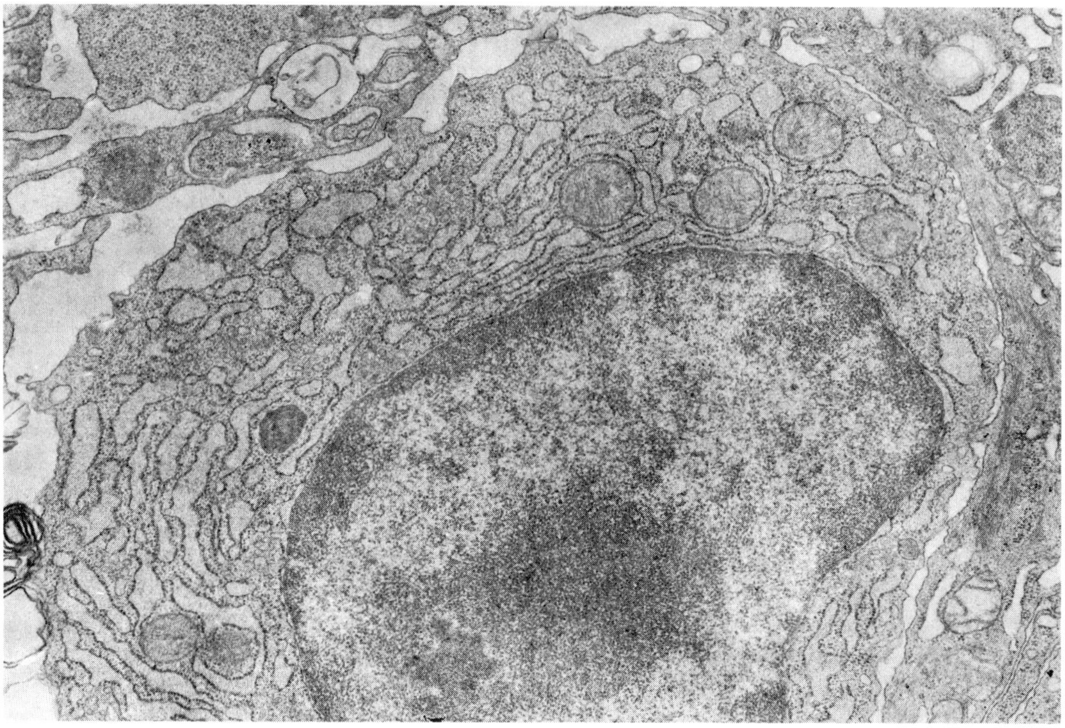

Fig. 9.12 Electron micrograph of a plasma cell from the golden hamster (*Mesocricetus auratus*). X14,000. (Courtesy of L. P. Merk.)

lation: *diffuse lymphoid tissue* and *lymph nodules.* The latter represent the organizational unit of the lymphoid organs.

In diffuse lymphoid tissue the lymphocytes form loose aggregates and there is no sharp line of demarcation between such aggregates and the surrounding connective tissue. Examples of diffuse lymphoid tissue are abundant in the submucosa of the gastrointestinal and respiratory tracts (Fig. 9.13).

In lymph nodules the lymphocytes are densely packed and hence appear as clearly delineated aggregations (Fig. 9.14), A nodule may contain a lightly stained core, the *germinal center* (Fig. 9.15), in which many medium-sized lymphocytes occur. It is because of the relatively greater abundance of cytoplasm provided by medium lymphocytes that the germinal centers appear lighter, there being proportionately fewer nuclei per unit of area in the microscopic field. It is presumed, but not established, that the medium-sized lymphocytes proliferate in the germinal centers, giving rise thereby to small lymphocytes. Lymph nodules may occur either singly, as in the intestine, or in unorganized clusters, as in the organ of Leydig and in Peyer's patches of the lower intestine (Fig. 9.15). They may also occur as partially encapsulated masses, such as in the tonsils (Fig. 9.16), or in the form of fully encapsulated organs, such as the lymph nodes, spleen, thymus, and bursa of Fabricius.

ORGAN OF LEYDIG

Characteristic of elasmobranchs, the organ of Leydig consists of a series of patches of lymphoid tissue in the esophageal mucosa. The organ includes not only lymphocytes of all sizes, but also hemocytoblasts, reticular cells, and all types of granulocytes in various stages of differentiation. Although normally the organ of Leydig is concerned only with the production of leukocytes, experimental extirpation of the spleen reveals that it can assume the function of producing erythrocytes and thrombocytes as well.

TONSILS

Tonsils are aggregations of lymph nodules associated with the pharynx in tetrapods. In

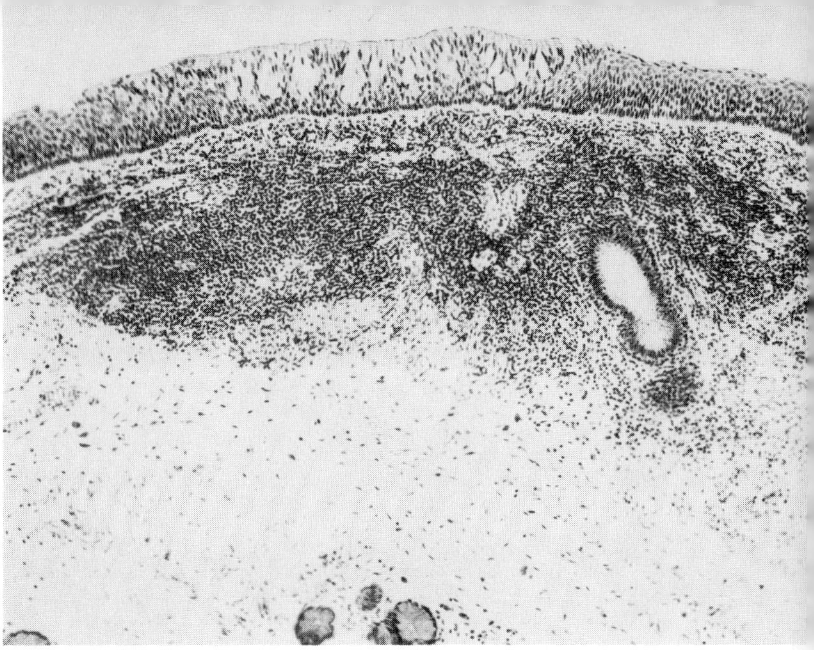

Fig. 9.13 Diffuse lymphoid tissue in the submucosa of the human uvula. X100.

amphibians they consist of little more than lymphoid masses in the roof of the pharynx that resemble the nodular concentrations encountered in the submucosa of the intestinal tract. In reptiles and birds and in some mammals these masses form better defined organs, the *pharyngeal tonsils.* In mammals the dorsal parts of the second pair of gill pouches give rise to the *palatine tonsils,* and associated with the lingual glands at the base of the tongue the lingual tonsils occur. *Tubal tonsils,* which are small masses of lymphoid tissue associated with the pharyngeal opening of the eustachian tubes have also been described in mammals.

The palatine tonsils of mammals (oddly lacking in some rodents) represent the most prominent and hence the most widely described tonsils (Fig. 9.16). They are covered by the stratified squamous epithelium of the pharynx, supported by a well-defined basement membrane. The epithelium is more or less deeply invaginated at various points to form characteristic pits or *crypts.* The epithelial lining of the crypts becomes progressively thinner with depth, and is infiltrated with lymphocytes. The ends of the deep crypts show more or less complete replacement of epithelium by lymphoid elements. Polymorphonuclear leukocytes which closely resemble heterophils, except that their cytoplasm is agranular, are frequently seen at the epithelial surface. They are believed to be derived from lymphoid cells, perhaps monocytes.

The bulk of the palatine tonsil consists of large numbers of lymph nodules segregated into clusters by crypts, on one side, and on the other by loose connective tissue *septa* derived from the underlying tonsillar *capsule.* The nodules are like those of lymph nodes, described below, and many of them have germinal centers. The greatest variety of cell types is to be encountered in the loose connective

Fig. 9.14 Lymph nodule in the intestinal wall of a mammal. The bases of the intestinal villi are visible in the upper part of the micrograph. X100.

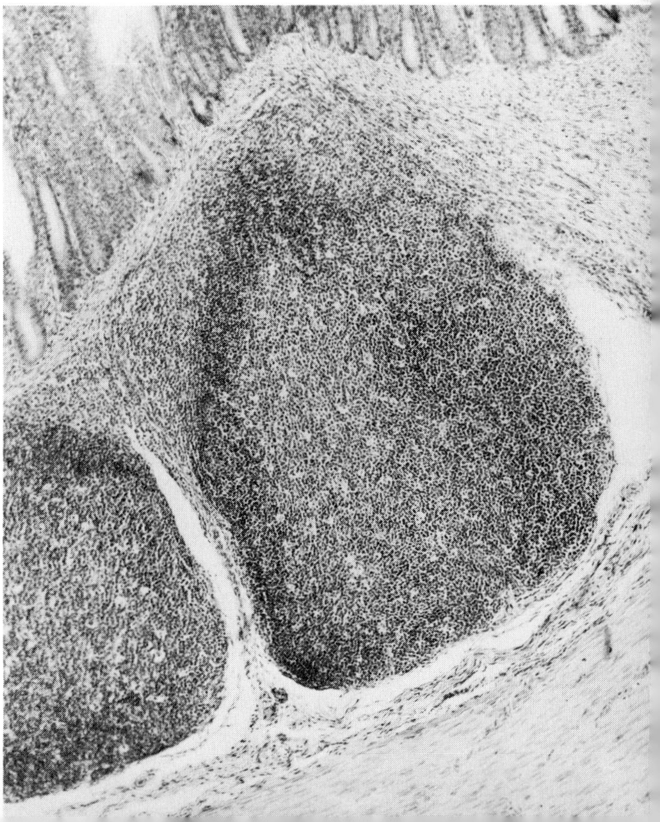

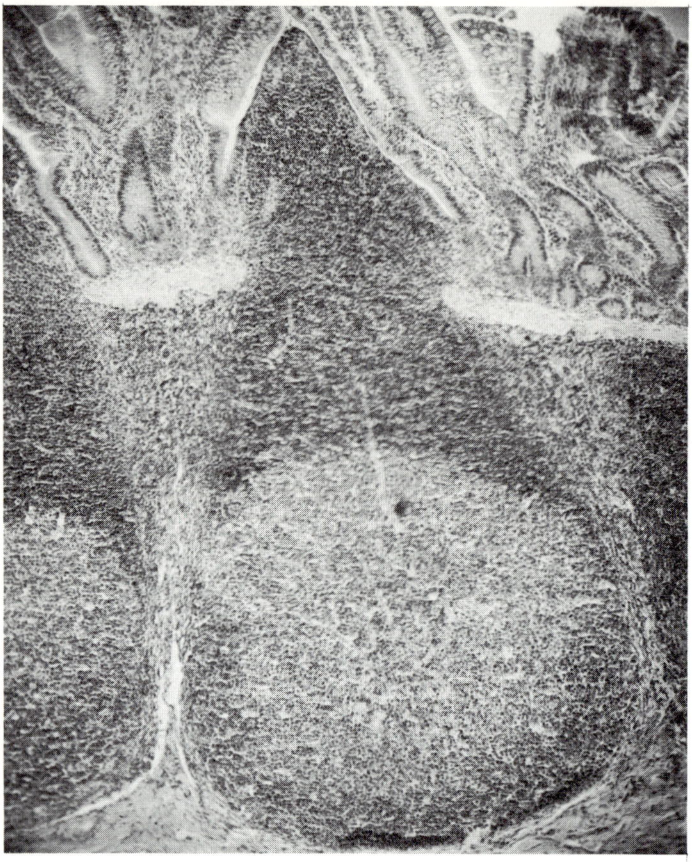

Fig. 9.15 A Peyer's patch in the human ileum protruding into the mucosa. Note the prominent germinal center. The loose connective tissue of the submucosa is restricted to a light band between the lamina propria and the lymphoid mass. X90.

Fig. 9.16 Vertical section through a palatine tonsil of a newborn rabbit. X150. (Courtesy of B. H. Harrison.)

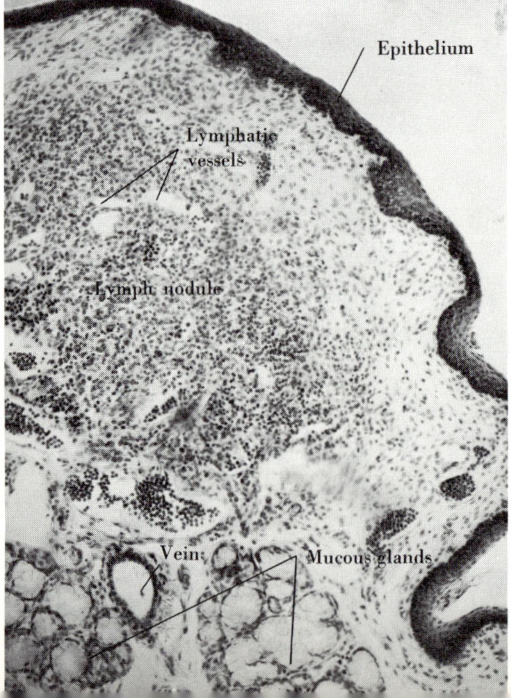

tissue of the septa, for there may be found lymphocytes of various sizes, tissue mast cells, plasma cells, and, occasionally, heterophils. Ducts connecting compound alveolar mucous glands in the submucosa with the pharyngeal lumen may be seen traversing the tonsillar tissue and its capsule.

The obvious function of tonsils is the production of lymphocytes, but whether this is the principal function is not known. The similarity in embryonic origin that exists between the palatine tonsils and the thymus, together with the facts that both organs undergo involution at about the time of puberty, and that in some species Hassal's corpuscles are present in both, are suggestive. The concentration of lymphoid tissue provided by the tonsils at the portal of entry into the entire gastrointestinal tract supports the postulate that the tonsils are the vanguard of the body's mechanisms of defense against infection by this route. Because of the very momentariness of contact between the tonsils and ingested materials, this relationship becomes a little hard to swallow! That their function more closely parallels that of the thymus is further supported by preliminary findings in the rabbit, which show in basal epithelial cells of the tonsil anlage fine structural characteristics very similar to those of primitive thymocytes and other lymphocytes (Fig. 9.17).

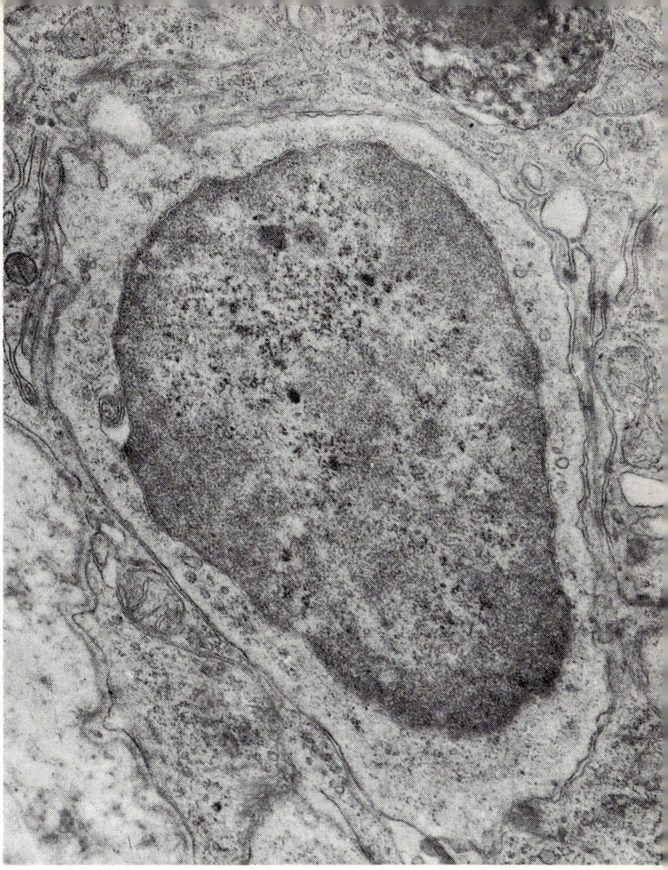

Fig. 9.17 Electron micrograph of a lymphoid cell differentiating in the tonsillar epithelium of a fetal rabbit. X11,000. (Courtesy of B. H. Harrison.)

LYMPH NODES

The lymph nodes are anatomically the most discrete of lymphoid organs, for reasons which soon will be evident. Although present in primitive form in the mesenteries of birds and crocodiles, they are best developed in and more characteristic of the mammals. A typical mammalian lymph node consists essentially of many clusters of lymph nodules enclosed within a fibroelastic capsule. Unlike any other lymphoid organ, the lymph node receives lymph from several *afferent lymphatic vessels* that penetrate the capsule along its convex surface. Just within the capsule is a *subcapsular sinus,* into which the lymph passes. The node is drained at a point marked by an indentation in the capsule, the *hilum,* by one or two *efferent lymphatic vessels* (Fig. 9.18). The hilum also marks the point of entry of an artery and nerve and emergence of a vein. (The functional relation of lymph nodes to the blood vascular system will be described below.)

The connective tissue of the capsule dips into the parenchyma of the node at more or less regular intervals as septa or trabeculae, accompanied by extensions of the peripheral sinus. In the well-defined outer zone, the cortex, the trabeculae and cortical sinuses appear as rather regular radii; but in the *medulla* and in the region of the hilum both trabeculae and sinuses are irregularly arranged, with extensive anastomosis (Fig. 9.18).

The lymphatic tissue of lymph nodes appears as more or less distinct nodules in the cortex, and most of these have germinal centers. In the medulla, however, because of the irregular arrangement of trabeculae and sinuses, the lymphatic tissue is compressed, as it were, into cords or strands (Fig. 9.19) with no nodular configuration. Such an arrangement of components enhances the efficiency of lymph nodes as filters of the lymph. The lymph is distributed around the periphery of the node by the subcapsular sinus and then it passes centripetally via the cortical and medullary sinuses along the trabeculae, coming into progressively more intimate contact with the reticular cells, macrophages, and lymphocytes along the way, until it emerges at the hilum through the efferent lymphatic.

Reticular cells form the stroma of the lymph node. They are stellate cells the protoplasmic processes of which freely anastomose along the reticular fiber network. The spaces between them are open in the sinuses, permitting the flow of lymph, but in the parenchyma the spaces are packed with lymphocytes and other free cells. Fixed macrophages occur in abundance along the walls of the sinuses. The free cells include lymphocytes of all sizes, as in the

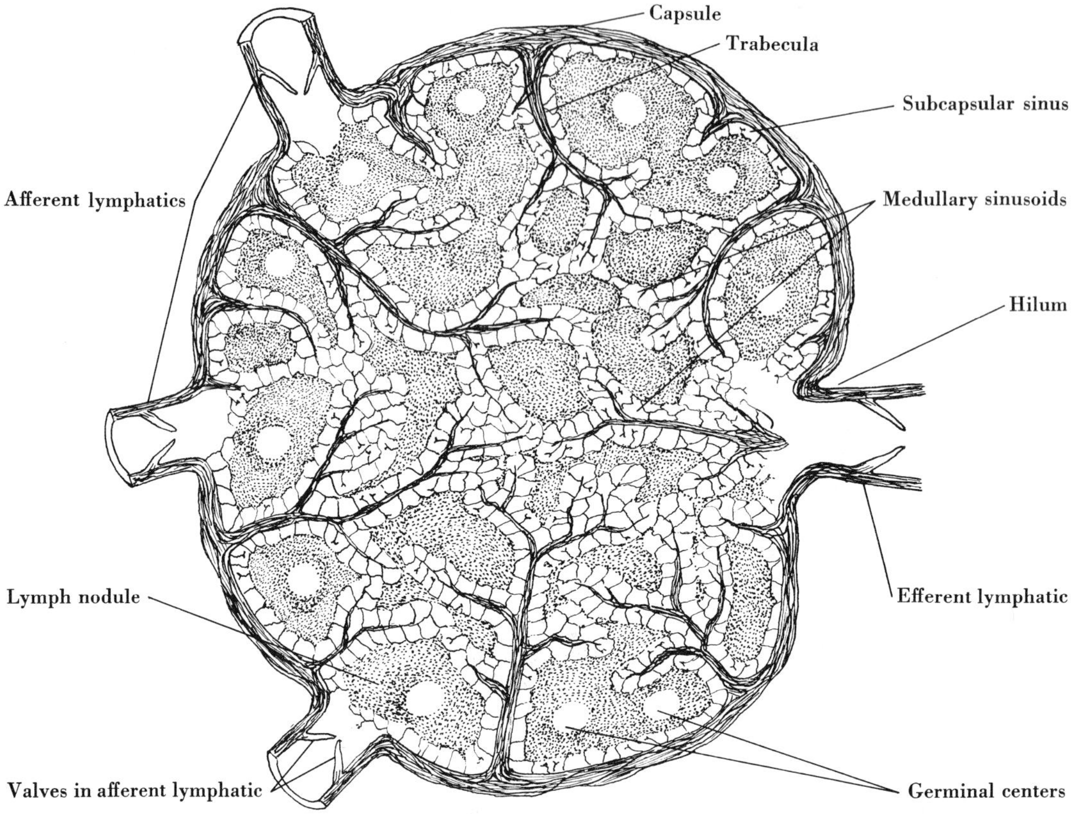

Fig. 9.18 Diagram of a typical lymph node.

lymph nodules described above, free macrophages, plasma cells, an occasional granulocyte, and, infrequently, some erythrocytes.

Lymph nodes are strategically located where the larger lymphatic vessels tend to be most numerous or to converge. Hence they are

Fig. 9.19 Photomicrograph of a section through a lymph node of a horse. Cortex and medulla, *a*, showing subcapsular sinus and lymph nodules with large germinal centers. X100. Medullary region under higher magnification, *b*, showing cords of lymphocytes and medullary sinusoids. X400.

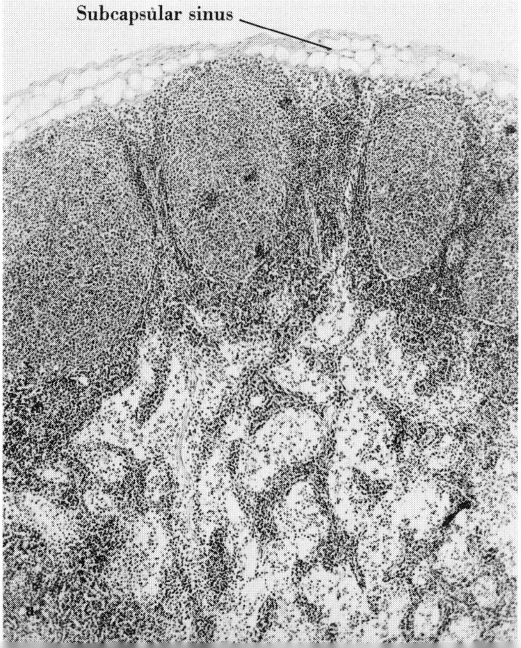

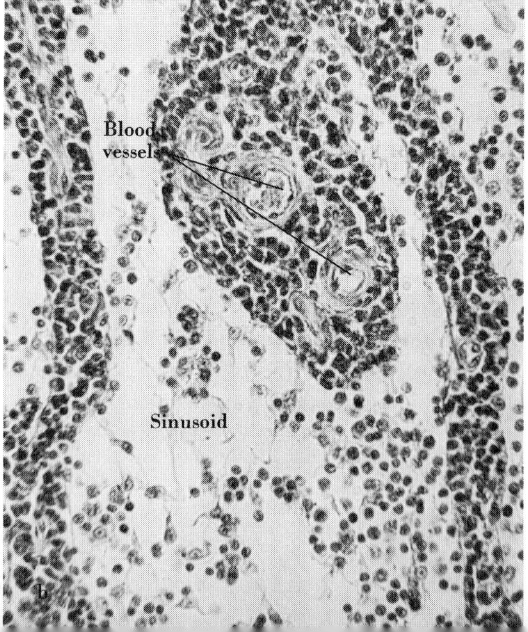

abundant in the mesenteries because of the rich lymphatic drainage of the gastrointestinal tract, and they occur in clusters in the neck, the axillae, the groin, and, to a less extent, in the popliteal area. Scattered nodes are found associated with the lungs and the heart. In some mammals, such as Rodentia, Lagomorpha, and Carnivora, fusion of certain mesenteric nodes takes place, resulting in a rather large organ, the *pancreas Aselli.*

The numbers, size, and certain structural details of lymph nodes vary widely among the mammalian genera. Individual nodes respond dramatically to infections distally by undergoing pronounced enlargement (hypertrophy). The response is usually reversible. However, the primary functions of the lymph nodes are uniform, namely, to filter the lymph before passing it on to the blood stream and to introduce lymphocytes into the circulation.

Small, encapsulated organs which resemble lymph nodes, except that they lack afferent lymphatic supply and their sinuses are filled with blood rather than lymph, occur in some mammals, notably the ungulates. Called *hemal nodes,* they are found in close association with large blood vessels adjacent to the vertebral column in the thoracic and lumbar regions, and occasionally near the spleen and the kidneys. They are structurally and functionally more akin to the spleen than to lymph nodes. Intermediate between hemal nodes and lymph nodes are the *hemolymph nodes,* found in pigs. They combine the structural and functional characteristics of the spleen and the lymph nodes, for their sinuses receive the contents of both blood vessels and lymphatic vessels.

THE SPLEEN

The evolution of the vertebrate spleen may be reconstructed from its appearance in the various orders of contemporary vertebrates. In hagfishes (*Myxine*) the splenic tissue is dispersed throughout the intestinal wall. In the lamprey (*Lampetra*) it is aggregated in the submucosa of the spiral valve. In the African lungfish (*Protopterus*) the splenic tissue forms a rather well-defined mass in the wall of the stomach. In elasmobranchs the spleen has separated away from the digestive tract as a distinct organ, retaining connection only by mesenteric attachment. It is drained by the hepatic portal vein, however, as are the stomach and intestine, and this relationship persists throughout the higher vertebrate classes.

In the cyclostomes and fishes the spleen and splenic tissue are primarily hemopoietic in function, producing all types of blood cells. Hence they are well populated with all of the primitive types of blood cells described above, as well as with macrophages, mast cells, plasma cells, and thrombocytes. In gnathostomatous fishes erythrocyte and thrombocyte production is segregated in cortical zones of *red pulp,* whereas the production of lymphocytes and some of the granulocytes takes place in the inner medullary or *white pulp* areas. Moreover, in higher fishes (Actinopterygii) the spleen has acquired an additional function, namely, blood cell destruction. Where blood cell destruction occurs in lower orders is not known.

The spleens of amphibians and reptiles in general continue to function in hemopoiesis, as mentioned earlier (see page 237), and histologically they have acquired a basic architecture similar to that which characterizes avian and mammalian spleens.

The avian or mammalian spleen is clearly the largest of the lymphoid organs. It is invested in a fibrous connective tissue capsule which is covered in turn by mesothelial cells of the visceral peritoneum. Fibroelastic trabeculae extend from the capsule into the splenic parenchyma, carrying blood vessels, lymphatics, and nerves with them. Both capsule and trabeculae incorporate smooth muscle elements. Unlike lymph or hemolymph nodes, no afferent lymphatic vessels are present. The spleen is indented by a more or less distinct hilum, where the vessels and nerves enter and leave the organ.

The microscopic structure of the spleen can be described best by tracing the course of its circulation from the hilum and back. The splenic artery or arteries branch within the thick capsular connective tissue of the hilum and pass via the trabeculae into the interior of

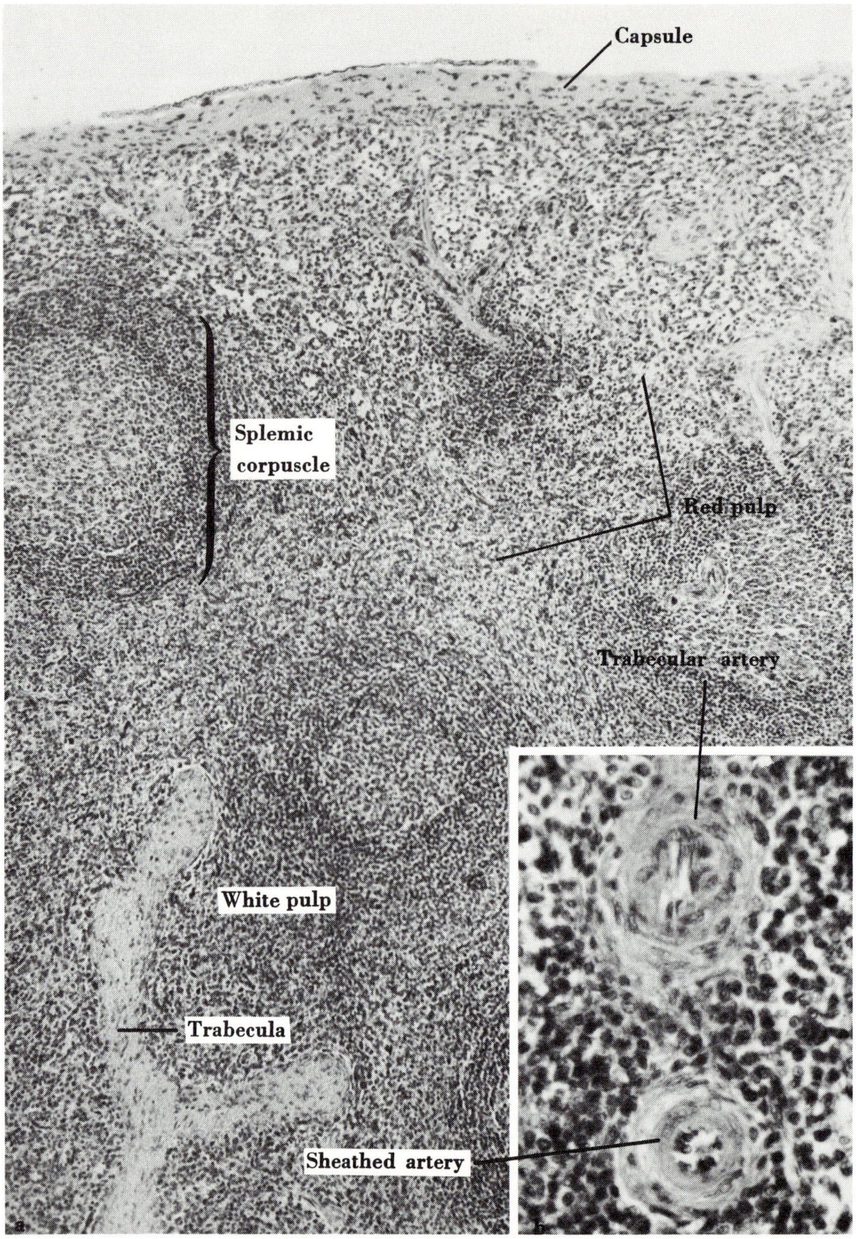

Fig. 9.20 Photomicrograph of the spleen of a monkey, *a*, showing capsule, trebeculae, splenic corpuscles, and red and white pulp. X100. An area in the white pulp of a human spleen, *b*. X450.

the spleen as the *trabecular* or *interlobular arteries*. Branches of the trabecular arteries which emerge from the trabeculae are called *central arteries*. They are characterized by a thick covering of lymphoid tissue whose lymphocytes penetrate the arterial walls. The lymphoid covering is intermittently thickened to form spherical masses, the *splenic* corpuscles (Fig. 9.20) which have the character of lymph nodules. The central arteries pass more or less (rather less, as a matter of fact) through the center of these corpuscles, which accounts for their name. Germinal centers are present in almost all of the splenic corpuscles of young animals, but become less numerous with age. The splenic corpuscles and the lymphoid tissue which surrounds the central arteries constitute the so-called white pulp. The cellular and connective tissue components of white pulp are in every respect like those described above for lymph nodules, although in some species monocytes may be relatively more numerous. The "white" appearance which these areas present in freshly cut tissue is attributable to the virtual absence of erythrocytes, in contradistinction to the red pulp.

After rather profuse branching through the white pulp, the central arteries give off numerous smaller branches which enter the red pulp. These branches, called the *penicilli,* are straight, like the bristles of a brush or *penicillus.* They appear successively as the *pulp arteries,* the *sheathed arteries,* and the *arterial capillaries.*

Pulp arteries, the longest of the three, are characterized by a thin but prominent layer of smooth muscle which is in direct contact with the red pulp. By contrast, their branches, the sheathed arteries, lack any conspicuous muscularis but are sheathed with thickened walls composed of concentrically arranged, spindle-shaped cells, which are presumably reticular cells, and longitudinally oriented reticular fibers. Erythrocytes are also frequently included among the sheath cells. The sheaths differ in thickness, appearing as prominent, ovoid bodies in the spleens of reptiles, birds, insectivorans, dogs, and pigs, but as very poorly developed structures in human spleens.

Two, three, or four arterial capillaries arise from each sheathed artery. Like other capillaries, they consist of only endothelium, supported scantily by reticular fibers and elongated pericytes. The arterial capillaries flare out as they terminate. The exact nature of the termination is controversial, for some authorities believe that the capillaries open into the intercellular spaces of the red pulp, thus creating an "open" system, and others believe that they open directly into the venous sinuses of the red pulp, thus maintaining a "closed" system. Compromise views are also held. Experimental evidence gained by the injection of India ink and other traceable materials into either the arterial or the venous supply of the spleen seems to favor the open theory of spleen circulation. In any event, blood cells of all kinds are abundant in the extravascular sites, the white and red pulp.

The venous sinuses should be considered as modified, expanded capillaries, for their walls consist of little more than a single layer of elongated cells supported by circularly disposed reticular fibers. The arrangement is reminiscent of barrel staves and hoops. As in red marrow and lymph node sinuses, the cells which constitute the splenic sinus wall are fixed macrophages, or, as they are customarily called in the spleen, reticular cells. They are not typical vascular endothelial cells. Their nuclei protrude unusually far into the lumen, particularly when the sinuses are constricted. Engulfed fragments of erythrocytes and leukocytes may be encountered in the cytoplasm of these cells, although they are not so actively phagocytic as their counterparts in the red pulp. The nature of the junction between adjacent cells has not been established yet, but one might anticipate that it is not very different from that which has been demonstrated for reticuloendothelial cells of the sinusoids of red marrow (see p. 238).

Lying between the venous sinuses is the red pulp, which has an anastomosing, cordlike arrangement. The framework of the red pulp, like that of the white pulp, is loose reticular connective tissue. The reticular cells and free macrophages are strongly phagocytic, so that they frequently contain fragmented erythrocytes and other blood cells in various stages of degradation. In addition to reticular cells and macrophages are a variable number of erythrocytes, all types of granulocytes, lymphocytes, monocytes, and, in many species, megakaryocytes.

The venous sinuses are drained by the *pulp veins,* which are characterized by an endothelium supported by a scanty network of reticular and elastic fibers and surrounded by rather compact red pulp. *In vivo* observations of splenic circulation suggest the presence of some smooth muscle fibers in the walls of the pulp veins near their origin from the sinuses, for they display sphincterlike activity. The pulp veins enter the trabeculae, where they coalesce to form the thin-walled *trabecular veins*, which, in turn, drain into the larger *splenic veins* in the vicinity of the hilum.

Lymphatic vessels occur in the trabeculae, coursing parallel to the trabecular arteries and veins; in some species they may even be found draining the white pulp. Unlike the lymph nodes, the spleen has no afferent lymphatic supply.

The innervation of the spleen is prominent, entering at the hilum and branching in the trabeculae. Consisting primarily of non-myelinated fibers, the nerves form periarterial plexuses throughout the white pulp and into the red pulp, where they innervate the vascular smooth muscle. They also innervate the smooth muscle elements of the capsule and trabeculae.

The function of the spleen has been primarily one of hemopoiesis from the lowest vertebrates to the highest. In most mammals the major part of the hemopoietic function is served only during embryonic and fetal life, namely, the production of erythrocytes, granulocytes, and platelets. The other part, the production of lymphocytes and monocytes, continues in the spleen throughout life. Yet the mammalian spleen reserves the capacity to resume erythropoiesis in the adult under certain conditions, such as in severe hemorrhage or anemia. There is also some evidence of a feedback relationship between the spleen and other lymphoid organs and the bone marrow, whereby the spleen influences the hemopoietic activities of these tissues.

Perhaps the most important function of the spleen in adult mammals is that of filtering the blood, as lymph nodes filter the lymph. By virtue of the particular architecture of the splenic sinuses and arrangement of its circulation, the probability that every component of the blood will come in contact with at least one reticular cell or free macrophage during the course of a single passage through the spleen is very high. Hence the spleen is quite efficient in removing wornout or moribund blood cells from the circulation. Because the extracellular fragmented erythrocytes in the red pulp of the spleen occur with greater frequency than in any other tissue or organ, the spleen is also believed to induce the physical breakdown of old blood cells in anticipation of phagocytosis. Even so, intact erythrocytes can be observed within the cytoplasm of splenic macrophages. The spleen is not indispensable in its role as the blood's housecleaner, for the red marrow and the liver take over the blood-destroying functions of the spleen if the spleen has been removed.

The spleen serves as a storage depot for blood. Its sinuses and pulp cords may be filled with blood, and the spleen as a consequence will be swollen. When there is demand for increased blood volume, the spleen contracts to furnish the increase. Contraction is effected by the smooth muscle and elastic fibers of the capsule and trabeculae. In addition, the amount of blood contained within a given sinus and adjacent pulp cords may be regulated by the contraction and/or relaxation of the vascular sphincters which have been described at either end of the sinus.

Finally, the role of the spleen in antibody production has been established, a function which it shares with the other lymphoid organs. It is of interest to note that, although the spleen may be removed with impunity because other tissues and organs readily take over its usual functions, it is almost indispensable for bone-marrow replacement in individuals who have been exposed to high doses of ionizing radiation, such as X-ray. This has been shown in experimental animals subjected to lethal doses of X-irradiation, but whose spleens have been protected from the rays by lead shielding. Such animals show a high survival ratio, unimpaired antibody production, and bone-marrow repopulation. Similar shielding of other lymphoid organs will preserve the antibody-producing capacity of the animal, but will neither result in bone-marrow replacement nor appreciably raise the survival ratio.

THYMUS GLAND

The thymus gland, which is found in all vertebrates except the cyclostomes, derives its name from the resemblance of its lobes in human beings to a leaf of thyme. It originates from dorsal outpocketings of the third and fourth visceral pouches during early embryonic life; thus for a time the anlagen of the thymus gland have lumina lined with epithelium of endodermal origin. The lumina soon disappear in most species, but the epithelial cells undergo extensive proliferation, invading the surrounding mesenchyme and forming a profusion of epithelial cords which impart a glandular appearance to the organ. In higher vertebrates the thymus primordium separates away from the pharyngeal region and moves down into the thoracic cage, coming to rest craniad to the heart and the aortic arch or arches. The primitive epithelial cells differentiate to form three different kinds of mature cells; thymocytes (thymic lymphocytes), phagocytic reticular cells, and nonphagocytic reticular cells. Other epithelial cells form whorls or laminated nests which become hyalinized and are known as *thymic* or *Hassall's corpuscles* (Fig. 9.21). Although their function, if they have one, is unknown, they are uniquely characteristic of the thymus and of the palatine tonsils of some species (e.g., the dog).

As the epithelial cords grow, they become arranged in lobes and lobules. These parts are separated from each other by connective tissue trabeculae, which arise from the surrounding capsule and introduce the organ's vascular, lymphatic, and nervous supply. Meanwhile, the epithelial cords expand and the thymocytes migrate centrifugally to become concentrated in the areas bordering the trabeculae and capsule, where the blood vessels are located. This results in the appearance of two zones, medullary and cortical (Fig. 9.21). Although not evident in random sections, the medullary zones are continuous with each other, as study of serial sections has shown. Because of the dense population of thymocytes in the cortex, the reticular cells are obscured, although they may be seen forming a more or less continuous separation between the blood vessels and thymic parenchyma. The completeness of the blood-thymus barrier so

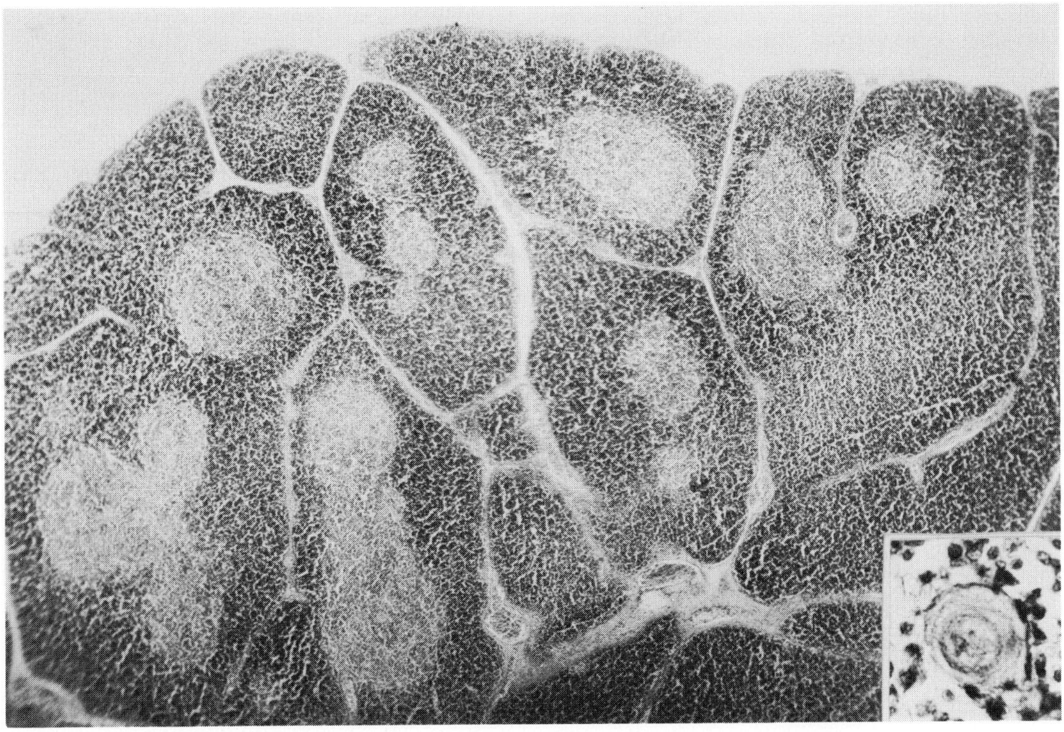

Fig. 9.21 Mammalian thymus, showing cortical and medullary zones. X35. Inset: detail of a medullary area, showing a Hassall's corpuscle. X600.

formed is open to question, but it does at least exist to an appreciable extent, and it may be of functional significance. The medulla, because of its lower thymocyte content, is more clearly seen to consist of a parenchyma of reticular cells and a loose protoplasmic mesh supporting thymic corpuscles and thymocytes; occasional plasma cells and eosinophilic myelocytes are also encountered.

The function of the thymus has been the subject of controversy and speculation for many years, and even now the question is not completely resolved. The thymus has long been believed to be an endocrine gland because of its glandular appearance and because of its response to other endocrine glands, particularly the gonads. Normally the thymus grows very rapidly during fetal life, reaching its relatively maximum size at about the time of birth in most placental mammals, or hatching in birds and reptiles. Although its growth continues after this, the rate is much reduced, and then at or about the time of puberty or onset of sexual maturity, the gland commences a more or less rapid involution, until it all but disappears. Involution is hastened by the administration of steroid hormones, such as the cortical hormones and the sex hormones, whereas the somatotropic or thyroid hormones have the opposite effect.

The most important function of the thymus, however, seems to be its role in immunogenesis and as a primary source of lymphocytes. To begin with, thymocytes are indistinguishable from lymphocytes. Moreover, lymphocytes are observed in the thymus (and also in the bursa of Fabricius, as described below) before they are to be found in either the blood or other lymphoid tissues in a wide variety of vertebrates (e.g., elasmobranchs, birds, marsupials, mice, and human beings). No evidence of immunological competence has been found in hagfishes (*Myxine*), which also lack any recognizable thymic tissue. Neonatal thymectomy of mice, hamsters, or rats is followed either by death or more or less complete suppression of immunological capabilities. As a consequence of some or all of the foregoing observations, among others, three different theories regarding the immunological function of the thymus have been proposed: (1) that the thymus populates all of the lymphoid tissues and organs of vertebrates above cyclostomes with their thymocytes, which become in fact the lymphocytes; (2) that the thymocytes contribute to the lymphocyte population, but do not represent the sole source; and (3) that the thymocytes are special lymphocytes that are concerned with antigen recognition, and that they impart information to the other lymphocytes in the other lymphoid tissues. The statement of the last hypothesis is oversimplified, to be consistent with the scope of this book, but one comment that should be made about it is that it accounts in part for some mechanism whereby the thymocytes are either coded to the individual's own classes of protein, or are coded according to certain "families" of antigens. In the former instance, the coding serves to inhibit the lymphocytes from producing autoantibodies, that is, antibodies which would act specifically against the individual's own tissues. Multiple sclerosis is believed to be an example of autoimmunity, reflecting a breakdown in some protective mechanism such as that just postulated. In the second instance, thymocytes coded for a certain family of antigens, when brought into action by exposure to an antigen belonging to this family, in turn stimulate the proliferation of clones of lymphocytes which in their turn produce the appropriate antibodies.

BURSA OF FABRICIUS

The bursa of Fabricius is a lymphoid organ which appears as a dorsal diverticulum or outpocketing of the cloaca in young birds. Its growth and subsequent involution, as well as its function, parallel those of the thymus gland in birds and other vertebrates. In addition it has been shown that its lymphocytes differentiate from epithelial cells of endodermal origin which line the lumen of the bursa. In every respect, therefore, thc bursa of Fabricius duplicates the thymus gland except in location, and even in this it is similar. It is almost as if the thymus, originating as it does from the pharynx, issues immunological passes at the front door, which the bursa verifies at the back door.

Histologically the bursa bears certain resemblances to the gut, for it is lined with a simple columnar or pseudostratified epithelium which is thrown into internal, longitudinal folds, or plicae. The tunica propria is included in the folds. Secondary plicae also occur as folds upon the primary plicae. The tunica propria in newly hatched birds is distended with lymphoid follicles which are characterized by an inner medulla and an outer cortex. Medulla and cortex are separated from each other by a basement membrane which has been shown to be derived from and continuous with the basement membrane which supports the luminal epithelial cells. The medullary cells which have differentiated from the epithelium include lymphocytoblasts and lymphocytes. The latter, and possibly undifferentiated epithelial cells as well, migrate into the lamina propria and contribute to the cortex. Lymphocytopoiesis takes place in both medullary and cortical zones of the follicles (Fig. 9.22).

The lymphoid follicles are separated from each other by loose connective tissue in which blood vessels and capillary loops may be seen as derivatives of a central artery and vein in each plica. Innervation is scanty and lymphatic vessels are not prominent. A thin muscularis mucosae and adventitia form the covering layers.

ORGANS OF PROPULSION AND CIRCULATION

The Heart

Both embryologically and phylogenetically the vertebrate heart represents a specialization of a large, ventral blood vessel. Even in adults, one finds that the transition from vascular endothelium to cardiac endothelium is anything but dramatic. The most pronounced difference is to be found in the microscopic structure and composition of the myocardium, which accounts for the major mass of the ventricular portions of the heart, and in the impulse-conducting system of the heart.

ENDOCARDIUM

The innermost layer of the heart, the endocardium, consists of an endothelium that is supported by a substantial layer of connective tissue of variable thickness. The single layer of cells forming the endothelium are irregular in outline in surface view, and polygonal or somewhat rounded in vertical section. Their nuclei are round or oval. Underlying the endothelium is a delicate and thin layer of collagenic fibers. Forming a distinct, middle layer of the endocardium is dense connective

Fig. 9.22 Section through the bursa of Fabricius from a chicken. X400. (Photomicrograph from a preparation by Samuel E. Moyer.)

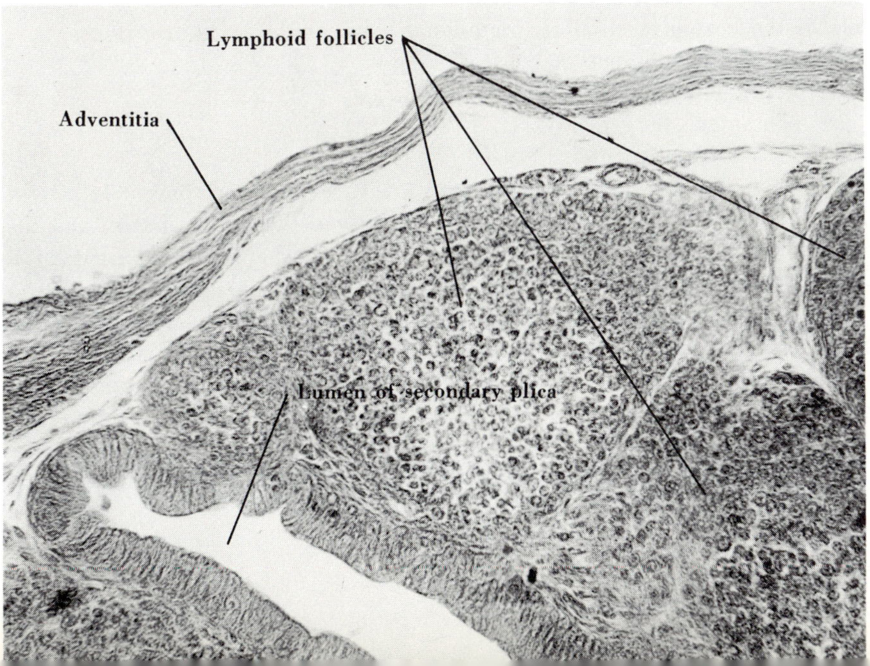

tissue comprised of elastic fibers, fibroblasts, and a variable amount of collagenic fibers. In the deeper parts of this layer some smooth muscle fibers may be seen. This layer is thickest where the myocardium is thinnest, which may be observed by comparing the atrial wall with that of the ventricle. The deepest layer of the endocardium consists of looser, fatty connective tissue, in which blood vessels and Purkinje fibers course. The endocardium forms not only the lining of the chambers of the heart, but also the covering of the valves and the chordae tendinae.

MYOCARDIUM

The middle, and in general the most prominent layer of the heart, is the myocardium—which is comprised primarily of bundles of cardiac muscle (see Chapter 5 for a definitive description of this kind of muscle tissue). In the atria and auricles, the bundles of cardiac muscle do not form a continuous layer, as in the ventricles. At the intervals between muscle bundles, the connective tissue of the deep endocardial layer joins and fuses with the connective tissue of the epicardium (described below). The endomysium of the myocardium is continuous with the connective tissue of both the endocardium, on the inside, and epicardium, on the outside.

EPICARDIUM

The covering of the heart, the epicardium, is connective tissue surmounted by a layer of mesothelium. Within the epicardium are found blood vessels of the coronary circulation, lymphatic capillaries and vessels, nerves, and varying amounts of fat. Where the large coronary vessels are contained, the epicardium is quite thick and loose, and contains much fatty tissue.

CARDIAC VALVES

The valves of the heart are leaflets composed of dense fibrous connective tissue covered on two sides by a fold of endocardium. In lower vertebrates (fishes and Amphibia) prominent valves occur at the junction of the sinus venosus and the atrium, in addition to the atrioventricular orifice and the aorta. The bases of the atrioventricular valves and of the valves which regulate the flow of blood out of the ventricle are supported by dense elastic connective tissue which is contiguous in the higher vertebrates, with similar connective tissue surrounding the atrioventricular and arterial foramina. Together, these dense, fibrous bundles form the *cardiac skeleton*, and serve not only to anchor the valves but also to reinforce the foramina against high internal pressures and to serve as areas of attachment of the cardiac muscle itself. It is in the mammals that the cardiac skeleton is most pronounced, and it may include chondroid (as in human beings) or hyaline cartilage (as in dogs) or bone (as in cattle). In old mammalian hearts of most any species, ossification is not infrequently encountered in the cardiac skeleton.

The atrioventricular valve leaflets are the only ones which are supported at their free ends by tendinous cords, the *chordae tendinae*, which attach to the papillary muscles of the ventricle. Like the cardiac valves, the chordae tendinae consist of connective tissue fibers covered with a thin layer of endocardium. The valves of the *conus arteriosus* are small and cup-shaped, which enables them to press against each other when they are in the closed position. The conus arteriosus of primitive fishes is characterized by numerous levels of such valves, but in terrestrial vertebrates there is only one level, comprising three semilunar valves. The minimum number is found in certain teleosts, which have only two semilunar valves in the conus arteriosus.

Blood Vessels

THE ARTERIES

Arteries are tubes of three concentric layers: the *tunica intima*, the *tunica media*, and the *tunica adventitia*. The intima consists of a single layer of endothelial cells supported by a basement lamina and connective tissue fibers. As with the media and adventitia, the intima varies in its structural details among the major categories of arteries. The media is the most important layer of arteries from the standpoint of determining the functional characteristics of the particular vessel at hand, as we shall see below. It shows the widest variation in thickness. The adventitia is generally the least dense of the arterial layers, and it merges with the surrounding connective tissue.

On the basis of structural and functional differences, arterial vessels may be considered to fall into one of three groups: *elastic arteries, muscular arteries,* and *arterioles.* In the first category are placed the large arteries, such as the aortae and the pulmonary arteries, which conduct the blood under relatively high pressure from the heart to their several branches, the muscular arteries. The muscular arteries distribute the blood to the various organs of the body, and, finally, the arterioles regulate the flow of blood to the capillary beds within the organs and tissues.

ELASTIC ARTERIES. Consistent with their job of conducting blood away from the heart to the main distributing channels, the elastic arteries are characterized by a large cross-sectional diameter and rather thin but strong elastic walls (Fig. 9.23). The walls are de-

Fig. 9.23 Transverse section through the aorta of a young turtle *Chrysemys elegans.* Embedded in Epon, stained with toluidine blue. X1500. (Courtesy of F. Craft.)

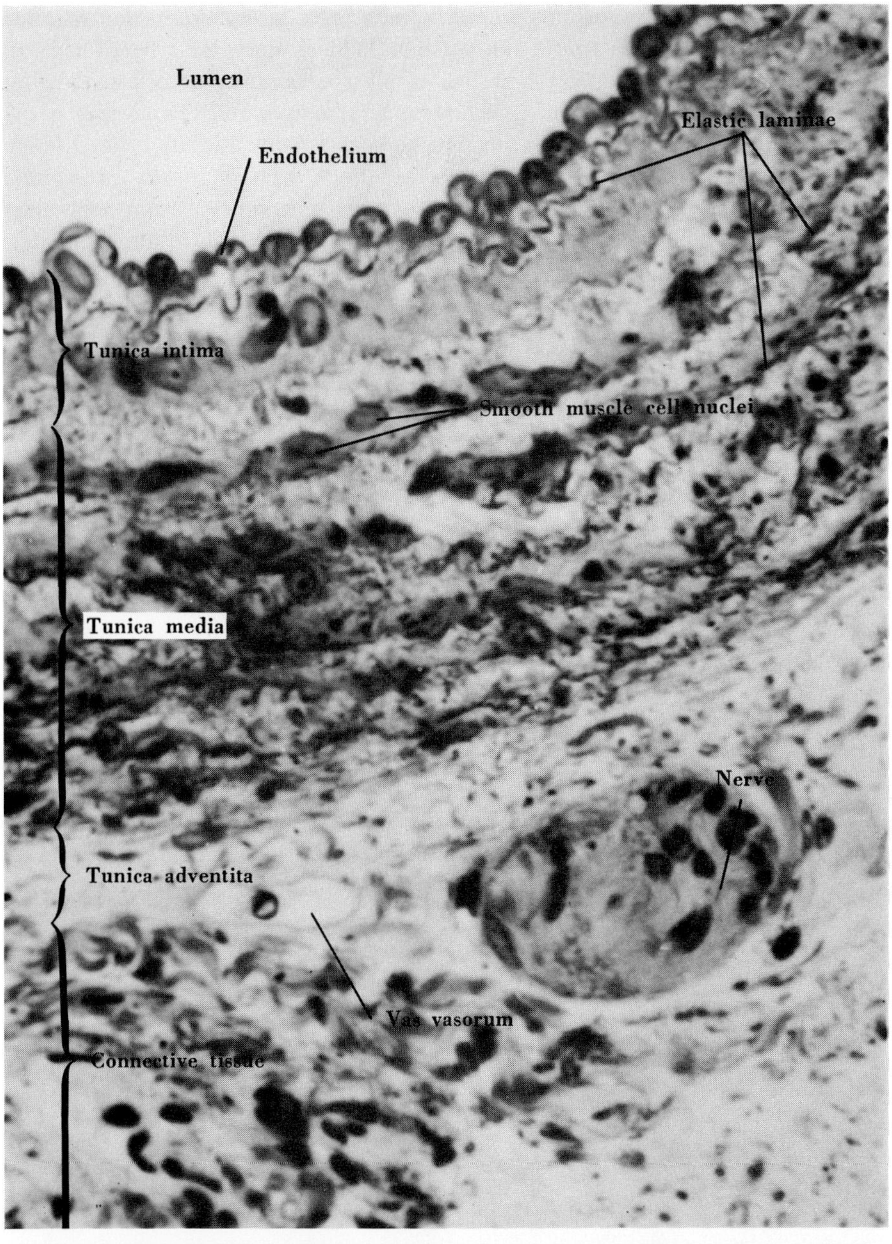

signed to absorb the shock of ventricular systole, so that the blood flow in the distributing arteries is smooth and pulsating rather than coming in the form of sharp, intermittent surges. The intima is made up of somewhat rounded (rather than flattened) endothelial cells and a poorly defined connective tissue layer that includes both collagenic and elastic fibers, fibroblasts, and small bundles of smooth muscle. The elastic fibers adjacent to the media often occur in fenestrated sheets. The fenestrations are round or oval openings which facilitate the movement of nutrient materials and metabolites across such otherwise-impermeable membranes as elastic fibers tend to form.

The media of elastic arteries is responsible for the name of this category. It consists primarily of elastic fibers which are in most species arranged in fenestrated, concentric cylinders alternating with thin layers of connective tissue. The latter are comprised of collagenic and elastic fibers, smooth muscle fibers, and fibroblasts, all embedded within an amorphous ground substance, probably chondroitin sulfuric acid and other mucopolysaccharides. The muscle fibers are arranged circularly, for the most part. Cardiac muscle may often be found in the media adjacent to the heart.

The tunica adventitia of the large, elastic arteries is thin and merges with the adjacent connective tissue. Blood vessels occur within the adventitia and send short branches into the media. The arterial wall is too thick to allow any significant transfusion of nutrients and metabolites directly between its tissues and the blood within its lumen, so it is vascularized by its own, considerably smaller system of vessels. They are called the *vasa vasorum*, which means literally "the vessels of the vessels." Lymphatic drainage also occurs in the larger vessels, and both myelinated and nonmyelinated nerves are found.

MUSCULAR ARTERIES. The transition from elastic to muscular arteries is generally gradual, so that vessels of intermediate or "mixed" character are neither unusual nor surprising. Typical muscular arteries have an intima which consists of flattened endothelial cells, basement lamina, and a connective tissue layer whose outer portion of elastic fibers forms a distinct *internal elastic membrane* (Fig. 9.24). This membrane contracts when the blood escapes, so that in routine histological preparations it appears as a prominent, undulating line separating the intima from the media. In the larger muscular arteries the internal elastic membrane is thick and fenestrated, and it may appear in two or more layers.

The media of muscular arteries is made up almost entirely of smooth muscle cells, which

Fig. 9.24 Transverse section through a muscular artery of a mammal. X100.

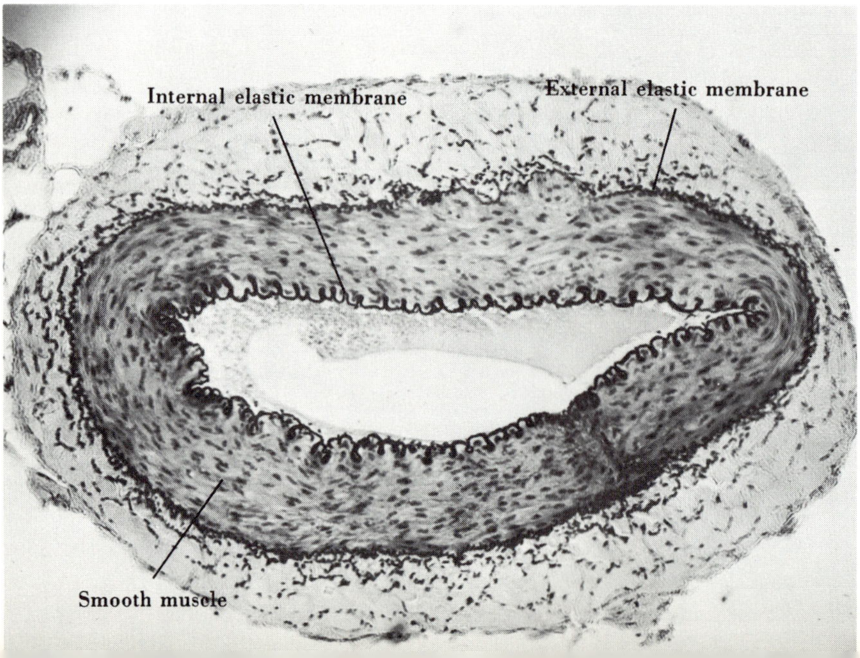

are circularly arranged in concentric layers (Fig. 9.24). They are supported by a delicate network of reticular fibers and sometimes loose networks of elastic fibers may be described weaving between the muscle layers. The media is usually the most prominent layer, although the adventitia may be as thick or thicker. It is rather profusely innervated by fibers of the autonomic nervous system. Impulses carried by these fibers mediate vasomotion, which is the alternating constriction and dilation characteristic of certain kinds of blood vessels. Vasomotion represents one of the major mechanisms for regulating blood pressure, as well as for regulating the blood flow to any given organ.

The adventitia of muscular arteries is characteristically thick, but it consists of loose connective tissue for the most part. Adjacent to the media it includes the external elastic membrane (Fig. 9.24), which is usually thick and fenestrated, and, like its counterpart in the intima, is usually contracted into a prominent undulating line in routine histological preparations. The remainder of the adventitia is very loose, providing considerable freedom to the artery for the activities of vasomotion. Included within its scaffolding of collagenous and elastic fibers are vasa vasorum, nerves, and fat cells. Its outer layers merge with the surrounding connective tissue.

ARTERIOLES. The smallest and structurally the simplest of the arterial categories are the arterioles. The intima includes a flattened endothelial pavement, basement lamina, a thin layer of collagenous and elastic fibers, and an internal elastic membrane. The last may be exceedingly thin or lacking in some arterioles. The media is also thin, and consists of little more than one or two layers of smooth muscle cells circularly disposed and supported by a very delicate reticular network. The adventitia is also thin and includes collagenous, reticular, and elastic fibers, nerves, and fibroblasts. Neither vasa vasorum nor an external elastic membrane is to be found (Fig. 9.25).

Other classifications of arteries include anastomoses of two sorts: (1) arterioarterial anastomoses, which connect collateral arteries

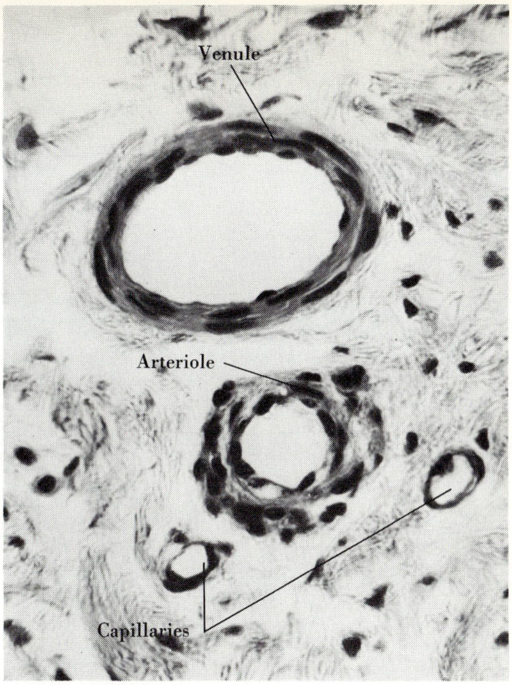

Fig. 9.25 Section through an arteriole and a venule in a human spermatic cord. X800.

of equivalent diameter (and, hence, equivalent blood pressure) or arterioles; and (2) arteriovenous anastomoses, of rare occurrence, which shunt blood from the arterial to the venular side of a capillary bed. Their histological characteristics are generally the same as those of the blood vessels which they interconnect. Still another class of arterial vessel is the *rete mirabile*, which is a specialized class of arteriole. (For details on the structure and distribution of retia mirabilia the student is referred to the chapters on the respiratory system and the excretory system.)

THE CAPILLARIES AND SINUSOIDS

The capillaries of vertebrates typically consist of an endothelial tube of one cell in thickness (Fig. 9.26) supported by a basement lamina and a very scanty network of collagenous fibers from the adjacent connective tissue. The diameter of the lumen is usually determined by the dimensions of the blood cells which must pass single file through it.

Associated with capillaries one finds "pericytes" or, in some locations, mast cells. At one time the pericytes were thought to be contractile cells and therefore responsible for capillary constriction. *In vivo* studies have demonstrated beyond much doubt that capillary con-

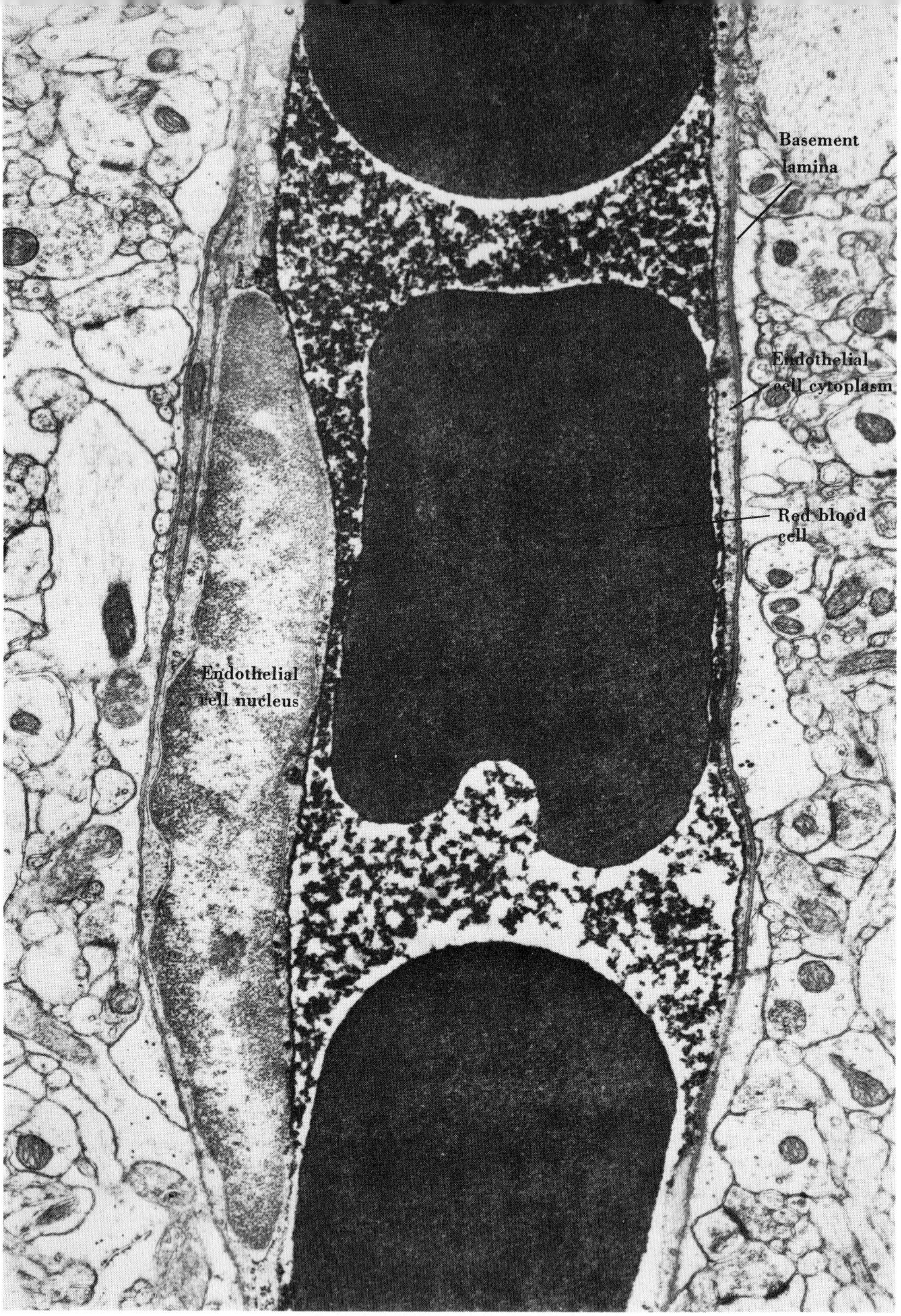

Fig. 9.26 Electron micrograph of a longitudinal section of a blood capillary. X20,000. (From T. S. Reese and M. J. Karnovsky. 1967. J. Cell Biol. 34:207–217.)

striction and dilatation are purely passive phenomena as far as the capillaries are concerned *per se*. When the arteriole leading into a capillary is dilated, the blood flows into and through the capillary, distending it. When the supplying arteriole constricts, blood flow ceases in the capillary and, providing that the venule draining the capillary is patent, the hydrostatic pressures outside the capillary cause it to collapse and discharge what blood may have remained within its lumen. Pericyte is a general term which defines the position of a cell rather than its function. Pericytes may be mesenchyme cells, macrophages, or even fibroblasts. Small arterioles are easily mistaken for capillaries, and their smooth muscle cells may be erroneously identified as pericytes. At the point where capillaries branch off from arterioles, a group of two or three smooth muscle cells may be found spiraling around the capillary origin, constituting a *precapillary sphincter* (Fig. 9.27). This device permits capillary constriction to occur independently of arteriolar constriction.

The patterns of capillary beds are quite variable, but they are a direct reflection of the function and architecture of the tissue or organ in which they occur. Capillaries tend to be long and nearly parallel to each other, or are arranged in elongated diamond-shaped nets in skeletal muscle, nerves, and tendons. They form flat networks in mesenteries and other thin membranes. They appear in basket-like shapes around the alveoli or secretory tubules of glands or within intestinal villi. In the lungs they form a three-dimensional mesh-work which is so dense that the interstices of the mesh may have smaller diameters than the capillaries themselves. Although capillary networks are quite profuse in almost all tissues, they become remarkably so in tissues of high metabolic rate. It is in the capillary beds that the major exchanges of materials between the circulating blood and the tissue fluids takes place.

Sinusoids are like capillaries in that they receive their fluid from a larger vessel and discharge it into a larger vessel. Examples of sinusoids which lie between arterioles and venules are found in bone marrow, the spleen, and the adrenal cortex; sinusoids between portal veins and venules are found in the liver and the hypophysis; and sinusoids between lymphatic vessels occur in the lymph nodes. Structurally alike, such sinusoids differ markedly from capillaries. Whereas the diameter of capillaries varies little from arterial to venous end and is remarkably uniform throughout the body, the diameters of sinusoids vary anywhere from capillary dimensions to six or eight times this value. The shape of sinusoids is typically irregular.

Fig. 9.27 Electron micrograph of a capillary origin from an arteriole, showing the precapillary sphincter. Note the points of attachment between the smooth muscle cells and the capillary endothelium. X8800. (From J. A. G. Rhodin. 1967. J. Ultrastruc. Res. 18:181–223.)

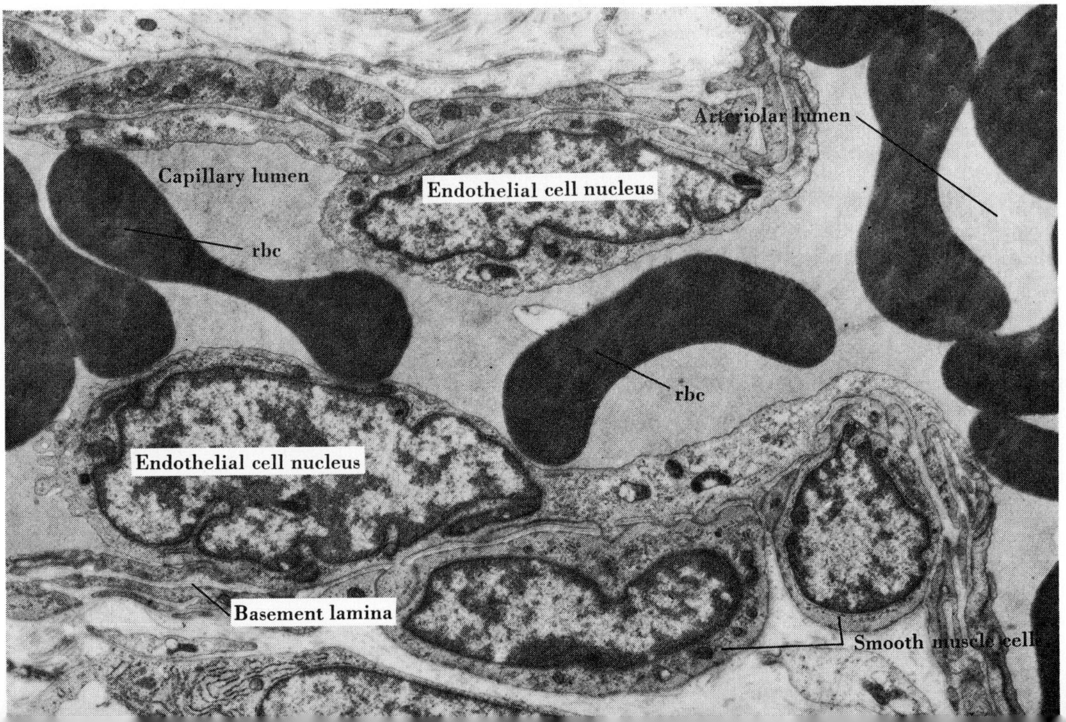

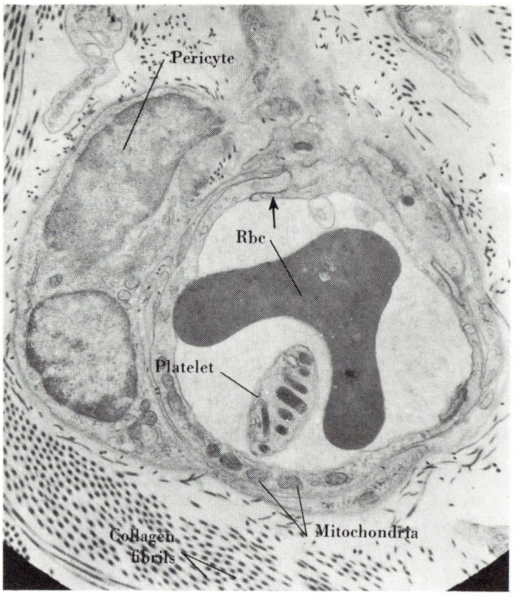

Fig. 9.28 Electron micrograph of a capillary in transverse section. Note the cell junction (*arrow*). X6000.

Whereas the capillaries are lined with continuous endothelium with strongly coherent cell junctions (Fig. 9.28), the sinusoids are lined by a loosely associated pavement of macrophages and reticular cells, the so-called reticuloendothelial system (Fig. 9.10). As already described, many of these cells are actively phagocytic. Finally, the sinusoids are supported by an extensive reticular connective tissue covering.

THE VEINS

In considering the dynamics of circulation, one appreciates the extent to which the structure of blood vessel walls is a function of the blood pressure. As the kinetic energy of the pulse diminishes from the time the blood leaves the heart to the time it reaches a value close to zero in the capillary beds, the vessel walls become progressively thinner and less resistant to pressure. The veins serve to return the blood to the heart from the capillary beds, and the only sources of mechanical energy remaining to effect this return are: (1) diffusion pressure from the tissues into the venous ends of the capillaries and venules themselves; (2) the pressure exerted incidentally on the veins by the contraction of skeletal muscles during locomotion and respiration and by smooth muscle during alimentation; and (3) in some cases gravity, such as in the venous return from the head and neck in human beings, giraffes, and other heads-up animals. The venous system is therefore a low-pressure system, characterized by vessels with a larger diameter than the corresponding artery, and a slow rate of flow. Accordingly the walls of veins are generally very much thinner and have less smooth muscle and considerably less elastic tissue than the corresponding arteries.

SMALL VEINS. The smallest veins are the ones derived from the conjunction of capillaries. They differ structurally from capillaries only by having a scattering of smooth muscle cells, more supporting collagenous fibers, and a larger lumen. Such small veins are known as *venules*. As the veins become larger by confluence, they acquire progressively more smooth muscle and more connective tissue, but the distinction between media and adventitia is not a clear one.

MEDIUM-SIZED VEINS. The veins of medium size have an endothelium of rounded or polygonal cells rather than flattened ones. In some cases the endothelium is thrown into a fold to form a simple flap valve, as in the veins of the legs of terrestrial vertebrates. Such valves help relieve the rather significant intrinsic pressure exerted by a column of blood on the vessel walls. Part of the tunica media is sandwiched in between the opposing endothelial layers.

The media of medium-sized veins is thin and contains a small amount of smooth muscle. The adventitia, which is not sharply delineated from the media, is the thickets part of the wall. Composed primarily of collagenous and elastic fibrous tissue, it often includes bundles of longitudinally oriented smooth muscle cells. Vasa vasorum are evident, in some cases reportedly draining directly into the lumen of the vein; nerves and lymphatic vessels are also present.

LARGE VEINS. The intima of the larger veins is more like that of arteries, because it has a more prominent subendothelial component

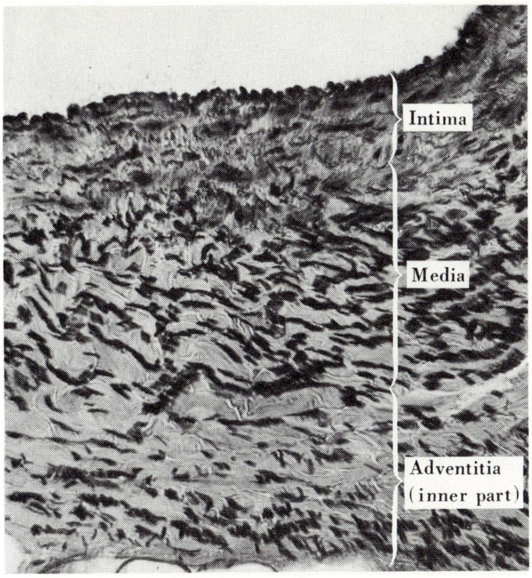

Fig. 9.29 A large vein in cross section. The lumen is at the top. X450.

than the smaller veins do—including in some cases a thin internal elastic membrane and a scattering of longitudinally oriented smooth muscle bundles. The media is still very thin, however, with only small amounts of smooth muscle. The adventitia is by far the thickest layer, and it usually contains prominent longitudinal bundles of smooth muscle (Fig. 9.29). Vasa vasorum, lymphatics, and nerves are also present, but there is no external elastic membrane.

The media of the venae cavae of mammals frequently includes some cardiac muscle adjacent to the heart.

Lymphatic Vessels

Lymphatic vessels differentiate from mesenchyme in the same manner as blood vessels, and in close association with the larger veins. In the Chondrichthyes the lymphatic vessels are so closely associated with the venous system that some authorities claim that in such forms it represents virtually a modification of the venous system. Curiously, the lymphatic system is well differentiated in the cyclostomes, and it is highly differentiated in the Osteichthyes and higher vertebrates.

The smallest vessels of the lymphatic system are the lymphatic capillaries. They are larger in diameter than blood capillaries, and they originate as small bulbs in the tissue spaces (Fig. 9.30). Lymphatic capillaries anastomose extensively with each other. They consist solely of an uninterrupted endothelium, with little or no evidence of supporting fibrous tissue of their own. The endothelial cells are more flattened than in blood capillaries and they cover a correspondingly larger area; thus they are difficult to resolve in optical sections (Fig. 9.16). Unlike blood capillaries, they lack associated pericytes. The mechanisms by which fluids pass from the tissue spaces through the endothelial cells and into the lumina of the lymphatic capillaries are not known, but they are presumed to be the same as apply to blood capillaries.

The lymphatic capillaries drain into larger lymphatic vessels, which are characterized by the addition of supporting tissue surrounding the endothelium. Attempts to distinguish three tunics, such as exist in blood vessels, are largely contrived, for there is rarely any clear-cut layering outside the endothelium. The supporting elements include bundles of collagenous fibers, elastic fibers, and varying amounts of smooth muscle. The last are ar-

Fig. 9.30 Lymph capillaries (lacteals) in a fold of the intestinal mucosa of lamprey (*Petromyzon*). X400.

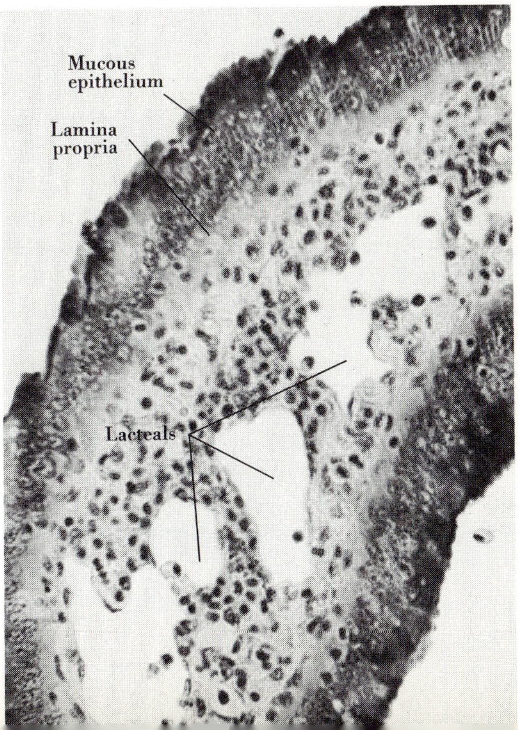

ranged circularly for the most part, except in the larger lymphatics. In tetrapods the lymphatic vessels of medium size have a characteristic beaded or moniliform appearance.

Valves are found in some of the medium-sized lymphatic vessels of tetrapods. They are similar histologically to those which occur in the veins, but they invariably occur in opposing pairs, frequently in groups, and they are very delicate. On the upstream side of the valves the vessel wall is slightly expanded and its musculature is more prominent.

Large lymphatic vessels are usually closely associated with large veins, in some species forming anastomosing networks about them, particularly in the mesenteries. These lymphatic vessels convene to form prominent trunks, namely: the subvertebrals, dorsal, and ventral lymph trunks in fishes; paired subvertebral trunks in Amphibia; a median subvertebral trunk in Reptilia, which divides anteriorly before it enters the precaval veins; paired thoracic ducts in birds; and a left thoracic duct and a right cervical duct in mammals. Such ducts sometimes possess an internal elastic membrane of longitudinally disposed elastic fibers. The muscular layers in what corresponds to the media of veins are well developed and include both longitudinal and circular bundles, the former predominating. Outside of this layer one finds fibroelastic connective tissue, occasional longitudinal muscular fibers, vasa vasorum, and a rich network of nerve fibers.

Lymphatic sinuses are encountered in the Osteichthyes and Amphibia, and they are especially large in the Anura. The lymphatic sinuses are flattened expansions of some of the larger lymphatic vessels, otherwise their histological characteristics are not significantly different from the vessels themselves. The wall of the sublingual lymphatic sinus of *Rana* has been used extensively as a transparent membrane for *in vivo* studies of the microcirculation.

LYMPH HEARTS. Lymph hearts (Fig. 9.31) are chambered, pulsatile modifications of certain lymphatic vessels which occur in various vertebrates at the points at which those vessels empty into veins. They have been described in all classes of vertebrates except Chondrichthyes and Mammalia. In the European catfish (*Silurus*), the eel (*Anguilla*), and trout (*Salmo*), paired caudal lymph hearts have been described in considerable detail. It is interesting to note that the pulse rate of the caudal lymph hearts in young eels is about 160/min, as compared with a cardiac pulse rate of only 60/min. The musculature of these hearts is striated and closely resembles cardiac muscle.

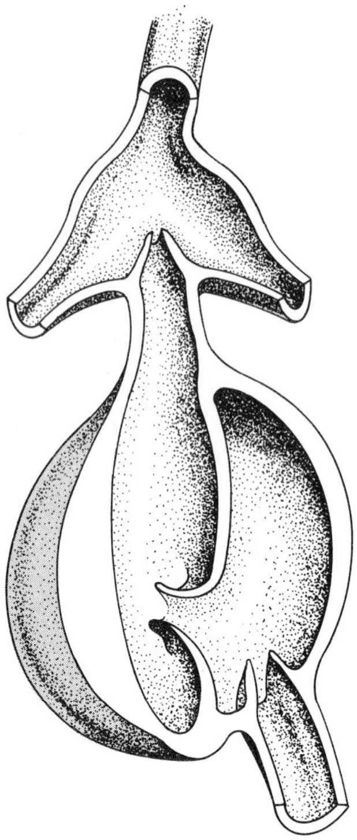

Fig. 9.31 Diagram of a lymph heart.

The number of lymph hearts is greatest in the amphibians, ranging from four in certain Anura to 14–20 in Urodela to about 200 in the Apoda. In frogs one pair of lymph hearts pumps into the vertebral veins, and another pair pumps into the transverse iliac vein. The lymph hearts of Apoda occur along the intersegmental veins. Reptiles have typically only one pair of lymph hearts, which pump into the iliac veins. Lymph hearts have been described in embryonic birds, but they are lacking in adults. True lymph hearts have not been described in mammals.

SELECTED REFERENCES

Ackerman, G. A., and R. A. Knouff. 1959. Lymphocytopoiesis in the bursa of Fabricius. Am. J. Anat. 104:163–205.

Bloom, W., and D. W. Fawcett. 1968. A textbook of histology. W. B. Saunders & Co., Philadelphia.

Eisen, G. 1897. Plasmocytes. Proc. Calif. Acad. Sci., Ser. III (Zoology) 1:2–72.

Emmel, V. E. 1924. Studies on the non-nucleated elements of the blood. II. The occurrence and genesis of non-nucleated erythrocytes or erythroplastids in vertebrates other than mammals. Am. J. Anat. 33:347–405.

Good, R. A., and A. E. Gabrielson, eds. 1964. The thymus in immunobiology. Hoeber Div. of Harper & Row, New York.

Miller, J. F. A. P., and D. Osaba. 1967. Current concepts of the immunological function of the thymus. Phys. Rev. 47:437–520.

Page, E. 1967. The occurrence of inclusions within membrane-limited structures that run longitudinally in the cells of mammalian heart muscle. J. Ultrastr. Res. 17:63–71.

Raviola, E., and G. Raviola. 1967. Striated muscle cells in the thymus of reptiles and birds: an electron microscopic study. Am. J. Anat. 121:623–646.

Reese, T. S., and M. J. Karnovsky. 1967. Fine structural localization of a blood-brain barrier to exogenous peroxidase. J. Cell Biol. 34:207–217.

Rhodin, J. A. G. 1967. The ultrastructure of mammalian arterioles and precapillary sphincters. J. Ultrastr. Res. 18:181–223.

Rund, J. Y. 1954. Vertebrates without erythrocytes and blood pigment. Nature 173:848.

Skutelsky, E., and D. Danon. 1967. An electron microscope study of nuclear elimination from the late erythroblast. J. Cell Biol. 33:625–635.

Weakley, B. S., D. I. Patt, and D. Shepro. 1964. Ultrastructure of the fetal thymus in the golden hamster. J. Morph. 115:319–354.

Wingstrand, K. G. 1956. Non-nucleated erythrocytes in a teleostean fish *Maurolicus mülleri* (Gmelin). Z. Zellforsch. 45:195–200.

Yoffey, J. M., and F. C. Courtice. 1956. Lymphatics, lymph and lymphoid tissue. Harvard University Press, Cambridge, Mass.

chapter 10
THE EXCRETORY SYSTEM

Excretion is the process whereby waste materials are eliminated from the blood or tissue fluids. Waste materials include (1) the waste products of metabolism, (2) nonmetabolites or substances of no use that have been introduced into the blood or tissues, and (3) water.

Metabolic Wastes

The principal metabolic wastes are the end products of the downgrading or catabolism of organic compounds. The deamination of proteins and the breakdown of nucleic acids produce residues known as the nitrogenous wastes such as urea, uric acid, hippuric acid, creatinine, creatine, ammonia, and ammonium compounds. The waste products of carbohydrate and fat metabolism are mainly carbon dioxide and water. Phosphates and sulfates are incidental waste products of metabolism, the former being derived mainly from phosphatides and the latter from proteins containing cystine. Several cations also belong in this category, such as potassium, sodium, calcium, and magnesium, which occur as urates, phosphates, sulfates, or chlorides, for the most part.

Nonmetabolites

The nature of the nonmetabolites varies enormously according to the environment. Marine animals absorb large quantities of inorganic salts, which must be excreted in order to maintain homeostasis and ionic equilibrium. Certain "inert" materials may be similarly absorbed by the intestine or respiratory epithelium and ultimately be excreted unchanged. For example, pentoses from many fruits may be absorbed and excreted in some species, such as human beings. Some potentially toxic material may become combined in the blood or tissues (usually the liver) with glycuronic acid and so rendered harmless; they may be excreted immediately as glycuronates. Examples are phenols formed from intestinal decomposition, or certain drugs, such as camphor, chloral, morphine, salicylates, or turpentine.

Water

Water is a metabolic byproduct of carbohydrate catabolism and of condensation processes. It may also be absorbed in large quantities (as it is in many fresh-water species) and be excreted promptly in order to maintain homeostasis. Water is atypical of excretions, however, because it is not entirely a waste product, even in excretion—through which it functions to hold other excretory products in solution and thus make excretion possible.

Various tissues and organs are concerned in one way or another with excretion in the vertebrates. The gills of fishes and amphibians function in excreting carbon dioxide, salts, and, in some cases, water. Similarly, the alimentary tract may and often does function in the excretion of salts and water, as may some of its associated glands, such as the salivary glands and the liver. The integument of mammals excretes to some extent via its sweat glands. The lungs excrete carbon dioxide as, one might say, half of its normal function. The kidneys are the principal excretory organs for nitrogenous wastes and water, however. By means of delicate control mechanisms, the kidneys also function in maintaining osmotic pressure, pH and ionic concentration, and blood sugar concentration within rather narrow and critical limits. In some species of vertebrates the kidneys also play a role in the regulation of blood pressure. In effect, then, the primary excretory organs are the respiratory organs and the kidneys, and the kidneys and their associated organs are considered to constitute the excretory system, *per se.* Because of the specialization in function shown by the rectal gland of elasmobranchs and the salt glands in birds and turtles, these organs will also be considered.

THE KIDNEYS

The kidneys develop embryonically from paired longitudinal ridges of mesoderm, the *nephrotomes,* which appear in or closely associated with the intermediate mesoderm. Differentiation of tubules commences at the anterior end of the nephrotome and continues caudad, with certain modifications along the way. The tubules forming at the very anterior end are segmentally arranged and open into the coelomic cavity by ciliated, funnel-shaped expansions, the *nephrostomes.* The opposite ends of the tubules turn caudad, each one effecting contact with the one posterior to it; in this way they form a duct on either side, the *archinephric duct,* which drains the constituent tubules. Three or four segments posteriorly, the first modification is seen. A knot of arteriolar capillaries (*rete mirabile*) becomes closely applied to the differentiating tubules, which respond to the contact by forming a cup-shaped enclosure around the blood vessel knot. The knot is known as the *glomerulus* and the cup as the *renal capsule.* The combination of the two is known as the *renal corpuscle.* The portion of the nephrotome described up to this point is transitory, generally degenerating as more caudal regions differentiate. Because of its primitive nature, its position, and its early differentiation, it is called the *pronephros;* with very few exceptions, it is not known to have any excretory function in adults, and has definitive form only in those vertebrates characterized by a larval stage.

Posterior to the pronephric portion, the nephrotome continues to differentiate tubules, but at this point they fail to open into the coelom in most species and the renal corpuscle becomes the structure of prime importance. Many tubules differentiate in each metameric segment, but they continue to join at the end opposite the glomerular capsule with the archinephric duct, which from this time on grows posteriorly by the proliferation of cells at its tip. In all anamniote vertebrates the rest of the nephrotome differentiates in essentially this way, except that in most species an accessory duct develops as an offshoot of the archinephric duct to drain the most posterior division. In contrast with the pronephros, these portions of the nephrotome function in excretion in adult anamniotes and constitute the *mesonephric* or *opisthonephric kidney.* The latter term is preferred in this context, because it implies inclusion of both the "middle" portion (mesonephros) and the posterior portion (metanephros) of the embryonic nephrotome. The archinephric duct in adult anamniotes is known variously as the nephric duct, mesonephric duct, and Wolffian duct. The accessory ducts are known either as such or, in recognition of their homologies in amniote vertebrates, as the *ureters.*

During differentiation, the kidneys of amniote vertebrates show striking similarities to those of lower vertebrates, including a somewhat aborted pronephros and the formation of a mesonephros. The latter is functional in reptiles and birds during embryonic or fetal life, and also in the embryos of those orders of mammals which do not have a well-developed placenta. However, only the portion of the nephrotome which differentiates in response to the developing ureters (homologous with the accessory ducts of anamniotes), the *metanephros,* is carried over into the adult. The metanephros varies enormously among the three classes of amniotes, and so it is not possible to generalize very far in comparing it to the opisthonephros—beyond the observation that it shows no evidence of metamerism and none of its tubules bear nephrostomes. The external shape of metanephroi may be correlated with body form in reptiles, i.e., it is long and slender in snakes, and compact and rounded in Chelonia. In birds the kidneys are lobulated but rather compact organs. In mammals they tend to be compact, "kidney-shaped" organs, having a characteristic indentation or *hilum* (Fig. 10.1). Mammalian kidneys may be *unilobar,* such as in monotremes, small marsupials, insectivores, chiroptera, and some rodents and carnivores; they may be *polylobar* but compact, such as in large marsupials, some carnivores, ungulates, and primates; or they may be distinctly *lobulated,* such as in Pinnipedia, Cetacea, and some ungulates. Moreover, in mammals the kidneys display a characteristic zonation internally; the renal capsules and convoluted tubules comprising an outer *cortex,* and the straight tubules and collecting ducts constituting the *medulla* (Fig. 10.1). In gross section the cortex appears granular, whereas the medulla is striated.

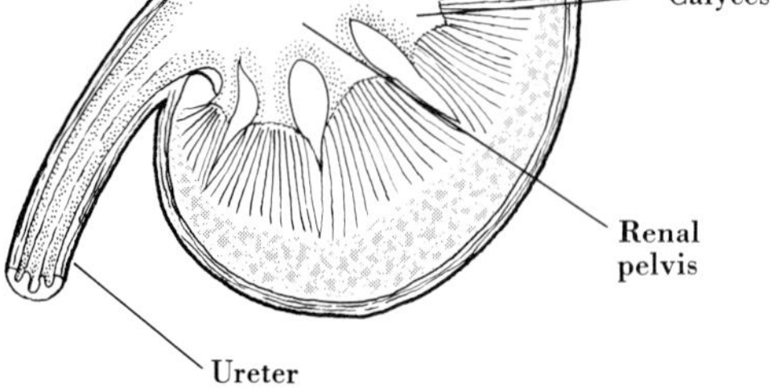

Fig. 10.1 Diagram of a mammalian kidney, excluding vascular and nervous supply.

Relation of the Vascular System

(1) The arterial supply of all vertebrate kidneys is derived directly from branches of the dorsal aorta, the *renal arteries,* which enter the kidneys and both nourish their tissues and supply blood via *afferent arterioles* to the renal glomeruli. Blood leaves the glomeruli by *efferent arterioles,* and is transported thence to capillary plexuses around the renal tubules. The blood is then collected by venules and small veins into the *renal veins,* by which it enters the posterior cardinal veins or the post cava, as the case may be (Fig. 10.2).

(2) In all vertebrates except the Agnatha and adult Mammalia, the kidneys receive blood from a second source, the *renal portal veins.* The renal portal system is extensively developed in most fishes and in urodelous amphibians, it is slightly reduced in the Anura, more reduced in most reptiles, and its functional significance in the birds is not definitely known. The degree of development of the renal portal system has been correlated with the degree of variability of the *milieu interieur.* Where they exist, the renal portal veins are generally formed by the confluence of most of the veins which drain the tail and, to a lesser extent, the segmental veins from the body wall. The renal portal veins then enter the kidneys and break up into small branches which join the capillary plexuses surrounding the renal tubules (Fig. 10.3). At this point the blood mixes with that which has come from the efferent arterioles, and together the two bloods leave the kidneys by way of the renal veins, as described above. It should be noted that the detailed anatomy of the renal

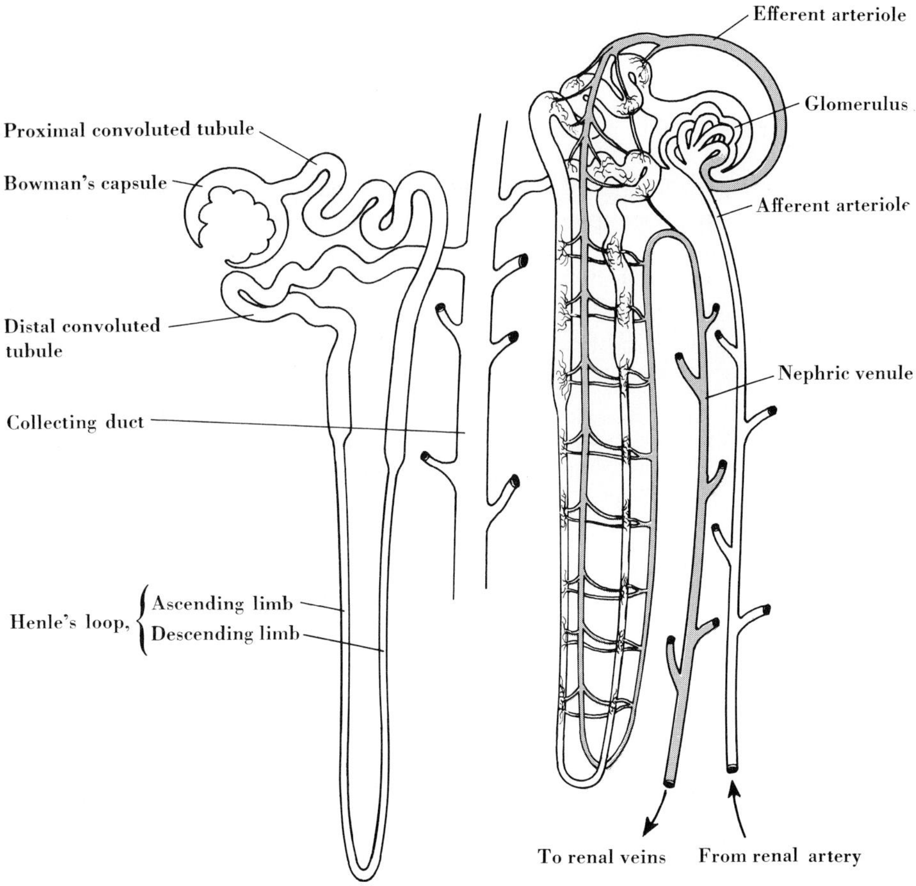

Fig. 10.2 Diagram of the nephron and of its vascular supply in amniotes.

portal system is highly variable among the fishes.

(3) In larval cyclostomes and in certain amphibians (*Rana*) nephrostomes convey body fluid from the coelom directly into venous spaces in the kidneys. In other species in which nephrostomes occur in the adult, the coelomic fluid is either swept into the renal tubules (such as in *Amia, Lepisosteus,* and urodelous amphibians) or into blind nephrostomatous pouches (such as in many elasmobranchs).

Renal Function

The morphological unit of renal function is the *nephron,* which consists of renal corpuscle and tubule (Fig. 10.2). All components of the blood except the corpuscles, macromolecules (plasma and serum proteins), fat droplets, and sufficient water to keep the former flowing are filtered from the blood through the walls of the glomerular vessels into the lumen of the renal capsule. Efficient filtration is achieved by virtue of the fact that the hydrostatic pressure of the blood in the glomerular capillaries (75 mm Hg in human glomerular capillaries) is approximately three times higher than in other capillaries in the body. Samplings of capsule fluid in frogs taken by means of micropipettes have verified that the fluid is neither more nor less than an ultrafiltrate of the plasma.

The volume of filtrate produced in a given period of time is amazingly high in fresh water vertebrates, as is the volume of their urine output. For example, *Petromyzon* excretes as much as 50 percent of its body weight at 18°C in 24 hr, with a normal fluctuation between 15 and 36 percent. Fresh water bony fishes excrete less than this, but still in the order of 5 to 12 percent. The filtra-

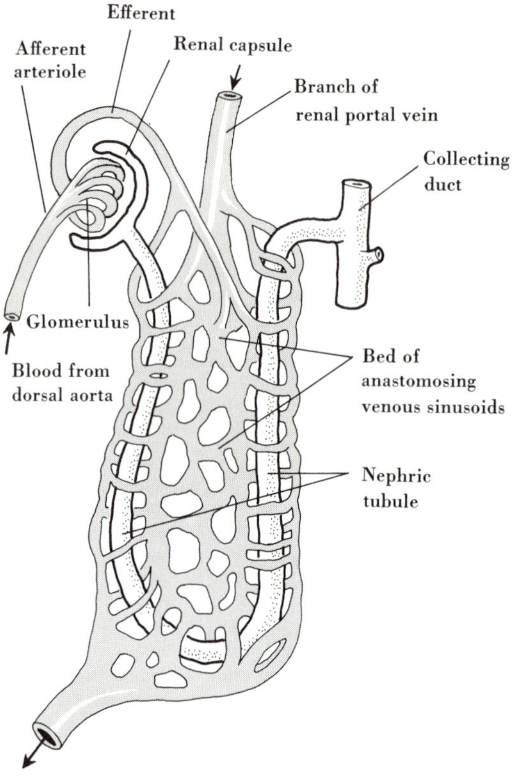

Fig. 10.3 Diagram of the relation of the renal portal vein to the nephron in anamniotes.

tion rate in human beings has been estimated at about 7.5L/hr, but because most of the volume is reabsorbed into the blood farther along the renal tubules, only about 60 ml of urine are excreted per hour. Expressed in another way, urine excretion in human beings amounts to 1–2 percent body weight per day.

Once the ultrafiltrate enters the kidney tubule, selective reabsorption of most of its contents—except urea, uric acid, and other nitrogenous wastes—takes place. Although much more is to be learned about the specific function of each segment of the nephron, a fairly complete accounting has been made both from some direct evidence, as cited above, and from a large body of indirect evidence. The proximal convoluted tubule is the site of reabsorption of most of the substances which are useful to the organism, viz., water, glucose, amino acids, ascorbic acid, and much of the sodium, chlorides, and bicarbonates. The proximal convoluted tubule is also the site at which other materials which escaped filtration in the capsule (presumably because they were either combined with or adsorbed on the macromolecular or corpuscular components of the blood) are added to the urine. These materials include creatinine, hippuric acid, and potassium. The segment of the tubule which extends between the proximal and distal convoluted segments is quite variable in length among the vertebrate species, as we shall see below. In birds and mammals it reaches its greatest length, and it makes use of the counter-current principle for reabsorbing water beyond the capacity of the proximal convoluted tubule. The distal convoluted tubule is the site of reabsorption of any remaining

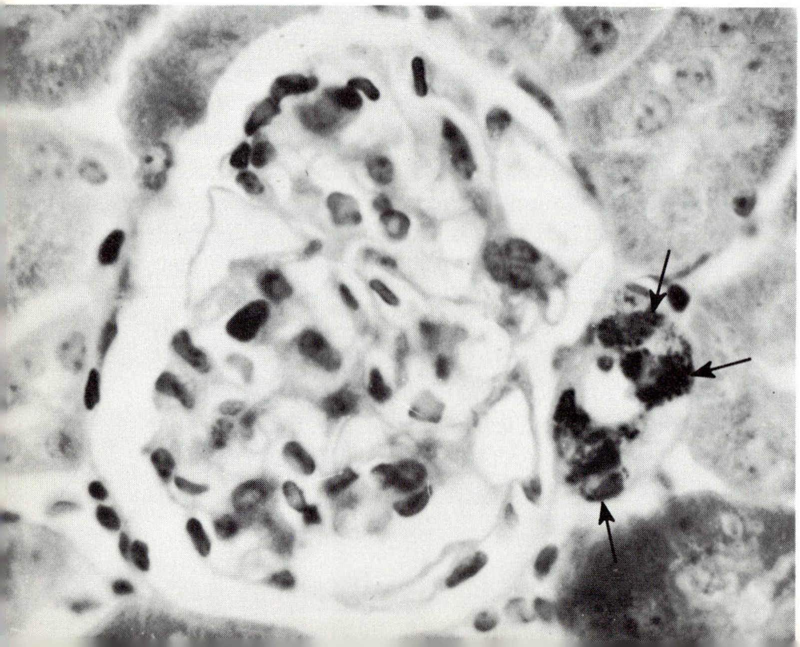

Fig. 10.4 Cross section of an afferent arteriole in an adult mouse, showing juxtaglomerular cells, *JG* (*arrows*), in the arteriolar wall. The *JG* cells are swollen with cytoplasmic granules. Approximately X3500. (Courtesy of C. T. Kaylor.)

sodium and water, consistent with the requirements of the body, and of phosphates. Here also the delicate acid-base balance of the blood is adjusted, which may involve the synthesis and/or excretion of ammonium hydroxide, in addition to shifts in sodium, phosphate, and hydrogen ions.

Urine passes from the nephrons into *collecting tubules* and *ducts* (Fig. 10.2) and thence into the Wolffian ducts or ureters, as the case may be. Whether the collecting ducts play a significant role in modifying the composition of urine is open to question. There is evidence that the antidiuretic hormone (see Chapter 4) affects water resorption by the collecting ducts as well as by the distal convoluted tubules, and some studies with rats implicate the collecting ducts in ion exchange.

The nephron incorporates a self-regulating mechanism which is poorly understood. By its action, certain glomeruli may be "closed down" because of constriction of their afferent arterioles; thus the rate of filtration and hence of urine production is correspondingly reduced. Two specialized structures which have been described in mammalian kidneys suggest a possible basis for this self-regulating mechanism, namely the *juxtaglomerular cells* of the afferent arterioles and the *macula densa* of each distal convoluted tubule.

The juxtaglomerular (JG) cells are a kind of myoepithelial cells which replace the smooth muscle cells in a finitely circumscribed area in the wall of the afferent arteriole just before it enters the renal capsule (Fig. 10.4). They stain lightly and bear some resemblance to the cells of the aortic body and the carotid body. It has been proposed that the JG cells excrete the humoral agent, *renin,* which is known to be produced in the kidneys. Renin, in turn, has been shown to transform *hypertensinogen* to *hypertensin.* Hypertensinogen is a humoral factor present in the blood. Its active counterpart, hypertensin (also known as *angiotonin*) stimulates vasoconstriction.

The function of the *macula densa* is less certain, but the fact that it is adjacent to or in contact with the JG cells suggests a functional relation. The *macula densa* is a group of slender, columnar cells in the wall of the distal convoluted tubules of mammals (Fig. 10.5). Because of the slenderness of the cells, the nuclei are close together, imparting a "dense" appearance to the cluster of cells in stained sections. It is conceivable that these cells impart information about the composition or osmolarity of the urine which is passing through them, and, on the basis of this information, the JG cells, via their secretion of renin and its subsequent effect on localized levels of hypertensin, might regulate the filtration rate at the capsule by simple vasodilation or constriction of the efferent arteriole. It must be emphasized, however, that this mechanism is speculative.

The microscopic anatomy of the various

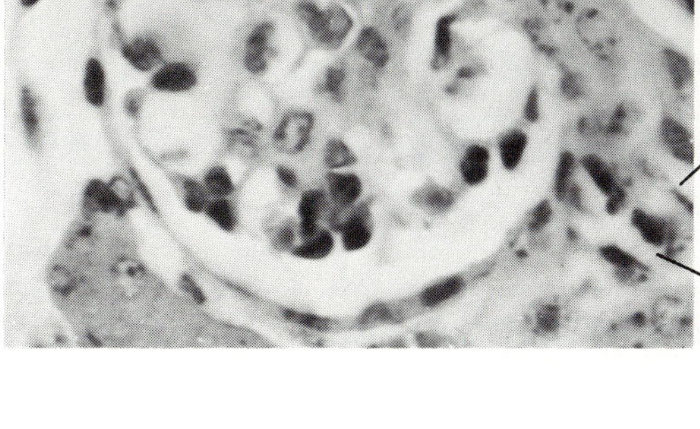

Fig. 10.5 Photomicrograph of a renal capsule in a fetal mouse, showing the macula densa cells of the distal convoluted tubule. These cells become thin and columnar in the adult. X900. (From C. T. Kaylor and J. M. Carter. 1967. Anat. Record 159:171–178.)

parts of the nephron is distinctive in most vertebrates, so that even in sections of the mammalian kidney that consist of more than a million nephrons, one may differentiate between the various parts (Fig. 10.6).

The Glomerulus

A word should be said here about aglomerular kidneys. Most marine teleost fishes are characterized either by very small glomeruli or by absence of a glomerulus. These modifications reflect their secondary adaptation to a hypertonic environment, where the problem of water retention within the body and tissues is a major one. In such forms, renal excretion is effected along the richly vascularized tubules by active secretion, rather than filtration. Interestingly, at least one species of fresh water fish, *Microphis boaja,* also has an aglomerular kidney. This species is a member of the Syngnathidae, which is an almost exclusively marine order of teleosts, and also exclusively aglomerular. Two species of desert anurans, *Cyclorana platycephalus* and *C. alboguttatus,* have been found to have degenerate or rudimentary glomeruli. Other groups of vertebrates, such as Chondrichthyes, marine turtles, marine birds, and marine mammals, have employed other means of adaptation to the salt-water environment, means which do not directly involve modification of kidney structure.

Fig. 10.6 Photomicrograph of mammalian kidney showing components of Bowman's capsule, the convoluted tubules, thick and thin limbs of Henle's loop, and collecting ducts. X500.

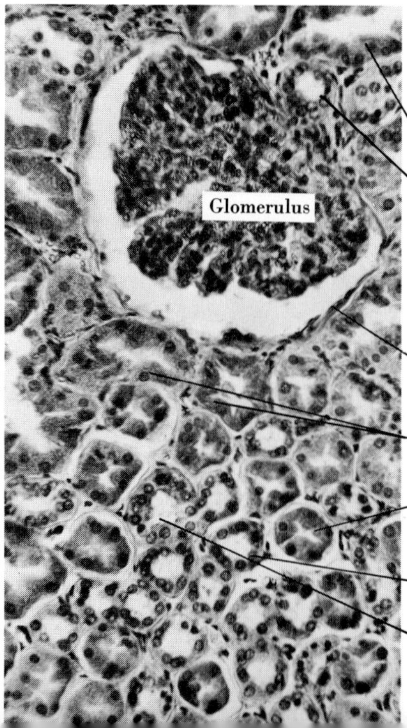

As stated above, the glomerulus is supplied by an afferent arteriole. This vessel is characterized by a thin endothelium which is supported by a basement lamina, an internal elastic membrane, and a thin but complete covering of circular muscle fibers. The adventitia is very thin or virtually lacking. As the afferent arteriole enters the renal capsule, marked by its cuff of JG cells (*see above*), it subdivides a number of times to give rise to the capillary loops of the glomerulus. These capillaries, a *rete mirabile,* consist of a single layer of very flattened endothelial cells supported by a basement lamina. Electron micrographs of glomerular capillaries reveal the presence of numerous *fenestrae* or holes in the endothelial pavement, which permit the blood to come into contact with the basement lamina (Fig. 10.7). Whether the fenestrae are complete openings or merely areas in which the cytoplasm is extremely attenuated is somewhat questionable. In either case, the spatial separation between the capillary lumen and the lumen of the renal capsule is sufficiently thin to account for the completeness and the rapidity with which the blood is filtered in the renal capsule.

The glomerular capillaries rejoin each other to form the efferent arteriole just before emerging from the capsule. The efferent arteriole is significantly smaller in cross-sectional diameter than is the corresponding afferent arteriole, a fact which probably accounts for the relatively high hydrostatic pressure within the glomerular capillaries. The efferent arteriole lacks an internal elastic membrane, but otherwise it is similar histologically to the afferent arteriole.

The Renal Capsule

Because the renal capsule is an involuted cup, like that obtained by pushing one's thumb into a soft tennis ball, it is double-walled. The inner wall with which the glomerulus makes

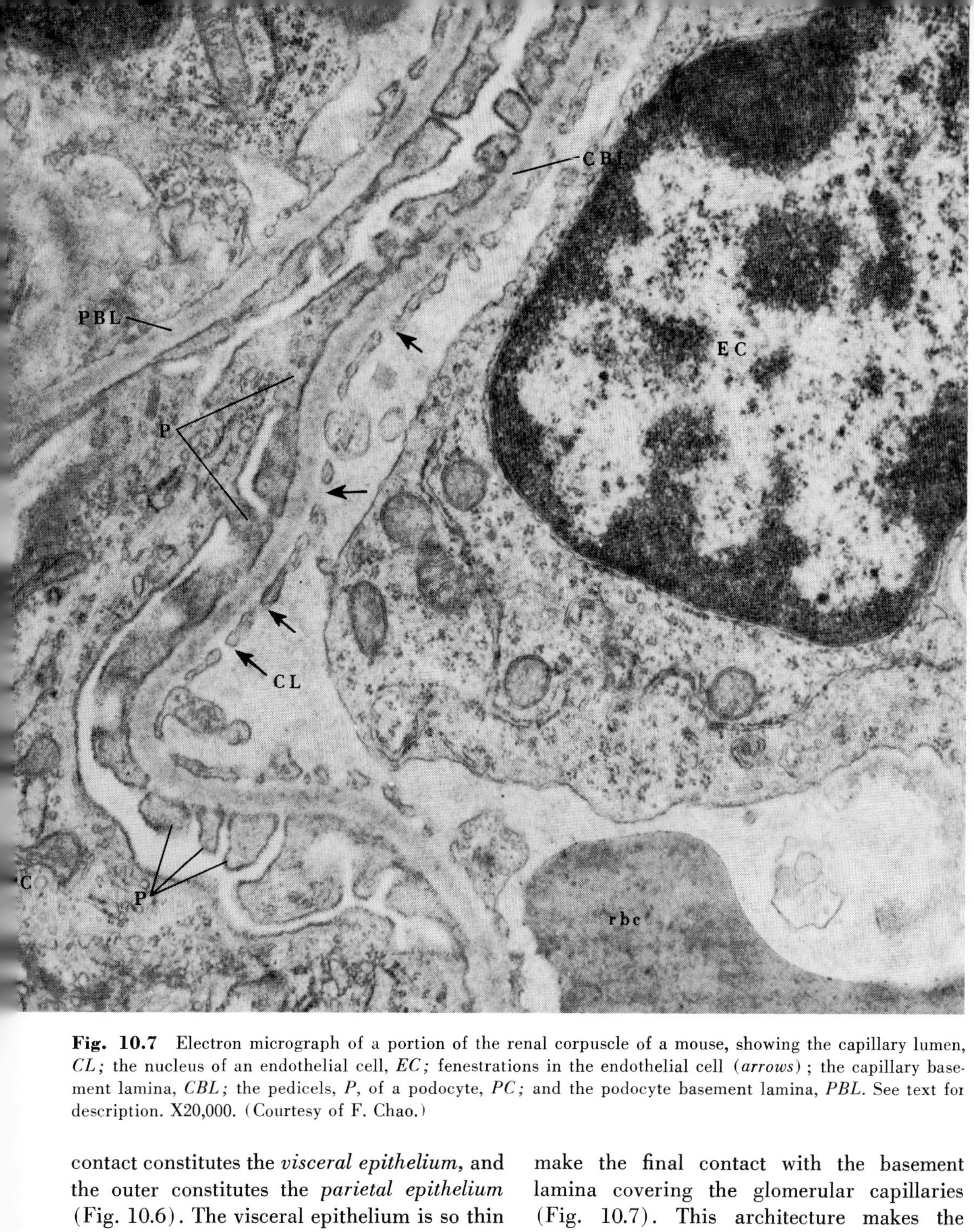

Fig. 10.7 Electron micrograph of a portion of the renal corpuscle of a mouse, showing the capillary lumen, *CL;* the nucleus of an endothelial cell, *EC;* fenestrations in the endothelial cell (*arrows*); the capillary basement lamina, *CBL;* the pedicels, *P*, of a podocyte, *PC;* and the podocyte basement lamina, *PBL*. See text for description. X20,000. (Courtesy of F. Chao.)

contact constitutes the *visceral epithelium,* and the outer constitutes the *parietal epithelium* (Fig. 10.6). The visceral epithelium is so thin it escaped the notice of microscopists for many years, but electron micrographs show not only that it exists but also that it has a complex architecture. Its cells touch each other only at intervals, leaving rather wide spaces or slits between them. In addition, extending inward toward the glomerulus, these cells have footlike processes upon which are even smaller processes, the pedicels, which make the final contact with the basement lamina covering the glomerular capillaries (Fig. 10.7). This architecture makes the visceral wall a sieve, which is held away from the capillaries by its pedicels in order to offer minimum impedence to the flow of the ultrafiltrate from the glomerular capillaries into the capsular lumen. Because of the presence of the footlike processes described above, the visceral epithelial cells are called *podocytes.* Because the epithelium of the capsule is supplied elsewhere with a prominent basement lamina, the basement lamina seen by electron microscopy separating the capillary endothelium from the pedicels is probably a fusion of the basement laminae of both the endothelium and the visceral epithelium. This is similar to the situation described in the alveoli of the lungs.

The parietal epithelium is clearly distinguishable as a single, continuous layer of squamous cells supported externally by a prominent basement membrane and separated from the visceral layer by a clear space of variable width. In cyclostomes the parietal cells may be ciliated or flagellated. In elasmobranchs they are almost always ciliated, as well as in some species of bony fishes and amphibians. In the higher vertebrates they are usually smooth-surfaced.

Proximal Convoluted Tubule

The proximal convoluted tubule is frequently constricted where it joins the capsule, in which case it is referred to as the "neck." The cells of the neck are equipped with flagella in the cyclostomes, and their motion in the living state is reminiscent of the flame cells of certain invertebrates. In most lower vertebrates above the cyclostomes these cells are ciliated and have a brush border formed by myriads of microvilli. Otherwise they are like the other cells of the proximal convoluted tubule.

Typically the cells of the proximal convoluted tubule have a brush border, they are cuboidal or low columnar, and the mitochondria and rough endoplasmic reticulum are concentrated in the basal zones, to which they impart a striated appearance in optical sections (Figs. 10.6 and 10.8). The lumina of the proximal convoluted tubules often appear starshaped in transverse section. In fishes and amphibians the cells are usually ciliated.

Distal Convoluted Tubule

In most nonterrestrial vertebrates, the proximal convoluted tubule merges almost imperceptibly into the distal convoluted tubule. The cells of the latter are usually low columnar and present a smooth surface to the lumen (Fig. 10.5). Cilia are rare or lacking, and the basal striations are not so pronounced as in the cells of the proximal convoluted tubule. The outside diameters of the two tubules are approximately the same, but the distal convoluted tubule has a larger internal diameter because of the evenness of the luminal surface of its epithelial cells. The structure of the distal convoluted tubule seems to have become fairly well established in the fishes, for it has essentially the same characteristics in higher vertebrates, except that the cells may be somewhat flatter.

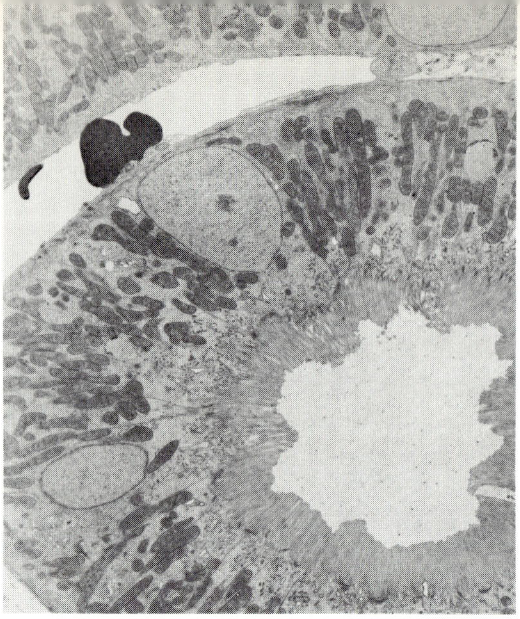

Fig. 10.8 Low power electron micrograph of a proximal convoluted tubule of a rat, showing that the brush border lining the lumen consists of microvilli. Note the characteristic orientation of the mitochondria. X4000. (From R. E. Bulger. 1963. Am. J. Anat. 116:237–255.)

Intermediate Segment

Interposed between the proximal and distal convoluted tubules may be found an *intermediate segment,* which is distinguished by a smaller cross-sectional diameter and more flattened epithelial cells without basal striations. Intermediate segments have been described in some of the more peripheral tubules in the kidneys of a few species of bony fishes. Urodeles have a long, ciliated intermediate segment, but in anurans it is either of moderate length or absent. Although generally conspicuous in reptiles, its length is variable in that group. In birds the intermediate segment

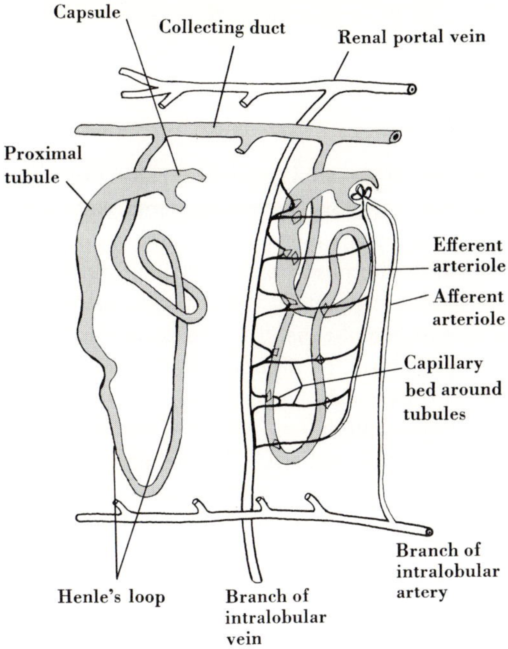

Fig. 10.9 Diagram of the nephron in the avian kidney.

is long and thrown into a loop (Fig. 10.9). In mammals it is similarly long and is known as the *loop of Henle* (Fig. 10.2). The *descending limb* of Henle's loop extends through the striated zone of the renal cortex, well into the medulla, before turning back to form the ascending limb. The descending limb is composed of a simple squamous epithelium and accordingly may be mistaken by an inexperienced eye for a blood capillary. The ascending limb has a thicker wall, because of its low cuboidal cells, and also a larger cross-sectional diameter.

Collecting Tubules and Ducts

Collecting tubules consist of a simple cuboidal epithelium when their diameter is small, which changes to simple columnar when, as collecting ducts, their diameter is larger (Fig. 10.6). The luminal surface is typically smooth (although occasional cilia and flagella have been described by electron microscopy), the cytoplasm is relatively clear and light-staining, the nuclei are round and highly basophilic, and the basement membrane is prominent. In fishes and amphibians the collecting tubules and ducts lead directly into the Wolffian duct in the mesonephric region of the kidney, and into the accessory duct in those species in which a clearly defined metanephric component applies.

Interstitial Tissue

The amount and kinds of cells interspersed among the uriniferous tubules, ducts, and blood vessels of the kidneys varies. In fishes, particularly in the teleosts, it is commonplace to find hemopoietic tissue comprising the bulk of what amounts to a considerable amount of interstitial tissue (Fig. 10.10). In general, however, the interstitial tissue is scanty and consists of reticular connective tissue, with some collagenous fibers, fibroblasts, and macrophages.

THE URINARY PASSAGES

The Wolffian Duct

The Wolffian or mesonephric duct courses along the dorsolateral aspect of the kidney in most anamniotes, but it is included within the capsule of the kidney up to the point where it emerges posteriorly. Thereafter it is covered by a serosa. The epithelial lining of the

Fig. 10.10 Section of kidney of the spiny dogfish, *Squalus acanthius*, showing a nodule of hemopoietic tissue among renal tubules. X500.

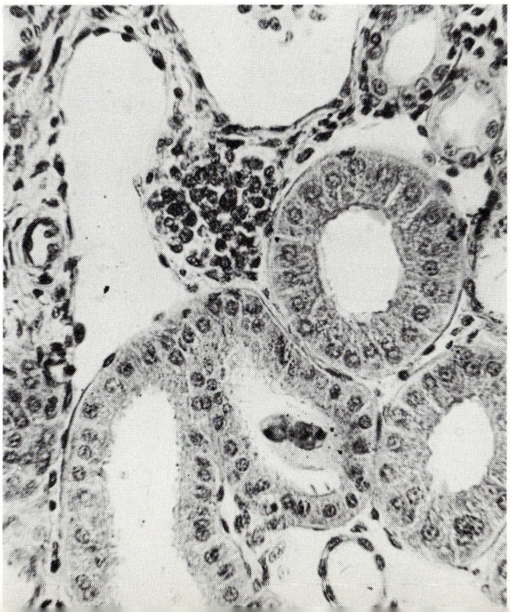

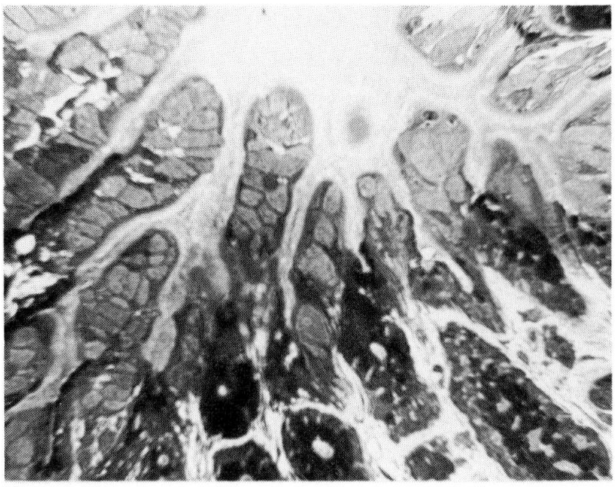

Fig. 10.11 An oblique section through the Wolffian duct of *Rana pipiens* showing clusters of mucous granules in the apical cytoplasm. X460.

Wolffian duct varies slightly along its length, consisting of a single layer of ciliated columnar cells anteriorly, which become progressively taller as one proceeds caudally. In freshwater fishes numerous goblet cells are present, but none is found in marine teleosts. It has been proposed that, because marine teleosts transform their waste nitrogen into trimethylamine, which is nontoxic, and because they excrete very little urine, there is no need of a protective or lubricating coat of mucin. In amphibians the epithelial cells contain granules of mucinogen in their apical cytoplasm and true goblet cells are few or lacking (Fig. 10.11). In both fishes and amphibians the epithelium is supported by a thin lamina propria, and together they are thrown into longitudinal folds which project into the lumen.

The mucosa thus formed by epithelium and lamina propria is surrounded by a delicate coat of collagenous connective tissue that contains a scattering of smooth muscle cells and, in many species, a thin layer of pigment cells. In the portion of the Wolffian duct that lies outside of the kidney, i.e., the extrarenal segment, the connective tissue coat is thicker and shows a progressive increase in smooth muscle elements. In fishes the smooth muscle is often segregated into two more or less distinct layers: an inner longitudinal layer and an outer circular layer. Interestingly, this arrangement, which is also characteristic of the ureter, is the opposite of that encountered in the digestive tract.

The Ureter

The ureters of amniote vertebrates are characterized by more definitive demarcation between mucosa, muscularis, and adventitia. The epithelium is simple columnar in lizards and snakes, and it is supported by a delicate basement membrane. In turtles and tortoises the epithelium is pseudostratified columnar, and no basement membrane is evident. In birds the epithelium is also pseudostratified columnar, but a distinct basement membrane is present. The mammalian ureter is lined with the transitional type of epithelium, but again a distinct basement membrane is lacking. Mucin-secreting cells occur interposed among the epithelial cells in amniotes, but they vary in relative numbers from species to species, depending apparently upon the chemical composition of the urine. Species which characteristically produce urine with high ammonia or uric acid content, such as birds, have a greater preponderance of mucin-secreting cells.

The lamina propria of the ureter is well defined and it accompanies the epithelium in its longitudinal folding. Fine elastic networks and smooth muscle fibers have been described in the lamina propria, although it is primarily collagenous in character (Fig. 10.12).

The muscularis of the ureter consists of two and sometimes three somewhat irregular layers of smooth muscle. The fibers of the innermost layer are invariably oriented longitudinally, and the fibers in the next layer circularly. The third layer occurs in the posterior segment of the ureter, and its fibers are longitudinally oriented and lie outside the circular layer. The coordinated contraction and relaxation of these muscles is manifest as peristalsis, by which the urine is propelled posteriorly through the ureter. It is interesting to note that only a small number of drugs is known to have an effect on ureteric peristalsis; these are histamine, the antihistamines, bradykinin, and serotonin.

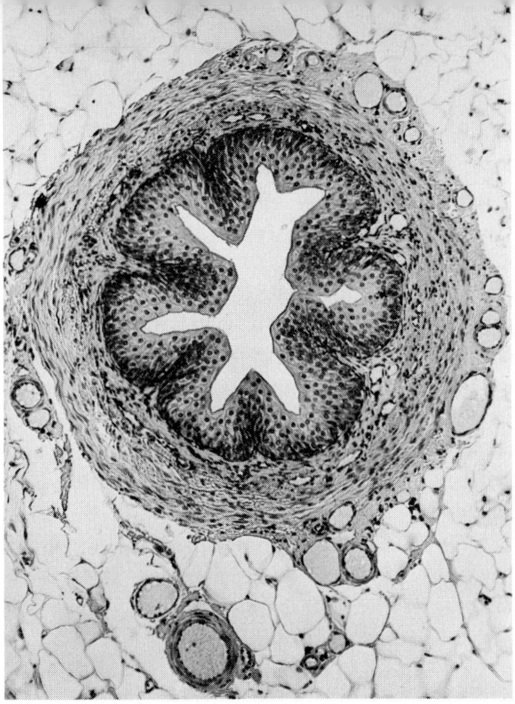

Fig. 10.12 Cross section of a mammalian ureter. Note the clear demarcation between the transitional epithelium, the lamina propria, the muscularis, and the adventitia. X150.

The adventitia of the ureter is very thin around the intrarenal segment and somewhat thicker in the extrarenal segment. It consists primarily of collagenous tissue, although a scattering of smooth muscle fibers may be found in some species.

The Urinary Bladder

Bladders which serve to store urine prior to micturition are found throughout the vertebrate subphylum. In fishes such bladders represent outpouchings of the fused distal ends of the Wolffian ducts, and so they are in every respect of mesodermal derivation. In amphibians the urinary bladder develops as a ventral outpocketing of the cloaca, and is presumed to be homologous with the so-called allantoic bladder of higher forms. For this reason, its epithelium is likely of endodermal origin. The urinary bladder of reptiles, birds, and mammals is derived, at least in part, from a portion of the foetal allantois; similarly, it is lined with cells of presumed endodermal origin. Among reptiles, the Ophidia and Crocodilia lack a urinary bladder. Among birds, only ostriches have a urinary bladder, all others having dispensed with them as excess baggage! The accessory bladders of some turtles are of particular interest, because they are reputed to be used as aquatic respiratory organs.

The urinary bladders of toads have been used in physiological studies of membrane transport and diffusion. They are lined by an epithelium which is simple squamous in the distended condition, but which consists of one or two layers of cuboidal or low columnar cells in the relaxed state. Four types of cells have been distinguished: granular, mitochondria-rich, mucous, and basal. The granular cells are the most numerous, and their luminal surface is covered with microvilli. The mitochondria-rich cells are flask-shaped, and extend from the basement lamina to the lumen. The basal cells are small and basophilic. The epithelial cells are firmly attached to each other laterally, presenting what has been described as a "leakproof membrane." Included within the epithelial layer are goblet cells and occasional polyhedral cells which may represent differentiating goblet cells. The lamina propria is collagenous, and includes blood vessels and irregularly arranged bundles of smooth muscle. It is covered by a thin serosa of squamous mesothelium.

The urinary bladder of the turtle, *Chrysemys picta,* is large and bilobed. Its epithelium is pseudostratified columnar with many mucinogen-secreting cells that bulge into the lumen. Occasionally eosinophils may be seen infiltrated among the epithelial cells. The lamina propria is loose and fibrous and accompanies the epithelium in its numerous folds. It contains small blood vessels, lymphatics, and nerve fibers. Eosinophils and lymphocytes are also seen scattered in the lamina propria. As in the mammalian bladder, no basement membrane is evident. The structure of the mucosa in *Chrysemys picta* is not consistent with a respiratory function. In *Pseudemys scripta* the luminal surface of the epithelium may appear by electron microscopy to be coated with branching, hairlike structures, and histochemical procedures have revealed a high concentration of polysaccharide to be associated with the surface.

The subepithelial connective tissue is somewhat thicker in the body of the urinary bladder of turtles than in the lateral lobes, otherwise the general characteristics are similar. It consists of fibroelastic connective tissue in which are included blood vessels, lymphatics, nerves, and prominent bundles of smooth muscle, as well as scattered eosinophils and loose aggregations of lymphocytes. The smooth muscle bundles, although irregularly arranged, are essentially oriented in three different axes, two perpendicular to each other and one oblique. Surrounding the connective tissue layer in a thin, fibrous adventitia which is covered over most of the organ by the squamous mesothelium of the visceral peritoneum.

The structure of the mammalian urinary bladder can be inferred from the structure of the ureters, notwithstanding the fact that their embryonic derivation is different. The transitional epithelium forming the lining of the urinary bladder, such as in the ureters, varies considerably in thickness and in the shape of its cells according to the extent to which the lumen is distended. When the bladder is empty, the surface cells are rounded and bulge into the lumen. Some of them are clearly binucleate. The cells underlying them progress from large and pear-shaped to small and polyhedral through four or five layers. As the bladder becomes distended with urine, first the folds of the mucosa level off and then the epithelial cells slide by one another—spreading out as they do so, ultimately presenting a stratified squamous epithelium of only two or three cells in thickness (Fig. 10.13). It is not because of this transition from one kind of epithelium to another that this kind of epithelium is called transitional, however, but rather because the tissue was believed by earlier histologists to be transitional between stratified squamous and columnar.

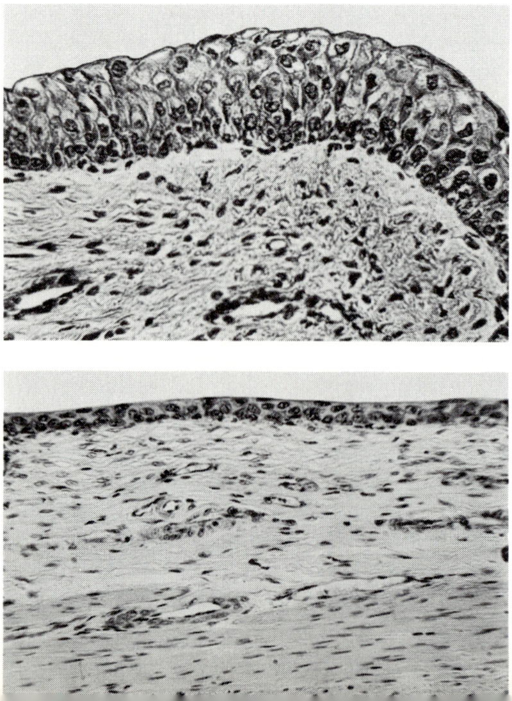

Fig. 10.13 Mammalian urinary bladder: relaxed, *above;* and distended, *below.* Some of the muscularis is included *below.* X600.

The lamina propria is sharply delimited from the epithelial layer, but not from the connective tissue surrounding it, for there is no basement membrane. Included within the lamina propria one may find diffuse lymph nodules or aggregations of lymphocytes. The epithelium and lamina propria are thrown into large folds in the undistended bladder.

Forming three irregular layers within the collagenous connective tissue stroma are inner longitudinal, middle circular, and outer longitudinal bundles of smooth muscle. In some areas it is evident that some of the bundles are diagonally disposed. Small lymphocyte nodules may also be encountered in this layer.

The anterior portion of the urinary bladder is covered by a serosa of simple squamous mesothelium; the rest is covered by a fibrous adventitia.

The Urethra

In all vertebrates except the Metatheria and the Eutheria the urine is voided into the cloaca, but in these two mammalian groups a new duct, the *urethra,* is present. In females the urethra serves solely a urinary function, transporting urine from the urinary bladder to the urogenital sinus or vestibule, i.e., the outside. In males only the very short segment extending between the urinary bladder and the openings of the prostate gland is solely urinary in function; the rest combines urinary and genital functions.

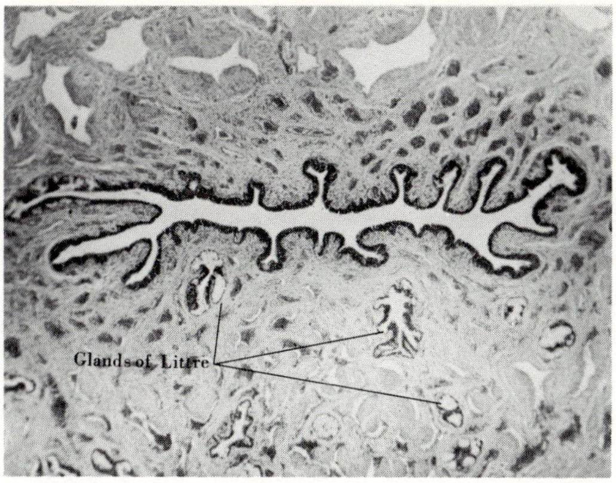

Fig. 10.14 Cross section of a human penile urethra. X60.

The female urethra is a short, thick-walled tube which consists of mucosa, submucosa, muscularis, and adventitia. The mucosal epithelium is transitional near the bladder, and then it changes through stratified or pseudostratified columnar to stratified squamous near the outside orifice. Mucus cells are common, occurring in clusters or in small diverticula which occasionally give rise to tubular mucous glands, the *glands of Littré*. Both the diverticular and the glands of Littré are more numerous and extensively developed in the male urethra (Fig. 10.14). A thin basement membrane is present, as well as a fibrous lamina propria, which, together with the mucous epithelium, is thrown into loose longitudinal folds, giving the lumen a crescentic shape in transverse section. The demarcation between the lamina propria and the submucosa is arbitrary in most species. The submucosa is rich in elastic fibers and networks of thin-walled venous channels, which make it semierectile.

The relatively thick muscularis of the female urethra consists of bundles of inner longitudinal smooth muscle interspersed with portions of the venous networks described above, and an outer circular layer of smooth muscle. The latter forms a prominent *involuntary sphincter* next to the urinary bladder. Surrounding the smooth muscle layers are bundles of circularly disposed skeletal muscle fibers, which form a prominent *voluntary sphincter* at the

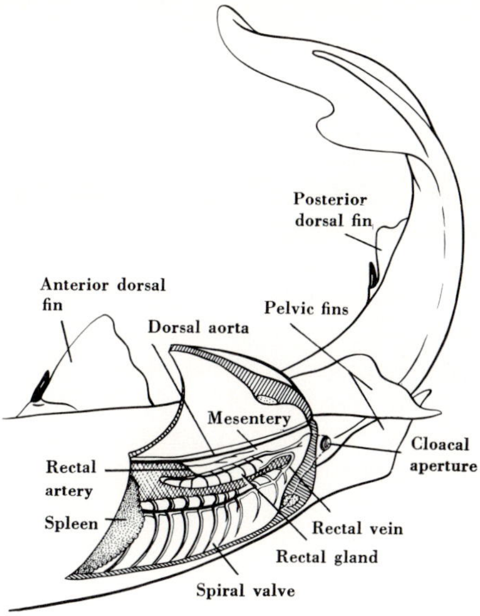

Fig. 10.15 Diagram of a dissection of the spiny dogfish, *Squalus acanthius,* showing the location of the rectal gland. (After R. E. Bulger. 1963. Anat. Record 147: 95–127.)

distal end of the urethra, where it opens into the vestibule. The adventitia is indefinite, being continuous with the connective tissue components of adjacent organs.

The urethra in males is long, extending from the neck of the urinary bladder to the *glans* of the penis. Only the proximal section, which lies between the bladder and the penis,

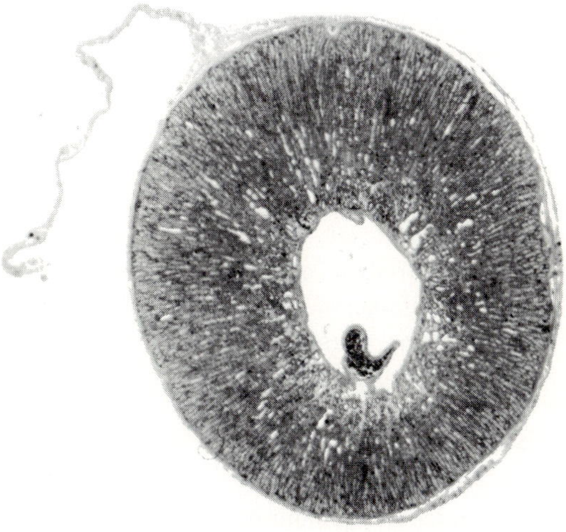

Fig. 10.16 Cross section of the rectal gland of *Squalus acanthius.* X24. (From R. E. Bulger. 1963. Anat. Record 147:95–127.)

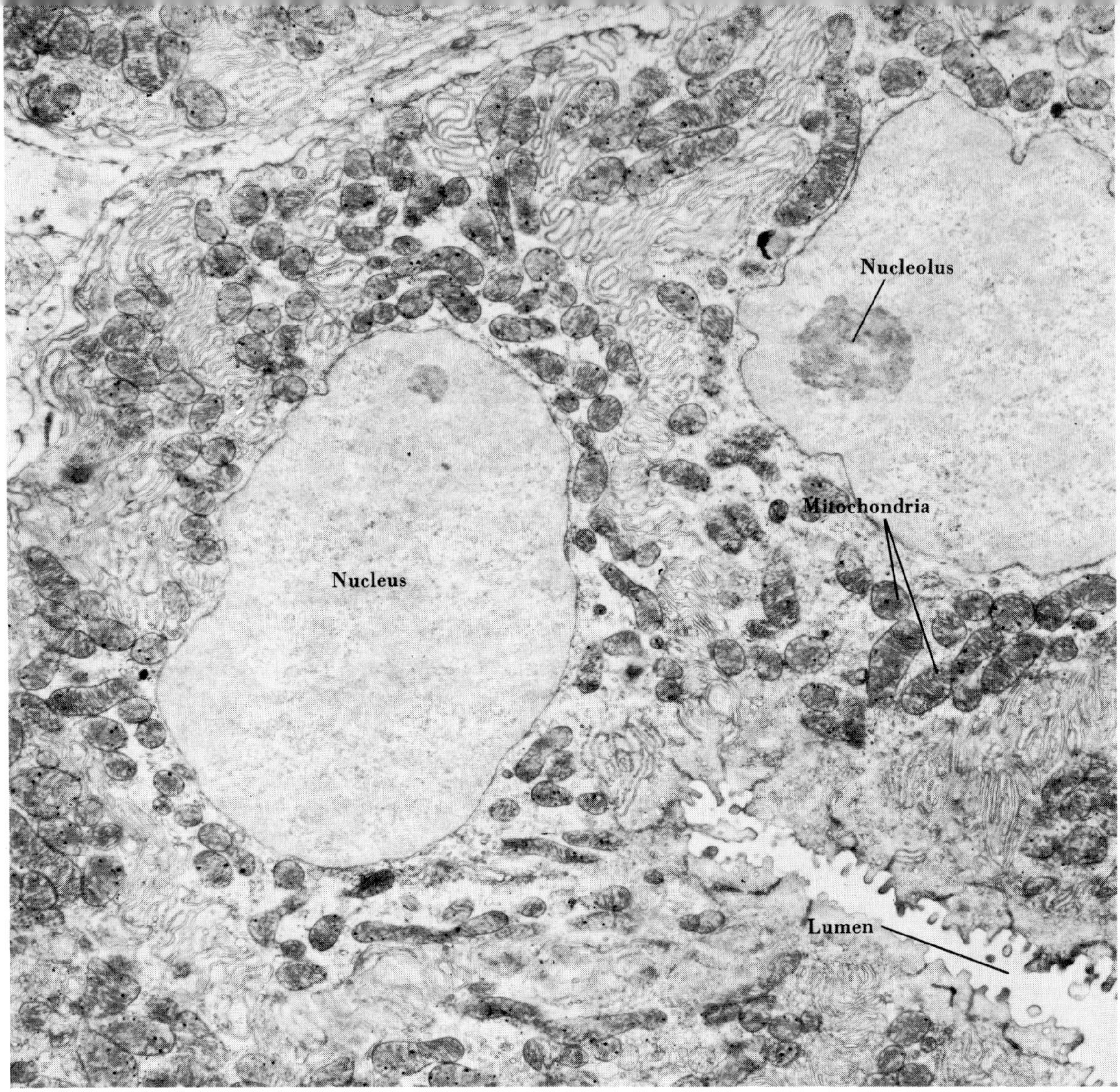

Fig. 10.17 Electron micrograph of cells of the outer tubular layer of the rectal gland of *Squalus acanthius*, showing the complicated interdigitations of cell membranes. X7600. (From R. E. Bulger. 1963. Anat. Record 147:94–127.)

is homologous with the female urethra. This section includes the *prostatic urethra* and the *membranous urethra*. The rest, and major part of the urethra forms from the folds of the urogenital sinus, the homologue of which in the female appears as the *labia minora*. In male development, the free edges of the urogenital folds come together and fuse, giving rise to the penile or *cavernous urethra* and the penis (see Chapter 11).

The mucosal epithelium of the male urethra is transitional in the prostatic section, and it varies between stratified and pseudostratified columnar from there to the *meatus* in the glans, where it becomes stratified squamous. Clusters of mucus cells, the *intraepithelial glands*, occur in irregular diverticula, the *lacunae of Margagni*. Branched tubular glands of mucus cells, the glands of Littré, described above, grow outward from the lacunae into the surrounding lamina propria and may even extend into the submucosa. (Other specialized glands of the urethra, the prostate and bulbourethral glands, are associated with its genital function and are considered in Chapter 11.) A thin basement membrane supports all of the epithelial cells.

As in the female urethra, there is no sharp

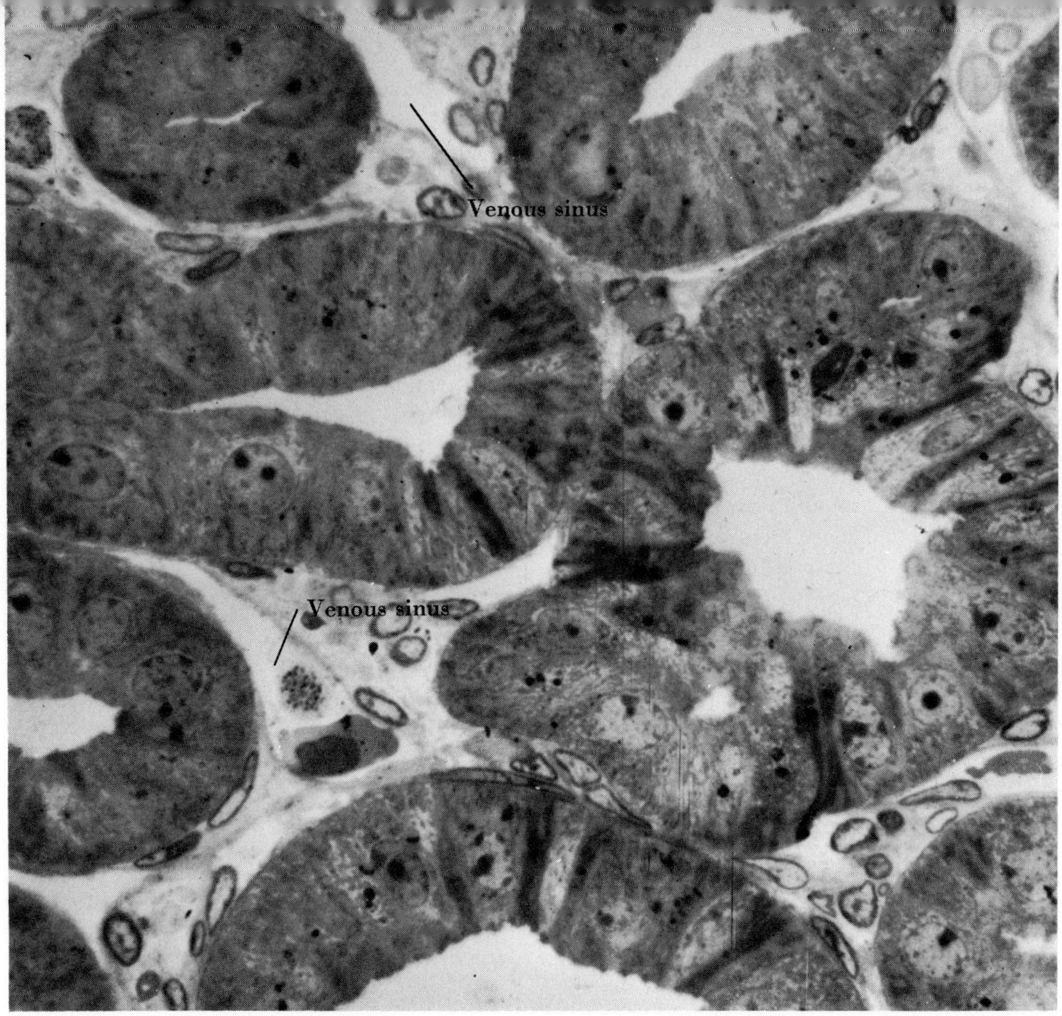

Fig. 10.18 Photomicrograph of excretory cells in the inner tubular layer of the rectal gland of *Squalus acanthius*. X1600. (From R. E. Bulger. 1963. Anat. Record 147:95–127.)

demarcation between the lamina propria and the submucosa. The submucosa and muscularis layers of the prostatic and membranous segments are essentially similar to their female-urethra counterparts as described above. No typical adventitia is present in these segments; instead, the prostatic urethra is embraced by tissues of the prostate gland, and the membranous urethra is encircled by skeletal muscle fibers of the voluntary sphincter. The cavernous urethra lacks either submucosa or muscularis, but is surrounded by erectile tissue and a dense outer sheath, as described in Chapter 11.

THE RECTAL GLAND

The rectal gland of elasmobranchs (Fig. 10.15) has been shown to be concerned with the active transport and excretion of sodium chloride. It is a compound and simple branched tubular gland organized around a central lumen which connects by a short duct to the intestine posterior to the spiral valve (Fig. 10.16). The central lumen is lined with a stratified cuboidal epithelium, although in some preparations the surface cells appear to be squamous. The most abundant surface cells are those which contain granules in their apical cytoplasm and have short microvillous projections on their luminal surface. In addition, rounded mucus cells and flask-shaped cells are found, and occasionally a cell which is distinguished in electron micrographs by the presence of large mitochondria. Beneath the surface layer is an intermediate layer of cells which are loosely associated and include macrophages, lymphocytes, and granulocytes in their intercellular spaces. The basal cells form a continuous layer resting on a thick basement lamina.

The tubular glands, radiating out from the central lumen, consist of a simple cuboidal epithelium the cells of which have been found by electron microscopy to have a complex form. The lateral surfaces of these cells connect with adjacent cells by interdigitating processes, and the apical surfaces are richly folded and project microvilli into the lumen of the tubule. The tubules are separated from each other by a loose connective tissue in which may be found wandering cells and venous sinuses (Figs. 10.17 and 10.18).

Between the glandular parenchyma of the organ and its fibrous connective tissue capsule is a layer of smooth muscle. The entire organ is covered by the cuboidal epithelium of the visceral peritoneum.

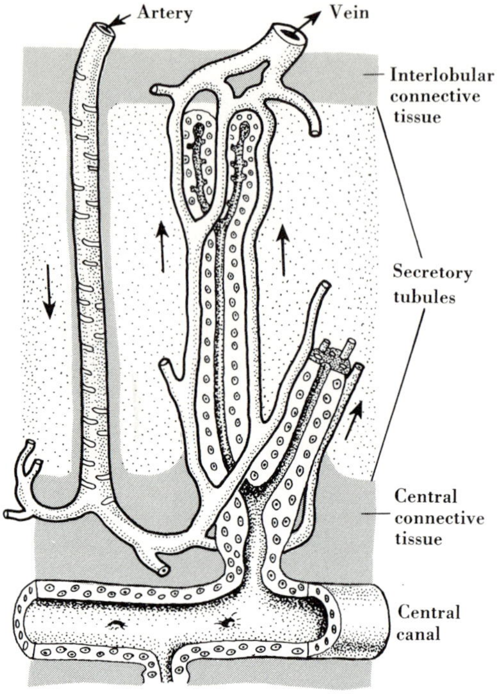

Fig. 10.20 Diagram illustrating the application of the countercurrent principle in the avian salt gland. (After R. Fänge *et al.* 1958. Biol. Bull. 115:162–171.)

THE AVIAN SALT GLAND

Many species of marine birds have prominent, paired glands lying in the suprorbital grooves of the frontal bones. As is the case with the elasmobranch rectal gland, they function in salt excretion, but they differentiate embryologically from outgrowths of the nasal cavity. Each gland consists of a series of parallel "lobes" alligned longitudinally, like bundles of rectal glands, as it were. Secretory tubules radiate from the central canal in each lobe (Fig. 10.19), and are simple or compound branched. The direction of blood flow through the capillaries surrounding the secretory tubules is opposite to the direction of flow of the excretion (Fig. 10.20). The cells of the secretory tubules are wedge-shaped columnar cells with a central nucleus and basal striations. Electron micrographs show numerous infoldings of the cell membrane at the basal end, which extend deeply into the cell, even reaching the apical surface. It has been suggested that such folding is a continuous process and is the mechanism by which salts are transferred from the subepithelial region to the lumen of the tubule. Fibrous connective tissue supporting a rich vascular network separates the tubules from one another. The gland is covered dorsally by a thin, tough, connective tissue membrane. Salt glands of similar architecture are also encountered in marine turtles (Fig. 10.21).

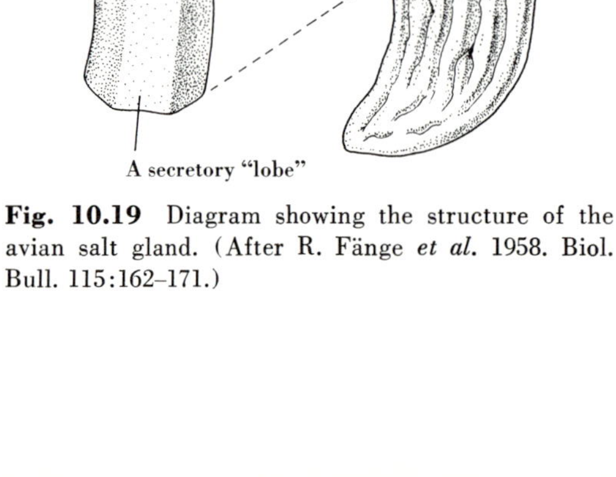

Fig. 10.19 Diagram showing the structure of the avian salt gland. (After R. Fänge *et al.* 1958. Biol. Bull. 115:162–171.)

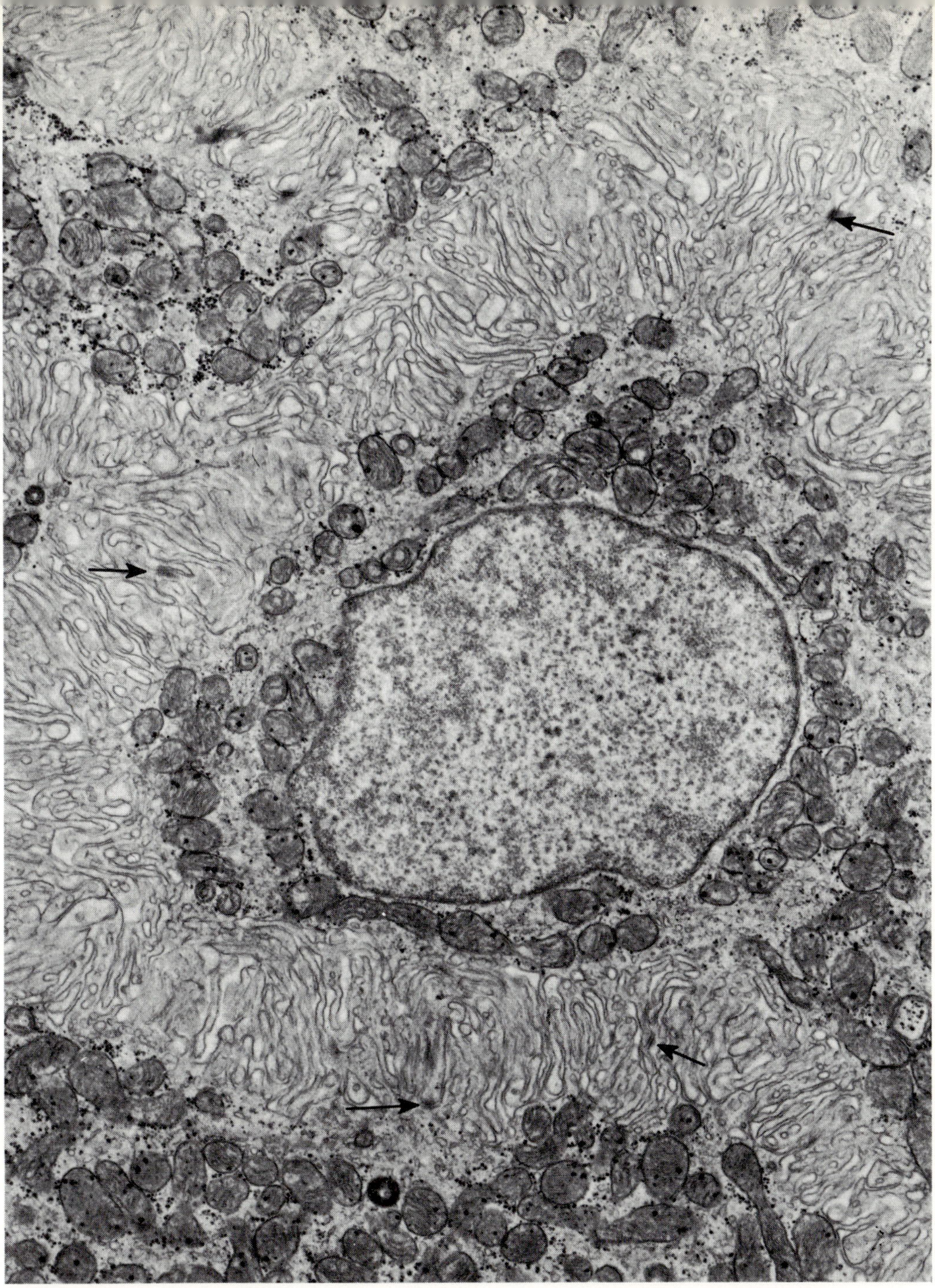

Fig. 10.21 Electron micrograph of a secretory cell in the salt-secreting lacrymal gland of the marine turtle, *Chelonia mydas,* showing the complicated interdigitations of the cell membranes. Compare with Fig. 10.17. Desmosomes (*arrows*) occur occasionally. X10,800. (From J. H. Abel and R. A. Ellis. 1966. Am. J. Anat. 118:337–358.)

SELECTED REFERENCES

Abel, J. H., Jr., and R. A. Ellis. 1966. Histochemical and electron microscopic observations on the salt-secreting lacrymal glands of marine turtles. Am. J. Anat. 118:337–358.

Bulger, Ruth E. 1963. Fine structure of the rectal (salt secreting) gland of the spiny dogfish, *Squalus acanthius*. Anat. Rec. 147:95–127.

Bulger, Ruth E. 1965. The fine structure of the aglomerular nephron of the toadfish, *Opsanus tau*. Amer. J. Anat. 117:171–192.

Bulger, Ruth E. 1965. Electron microscopy of the stratified epithelium lining the excretory canal of the dogfish rectal gland. Anat. Rec. 151:589–607.

Catacunan-Labay, P., and S. Boyarsky. 1966. Bradykinin: effect on ureteral peristalsis. Science 151(3706):78.

Choi, J. K. 1963. The fine structure of the urinary bladder of the toad, *Bufo marinus*. J. Cell Biol. 16:53–72.

Dawson, A. B. 1951. Functional and degenerating or rudimentary glomeruli in the kidney of two species of Australian frog, *Cyclorana* (*Chiroleptes*) *platycephalus* and *alboguttatus* (Gunther). Anat. Rec. 109:417–429.

Doyle, W. C. 1960. The principal cells of the salt gland of marine birds. Exptl. Cell Res. 21:386–393.

Fänge, R., K. Schmidt-Nielsen and H. Osaki. 1958. The salt gland of the herring gull, *Larus argentatus*. Biol. Bull. 115:162–171.

Liu, Hin-Ching. 1962. The comparative structure of the ureter. Am. J. Anat. 111:1–15.

Pitts, R. F. 1959. The physiological basis of diuretic therapy. Charles C Thomas, Springfield, Ill.

Saladino, A. J. 1966. Light and electronmicroscopic observations on the cell surface coat of the turtle bladder. Proc. Fed. Am. Soc. Exptl. Biol. 25 (2):599. (Abstract)

Smith, H. W. 1951. The kidney: structure and function in health and disease. Oxford University Press, New York.

Sturkie, P. D. 1965. Avian physiology. 2nd ed. Comstock Publishing Associates Div., Cornell University Press, Ithaca, New York.

chapter 11
THE REPRODUCTIVE SYSTEM

The reproductive system of vertebrates consists of the gonads and the genital ducts, as well as of the glands and accessory structures functionally associated and anatomically contiguous with them. As in the case of the urinary system, the evolution of the reproductive system is revealed to some extent both by its anatomy in modern, adult species, and by its embryonic development.

The "Indifferent" Gonad

Although the sexes are typically separate in vertebrates, a single, sexually indifferent pair of gonads appears in the embryo, which later acquires the requisite modifications to identify it as a pair of ovaries or as a pair of testes. The gonads are first recognized in the early embryo as a pair of longitudinal thickenings of the visceral mesoderm, the *genital ridges,* which appear medial to the nephrotomes. They are covered by a simple cuboidal epithelium, the *germinal epithelium,* in which large, rounded cells, the *presumptive germ cells,* soon become evident. The precise site of origin of the presumptive germ cells is open to question, for in some species they appear to arise *in situ,* whereas in others they have been shown quite conclusively to be derived from the outlying yolk sac membrane, from which they migrate to their final location in the germinal epithelium. Oocytes and spermatogonia are believed to be derived directly from the presumptive germ cells.

The germinal epithelium proliferates cords of cells centripetally, the *primary sex cords,* which soon separate away from the epithelium. They induce simultaneously the underlying cells to form the *rete cords,* with which they become continuous. Together, the primary sex cords, the rete cords, and the interstitial cells constitute the gonadal *medulla.* Meanwhile a thin layer of cells, the incipient *tunica albuginea,* has commenced differentiating between the germinal epithelium and the detached primary sex cords. From this point, further differentiation follows one of two lines, leading to the formation of the testis, on the one hand, or to the ovary, on the other.

Testicular Differentiation

The germinal epithelium becomes nothing more than a covering epithelium if the indifferent gonad proceeds in the direction of testicular differentiation. The tunica albuginea becomes collagenous and forms a rather tough connective tissue capsule surrounding the testis. The primary sex cords in the medulla become hollowed out and appear as seminiferous tubules, the primary function of which becomes cytogenic, i.e., the production of sperm cells. The interstitial cells which lie between the seminiferous tubules are stimulated by secretion of the anterior lobe of the pituitary gland to elaborate the male hormones, principally testosterone. The rete cords also become hollowed out and, joining certain nephric tubules which have not been included in the renal function of the kidney, they provide the avenues of exit for the mature spermatozoa to the Wolffian duct. In amniotes the Wolffian duct loses its embryonic urinary function, if it had one, and serves solely the transportation of sperm; accordingly, it is known as the spermatic duct or *vas deferens.* In cyclostomes and one family of teleosts (Salmonidae) no such connections are established, for the sperm are shed directly into the body cavity, from which they escape via genital pores in the body wall.

Ovarian Differentiation

The first event marking the differentiation of the indifferent gonad into an ovary is the establishment of a thick gonadal cortex by proliferation of cells of the germinal epithelium. Secondary sex cords arise from the inner aspect of the cortex and push inward toward the medulla. The cells of the secondary sex cords form primary follicles, which consist of an oocyte surrounded by a single investing layer of cuboidal cells.

Hence it is clear that the indifferent gonad which is developed by the early embryo consists essentially of two parts: the cortex and the medulla. The former has the capacity to differentiate into an ovary, and the latter into a testis. The factors involved in determining which part will develop are not completely understood, but the primary factor is doubtless genetic. It is interesting to note that in some cases temperature can be a decisive factor. If frog tadpoles are transferred to warm water, differentiation of the gonadal medulla is favored and most of the tadpoles develop into males. On the other hand, low temperatures favor cortical and hence ovarian development, leading to a preponderance of females. In mammals there is some indication of an opposite temperature effect, for in many species the testes do not develop fully or normally unless they are at a temperature lower than body temperature, as provided for by their descent into the scrotal sac. In some teleosts and amphibians the undifferentiated component retains its potential into adult life and sex reversal can occur spontaneously or can be induced either by extirpation of the functional component or by exogenous administration of the appropriate sex hormone. In the female of most birds (all except the Falconiiformes), only the left gonad becomes an ovary; the right remains undifferentiated but consists chiefly of medullary tissue. If the ovary is removed or is destroyed by disease, the right gonad may differentiate into a functional testis with consequent complete reversal of the sex of the individual.

Although pseudohermaphroditism and sex reversal are anomalously encountered in nearly all classes of vertebrates, hermaphroditism as a normal condition appears to be restricted to the lower forms. Cyclostomes and certain groups of teleosts (such as Menidae, Sparidae, and Serranidae) have been reported to display protandrous hermaphroditism, i.e., to function as males early in life, and then as females later. Cases of self-fertilization have allegedly occurred in the *Serranidae,* which implies true or functional hermaphroditism.

THE MALE REPRODUCTIVE SYSTEM

The Testis

The testes of adult vertebrates are cylindrical, ovoid, or spherical compound tubular glands covered by coelomic epithelium and a fibrous capsule, the tunica albuginea. At its point of attachment to the dorsal body wall (or its equivalent position in mammals) the capsule

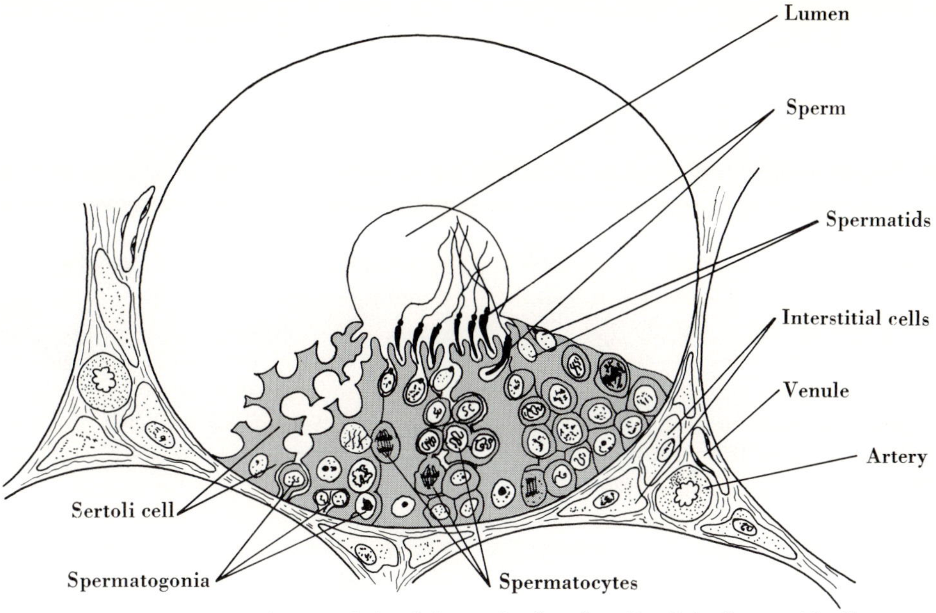

Fig. 11.1 Diagram of a seminiferous tubule of the testis. One Sertoli cell is shown with all spermatogenic cells removed.

is thickened and radiates a number of projections or septa into the gland, dividing it into lobules. Some, but not all, of the septa extend to and join the capsule on the opposite side. They consist of fibroelastic tissue. The point of attachment also serves as the point of entrance or issue of blood vessels, nerves, and efferent sperm ducts.

Within the lobules are distributed the more or less tortuously twisted seminiferous tubules, separated from each other by loose connective tissue and interstitial tissue. Two basically different kinds of cells make up the seminiferous tubules: spermatogenic cells and Sertoli cells. The latter are long, highly irregular cells the bases of which are spaced at intervals around the periphery of each tubule and project into the lumen. Their bases rest on the sharply defined basement membrane which encircles the tubule, but elsewhere their limits are poorly defined because of their exceedingly irregular shape. They are most readily distinguished by their nuclei, which are generally ovoid, deeply grooved, and possess a prominent nucleolus. In some species, such as human beings, a rod-shaped or needle-shaped crystalloid inclusion may be found adjacent to the nucleus. Also called *nurse cells,* the Sertoli cells hold the spermatogenic cells within numerous invaginations of their cell membrane until spermatogenesis is complete, whereupon the sperm cells are released (Figs. 11.1, 11.2, 11.3).

The spermatogenic cells are by far the most numerous cells in the seminiferous tubules, and they include four clearly distinguishable types, as described below. (For more complete details concerning chromosomal behavior and transformations, Chapter 2 should be reviewed.)

SPERMATOGONIA

The spermatogonia are relatively small cells, and they are most abundant around the periphery of each tubule, constituting two or more cell layers resting on the basement membrane between the bases of the Sertoli cells (Figs. 11.1, 11.3). Immature spermatogonia may also possess a small crystalloid adjacent to the nucleus. These cells multiply by mitosis —giving rise to the second type of cell, the primary spermatocytes.

SPERMATOCYTES

Spermatocytes are generally significantly larger than the spermatogonia and are located

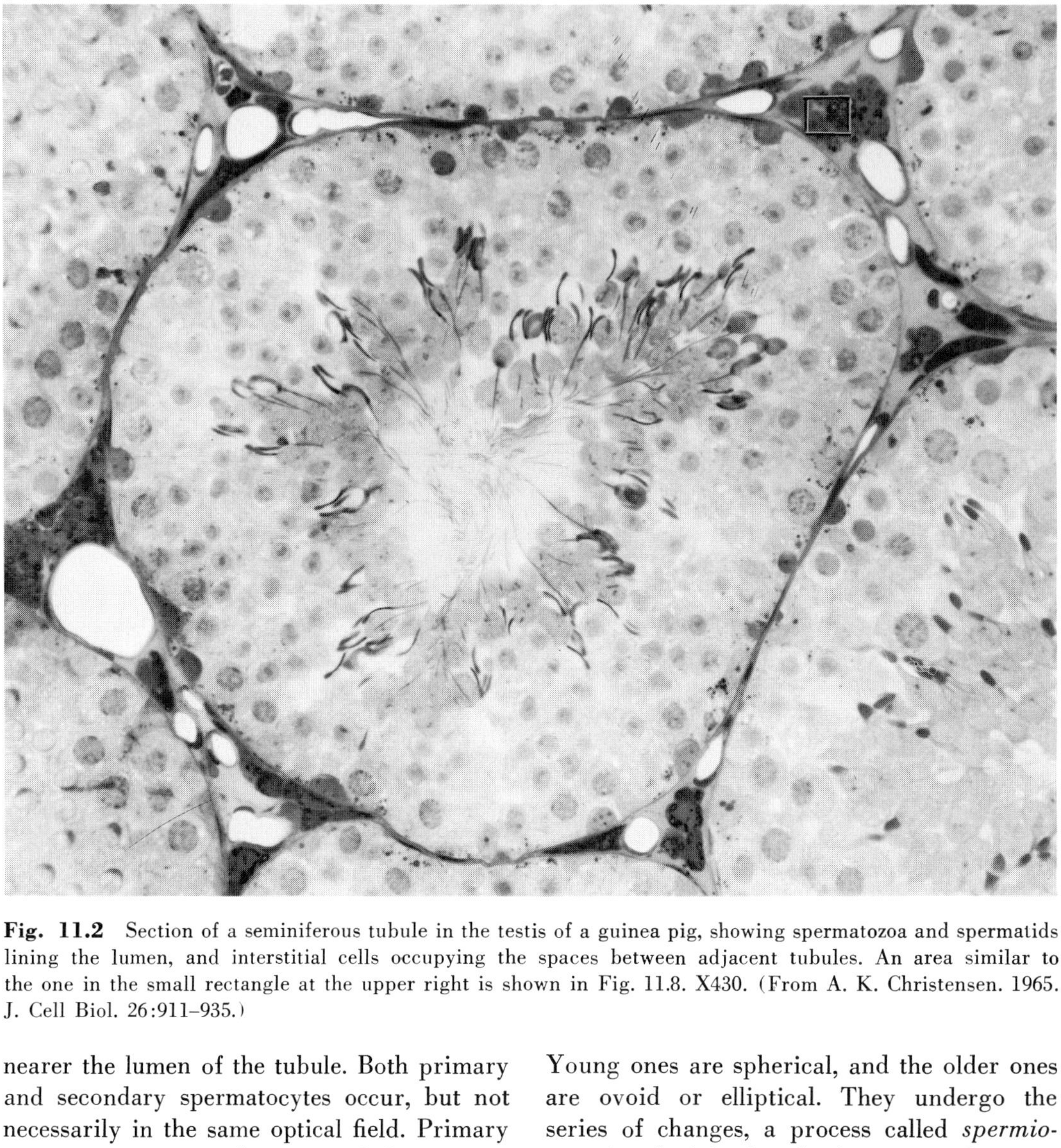

Fig. 11.2 Section of a seminiferous tubule in the testis of a guinea pig, showing spermatozoa and spermatids lining the lumen, and interstitial cells occupying the spaces between adjacent tubules. An area similar to the one in the small rectangle at the upper right is shown in Fig. 11.8. X430. (From A. K. Christensen. 1965. J. Cell Biol. 26:911–935.)

nearer the lumen of the tubule. Both primary and secondary spermatocytes occur, but not necessarily in the same optical field. Primary spermatocytes are characterized by a prolonged prophase, and so they appear more numerous at any given instant and most of them are in prophase. The time span of the secondary spermatocytes is shorter, and so they are not seen in such abundance. Electron microscope studies have revealed that cytoplasmic separation of spermatocytes during meiosis is not complete, and so the daughter cells remain together in clusters or clones.

SPERMATIDS

Spermatids are haploid cells with a spherical nucleus and scanty cytoplasm (Fig. 11.4). Young ones are spherical, and the older ones are ovoid or elliptical. They undergo the series of changes, a process called *spermiogenesis,* which lead to the formation of mature spermatozoa. These changes involve the formation of the highly basophilic and refractile *acrosome* at one end and the *tail piece* at the other end, the consolidation of mitochondria into the middle piece, and the sloughing off of all excess cytoplasm. The sequence of these changes is depicted in Fig. 11.5.

SPERMATOZOA

The mature *spermatozoa,* or *sperm,* occupy a central position in the tubule. Their heads are embedded in the pockets of the Sertoli cells, and their tails lie in parallel arrays in the

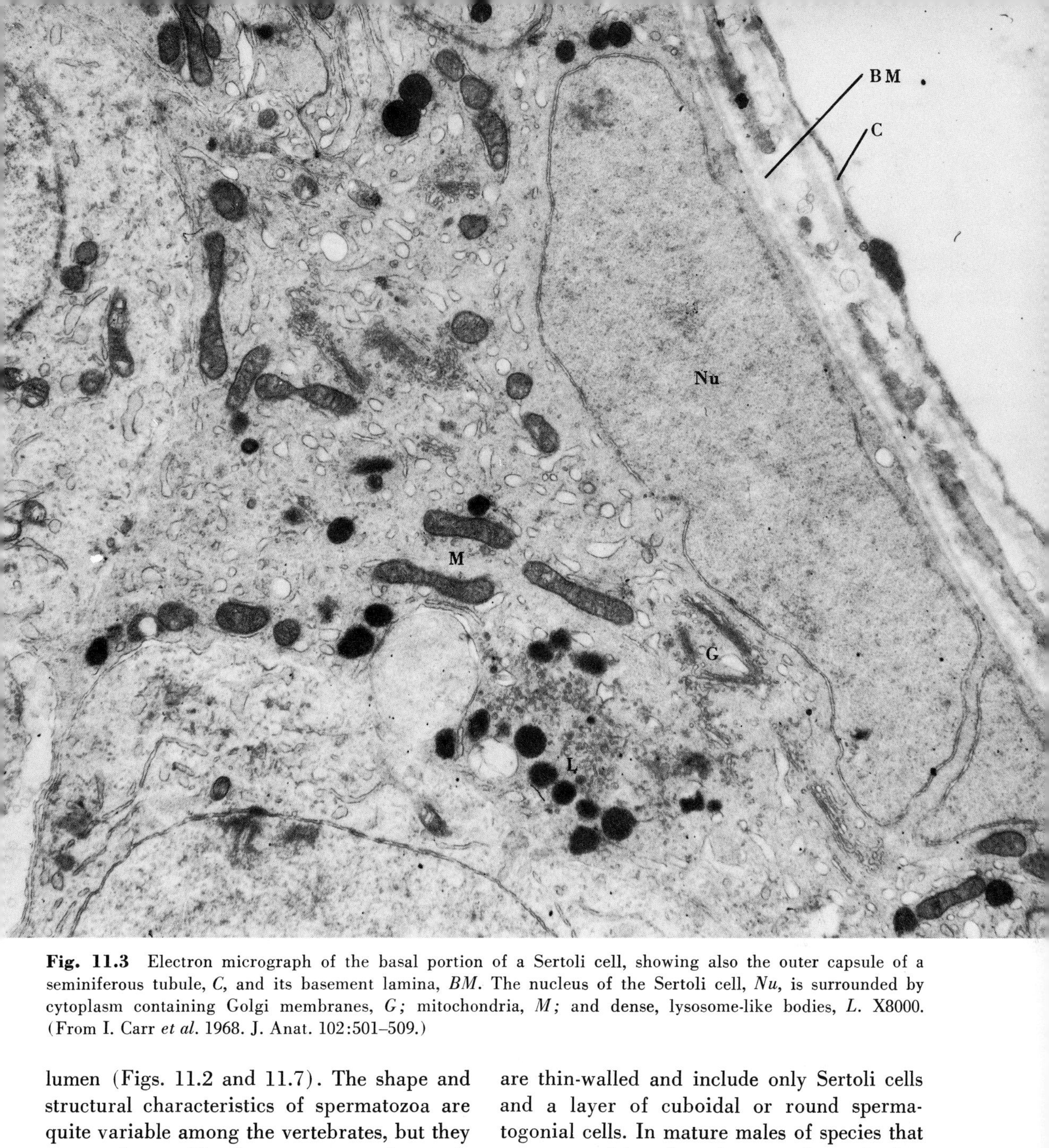

Fig. 11.3 Electron micrograph of the basal portion of a Sertoli cell, showing also the outer capsule of a seminiferous tubule, *C*, and its basement lamina, *BM*. The nucleus of the Sertoli cell, *Nu*, is surrounded by cytoplasm containing Golgi membranes, *G*; mitochondria, *M*; and dense, lysosome-like bodies, *L*. X8000. (From I. Carr *et al.* 1968. J. Anat. 102:501–509.)

lumen (Figs. 11.2 and 11.7). The shape and structural characteristics of spermatozoa are quite variable among the vertebrates, but they are species-specific. An assortment of types is shown in Fig. 11.6. Not infrequently two-headed or two-tailed sperm may be encountered, or sperm which have not successfully cast off all of their excess cytoplasm. The spermatozoa are quiescent while in the testes, becoming motile only during or after emission from the reproductive tracts.

The proportion and distribution of the various cell types is highly variable and merits some comment. In immature testes the tubules are thin-walled and include only Sertoli cells and a layer of cuboidal or round spermatogonial cells. In mature males of species that are seasonal breeders the testes have the immature appearance between times of "rut." In fact, the number of spermatogonial cells is cut back so drastically in some of these species at these times that they are difficult to find. In other species a large number of spermatogonia in a given part of a tubule may differentiate into primary spermatocytes, which in turn will become secondary spermatocytes, all at nearly the same time, and so on. As a result of such synchronous behavior, sections of the

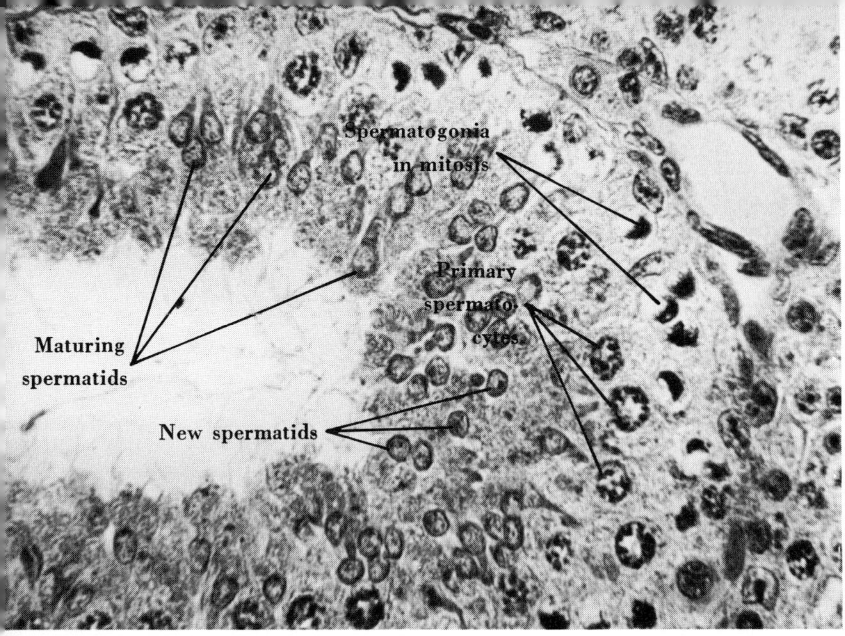

Fig. 11.4 A portion of a seminiferous tubule of a guinea pig. The dense and flocculent material that surrounds the spermatids is Sertoli cell cytoplasm. Note sperm tails in the lumen. X450.

testis will show one tubule in which the predominant cells are primary spermatocytes in leptotene, while next to it is a tubule consisting mainly of secondary spermatocytes, and next to this one that includes great numbers of spermatids, and perhaps next to this another tubule of primary spermatocytes in the diplotene stage of meiotic prophase. This kind of synchrony is common in fishes and amphibians (Fig. 11.7). In teleost fishes the tubules

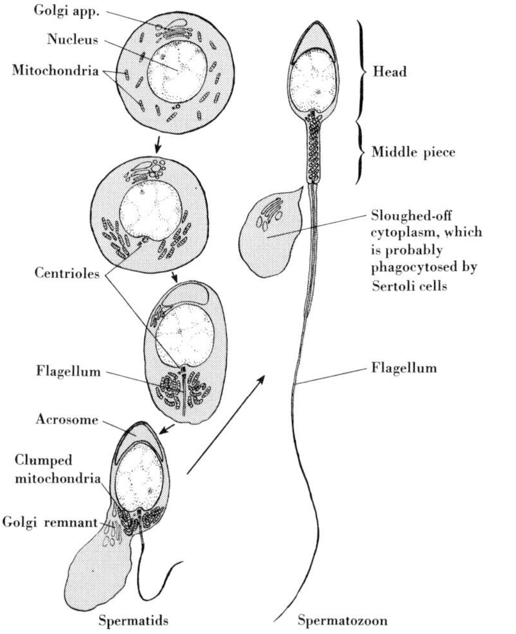

Fig. 11.5 A diagram showing metamorphosis of a spermatid to a mature spermatozoon.

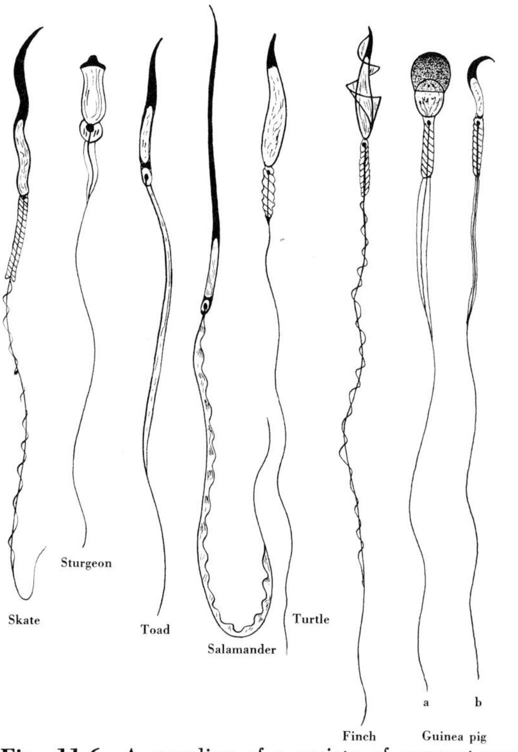

Fig. 11.6 A sampling of a variety of spermatozoa of vertebrates. The guinea pig sperm is shown in both "top" view, *a*, and side view, *b*.

lose their lumen, owing to extensive cell proliferation, and acquire a cystic organization. When the sperm mature, the walls of the cysts burst, releasing the cells into a central space in each lobule. After each breeding season, consequently, extensive regeneration of the

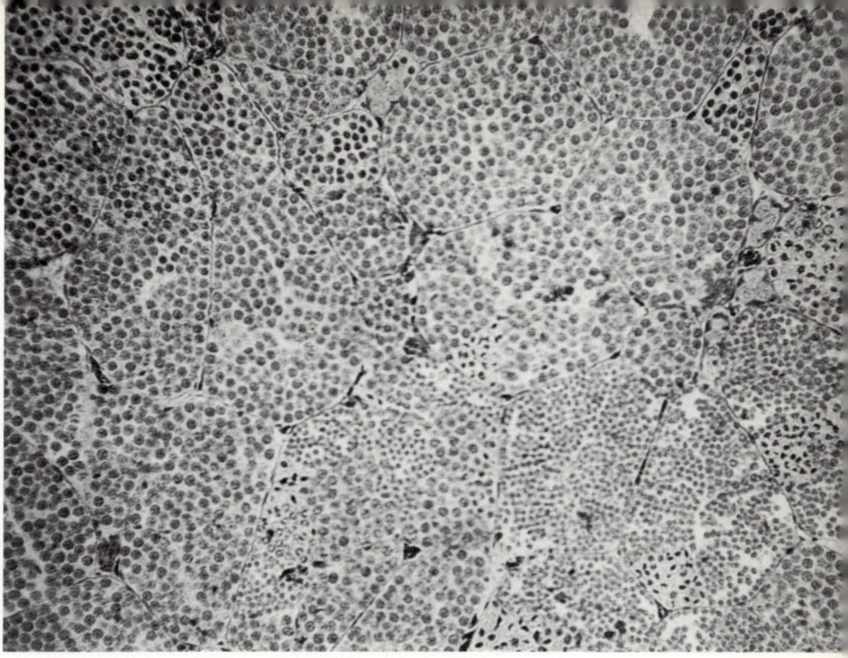

Fig. 11.7 Testis of the urodele *Ambystoma tigrinum*, showing cystic organization. Note that all the cells in any given cyst are in the same stage of meiosis. X100.

testes takes place, except in migratory salmon, in which regeneration fails to occur and death ensues.

In other species, primarily among the mammals, spermatogenic activity progresses in waves initiated at the rete testis, so that serial sections starting at this point will show levels of intense spermatogenic activity alternating through transitional zones with levels of inactivity. In the active zones all stages of spermatogenesis are observable and usually these stages are arranged in ordered layers from the spermatogonia on the basement membrane to the mature spermatozoa in the lumen. In the inactive zones spermatogonia are the only spermatogenic cells. In other mammals there seems to be little regularity between active and inactive areas of spermatogenesis.

The interstitial tissue lying between the seminiferous tubules consists of fibroblasts, blood vessels, lymphatic capillaries, nerve plexuses, and compact clusters of cells called the interstitial cells or *cells of Leydig* (Figs. 11.1, 11.2 and 11.8). The last are specifically implicated in the production of testosterone. They are rather large cells, somewhat irregular in shape, and they have a round nucleus containing one or two dense nucleoli. In some species binucleate cells of Leydig may be found. The cytoplasm is rich in mitochondria, Golgi substance, and granules. Some of the granules are lipid and are dissolved out by standard histological procedures, which produces a vacuolated appearance in the peripheral cytoplasm. Other granules are pigment, largely lipofuscin. In addition, numerous rod-shaped crystalloids of an albuminous substance are found in some species, notably human beings. The number of cells of Leydig is influenced strongly by the interstitial cell stimulating hormone (ICSH) of the anterior pituitary, and also by levels of cortical steroids.

The propulsion of sperm cells and the detritus which represents the fragments of sloughed-off cytoplasm is effected in three ways: (1) cell pressure (i.e., one cell pushing against another as a consequence of growth and multiplication); (2) transfer of fluid from the interstitial tissues into the tubules, thereby creating a testicular effluent which carries the sperm along; and (3) contraction of smooth muscle in the testicular capsule. The last is more prominent and probably more significant in the lower vertebrates. The testicular effluent is reabsorbed by cells of the rete testis and epididymis, thereby effecting a kind of intratesticular circulation.

The rete testis is formed of anastomosing tubules and spaces the lining cells of which are continuous with and probably represent modifications of the Sertoli cells—except that they have a regular columnar or cuboidal shape, often are flagellated, and the nuclei are more basophilic. The transition from semi-

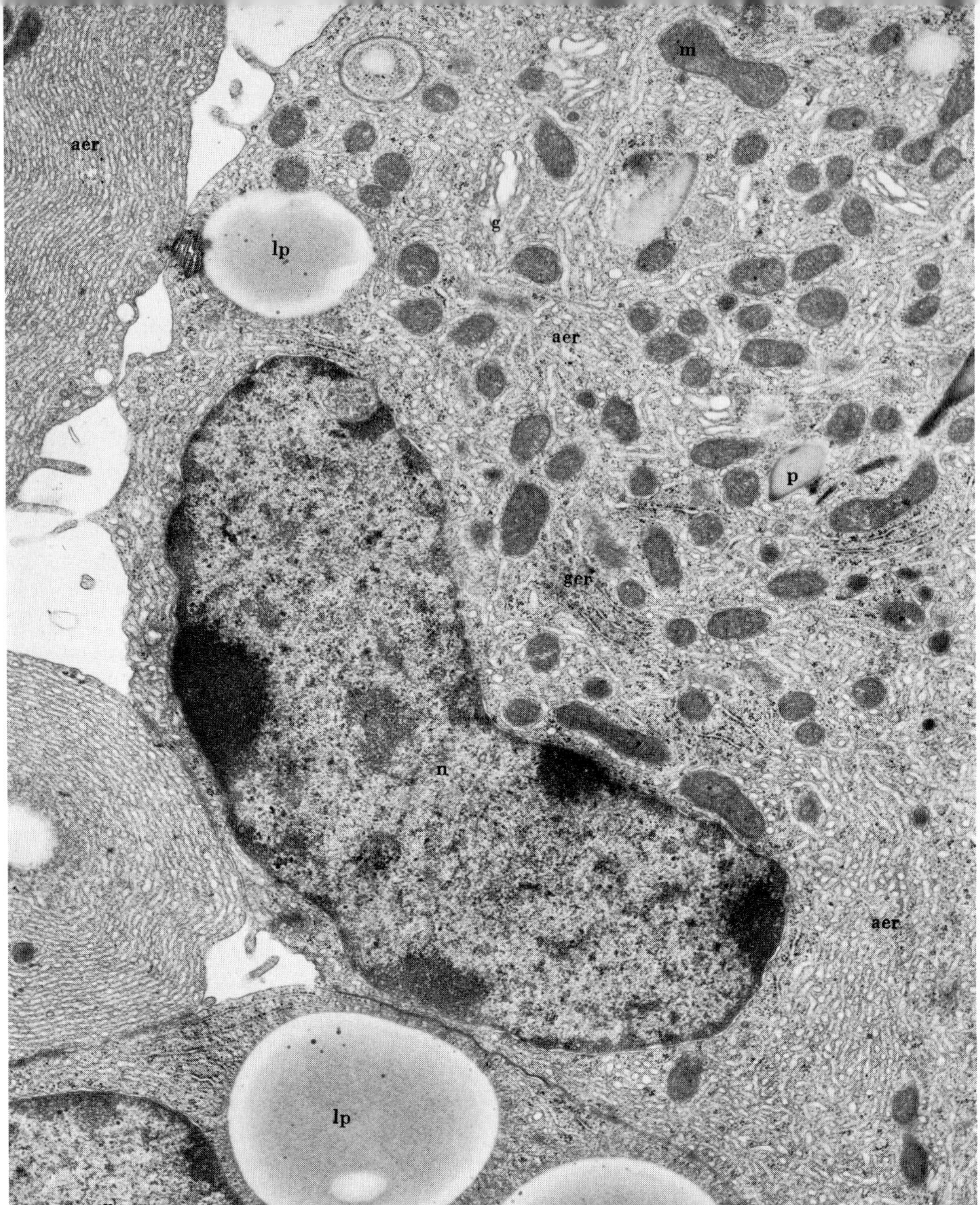

Fig. 11.8 Interstitial cells in the testis of the guinea pig, showing nuclei, *n;* smooth-surfaced ER, *aer;* rough-surfaced ER, *ger;* Golgi vesicles, *g;* mitochondria, *m;* lipid droplets, *lp;* and lipofuscin pigment, *p.* X14,000. (From A. K. Christensen. 1965. J. Cell Biol. 26:911–935.)

niferous tubule to rete is typically abrupt. In elasmobranchs and bony fishes the rete tubules are invested in a fibroelastic sheath which includes some smooth muscle elements, and so they are able to provide some propulsive force to the semen. In higher forms no contractile properties have been described, and in many the epithelium is simple squamous.

The Reproductive Tracts

In all vertebrates except the Agnatha and most teleosts, the male genital ducts are derivatives of the urinary system. Both cyclostomes and the Salmonidae are lacking in genital ducts. Many other teleosts have developed sperm ducts independent of the urinary system. With respect to gnathostome fishes in general, the variations in genital ducts from species to species are too extensive to fall within the scope of this book. Nevertheless, the affinities of the male reproductive ducts to the urinary ducts are very clear both in those fishes in which such affinities do exist and in the amphibians. Even in adults one may see certain tubules of the kidney, most commonly those at the cranial end, connecting the testis directly with the Wolffian duct. Called the vasa efferentia, these tubules are formed of a simple cuboidal epithelium supported by encircling connective tissue which may contain scattered smooth muscle cells. In the fishes the *vasa efferentia* come together to form larger, twisted ducts, the epididymal tubules, which are closely adherent to the ventral surface of the kidney. These tubules are lined with a simple ciliated columnar epithelium which is thrown into longitudinal folds, like those of the Wolffian duct, which they join. Numerous goblet cells occur among the epithelial cells. In amphibians the number of epididymal tubules is reduced, but they have the same essential characteristics as those of fishes.

The vasa efferentia of amniotes are lined with ciliated cuboidal and/or low columnar cells. In mammals the columnar cells often are not ciliated and the distribution of cuboidal and columnar types is such as to impart an undulant or folded appearance to the inner wall (Fig. 11.9). The epithelial cells are glandular, as indicated by their granular and acidophilic cytoplasm. The basement membrane is well defined, and smooth muscle cells are scattered in the supporting connective tissue. In mammals the epithelium shows a transition to pseudostratified columnar as it approaches the epididymis.

In amniotes the tortuous, anastomosing epididymal tubules are closely bound together by a collagenous network into a discrete body, the *epididymis*. In reptiles the epididymis may be as large as or larger than the testis; in birds it is relatively small; and in mammals, although varying considerably in size relative to the testis, it is always prominent. The epithelium of the epididymal tubules in reptiles consists of tall columnar cells with basal nuclei and cytoplasm filled with acidophilic granules. In birds the epithelium is ciliated low columnar. In mammals the epididymis consists of a single, tortuous duct the epithelium of which is pseudostratified, consisting of slender columnar cells and rounded basal cells, both of which are in contact with the basement membrane. The columnar cells bear nonmotile processes, *stereocilia,* on their luminal surface (Fig. 11.10), their nuclei are elongate, and their cytoplasm includes droplets and granules of various kinds. The connective tissue surrounding the epididymal duct contains a small amount of circular smooth muscle, which increases distally and is joined by some longitudinal bundles near the *vas deferens*.

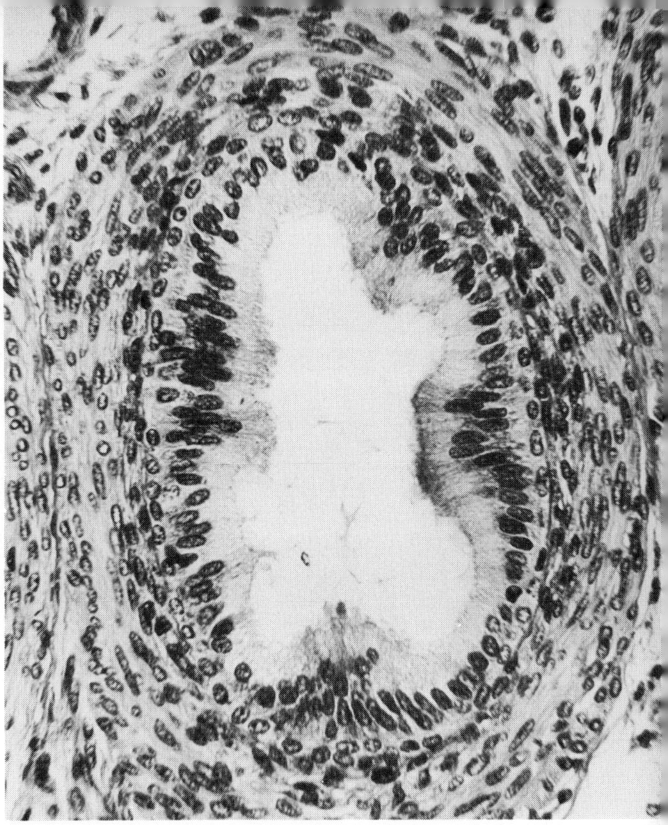

Fig. 11.9 Cross section of a vas efferens in the testis of the guinea pig. The epithelium lining this specimen varies from low to tall columnar and includes some pseudostratified areas. Note the circumferential layer of smooth muscle cells. X500.

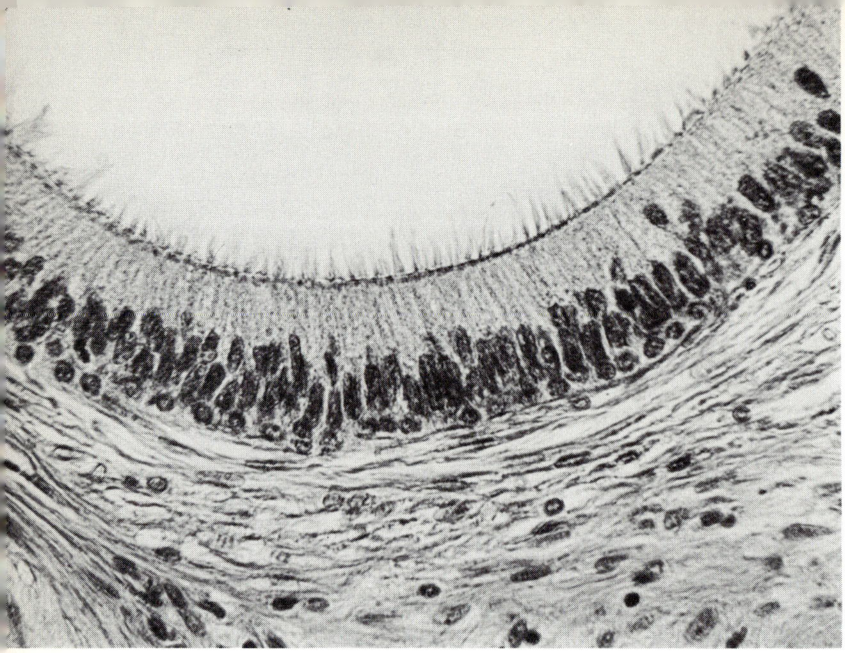

Fig. 11.10 Part of the vas deferens in the region of the epididymis. Note the pseudostratified columnar epithelium, the stereocilia, and the subepithelial muscle and connective tissue. X500.

The main genital ducts in male vertebrates are the Wolffian ducts, except in those teleosts which have developed a sperm duct independently of the urinary system, and in the cyclostomes and members of the teleost family Salmonidae (such as salmon and trout) that release their sperm directly into the body cavity from whence it emerges via genital pores in the posterior body wall. In most anamniotes the Wolffian ducts, described in Chapter 10, may be considered to be primarily excretory in function. In the amniotes, however, the ureters have taken over the urinary function and the Wolffian ducts are left with only their genital function; thus they are known as the sperm ducts per se, or, technically, as the *vasa deferentia* (sing., *vas deferens*).

The vas deferens of reptiles and birds is lined with columnar cells, usually ciliated, supported by a thin basement membrane and elastic lamina propria. Smooth muscle fibers, both circularly and longitudinally arranged, lie in a fibroelastic connective tissue covering which is continuous with that of the ureter. In many species blood vessels supplying the cloaca are also included within the same connective tissue bundle. Fusion of the distal ends of the vas deferens and the ureter may occur just before they emerge into the cloaca through slitlike openings. The vas deferens of mammals is similar except that longitudinal folds of the mucosa are more pronounced, the epithelium is pseudostratified with some stereocilia, and the musculature is better developed, consisting of a thin inner longitudinal layer, a rather thick middle circular layer, and an outer longitudinal layer. A few bundles of smooth muscle may also be encountered in the adventitia. There is no distinct submucosa in the vas deferens of any of the amniotes (Fig. 11.11).

The testes of Proboscidea and of the large marine mammals, such as Cetacea and Sirenia, are permanently located within the body cavity. The testes of most insectivores, bats, most marsupials, and rodents are also located within the body cavity, but during the breeding seasons they lie in the scrotal sacs. In the other orders of placental mammals (notably the primates, ungulates, and most of the carnivores) the testes descend into the scrotal sacs early in development and remain there throughout life. While descending into the scrotal sacs, each testis is accompanied by four conspicuous structures: the *spermatic artery*, the *pampiniform plexus* of veins, nerves of the *spermatic plexus*, and the vas deferens. The four are ensheathed together in loose connective tissue in which some longitudinal smooth muscle bundles may be observed, forming in this way a sort of conduit, the *spermatic cord* (Fig. 11.12). The spermatic cord extends from the testis to the point at which the vas deferens crosses the ureter and swings caudad to enter the urethra.

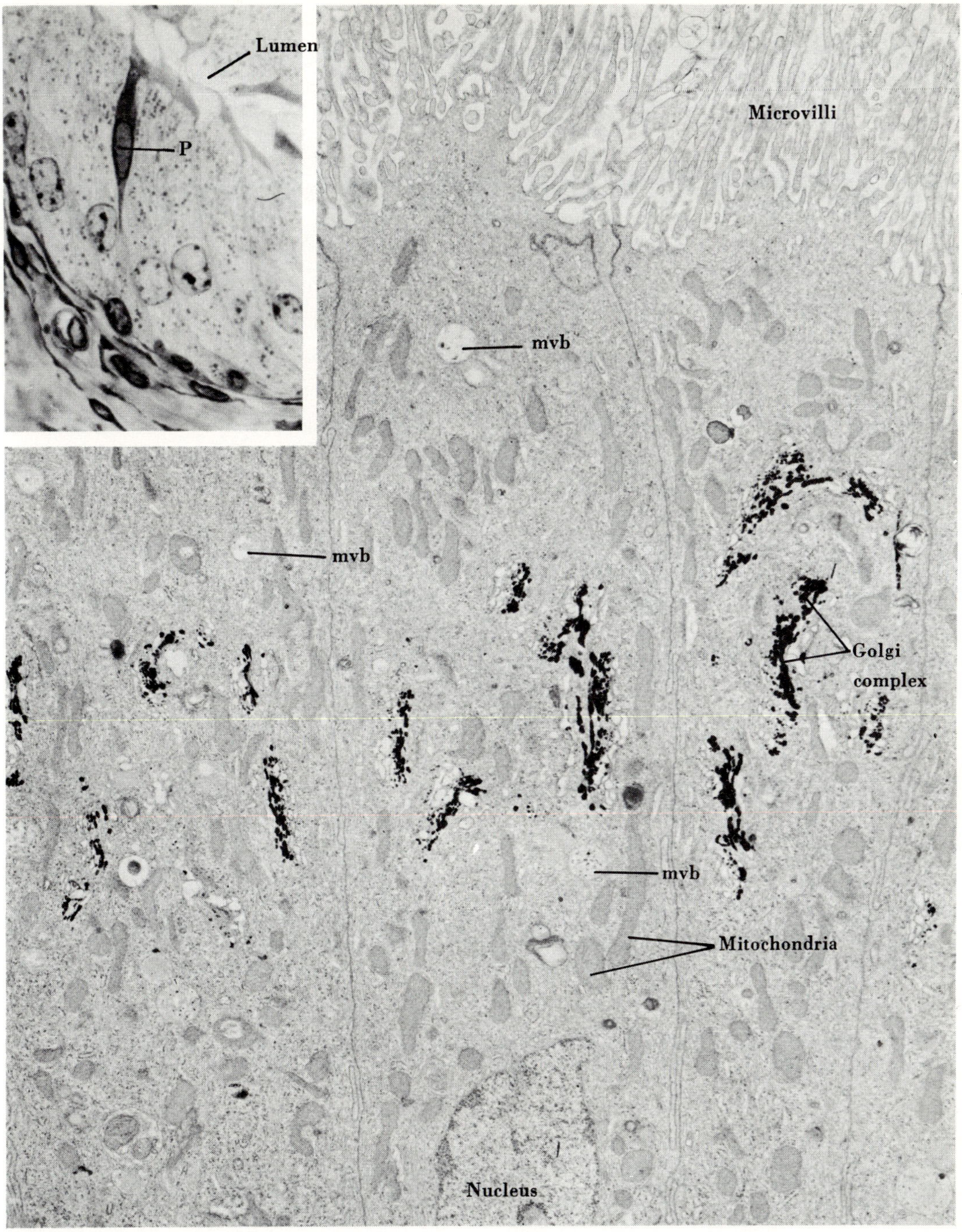

Fig. 11.11 Low-power electron micrograph of several epithelial cells in the vas deferens of a rat, showing microvilli on the luminal surface. The cisternae of the Golgi complex are filled with dense deposits of lead phosphate which were introduced by the preparative procedures. Multivesicular bodies, *mvb*, are characteristic of the cytoplasm of these cells, presumably functioning in protein absorption. X800. Inset: photomicrograph of vas deferens epithelium, showing the tall columnar cells with long microvilli (stereocilia?) that project into the narrow lumen. Occasionally a deeply stained "pencil" cell, *p*, is encountered. The lower layer of cells includes principally smooth muscle cells and fibroblasts. X700. (Both photomicrographs from D. S. Friend and M. G. Farquhar. 1967. J. Cell Biol. 35:357–376.)

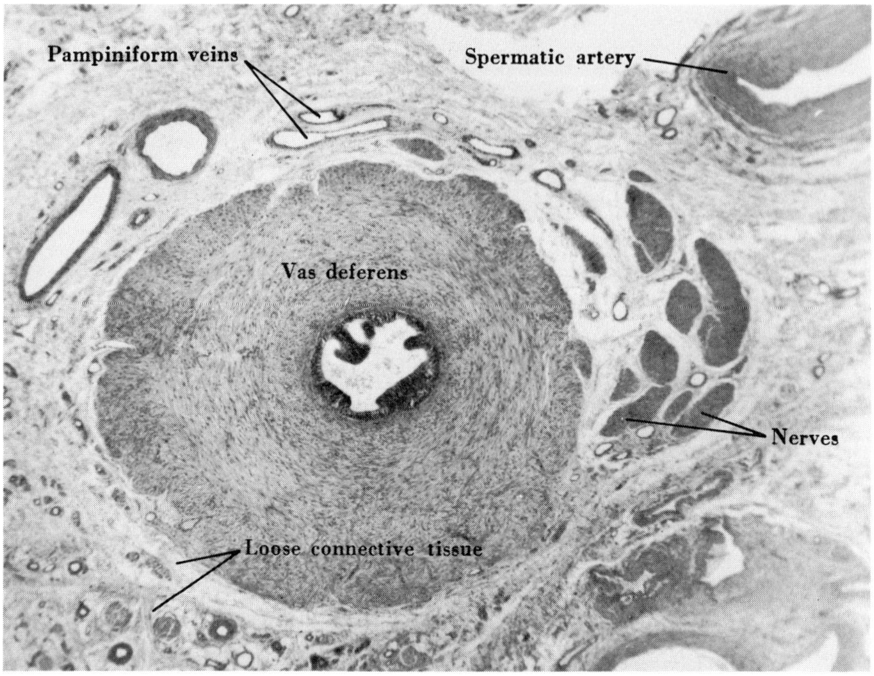

Fig. 11.12 Spermatic cord of a mammal. The smooth muscle wall of the vas deferens is characteristically very thick, as here. Arteries and blood vessels are supported within a matrix of loose connective tissue. X30.

Just before joining the urethra, the vas deferens of certain mammals (such as primates, some carnivores, and most rodents) becomes dilated to form the spindle-shaped *ampulla,* the convoluted epithelial lining of which is often rich with mucus-secreting cells. The short, constricted section of vas deferens between the ampulla and the urethra is known as the *ejaculatory duct.*

With the exception of the carnivores, placental mammals have an elongate, saccular gland, the seminal vesicle, which opens into the ejaculatory duct. The epithelial lining of the seminal vesicle is so extensively folded and convoluted that in transverse section it appears like a sponge (Fig. 11.13). The epithelium is pseudostratified or stratified columnar, and the apical cells frequently are characterized by a single flagellum. Numerous acidophilic granules are found in the epithelial cells, as well as varying amounts of globules of fatty, yellow pigment. The secretion is thick and viscous and contains fructose in some species, and so presumably it helps nourish the spermatozoa.

INTROMITTENT ORGANS

Intromittent organs are organs which facilitate the transfer of spermatozoa from the male reproductive tracts to the cloaca or lower reproductive tracts of the female. In the Chondrostei they are known as *claspers,* and generally represent modifications of the medial portions of the pelvic fins of the male. Each clasper has a medial longitudinal groove, so designed that when the two claspers are approximated during copulation they form a tube extending from the cloaca to the distal ends of the claspers. In the abdominal wall of sharks there is a muscular sac, the *siphon,* which communicates with the proximal portion of the claspers. During copulation, the siphon fills with water, the claspers come together, and the tube they form fills with spermatozoa. The claspers are then inserted into the cloacal aperture of the female and in the process they are bent in such a way that the proximal part lies opposite the siphon. The siphon then contracts, forcing the spermatozoa into the female. In skates and

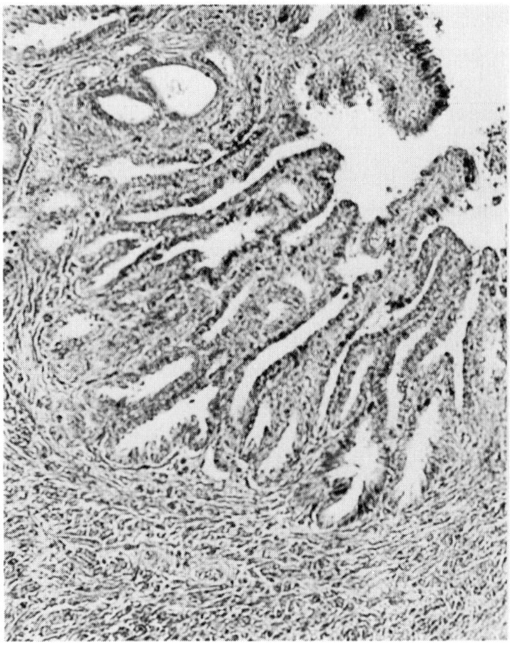

Fig. 11.13 Transverse section of a human seminal vesicle. X100.

rays no siphon is present, but in its place is glandular tissue (the pterygopodial glands), the function of which is not known.

In some teleosts an intromittent organ, the *gonopodium,* appears as a modification of the anterior border of the anal fin. In others the hemal spines of certain caudal vertebrae may serve as intromittent organs.

Although amphibians lack specific intromittent organs, two interesting adaptations deserve mention. In some urodeles such as *Triturus,* the spermatozoa are packaged in mucus envelopes, the *spermatophores,* which are deposited on the bottom of ponds or streams or even on land in some species, and are subsequently picked up by the cloacal lips of the female to effect internal fertilization. Male Apoda are able to protrude their cloaca, fitting it into the cloacal aperture of the female during copulation.

Two kinds of intromittent organs are found in reptiles and birds. Snakes and lizards (excepting *Sphenodon,* which has no copulatory organs) have paired, evertible structures, called *hemipenes,* located one on either side of the cloacal aperture. The epithelium of these organs has been described as being arranged in "pleats, calyces, and flounces" and equipped with sharp, sloping spines or hooks. Although erectile tissue is not usually present in hemipenes, an exception has been noted in pythons.

Crocodilians, chelonians, ostriches, and members of the Anseriformes (such as ducks and geese) have an unpaired, spatulate *phallus.* It is formed by the fusion of parallel ridges in the ventral wall of the cloaca and it incorporates a special kind of vascularized tissue, *cavernous* or *erectile tissue,* as described below. During sexual excitement, the phallus becomes engorged with blood, it extends, and the sides curl dorsally toward each other so that the organ forms a more or less closed tube for the conduction of the semen. In females the organ is present in reduced form as the *clitoris.* Vestiges of such organs can be found in other species of birds. There is little doubt that the reptilian phallus is the prototype or progenitor of the mammalian penis, with which it is homologous to a marked degree.

The intromittent organ of mammals is a closed, erectile tube, the *penis.* In monotremes it lies in the ventral wall of the cloaca, as in reptiles, but it differs from the reptilian phallus in being a closed tube the canal of which is surrounded by a mass of erectile tissue, the *corpus spongiosum,* in addition to having the two masses of cavernous tissue already encountered in reptiles. Unlike that of marsupials and placental mammals, the penile canal of monotremes probably does not conduct urine, for the urinary bladder connects directly with the cloaca farther anteriorly by a short urethral duct.

In marsupials and placental mammals the penis embraces the urethra, which conveys both urine and semen. It includes what may be considered to be three bodies of erectile tissue. One of these, the corpus spongiosum (*corpus cavernosum urethrae*), as in the monotremes, completely surrounds the penile urethra. It consists of a compact mass of venous sinuses of rather uniform size, separated by networks of fibrous and elastic tissue. It is surrounded by a fibrous membrane, the tunica albuginea, which includes circularly arranged smooth muscle fibers (Fig. 11.14).

Lying above the corpus spongiosum are the more massive, paired *corpora cavernosae* (*corpora cavernosae penis*). They consist also of anastomosing venous sinuses, which progress from small, slitlike cavities peripherally to large cavernous spaces toward the center. The tunica albuginea surrounding the paired corpora is very thick and is composed of an outer layer of longitudinally arranged and an inner layer of circularly arranged collagenous fibers, in addition to networks of elastic fibers. The tunica albuginea also forms a septum which partially separates the two halves from each other, as well as trabeculae which separate the cavernous spaces from each other. In addition to collagenous and elastic fibers, the trabeculae contain smooth muscle cells surrounding the sinuses. Blood enters the three sets of erectile tissue by way of the trabeculae through looped arteries, the *helicine arteries,* which are in a state of partial contraction (tone) except during sexual excitement. Sexual stimulation produces some interesting hydrodynamic changes. It causes the arterial smooth muscle to relax, resulting in dilation and greatly increased blood input. The more centrally located venous sinuses become engorged with blood to the extent that they compress the peripherally located draining veins, thereby facilitating a pronounced buildup of arterial pressure within the sinuses. Because of the restriction imposed by the tough tunica albuginea, the cavernous tissue becomes extremely hard and the penis becomes extended or, in the case of primates, erect. After the sexual climax (orgasm) the helicine arteries constrict. The intima of many of these vessels has cushionlike thickenings which, on vasoconstriction, occlude the lumen, thereby virtually shutting off the input into the lacunae. Constriction of the trabecular smooth muscle surrounding the sinuses expresses the blood slowly from the sinuses, and the organ becomes flaccid.

The distal end of the corpus spongiosum extends beyond the ends of the corpora cavernosae, and expands to form the *glans penis* (glans means "acorn," which describes the shape of the end of the human penis). The glans is not surrounded by tunica albuginea, however, but is in direct contact with the integumental covering. The glans varies in form among the different mammalian species, but its histologic appearance is quite uniform. Its epidermis is thin, richly vascularized, and devoid of glands, pigment, or hair follicles. Because the spongy tissue of the glans is not enclosed in the tunica albuginea, it swells considerably during sexual excitement.

Elsewhere the penis is covered by a rather thick layer of loose connective tissue in which may be found smooth muscle bundles, lymphatic vessels, blood vessels, and nerves, but a paucity of fatty tissue. The connective tissue is surmounted by a thin integumentary layer with tall dermal papillae, which is reflected over the glans as the *prepuce.* In most species of mammals the skin is continuous dorsally with the skin of the abdomen and is of rather similar character; however, in the higher primates, which have a pendulous penis, the skin does form a cylindrical covering and the prepuce is therefore more readily delineated. Except for the undersurface of the prepuce, the skin is characterized by hair follicles, pigment, and, according to species, occasional small sweat glands. The undersurface of the prepuce is hairless, but it has modified sebaceous glands, the *preputial glands* (*glands of Tyson*) which secrete a waxy substance called *smegma.*

The skin of the glans and of the prepuce is liberally supplied with free sensory nerve endings and, in deeper areas, with encapsulated corpuscles of various types, such as Meissner's corpuscles, "genital" corpuscles, and corpuscles of Vater-Pacini (see Chapter 12).

Associated Sex Glands

In addition to the secretions of the lining epithelium of the genital ducts, of the seminal vesicles, and of the glands of Littré, already described, there are those of two major glands in mammals which are primarily concerned with the ejaculation of seminal fluid. These are the *prostate gland* and the *bulbourethral* (*Cowper's*) *glands.*

THE PROSTATE GLAND

The prostate gland is an aggregation of branched tubuloalveolar glands surrounding the prostatic portion of the urethra into which it opens by numerous ducts. It is frequently lobulated. It is encapsulated by a fibroelastic membrane, richly vascularized, in the inner layers of which may be found autonomic ganglia and bundles of smooth muscle fibers. The gland has a considerable stroma formed by trabecular penetrations of the capsule, which also include collagenous and elastic fibers, numerous blood vessels, lymphatic clefts and vessels, myelinated and nonmyelinated nerves, and smooth muscle cells scattered in various directions.

The epithelium of the alveoli is mainly simple cuboidal or low columnar, although basal cells may be present as in pseudostratified epithelium (Fig. 11.15), and it tends to be folded. No basement membrane is present, but rather the epithelium is supported directly by the connective tissue of the stroma. Secretion granules and lipid droplets are abundant in the apical cytoplasm and in blebs protruding into the lumen. The tubules are lined with simple cuboidal epithelium, which changes to simple columnar in the ducts, merging in turn with transitional epithelium near the junction of the latter with the urethra.

In some alveoli one may find prominent, spherical bodies, the corpora amylacea, which are formed of concentric layers of an albuminous protein and nucleic acids. Normally they are discharged with the ejaculum, in which they appear as soft, yellowish globules. As sexual activity decreases with age, the corpora amylacea increase in number and size and may become calcified.

The utriculus prostaticus is a small sac or system of tubules of folded mucous membrane found within the prostate gland. Its epithelium is simple or pseudostratified columnar. Although a remnant of the fetal Müllerian duct, it is believed to be functional, contributing to the prostatic secretion.

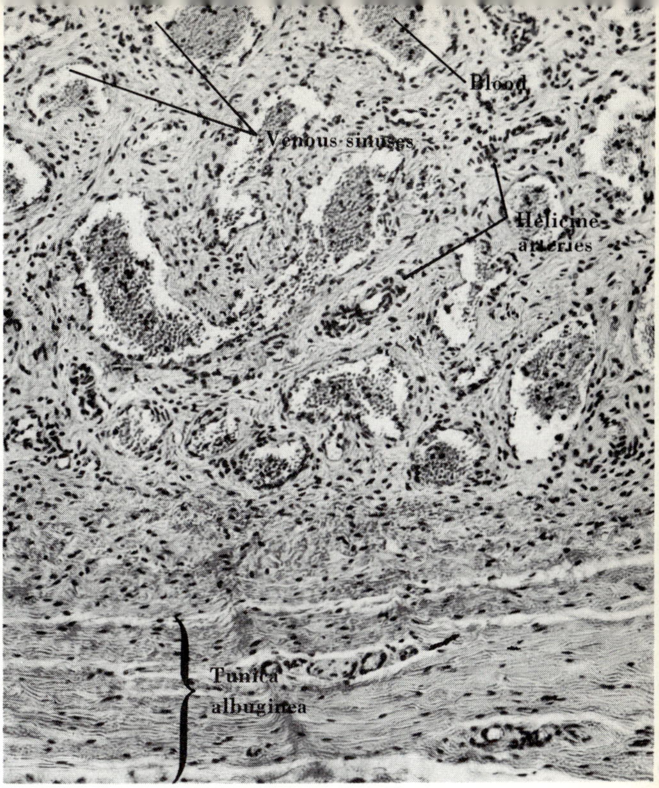

Fig. 11.14 Corpus spongiosum of a human penis. The numerous venous sinuses are partly filled with blood. X150.

The prostate gland is present only in marsupials and placental mammals, although it is absent in edentates and cetaceans. Its secretion, which is slightly alkaline, includes appreciable quantities of citric acid, acid phosphatase, eosinophilic lipid bodies, and fructose. The prostate gland contributes most of the bulk of the seminal fluid in most species, including man. It is forcibly expressed during ejaculation, carrying with it the sperm which have been simultaneously forced out of the epididymis by strong contractions of the smooth muscle in its walls. The prostatic fluid dilutes the semen, facilitating and augmenting sperm motility. In some mammals, such as rodents, part of the prostatic secretion coagulates on mixing with the secretions of the seminal vesicles and forms the copulation plug characteristically found in the vagina of the female after mating. This plug is believed to aid in the retention of sperm in the female reproductive tract and thus to enhance the probability of fertilization. The prostate of the opossum, *Didelphis virginiana,* shows an interesting zonation into three morphologically and physiologically distinct segments (Fig.

11.16). The anterior segment is apocrine and secretes several enzymes (as acid phosphatase and aliesterase).

THE BULBOURETHRAL GLANDS

The bulbourethral or *Cowper's glands* are paired, compound tubuloalveolar glands with ducts and tubules of varying diameters. Each gland is invested in a thin, fibrous capsule, from which septa protrude, dividing the gland into lobules. The septa contain both skeletal and smooth muscle fibers, the former predominating. The glandular epithelium is simple squamous in the distended alveoli, and elsewhere it varies between simple cuboidal and simple columnar. The columnar cells are mostly lightly staining, mucin-secreting cells with basal nuclei. The other cells are mostly acidophilic and often contain fusiform, acidophilic inclusion bodies and droplets of "colloid." The epithelium is supported by a definitive basement membrane. The main ducts are lined with simple columnar epithelium, which become stratified columnar near their junction with the lumen of the cavernous portion of the urethra. Some of the lining cells are secretory. The main ducts are surrounded by circular smooth muscle fibers.

The secretion of the bulbourethral glands is clear, viscid, and mucilaginous. Its slight alkalinity serves to render the pH of the urethra and of the vagina more favorable for sperm survival, as well as to aid in lubrication. Secretion accompanies sexual arousal, and so it is preparatory to copulation and ejaculation.

THE FEMALE REPRODUCTIVE SYSTEM

The detailed anatomy of the female reproductive system reflects faithfully the mechanisms by which the reproductive process assures survival of the species. Although it is not the purpose of this text to classify and catalogue the various mechanisms employed by the vertebrates, a sampling of some of them supplemented by well-known examples will be given.

Doubtless the most straightforward and simplest mechanism is predicated on probability theory. If the probability that a given unprotected ovum cast into the sea will success-

Fig. 11.15 Photomicrograph of the outer part of the prostate of a squirrel (*Sciurus carolinensis*), showing the capsule (*top*) and the glandular alveoli. X150.

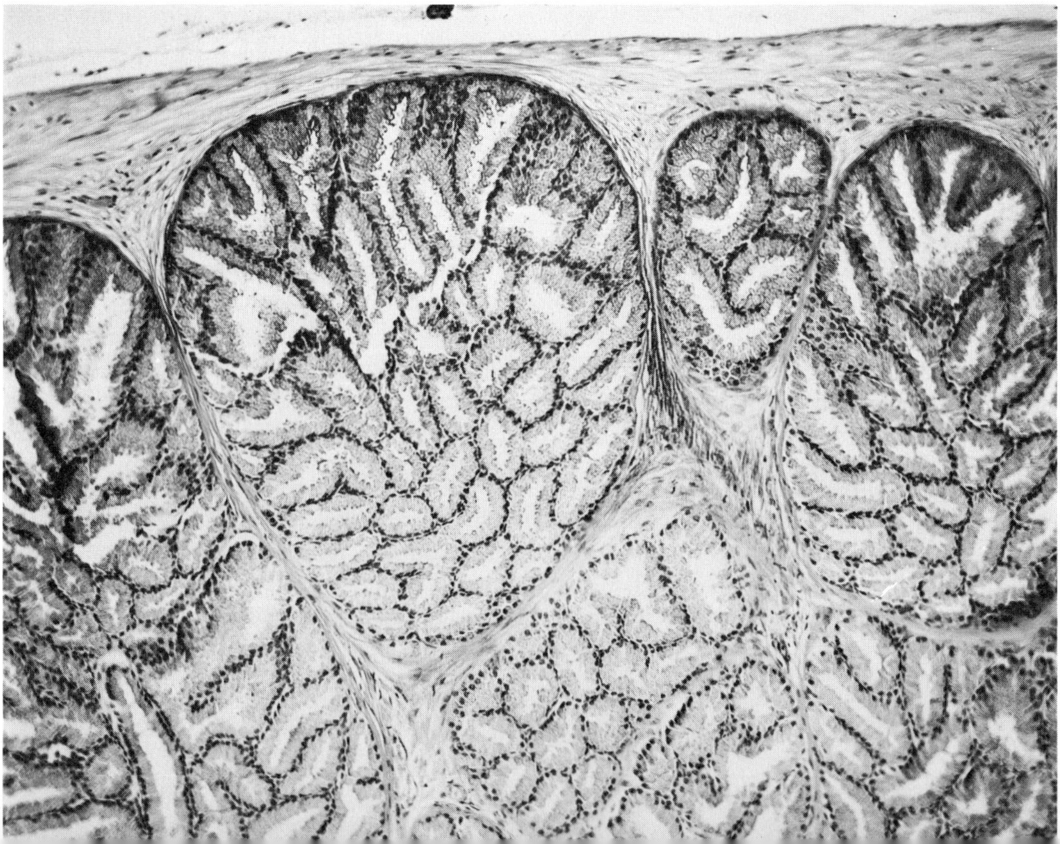

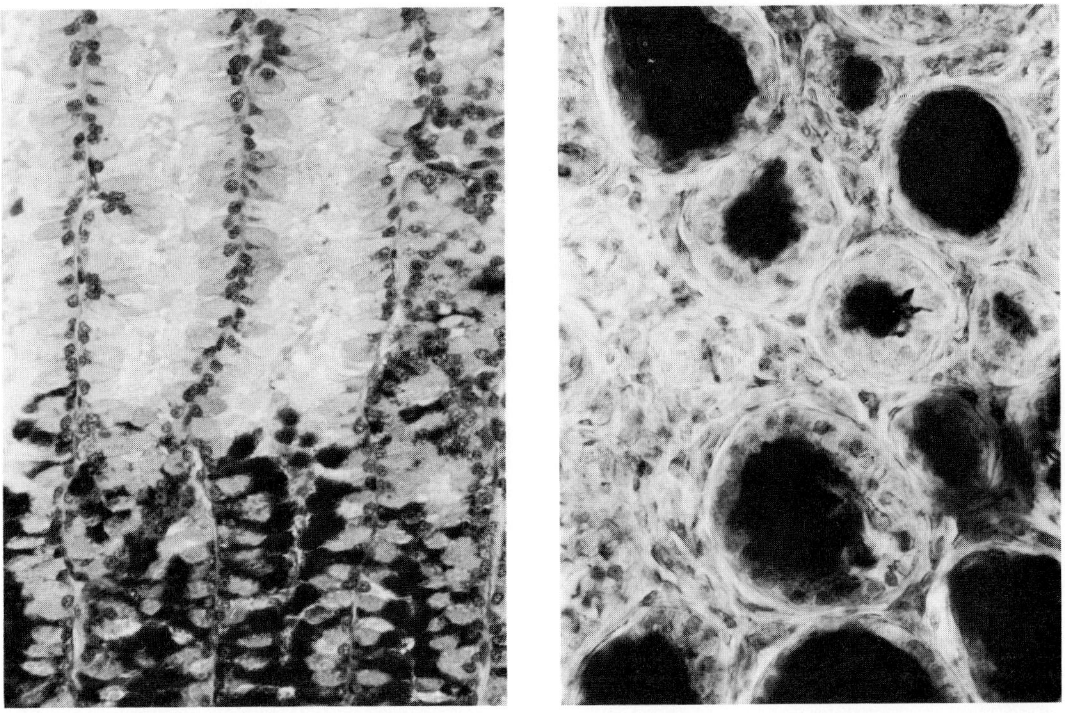

Fig. 11.16 The prostate of the opossum *Didelphis virginiana.* Section at the transition between outer and middle zones of one segment, longitudinal to the secretory tubules, *a.* Stained by the PAS reaction. Section of another segment, transverse to the secretory tubules, *b.* Stained by the acid-phosphatase reaction. Both sections X300. (From A. Hruban *et al.* 1965. J. Exp. Zool. 160:81–106.)

fully complete development and survive to maturity is 1×10^{-6}, it follows that a minimum of 1,000,000 eggs must be so cast in order to assure *minimum* survival for one more generation. Twice this number is required to provide for a representative of each sex to make survival for additional generations possible. Moreover, the probability of an encounter between a male and a female during the breeding season must be considered, as well as the probability that both individuals will be fully fertile. In view of only these simple conditions, it is therefore not surprising to learn that a fully grown ocean sunfish, *Mola mola,* may shed 28,000,000 eggs in one breeding season, and thereby just manage to contribute its share to the maintenance of a fairly stable population of sunfish from generation to generation. The test of the feasibility of such a mechanism is not its economy, but whether it works. It works.

Nevertheless, more economical methods have been devised, but their workability depends on ways of increasing survival probability. One way is to make the eggs unpalatable to predators, another is to provide the eggs with a protective covering, or to deposit them in large coherent masses. Eggs that float will be beyond the range of some predators. Eggs that sink will be beyond the range of others. Eggs buried in the sand or mud have a better chance of survival, assuming they are adapted to such environments. Eggs that stick to things along a river bank or bed are less likely to be washed out to sea. Eggs containing large amounts of yolk enable the young to foreshorten or bypass a susceptible larval stage and arrive more quickly at a self-sufficient stage of development. All of these mechanisms have been exploited by fishes, and some of them by amphibians, reptiles, and birds.

Many vertebrates provide direct protection to their eggs and embryos, and in a variety of ways. For example, the sticklebacks (*Gasterosteus* and *Eucalia*) build nests for their eggs and guard them and care for the fry for a time after they have hatched. The male seahorse (*Hippocampus*) and pipefish (*Syngnathus*) house the fertilized eggs in a brood pouch on the ventral body wall. Many catfishes (e.g., *Loricaria* and *Galeichthys*) carry their eggs and fry in the buccal cavity. The male Chilean frog (*Rhinoderma*) carries the eggs in its vocal sacs. The male surinam toad (*Pipa*) spreads the fertilized eggs on the female's back; the eggs sink into small pits in the skin, where they remain until metamorphosis. The female marsupial frog (*Gastrotheca*) has a permanent brood pouch on her back.

The most generally efficacious way of increasing survival probability is viviparity, that is, retaining the fertilized eggs in the mother's reproductive tract for a certain critical period of time. Examples of this are found in all classes of vertebrates except the Agnatha and the Aves. In some species the egg is supplied with sufficient stores of yolk to nourish it during gestation (ovoviviparity) as in many elasmobranchs, some teleosts (e.g., *Lebistes*), some Apoda (*Typhlonectes*) and salamanders (*Salamandra atra* and *S. maculosa*), and the ovoviviparous reptiles, of which there are many. (In the live-bearing teleosts, embryonic development takes place not in the genital tract, however, but in the ovary itself.) In other species the developing egg depends upon the mother for some nourishment and respiratory exchange, as evidenced by specialized structures for these purposes (e.g., *Mustelus* and *Sphyrna* among the sharks). True viviparity, wherein the egg is virtually devoid of yolk and almost entirely dependent upon direct maternal sources for nourishment and respiration is first successfully achieved by the marsupials, whose uterine glands secrete a nourishing "milk," and perfected by placental mammals, as is well known.

As a general rule, the more effectively the eggs are protected, the fewer eggs are produced. But egg number may also be correlated to some extent with the characteristic size of the egg, the correlation being an inverse one. On the other hand, in mammals, whose eggs are uniformly microscopic in size and relatively yolk-free, the number of eggs has in general an inverse relation to the adult size characteristic of individuals of the species; i.e., large mammals (cetaceans, elephants, and ungulates) typically have single births, whereas rodents and insectivores have quite sizeable litters of 10 or more. To be sure, such generalizations are influenced by other factors, such as length of postnatal care, of which the human species is an extreme example.

According to the conditions outlined above, the ovaries vary in size, in the number of active follicles, in the role played by the follicle or "nurse" cells, and in the relative size and duration of corpora lutea. Similarly, the oviducts vary in respect to the number, location, and nature of their glands, musculature, and regional specializations of other kinds, as uterus, vagina, and seminal receptacles.

The Ovary

The vertebrate ovary, as pointed out above, consists of a medulla and a cortex invested in an epithelium which is continuous with the mesothelium of the suspending mesentery or ligament. The medulla appears as a richly vascularized stroma of collagenous and reticular fibers and fusiform fibroblasts, with nervous and lymphatic elements also represented. In lower forms myeloid tissue is frequently also a component. Elastic fibers are not common in most species, and smooth muscle is usually confined to the walls of the blood vessels, although bundles of smooth muscle may be found in the medullary stroma adjacent to the point of attachment of the mesovarium in many species. In the ovaries of frogs, smooth muscle is included in the theca surrounding each maturing ovum, as described below. In species characterized by the production of large numbers of ova, the medulla is evident only as a rather thin and loose connective tissue membrane which follows the folding of a more prominent cortex. The medulla of the mammalian ovary is considerably more prominent.

The ovarian cortex in higher forms is sepa-

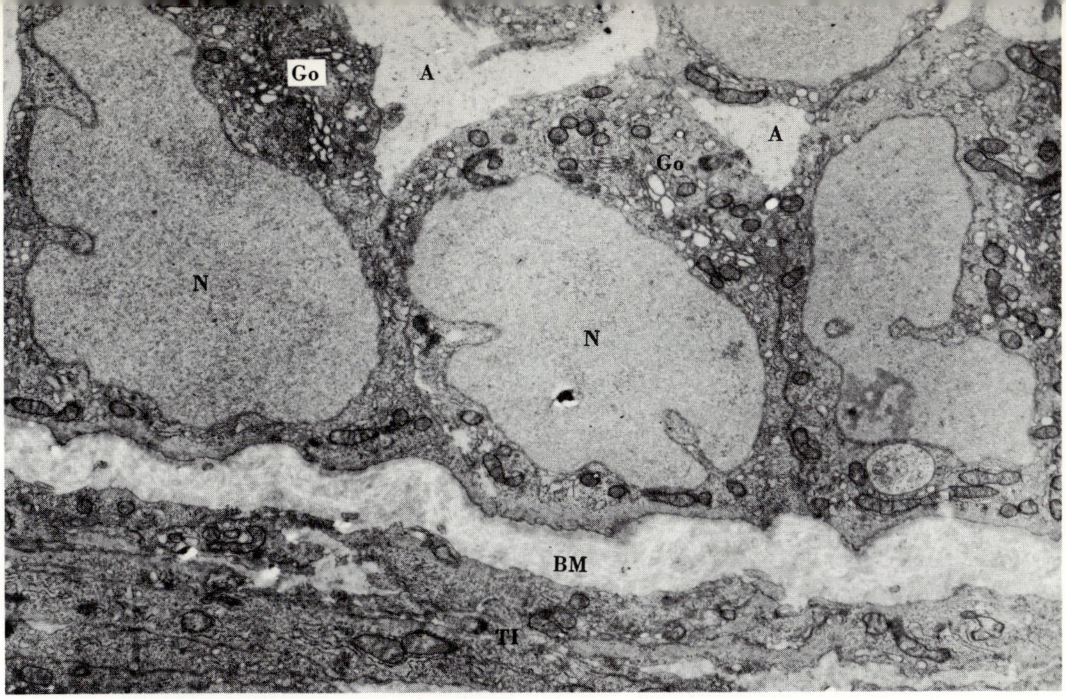

Fig. 11.17 Electron micrograph of a portion of the wall of a mature ovarian follicle in a rabbit, showing a row of granulosa cells. The Golgi vesicles, *Go*, lie between the granulosa cell nuclei, *N*, and the antrum, *A*. Note the well-defined basement membrane, *BM*, that lies between the granulosa cells and the cells of the theca interna, *TI*. X5000. (From E. Joan Blanchette. 1966. J. Cell Biol. 31:501–516.)

rated from the erstwhile germinal epithelium on the outside by a more or less dense layer of connective tissue, the *tunica albuginea*. In other species the germinal epithelium and cortex are contiguous.

The ovarian cortex is comprised of cords of cells derived from the germinal epithelium and includes the primordial germ cells. The latter appear very early in embryonic development and are generally held to become transformed directly into the primary oocytes (*see p. 284*). As each primary oocyte grows in size, a single layer of flattened or cuboidal cortical cells forms a follicle about it. In vertebrates whose eggs accumulate large quantities of yolk, the follicle cells have been shown to be the source of the yolk, and so the term *nurse cells* has been applied to them. Because of their granular appearance, they are also called *granulosa cells* in mammals. In the rabbit they vary in shape from cuboidal to columnar. The nucleus is large and of irregular shape, and the cytoplasmic organelles appear to be polarized toward the centrum (Fig. 11.17). There is some evidence that even in alecithal (relatively yolk-free) eggs, some of the follicle cells provide nourishment to the developing ovum. In either case, the follicle so formed is called a *primary follicle*, numbers of which may be seen concentrated in the peripheral areas of the cortex.

As the ovum increases in size, its nucleus also enlarges, the mitochondria become more uniformly distributed throughout the cytoplasm, and yolk granules and lipid droplets accumulate. In all vertebrates except the marsupials and placental mammals, the yolk and lipid droplets become so dense that they obscure other cytoplasmic components. In animals which produce very large numbers of eggs within one breeding season, all of the primary follicles grow and mature synchronously so that the cortex comes to consist of little else than eggs in the same stage. In species which produce only a small number of eggs, such as elasmobranchs, reptiles, birds, and mammals, eggs in almost all stages of maturation may be found in the same ovary at the same time (Fig. 11.18).

The maturing ovum soon acquires a *vitelline membrane*. This is a tough, refractile film secreted by the egg, which lies immediately outside the egg cell membrane. In most nonmammalian species, the egg cytoplasm adjacent to the vitelline membrane has a striated appearance, so that it is called the *zona radiata.* In mammals the zona radiata is extracellular, lying outside the vitelline membrane, and its origin is uncertain. In many mammals, such as human beings, it is not striated, but appears as a clear, refractile, deeply staining area rich in glycoprotein, the *zona pellucida.* When the zona radiata or zona pellucida appear, the follicle is said to be a *secondary follicle* (Fig. 11.19). Concomitantly the follicle pushes centripetally and the follicle cells proliferate, establishing several layers around the enclosed egg. Spindle-shaped fibroblasts and reticular fibers from the cortical stroma surround the follicle in several concentric layers, the theca folliculi, which may incorporate some smooth muscle fibers, as occurs in amphibians. In mammals the theca later becomes subdivided into a vascular *theca interna* and a more compact and fibrous *theca externa* (Fig. 11.19). The interna is separated from the granulosa cells of the follicle by a distinct basement membrane, which has a glassy appearance in mammals. The externa merges gradually with the surrounding stroma.

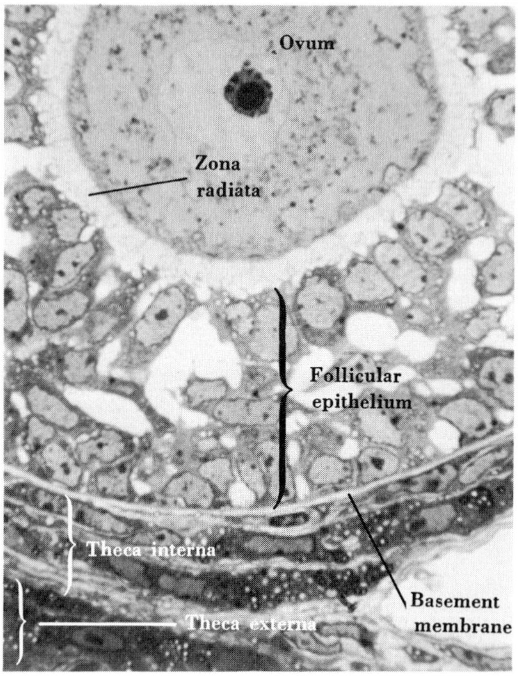

Fig. 11.19 Secondary follicle of a rat. This specimen was embedded in Epon and stained with toluidine blue. X1000. (Courtesy of F. B. Merk.)

Fig. 11.18 Cortical region of a cat's ovary showing ova in various early stages of growth and differentiation. X400.

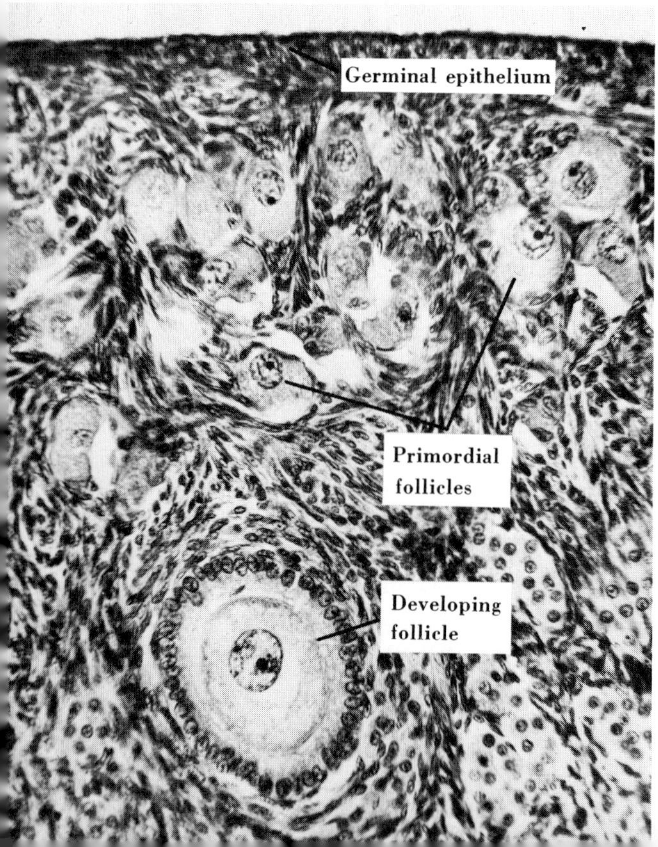

The ova of reptiles and birds show eggs of progressive stages of maturation arranged longitudinally in a sort of assembly-line fashion. In mammals the various stages appear to be randomly distributed. Mature follicles occupy the width of the cortex and bulge conspicuously outward. Under the influence of a critical balance between levels of the follicle-stimulating hormone (FSH) and the luteinizing hormone (LH), the attenuated ovarian capsule breaks open and the eggs are released, a process called ovulation. The posterior-pituitary hormone, oxytocin, is believed also to be implicated in this process, but its precise role is obscure.

Maturation of the egg in mammals is accompanied by some additional events and structural changes which have not been so clearly described in other species, but are nonetheless fundamental and deserve some detailed account. After the secondary follicle is established in the mammalian ovary, clefts appear among the granulosa cells of the follicle. The clefts then coalesce, ultimately forming a single, fluid-filled space, the *follicular cavity* or *antrum.* At this juncture the follicle is called a *tertiary follicle.* The ovum of the tertiary follicle lies in an eccentric position, embedded in a cluster of follicle cells called the *cumulus oophorous,* which protrudes into the antrum (Fig. 11.20). Among the granulosa cells an occasional round, dense-staining body may be found (*body of Call-Exner*). Such bodies are believed to represent centers of secretion of follicular fluid. The cumulus cells in contact with the zona pellucida of the egg become tall columnar in shape, and fine processes may be seen extending from them through the zona pellucida, presumably indicative of a nutritive function. They remain rather firmly adherent to the egg as a crown of radiating cells, hence they are known as the *corona radiata* (Fig. 11.20).

The so-called *mature* or *Graafian follicle* is achieved when clefts appearing in the cumulus weaken the attachment of the ovum to the follicular wall and the ovum attains its maximum volume. In many species the first polar body is produced at about this time, and may be seen insinuated between the egg cell membrane and the vitelline membrane. The second polar body is not produced until after ovulation and, in many cases, only after the specific stimulus of sperm entry.

Follicle degeneration is a common event in vertebrates, from fishes to mammals. For reasons not well understood, some oocytes disintegrate and thus never succeed in perforating the ovarian wall (*atresia*). Atretic degeneration can occur at any time during follicular growth, but it is most common among primary follicles.

Postovulatory changes in the ovary fall into three general classes. In seasonal breeders with large numbers of eggs, such as most fishes and amphibians, the postovulatory ovaries appear as shrivelled membranes, and consist of little more than the connective tissue of the already skimpy medulla and a thin germinal epithelium. A new cortex soon develops by proliferation of the germinal epithelium, however, and progressive changes leading to the formation of new follicles in preparation for the next breeding season can be observed. In most other nonmammalian vertebrates, the postovulatory ovaries show no pronounced changes. The empty follicles first appear as clumps of cells, which become reorganized as part of the general stroma.

The third class of postovulatory change is that wherein the *corpus luteum* is formed. The corpus luteum has been found to have a rather wide distribution among vertebrate animals, having been described in several ovoviviparous fishes (e.g., *Mustelus canis* and *Raja batis*); in a variety of both viviparous and ovoviviparous reptilian genera (e.g., *Chelydra, Terrapene, Hoplodactylus,* and *Lacerta*); in birds (*Colaeus* and *Gallus*), and in all mammals. Named for its yellowish appearance in human beings, the corpus luteum develops from the residual follicle after ovulation. After the egg is released from the ovary, the antrum fills with blood and the perforation heals. The granulosa cells enlarge, their nuclei become vesiculated, and lipid droplets appear in the cytoplasm. There is a great increase in the smooth endoplasmic reticulum, and an accumulation of membrane-bound glycogen granules, often in a rosette pattern (Figs. 11.21 and 11.22). In some species, such as primates and certain ungulates, a yellow or orange carotenoid pigment also accumulates in these cells, accounting for the characteristic color of the corpora lutea in these forms. Some of the cells of the theca interna contribute to the corpus luteum, and so two kinds of cells, the *granulosa-lutein* and *theca-lutein* cells can be distinguished. The latter are smaller and their nuclei are more basophilic than in the former, except in the box tortoise (*Terrapene carolina*), in which the opposite is true. In many ovoviviparous reptiles the theca contributes nothing to the corpus luteum. The corpus luteum has been shown to be of endocrine function in mammals. It secretes the ovarian hormone, progesterone, which inhibits ovulation, prepares the uterus for implantation, and maintains the gravid uterus. The luteal hormone is believed to be implicated in regulating the secretion of tertiary membranes of the oviduct in oviparous species (*see below*) and of maintaining the egg in the oviduct or the ovary (as the case may be) in both ovoviviparous and vivip-

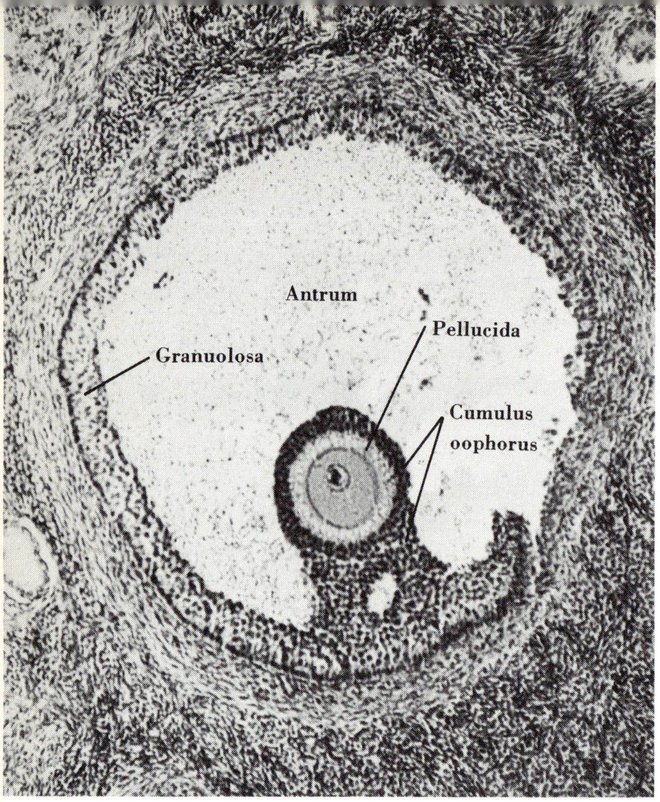

Fig. 11.20 Tertiary follicle of a cat. X95.

arous nonplacental species. In all species it appears to inhibit ovulation and thus "space" successive ovulations. In the snakes *Thamnophis* and *Natrix,* it has been claimed that the corpus luteum "antedates" its possible use as an endocrine organ.

There is some overlap of secretion between the follicle and the corpus luteum in mammals, for although the granulosa cells of the follicle secrete mainly estrogen, they also secrete small amounts of progesterone just before they become transformed into luteal cells (i.e., prior to ovulation), and the corpus luteum also secretes small quantities of estrogen. (The role of the pituitary gland in regulating the endocrinal function of the ovary is discussed in Chapter 14.)

The Oviducts

The oviducts are paired tubes that ultimately convey the eggs to the outside. They are lacking in cyclostomes and salmonoid fishes, as previously mentioned, and they are unpaired in certain species, notably most birds. Except in teleosts, the oviducts arise embryonically from infolding of the coelomic epithelium alongside the Wolffian duct or, as is the case with elasmobranchs and urodeles, by lengthwise splitting of the Wolffian duct itself. The

Fig. 11.21 Electron micrograph of a part of a lutein cell 20 hours after ovulation. The smooth-surfaced ER appears as whorls of fenestrated cisternae surrounding small clusters of mitochondria. X17,000. (From E. Joan Blanchette. 1966. J. Cell Biol. 31:517–542.)

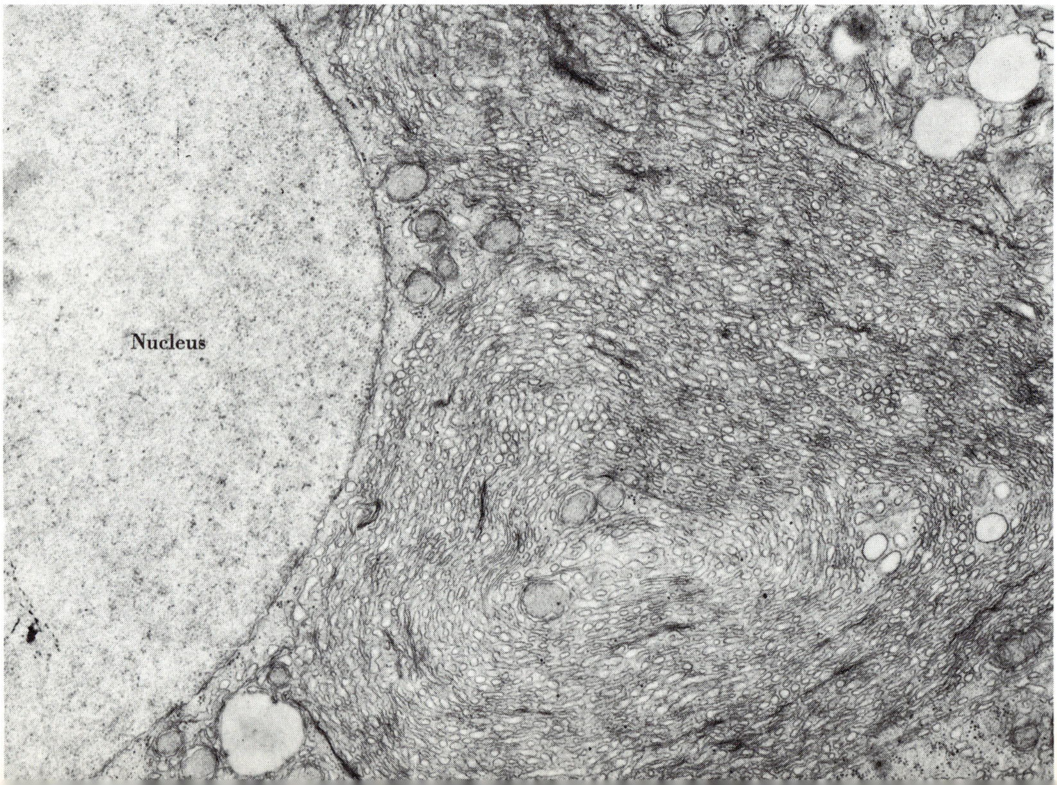

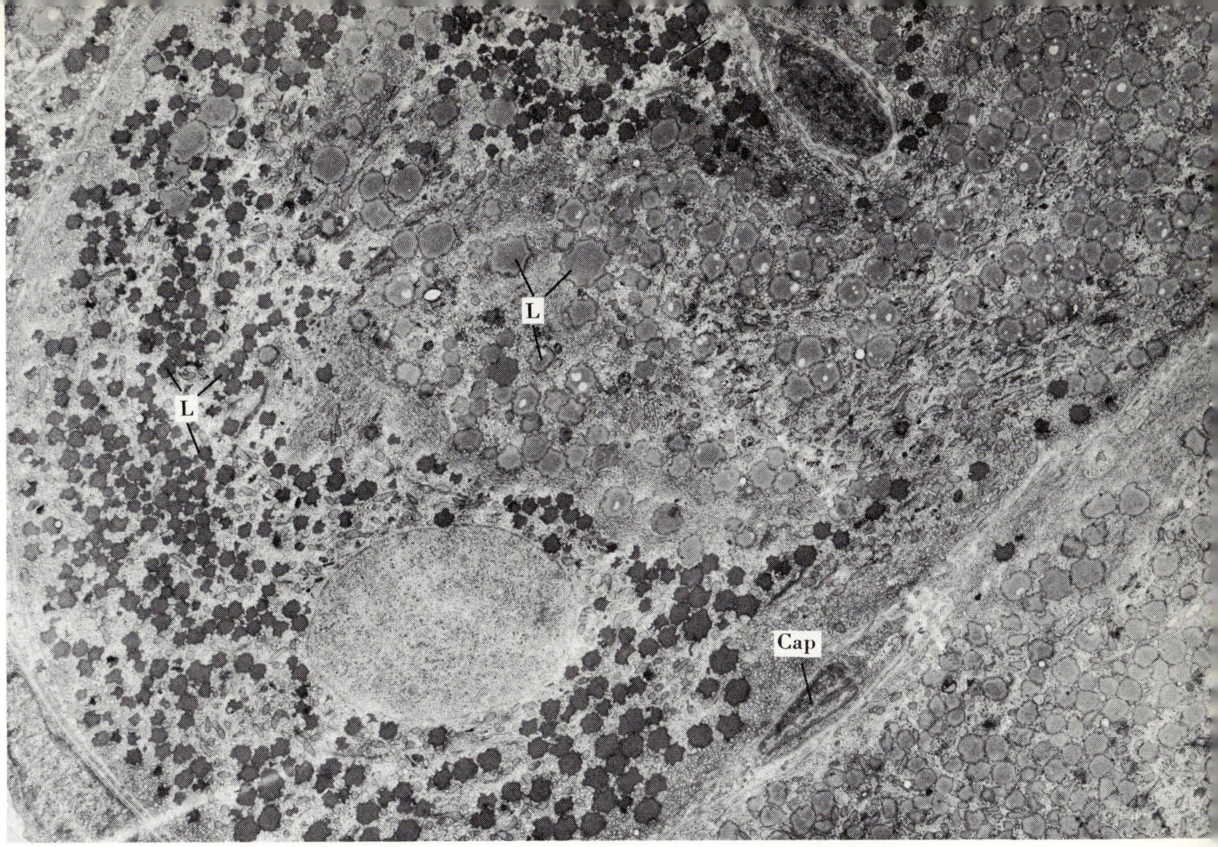

Fig. 11.22 Electron micrograph of a rabbit lutein cell 25 days after ovulation. Two kinds of lipid droplets, *L*, occupy most of the cytoplasm. Parts of capillaries, *Cap*, are present. X5000. (From E. Joan Blanchette. 1966. J. Cell Biol. 31:517–542.)

opening into the oviduct, the ostium, is usually flared or trumpet-shaped and it may be equipped with a fringe of tentacular processes, the *fimbriae*. The paired oviducts of elasmobranchs are joined anteriorly and thus they share a single, median ostium. In teleosts the peritoneum folds over the ovaries, isolating them from the coelom, and the folds continue caudally as special oviducal tubes leading into the cloaca. In all cases the histological characteristics of the oviduct reflect the treatment accorded the egg between the time it is received by the oviduct and the moment it is released to the outside. The appearance of the oviduct also depends on degree of sexual maturity and the stage of the reproductive cycle.

The simplest oviducts are found predictably in oviparous fishes and amphibians, as exemplified by the frog. They consist of a mucosa, a sort of muscularis, and a serosa. Starting at the ostium, one finds that the oviduct is thin-walled. Its mucosa is composed of simple, ciliated, columnar epithelium which is thrown into longitudinal folds, and a thin and cellular lamina propria. The muscularis, hardly a distinct coat at this level, consists of a thin encircling layer of connective tissue with a scattering of circulary oriented smooth muscle cells. Proceeding posteriorly one finds the wall thickens, owing to the appearance of tubular glands in the mucosa and the accumulation of smooth muscle in the muscularis. The secretory cells of the glands are also columnar, and their secretion is a thick mucus which forms a gelatinous coating about the eggs as they pass through. Between breeding seasons the glands are greatly reduced in number and extent.

The oviducts of oviparous vertebrates which lay large, yolky eggs, are somewhat more complex. This is not because of the quantity of yolk, per se, but because of the nature and quantity of the tertiary membranes added to the eggs in the oviduct. In some elasmobranchs, such as skates and some species of sharks (e.g., *Scyllium*, *Heterodontus*, and *Raja*) the tertiary membranes include both ovalbumin and a keratinlike shell. In oviparous reptiles a leathery shell, also of a keratin nature, and in birds a calcareous shell surmounting a pair of keratinous membranes (the inner and outer shell membranes) are formed in addition to a covering of ovalbumin.

In most elasmobranchs the oviducts have a highly differentiated *nidimental* or *shell gland,* and in ovoviviparous species a clearly defined uterus as well. The histologic characteristics of the oviduct itself are similar to those described above for the frog.

The nidimental gland of elasmobranchs appears as a rounded expansion of the oviduct wall in mature individuals, and it is considerably larger in oviparous (e.g., *Raja* and *Scyllium*) than in ovoviviparous species (e.g., *Mustelus* and *Squalus*). The anterior part of the gland consists of a mass of branched tubules of epithelial cells that secrete the albuminous covering of the egg. Secretory granules of mucoprotein fill the cytoplasm of these cells during the height of their secretory phase. The mucosa of the posterior part of the nidimental gland is richly folded to form villi that protrude into the lumen of the oviduct. The crypt of the villi give rise to branched tubular glands of simple columnar epithelium. Most of the cells are glandular, and their cytoplasm is charged with granules of "prokeratin" (*ovokeratin*) and/or mucoprotein. Interspersed among these cells are smaller, ciliated epithelial cells the nuclei of which are located in a more apical position. The lamina propria which supports the epithelial cells forms a richly vascularized stroma between the glandular channels. Within the lumina of the tubules fibers of ovokeratin may be seen during the most active phases of secretion. These fibers are deposited in several layers around the egg as it passes through the central lumen of the gland. The posterior-most part of the gland, which is tubuloalveolar, secretes mainly mucin to lubricate the eggs and facilitate continued passage through the oviduct and uterus.

The oviducts of birds and reptiles (Fig. 11.23) are similarly equipped with glandular components for the secretion of ovalbumin and ovokeratin, and in addition it elaborates a calcareous shell for the final covering of the egg in birds. Four more or less distinct regions can be differentiated in the avian oviduct: the funnel or *infundibulum,* the albumin-secreting region, the isthmus, and the uterus. The anterior part of the infundibulum has essentially the typical vertebrate oviduct structure, being lined with an epithelium of ciliated columnar cells and mucus-secreting cells which is thrown into longitudinal folds by the supporting lamina propria. Simple tubular glands appear in the lamina propria of the posterior part of the infundibulum, however, and they secrete the dense albumin of the egg's chalaza.

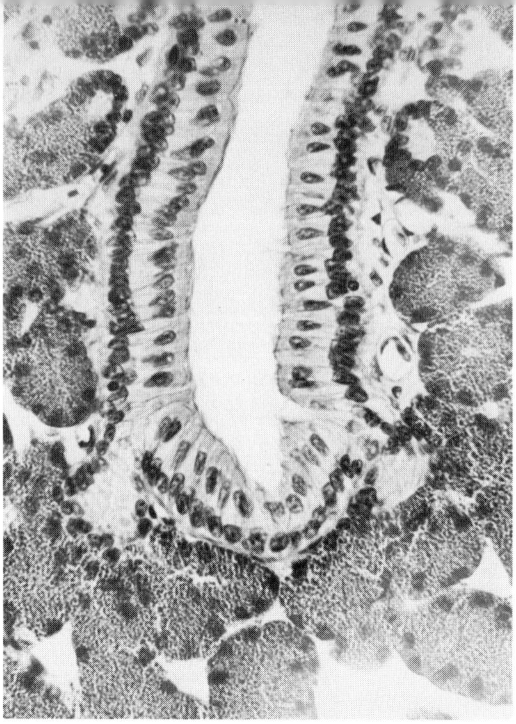

Fig. 11.23 Secretory epithelium of the oviduct of turtle. Involution of the basal layer of epithelial cells is seen at two points, where they connect with the subepithelial glands (dark and granular) that secrete ovalbumin. X500.

Tubular glands also appear in the next three regions of the avian oviduct, but the character of their secretions differ. In the albumin-secreting region, the tubule cells are highly acidophilic during their active phase, owing to the dense accumulation of granules of the precursor of ovalbumen. The granules fuse into masses within cytoplasmic vacuoles before being secreted into the lumen. The tubular gland cells of the isthmus show dense apical granules of varying sizes, which do not

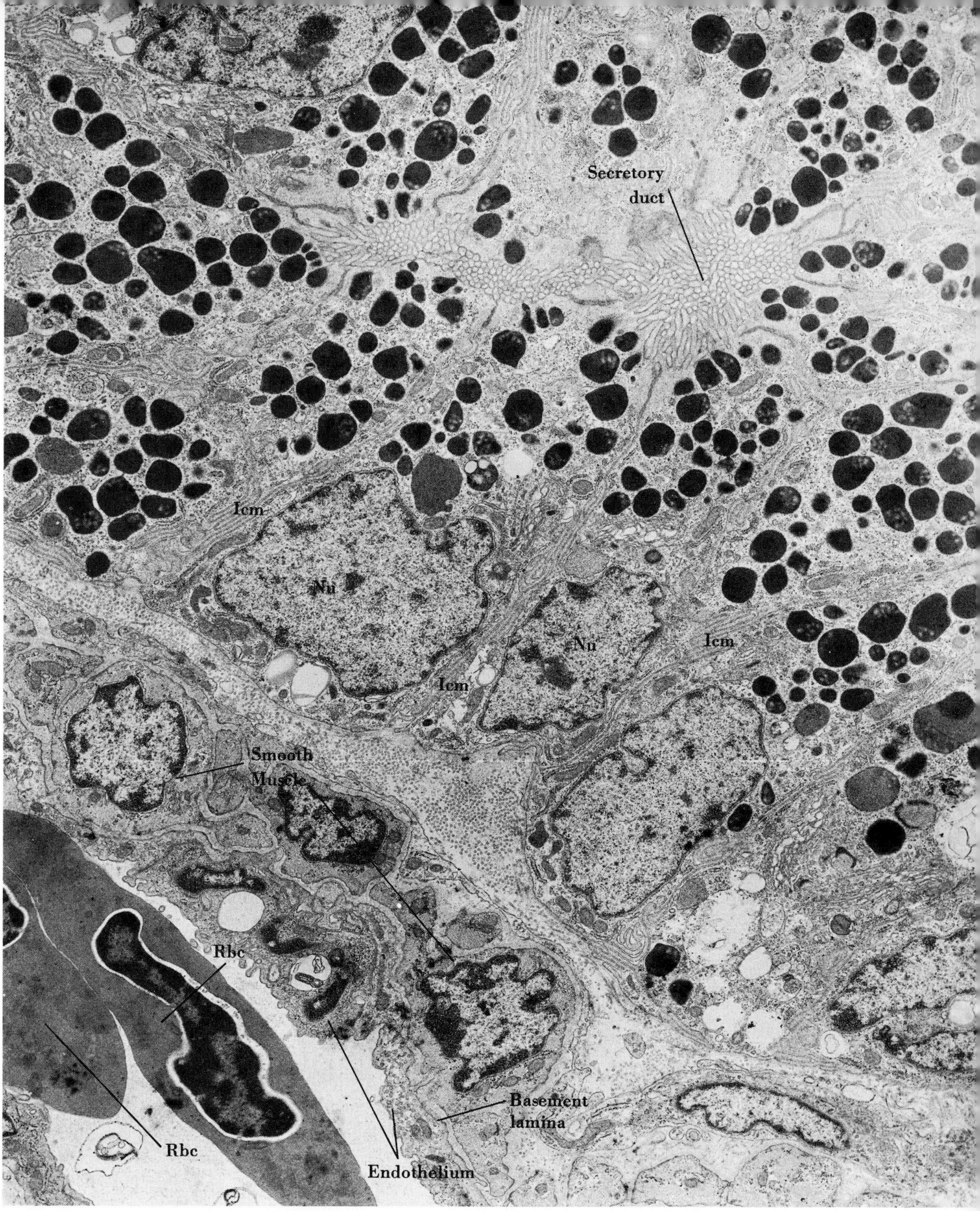

Fig. 11.24 Electron micrograph of the secretory epithelium of the uterine isthmus of the hen. The lumen of the secretory duct is filled with microvilli. The dark granules are believed to be precursors of ovokeratin. The cell membranes of adjoining cells, *Icm*, show complicated interdigitations. A venule cuts across the lower left corner of the micrograph. X8000. (Courtesy of F. Craft.)

fuse until after they have been secreted. These granules have been shown to be composed of neutral mucopolysaccharides and sulfur-containing proteins and to have a rather complex ultrastructure in the hen (Fig. 11.24). Such glands are of special interest because they represent a situation wherein keratin becomes an extracellular product, for in all other cases this widespread protein is entirely intracellular.

The epithelium of the uterus of the bird is pseudostratified columnar with coiled tubular glands extending into the lamina propria. The apical gland cells stain uniformly and lightly, without cytoplasmic granules but with small vacuoles which contain dense particles demonstrable by electron microscopy. During active shell formation the vacuoles migrate to the tips of microvilli protruding into the lumen and discharge their contents. There is little doubt that these cells effect the transport and concentration of calcium, which becomes deposited on the surfaces of the keratin fibers of the external shell membrane. The basal cells of the uterus are not ciliated, but they do have microvilli. Those in the anterior part of the uterus secrete a sulfated mucopolysaccharide, whereas those in the posterior part secrete mucoproteins.

The mammalian oviduct is structurally fairly uncomplicated. Most of the cells of its simple columnar epithelium are ciliated, the others are mucin-secreting. The lamina propria is loosely fibrous and contains fusiform fibroblasts, mast cells, and occasional macrophages. The two layers of the mucosa are thrown into numerous longitudinal folds that all but occlude the lumen (Fig. 11.25). The lamina propria of the fimbriae is unusual because it contains numerous large blood vessels, mostly veins, and smooth muscle cells. During ovulation the veins become engorged with blood, much as the sinuses of erectile tissue do, and the smooth muscle contracts rhythmically. The result of the two actions is that the fimbriae become closely applied to the ovary and stroke it, "egging it on," as it were—thereby assuring that when the ova are released they will be picked up and swept by the cilia into the oviduct. The muscularis of the oviduct consists of two layers of smooth muscle bundles, inner circular and outer longitudinal, which intermingle in the middle. They undergo rhythmic contractions during ovulation, generating peristaltic waves sweeping toward the uterus.

Fig. 11.25 Transverse section of the oviduct of a cat. See text for description. X100.

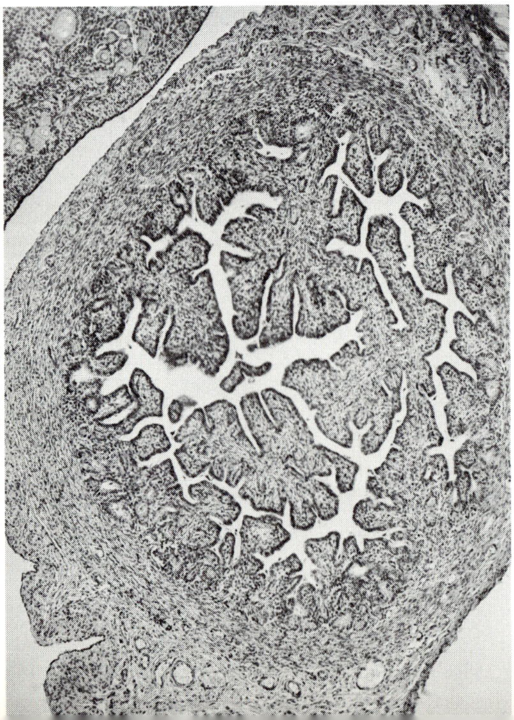

The uterus represents the posterior, expanded portion of the oviduct in most species, such as in the bird, as described above. In mammals characterized by the *uterus simplex*, as in man and other higher primates, it represents the lower portions of both oviducts which have united to form a single, expanded chamber. In amphibians and other oviparous forms, the uterus is very similar to the oviduct histologically—save for increased amounts of smooth muscle and connective tissue components.

In ovoviviparous elasmobranchs and reptiles, the uterine wall is considerably thicker and its lining epithelium is often stratified cuboidal and rich in gland cells with a granular, acidophilic cytoplasm. The lamina propria is quite vascular and, together with the epithelium, is thrown into an intricate network of folds. A distinct submucosa of loose connective tissue is present, and in elasmobranchs it, too, is included in the major folds of the lining, carrying with it very prominent blood vessels. In such forms the egg comes into intimate contact with the uterine epithelium and significant exchange of respiratory gases, excretory products, and probably nutritional substances takes place. In certain genera of sharks (*Scoliodon* and *Sphyrna*)

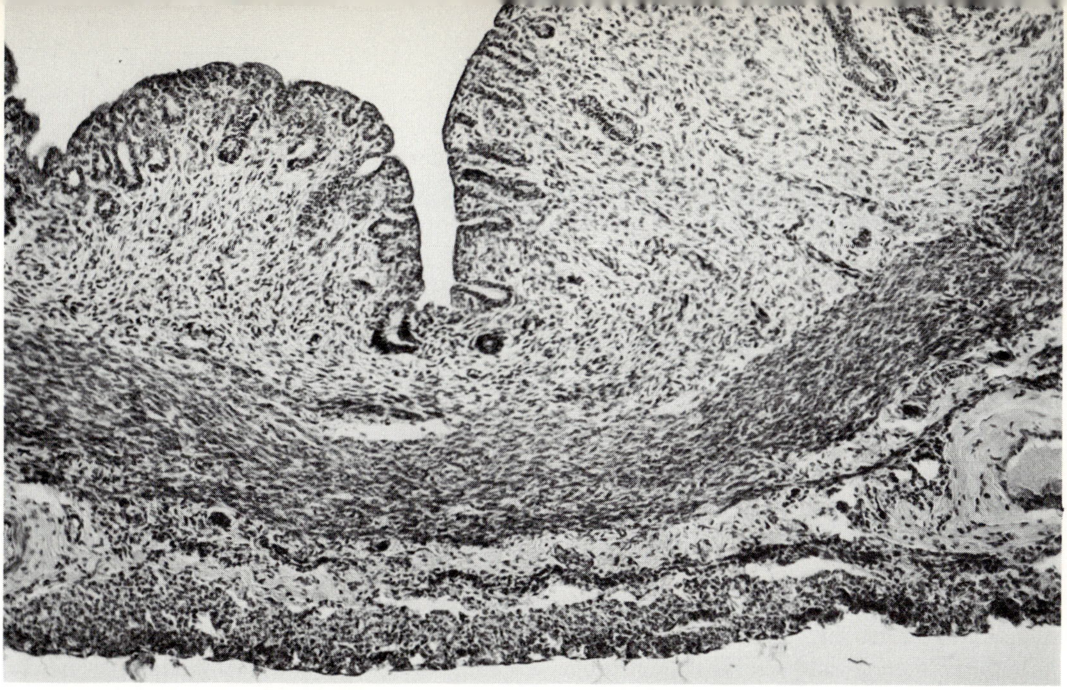

Fig. 11.26 Section of the resting uterus of a cat, showing the glandular myometrium and the layers of smooth muscle. X100.

the folds of the uterine mucosa are especially well developed where they make contact with the yolk stalk of the embryo. In the smooth dogfish, *Mustelus canis,* there is a sort of yolk-sac placenta formed, which effects a very intimate relation between the yolk sac and the uterine wall. In the blue shark, *Prionace glauca,* processes of the yolk sac interdigitate with processes of the uterine lining.

Even more intricate mechanisms are employed in certain teleosts, but there development is within the ovary rather than in the uterus, and parts of the embryo other than the yolk sac may be involved. For example, in the Jenynsidae, villous projections of the ovarian wall grow into the gill chambers of the embryos. In the *Brotulidae* not only are the yolk sac, body wall, and gills involved in absorbing nutrients from the ovary, but the embryos also ingest intraovarian fluids and feed on ovarian tissues and other eggs!

The mammalian uterus (Fig. 11.26) is composed of three layers of tissue: a mucosa (*endometrium*), a muscularis (*myometrium*), and a serosa. The endometrium consists of a simple columnar epithelium, some of the cells being ciliated, and a lamina propria that is subject to extreme variation during the course of the estrous cycle or during pregnancy. The variation is attributable primarily to changes in the length and complexity of the tubular epithelial glands lying within it and in the richness of its blood supply. In the resting uterus (anestrus or intermenstrual) the tubular glands are short and fairly straight, the blood supply is minimal, and the endometrium is at its thinnest. Scattered throughout the lamina propria (stroma) are fibroblasts, mesenchyme cells, lymphocytes, and occasional macrophages. Corresponding with follicular growth in the ovary (proestrus or premenstrual), the tubular glands lengthen and start to coil. Some distal branching may occur as well. The arterioles lengthen and spiral, the veins enlarge, and capillaries become more numerous. The gland cells accumulate glycogen and lipid droplets, and mucus cells become more numerous and active so that the lumina of the tubules become distended with secretion. This endometrial hypertrophy reaches a climax at about the time of ovulation (corresponding with estrus in most nonanthropoids) if fertilization does not occur, whereupon the process is reversed either gradually (metestrus and postestrus) or suddenly (menstruation). If fertilization of the ovum does occur, endometrial hypertrophy continues, climaxing in placentation.

Certain giant cells, the so-called decidual cells, appear in the endometrium during pregnancy and may also be observed in the human premenstrual uterus. They are thought to be derived from the mesenchyme cells of the lamina propria. Their cytoplasm contains many fine, acidophilic granules, and two dense-staining nuclei are usually present. For details of the differences between the estrous cycle characteristic of most mammals and the menstrual cycle, which is characteristic of the higher primates, and for details of placentation, the reader is referred to works on embryology and endocrinology.

The myometrium also undergoes corresponding hypertrophy, involving an increase in both the number and size of its constituent cells and the amount of collagenous fibers and intercullular fluid. In the resting uterus the smooth muscle fibers are arranged in bundles or sheets in several layers separated by loose connective tissue in which may be found blood vessels, fibroblasts, mesenchyme cells, mast cells, and individual smooth muscle cells. Elastic fibers are abundant between the outer muscle bundles. In the human uterus the myometrial muscle is distributed in four more or less distinct layers: (1) a thin *stratum submucosum* of predominantly longitudinal fibers, but also some circular and oblique, lying immediately under the lamina propria; (2) a thick *stratum vasculare* of predominantly circular and oblique fibers, but also exceedingly rich in large blood vessels; (3) a *stratum supravasculare* of circular and longitudinal fibers; and (4), as the outermost layer just inside the serosa, a *stratum subserosum,* the fibers of which are oriented longitudinally. The fibers of the supravasculare and subserosum are continuous with those of the oviducts and suspending ligaments. As the uterus increases in thickness in response to estrogen and progesterone, the muscle increases both in amount and in the length of individual fibers. The increase in amount is achieved by both division of muscle cells and by the differentiation of mesenchyme cells, particularly in the stratum vasculare. The increase in length which the smooth muscle cells undergo is remarkable, amounting in the human pregnant uterus to a change from approximately 50 to 500μ or more.

Before joining the vagina, the uterus decreases sharply in diameter to form the tubular *cervix.* The cervical mucosa includes many mucus-secreting cells the secretions of which lubricate the vagina during sexual excitement; tubular glands are present but they are reduced in number and length. Surrounding the mucosa is a dense layer of fibrous connective tissue in which elastic elements predominate; smooth muscle fibers are variable in number and randomly distributed, and the arterioles have lost their characteristic spiral shape. The cervix protrudes a short way into the lumen of the vagina as the *portio vaginalis,* abruptly changing its mucous epithelium from tall columnar to stratified squamous.

The vagina is the last major modified segment of the oviduct. It represents an adaptation for reception of the penis in marsupials and placental mammals. Vaginae have been described in lower classes of vertebrates, but there they represent little more than the constricted terminus of the oviduct or uterus. The true vagina of mammals is thinwalled; its mucosa consists of stratified squamous epithelium and a glandless but well-vascularized though fairly dense lamina propria, both of which are arranged in transverse ridges or rugae. The more superficial cells of the epithelium exhibit characteristic changes that correspond with the ovulatory cycle in most mammals. Moreover, lymphocytes and neutrophils tend to accumulate in the lamina propria and to invade the epithelial layers. Microscopic examination of washes or lavages of the vagina can be useful in determining the stage of the ovulatory cycle in some mammals, such as most rodents. For example, during proestrus the vaginal lavages of laboratory rats reveal mainly nucleated epithelial cells; during estrus both nucleated and desquamated, enucleate cells ("plates") are evident; in metestrus large numbers of plates and many leukocytes are found; and, finally, in diestrus nucleated cells reappear, squamous plates are still present, and leukocytes are quite numerous. Although not normally cornified, the more superficial cells of the vaginal epithelium contain keratohyaline granules and are rich in glycogen. The latter substance has been shown to increase with estrogen administration.

The smooth muscle fibers of the vaginal muscularis are arranged in intertwined circular and longitudinal bundles, predominantly the latter, which are continuous with those of the myometrium. The bundles are separated by dense fibroelastic connective tissue. Skeletal muscle fibers may be found in the muscularis at the vaginal orifice. Surrounding the muscularis is an adventitia or fibrosa containing many coarse elastic fibers. A thin fold of the mucosa, the *hymen,* may be found partially occluding the external orifice in virgin females of some species.

SELECTED REFERENCES

Aitken, R. N. C., and H. S. Johnston. 1963. Observations on the fine structure of the infundibulum of the avian oviduct. J. Anat. 97:87–99.

Blanchette, E. J. 1966. Ovarian steroid cells. I. Differentiation of the lutein cell from the granulosa follicle cell during the preovulatory stage and under the influence of exogenous gonadotrophins. J. Cell Biol. 31:501–517.

Blanchette, E. J. 1966. Ovarian steroid cells. II. The lutein cell. J. Cell Biol. 31:517–543.

Filhol, J., and H. Garrault. 1938. La secretion de la prokeratine et la formation de la capsule ovulaire chez les Selacians. Arch. Anat. microsc. 34:105–145.

Harrison, R. J. 1948. The development and fate of the corpus luteum in the vertebrate series. Biol. Rev. Camb. Phil. Soc. 23:296–351.

Hett, J. 1923. Das Corpus luteum der Dohle. Arch. für mikr. Anat. 97:718–836.

Hruban, A., J. Martan, A. Slesers, D. F. Steiner, M. Lubran, and M. Rechcigl. 1965. Fine structure of the prostatic epithelium of the opossum (*Didelphis virginiana* Kerr). J. Exp. Zool. 160:81–106.

Johnston, H. S., R. N. C. Aitken, and G. M. Wyburn. 1963. The fine structure of the uterus of the domestic fowl. J. Anat. 97:333–344.

Lagler, K. F., J. E. Bardach, and R. R. Miller. 1962. Ichthyology, the study of fishes. John Wiley & Sons, New York.

chapter 12
NERVOUS TISSUE AND ORGANS

THE NERVOUS SYSTEM reaches its highest degree of differentiation and complexity in vertebrate animals. Even in those contemporary species which are phylogenetically closest to primitive ancestral vertebrates, namely, the cyclostomes, one finds nervous systems that are far more extensive and advanced than in any invertebrate phylum. Yet the progressive adaptations which have characterized the evolution of the vertebrate classes include increases in complexity of the nervous system so far beyond that achieved by the cyclostomes as to make even their great advance seem insignificant.

The mechanisms by which the so-called higher vertebrates have developed the fantastic "brain power" shown by some human beings do not include anything mysterious or, indeed, revolutionary, as far as can be determined. Basically the way in which the nervous system works is the same in all animals that have anything at all by way of nervous tissue. To be sure, an interesting evolutionary dichotomy is noted in the fact that in invertebrates the endocrinal role of the nervous system has been stressed whereas in the vertebrates this role has been limited by and large to the supraoptic and paraventricular nuclei of the brain, and emphasis has been placed instead on the development of fiber tracts and association centers. This is especially pronounced in the higher vertebrates, in which the advances represent essentially an increase in the total number of nerve cells involved, in the speed and efficiency of impulse conduction and transmission, and, most of all, in the variety and complexity of synaptic combinations effected. New mechanisms of intercullular transmission may also be involved, inasmuch as structural modification of the nerve synapse is also evident in the higher vertebrates.

The nervous system of all vertebrates may be divided conveniently into two parts, the *central nervous system* and the *peripheral nervous system.* Although vascular, connective, and glandular tissues are involved, the essential component of both parts is *nervous tissue.* Nervous tissue in turn may be considered to have a dual composition, *neuron cells* (*neurons*) and *nervous interstitial cells.* The former are concerned with nervous function *per se,* that is, with the generation, conduction, and transmission of the nerve impulse. Cells of the latter type are considered to have primarily a supportive function, and their broad categories include (1) the *neuroglial cells;* (2) the *ependymal cells,* which form the lining of the cavities of the brain and spinal cord; and (3) the Schwann cells of peripheral nerves and peripheral ganglia. The kinds of supportive functions attributed to nervous interstitial cells are both trophic and mechanical, as discussed below. It has been proposed that the neuroglial cells may also play a role in memory.

Embryonic derivation of nervous tissue. Nervous tissue arises in vertebrate embryos from dorsal elements of ectoderm which become delineated as a thickened *neural plate* after gastrulation, in direct response to the establishment of the so-called chordamesoderm. The neural plate invaginates along its longitudinal axis, forming a transitory *neural groove,* the lateral edges of which soon meet and join to form the *neural tube* (Fig. 12.1). The marginal cells of the neural plate meanwhile have been drawn toward the midline and consolidated into a pair of dorsolateral strips, the neural crests. Subsequently the neural crests migrate ventrally and come to lie between the neural tube and the bilateral masses of mesoderm. As the mesoderm arranges itself into segmental blocks, the *myotomes,* it induces the neural crests to pinch into segments which alternate with the myotomes.

The anterior part of the neural tube undergoes pronounced modifications in form at a very early stage. These modifications are brought about by a series of invaginations, evaginations, thickenings, thinnings, and flexures which lead to the formation of the various regions of the brain and major sense organs (see Chapter 13). The rest of the neural tube differentiates into the spinal cord with little change in external form. The processes of cellular differentiation are identical for the most part in both the brain and the spinal cord, but they are more easily followed in the latter.

The ancestry of the cells lining the neural canal, the *ependymal cells,* can be traced back directly to the superficial ectodermal cells of

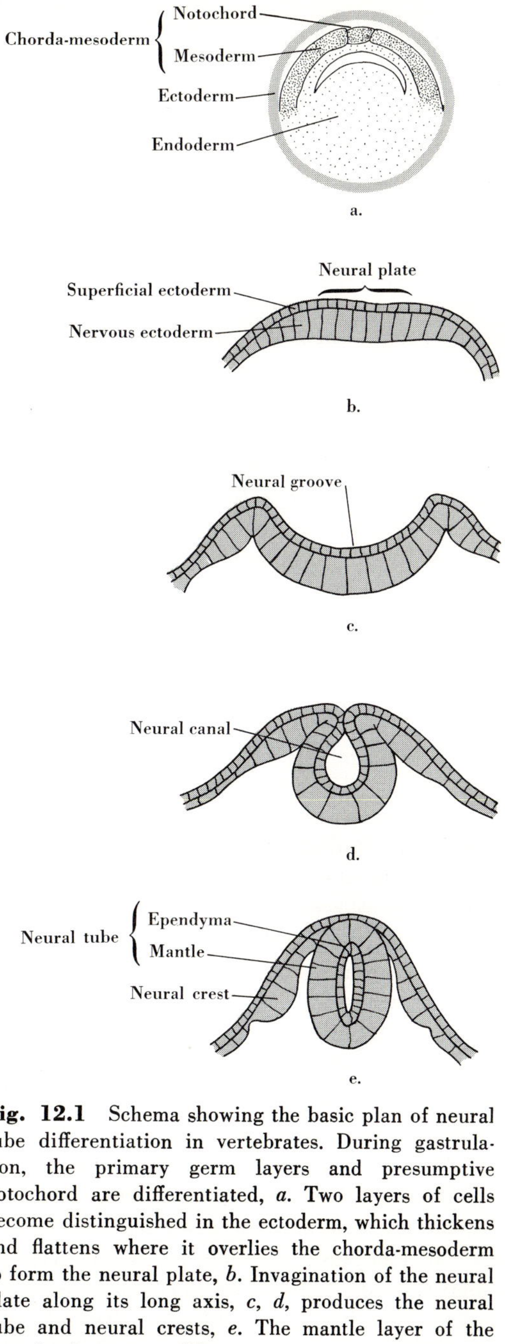

Fig. 12.1 Schema showing the basic plan of neural tube differentiation in vertebrates. During gastrulation, the primary germ layers and presumptive notochord are differentiated, *a.* Two layers of cells become distinguished in the ectoderm, which thickens and flattens where it overlies the chorda-mesoderm to form the neural plate, *b.* Invagination of the neural plate along its long axis, *c, d,* produces the neural tube and neural crests, *e.* The mantle layer of the neural tube soon proliferates an outer marginal layer, not shown here.

the neural plate which, in turn, were continuous with the other superficial cells surrounding the embryo and eventually give rise to epidermis. It is not surprising, therefore, to learn that the ependymal cells appear as a kind of epithelium and are usually ciliated. The layer of cells which surrounds the ependymal layer and which is derived directly from the thickened portion of the neural plate is called the *mantle layer*. Protoplasmic processes from certain cells of both layers move centrifugally to form the thin, outer, investing *marginal layer*. Most of the cells of the mantle layer soon become identified as *neuroblasts*, from which the neurons of the gray matter differentiate. The rest of the cells of the mantle layer become *spongioblasts*, which, together with the cells of the ependyma, differentiate into various kinds of nervous interstitial cells. By differential growth the neuroblasts of the spinal cord become arranged into dorsal and ventral columns whose appearance in sections of freshly excised spinal cord is gray, whence the terms "gray matter" and "gray columns." Gray matter contrasts sharply with "white matter," a term which applies to those areas of the brain and spinal cord in which myelinated nerve fibers predominate. The white appearance is directly attributable to the myelin.

While the cells of the central nervous system are differentiating, the neural crests give rise to a number of different structures. The major portion of each neural crest segment differentiates into the cranial and spinal ganglia. Within the ganglia two principal types of stem cells appear: neuroblasts and nervous interstitial cells. In most vertebrate species the neuroblasts differentiate into sensory neurons and the interstitial cells differentiate into *Schwann cells*, the structure and functions of which are described below. Other elements of the neural crests become detached from the segmental clusters and migrate to a wide variety of locations, ultimately giving rise to ganglia of the autonomic nervous system, to suprarenal tissue, to chromatophores, and even—according to some experimental evidence—to components of the visceral skeleton and of the teeth.

The manifold changes undergone by the neural tube and neural crests lead to the formation of the mostly highly integrated system of the body, a condition which is implicit when one considers that the *raison d'être* of the nervous system is functional integration of all of the other systems of organs and tissues. As stated above, the major anatomical divisions of the nervous system are the central nervous system and the peripheral nervous system. The central nervous system in turn is divided into the brain and the spinal cord. As we shall see, this division is not entirely arbitrary, as it might seem to be in view of the identity of embryonic origin and the anatomical contiguity of the two parts.

The peripheral nervous system includes the cranial and spinal nerves and their branches. It is comprised of two more or less equal parts, afferent and efferent, on the basis of the direction in which the nerve fibers normally conduct their impulses; that is, the *afferent* portion includes those elements concerned with conducting impulses *toward* the central nervous system, whereas the *efferent* portion includes the elements concerned with conducting impulses *away from* the central nervous system. Synonymous with afferent and efferent therefore are the functional terms *sensory* and *motor*, respectively.

An additional subdivision of the peripheral nervous system then presents itself on the basis of the nature and location of the peripheral organs in which the afferent and efferent fibers terminate. Those which have their terminus in the so-called "visceral" organs, such as organs of the circulatory, digestive, respiratory, endocrine, and urogenital systems, are accordingly called *visceral* components, whereas those which innervate the remaining parts of the body are called *somatic* components. On reflection, one will note that the somatic components are concerned with the physical relation of the organism to its outside environment, for one thing, and for another that this relation is a "conscious" one. Hence the somatic afferent (sensory) components have their termini of origin largely in the integument and in the major sense organs, and the somatic efferent (motor) components terminate in skeletal muscles. Similarly one will note that the visceral components are concerned with the "housekeeping" functions of the organism, i.e., the maintenance of the functions that must go on irrespective of changes in the outside environment, functions which are more or less directly related to metabolism and reproduction. This is not to say that they are unaffected by the external environment. Within limits these functions are so routine that not only have they become "automatic" in the sense that they go on without conscious effort, but they do so regardless of what the individual may will; i.e., they are involuntary.

One may see that sensory impulses require no alternatives beyond being visceral or somatic, for they are concerned with providing information to the central nervous system, and that is all. Motor impulses, on the other hand, do require alternatives beyond visceral versus somatic. In the case of somatic motor, alternatives such as motion forward or backward, upward or downward, left or right, etc., obtain. Most of these can be resolved into the alternative functions, flexion and extension, and accordingly most muscles are arranged in antagonistic pairs. Because there is no anatomic or physiologic difference between nerve fibers which carry impulses to flexor muscles and those which carry impulses to extensor muscles, no special distinction between them is justified.

In the case of visceral motor functions, skeletal members are not involved and so the muscles cannot be arranged in opposing pairs. Furthermore, forces of other kinds may be involved, such as hydrostatic pressure in the case of blood vessels and parts of the gastrointestinal tract. A basic characteristic of visceral motor function is *tone* or *tonic contraction,* which is a more or less prolonged state of partial contraction exhibited by smooth muscle fibers. Tone is achieved by a steady flow of impulses to the muscle cells via excitatory visceral motor nerves. Regulation of function is then achieved by the alternatives of either increasing muscle contractility above the tonic level, or by decreasing it below this level. The increase is brought about by impulses carried by excitatory nerves, and depressed activity is effected by impulses carried by inhibitory nerves. Although the physicochemical nature of impulses in the two kinds of nerve is identical, the chemical substances released at the endings in response to an impulse differ. Therefore, in order to achieve this kind of control, most visceral organs have *dual* innervation by both excitatory and inhibitory nerves. The chemical substances by which the two kinds of nerves mediate their function are primarily *acetylcholine* and either *adrenalin* or *noradrenalin.* Visceral motor nerves which work by the former substance are called *cholinergic* and the others are called *adrenergic.* Because of rather distinct differences in anatomical relations to the central nervous system, visceral motor nerves are grouped into two divisions: the *sympathetic division,* which includes all adrenergic nerves, and the *parasympathetic division,* which includes all visceral motor nerves that are cholinergic. Because of its uniqueness in these respects, and because of its involuntary nature, the visceral motor system is also distinguished by the name *autonomic nervous system.*

It should be made clear at this point that the parasympathetic division is not alone in being cholinergic. Somatic motor nerves are also cholinergic, and evidence is accumulating to indicate that acetylcholine is the most widespread synaptic transmitter, inasmuch as cholinesterase and/or acetylcholine esterase are found at synaptic junctions in the central nervous system. Another important distinction should be made between preganglionic and postganglionic fibers of the sympathetic division: namely, that the preganglionic fibers are cholinergic and that it is specifically and only the postganglionic fibers that are adrenergic.

Somatic motor neurons in mammals have all their cell bodies within the central nervous system and their fibers pass directly without any discontinuity to the skeletal muscles which they innervate. Visceral motor nerves, on the other hand, whether sympathetic or parasympathetic, have their cell bodies in two locations. The primary cell station, like that of the somatic motor neurons, is in the gray matter of the central nervous system. From there the axonal fibers pass peripherally out of the central nervous system to secondary cell stations, the *autonomic* ganglia, where they synapse with the dendritic processes of secon-

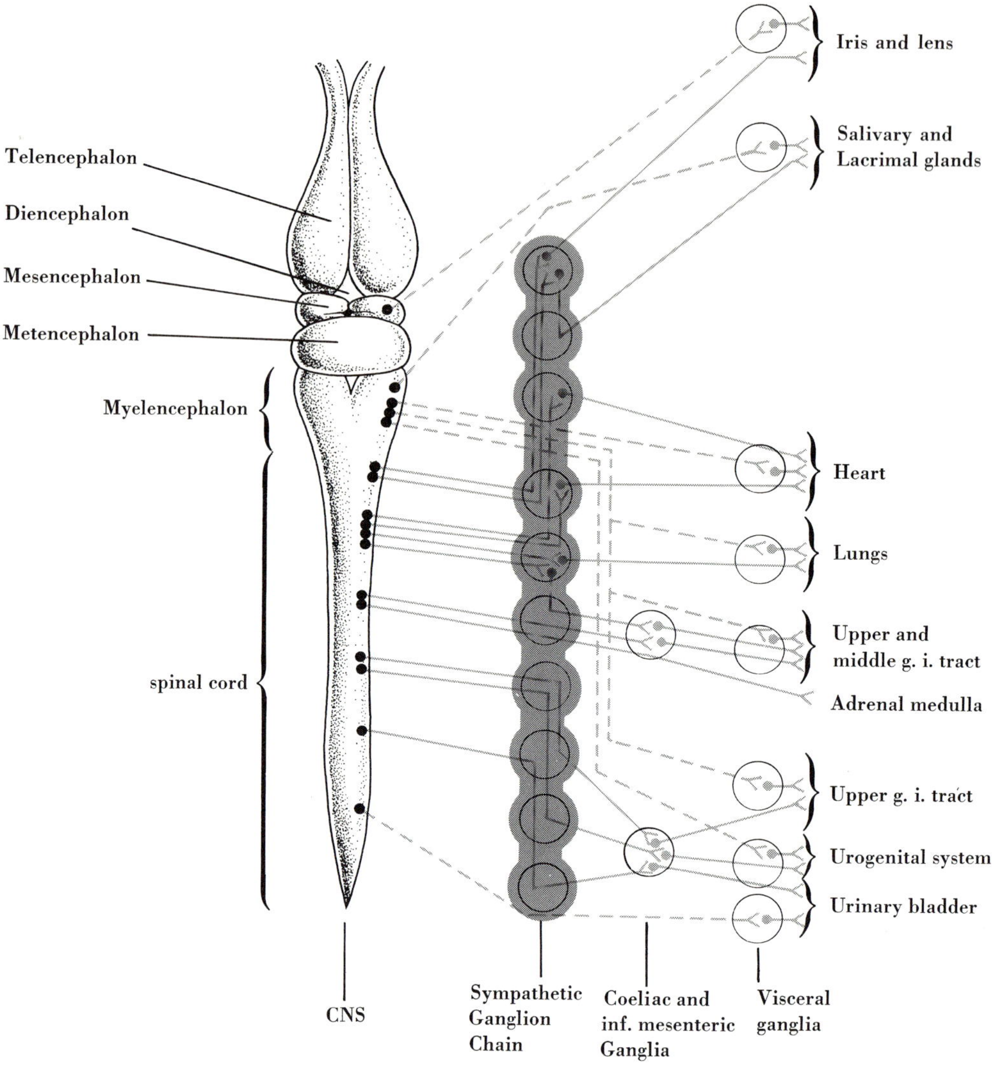

Fig. 12.2 Diagrams showing the relations between the *CNS*, the autonomic nervous system, and the major outlying ganglia. Preganglionic fibers of the parasympathetic division are shown in broken line; all other autonomic fibers are shown in solid lines.

dary neurons. The axonal processes of the latter then pass more or less directly to the muscle (smooth or cardiac) that they innervate (Fig. 12.2). The autonomic fibers that originate in the central nervous system are called preganglionic fibers and those that originate in the autonomic ganglia are called postganglionic fibers. A unique characteristic of postganglionic fibers is that they are not myelinated; thus they appear gray rather than white.

It is the postganglionic components that are derived from migrating cells of the neural crests. The cell bodies usually occur in clusters, or ganglia, which are distributed in various sites in the body. Except in cyclostomes and elasmobranchs, most of the sympathetic ganglia occur in a pair of chains that lie parasagittally between the dorsal aorta and the spinal column in the cervical, thoracic, and lumbar regions. The preganglionic fibers leave the spinal cord via the spinal nerves, then branch off as the *white rami*, which lead directly to the ganglionic chain (Fig. 12.2). Most of the preganglionic fibers synapse with the secondary neurons in these ganglia, but a number pass through and synapse in outlying ganglia in the viscera instead. The postganglionic fibers that originate in the ganglia of the sympathetic chain may either pass directly to their target organ (e.g., the heart) or they may rejoin the spinal nerves via the *gray rami* (Fig. 12.2) and thus accompany other nerves for their peripheral distribution to blood vessels and glands. In any case, as a general rule the preganglionic fibers of the sympathetic division are relatively shorter than the postganglionic.

The ganglia of the parasympathetic division are not connected in chains and often are more diffuse. For the most part they lie immediately adjacent to the target organ, or actually within its walls. As a consequence the postganglionic fibers are usually very short, and the preganglionic correspondingly long (Fig. 12.2).

NERVOUS TISSUES

The Neuron

The primary functional cells of the nervous system, regardless of location, are the neurons. Typically a neuron consists of a cell body, in which is located the nucleus, and two processes, a dendritic process or processes and an axonal process. Dendrites are characterized by terminal arborizations of various kinds, and the axonal process terminates in a spray of fine fibers, the axonal *telodendria* (Fig. 12.3).

NERVE CELL BODY

The neuron cell body occurs in various sizes and shapes, although a round or oval form is prevalent. The nucleus is relatively large with a regular contour, lightly basophilic, and characterized by a large nucleolus, usually solitary. In addition to Golgi substance and mitochondria, the nerve cell cytoplasm shows distinct basophilic condensations, the *Nissl substance* (Fig. 12.4). By electron microscopy the Nissl substance appears as stacks of rough-surfaced endoplasmic reticulum that sandwich an abundance of polyribosomes. Nissl substance may be found in all parts of the nerve cell cytoplasm except for the areas immediately adjacent to the cell surface, the nuclear surface, and the vicinity of the origin of the axonal fiber (the axon hillock). The aggregations of Nissl substance are large and numerous in somatic motor and other large neurons, but they are small and diffuse in sensory cells and in autonomic ganglion cells. The observation that the Nissl substance increases with moderate activity of its parent cell but decreases on overstimulation (*chromatolysis*), as well as the accumulating evidence that RNA is implicated in memory processes, suggest that the Nissl substance is intimately involved in neurological function *per se*. The

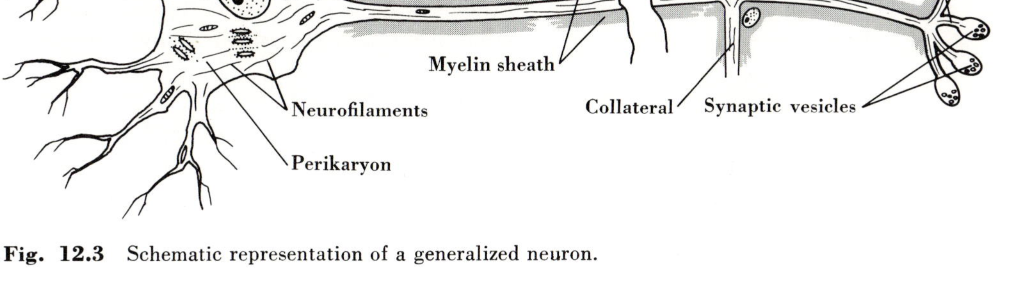

Fig. 12.3 Schematic representation of a generalized neuron.

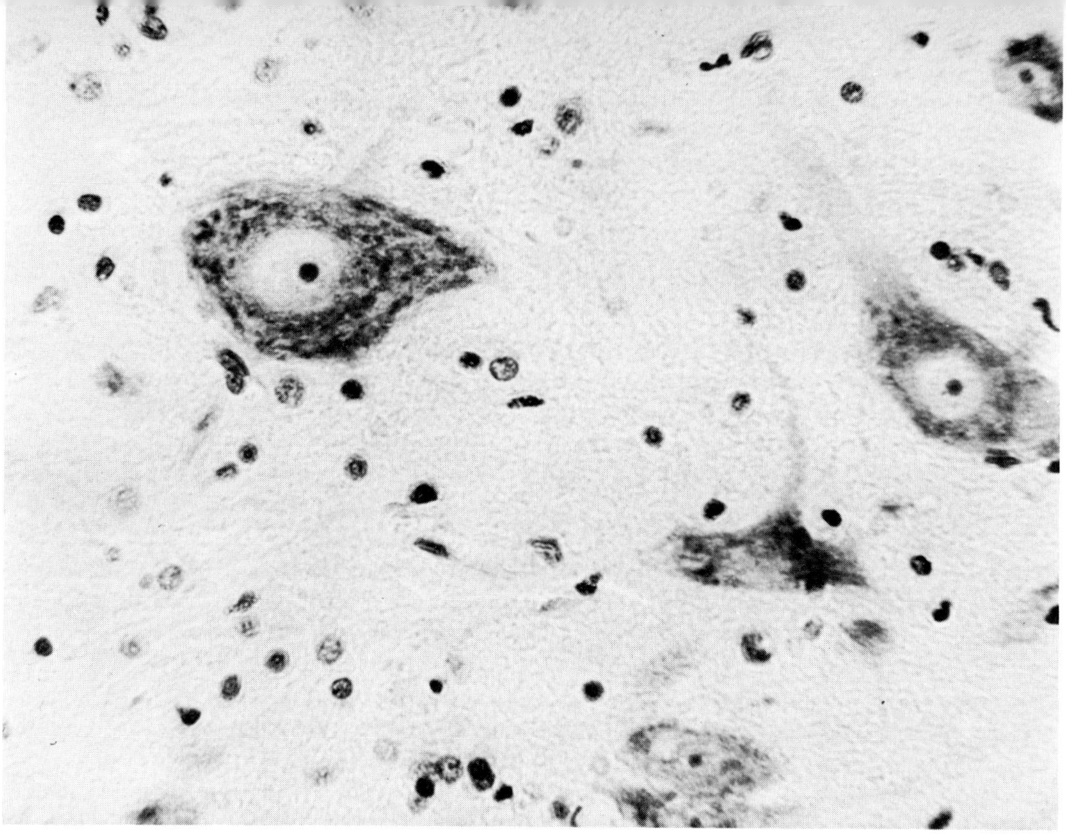

Fig. 12.4 Photomicrograph of motor neurons in the spinal cord, stained to demonstrate the Nissl substance. X450.

mechanisms of selective dendritic stimulation, such as that implied in complex behavioral responses, are not known, but it is suspected that products of Nissl substance activity are directly implicated.

A second cytoplasmic inclusion characteristic of neurons, especially when they are prepared by silver impregnation, is the *neurofibrils,* which occupy much of the perikaryon and are visible in the terminal processes. They are believed to be fixation artifacts, perhaps representing aggregations of the neurofilaments that have been described by electron microscopy. Neurofilaments are tubular fibrils that measure about 100A in diameter and are found throughout the neuron cytoplasm. They are believed to be protein in chemical composition, but their function is not understood. Microtubules, measuring about 200A in diameter, are also demonstrable by electron microscopy in the neuronal cytoplasm.

Pigment inclusions are not infrequently found in the cytoplasm of nerve cell bodies. Granules of melanin are common in cells of the olfactory bulb, in certain areas of the midbrain and hindbrain, and in some cells of the spinal and sympathetic ganglia. Lipofuscin is another kind of pigment found in nervous tissue. It occurs as brownish-yellow granules dispersed in the cytoplasm, but occasionally as large rounded masses. They are most common in nerve cells of aged individuals.

NERVE CELL PROCESSES

Neurons typically have a single axon, which arises from an area of clear cytoplasm, the axon hillock, which was mentioned above. Under usual conditions the axon conducts impulses only in one direction. Terminal arborizations, on the other hand, vary in number from one to many. Receptor or dendritic arborizations typically convey impulses toward the perikaryon and in most neurons (*see below*) they arise directly from the perikaryon. Axonal telodendria distribute nervous impulses to other neurons, muscle, or glands, and always arise from the end of an axon or collateral (Fig. 12.3). Neurons may be classified according to the number and disposition of their axonal and dendritic processes as unipolar, bipolar, and multipolar.

True unipolar neurons, those with only one process, are generally embryonic and transitional, although some sympathetic ganglion cells in the adult frog have been shown to be unipolar. Pseudounipolar neurons are neurons which were at one time bipolar, having had two apposed processes, but the processes have converged and fused adjacent to the perikaryon so that they bifurcate sharply from each other, thus acquiring a T- or Y-shape (Fig. 12.5). Most neurons of the cerebrospinal ganglia are of this type. The long process that extends from the point of divergence peripherally carries impulses toward the cell body. It has its terminal arborization, in this case a receptor arborization, in or associated with a sensory area. The other process carries impulses away from the cell body to the spinal cord or brain, where it terminates in a synaptic spray of axonal telodendria.

Many descriptive works in histology and physiology apply the term "dendrite" to all processes that convey impulses toward the perikaryon, and restrict the term "axon" to that process that conveys impulses away from the perikaryon. Such terminology implies that structural and functional differences exist between the two major processes of sensory neurons, when in fact there is apparently no difference at all. The sites of functional difference are the terminal arborizations, as stated above; they will be more fully described below.

Bipolar neurons have two axonal processes that emerge opposite each other from the cell body. Bipolar neurons are found in cerebrospinal ganglia of embryos, in the ganglia of the eighth cranial (auditory) nerve, and, in modified form, in the retina of the eye (Fig. 12.5).

Multipolar neurons are the most numerous. They are characterized by a single axon arising from one side of the cell body, and several dendritic or receptor processes (Fig. 12.5). The axon is generally long, with few collateral processes, and it has a small and uniform cross-sectional diameter. The receptor processes are usually short and branch profusely

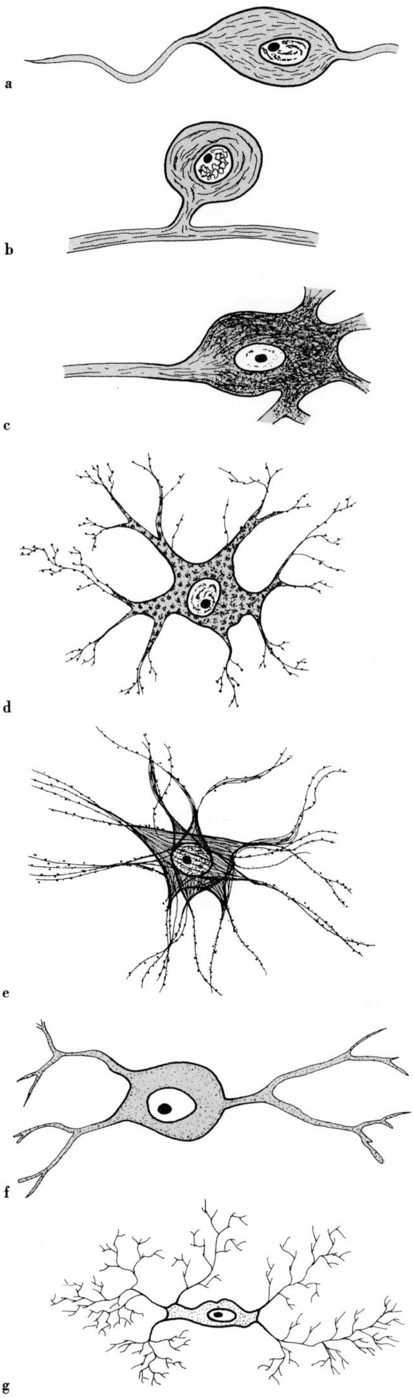

Fig. 12.5 The variety of cell types that characterize nervous tissue. There are basically three forms of neurons—bipolar, *a;* pseudo-unipolar, *b;* and multipolar, *c.* There are five types of nervous interstitial cells—protoplasmic astrocytes, *d;* fibrous astrocytes, *e;* oligodendrocytes, *f;* microglial cells, *g;* and Schwann cells. The last are figured elsewhere.

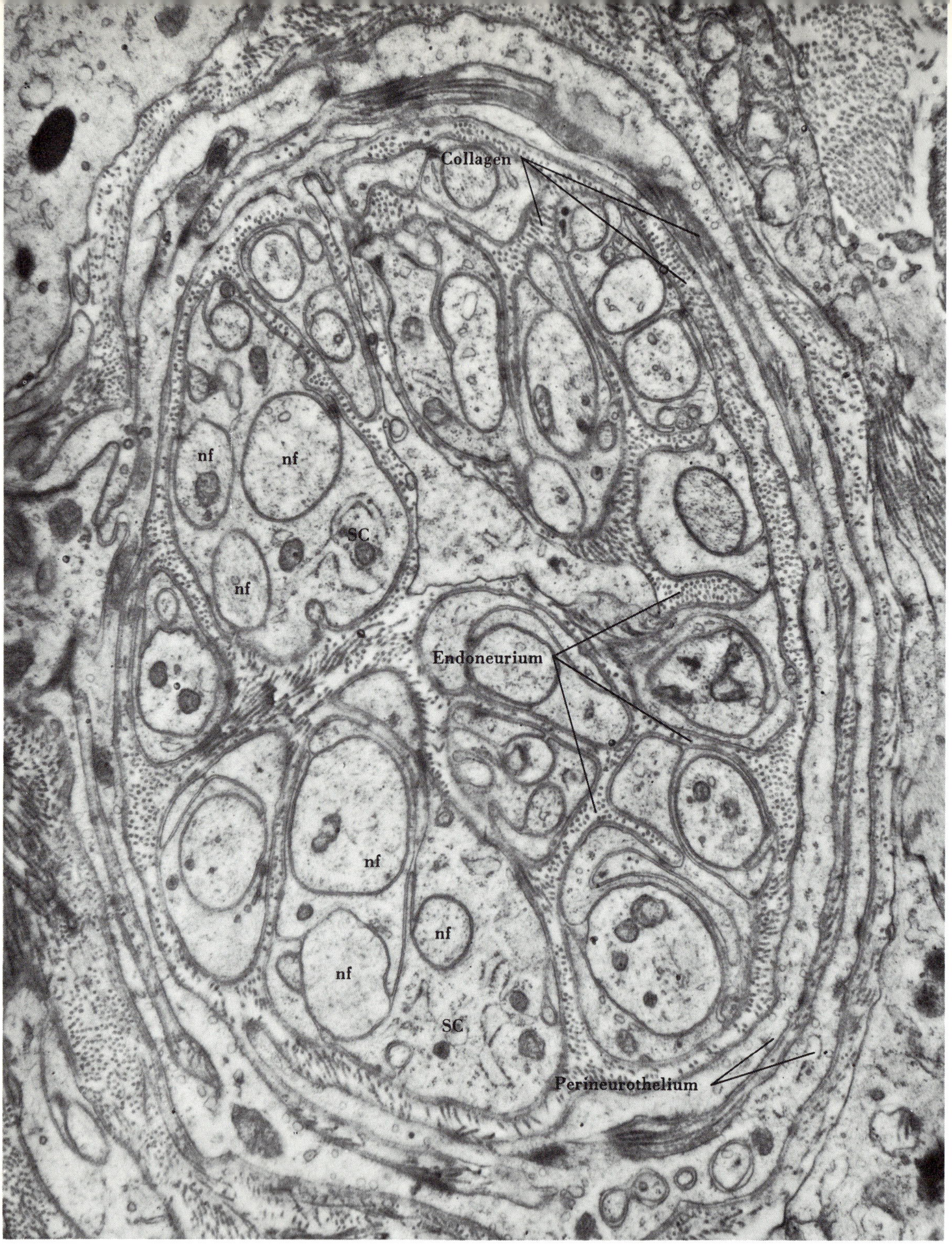

Fig. 12.6 Electron micrograph of a nonmyelinated nerve in a rat showing Schwann cells, *SC*, embracing multiple numbers of nerve fibers, *nf*. (Courtesy of Alan Peters.)

close to their origin from the perikaryon into slender, irregular fibers. Multipolar neurons are the most abundant type, and therefore represent the "typical" neuron. They include most of the neurons of the central nervous system and autonomic ganglia.

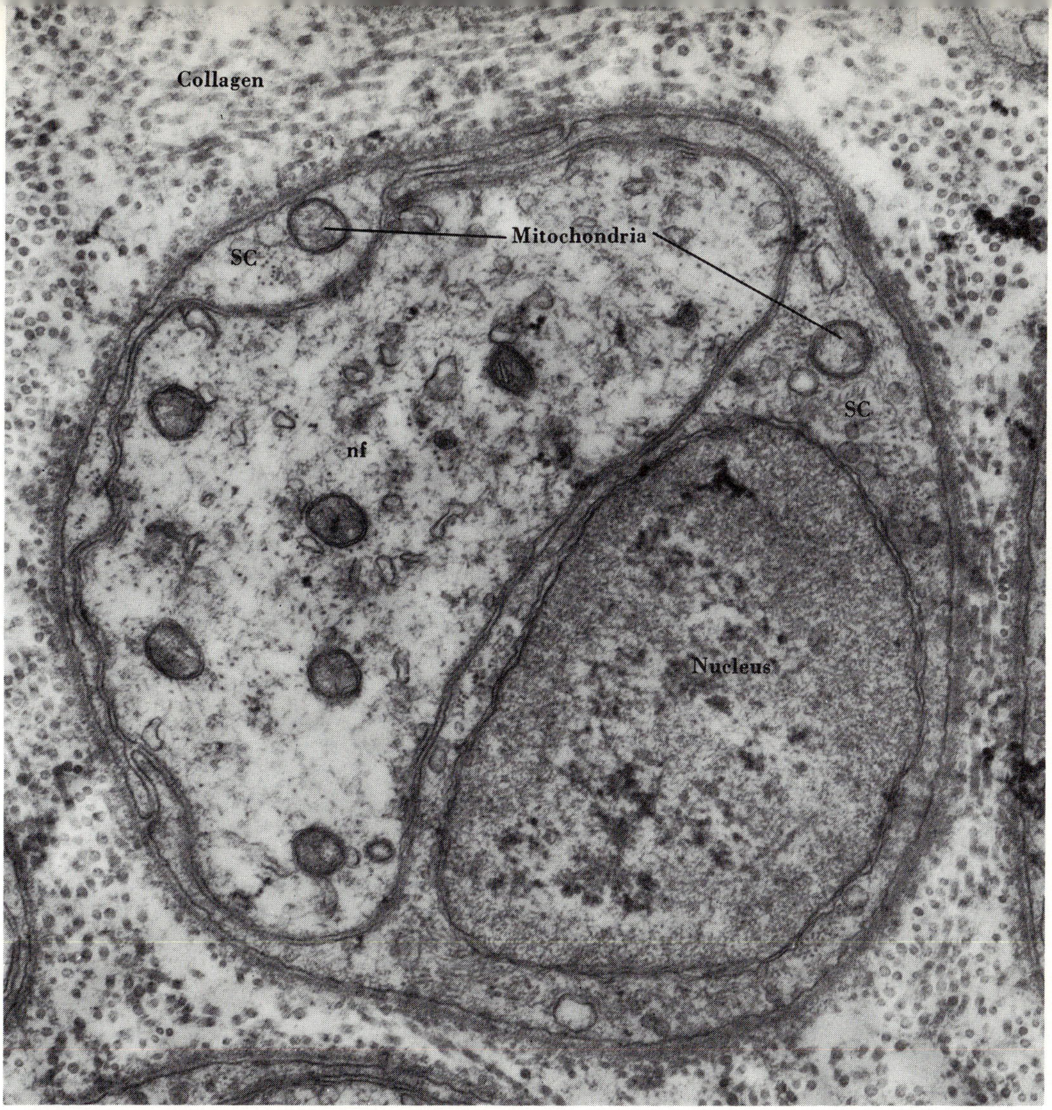

Fig. 12.7 Electron micrograph of a single nerve fiber, *nf*, in the hagfish, *Myxine glutinosa*, embraced by parts of two Schwann cells, *SC*. (Courtesy of Alan Peters.)

THE AXON. The axons of neurons are characterized by a rather uniform diameter and regular contour. The cytoplasm consists of a gelatinous neuroplasm in which are dispersed longitudinally oriented neurofibrils, microtubules, and occasional mitochondria. The limiting plasma membrane is called the *axolemma.* Axons are surrounded by nervous interstitial cells except at their origin from the perikaryon, at their termination, and at their points of branching. The interstitial cells surrounding axons of the peripheral nervous system are the Schwann cells, whereas those in the central nervous system are the glial cells, particularly the oligodendrocytes.

In nonmyelinated nerves of the peripheral nervous system, specifically the postganglionic fibers of the autonomic nervous system, several axons are embraced by invaginations of a singly arranged series of Schwann cells running along the length of the axons (Fig. 12.6). Nonmyelinated nerves of the central nervous system have no covering of such an intimate nature; rather, they course through a matrix of neuroglial cells.

Myelinated axons of the peripheral nerves are surrounded individually by a single, continuous row of Schwann cells. The most primitive condition occurs in Myxine (Fig. 12.7), in which one or two Schwann cells make a single layer around the axon. Actually this does not constitute myelin, and the cyclostomes are unique among vertebrates for their total lack of myelinated nerve sheaths. In

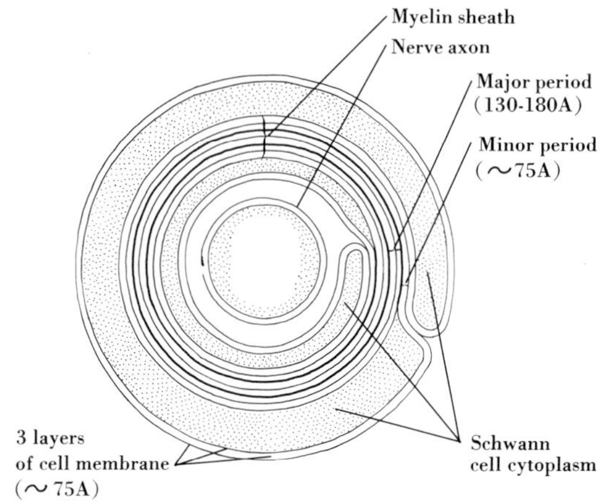

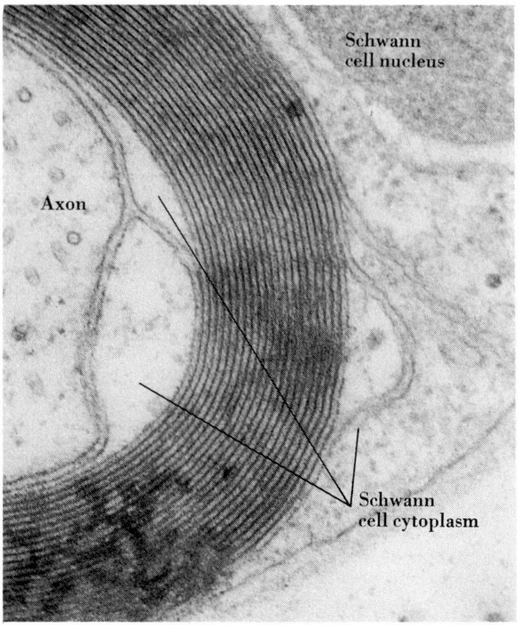

Fig. 12.8 The myelin sheath of peripheral nerve fibers. Diagram of a cross section of a myelinated nerve fiber, showing how the myelin sheath is formed by an encircling Schwann cell, *a*. (Courtesy of Alan Peters.) Electron micrograph of a portion of a myelinated nerve fiber, *b*. X135,000. (Courtesy of Humberto Cravioto.)

Fig. 12.9 Electron micrograph of a longitudinal section of a nerve fiber at a node of Ranvier. X64,000. (Courtesy of Alan Peters.) Compare with Fig. 12.3.

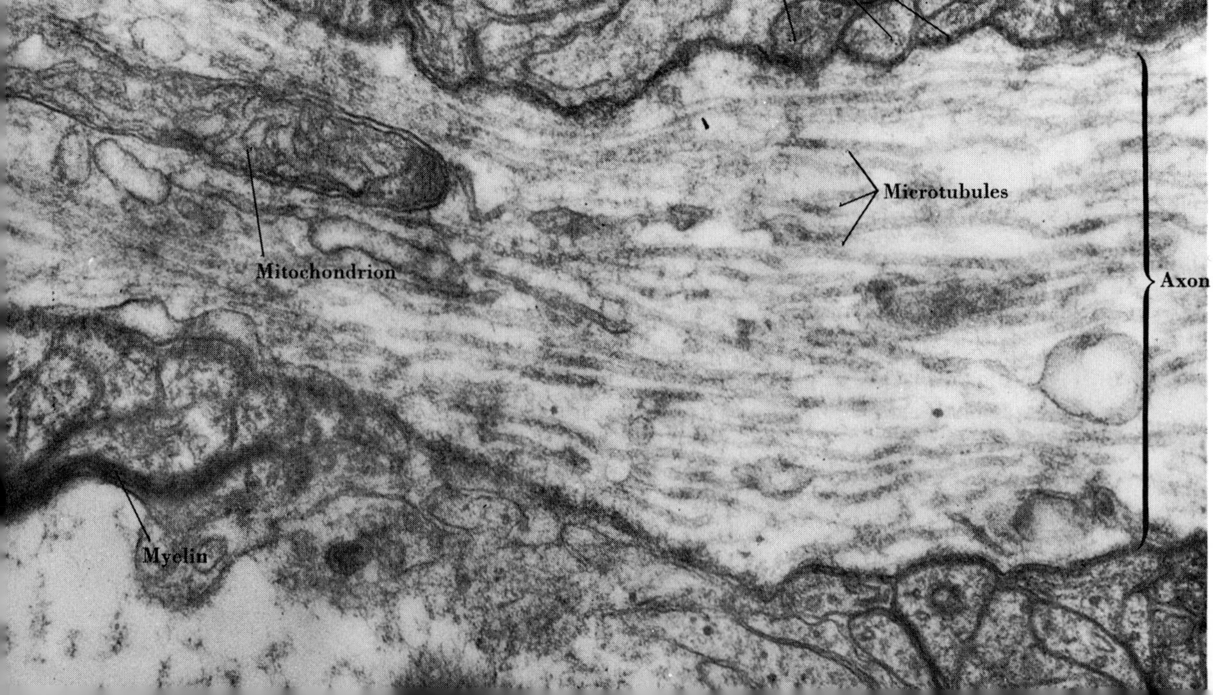

higher vertebrates a given Schwann cell is rolled several times (as many as fifty times in some nerves) around each axon so that the axon is invested in concentric layers of Schwann cell membrane (Fig. 12.8). In the process of rolling around the axon, the cytoplasm of the Schwann cell is squeezed out from between its layers of plasma membrane, so that the wrapping consists of concentric layers of plasma membrane sandwiched between two layers of Schwann cell cytoplasm (Fig. 12.8). The nucleus lies in the peripheral layer of cytoplasm. It will be recalled (see Chapter 2) that the plasma membrane consists typically of a bimolecular leaflet of lipid molecules sandwiched between two protein layers with a total thickness of approximately 75A. Accordingly, cross sections of the myelin sheath show it to be made of concentric layers of membranes with a minor period of approximately 75A between protein layers and a major period of about twice this (130–180A) between the lines formed by the fusion of the *inner* protein layers of the plasma membrane after expression of the cytoplasm (Fig. 12.8).

The lipid component of the Schwann cell membrane constitutes the myelin, and it includes a variety of lipid substances, such as cholesterol and various cerebrosides and phospholipids. The conjunction of two adjacent Schwann cells of the same axon is marked as a constriction, the *node of Ranvier* (Figs. 12.3 and 12.9). At the node the myelin sheath is discontinuous, but the outer membranes of the neighboring Schwann cells approximate each other so that not much of the axon is uncovered. The period between successive nodes is generally quite uniform for a given nerve fiber, but it is usually longer in axons of large diameter, with the consequence that internodal distances enjoy an enormous range, (about 75μ to more than 1000μ) from one axon to another.

The outermost layer of the encircling Schwann cell contains most of the cell's cytoplasm and it appears in optical sections as a cytoplasmic limiting membrane called the *neurolemma.* In some longitudinal sections of nerves the neurolemma may be seen intruding oblique clefts through the underlying myelin sheath to the axon. These intrusions are known as the *incisures of Schmidt-Lanterman.* Their significance is probably metabolic, for they are filled with a homogeneous material which is rich in carbohydrate and polypeptides, or they may be artifacts.

Myelinated fibers of the central nervous system derive their sheaths not from Schwann cells but from processes of glial cells, probably oligodendrocytes, which are believed to encircle the axons in a similar manner as Schwann cells (Fig. 12.10). Processes of a given glial cell may contribute to the myelin sheath of more than one axon, however. Because the sheath is formed by cell processes rather than by whole cells, the cytoplasm extruded from concentric layers is transferred to the parent cell, which may be removed some distance from the axon. Consequently a "neurilemma" is usually not found (compare Figs. 12.8 and 12.10), and nodes of Ranvier, although present, are considerably less conspicuous than they are in peripheral nerves.

Modified sheaths of various kinds have been described in the sympathetic ganglia of frogs. The endings of the preganglionic fibers are covered not by myelin, but by enveloping satellite cells (*see below*), and other fibers are covered by interstitial cells that form a protoplasmic sheath. In some cases the sheath

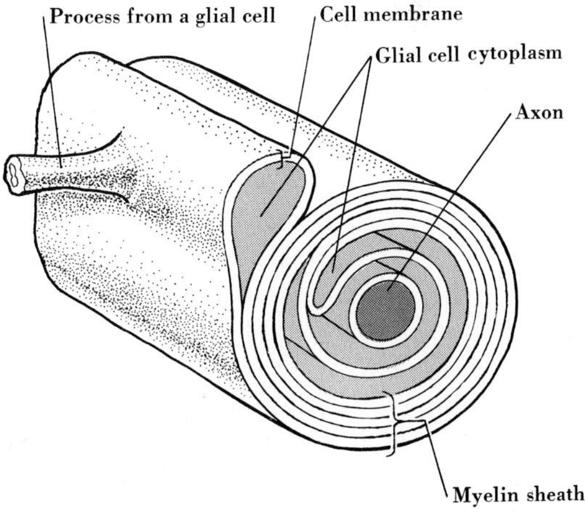

Fig. 12.10 Diagram showing that the myelin sheath in the CNS is formed by an expanded process of an interstitial cell that is wrapped around the nerve fiber. (Courtesy of Alan Peters.) Compare with Fig. 12.8.

cells overlap, thereby contributing more cell membranes to the sheath.

The function of the myelin sheath has not been definitely determined. Because of the high dielectric properties of lipid, it seems reasonable to assume that the sheath has an insulating function and prevents action currents and therefore nervous impulses from jumping across from one axon to another. On the other hand, there is a direct correlation between the velocity at which the impulse is conducted and the thickness of the myelin sheath. This may be a spurious correlation, because the thickness of the sheath also seems to be a function of the thickness of the axon, which in turn is directly correlated with conduction velocity. Biochemical data on neuroglia and Schwann cells support the hypothesis that they serve a trophic function for maintaining the structural and functional integrity of white matter and peripheral nerves. Presumably the myelin sheath is implicated in this function.

The myelin sheath may also be a device for facilitating extra-axonal conduction of nervous impulses, although this concept seems to conflict with the idea that the sheath is a kind of insulator. There is some evidence, however, that the action current which accompanies the nervous impulse "leaks out" at the node of Ranvier and is propagated as an electrotonic current through the extracellular fluid to the next node. This kind of jumping from node to node is called *saltatory conduction,* and it is much more rapid than membrane conduction. Such conduction is decremental, and so it must be reinforced at each node in order to be maintained. Although the theory of saltatory conduction helps to explain the high conduction velocity of myelinated fibers, it does present many obvious inconsistencies that are difficult to explain.

In addition to the morphological characteristics described above, axons are distinguished by a unique physiological characteristic, the so-called *action* or *spike potential.* At rest the concentration of sodium ions is much higher outside the axolemma than it is within the axon, so that an electric potential exists across the membrane, the so-called resting potential. A nervous impulse is generated when the axolemma becomes depolarized, that is, when it becomes permeable to sodium ions so that they may flow into the axon. The resultant influx of positively charged ions sets up an electrotonic current which depolarizes adjacent parts of the axolemma and thus propagates a wave of depolarization which can be measured electrically and is known as the action potential. In electrical recordings the action potential of nerve axons has the shape of a spike. The action potential is self-propagating and it advances along the membrane without decrement like a row of falling dominoes. A stimulus which is strong enough to initiate an action potential will elicit a maximum response, which is the basis of the "all or none" concept of nervous conduction. For further information on this subject the reader is referred to a work on neural physiology.

TERMINAL ARBORIZATIONS AND THE SYNAPSE.

Receptor Processes. The receptor or dendritic processes of neurons of the central nervous system are usually rather broad at the base and incorporate the same diversity of cytoplasmic organelles as does the perikaryon (Fig. 12.11). Some end bluntly, whereas others taper quickly into a variable number of more or less short, slender branches with numerous terminal twigs (Fig. 12.5) which usually end not far from the main process. The branches and twigs are often contorted and have a thorny appearance. The twigs often have swollen or knobbed "gemmules" at their ends where they make synaptic connection with the axonic arborizations of other neurons. The function of the dendritic processes is to receive impulses from the axonal telodendria of the synapse and pass them on to the conducting fiber by way of the perikaryon. In some instances, such as in the sympathetic ganglia of frogs, dendritic processes are reduced or lacking and impulses are picked up directly by the membrane of the cell body.

In contrast to the spike potential characteristic of nerve axons, dendritic processes are characterized by a graded response to stimulation. The response is in the form of an electrotonic current which is conducted with decrement to the axonal process.

The receptor processes of afferent fibers of

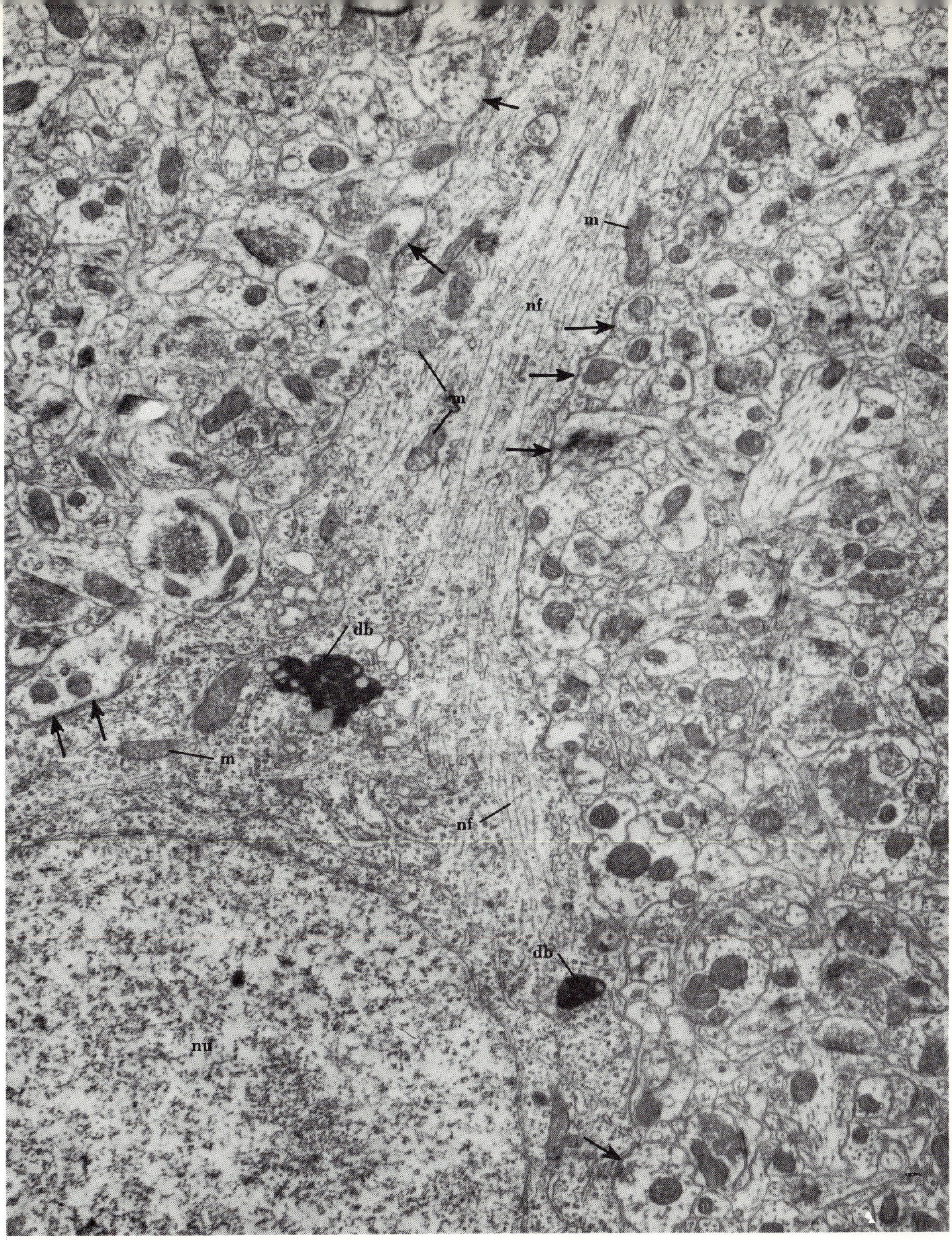

Fig. 12.11 Electron micrograph of the base of a dendritic process, showing neurofibrils, *nf;* mitochondria, *m;* and irregular dense bodies, *db* (lipofuscin?); and nucleus, *nu.* The process is embraced on both sides by numerous terminal twigs of nerve fibers in which may be seen neurofilaments, mitochondria, and synaptic vesicles. Note the tight junctions where the twigs synapse (*arrows*) with the cell membrane of the dendritic process. (Courtesy of Alan Peters.)

the peripheral afferent nerves are highly varied and often quite complicated, but they show either one of two arrangements. Either they occur as *free nerve endings* or as encapsulated nerve endings. Free nerve endings are found in epithelia, serous membranes, connective tissue, and muscles. They appear as slender, nonmyelinated tendrils insinuated between or wrapped around the cells of the tissue which they innervate. In epithelia they often end in little knoblike swellings between the epithelial cells. In other cases they may

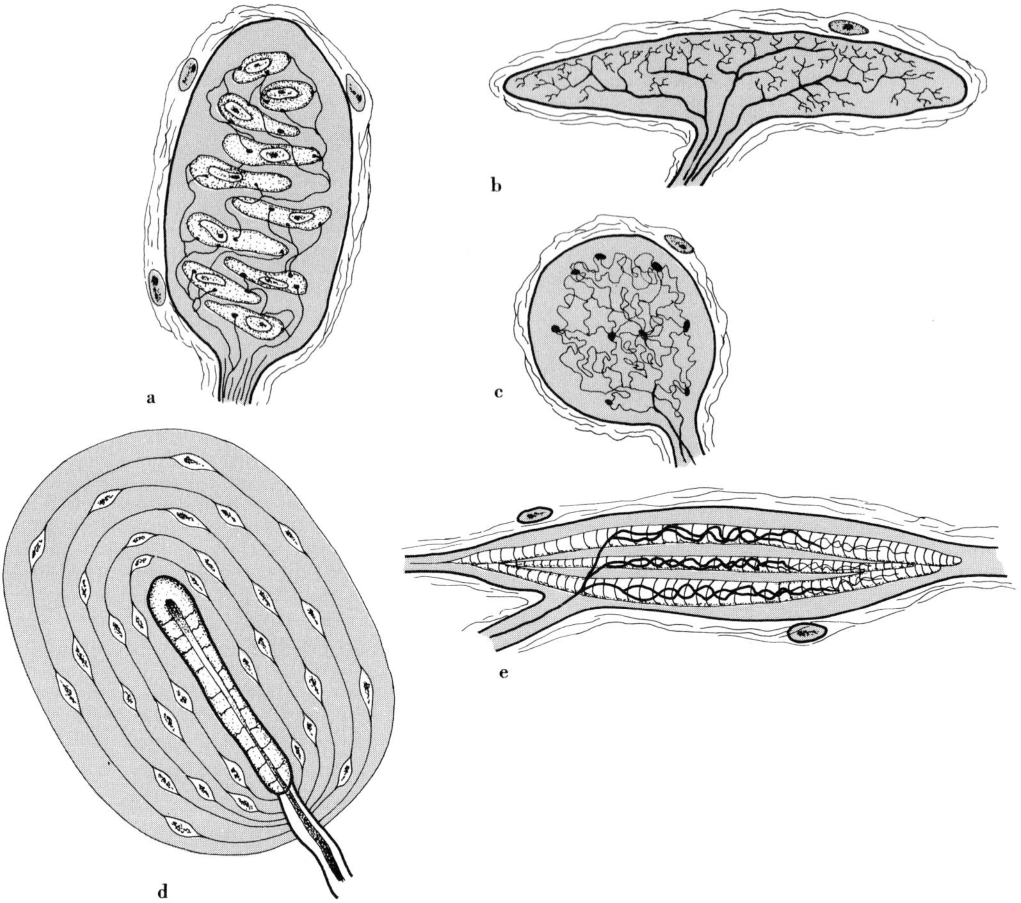

Fig. 12.12 Drawings of encapsulated nerve endings: *a*, Meissner's corpuscle (touch); *b*, end bulb of Ruffini (heat); *c*, end bulb of Kraus (cold); *d*, Pacinian corpuscle (pressure); and *e*, muscle spindle (stretch).

form a plexus or basket around a specialized sensory receptor cell, as will be seen below (see Chapter 13). In muscle tissue the afferent nerve endings may be found twined around individual muscle fibers. In connective tissue the free sensory endings may form rather complicated plexuses of anastomosing fibers.

Encapsulated sensory endings are characteristic of amniotes and represent an adaptation to terrestrial life. Such endings are invested in specialized connective tissue capsules in various ways. The major types are characterized in mammals as *end bulbs, Meissner's corpuscles, Pacinian corpuscles, muscle spindles,* and *muscle-tendon spindles* (Fig. 12.12). In end bulbs, such as the end bulbs of Krause or Ruffini, the nerve endings form a skein which is enclosed by a thin-walled connective tissue capsule of flattened cells. Such capsules are widely distributed in the skin and mucous membranes, in the conjunctiva of the eye, and in the heart, and they are sensitive to thermal stimuli. Meissner's corpuscles are large, ellipsoid, tactile receptors located in the skin of the digits and palm and sole. They may measure up to 180μ in length. The capsule is fibrous and encloses a core of flattened tactile cells separated by fibrous septa. The sensory nerve enters at one end of the capsule and sends branches spiralling through it, each branch terminating in a plexus of fibers adjacent to a tactile cell.

Pacinian corpuscles represent a lamellar type of corpuscle of macroscopic dimensions (up to 4 mm in length) and are sensitive to mechanical pressure. The capsule consists of several to many layers of fibrous connective tissue investing a cylindrical core of rather gelatinous intercellular fluid. The sensory nerve enters at one end of the capsule, gives off branches laterally within the core, and terminates in an expansion or knob at the far end of the core. Pacinian corpuscles occur in the deep subcutaneous connective tissue of the feet, in the connective tissue associated with the joints, tendons, and ligaments, in the penis, clitoris, mammary gland and nipple, and in certain visceral organs such as the pancreas, urethra, mesenteries, and parietal peritoneum.

Muscle spindles (neuromuscular spindles) are cylindrical capsules which enclose several small skeletal muscle fibers of the red variety, connective tissue, blood vessels, and both motor and sensory nerve fibers that enter at various points. The sensory fibers branch profusely within the capsule into receptor arborizations that form a plexus or a spiral around the enclosed muscle fibers. Muscle-tendon spindles are rather similar in structure, except that bundles of tendon instead of muscle fibers are present within the capsule. The tendons are innervated by complex arborizations of sensory nerve endings. They derive their name from their location, which is at the muscle-tendon junction.

Axonal Telodendria. In terms of function, the axonal telodendria are the opposite of the receptor arborizations, that is, they are concerned with the transmission of the nerve impulse rather than with reception of it. They occur generally as a spray of branches from the end of the axonal fiber and are nonmyelinated. Their tips are often expanded somewhat as the terminal boutons, in which mitochondria are numerous and the neurofibrils appear to form terminal loops. In most cases numerous small vesicles, the *presynaptic vesicles,* are observable by electron microscopy in the endings or boutons, and are especially abundant adjacent to the point of synaptic contact (Fig. 12.13). How the vesicles are formed is not understood, but they are believed to be secretory vesicles, probably containing acetylcholine or its precursor substance (or, according to some investigators, acetylcholine esterase), and to be concerned with transmission of the nerve impulse.

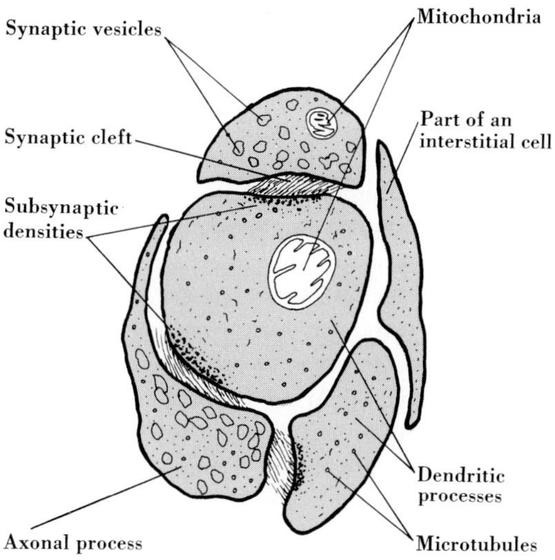

Fig. 12.13 Drawing of synaptic junction between two axonal (presynaptic) processes, and two dendritic (postsynaptic) processes. The synaptic vesicles are enclosed in a unit membrane and are generally concentrated adjacent to the synapse. The synaptic cleft contains an amorphous substance that is slightly electron-dense.

Axonal telodendria are concerned, therefore, with transmitting nervous impulses either to other neurons (by synapses) or to effector organs (by motor endings). As is the case with receptor dendrites, axonal telodendria conduct with decrement.

The Synapse. Although neurons are capable of transmitting nervous impulses along their fibers in either direction, ordinarily they do not—primarily because of the valvelike action of the nervous synapse. The synapse is the physical conjunction of the axonal telodendria and the receptor sites of two neurons. The axonal telodendria constitute the presynaptic fibers and the receptor sites are represented either by the perikaryonal membrane or by the receptor arborizations. The latter may also be referred to as the postsynaptic fibers. When a nervous impulse reaches the presynaptic fibers, acetylocholine or some other chemical mediator is liberated by them. This substance alters the permeability of the juxtaposed receptor site, thereby depolarizing its membrane and generating a new nervous impulse, which is promptly propagated by the second neuron to its axonal telodendria. The action of the acetylcholine liberated by the presynaptic fibers is short-lived, however, for the enzyme, acetylcholine esterase, which is found at synaptic junctions, hydrolyzes and thereby inactivates acetylcholine almost immediately. This permits the membrane of the postsynaptic site to become repolarized quickly and thus ready to receive another impulse. When acetylcholine esterase is experimentally blocked in some way, the lingering presence of acetylcholine prevents recovery of the postsynaptic fiber, thereby rendering it refractory to subsequent stimulation. Synaptic thresholds, i.e., the levels of intensity of stimulation required to generate a new impulse, are profoundly affected by a number of drugs and other chemical substances which act either on the permeability of the synaptic membranes themselves or on the choline-cholinesterase system. Such neurological drugs include reserpine, esserine, serotonin, histamine, and the catecholamines, among others.

When the action potential of the axon reaches the telodendria, it is transformed into a graded or decremental potential, as mentioned above. The effect that this potential has on the postsynaptic site depends upon a number of factors, some of which may vary from site to site. The greater the frequency of action potentials reaching the presynaptic fibers, the greater the presynaptic potential and, presumably, the greater the amount of chemical mediator released. Depending thus on the strength of the presynaptic potential, and upon the kind of chemical mediator released, the effect on the postsynaptic site may be excitatory, facilitory, or inhibitory. An excitatory effect is one which results in the generation of a new action potential in the postsynaptic axon. A facilitory effect is one which hypopolarizes the postsynaptic neuron, thus lowering its threshold and rendering it more easily excitable without actually generating an action potential. An inhibitory effect is one which hyperpolarizes the postsynaptic site, thereby raising its threshold and rendering it more difficult to excite. If one visualizes the axonal telodendria of one neuron synapsing with receptor sites on several other neurons, one may see how this mechanism of modulation can determine how many of these secondary neurons will be excited (or inhibited) by impulses reaching them from the primary neuron. "Although the axons may be considered relatively simple message-relaying cables, the synaptic regions are the *choice points* at which complex 'decisions' are made concerning how and along what pathways messages will be relayed. The great complexity of central nervous control functions must be caused by the delicate and highly variable discharging and generator processes occurring at the synaptic regions" (Altman, p. 197).

Two types of synaptic associations are commonly found in vertebrates. In *axosomatic synapses* the axonal telodendria make synaptic contact with the perikaryon of another neuron (Fig. 12.14). In *axodendritic synapses* contact is between the axonal telodendria of one neuron and the receptor arborizations of another (Fig. 12.14). Combinations of the two types also occur, wherein the axonal telodendria synapse with both the cell body and the receptor arborizations. In all cases the contact is characterized by very close approximation of the membranes of both terminals, with a space of only about 200A between them. Small densities reminiscent of desmosomes are often observed by electron microscopy on opposite sides of the apposed membranes (Fig. 12.15). So-called "tight junctions" or "electrotonic junctions" are observed in some synapses. Here there is no intervening space between apposed cell membranes; indeed, the outer component of each membrane is fused. At such sites it is believed that transfer of the nervous impulse may be a direct electrical phenomenon, and therefore extremely rapid.

MOTOR ENDINGS. The motor endings of postganglionic fibers of the autonomic nervous system arise generally from complicated plexuses in close association with the effector organ. The endings may be simple terminations which make contact with the surface of the muscle fibers or gland cells, as the case may be, or they may show one or more terminal swellings, called *visceral motor endings*. The motor endings of myelinated nerves on skeletal muscle form clusters of terminal

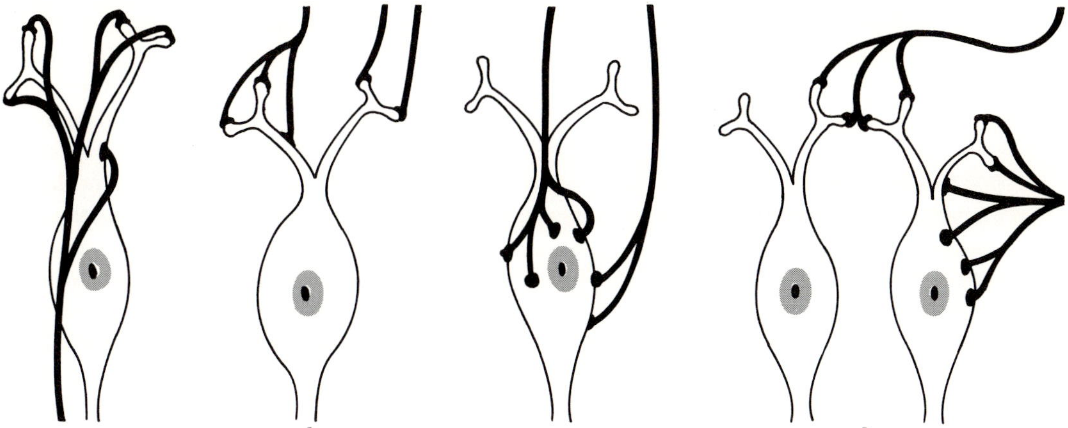

Fig. 12.14 Diagram of various synaptic relations between neurons. The axonal process of one neuron may lie antiparallel to the cell body and dendritic processes of another neuron, *a*; or be oppositely apposed, *b*; or approached from one side, *d*. Axodendritic synapses are shown in *a*, *b*, and *d*. Axosomatic synapses are shown in *c* and *d*. Both kinds of synapses are shown in *d*. Note that one neuron may receive impulses from more than one axon (*b*, *c*, *d*), and one axon may transmit an impulse to more than one neuron (*d*).

swellings, the *motor end plates,* which lie in shallow grooves or depressions of the sarcolemma (Fig. 12.16). The sarcolemma of these grooves may also be richly infolded, presumably in order to provide more surface for the action of the chemical transmitter released by the nerve ending. In no case, however, is there any evidence of protoplasmic continuity between the end plate and the muscle cell. The nuclei of the muscle cells tend to be concentrated in the vicinity of motor end plates. It should also be noted that the neurilemma is absent in the zone of immediate contact between the end plate and the sarcolemma.

The number of muscle cells which are innervated by a single neuron is directly related to the refinement of muscular function. In large muscle organs characterized by relatively gross movements, as in the extrinsic appendicular muscles (e.g., *m. pectoralis*), branches of a single neuron innervate several muscle fibers. In contrast with this is the nearly one-to-one relation found in the extrinsic muscles of the eye.

The ultrastructure of motor endings is similar to that of presynaptic nerve endings. Mitochondria and synaptic vesicles are both numerous. Except for the sympathetic nerves, the motor endings appear to be cholinergic. Sympathetic endings liberate adrenalin or noradrenalin to effect motor response.

Fig. 12.15 Electron micrograph of axodendritic synapses. The long dendritic process contains microtubules, *mt,* and shows cytoplasmic densities (*arrows*) at the cell membrane where it synapses with presynaptic fibers, *pf*. The latter, identified by their numerous synaptic vesicles, *sv,* also synapse with postsynaptic fibers, *ppf,* some of which are doubtless branches of the dendritic process. Part of a myelinated nerve fiber appears in one corner. (Courtesy of Alan Peters.)

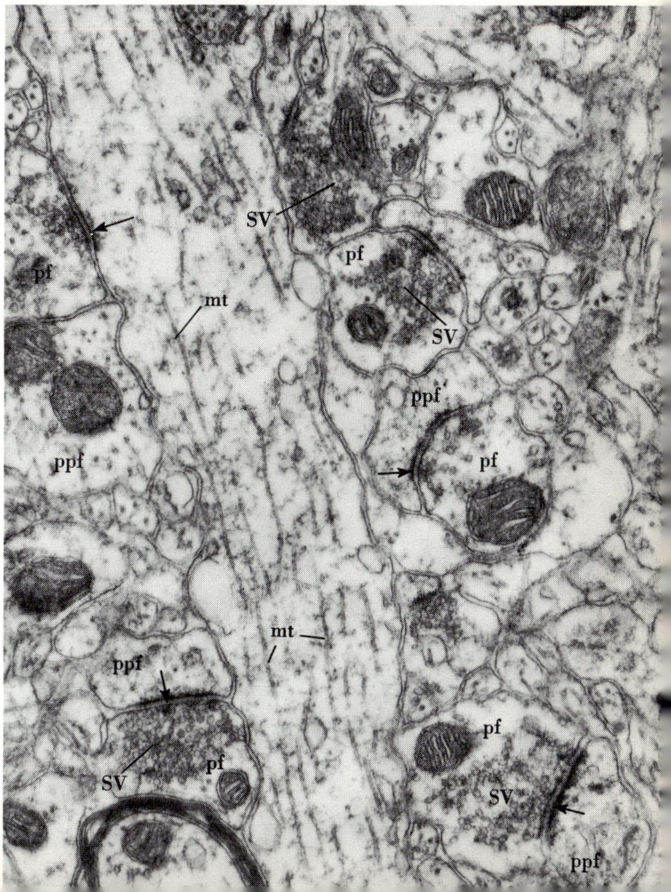

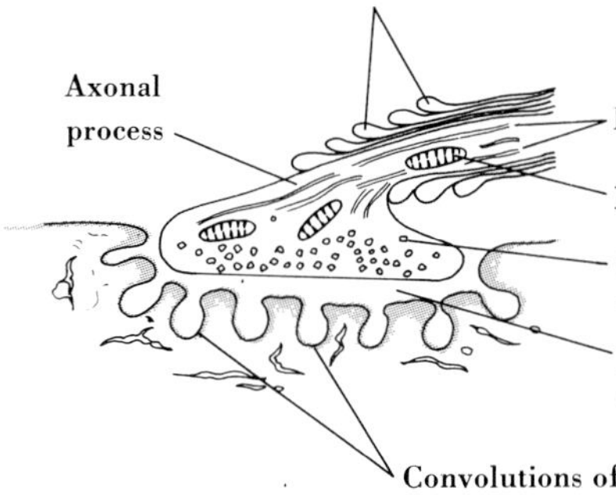

Fig. 12.16 Diagram of a motor end plate.

Nervous Interstitial Cells

The major categories of nervous interstitial cells are the ependymal cells and the neuroglial cells (glia) in the CNS, and the Schwann cells in the PNS.

EPENDYMA

Ependymal cells are derived initially from the superficial layer of the neural plate and they form the epithelial lining of the ventricles of the brain, the epithelial lining of the neural canal of the spinal cord, and the *lamina epithelialis* of the *telae choroideae.* Except in the last, the ependymal cells tend to be tall cells which taper at their bases. The tapered ends may become fibrous, similar to the fibers of neuroglial cells, and similarly branched. The functions of the fibrous processes are not known, but they may serve as channels for the movement of materials required for the metabolism of the parent cells. Ependymal cells are ciliated in all vertebrate embryos, and in the adults of fishes and amphibia. Cilia persist on some of the ependymal cells in higher forms as well.

The telae choroideae are richly vascularized membranes that differentiate from the roof of the third and fourth ventricles of the brain. They consist of rather extensive intrusions of the pia mater (*see below*) into the ventricles, intrusions that are covered by a continuous layer of modified ependyma, the *lamina epithelialis.* The loose connective tissue of the pial component contains a rich vascular network, the *choroid plexus* (Fig. 12.22), which nourishes the cells of the lamina epithelialis. The tela choroidea derived from the roof of the third ventricle is generally the more extensive, for it bifurcates and passes through the foramina of Monro into the lateral ventricles, where it fans out along the dorsal and lateral walls. In fishes and amphibians the tela choroidea derived from the fourth ventricle is also a rather prominent membrane, but in amniote vertebrates it is considerably less conspicuous. The cells of the lamina epithelialis are cuboidal in shape, and they are supported by a basement lamina. Their luminal surface, which is ciliated in most species, forms an irregular brush or striated border, which is consistent with their secretory function. The basal portions of the cells are joined to each other by interdigitations of the cell membrane (Fig. 12.17). The nucleus is round and basophilic. The cytoplasm contains many mitochondria of both filamentous and granular form, free ribosomes, lipid droplets, and clear vacuoles. In mammals extensive rough-surfaced endoplasmic reticulum occurs, often arranged in stacks or whorls.

The telae choroideae represent the principal source of the cerebrospinal fluid. The ependymal cells of the neural canal are believed to make a minor contribution. Cerebrospinal fluid is clear and slightly viscous, with chemical and physical characteristics similar to those of the aqueous humor of the eye. It is sparsely populated by lymphocytes which have migrated from the capillary networks of the choroid plexuses through the lamina epithelialis. The circulation of the cerebrospinal fluid is mainly from anterior to posterior in the brain, and the reverse in the spinal cord. Its principal avenues of exit are the foramina of Magendie and of Luschka in the roof of the fourth ventricle. Some is probably absorbed by the ependymal cells. From the foramina of Magendie and Luschka the cerebrospinal fluid passes into a sinus, the cerebellomedullary cistern, and thence into the subarachnoid spaces (*see below*). From the subarachnoid spaces it is largely absorbed via the arachnoid villi into endocranial venous sinuses, although it may pass into certain lymphatic vessels which drain the central nervous system, or it

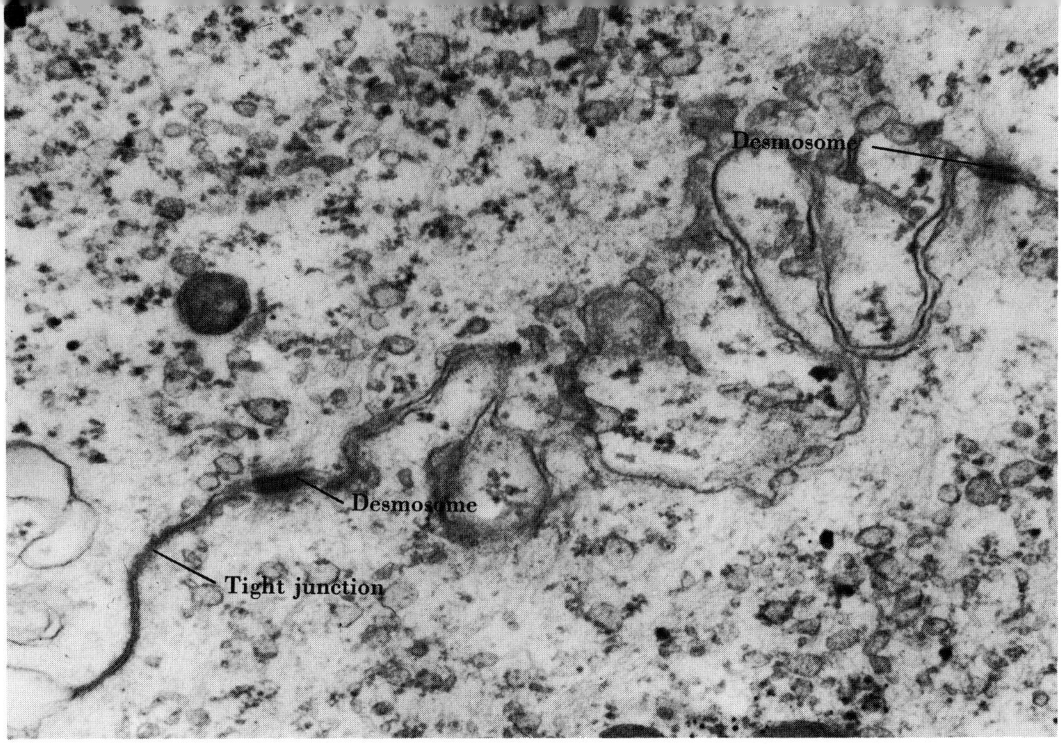

Fig. 12.17 Electron micrograph of cell junctions in the choroid plexus of the 3rd ventricle of *Necturus*. Approximately X43,000. (From S. J. Carpenter. 1966. J. Comp. Neurol. 127:413–434).

may issue through spaces between the cranial nerves and their epineural sheaths and be picked up by lymphatic vessels. The cerebrospinal fluid that flows through the subarachnoid spaces enveloping the spinal cord seeps out through spaces between the roots of the spinal nerves and their dural membranes into venous sinuses strategically located for this purpose. The function of the cerebrospinal fluid is believed to be twofold: nutritive and hydrostatic. The latter is probably the primary function. Because of the very limited amount of intercellular space in nervous tissue (3–5 percent total volume in nervous tissue as compared with 20 percent in skeletal muscle, for example) some kind of "expansion tank" must be provided to compensate for changes in volume occasioned by vasomotion within the central nervous system. The ventricles are well situated for this function. One may visualize that as the brain swells with vasodilatation, the ventricles passively contract and the cerebrospinal fluid is expressed through its escape ports. Also along the lines of hydrostatic function, a shock-absorbing role has been proposed for the cerebrospinal fluid. Any localized pressure applied to the central nervous system, as by a blow, for example, would be transmitted fairly uniformly by the cerebrospinal fluid throughout the central nervous system, with consequent widespread dissipation of the mechanical energy involved, thereby diminishing any injurious effect.

NEUROGLIA

Neuroglial cells are generally classified into four major types in the higher vertebrates: protoplasmic astrocytes, fibrous astrocytes, oligodendrocytes, and microglial cells. The first three types differentiate from spongioblasts, which are derivatives of the embryonic neural ectoderm. The microglial cells are believed to be mesenchymal derivatives.

PROTOPLASMIC ASTROCYTES. Astrocytes, or *astroglia,* are stellate cells with branching processes. The processes of protoplasmic astrocytes are rather numerous, profusely branched and folded, and they contain many round or oval granules, the *gliosomes.* Gliosomes are also present throughout the rather scanty perinuclear cytoplasm. The nucleus stains lightly and lacks a nucleolus. Protoplasmic astrocytes are most abundant in the gray matter of the central nervous system, where they are associated with neuron cell bodies or,

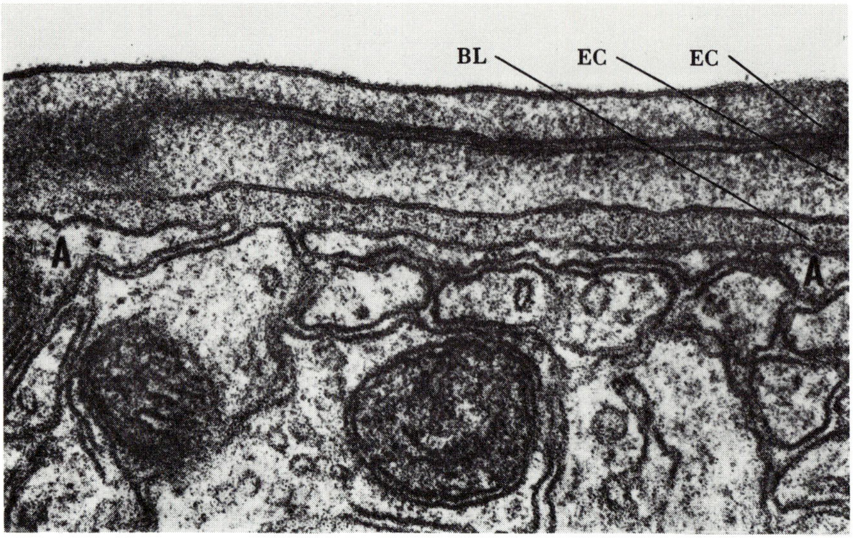

Fig. 12.18 Electron micrograph showing the perivascular end-feet of two astrocytes, *A*, next to the basement lamina, *BL*, of two overlapping endothelial cells, *EC*, of a capillary. (From T. S. Reese and M. J. Karnovsky. 1967. J. Cell Biol. 34:207–217.)

more commonly, with blood vessels. Where their protoplasmic processes touch the blood vessel walls, they are expanded as *perivascular feet* (Fig. 12.18). Oligodendrocytes are also found in small numbers adjacent to small blood vessels, and they, too, have been observed to form perivascular feet. The capillaries and small blood vessels in the central nervous system are invested in a sort of sheath formed in this way, and this is proposed to be the morphological basis of the so-called "blood-brain barrier," which is discussed more fully below.

FIBROUS ASTROCYTES. These structures resemble the protoplasmic astrocytes except in certain respects. For one, their cell processes are less numerous and often broader. The cytoplasm of both the cell body and its processes contain characteristic, fine, unbranched fibers, which by electron microscopy appear as delicate tonofibrils. Finally, the fibrous astrocytes are most abundant in the white matter of the central nervous system and at the pia-glial membrane. As do protoplasmic astrocytes, they form perivascular foot plates around small blood vessels (Fig. 12.5).

OLIGODENDROCYTES. Oligodendrocytes are somewhat smaller than astrocytes and have a correspondingly smaller nucleus, which is round or oval. Their cytoplasmic processes are less numerous and show few branches. Oligodendrocytes of gray matter are found most commonly in close association with neuron cell bodies, where they are known as *perineuronal satellites*, or closely applied to small blood vessels, as mentioned above. They have relatively few mitochondria and rather scanty rough-surfaced endoplasmic reticulum. Cells which are intermediate between oligodendrocytes and astrocytes are not uncommon. The oligodendrocytes of white matter, on the other hand, have more extensive endoplasmic reticulum, free ribosomes, and more mitochondria. They are arranged in rows in the white matter, where they function as neurolemmal cells, as described above (Figs. 12.5 and 12.10).

MICROGLIA. Microglia are the smallest of the neuroglial cells (Fig. 12.5). Their nuclei are small, strongly basophilic, and usually elongated, although at times they are irregular in shape. Cytoplasm is scanty and formed into numerous processes with short, irregular twigs or spines. Neither fibrils nor gliosomes are present. In situations of inflammation or trauma, the microglia become actively amoeboid and phagocytic. Although more numerous in gray matter than in white, they occur throughout the central nervous system, where they are most commonly associated either with blood vessels or with neuronal cell bodies.

FUNCTIONS OF NEUROGLIA. As mentioned above, biochemical data on neuroglia and Schwann cells suggest that they have a trophic function. Part of this role seems to consist of providing respiratory and other enzymes to the conducting pathways which are remote from enzyme-forming sources in the perikaryons of the neurons. This idea is suggested by the observations that oligodendrocytes and, to a considerable extent, Schwann cells and satellite cells (1) possess considerable RNA; (2) show active incorporation of RNA and protein precursors; (3) have a high cytochrome exidase activity and high QO_2; and (4) are rich in DPN and TPN diaphorases and in malic and lactic dehydrogenases. As a matter of fact, these properties also are consistent with the idea that nervous interstitial cells, particularly the oligodendrocytes, may be concerned with transmission of nervous activity.

After the discovery of Ehrlich in 1885 that certain aniline dyes fail to diffuse from the blood to the brain, the concept that some sort of "blood-brain barrier" exists was explored and, until rather recently, has been widely accepted. The neuroglia were believed to represent the morphological basis of the blood-brain barrier. Electron microscopy has failed to provide any convincing evidence that any such structural barrier exists, for the capillary endothelium in the brain does not differ significantly from that in most other tissues. Moreover, the pericapillary sheath formed by astrocytes and, to a less extent, by oligodendrocytes and microglia does not appear to be complete, covering no more than approximately 85 percent of the capillary wall. Finally, the majority of physiological phenomena that have been ascribed to a blood-brain barrier can be explained either as well as or better in terms of central nervous system metabolism and special physiological properties of the capillary endothelium.

SCHWANN CELLS

The neuroglia find their counterpart in the peripheral nervous system as the Schwann cells. These cells have already been described in connection with the myelin sheaths of the axonal fibers of peripheral nerves (*see above*). They are also clearly discriminated as small cells clustered around the multipolar neurons of the cerebrospinal ganglia and autonomic ganglia (Fig. 12.19). Like their counterparts in the central nervous system, the Schwann cells are believed to be concerned primarily with supporting the metabolic functions of the neurons with which they are associated.

ORGANS OF THE NERVOUS SYSTEM

The detailed histologic structure of the organs of the nervous system (especially of the CNS) of vertebrates is far too extensive for a work of this scope; thus only the more generalized aspects will be considered. Students are referred to the three-volume work of Ariens Kappers *et al.*, cited at the end of this chapter, for details.

Any understanding of the structure of the organs of the nervous system depends upon some understanding, however rudimentary, of the mechanics of nervous function. In essence the nervous system is that system of organs which, primarily by means of nervous im-

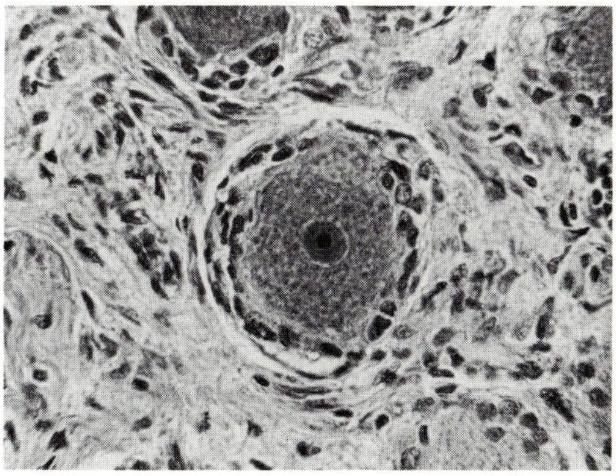

Fig. 12.19 Photomicrograph of a neuron in a spinal ganglion of a cat, encircled by the nuclei of Schwann cells. X500.

pulses, integrates the activities of all the diverse parts of the body in relation to the environment. To do this, the nervous system utilizes three components: a receiver component, an effector component, and, between the two, a selector or integrator component. The receiver component is represented by afferent nerves, which transmit signals from the environment in the form of nervous impulses. The signals may originate from any kind of environmental change, such as temperature, light intensity or movement, chemical composition of the environment, or some physical factor that might be interpreted as pressure, friction, sound, etc. The integument and/or organs of special sense are equipped with receptor organs which are selectively sensitive to specific kinds of environmental changes. Hence there are thermal receptors, photoreceptors, chemoreceptors, tangoreceptors, etc.

The signals from the environment are received by the selector or integrator component, which is represented by the central nervous system. Here the decisions are made as to where the signals should be sent, i.e., to what effector organs. To a considerable degree, if not entirely so, the central nervous system is "programed" for its decision-making operation. Where it transmits the signals depends upon the sensory nerves or combination of sensory nerves from which it has received such signals. Once the decision is made, the signals are passed on to the appropriate effector organ, which brings about an adjustment appropriate to the environmental change in question.

One of the simplest kinds of nervous function is the disynaptic spinal reflex arc. If one lightly pinches the toe of a frog, a nervous impulse is generated in the free nerve endings of sensory nerve fibers in the skin of that toe. The impulse passes via the corresponding nerve axon to the spinal cord. In the gray matter of the cord the impulse is transferred across a synaptic junction to a relay or *internuncial neuron,* then across another synapse to a motor neuron. The last passes the impulse out of the spinal cord via its axon to certain flexor muscles in the leg on which the pinched toe is appended. The flexor muscles respond by contracting, and the foot *plus* its offended toe are withdrawn from the source of irritation, the pinch. If the pinch is somewhat stronger, the sensory impulses excite a greater number of synaptic junctions in the spinal cord, some of which involve neurons that carry impulses across to the opposite side of the cord. Such *commissural neurons* excite motor neurons in this opposite side, motor neurons which carry impulses to extensor muscles in the leg contralateral to the one whose toe was pinched. The effect of such a pinch is therefore twofold: not only is the offended foot withdrawn as before, but the opposite foot kicks out, thereby pushing the whole frog away from the source of irritation. The frog could be "asleep" while all this was going on and not even awaken. The same responses can be elicited in a frog whose brain has been removed, destroyed, or severed from the spinal cord.

If a *very strong* pinch is applied to the toe of a frog with brain intact, a burst of sensory impulses is transmitted not only to ipsilateral (same side) and contralateral (opposite side) motor neurons of the corresponding segmental level but also to other levels of the central nervous system, including the brain. The synaptic patterns then understandably become quite complex, but, as far as can be determined, they follow precisely predetermined pathways which ultimately find expression in complex behavioral responses. If the frog was asleep before this, it will certainly awaken now. It will go hopping off, if on land, or, if its receptor organs "tell" it that it is in the water, it will swim away; for in the brain the sensory impulses that originated from the pinch are correlated with sensory impulses coming in from the eyes, the semicircular canals of the inner ear, and many other receptor sites, and all are integrated into a pattern of meaningful motor and effector responses.

From the foregoing highly simplified description of nervous function it becomes evident that the spinal cord is programed to integrate rather uncomplicated reflex activities, whereas the brain is used to integrate activities which involve many parts of the body—but especially those activities which require information from the major sense organs.

The Spinal Cord

The entire central nervous system of Amphioxus is similar to the spinal cord of vertebrates. For this reason, among others, it is thought that the vertebrate spinal cord presents the basic features of the central nervous system, while the brain presents modifications related initially to the reception and correlation of sensory impulses originating in the cranially concentrated organs of special sense (olfaction, vision, equilibrium, and hearing). Advances in vertebrate evolution have been accompanied by the brain's assuming more and more the functions of correlating information received from other regions of the body, while the spinal cord has undergone relatively minor changes, however significant and progressive.

The basic feature of tissue organization in the spinal cord is the segregation of white matter into longitudinal strips surrounding the gray matter. As noted above, white matter consists mainly of myelinated nerve axons and nervous interstitial cells, whereas gray matter consists of neuronal perikaryons, synapses, and interstitial cells. The progressive changes seen in the vertebrate classes are in (1) the degree of consolidation of white matter into clearly defined tracts (*funiculi*); (2) the zonation of gray matter into areas of common function (*columns* or *nuclei*); and (3) the appearance of longitudinal grooves (*sulci* or *fissures*).

In cyclostomes the spinal cord extends nearly the full length of the vertebral column. It is quite flattened and rather uniform in diameter throughout its length, and it has no sulci or fissures. Although the demarcation between gray and white matter is not clear, the gray matter is concentrated in the middle around the central canal (Fig. 12.20). In elasmobranchs and bony fishes the demarcation becomes more pronounced, with some evidence of functional zonation as well. The dorsal column of gray matter is largely sensory in function and is characterized by the so-called *substantia gelatinosa* Rolandi. This is a complex of dendritic processes, nonmyelinated and thinly myelinated fibers, and synaptic junctions formed between incoming somatic sensory axons and secondary or intercalated neurons. The substantia gelatinosa is most highly developed in mammals, particularly in the ungulates, where it is concerned to a considerable degree with the distribution of impulses coming in from integumentary sensory receptors. The ventral gray matter of fishes is in the form of paired columns that consist mainly of somatic motor neurons. The dendritic processes, however, tend to be long—reaching up into dorsal or dorsolateral areas to synapse with other neurons. The distribution of visceral neurons is not clear in the lower vertebrates, but presumably they are most abundant in the zone between dorsal and ventral columns. A dorsal median sulcus is introduced and is present in the spinal cords of most fishes (Fig. 12.20).

The spinal cords of amphibians possess both a dorsal median sulcus and a ventral median fissure. White matter is more sharply segregated into longitudinal tracts, the funiculi and the gray matter have assumed the shape of a butterfly, a characteristic of higher vertebrates (Fig. 12.20). Correlated with the extensive development of paired appendages in anuran amphibians are cervical and lumbar enlargements of the spinal cord. They are also present in urodeles, but not nearly so prominently. These enlargements are most pronounced in amniotes (except for snakes and limbless lizards), and are believed to have reached their maximum dimensions in the giant dinosaurs. (The enormous lumbar enlargement of some dinosaurs has been said to enable them to reason *a posteriori* as well as *a priori.*) A special modification of the lumbar enlargement occurs in birds. The dorsal halves of the spinal cord are separated by a broad, rhomboid fissure, the *sinus rhomboidalis.* This is filled with nervous interstitial cells and fibers, which form a gelatinous mass (Fig. 12.20). The functional significance of the sinus rhomboidalis is not known.

In amniote vertebrates the dorsal sulcus and

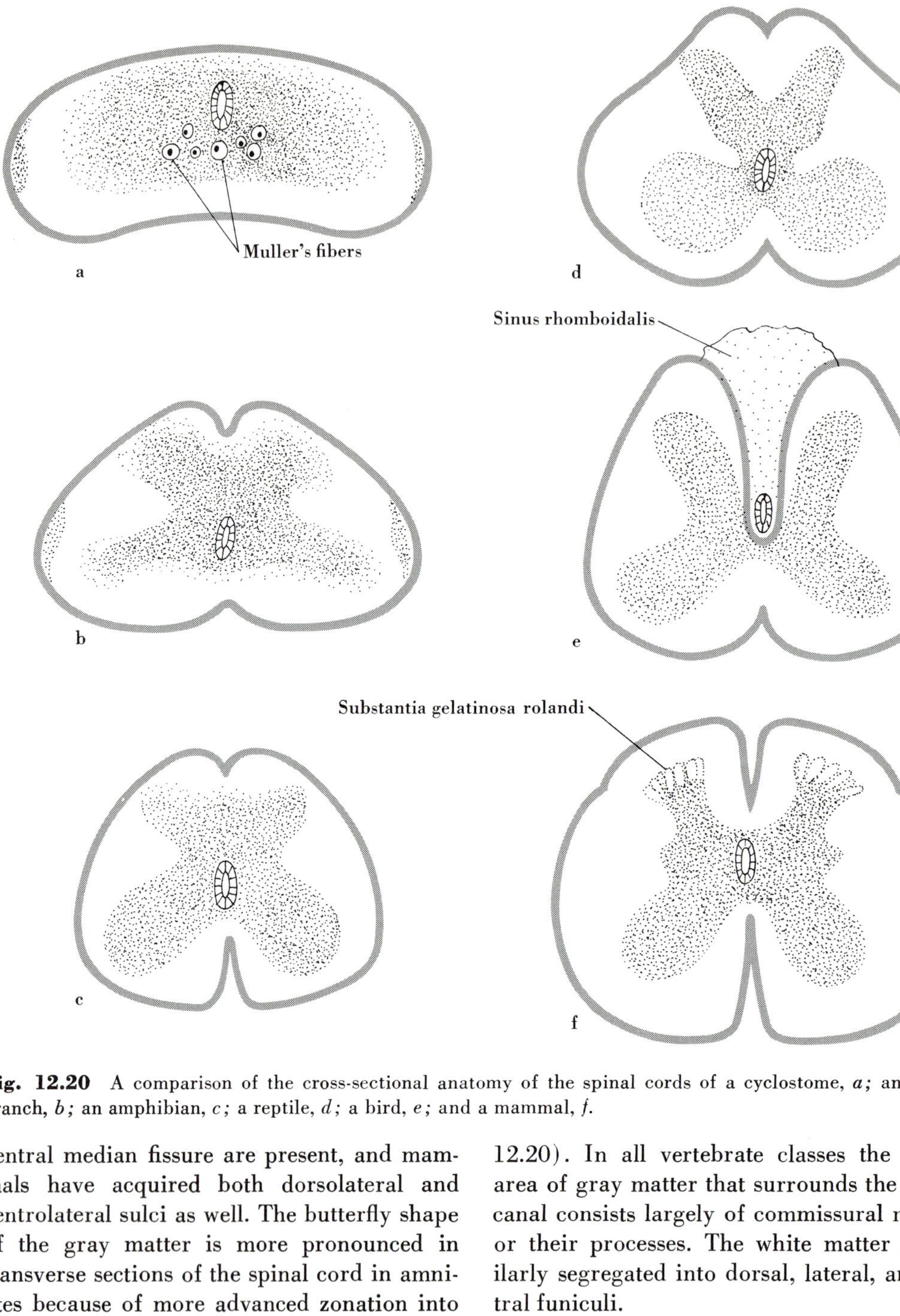

Fig. 12.20 A comparison of the cross-sectional anatomy of the spinal cords of a cyclostome, *a;* an elasmobranch, *b;* an amphibian, *c;* a reptile, *d;* a bird, *e;* and a mammal, *f*.

ventral median fissure are present, and mammals have acquired both dorsolateral and ventrolateral sulci as well. The butterfly shape of the gray matter is more pronounced in transverse sections of the spinal cord in amniotes because of more advanced zonation into four columns on either side. The dorsomedial column is concerned with somatic sensory function, the dorsolateral with visceral sensory. The ventrolateral column is concerned with visceral motor function, and the ventromedial with somatic motor (Fig. 12.20). In all vertebrate classes the middle area of gray matter that surrounds the central canal consists largely of commissural neurons or their processes. The white matter is similarly segregated into dorsal, lateral, and ventral funiculi.

The dorsal funiculus of the vertebrate spinal cord consists of both ascending and descending tracts of fibers, most of which connect with higher centers, except in cyclostomes and elasmobranchs, in which few if any extend very far cranially. It is significant to note that the dorsal funiculus is extremely limited in size in cyclostomes and fishes, however, and that it becomes increasingly more expanded with each of the higher vertebrate classes as more and more ascending tracts are incorporated in it. This correlates with the increasing role played by the brain in integrating functions of parts of the body posterior to the head. One cannot help but conclude that a fish has little knowledge of what goes on behind its eighth cranial nerve!

The ventral funiculus of the spinal cord consists chiefly of the axons of motor neurons the cell bodies of which are located at other levels, mostly at anterior levels, as the medulla oblongata and cerebellum. Consequently most of the fibers represent descending tracts. In *Petromyzon* giant motor axons, *Müller's fibers*, are a distinctive component of the ventral white substance of the cord. Numbering six to eight on a side, these fibers originate from cell bodies located in the mesencephalon and the medulla oblongata and terminate without crossing over (i.e., without *decussation*) in synapses with motor neurons in the caudal regions of the spinal cord (Fig. 12.20).

The lateral funiculi include both descending and ascending fibers the cell bodies of which are located for the most part at other levels of the spinal cord and are both motor and sensory in function. In cyclostomes, fishes, and amphibians, the motor and the visceral neurons effect extensive dendritic arborizations along the edge of the lateral funiculi, which are seen as the *marginal dendritic net*. Such nets also occur in reptiles, but they are considerably reduced. Instead one finds in some higher reptiles (Crocodilia) the appearance of segmented clusters of neuron cell bodies (the marginal nuclei) in corresponding positions on either side. Marginal nuclei are quite prominent in the spinal cords of birds, even to the point of producing slight elevations just dorsal to the position of the ventral roots of the spinal nerves. The function of the marginal nuclei is not known, but they are believed to be intercalated neurons which decussate (*see p. 338*).

At this juncture some distinction should be made among the various functional categories of neurons observable in sections of the spinal cord. Although there is some disagreement about lower forms of vertebrates (in cyclostomes and some fishes intramedullary sensory neurons have been described), it seems that the cell bodies of the sensory neurons of the peripheral nervous system lie outside of the spinal cord. Their proximal axons enter the cord dorsolaterally and undergo synapsis in the dorsal gray columns with either motor neurons or, more commonly, with *intercalated neurons,* such as the commissural and internuncial neurons mentioned above (Fig. 12.21).

SPINAL MOTOR NEURONS

The cell bodies of somatic motor neurons are distinguishable by their large size, their copious Nissl substance, and the fact that they are located invariably in the ventral gray columns. In cyclostomes and elasmobranchs many motor neurons have very long receptor dendrite processes that reach up into the dorsal gray substance to synapse directly with incoming sensory telodendria. In higher forms intercalated neurons are usually interposed between the motor neurons and the sensory telodendria, exceptions being in primitive monosynaptic reflex arcs, such as the extensor stretch reflexes.

The cell bodies of visceral motor neurons are generally rather smaller; their Nissl substance consists of a dusting of fine particles, and their location is somewhat variable in lower forms. In cyclostomes and some fishes the location is uncertain. In other fishes and in amphibians the visceral perikaryons are distributed mainly throughout both the dorsal and ventral portions of the lateral gray substance, but primarily in the latter. Visceral motor fibers have been shown in both cyclostomes and fishes emerging by way of the dorsal roots of the spinal nerves. This is not the case in higher forms, as we shall see below, although birds have the curious fibers of von Lenhossek, which emerge by way of the dorsal roots of certain cervical spinal nerves. The cell bodies of the fibers of von

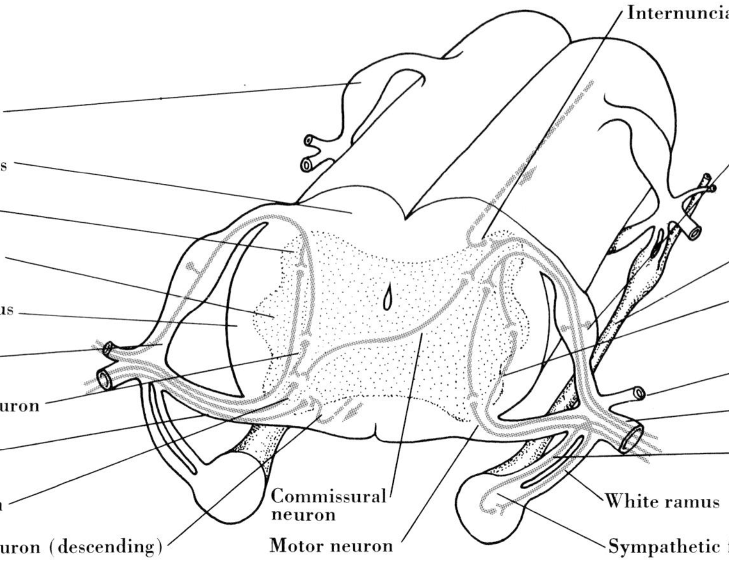

Fig. 12.21 A schematic representation of the spinal cord of a tetrapod.

Lenhossek are situated in the ventral columns of gray matter and have the morphological characteristics of motor neurons. Whether they are somatic or visceral is not clear, however.

INTERCALATED NEURONS

Intercalated neurons are those neurons which are interposed between sensory and motor neurons in the spinal cord. They may conveniently be classified into three kinds on the basis of where they send their axons. Commissural neurons have short axons that cross the spinal cord to the opposite side at essentially the same segmental level. Their cell bodies are concentrated generally in the dorsal column, except in fishes, where they are abundant in the ventral column. Internuncial neurons have extremely short axonal processes and both perikaryon and processes are confined to the same side and segmental level of the spinal cord. Funicular neurons are intercalated neurons which send more or less long axonal processes up or down the funiculi of the cord. The perikaryons of those that form ascending tracts are located in the dorsal column for the most part, whereas those of the descending fibers are primarily either in the ventral column or in the brain.

The Brain

The gross anatomic and the histologic structure of the brain reach a very high degree of complexity in the higher vertebrates, and every part plays a vital role. It must be borne in mind that any attempt to present a condensed description of the histologic characteristics of the brain entails compromises and subjective judgments of what should be included and what should be omitted. The treatment that follows is intended primarily to give the student an idea of evolutionary trends and some sense of the organization of the brain.

The brain first developed in primitive vertebrates as a modification of the anterior end of the nerve cord in response to the appearance, one after the other, of specialized organs for smell, sight, equilibrium, and hearing in a head that heretofore had been concerned with little more than supporting a mouth and a series of gill pouches. The situation was further complicated by several modifications in the pharyngeal region that led to the development of jaws and reduction in the number of gill pairs, with consequent foreshortening of the head and flexion of the brain. As with the rest of the neural tube, the brain is segmented —reflecting the myomeric organization of the

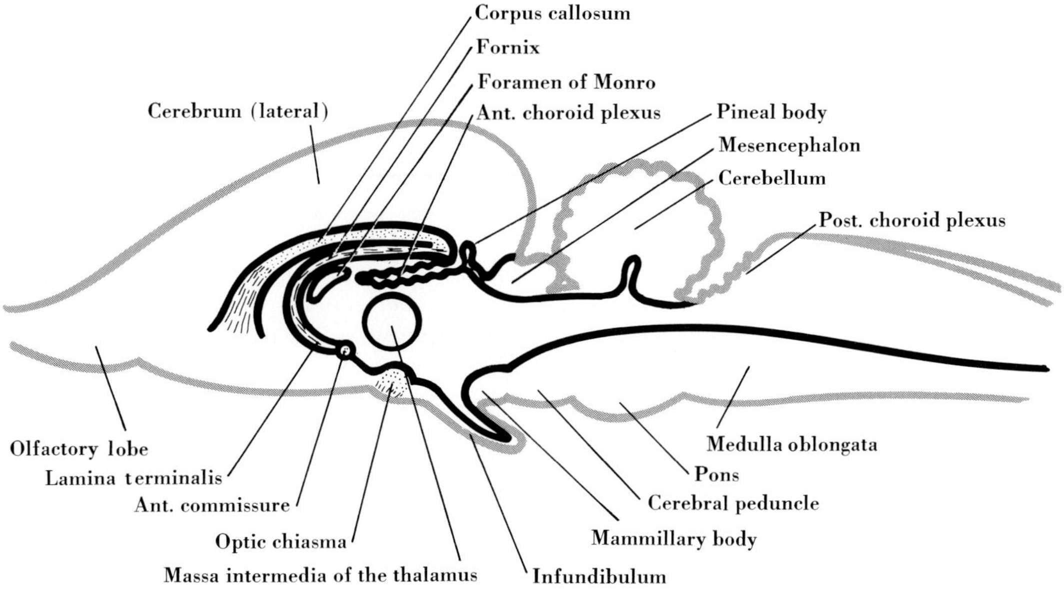

Fig. 12.22 A diagram of a generalized vertebrate brain to show the anatomic relations of its major parts.

body. Consolidation of segments has occurred, however, and there is superimposed during embryonic development a secondary segmentation which has been retained throughout the vertebrate classes. Each unit of secondary segmentation serves a group of functions which have remained rather permanently identified with it throughout the adaptive changes which have characterized vertebrate evolution; thus the description that follows is organized according to these embryonic segmental units from spinal cord end to the anterior end of the brain.

MYELENCEPHALON

At a level which lies just anterior to the first pair of cervical spinal nerves, the columns and funiculi of the nerve cord show differences in their arrangement and the central canal is somewhat enlarged. This level represents the beginning of transition from the spinal cord to the brain and arbitrarily marks the posterior limit of the *myelencephalon* (Fig. 12.22). Transverse sections through the middle of the myelencephalon of nonmammalian vertebrates reveal its basic organization (Fig. 12.23). The central canal is a broad and, in cross section, V-shaped chamber, the *fourth ventricle* of the brain. The dorsal wall is very thin and richly vascularized and forms the posterior choroid plexus or tela choroidea, the functions and microscopic structure of which

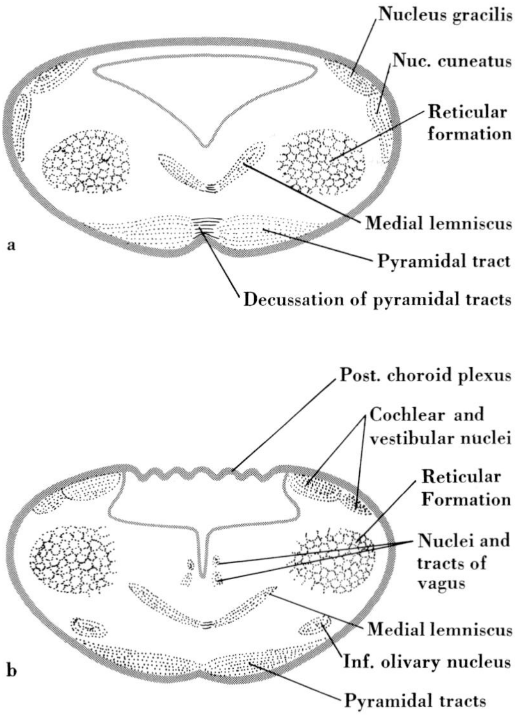

Fig. 12.23 Diagrams of the medulla oblongata at a posterior level, *a*, and at an anterior level, *b*. See text for description.

have already been described. The lateral and ventral walls are thick and constitute the *medulla oblongata* of the brain. In mammals the posterior choroid plexus is replaced by a thin, avascular membrane, the *medullary velum,* which separates the greatly compressed fourth ventricle from the overlying cerebellum.

MEDULLA OBLONGATA. The medulla oblongata, or more simply *medulla,* is comprised of both tracts of fibers and columns or aggregations of gray matter. Concentrations of gray matter in the central nervous system are usually called *nuclei.* Ascending fiber tracts of the dorsal funiculus of the spinal cord terminate in two dorsal pairs of nuclei, the nuclei *cuneatus* and *gracilis.* Here they synapse with other neurons that give rise to a new pair of fiber tracts, the medial lemniscus (Fig. 12.23), whose fibers decussate and then proceed anteriorly to the thalamus. In tetrapods the nucleus cuneatus receives the axons the cell bodies of which are located in the spinal ganglia of the nerves of the cervical plexus. The sensory components of these nerves are mainly discriminative; that is, they are concerned ultimately with sensory synthesis, which to human beings results in such sensations as recognizing texture, shape, size, and weight of external objects. The nucleus gracilis, on the other hand, receives sensory fibers from the lumbosacral plexus, which supplies analogous regions of the posterior appendage.

Prominent *pyramidal tracts* may be seen in the ventral part of the medulla. They consist of descending bundles of fibers which decussate in the posterior levels of the medulla. They include motor fibers whose cell bodies are located in motor nuclei of the cerebellum and mesencephalon, as described below. Giant motor neurons, characteristic of cyclostomes, fishes, and amphibians (e.g., the Müller's fibers mentioned above) are located in the medulla and midbrain; they send their fibers along the pyramidal tracts on their way to the skeletal muscles of the trunk and tail.

The anterior part of the medulla is even less like the spinal cord in its organization. The dorsolateral portion becomes the *acousticolateralis* area in cyclostomes and fishes. This is a sensory area in which are located nuclei for the lateral line and inner ear. In terrestrial vertebrates this area of the medulla is resolved into two major nuclei, the *cochlear* and *vestibular nuclei* (Fig. 12.23), which include cell bodies and synapses of neurons for the transmission of impulses coming in from the cochlea and the semicircular canals of the inner ear, respectively, via the eighth cranial nerve. Fibers originating in these nuclei pass up to the cerebellum and mesencephalon.

In the anterior part of the medulla the paired *inferior olivary nuclei* also appear, lying ventrally on either side of the pyramidal tracts. The olivary nucleus is conspicuous because of its shape in mammals, which is like that of a crumpled sac. It is formed by (1) the axonal telodendria of neurons the cell bodies of which are located in the lenticular nucleus of the cerebrum and (2) the dendritic processes and cell bodies of neurons the axons of which emerge from the mouth of the sac and pass up into the cerebellar cortex.

The rest of the medulla oblongata consists in most vertebrates of less-well-defined ascending and descending tracts and nuclei, the majority of which form an irregular system of cells and fibers called the *reticular formation* (Fig. 12.23). The functions of these elements are nonetheless important, for the reticular formation is concerned with mediating vital reflexes of the respiratory and cardiovascular systems, as well as mediating postural reflexes.

Finally, the anterior-most part of the medulla oblongata forms prominent *vagal lobes* in some fishes, where they represent expansions of visceral sensory areas associated with the sense of taste.

METENCEPHALON

The metencephalic region of the embryonic brain becomes identified in the adult as the *cerebellum* and, in endothermic vertebrates, the *pons.*

CEREBELLUM. One could tell a good deal about a vertebrate animal just by looking at its cerebellum. There is a direct correlation between physical activity and the complexity and size of the cerebellum, for this part of the brain includes the major centers for correlating proprioceptive impulses from muscles and impulses from the organs of equilibrium. The cerebellum is poorly developed in sedentary forms, such as cyclostomes, some fishes, amphibians, and most reptiles. It is well developed in active fishes, such as sharks and trout. It is most highly developed in the birds and mammals.

The cerebellum has developed in three stages during the adaptive evolution of the vertebrates. The first stage is seen in fishes and amphibians, in which the cerebellum serves its basic function of regulating body movements according to information received primarily from the inner ear and from the lateral line system. To a considerably less extent some information is also received from olfactory and visual organs. The vestibular nerve, a part of the eighth cranial nerve, brings in impulses from the semicircular canals of the inner ear and provides information about spatial orientation. If this nerve is cut, a fish literally doesn't know which way is up! The nerves coming in from the lateral line system (sensory elements of cranial nerves VII, IX, and X) supply information about depth, water movement (such as currents or waves), and vibrations emanating from or reflected by objects nearby. The part of the cerebellum that receives fibers from these nerves represents the posteriodorsal metencephalon. In fishes its expansion is commensurate with the physical activity characteristic of the species and constitutes the *vestibulocerebellum.* In elasmobranchs, lungfishes, and other primitive fishes, bilateral projections of the vestibulocerebellum, the *restiform bodies,* are present. They are contiguous with the medulla oblongata and they include most of the axonal telodendria of the vestibular nerve.

The second stage of cerebellar development reflects the development of paired limbs and adaptation to terrestrial life. Accordingly there is more extensive involvement of proprioceptive impulses from muscles of the limbs, as well as impulses from higher centers by way of the olivary and other nuclei. These afferents enter the anterior part of the cerebellum, which becomes correspondingly expanded to form the *somatocerebellum.* Hence, through the somatocerebellum a terrestrial vertebrate is able to execute the rather complex muscular coordination required to walk across uneven surfaces of varying textures under conditions of gravitational force that are considerably different from those encountered in the water.

The third stage of cerebellar development is found in the mammals and represents a response to increased cerebrocortical control of physical activity, which is manifested as deliberate and highly coordinated movement. The terminations of the afferent fibers involved are in the dorsolateral portions of the metencephalon, which undergoes extensive expansion to form the *cerebrocerebellum.*

The distribution of white and gray matter in the cerebellum is the opposite of that encountered in the spinal cord. The gray matter forms the outer covering, the *cerebellar cortex,* whereas the white matter is in the middle. Moreover, as the gray matter increases in amount in the cerebrocerebellum, it becomes convoluted, forming long parallel folds, the *foliae.* The underlying white matter is also involved in the convolutions, and appears in sections of the cerebellum as a spectacular spray of branching white tracts, called the arbor vitae.

The microscopic organization of the cerebellar cortex (Fig. 12.24) is uniform over the entire cerebellum. It consists of two distinct layers, an outer *molecular* or *plexiform layer* and an inner *granular* or *nuclear layer,* which are separated from each other by a single layer of large cells, the *Purkinje cells.* The molecular layer is uniform in thickness and it consists mostly of synapses between incoming axons and the dendritic processes of Purkinje cells. The few cells that are present in the molecular layer are the *stellate cells* and the *basket cells,* although these are lacking in fishes and amphibians. The stellate cells are superficial in position. Their dendritic processes pick up impulses from axons which reach up into the molecular layer from the granular layer and convey them across the

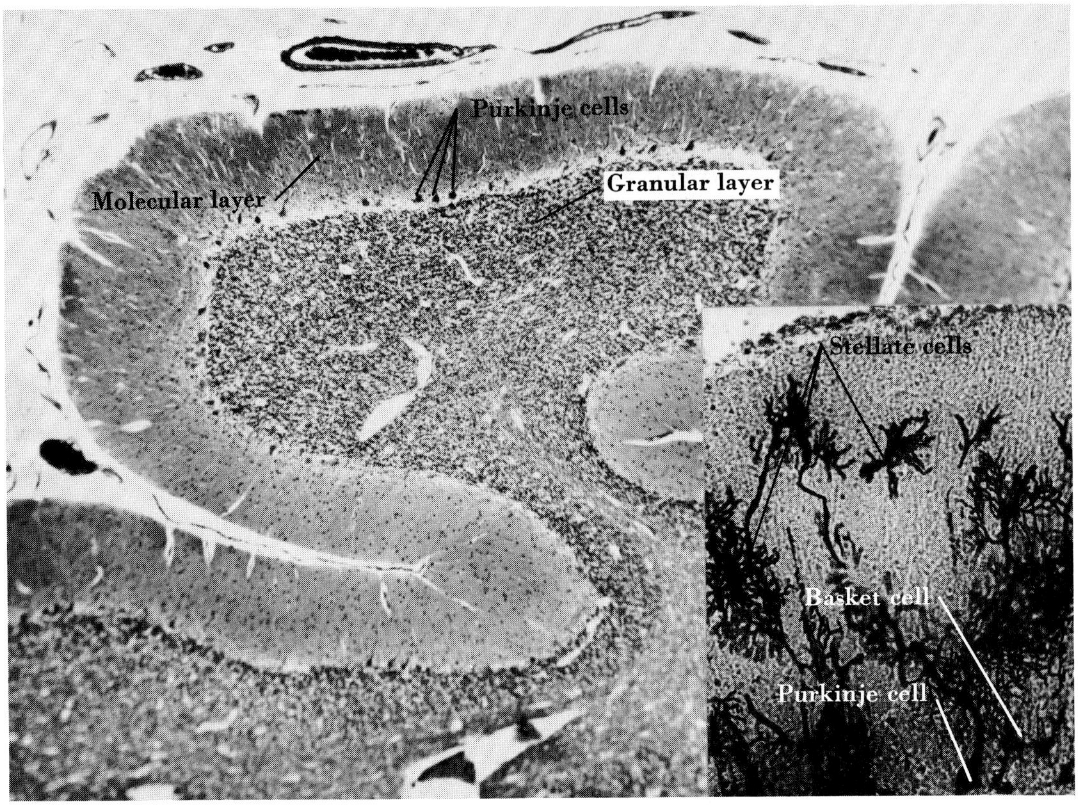

Fig. 12.24 Photomicrograph of a human cerebellar convolution. X35. Insert: A Golgi preparation of the molecular layer, showing parts of Purkinje cells, stellate cells, and basket cells. X100.

folia to the dendritic processes of Purkinje cells. Basket cells, on the other hand, lie deeper in the molecular layer. Their dendrites also branch out profusely and their axons pass across the folia and form baskets of intertwining axonal processes around the Purkinje cell bodies.

The Purkinje cells have a most remarkable appearance and arrangement (Fig. 12.5). The cell body is large (35–65μ in diameter in man) and flask-shaped and contains a number of Nissl bodies, which tend to a concentric arrangement. The dendritic processes branch off rather sharply and fan out in one plane through the entire thickness of the molecular layer. In convoluted cerebellums, this plane is at right angles to the long axis of the folia. The single axon passes straight down through the granular layer and either continues to central cerebellar nuclei or turns and runs into other foliae. Collaterals branch off from the axon in the granular layer and return to the molecular layer to synapse with other Purkinje cells. The Purkinje cell axons constitute the *only efferents from the cerebellar cortex.*

True Purkinje cells are not present in cyclostomes, but giant cells which are believed to be their forerunners are found scattered in the granular layer in lampreys. Purkinje cells are present and well differentiated in elasmobranchs and all higher forms, however. In many teleosts and amphibians they are situated in more than one layer and are less regularly distributed between the molecular and granular layers.

The granular layer of the cerebellar cortex varies widely in its thickness and distribution in different areas of the cerebellum in lower vertebrates, but it is fairly uniform in these respects in higher forms. It is densely populated by small neuron cells, the *granular cells,* and has an abundance of axonal and dendritic processes. Neuroglial cells and occasional large neuron cells of unknown function are also found. Granular cells have three to six short dendritic processes which synapse with the terminations of axons coming in from the underlying white matter. The axons of the granular cells are poorly myelinated or non-myelinated. They pass up into the molecular layer, where they bifurcate sharply. The two branches run in opposite directions parallel to the long axis of the folia and consequently, by synapsing with the terminal twigs of Purkinje cell arborizations, they interconnect Purkinje cells in different rows. The arrangement is similar to that of telephone wires strung from pole to pole. Transverse connections are made by synapses with the dendrites of stellate or basket cells. In this way, incoming impulses can pass both along and between the rows of Purkinje cells, thereby effecting almost limitless combinations of circuits.

The fibers in the granular layer include also the *mossy fibers* and the *climbing fibers* (Fig. 12.24). The mossy fibers are coarse myelinated fibers whose sprays of telodendria synapse with the granular cells. Climbing fibers pass straight up to the Purkinje cells, where they branch profusely. The branches climb up the dendritic processes of the Purkinje cells, like a vine, synapsing with them along the way, but they do not extend as far as the terminal twigs of the Purkinje cells. The interrelations of the various cells and fibers are shown in the stereogram in Fig. 12.25.

PONS. The ventral part of the metencephalon consists largely of tracts of fibers that connect the medulla and spinal cord with the cerebellum and higher levels of the brain. These structures include anterior extensions of the pyramidal tracts and medial lemniscus already mentioned in connection with the medulla, and others. In birds and in mammals, however, this part of the metencephalon is more conspicuous, and is known as the pons. The appearance and complexity of the pons is a direct function of the increased size and complexity of the cerebellum. Extensive decussation of fibers occurs here, so that complete interrelation of all cerebellar afferents can be effected. More ventrally located in the pons is the reticular formation, which is continuous posteriorly with that described in the medulla and continues anteriorly through the mesencephalon into the diencephalon.

MESENCEPHALON

The mesencephalon or midbrain consists of three regions which differ in structure and function. The dorsal region is the *tectum*, the

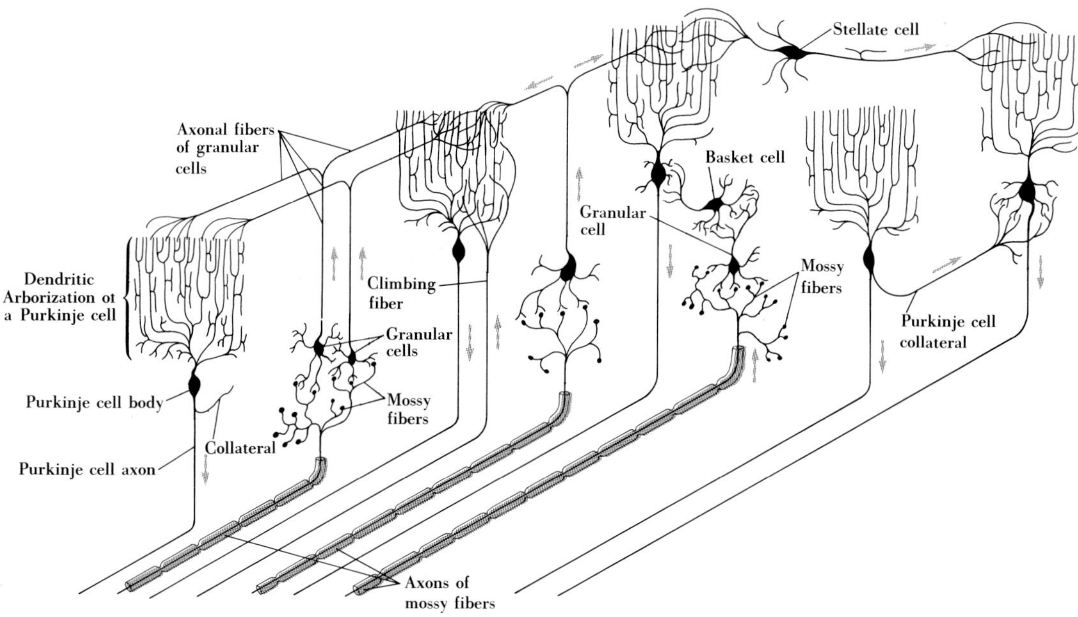

Fig. 12.25 Stereogram of the cerebellar cortex showing the three-dimensional hook-up of component neurons.

ventral is the *tegmentum,* and the area between the two is the *central gray of the mesencephalon.*

TECTUM. Except in mammals and ophidians (snakes), the tectum of vertebrates appears as a pair of dorsolateral *optic lobes* (*corpora bigemina*) which are quite conspicuous in some groups of vertebrates, such as anurans and birds, and barely perceptible in others, such as cyclostomes. In ophidians and mammals the tectum is further divided to make four areas (*corpora quadrigemina*): a pair of *anterior colliculi* and a pair of *posterior colliculi.* The tectum plays a more important role in lower forms, for it is the chief center for the termination of the optic tracts, as well as for its receiving ascending secondary fibers from the spinal cord and medulla and fibers from the trigeminal nuclei. Hence it is an important center for correlating optic and other exteroceptive impulses. In higher reptiles and in mammals the ascending spinal components tend to bypass the tectum and go more or less directly to the thalami. Moreover, in these forms the auditory sense has achieved greater importance and its centers have moved in part up into the posterior colliculi, thereby effecting better correlation with visual impulses, which enter the anterior colliculi. Despite this, the major auditory and visual centers of mammals have been removed to the cerebral cortex, so that the tectum is relegated essentially to the role of reflex center for light and sound.

In microscopic structure the tectum of cyclostomes is largely in the form of a choroid plexus, but part of it is thickened and organized into roughly three to four layers. The outermost layer is made up of fibers of the optic tract with *horizontal cells* dispersed among them (Fig. 12.26). The next layer is thicker and it includes numerous *vertical cells,* the dendritic processes of which extend toward the surface while their axons are directed centrally. Other cells in this layer are intercalated cells which correlate impulses coming in from the optic tracts with impulses coming in from the medulla oblongata. Efferent fibers from the tectum form a poorly defined third layer. These fibers pass caudally for the most part and synapse in the medulla with motor neurons for medullary and spinal outflow.

In elasmobranchs the lamellar structure of the tectum is more distinct, and it becomes even more so in higher forms. As in cyclostomes, the outermost layer of the tectum of

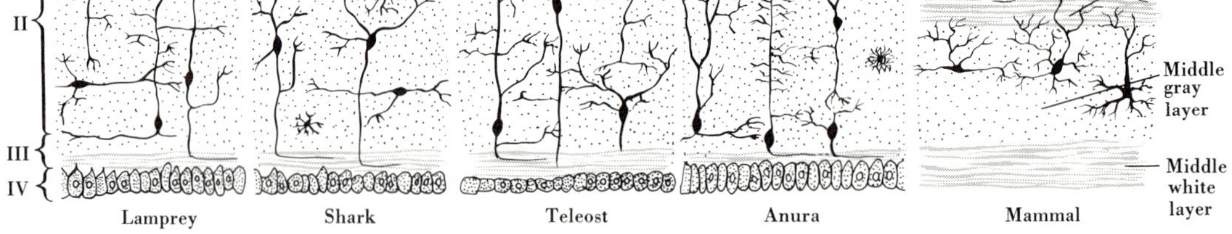

I = Optic tract (except in mammals)
II = Zone of intercalated cells and efferent neurons
III = Periventricular fibrous layer
IV = Ependyma (except in mammals)

Fig. 12.26 A diagram of neuronal arrangements in the optic tectums of representatives of the major vertebrate classes. (After Ramon y Cajal, ca. 1898, and P. Ramon. 1890.)

nonmammalian vertebrates is relatively acellular, consisting mainly of fibers of the optic tracts and occasional horizontal cells (Fig. 12.26). The rest of the tectum is arranged in a series of alternating layers of gray and white matter that represent, respectively, the various kinds of cells and fibers that correlate visual impulses with other exteroceptive impulses, and of efferents that pass these impulses on to motor centers in other parts of the brain and spinal cord.

The optic tectum of mammals is also layered, but the optic tract components have migrated to a deeper position (Fig. 12.26). Indications of an inward migration have been described in various lower forms, such as fish, frogs, and turtles, so the shift in mammals is not unheralded. The innermost layer of gray matter in the mammalian tectum, the periventricular layer, has neurons of one of the roots of the trigeminal nerve distributed in its outer border. As mentioned above, the main visual centers in mammals have shifted to the cerebral cortex and the tectum receives only a small representation of the optic tracts. This representation involves a rich heritage of reflex connections, however. Among these are connections to the iris, for direct tropic response of the eyes, and to motor nuclei of the head and neck, as well as to spinal motor nuclei, for reflex bodily responses. When one is "startled," as by the sudden appearance of a face at the window, or a flash of lightning, or an unexpected footfall, it is tectal reflexes that are involved.

TEGMENTUM. A significant portion of the tegmentum is occupied by reticular formation, part of which is consolidated into a variable number of motor and correlating nuclei. The rest of the tegmentum consists of ascending and descending fiber tracts and decussations of fibers to and from the tectum. In elasmobranchs a pair of *lateral* and a pair of *medial tegmental nuclei* are evident. Between the latter and the tectum on either side is a rather inconspicuous area called the *torus semicircularis.* In bony fishes this becomes quite prominent, especially in those species that have well-developed lateral line systems, such as *Mormyrus.* The function of the torus semicirculrais is gravistatic, for it receives impulses from parts of the acousticolateralis system. It is also present and well developed in higher vertebrates, becoming very closely associated both spatially and functionally with the overlying tectum. In birds, the homologue of the torus semicircularis, the so-called *lateral mesencephalic nucleus,* constitutes the major mass of the tegmentum, even rivaling the tectum in size.

In elasmobranchs the *interpeduncular nucleus* becomes well defined. Its homologue is present in cyclostomes, but there it is poorly differentiated. The interpeduncular nucleus persists with amazing constancy throughout the other vertebrate classes, maintaining its position in the midventral wall of the midbrain. It receives terminal fibers from the habenulopeduncular tract, the hypothalamus, and the reticular formation; it is functionally associated with the sympathetic nervous system.

Introduced by elasmobranchs, but better delineated in bony fishes, is the *nucleus isthmi,* which represents a consolidation of a portion of the lateral tegmental nucleus mentioned above. It appears in the caudal part of the tegmentum, adjacent to the cerebellum. In anurans the nucleus isthmi becomes a more conspicuous structure by virtue of its outer layer of deep-staining cells that cover a medulla of scattered cells and fibers. In reptiles it is even more prominent as a pair of lateral bulges. The nucleus isthmi in birds is reduced to a flattened area compressed between the lateral mesencephalic nucleus and the skull. The major connections of the nucleus isthmi is with the optic tectum and the acoustic nuclei, and so it serves as a correlation center for optic and acoustic impulses. In reptiles and birds it also has some connections with the cerebellum.

Perhaps the most conspicuous nucleus in the tegmentum of amniotes, especially mammals, is the red nucleus (Fig. 12.27). It has a rich vascular supply, which accounts for its characteristic color, and some of its motor neurons are very large. It is derived phylogenetically from poorly defined reticular formation in the

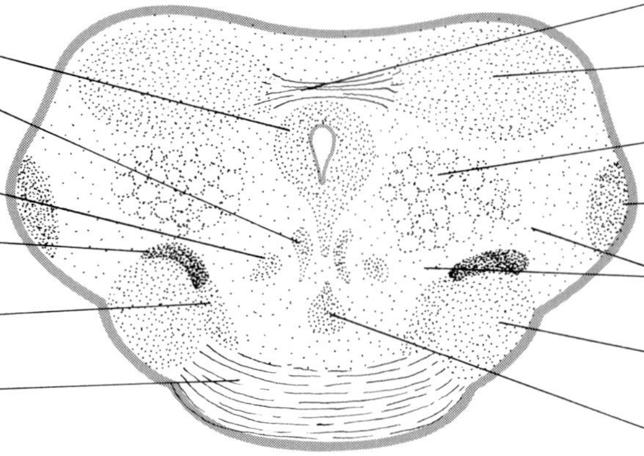

Fig. 12.27 A diagram of a cross-section through the mesencephalon of an amniote.

ventral tegmentum and it has two distinct sizes of motor neurons, small and giant. The giant neurons are similar to those that are characteristic of the reticular formation in the cerebellum and medulla of amphibians and fishes. Efferents from the giant cell portion of the red nucleus decussate in the tegmentum and traverse the pons and the medulla on their way to the spinal cord, giving off collaterals and synaptic terminals to motor neurons of axial musculature *en route*. The efferents from the small cells of the red nucleus pass up to the dorsal thalami and effect connections with the cerebral cortex. They also send representatives to the inferior olivary nucleus and to reticular formation of the metencephalon and myelencephalon. Afferents to the red nucleus come mainly from the cerebellum, and to a lesser extent from the cerebral cortex. The function of the red nucleus is in righting reflexes, as aptly illustrated when a blindfolded cat is dropped back-downward and rights itself in midair to land squarely on all four feet.

Present in reptiles, but more conspicuous in mammals, is another consolidation of reticular formation, the paired *substantia nigra* (Fig. 12.27), lying on either side of the red nuclei. Some of the cells of the nigra are large and stellate and contain a characteristic dark pigment. Although considerable attention has been devoted to the nigra, its functions are poorly understood. It is believed to be primarily a center for coordinating movements for reflex walking and, in conjunction with certain subthalamic nuclei, with smoothing out tonic adjustments.

Two pairs of important nuclei in the tegmentum are also the nuclei for the third (oculomotor) and fourth (trochlear) cranial nerves. Except in cyclostomes, these are located one pair in front of the other in the medial part of the tegmentum at the margin of the middle gray. The trochlear nucleus of cyclostomes is uniquely situated in the cerebellum.

Other nuclei are found in the tegmentum, but of them the major ones of regular occurrence that have not already been mentioned are the medial and lateral geniculate bodies. These properly lie in the precincts of the diencephalon, but they often extend back into the midbrain.

The white matter of the midbrain is confined largely to the tegmental portion and appears as fascicles, tracts, and commissures. Among the more consistently prominent ones are the medial lemniscus, described above in connection with the medulla oblongata, the tracts of cranial nerves III and IV, and the *posterior commissure*. The last lies in the torsal tegmentum at the line of transition from mesencephalon to diencephalon. Its fibers arise in various nuclei of the diencephalon and mesencephalon, principally the lateral geniculate, the tectal and pretectal, and the reticular nuclei. Finally, in the ventral margin of the tegmentum in amniotes, are the *cerebral peduncles*. By no means prominent structures in reptiles or birds, they appear as distinct pyramidal tracts in mammals. In primates, especially human beings, they constitute a major bulk of the midbrain region. The origin of the fibers that make up the cerebral peduncles is in the cerebral cortex, and their termini are chiefly in the pons, the medulla, and the spinal cord.

DIENCEPHALON

The vertebrate diencephalon, like the mesencephalon, is short in its anterior-posterior dimension, and it is an ancient and vital reflex region. It is readily divisible into dorsal, lateral, and ventral portions which constitute respectively the epithalamus, the thalamus, and the hypothalamus. The thalamus in turn is divisible into dorsal thalamus and ventral or subthalamus.

EPITHALAMUS. This portion of the diencephalon is the least variable from cyclostomes to mammals. Its centers are primarily for olfactosomatic correlation, but it is also the location of the epiphyseal apparatus and the anterior choroid plexus. The histology of the latter (telae choroidea) has already been considered. It is worth noting that in many fishes the anterior choroid plexus bulges outward, rather than inward, and may form large saclike structures of glandular nature, as in *Lepisosteus* and *Amia*.

The epiphyseal apparatus includes two organs, the *parietal body* and, behind this, the *pineal body* or *epiphysis*. Both are present in lampreys and in many fishes and reptiles, but only the pineal body is represented in other vertebrates. The parietal body forms a small median eye in lampreys, *Sphenodon*, and certain lizards; in some cases it is equipped with a fairly well-differentiated retina and lens (Fig. 12.28). It is doubtful if it can discriminate more than between light and dark in most cases.

The pineal body also appears to be a photoreceptor in cyclostomes, but in other vertebrates it is a glandlike organ the function of which has not been elucidated. Two types of cells have been identified in bovine pineal glands: one includes the so-called pinealocytes, and the other is glial. Unique to the

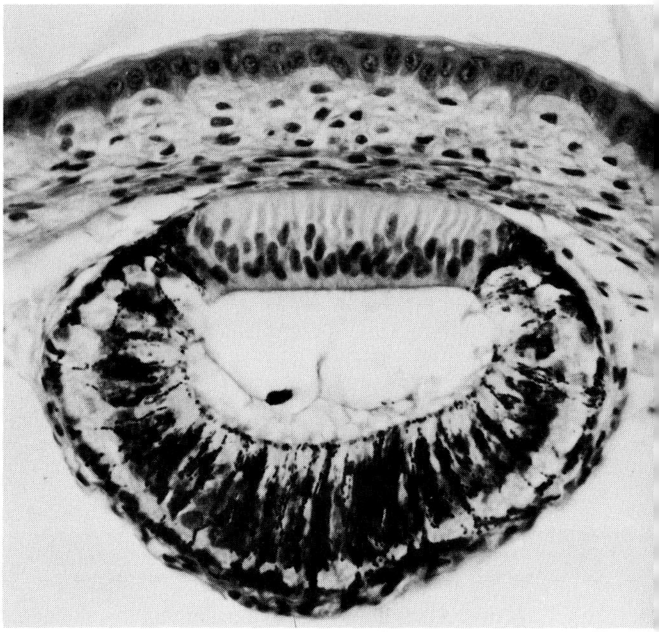

Fig. 12.28 A vertical section through the median eye of the lizard *Sceloporus occidentalis*, showing well-differentiated lens and retina. The skin opposite the lens is transparent in life. X590. (From R. C. Stebbins and R. M. Eakin. 1958. Am. Mus. Novitates No. 1870:1–40.)

former are the large number of smooth-surfaced vesicles and striated microtubules seen in electron micrographs. A number of compounds have been extracted from mammalian pineal glands; the principal ones are serotonin, which is found in higher concentration in the pineal gland than in any other part of the brain, and a compound known as *melatonin* (5-methoxyindole-3-acetic acid). Acetone extracts of mammalian pineal glands elicit melanophore contraction in frogs, and when the substance is injected into rats it appears to inhibit estrus and to result in reduced ovarian weight.

Associated primitively with the epiphyseal apparatus are the *habenular nuclei*, which receive afferents from the epiphysis, from the dorsal thalamus, and, after decussation in the *habenular commissure*, from the rhinencephalon. They send efferents to the interpeduncular nucleus of the tegmentum by way of the habenulopeduncular tracts. The function of the habenular nuclei is not known, but in view of their connections, they are believed to be concerned with correlating olfactory and proprioceptive impulses. Certainly they reach their highest development in species characterized by a well-developed olfactory sense.

DORSAL THALAMUS. This region of the diencephalon shows the greatest response to the evolutionary development of higher centers. It is primarily a relay station for ascending impulses and is responsible for sending them either to efferent centers of the ventral thalamus and tectum, particularly in the lower vertebrates, or to the cerebral cortex. In mammals it receives all exteroceptive and proprioceptive impulses for relay to the cortex. Because there are few exteroceptive ascending pathways in cyclostomes and fishes, the dorsal thalamus is not well developed in these forms. In cyclostomes, for example, the dorsal thalamus appears as the primitive *median lobe* of the diencephalon. Although the connections of this lobe have not been thoroughly determined, the structure does not seem to receive any fibers from the forebrain and it is not related to the olfactory system. A small group of cells is present, however, which effect synapsis with some fibers of the optic tract, probably collaterals, and so may be a forerunner of the lateral geniculate nucleus of higher forms.

A definitive lateral geniculate nucleus first appears in the elasmobranchs, where it receives collaterals from the optic tract and scattered bundles of ascending fibers from the trigeminal nerve and the acousticolateralis system. Otherwise the dorsal thalamus in these forms is small and primitive. In bony fishes the lateral geniculate nucleus is more advanced, for in some species (e.g., species of Pleuronectidae) its optic components are laminated.

The dorsal thalamus of urodeles is similar to that of bony fishes, but in anurans it shows an appreciable increase in relative size. In frogs projection fibers from the lateral geniculate nuclei to the forebrain occur, and a mass of cells appears medially which receives fibers from lower centers and sends efferents to the dorsolateral wall of the forebrain.

In reptiles and birds a number of nuclei appear in the dorsal thalamus, all of which have intimate connection with the forebrain and anticipate the extensive expansion of the dorsal thalamus characteristic of the mammals. This advance is particularly evident in the Crocodilia and the birds. The most prominent of these is the *nucleus rotundus,* which is believed by some to be homologous with certain components of the medial group of nuclei of mammals described below.

It is primarily through the dorsal thalamus that the cortex of the highly expanded cerebrum has become integrated with the rest of the nervous system in mammals. Consequently the dorsal thalamus is very highly developed and includes a large number of nuclei. These nuclei are arranged in four anatomically distinct groups: the anterior, the medial, the lateral, and the posterior groups of nuclei (Fig. 12.29). The anterior group is usually the smallest. Its nuclei connect by afferent and efferent fibers with the mammillary bodies of the hypothalamus and with the olfactory portions of the forebrain, and they function as olfactovisceral and olfactosomatic correlation centers. The medial group includes several nuclei that connect with the hypothalamus, with other thalamic nuclei, and (some of them) with parts of the prefrontal and temporal cortex. Although the functions of this group are largely unknown, it has been found that surgical destruction of its dorsomedial component in human beings is effective in modifying severe personality disorders. In general the medial group is believed to be concerned primarily with intrathalamic correlation.

The lateral group of dorsal thalamic nuclei in mammals includes several nuclei. Those in the dorsal portion connect with other thalamic nuclei and send fibers to the cerebral cortex. Those in the ventral portion serve as relay stations for proprioceptive and exteroceptive impulses to be projected onto sensory and motor areas of the cortex.

Finally, the posterior group of dorsal thalamic nuclei in mammals includes three major gray masses: the lateral and medial geniculate nuclei and the *pulvinar nucleus.* The former

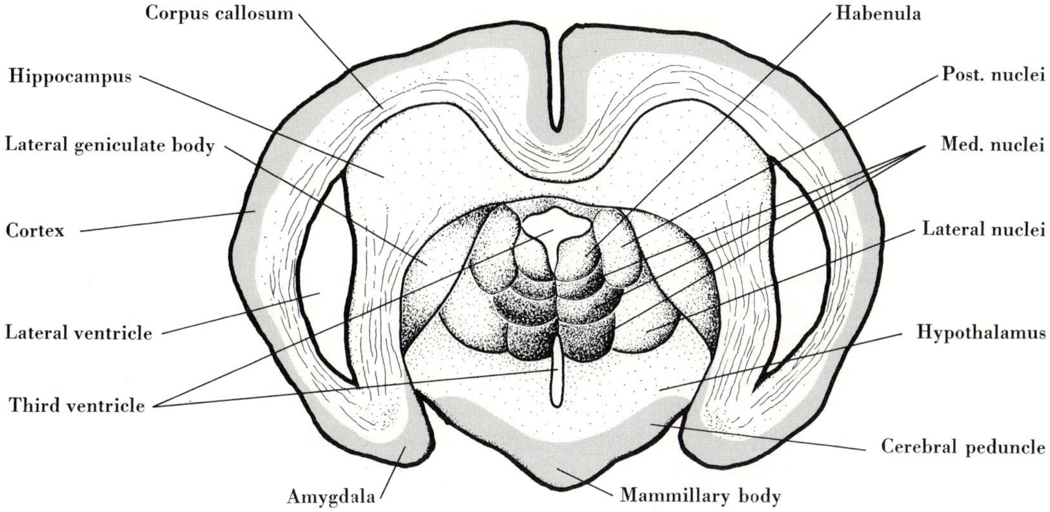

Fig. 12.29 Drawing of a vertical section through the thalamus of a generalized mammal.

are often found intruding into the mesencephalic tegmentum, with which they have close functional correlations. The lateral geniculate nucleus receives fibers from the optic tract and projects to the cortex. In mammals the laminate structure of the lateral geniculate is particularly striking. At the optic chiasma, decussation of half of the fibers of the optic tracts takes place, so that each eye is represented on each side of the brain. These fibers sweep up along the sides of the diencephalon and terminate in the lateral geniculate nucleus in alternate layers for each eye, with a thin sheet of fibers between each layer. The position of the fibers in any layer is directly related to the position of their corresponding cell bodies in the retina. The medial geniculate nucleus receives auditory impulses and connects with auditory areas of the temporal cortex. The pulvinar nucleus is believed to be concerned with interthalamic correlation, for it connects with adjacent nuclei and projects to and receives fibers from the cerebral cortex.

VENTRAL THALAMUS. The functions and connections of the ventral thalamus or subthalamus are rather obscure. In mammals, in which the most is known about the structure, the subthalamic nuclei in conjunction with the substantia nigra seem to be concerned mostly with smoothing out reflex tonic adjustments, particularly those related to postural and locomotor reflexes. Hence the expression of their activity is often subtle and more qualitative than quantitative. The major associations of its cells appear to be with the basal ganglia, the substantia nigra, and possibly the red nucleus.

In cyclostomes and elasmobranchs the subthalamus shows little by way of discrete nuclear arrangement. In bony fishes a variety of nuclei with varied connections have been described, but their homologies with subthalamic nuclei in higher forms are uncertain and conjectural. In frogs an *entopeduncular group* of nuclei is introduced. Although little more than clusters of cells with lateral forebrain associations, this group is believed to carry over into higher vertebrates and to contribute to the development of the substantia nigra of amniotes. In reptiles and birds, in addition to paired entopeduncular nuclei, clusters of large cells have been described which contribute fibers to pretectal and dorsal thalamic regions.

In mammals the subthalamic gray matter is concentrated into three or four pairs of nuclei: a pair of small entopeduncular nuclei; a bilateral and diffuse *zona incerta,* part of which may occur separately as *Forel's fields;* and a pair of *subthalamic* (*Luys's*) nuclei. The last is characterized by large multipolar cells with coarse granules in their cytoplasm. In many mammals the substantia nigra and red nucleus may extend into the subthalamic region.

HYPOTHALAMUS. The secretory role of the diencephalon is of long standing phylogenetically, particularly in its dorsal and ventral areas. Dorsally this role is seen in the anterior choroid plexus and the epiphyseal system. Ventrally it is manifest in the ependymal cells lining the floor of the diocoele, especially in the portion that lines the *infundibulum,* and in the cells that constitute some, if not most, of the hypothalamic nuclei. In addition, at least in higher vertebrates, hypothalamic centers regulate the release of tropic hormones by another gland, the *adenohypophysis* (see Chapter 14). The principal hormones that mammalian hypothalamic nuclei have been proved to secrete are *vasopressin* and *oxytocin,* which are also discussed in Chapter 14. Although few species of lower vertebrates have been examined for these compounds, oxytocic activity has been found in all classes except the Agnatha. Substances similar to vasopressin have been demonstrated in frogs, toads, fowl, and a teleost (*Pollachius virens*).

The hypothalamus, including the mammillary bodies, is derived from the floor of the diencephalon, along with the *optic chiasma, supraoptic commissure,* and *neurohypophysis;* it is represented in all vertebrates. In cyclostomes and elasmobranchs it is the largest subdivision of the diencephalon and is the principal center for coordination of sensory impulses—it receives fibers from the olfactory tract, from tectal centers, and from the acousticolateralis areas of the medulla. Its principal efferents go to visceral centers in the medulla oblongata. In addition to the foregoing, the elasmobranch hypothalamus also receives fibers from what is construed to be the primordium of the hippocampus of the forebrain and sends efferents to the cerebellum. Extending as a prominent ventral outpocketing of the hypothalamus in these lower forms is the infundibulum, the caudal part of which comprises the neurohypophysis and, in elasmobranchs, a richly vascularized epithelial membrane, the *saccus vasculosis.* In elasmobranchs the infundibulum is also characterized by bilateral, thick-walled projections, the *lobi inferiores,* which are evident also in bony fishes.

The hypothalamus of bony fishes is well developed in terms of nuclear formation, although these structures are variable in number, size, and distribution among the many groups. The nuclei include *preoptic, lateral hypothalamic, anterior hypothalamic,* and *posterior hypothalamic nuclei.* The preoptic nucleus has a *magnocellular* component of large (12–30μ in diameter) flask-shaped cells with a rich capillary supply, and a *parvocellular* component of small cells that embraces the magnocellularis on three sides. The magnocellularis is homologous with the supraoptic and paraventricular nuclei of higher vertebrates and in some bony fishes, such as *Anguilla,* its cells are among the largest in the central nervous system. The lateral hypothalamic nuclei, located at the base of the infundibular stalk, represent ventrocaudal extensions of the preoptic nuclei and are characterized by small cells. The anterior hypothalamic nucleus, although quite large in some species (e.g., *Periophthalmus*), also consists of rather small cells. The posterior hypothalamic nucleus is located posterior to the infundibular stalk and is also quite large in some species, such as the teleost *Periophthalmus.*

The saccus vasculosis of elasmobranchs and bony fishes is very highly developed. It consists of cuboidal or columnar neurosecretory cells equipped with knobbed processes that project into the ventricular lumen, and of small secretory cells. Nerve fibers from the former appear in distinct tracts that pass dorsally to the dorsal thalamus and to the posterior hypothalamic nucleus. The function of the saccus vasculosis is uncertain, but it is believed to be concerned with sensing pressure changes in the ventricle.

As in cyclostomes and elasmobranchs, the hypothalamus in bony fishes is primarily concerned with correlating olfactory impulses with gustatory and other sensory impulses; thus it is important in feeding reflexes. By way of comparison, the hypothalamus of amphibians is reduced in size, which is related to the fact that increased skin sensibility in these forms has made the olfactory and gustatory senses less important. However, despite its smaller relative size, the hypothalamus of amphibians has fairly well-defined nuclei, which include large preoptic nuclei, a pair of posterior nuclei that are believed to be homologous with the mammillary nuclei of amniotes, and a group of ventromedial nuclei that are probably the forerunners of the tuber cinerium of amniotes. The saccus vasculosis, when present, is small and its connections are uncertain.

In reptiles and birds the hypothalamus shows even more reduction in relative size, but its nuclei are well differentiated. The infundibular base in snakes has thick walls reminiscent of the lobi inferiores of fishes. A saccus vasculosis is lacking. The hypothalamic nuclei include preoptic, anterior hypothalamic, periventricular, lateral hypothalamic, and ventral hypothalamic nuclei and a pair of mammillary bodies. The anterior and lateral nuclei are not sharply separated from each other, and both receive fibers from olfactory areas of the forebrain via the *medial forebrain bundle.* The periventricular nuclei lie, as their name implies, along the ventricle and they include rows of deeply staining neurons and interstitial cells of the olfactory tract. The mammillary bodies connect with the hippocampal region of the forebrain by way of prominent bundles of the *fornix.*

The hypothalamus of mammals is characterized by considerable variability in nuclear arrangement. In general three nuclear groups may be distinguished: (1) the anterior nuclei, which include *anterior hypothalamic, paraventricular,* and *supraoptic nuclei;* (2) the medial or *tuberal* nuclei, which include *ventromedial, dorsomedial, periventricular* (*arcuate*), and *lateral tuberal nuclei;* and (3) the posterior group, which includes the *posterior nuclei* and the mammillary bodies. The tuberal nuclei and tracts constitute the *tuber cinereum* of the mammalian brain. The mammalian hypothalamus is further characterized by several fiber systems and decussations, the most prominent of which are the medial forebrain bundle and the fornix, mentioned above; the *mammillary peduncle,* the *mammillothalamic tract,* and its associated *mammillotegmental tract;* and the *hypothalamohypophyseal tract.* The connections of these tracts are implicit in their names.

More is known about certain functions of the hypothalamus in mammals than in lower forms, aside from the ones already described, but this does not imply that similar functions may not be performed by the hypothalamus in all vertebrates. The mammalian hypothalamus is a vital center for the regulation of metabolic activities both through its neurosecretions and its control of the autonomic nervous system. Its role as a neurosecretory organ are discussed more fully in Chapter 14. It is sufficient to note here that both vasopressin and oxytocin are secreted by neurons of the supraoptic and paraventricular nuclei and are transported via the hypothalamohypophyseal tracts to the neurohypophysis. The autonomic regulatory and control parameters of the hypothalamus include such vital functions as vasomotion; cardiac acceleration and depression; regulation of body temperature, sleep and wakefulness, hunger, thirst, micturition, and gastrointestinal peristalsis; and the expression of certain emotions, such as rage and fright.

TELENCEPHALON

The vertebrate telencephalon or forebrain develops from the anterior-most portion of the embryonic neural tube and is contiguous with the diencephalon. In fact, it represents the anterior extension of dorsolateral components of the diencephalon. Primitively the telencephalon is associated with the olfactory sense, for in lower vertebrates all but a small dorsal portion is solely concerned with the reception and transmission of olfactory impulses. The small dorsal portion, the *neopallium,* receives a meager representation of fibers from the lemnisci, which bring in sensory impulses from lower levels. While remaining diminutive in fishes and amphibians, the neopallium shows an increase in size and functional significance in reptiles, and in mammals it is expanded to the extent of becoming the largest part of the brain. In the primates, particularly in the hominoids, the neopallium comprises more than three-quarters of the total brain mass. This expansion has been correlated with increased input from nearly all sensory pathways, the arrangement of constituent neurons into complex integrating and correlating centers, and the increase in efferent pathways to all other regions of the central nervous system. The basic connections already existed in the lowest forms, but in them their function was little more than that of secondary correlation of certain sensory impulses with the dominating olfactory function.

The two major regions of the vertebrate forebrain are the *pallium* and the *corpus striatum.* The pallium comprises the dorsal wall of the forebrain and thereby the roof of the *telocoele* or forebrain ventricle. The corpus striatum represents the thickened floor of the forebrain and it consists largely of reticular formation in lower forms but becomes consolidated into the so-called *basal ganglia* in higher vertebrates.

In cyclostomes the forebrain consists of paired *olfactory bulbs* or *rhinencephalon* and, behind them, paired *olfactory lobes* or *cerebral hemispheres.* The former are prominent and develop in conjunction with the olfactory epithelium in which the primary neurons are located. In lampreys the olfactory bulbs show little differentiation, but in the myxinoids they show a laminar arrangement of components. Incoming olfactory fibers form an outer fibrous layer. A more or less distinct glomerular layer appears as a region where several secondary neurons synapse with each primary fiber coming in from the olfactory epithelium, an arrangement that has a pronounced ballooning and summation effect on olfactory input. Finally, in myxinoids an interglomerular layer is seen that includes large bipolar and multipolar cells believed to be homologous with the *mitral cells* found in the rhinencephalons of higher forms.

The olfactory lobes of lampreys consist largely of reticular formation, with efferents going primarily to habenular and hypothalamic areas. In myxinoids the cells and fibers of the olfactory lobes are arranged in a number of layers surrounding an area that is believed to be primordial to the hippocampus of higher forms. It may be significant that some of the cells of the myxinoid olfactory lobe are pyramidal in shape and their laminar arrangement is reminiscent of that of the cerebral cortex.

The basal part of the cyclostome telencephalon, the forerunner of the striatum in higher forms, includes cells and fibers that connect with dorsal and ventral thalamic areas, as well as a poorly defined concentration of cell bodies that constitute the preoptic nucleus described above.

The organization of the telencephalon of elasmobranchs and bony fishes is essentially like that of myxinoids except for separation of the olfactory bulbs from the olfactory lobes by attenuated olfactory tracts. Because of this separation, the olfactory lobes are generally identified as cerebral hemispheres. The hippocampal primordia are somewhat better differentiated as parts of the dorsolateral and dorsomedial walls of the cerebral hemispheres and they have more clearly defined commissural interconnections. Moreover, there are connections of the telencephalon with the cerebellum, which cannot be discerned in cyclostomes. Bony fishes show more nuclear consolidations than are evident in elasmobranchs, but these are highly variable among the various orders. It is sufficient to note that they all have olfactory connections.

The telencephalon of amphibians presents some significant advances over that of fishes, advances that can be correlated with adaptation to terrestrial life with consequent shifts from a dominant olfactory sensibility to increased visual, auditory, and cutaneous sensibility. There is no sharp demarcation between olfactory bulbs and cerebral hemispheres. The cerebral hemispheres are expanded by lateral evagination so that paired ventricles are present. The fiber connections between hippocampus and the mammillary nuclei are consolidated into a more or less definitive fornix, and more extensive interconnections are established between telencephalic nuclei by tertiary tracts. Of particular interest are connections which relate the pallium with the dorsal part of the corpus striatum and with the hippocampus and fibers that form a ventricular fiber system; these connections may be considered to be forerunners of the cortical association systems of higher vertebrates. Finally, ascending fibers are introduced into the medial and lateral forebrain bundles—carrying visceral and possibly gustatory impulses anteriorly for correlation with olfactory impulses in the telencephalon, thereby setting the stage for the development of telencephalic areas as important correlation centers.

The telencephalon of reptiles shows considerable advance over that of amphibians, owing primarily to a more highly developed striatal complex, the so-called basal ganglia, and to the appearance of cortical differentiation in the hippocampus, the piriform lobe, and the pallium. The dorsolateral area of the telencephalon is striated. It is quite large and anteriorly it intrudes into the lateral ventricle as a ridge, which is particularly prominent in alligators. The cells in the dorsolateral area have a cortical arrangement, and those in the anterior ridged portion have connections with ascending tracts from lower centers and are believed to be homologous with the mammalian caudate nucleus and putamen. The posterior portion, which retains olfactory connections, is believed to be homologous with the amygdaloid complex of mammals.

In birds considerable reduction of the olfactory lobes is encountered. This is correlated with the reduced importance of olfaction in birds. In sparrows, for example, the olfactory bulbs are minute and fused together to form a median appendage on the anterior ventral aspect of the cerebrum. There is also a reduction in the development of secondary olfactory centers in the cerebral hemispheres as compared with reptiles. The striatal or basal nuclei are well differentiated, as in the higher reptiles, although they differ somewhat in distribution. These nuclei are responsible for the stereotyped or instinctive behavior of birds, such as nesting and migration. Laminated cortical areas have been described in various birds, such as hummingbirds and the kiwi, but it is questionable if they are homologous with mammalian cortical laminations. There is little question, however, that they represent the same kind of response to increased demand for intricate functional correlation. Increased functional complexity is also reflected in the intricate fiber connections found in the cerebral hemispheres. The lateral forebrain bundles are highly developed, carrying impulses both to and from tegmentum, cerebellum, tectum, and dorsal thalamus, as well as to the hypothalamus.

The mammalian telencephalon is characterized by extreme development of a dorsolateral portion of the pallium called the neopallium. This has become so voluminous on either side that it has overgrown the diencephalic and mesencephalic divisions as well as, to a considerable extent, the cerebellum. The neopallium is represented in lower vertebrates as a small, ill-defined area which receives afferents from the thalami. In bony fishes the afferents carry visceral and perhaps gustatory impulses. In amphibians, visceral, gustatory, and perhaps a small representation of somatic sensory impulses are relayed to the neopallium and the framework for extensive intercommunicating tracts within the telencephalon is first suggested.

In mammals progressive development of the neopallium is evident from what is essentially a reptilian condition in prototherians, through moderate development in metatherians and the small eutherians (e.g., insectivorans and rodents), pronounced development in the ungulates, advanced development in carnivorans and primitive primates (e.g., lemurs and tarsiers) and cetaceans, and, finally, to extreme complexity in the higher primates. The gross criteria for degree of development include the size of the cerebrum relative to the rest of the brain and the presence and complexity of convolutions of the cerebral surface. Expansion of the cerebrum in mammals was facilitated by the extension of the three major sensory systems, visual, auditory, and somatic sensory, from cell stations in the thalami to the neopallium. In response, the cells of the neopallium became arranged in a series of layers concentrated toward the surface, thereby forming the *neocortex*, and efferent connections were developed to lower centers. Cortical afferents and efferents converged to form a distinctive fibrous lamina, the *internal capsule.* Intercommunication between right and left cerebral cortices, already evident as commissural fibers in reptiles, became extensively developed as the prominent *corpus callosum.*

THE CEREBRAL CORTEX. In the development of the neopallium the gray matter became concentrated along the surface as the neocortex or, as it is known in mammals, the *cerebral cortex.* The cerebral cortex is rather uniform in thickness and histologic appearance over the entire cerebrum. Seven morphologically different types of neurons characterize the layers of the cortex (Fig. 12.30). *Horizontal cells* have a fusiform cell body with its long axis tangentially oriented. *Golgi Type I* cells are medium-sized to large, with short dendritic processes and a long axon. *Golgi Type II* cells are small neurons with a very short axon. *Pyramidal cells* have a distinctive triangular form with the apex directed peripherally, and they vary in size from small to very large. *Star cells* are medium to large multipolar neurons with relatively short dendritic processes that radiate out uniformly from the cell body. *Star pyramidal cells* are large multipolar cells with an irregular triangular outline. The dendritic process that emerges from the apex is long, like that of

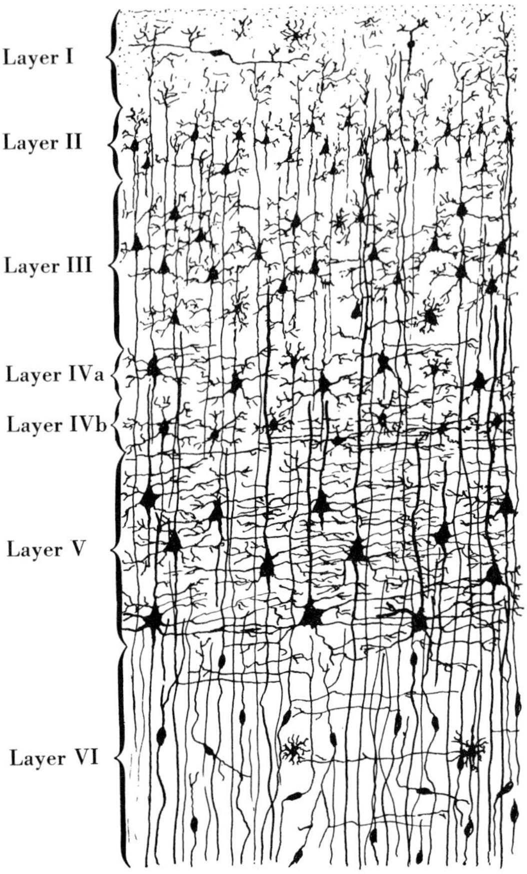

Fig. 12.31 A diagram to show the layers of the mammalian cerebral cortex.

Fig. 12.30 Photomicrograph of a Golgi preparation of the cerebral cortex of a rabbit. X100.

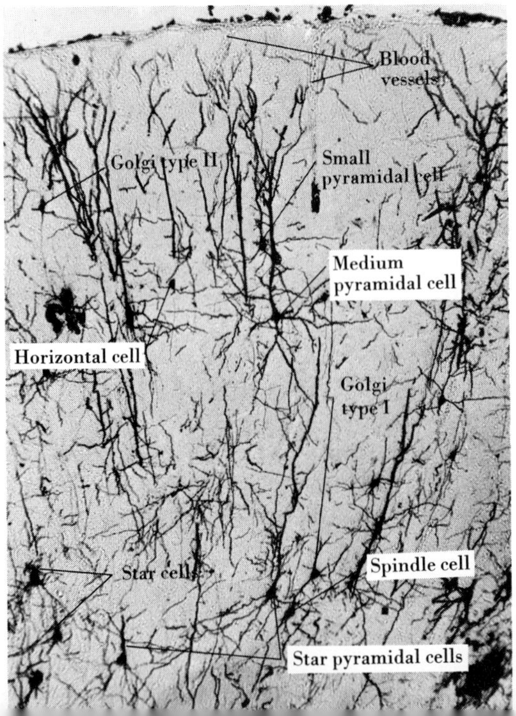

pyramidal cells. *Spindle cells* are small- to medium-sized bipolar cells with a fusiform body. These cells and their processes are arranged in seven layers in most cortical areas. as follows (Fig. 12.31).

Layer I. The molecular or plexiform layer is the outermost layer. It consists mainly of axons and dendritic processes from underlying layers, and it has a sparse population of horizontal cells and Golgi Type II cells.

Layer II. The small pyramidal cell layer consists mostly of small pyramidal neurons the dendritic processes of which are short and are confined primarily to this layer. Their axons pass centripetally to lower layers or form association or commissural connections with other parts of the cortex.

Layer III. The medium pyramidal cell layer contains medium-sized pyramidal cells whose axons, which originate from the base of the triangle, pass centripetally through underlying layers to the corpus callosum and internal capsule, giving off collaterals along the way. Their dendritic processes receive impulses primarily from thalamic centers.

Layer IVa. The star pyramidal cell layer, composed mainly of star pyramidal cells, with some Golgi Type I cells, makes dendritic contact with Layer I and, by collaterals from the apical dendrite, with other cells in Layers IVa and IVb.

Layer IVb. The star cell layer has dendrites that are confined entirely to Layers IVa and IVb, both of which include dense fibrillar plexuses with incoming axons from thalamic centers. Layers IVa and IVb are sometimes considered as a single layer, called the *internal granular layer.*

Layer V. The large pyramidal cell layer contains principal neurons that are large pyramidal cells which are especially numerous and large (Betz cells) in motor areas of the cortex. The apical dendrites of the cells of this layer reach the plexiform layer, with collaterals and basilar dendrites distributed exclusively within Layer V. Most of the cortical efferents originate in this layer, passing centripetally into the internal capsule.

Layer VI. The spindle cell layer has long, medium, and short spindle cells. The long spindles give off collaterals from their dendritic process close to the perikaryon. The dendritic process then passes directly to the plexiform layer. The dendritic process of the medium spindles end in Layer IV in synapsis with thalamic afferents, while those of the short spindle cells end in Layer V. The axons of spindle cells are distributed largely to other cortical areas as association fibers.

The connections between the various cells of the cortical layers are very complex, and the number of cells involved is fantastically large; their number has been reckoned at about 2.6×10^9 in the human cerebral cortex. Reverberating circuits are set up whereby a given impulse arriving in the cortex from a thalamic center may pass through a series of cortical synapses and reverberate in cycles either within the cortex or, by way of cortical efferents, between cortex and thalamus. Such reverberating circuits may continue to function as long as they are metabolically supported or until other entering impulses cause them to be channeled off or otherwise modified. Certain synapses which may not fire the first few times around may become successively hypopolarized and eventually fire (summation), in this way introducing new chains of connections and even additional reverberating circuits. Mental processes such as thinking, remembering, or deliberating are believed to involve such phenomena.

THE BASAL GANGLIA. The mammalian counterpart of the corpus striatum of lower forms, plus an ancient portion of the rhinencephalon, comprise a series of paired nuclei called the basal ganglia. These are the *caudate, lentiform,* and *amygdaloid nuclei* and the *claustrum* (Fig. 12.32). The caudate nucleus is shaped like a tail which starts broadly at the anterior end of the lentiform nucleus, arches dorsocaudally over it, swings around until it is directed anteriorly again, and terminates in the amygdaloid nucleus ventrally. The most significant of its neurons are small stellate cells and large multipolar cells that connect by their axons with the lentiform nucleus.

The lentiform nucleus is an ovoid mass consisting of two parts, a lateral *putamen* and a medial *globus pallidus.* The former includes small polygonal cells and larger multipolar cells. Axons of the larger cells join those of the globus pallidus to form the efferent pathway of the lentiform nucleus. The globus pallidus contains large motor neurons, which may be pyramidal or spindle-shaped, and it represents the main efferent center of the basal ganglia.

The amygdaloid nucleus is phylogenetically very old, being a representative of the primitive rhinencephalon. Unlike other components of the basal ganglia, it has no motor function but, along with the hippocampus, is olfactory. The claustrum is poorly understood. It has

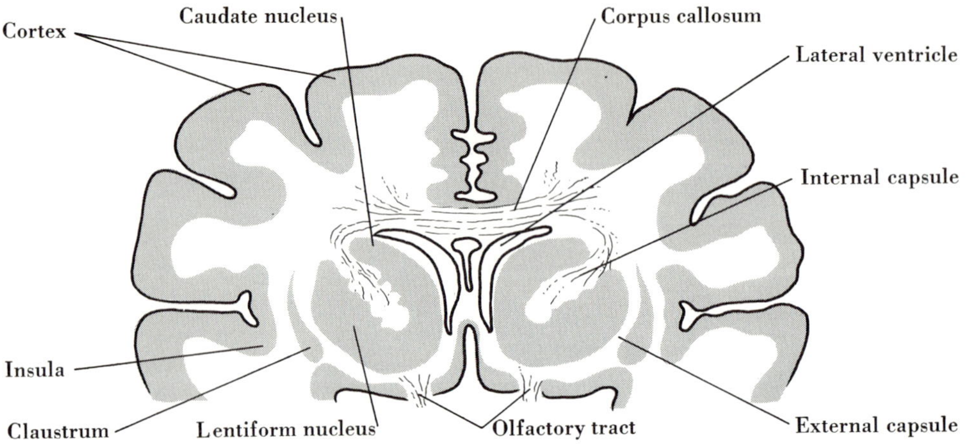

Fig. 12.32 A diagram of a vertical section through the mammalian forebrain to show the basal ganglia.

direct connections with the cerebral cortex and indirect connections with the olfactory bulb and the thalamus.

Very little is known about the function of the basal ganglia in mammals. The caudate-lentiform component appears to be tied functionally to the red nucleus, the substantia nigra, and the reticular formation of the brain stem, and to be concerned with stereotyped movements and with coarse motor coordination. The extent to which the cerebral cortex has dominated the functions of other parts of the brain is nowhere more evident than it is in the effect it has had on the apparent functional importance of the basal ganglia.

The Peripheral Ganglia

The peripheral counterparts of the nuclei and ganglia of the central nervous system are the peripheral ganglia, for these are concentrations of nerve cell bodies which lie outside of the brain and spinal cord. These include the *cerebrospinal ganglia,* which are sensory in function, and the *autonomic ganglia,* which are visceral motor.

CEREBROSPINAL GANGLIA

The ganglia of the cranial nerves and of the spinal nerves are essentially alike and thus they are known collectively as the *cerebrospinal ganglia.* They appear as prominent swellings of the sensory components of peripheral nerves adjacent or very close to the points where these nerves enter the medulla oblongata or the spinal cord. Such ganglia are lacking in cranial nerves I and II because the cell bodies of these nerves are situated respectively in the olfactory epithelium and the retina, and they are lacking in cranial nerves III, IV, and VI because these tracts include only motor components.

Each cerebrospinal ganglion is invested within a fibrous connective tissue capsule which is continuous with the connective tissue covering (*epineurium*) of its corresponding peripheral nerve. Trabecular extensions of the capsule invade the ganglion, dividing it into irregular compartments of nerve cell bodies and associated nerve bundles. The connective tissue of the trabeculae is continuous with the *perineurium* of the nerve bundles as they issue from the confines of the ganglion into the peripheral nerve.

The ganglionic neurons are rounded cells with a smooth contour. In cyclostomes both bipolar and unipolar neurons are found, but in the adults of other vertebrates they are, with few exceptions, unipolar. These cells vary in diameter from about 5 to 100μ or more. The nuclei are round and light staining, they generally occupy a central position, and they have a prominent nucleolus (Fig. 12.33). The cytoplasm contains the usual complement of organelles, plus neurofibrils and Nissl bodies which are more diffusely distributed than in motor neurons of corresponding size in the central nervous system. Pigment inclusions of *lipofuscin* are common in the cytoplasm, particularly with advanced age.

Except in cyclostomes, in which myelinated fibers are unknown, the larger ganglionic neurons give rise to myelinated axons. Some of the smaller, dark-staining neurons are believed to be visceral sensory and to give rise to unmyelinated fibers. The single myelinated axon of the larger cells may pass directly to the adjacent nerve bundle, where it bifurcates, sending one branch peripherally to a receptor site and the other to enter the central nervous system. On the other hand, after the axon leaves the cell body it may form a loose coil or "glomerulus" around the cell body before passing to the nerve bundle.

Each nerve cell body is surrounded by a single layer of small nervous interstitial cells, the Schwann cells, which are believed to serve a nutritive function. These cells are surrounded in turn by a single layer of fibroblasts. In cyclostomes the neuron cell bodies are not confined to ganglia, for they also occur within the trunks of certain peripheral nerves.

AUTONOMIC GANGLIA

The ganglia of autonomic nerves vary considerably in size. The ganglia of the sympathetic chain are fairly uniform in size, but the coeliac or anterior mesenteric ganglia of mammals tend to be very large. Ganglia of the parasympathetic division are usually diffuse and incorporated within or form a plexus on the visceral organs which they innervate. In elasmobranchs the sympathetic fibers leave the spinal cord by way of the dorsal roots of the spinal nerves and then many of them separate near the point of junction of dorsal and ventral roots as visceral efferent branches that go directly to the digestive tract and blood vessels. The situation in cyclostomes has not yet been clarified, except that sympathetic nerves have been shown to be incorporated in the vagus nerve. Definitive chains of sympathetic ganglia first appear in teleosts and are continued in all higher vertebrates.

When discrete autonomic ganglia are found, they have essentially the same histologic organization as cerebrospinal ganglia with respect to capsule and trabeculae. The neurons, however, are multipolar and hence irregular in outline and they are generally smaller. The nucleus is usually eccentric in position. Some but not all of the neurons, including their dendritic processes, are surrounded by Schwann cells.

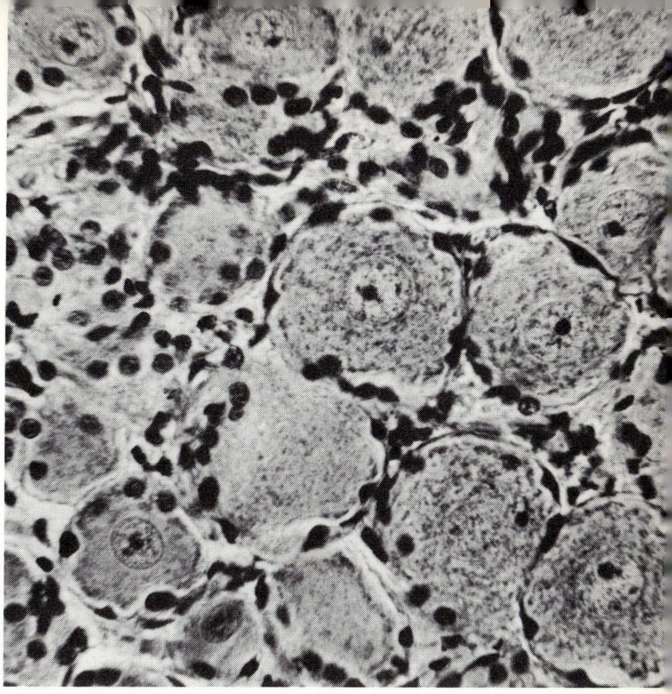

Fig. 12.33 Photomicrograph of a section through a spinal ganglion of a dog. X500.

The Peripheral Nerves

As noted earlier, the axons of peripheral nerves, whether myelinated or unmyelinated, are ensheathed by Schwann cells. The axons of myelinated fibers or groups of unmyelinated fibers are bound together in parallel array by fine collagenous fibers into fascicles. Each fascicle is invested in an *endoneurium* of connective tissue that includes collagenous fibers, fibroblasts, fixed macrophages, and blood capillaries (Fig. 12.34). Groups of fascicles are bound together as nerves by the *perineurium*. The perineurium is rather distinct from either the epineurium or endoneurium, because it is made up of concentric layers of cells that are endothelial rather than fibroblastic in character. These cells and their processes are joined to each other to form a continuous multilaminar cellular sheath around the nerve bundle. A thin but nonetheless distinct basement lamina separates each

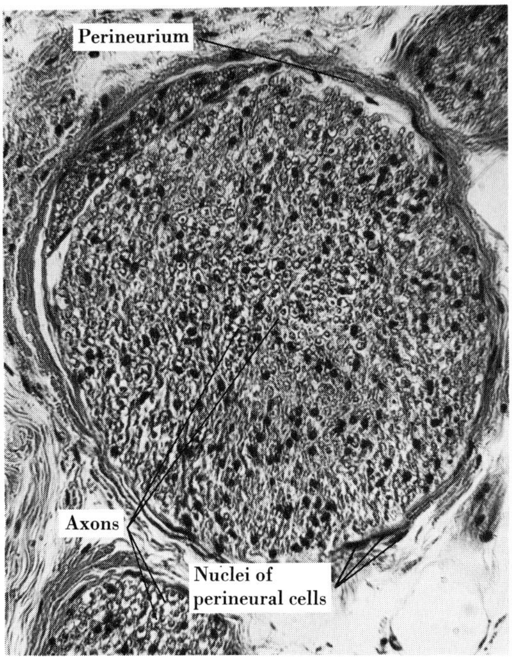

Fig. 12.34 Photomicrograph of a cross section of a peripheral nerve in a monkey. X450.

cell layer from the adjacent collagenous fibers. Because of its endothelial character, the term *perineurothelium* has been proposed as more appropriate than perineurium, and it has also been proposed that it functions as a diffusion barrier, thereby physiologically isolating the nerve bundles. Nerve trunks include numbers of such nerves which are ensheathed in a rather dense connective tissue membrane, the epineurium. The epineurium includes coarse bundles of collagenous fibers, a scanty supply of fibroblasts, and blood vessels and lymphatics. The epineurium may extend septa into the nerve trunk, separating groups of nerves.

Sections of peripheral nerves reveal axons of several kinds, which in mammals can be grouped into three types on the basis of rate of conduction. Type A fibers are those with a large diameter, a thick myelin sheath, long internodes, and rapid rate of conduction (15–100 m/sec). This type includes skeletal motor fibers and some somatic sensory fibers. Type B fibers have a smaller diameter, a thin myelin sheath with short internodes or no myelin sheath at all, and an intermediate conduction rate (3–14 m/sec). They consist mainly of visceral and somatic sensory fibers. Type C fibers are largely unmyelinated visceral motor and visceral sensory fibers that conduct at rates less than 3 m/sec.

The Meninges

The brain and spinal cord are enclosed by continuous connective tissue membranes which serve the dual functions of providing mechanical protection and of supporting the blood vessels that supply the central nervous system. In cyclostomes and fishes these functions are met by a single membrane, the *meninx primitiva.* This is a fibrous connective tissue membrane with a rich vascular supply that lies in intimate contact over the entire surface of the brain and spinal cord. In the regions of the telae choroidea it forms the vascular supporting layer. Where larger blood vessels enter and leave the brain, the meninx primitiva involutes a short distance with them—forming a cuff or sleeve around the proximal portion of the blood vessel. It also follows nerve roots that emerge from the central nervous system and continues peripherally as or continuous with the epineural membrane.

Between the meninx primitiva and the connective tissue lining of the brain case or spinal canal (perichondrium or periosteum, as the case may be) is a space that varies in width in different locations and according to slight changes in the volume of the brain or spinal cord. Such changes, it should be recalled, would be occasioned by the rates of secretion or absorption of the cerebrospinal fluid. This space, the *perimeningeal space,* is traversed by a loose meshwork of fine collagenous fibers and is filled with cerebrospinal fluid. It will also be recalled that the cerebrospinal fluid is secreted into the ventricles and central canal of the central nervous system and then it passes through foramina into the meningeal spaces. From these spaces it diffuses into venous sinuses and vessels of the meninx and thus is returned to the blood.

In amphibians, reptiles, and birds the

meninges consist of two distinct layers, the *leptomeninx* that immediately covers the brain and spinal cord and, outside of this, a tougher fibrous layer, the *pachymeninx* or *dura mater*. The leptomeninx is doubtless the homologue of the meninx primitiva, for it has a similar embryonic origin and it has similar physiologic and histologic characteristics. As its name signifies, the leptomeninx is thin, and it is well vascularized. The space which separates it from the dura mater, the *subdural space,* is essentially similar to the perimeningeal space of lower forms. Another space, the *epidural space,* separates the dura from the periosteal lining of the cranium and spinal cord. Filled with lymph or tissue fluid, this space is most evident around the spinal cord. The dura includes some blood vessels and nerves and it is continuous with the epineurium of peripheral nerves, whereas the leptomeninx probably is not—except that it does accompany at least part of the optic nerve.

The leptomeninx of mammals consists of two membranes, an inner *pia mater* and an outer *arachnoid membrane,* which are separated from each other by the *subarachnoid space* (Fig. 12.35). The pia is closely adherent to the brain and spinal cord, even dipping down into their sulci and fissures. Homologous with the leptomeninx of lower vertebrates, it is a richly vascularized membrane and contributes most of the blood vessels that nourish the central nervous system, as well as forming the choroid plexuses of the telae choroidea. The blood vessels are supported by a network of interlacing collagenous fiber bundles and elastic fibers and are characterized by perivascular plexuses of sympathetic nerves as well as sensory nerve endings. Fibroblasts, fixed macrophages, perivascular mast cells and lymphocytes, and a surface layer of squamous cells form the cellular components of the pia.

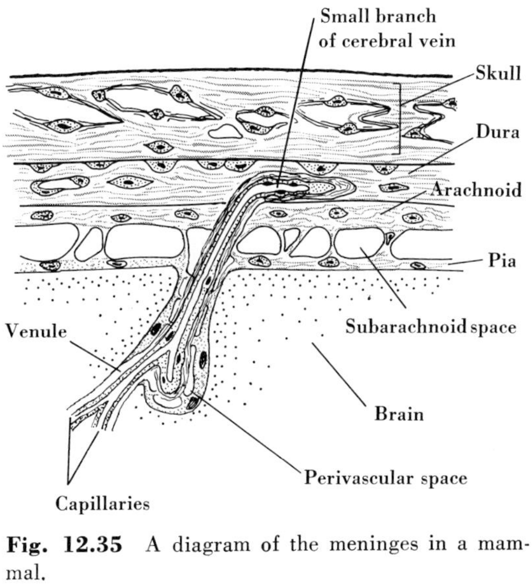

Fig. 12.35 A diagram of the meninges in a mammal.

The arachnoid membrane is avascular. Its connective tissue composition is similar to that of the pia, however, and its outer surface also consists of a single layer of squamous cells that form a kind of serous membrane. From the inner surface delicate cobwebs and strands of collagenous fibers arise and traverse the subarachnoid space, attaching to the pia. Unlike the pia, the arachnoid membrane does not dip down into fissures and sulci, although the subarachnoid space does.

The dura mater of mammals is a tough, fibroelastic membrane that covers the other meninges of the brain and the spinal cord. In the brain region it is fused with the periosteum of the inside of the cranium so that there is no epidural space. It is involuted to form a prominent longitudinal septum, the *falx cerebri,* between left and right cerebreal hemispheres, and a transverse septum, the *tentorium,* between the cerebrum and the cerebellum. The two layers of dura thus apposed in these septa are separated by prominent venous sinuses. Moreover, the dura of these septa is made up of a looser connective tissue than elsewhere and is perforated by villous projections of the arachnoid membrane into the venous sinuses. Such arachnoid villi provide a way of escape for the cerebrospinal fluid into the blood.

The spinal dura does not fuse with the periosteum of the vertebral canal, but is separated from it by a rather wide epidural space containing loose, fatty, connective tissue, and blood vessels, and being bridged on either side by a series of denticulate ligaments.

SELECTED REFERENCES

Altman, J. 1966. Organic foundations of animal behavior. Holt, Rinehart and Winston, New York.

Anderson, E. 1965. The anatomy of bovine and ovine pineals: Light and electron microscopic studies. J. Ultrastruct. Res., supplement 8.

Ariens Kappers, C. V., G. C. Huber, and E. C. Crosby. 1960. The comparative anatomy of the nervous system of vertebrates, including man. Vols. I–III. Hafner, New York.

Bairati, A., and F. Maccagnani, 1950. Richerche sulla glioarchitettonica dei vertebrati. I. Anfibi. II Ucelli. Monit. Zool. It., supplement 58.

Bennett, M. V. C., Y. Nakajima, and G. D. Pappas. 1967. Physiology and ultrastructure of electrotonic junctions: I. Supramedullary neurons. J. Neurophysiology 30:1–179.

Brodal, A., and Fänge, R., eds. 1963. The biology of Myxine. Universitetsforlaget, Oslo.

Cohen, M. M., and R. S. Snider, eds. 1964. Morphological and biochemical correlates of neural activity. Hoeber Div. of Harper & Row, New York.

Cravioto, H. 1966. Filamentous and tubular structures in mammalian axoplasm: An electron microscope study. J. Comp. Neurol. 126(3):453–461.

Cravioto, H. 1966. The perineurium as a diffusion barrier–ultrastructural correlates. Bull. Los Angeles Neurol. Soc. 31(4):196–208.

Glees, P. 1955. Neuroglia; morphology and function. Blackwell Scientific Publications, Oxford.

Horstmann, E. 1954. Die Faserglia des Selachiergehirns. Z. Zellforsch. 39:588–617.

Kruger, L., and D. S. Maxwell. 1966. The fine structure of ependymal processes in the teleost optic tectum. Am. J. Anat. 119: 479–498.

Lehman, H. J. 1957. Über Struktur und Funktion der Perineuralen Diffusion Barriere. Z. Zellforsch. 46:232.

McMahon, V. J. 1967. Fine structure of synapses in the dorsal nucleus of the lateral geniculate body of normal and blinded rats. Z. Zellforsch. 76(1):116–146.

Nachmansohn, D. 1961. Chemical factors controlling nerve activity. Science 134:1962.

Palay, S. L. 1960. The fine structure of the secretory neurons in the preoptic nucleus of the goldfish (*Carassius auratus*). Anat. Rec. 138:417–425.

Peters, A. 1961. Anatomical considerations of the site of the blood-brain barrier. J. Anat. 95 (suppl.):20–22.

Pick, J. 1963. The submicroscopic organization of the sympathetic ganglion in the frog (*Rana pipiens*). J. Comp. Neurol. 120:409–462.

Ramon y Cajal, S. 1952. I. Histologie du systeme nerveux de l'homme et des vertèbres. Consijo Superior de Investigaciones Cientificas, Madrid.

Scharrer, E., and B. Scharrer. 1963. Neuroendocrinology. Columbia University Press, New York.

Shantaverappa, T. R., and G. H. Bourne. 1967. Nature and origin of perisynaptic cells of the motor end plate. International Review of Cytology. G. H. Bourne and J. F. Danielli, eds. 21:353–364.

Shauklin, W. M., and N. A. Azzam. 1964. Histological and histochemical studies on the incisures of Schmidt Lanterman. J. Comp. Neurol. 123:5–9.

chapter 13
SENSE ORGANS

THE CENTRAL NERVOUS SYSTEM of vertebrates is physically removed from the outside environment, and so it is ordinarily not directly affected by most kinds of environmental changes. With the general exceptions of the olfactory organ and free sensory nerve endings (see Chapter 12), stimulation of the nervous system is effected *indirectly* through intermediary components, the *sensory receptors.* Sensory receptors are functionally discriminative. As described in the preceding chapter, some receptor organs are specifically sensitive to touch or pressure, others to heat, others to cold, and still others to stretch. Even more discriminative are receptors for the organs of special sense, which will be described in this chapter.

Sensation *per se* is probably entirely a function of certain sensory centers of the brain. In such matters human beings can speak only for themselves, for no way has yet been devised whereby either the nature or intensity of sensation in other animals can be determined, if indeed the word sensation is even appropriate to nonhuman species. In fact, one cannot be certain that the sensation perceived by one person is identical to that perceived by another, even when all possible variables are controlled. Uncertainties about the universality of sensation, as well as the correspondence between sensation and the physical environment, have stimulated philosophers since at least the time of Plato to consider the nature of reality itself.

Although the ultimate nature of sensation is essentially unknown, the mechanisms and phenomena necessary to it are quite well understood. It is known, for example, that specific receptor cells translate correspondingly specific environmental changes or conditions into nerve stimuli at the sensory endings of corresponding sensory nerves. There the stimuli are transformed into nerve impulses, the physicochemical nature of which is the same for all nerves and for all stimuli. The nerve impulse is then transmitted, usually through a synaptic series of neurons, to sensory areas of the brain. It is known that in human beings all of the areas of conscious sensation reside in the cerebral cortex, and that it is in these areas that the actual phenomenon called sensation occurs. Thus for human beings it can be said that the cerebral cortex sees, hears, smells, and feels and that the eyes, ears, and other sense organs are the instruments for these respective functions. There is scientific justification for the assertion made many years ago that if one could functionally graft the optic nerves to the inner ears and the acoustic nerves to the eyes, one could hear lightning and see thunder. The anatomic basis for sensation is clearly the precise areas of the cerebral cortex that receive the nervous impulses in question.

Because other mammals have essentially the same kind of neurological equipment as human beings, including similar projection pathways to the cerebral cortex from all exteroceptive sense organs, and similar—albeit not so extensive—association pathways between cortical areas, there is every likelihood that these mammals have sensations and that their sensations are like those of human beings. As for nonmammalian vertebrates, which, with the likely exception of the birds, largely lack any sort of neurological organ functionally comparable to the mammalian cerebral cortex, there is serious question if there is any such thing as conscious sensation as we know it.

From the standpoint of function, the organs of special sense can be grouped into three systems; chemoreceptive, mechanoreceptive, and photoreceptive.

THE CHEMORECEPTIVE SYSTEM

Most epithelia that are normally moist are sensitive to chemical irritation via their free sensory nerve endings. Thus the entire integument of most aquatic mammals and the mucous membranes of terrestrial animals, such as the linings of the mouth and nostrils, of the eyelids, and of the anus, are sensitive to chemical irritants. In addition to such nonspecific chemical sensitivity is a system of receptor cells or organs that have specific chemical sensitivities. Included within this system are the *olfactory organs* and the *taste buds.* Although both are chemoreceptive, they differ from each other anatomically and physiologically.

Olfactory Organs

Perhaps the most primitive of the vertebrate senses and the least understood is the sense of smell or olfaction. The variety of chemical substances that can be discriminated is, of course, not known, but it is obviously very large when one considers the familiar, highly developed olfactory abilities of dogs. Perhaps even more astonishing is the sensitivity of the olfactory epithelium to extremely low concentrations, actually trace quantities, of some kinds of chemical substances, such as the mercaptans (the odor of skunk, for example). It is inconceivable that there is one kind of receptor cell for every chemical substance or molecular configuration capable of being smelled. It is more likely that certain cells are sensitive to a "family" of substances, and certain other cells to a different family, and so on, with sufficient overlap between families to provide a mechanism for discrimination within families. On the other hand, experiments with frog olfactory epithelia and nerves have indicated that any olfactory cell will respond to a range of odors sufficiently broad to suggest that the neurophysiological basis of olfactory specificity is the nature of the nervous impulse generated by the odor. It is known that nervous impulses, while not differing in their electrochemical nature, can differ in frequency and in patterns of spacing between successive spikes or barrages of spikes. Attempts are being made to correlate such patterns with specific odors.

Most of the 10,000 or more nuances of odor distinguished by "educated" human noses can be grouped into seven classes of common odors: ethereal, camphoraceous, musky, floral, minty, pungent, and putrid. The determining characteristics of the first five classes have been shown to be the size and the shape of the molecule that produces the odor. The last two depend upon particular electronic characteristics of the molecule.

The olfactory epithelia of vertebrate animals are basically alike, and their sensory receptor components bear many striking similarities to olfactory receptors in invertebrates as well. In vertebrates the olfactory epithelium consists of bipolar neurons, sustentacular cells, and basal cells (Fig. 13.1).

The bipolar neurons are the sensory receptors and are distinguished by a round nucleus. They occupy the middle zone of the olfactory epithelium, although their cell bodies are distributed in several layers. The axonal processes pass into the underlying connective tissue, where they form the nonmyelinated olfactory tracts that enter the olfactory lobe of the telencephalon. The single dendritic process extends from the cell body to the free surface of the olfactory epithelium, where it expands slightly as the *olfactory vesicle*. Individual bipolar neurons are separated from each other throughout the length of their cell bodies and dendritic processes by the sustentacular cells. The olfactory vesicle is equipped with several nonmotile cilia (average, 6) that measure up to 200μ in length, and it incorporates a corresponding number of basal bodies, ciliary rootlets, and centrioles (Fig. 13.2). The proximal portion of the cilium measures about 250mμ in diameter in frogs and includes the typical arrangement of 9+2 double filaments (see Chapter 2). The distal portion of the cilium, accounting for about 80 percent of the

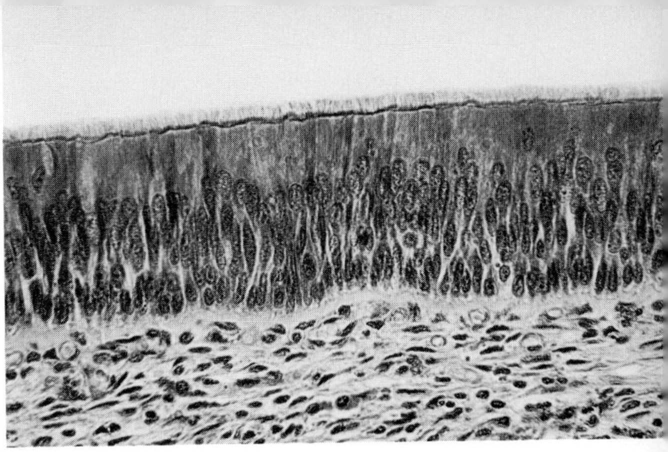

Fig. 13.1 Mammalian olfactory epithelium. X500.

length, is about 150mμ in diameter and includes a 9+2 arrangement of *single* filaments. The transition from double to single filaments is sharp, occupying a zone only about 0.5μ in length. The distal portion is easily broken off, whereupon the remaining cilium may assume an uncoordinated motility. Normally the cilia are protected by a continuous coating of mucus secreted by the sustentacular cells, Bowman's glands, and other cells of the adjacent mucous epithelium.

The sustentacular cells have an oval nucleus and are the most numerous cell type. The distal part of the cell is broad and cylindrical and extends to the free surface, where it bears

Fig. 13.2 Electron micrograph of the olfactory mucosa of a mouse. (From Donald Frisch. 1967. Am. J. Anat. 121:87–120.)

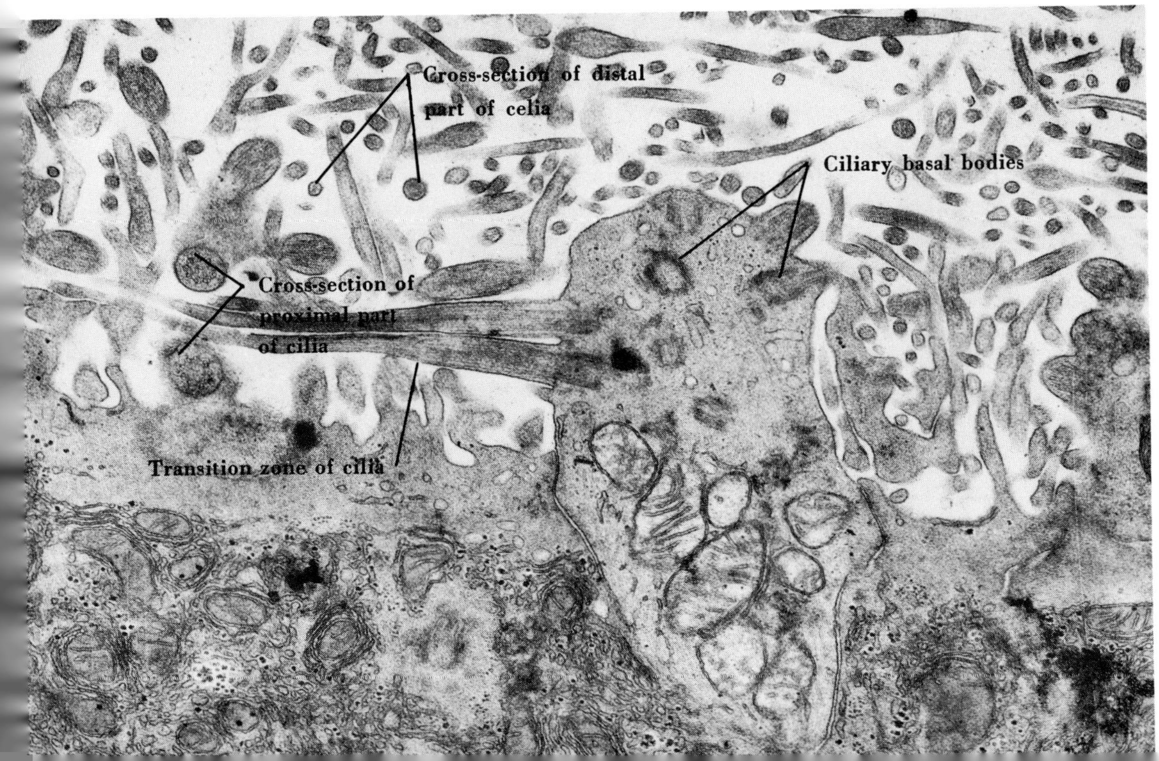

a striated border formed by a dense carpet of microvilli. The cytoplasm contains pigment granules and many mucous granules similar to those of goblet cells, except that they are extruded one-by-one rather than *en masse.* A filamentous process extends from the cell body down through the layer of bipolar neurons and the layer of basal cells to the basement lamina.

The basal cells are small and irregular in shape, with a rounded or oval nucleus, and they lie on the basement lamina. The submucosal connective tissue consists of a loose network of collagenous and elastic fibers and fibroblasts and includes clusters of simple branched tubular *glands of Bowman.* The secretory cells of the glands of Bowman contain yellowish pigment granules and mucous granules similar to those of the sustentacular cells.

Vomeronasal Organs

Located in the oral cavity of amphibians, reptiles, and certain mammals are paired inpocketings of mucous membrane that are presumed to have an olfactory function. These are the *vomeronasal organs* or Jacobson's organs. In amphibians in which they have been described, they appear less as an organ than as an area of sensory epithelium located on the ventral surface of the nasal cavity. The vomeronasal organ is perhaps most highly developed in lizards and snakes, in which it appears as a thickened epithelium that lines paired chambers in the roof of the mouth that communicate with the mouth by a pair of short canals (Fig. 13.3). When present in mammals the vomeronasal organ is a tubular formation on either side of the nasal septum and it opens anteriorly either into the nasal cavity directly or into the nasopalatine canal. The mammalian orders in which it has been described are the insectivorans, rodents, carnivorans, the ungulate orders, and the marsupial orders.

Two types of cells are distinguished in the sensory epithelium of the vomeronasal organ: sustentacular cells and sensory cells. The former are elongate cells the nuclei of which are also elongate; they are basophilic and comprise the outer nuclear layer of the epithelium (Fig. 13.3). The apices of the sustentacular cells join each at the lumen by terminal bars to form an outer limiting, nonciliated membrane. The sensory cells form the basal layer of the epithelium. They have round or oval, lightly staining nuclei that are distributed in several levels. Each sensory cell sends a slender apical process luminally between the sustentacular cells to the limiting membrane, where it expands as a terminal bulb. It therefore does not reach the luminal surface and, unlike olfactory sensory cells, it is devoid of cilia.

Taste Buds

Taste buds have a diversified distribution among the vertebrates. In fishes they are found mainly buried in the epithelium of the gill rakers and gill arches, but they may also occur in the mouth and pharynx and, in some species, on the fins or on the integument of the head and body. Bottom foragers, such as the sturgeon (*Acipenser*) and catfish (*Ictalurus*) have numerous taste buds or barbels associated with the snout or mouth. The taste buds of terrestrial vertebrates are restricted to the tongue and, to some extent, the pharynx.

The kinds of chemical substances to which taste buds are sensitive is rather large but, speaking again from the human standpoint, they may be grouped into five classes on the basis of the kinds of sensation with which they are correlated: namely, substances that taste sweet, sour, salty, metallic, or bitter. The taste buds of carp (*Cyprinus*) respond to a similar grouping and also to sapid substances that elicit no human response. Sea robins (*Prionotus*) and hake (*Urophycis*) apparently can taste sapid and sour substances, but not salty, sweet, or bitter ones.

The microscopic structure of taste buds is rather straightforward (Fig. 13.4). Each bud consists of an ovoid cluster of cells of two types, *gustatory receptor cells* and *sustentacular cells.* The receptor cells are generally oval or fusiform and have basophilic nuclei and densely granular cytoplasm. The apical end of

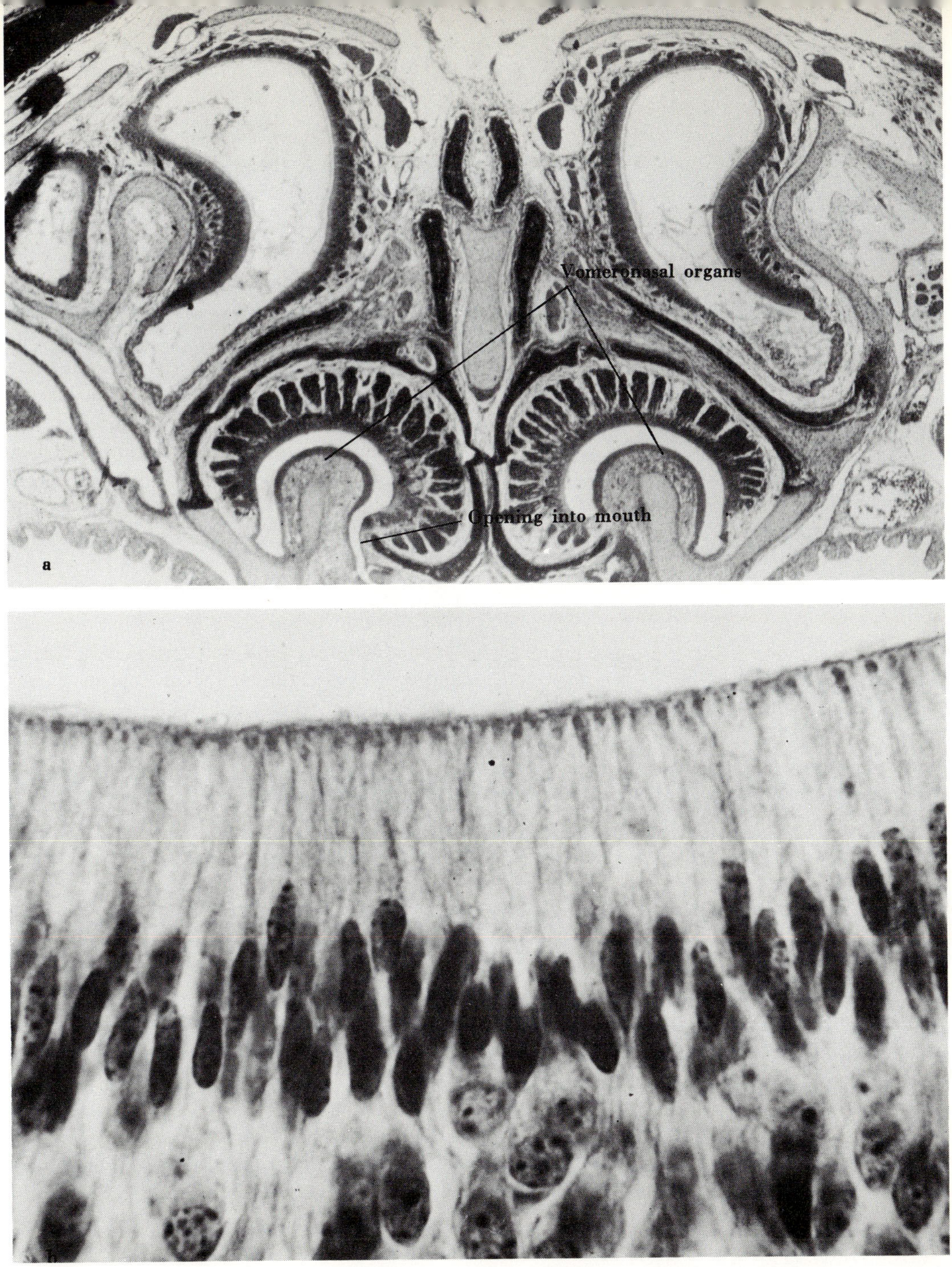

Fig. 13.3 The vomeronasal organ. A low-power micrograph of a transverse section of the head of a snake (*Vipera*) is shown in *a*. This section includes the opening of the left vomeronasal organ into the mouth. A high-power micrograph of the vomeronasal epithelium in a rat is shown in *b*. The sensory cells have round nuclei in the basal layers and extend up to the superficial limiting membrane in clublike swellings. The cells with elongate nuclei are the sustentacular cells. Their processes form the limiting membrane. (Both figures from R. Cordier. 1964. *In* The cell, J. Brachet and A. E. Mirsky, eds., vol. VI, Academic Press, New York.)

each receptor cell appears as a process that protrudes into a small pitlike depression in the epithelium, the *taste pore* (Fig. 13.4). These apical processes bear numerous microvilli which provide a large amount of surface for contact with the stimulus. Apical processes are

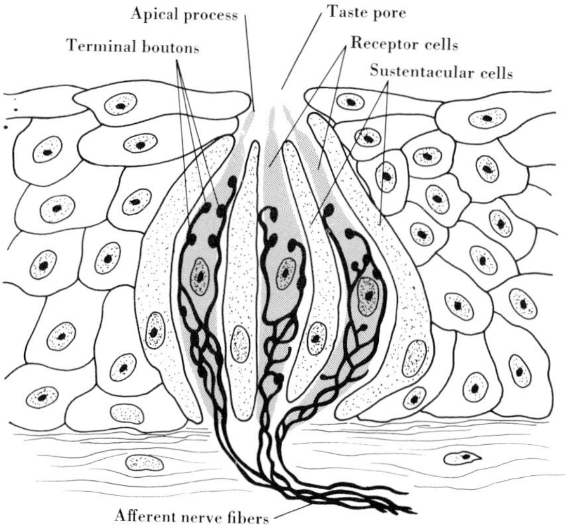

Fig. 13.4 Structure of a taste bud.

lacking in some marine teleosts, however. Whether stimulation occurs on the microvillous surface or within the apical process is not known. The body and basal process of the receptor cell are surrounded by a plexus of myelinated sensory nerve endings whose processes terminate as amyelinated boutons, mainly in the apical region of the receptor cell. Contact between bouton and receptor is effected by interdigitation of their cell membranes.

The sustentacular cells are large, fusiform cells with oval, pale nuclei and lightly staining cytoplasm. They extend from the basement lamina to the taste pore and they both embrace the centrally located receptor cells peripherally and separate them from each other internally.

THE MECHANORECEPTOR (ACOUSTICOLATERALIS) SYSTEM

This system includes the special sense organs that are concerned with providing the central nervous system with information about certain kinds of changes or conditions in the physical nature of the environment, such as vibrations (sound), pressure (depth), flow (rheoscopy), and gravity (attitude or orientation). All of the components of this system in vertebrates are believed to have been derived from a common, ancestral neuroepithelial type of cell similar to those found in modern fishes as the *neuromast cells* of the lateral line organ.

A typical neuromast cell consists of a nucleate cell body that supports a motile cilium and a cluster of elongate stereocilia on its apical end (Fig. 13.5). These processes usually appear embedded in an extracellular gelatinous substance that clings tenaciously to them—forming a clear, knoblike body, the cupula. The position of the cilium in relation to the stereocilia is precise and is correlated with the direction of movement of the stimulus (Fig. 13.5). Whereas there is evidence that some neuromasts may respond to chemical, thermal, or osmotic stimuli, in general they respond to shearing forces that act upon their protruding cilia and stereocilia. Neuromasts comprise the sensory receptors of the lateral line system and of the inner ear, which is an evolutionary derivative of the lateral line. Together these mechanoreceptors constitute the acousticolateralis system, the central connections of which are described in Chapter 12.

The Lateral Line System

The lateral line system is found in fishes and in the aquatic stages of amphibians. Although concerned primarily with perception of low-frequency vibration, turbulence, and the flow of the surrounding medium, certain components of the lateral line system are adapted to respond to other kinds of stimuli. One especially interesting adaptation is seen in the Mormyridae and some of the Gymnotidae, fishes that generate a weak electrical field about them. Distortions of the electrical field by the presence of other fishes or objects are sensed by these animals, enabling them thereby to inhabit murky or turbid waters. Parts of the lateral line system located in the skin of the head function as the electrical receptors for this activity. The ampullae of Lorenzini (Fig. 13.6) are isolated lateral line components scattered about the head regions of elasmobranch fishes. Although their function is uncertain, they have been shown to be sensitive to electrical, thermal, and salinity

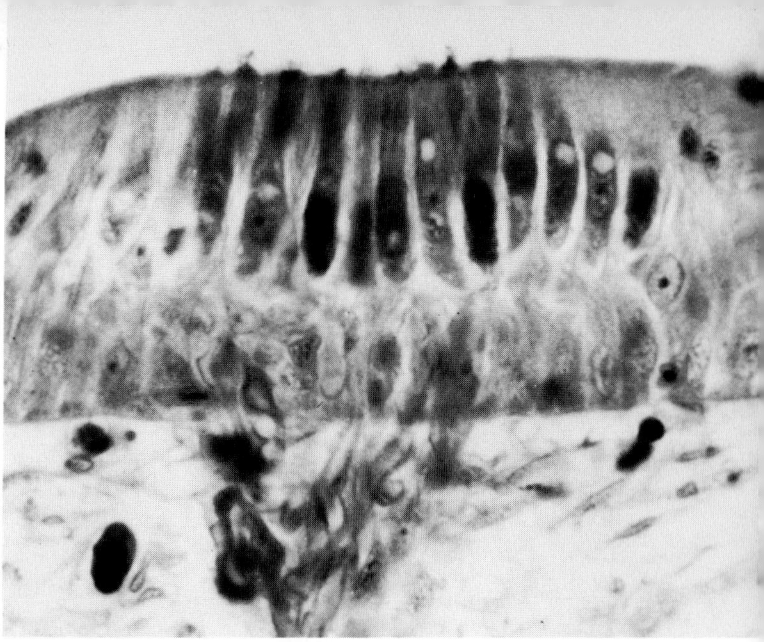

Fig. 13.5 Structure of the neuromast. A neuromast cluster in the skin of the teleost *Xiphophorus* is shown *right.* The sensory cells have apical hairs and supranuclear globules. (From R. Cordier. 1964. *In* The cell, J. Brachet and A. E. Mirsky, eds., vol. VI, Academic Press, New York.) Schema of a neuromast cell, showing the polarity implicit in the arrangement of stereocilia and kinetocilia, is shown *below.* (After Å. Flock. 1965. Symp. Quant. Biol. 30:133–145.)

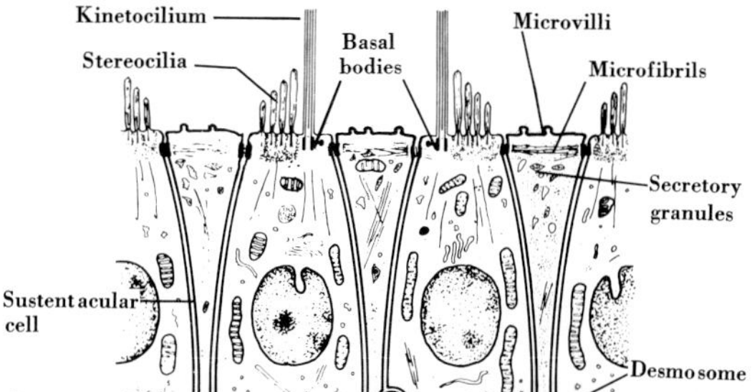

stimuli, as well as to mechanical stimulation.

Other isolated lateral line components described in various fishes are the *vesicles of Savi* (Fig. 13.6) of several genera of rays, the *organs of Fahrenholtz* of larval dipnoans (Fig. 13.6), and the *spiracular organs* of the Chondrichthyes, Holostei, and Chondrostei; but their functions are unknown.

The lateral line organ *per se* consists of clusters of sensory receptors arranged in definite lines in the integument of the head and trunk. The individual clusters may lie in pits or depressions in the skin or in canals that are either closed or partially open. The canals have been shown to develop initially from depressions. Canals and pits alike are filled with a gelatinous or semigelatinous mucus substance through which the appropriate stimulus is transmitted from the environment to the receptor cells.

The cluster of receptor cells is supported by sustentacular cells and, in some cases, by basal cells that attach to the basement lamina. Separating these components from the cells of the epithelium is a fourth class of cells called the *mantle cells* (Fig. 13.7).

The receptor cells are club-shaped or columnar and bear three types of projections on the exposed apical end. These include the motile cilium (kinetocilium) and the long stereocilia mentioned above, as well as small irregularly situated microvilli (0.1μ diameter). The cilium is always located on the periphery of the patch of stereocilia and in the axis of movement of the stimulus (Fig. 13.8), but in some clusters it is on the "windward" and in others it is on the "leeward" edge of the patch, to borrow nautical terminology. The functional significance of this arrangement has been investigated and it has been shown that a surge of movement along the axis in one direction causes depolarization of receptors whose cilium is on the far (leeward) side of the patch of stereocilia and hyperpolarization of the receptors whose cilium is on the near (windward) side of the patch. The

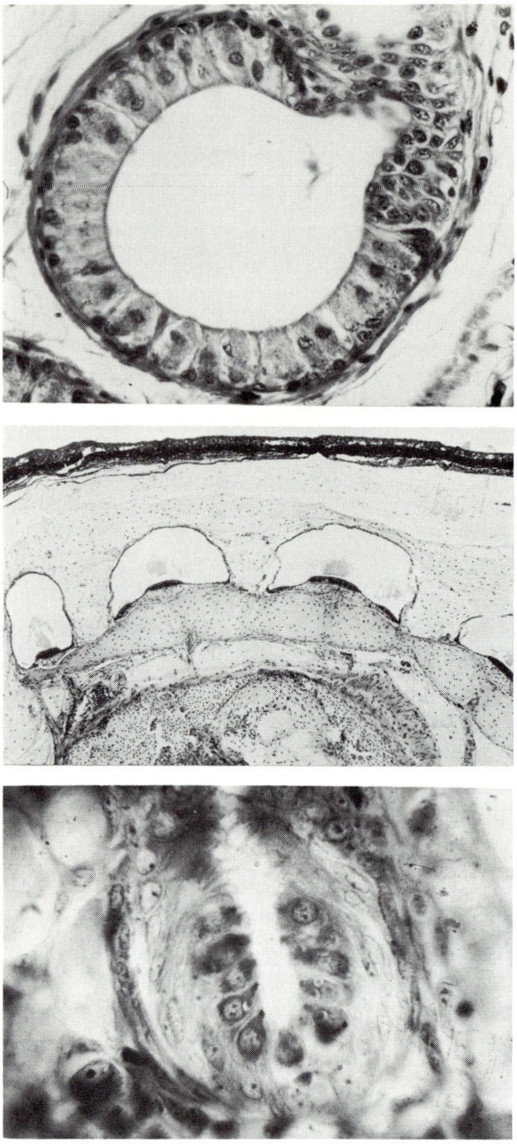

Fig. 13.6 Some isolated components of the lateral line system. The ampulla of Lorenzini of *Plotosus anguillaris* is shown at *top*. Ovoid sensory cells alternate with narrow supporting cells. Vesicles of Savi in *Torpedo marmorata* located in a row on a plate of chondroid are shown at *middle*. Note the afferent nerve fibers and the cupulae. Organ of Fahrenholz in *Polypterus weeksi* is shown *directly above*. The sensory cells are dark and pear-shaped. (All figures from R. Cordier. 1964. *In* J. Brachet and A. E. Mirsky, eds., The cell, vol. VI, Academic Press, New York.)

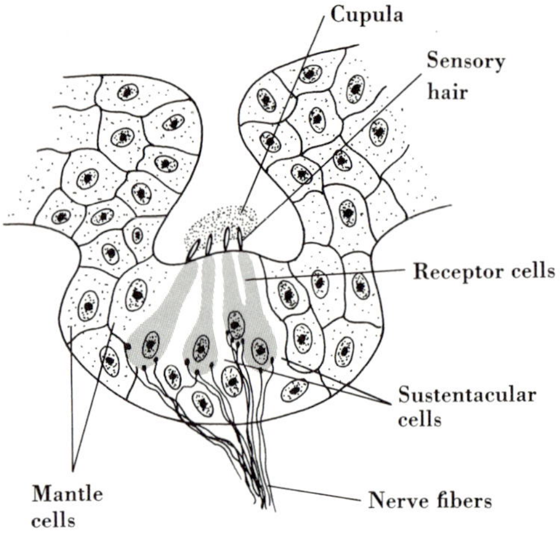

Fig. 13.7 A diagram of a lateral line cluster.

former results in excitation and the latter in inhibition, and so together they provide a mechanism for discriminating between motion in either direction, with each having an excitatory and an inhibitory component. The amount of depolarization always exceeds that of hyperpolarization, however. Although the acousticolateralis system is entirely sensory in function, it is interesting to note that electron micrographs reveal the presence of both afferent and efferent nerve endings. The efferent nerves are believed to provide feedback for modifying the sensory response.

The sustentacular cells of the lateral line organ are columnar or spindle-shaped and they are distributed among the receptor cells. They have a well-developed endoplasmic

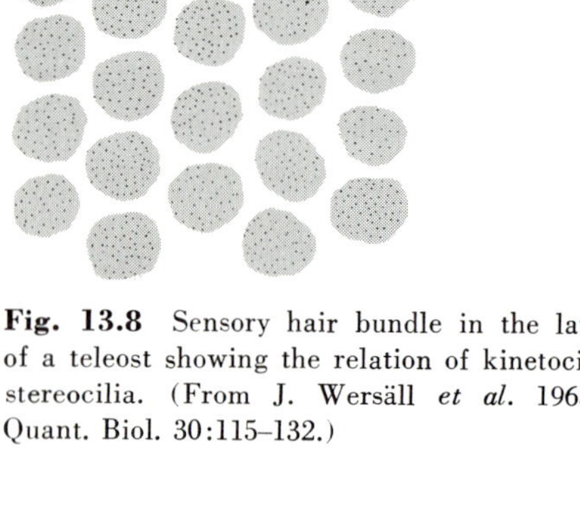

Fig. 13.8 Sensory hair bundle in the lateral line of a teleost showing the relation of kinetocilium and stereocilia. (From J. Wersäll *et al.* 1965. Symp. Quant. Biol. 30:115–132.)

reticulum with large cisternae, which gives the cells a vacuolated appearance. The Golgi substance is also conspicuous. The basal cells have a similar ultrastructure as the sustentacular cells, but they are rounded or irregular in shape and are confined to the base of the cell cluster.

The Inner Ear

The inner ear traces its origins back to the portion of the lateral line system that was innervated by the eighth cranial nerve, thereby leaving the rest of the lateral line system innervated by cranial nerves VII, IX, and X. Like other components of the lateral line or acousticolateralis system, the inner ear differentiates from inpocketing ectoderm, but in this case the closure is complete so that a pair of vesicles, the *otic vesicles,* is formed. In myxinoid cyclostomes the otic vesicle gives rise to a vertical semicircular canal, and little else. In the petromyzonts two vertical semicircular canals are formed. In the Chondrichthyes and all higher vertebrates three semicircular canals are present, two vertical and one horizontal. In the fishes a horizontal constriction also appears around the differentiating otic vesicles, partially dividing them into a *utricular* chamber dorsally and a *saccular* chamber ventrally. The semicircular canals develop as foldings of the wall of the utriculus. The sacculus forms a ventromedial or posteroventral outpocketing, the *lagena,* in fishes and amphibians, which becomes elongated and curved in reptiles and birds, and appears as the coiled *cochlear duct* in mammals. The number of gyres in the cochlear spiral reaches a maximum of five in alpacas.

The early division of the otic vesicle into utriculus and sacculus is significant, because the utriculus and its derivatives are concerned primarily and almost exclusively with the function of equilibration, whereas the sacculus and its important derivative, the lagena or cochlear duct, are concerned primarily with the auditory sense. All of the foregoing structures are hollow and are filled with a viscous fluid secretion, the *endolymph.* The major part of the walls of the utriculus and sacculus appears as a thin epithelium that is supported externally by a dense feltwork of collagenous fibers. Because of the complexity of its folds, outpocketings, and canals, the entire inner ear structure, with its supporting dense connective tissue, has become known as the *membranous labyrinth.* It lies embedded in the skeletal tissue of the skull. The remaining cells of the utricular and saccular walls appear as rather sharply circumscribed and elevated patches of sensory epithelium known as *cristae* in the ampullae of the semicircular canals, *maculae* in the utriculus and sacculus (and in the lagena in fish and amphibians), and as the *organ of Corti* in the cochlear duct. The sensory epithelium has essentially the same histologic characteristics as the neuromasts of the lateral line system, except that in adult mammals the kinetocilia of the organ of Corti degenerate and disappear (Fig. 13.9).

Concretions of various sorts (*otoliths*) are

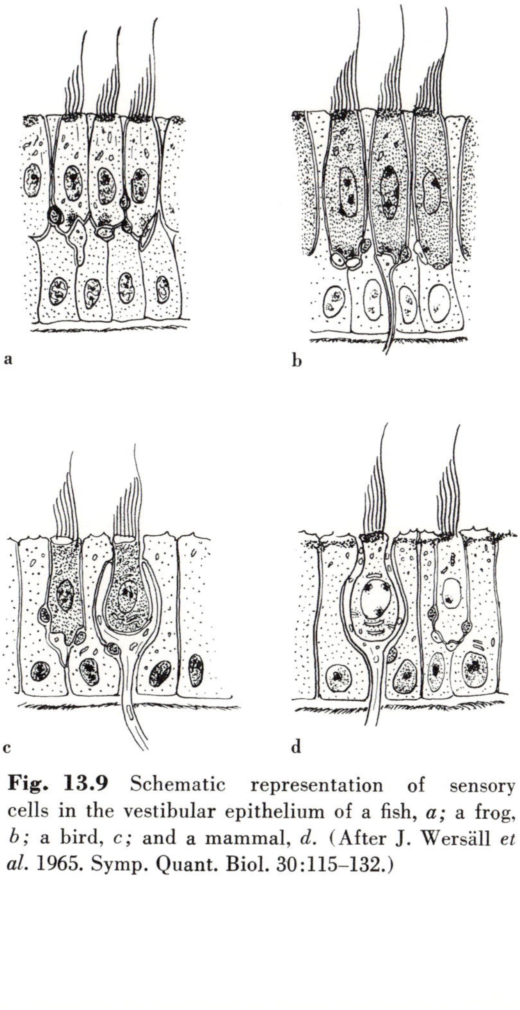

Fig. 13.9 Schematic representation of sensory cells in the vestibular epithelium of a fish, *a;* a frog, *b;* a bird, *c;* and a mammal, *d.* (After J. Wersäll *et al.* 1965. Symp. Quant. Biol. 30:115–132.)

found associated with the cristae and maculae, particularly in the lower vertebrates. In fishes they may be quite large, nearly filling the utriculus and sacculus. The inner ear of many elasmobranchs, on the other hand, communicates with the outside by an invaginated duct from the integument, and grains of sand find their way via this duct to the inner ear, where presumably they function as otoliths. Otoliths of the utriculus and sacculus normally rest on the sensory epithelium, and their weight elicits a rhythmic firing of impulses to the central nervous system via the vestibular branch of the eighth cranial nerve. If the animal is tilted from its normal attitude, the weight of the otolith produces a shearing force on the sensory hairs, and this causes an increase in the rate of impulses being fired. A sense of dynamic equilibrium, that is, of changing attitudes during motion, is achieved in a similar way, but it appears to depend on the surging or displacement of the endolymph in the semicircular canals.

The sense of hearing is believed to have developed as a highly discriminating sense with the evolution of the tetrapods, representing primarily an adaptation to a gaseous environment. Inasmuch as both water and protoplasm are incompressible and the density of fish is nearly that of water, fish might be considered relatively "transparent" to sound. Sounds of low frequency or sharp sounds of high intensity can certainly be perceived, probably because such sounds produce turbulence—to which the lateral line system is very sensitive. In some families of fishes, such as the cods (*Gadidae*) and herring (*Clupeidae*), an anterior extension of the swim bladder reaches the inner ear, thus providing a mechanism whereby sound waves can be intercepted and transmitted to the ear and better perceived. In the Cypriniformes, which includes a large number of fresh-water fishes, a paired series of three or four bones (*Weberian ossicles*) that are derived from the first three or four vertebrae, connect the swim bladder with the inner ear, much as do the otic ossicles in higher vertebrates. Doubtless the sound-discriminating abilities of these forms is greatly enhanced.

Amphibians, reptiles, and birds have a pistonlike bone, the *columella,* that transmits sound vibrations from the eardrum (*tympanum*) to a membrane-covered opening in the inner ear capsule. This opening communicates with a fluid-filled space between the capsule and the lagena portion of the membranous labyrinth. The fluid, called *perilymph,* is derived directly from the cerebrospinal fluid along the route of the auditory nerve. Oscillation or surging of the perilymph by the piston action of the columella therefore transfers pressure waves to all parts of the lagena. The tip of the columella that touches the ear drum is cartilaginous and able to bend if a loud noise strikes the eardrum, thus protecting the inner ear from damage.

Mammals have a series of three otic ossicles (*malleus, incus,* and *stapes*) lying between the tympanic membrane and the inner ear. The stapes is homologous with the columella. These bones are arranged in such a way that the amplitude of movement of the stapes against the inner ear is about two-thirds less than that of the malleus at the tympanum, but the force is correspondingly two-thirds greater —which results in increased sound pressure.

Mammals have also added to the auditory apparatus an *external ear* (*pinna*) which acts as a trumpet for concentrating sound waves. In many species the external ear can be moved so it is directed toward the sound. The relation of the perilymphatic space and the cochlear duct is also more elaborate in mammals. The perilymph is contained within a well-defined canal that follows up along one side of the coiled cochlear duct, recurves over the end of it through a small opening, the *helicotrema,* and then spirals down along the other side. A vertical section through the labyrinth at an intermediate level of the coiled cochlea shows the canals in cross section (Fig. 13.10) as the ascending perilymphatic canal (*scala vestibuli*), the descending perilymphatic canal (*scala tympani*) and, between them, the cochlear duct (*scala media*). The separation between vestibuli and media is achieved by the thin *Reissner's membrane,* and between media and the scala tympani by the thick *basilar membrane.* Pressure waves are introduced by

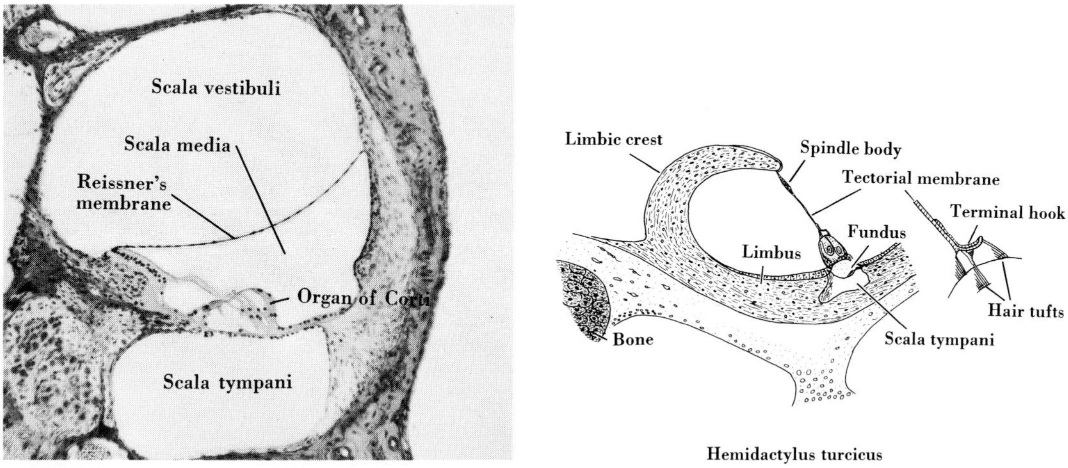

Fig. 13.10 The cochlear duct and organ of Corti. A vertical section of the cochlear duct of a mammal, showing the scalae and the organ of Corti, is shown in *a*. X100. The organ of Corti in the warty gecko, *Hemidactylus turcicus*, is shown in *b*. This is one of several interesting variations found in lizards. (Part *b* from E. G. Wever. 1967. J. Morphol. 122:307–320.)

the piston action of the stapes to the perilymph of the scala vestibuli through the *fenestra ovalis;* they pass up the scala vestibuli, then across the scala media to the scala tympani, and are dissipated by the movement of the membranous covering of the counterpart of the fenestra ovalis, the *fenestra rotunda.* This window, as the ovalis, faces the air-filled cavity of the middle ear chamber.

The sensory epithelium of the cochlea duct is incorporated into a complex structure called the *organ of Corti* (Fig. 13.11) that lies in the basilar membrane. Two kinds of receptor cells are evident in mammals, the *inner hair cells* and the *outer hair cells* (Fig. 13.12). These cells are separated from each other by the *tunnel of Corti,* which extends the length of the cochlear duct. Overlying the hair cells is a gelatinous membrane, the tectorial membrane, which is readily distorted by fixation procedures. Normally it is in continuous contact with the stereocilia of the inner and outer hair cells. Although it is generally held that pressure waves cause the stereocilia to brush or press against the tectorial membrane, thereby directly inducing nervous stimulation, there is some indication that the role of the tectorial membrane may not be so passive. It has been suggested that because of the crystalline arrangement of the water molecules in the protein-water gel of the tectorial membrane, piezoelectricity may be generated when the tectorial membrane undulates and that it is this electrical effect, amplified by the semiconductor characteristics of the tectorial membrane, that stimulates the receptor cells. In essence this theory proposes that the tectorial membrane and the hair cells function like a solid-state electronic pickup.

The histologic features of the cochlea and of the organ of Corti are presented in Fig. 13.10.

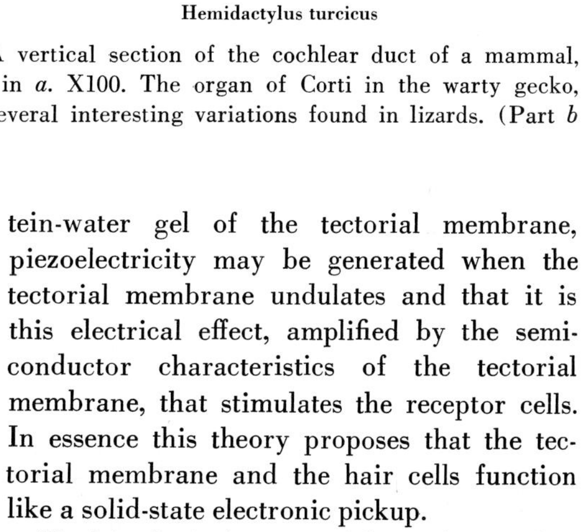

Fig. 13.11 A detailed stereogram of the organ of Corti. The sensory epithelium is shown "through" the tectorial membrane. (After J. Wersäll *et al.* 1965. Symp. Quant. Biol. 30:115–132.)

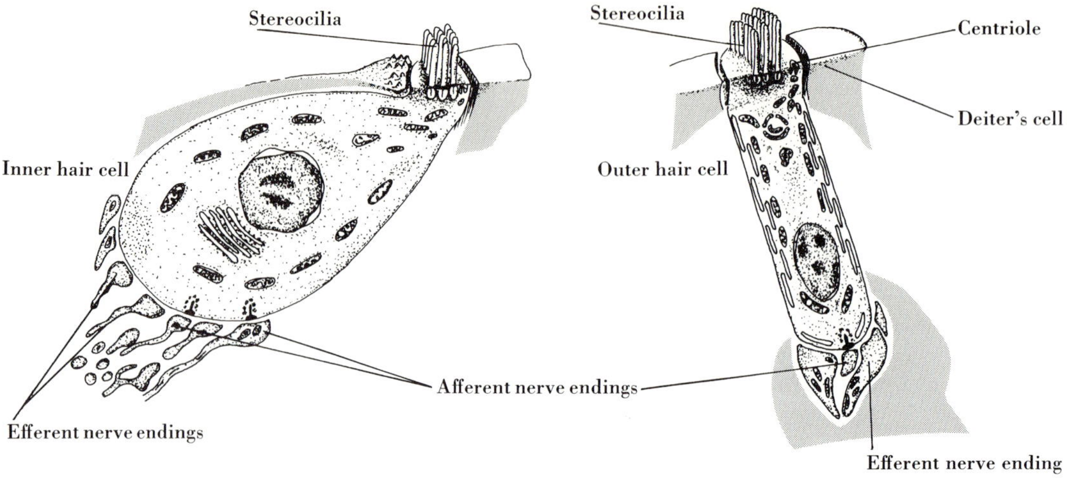

Fig. 13.12 Schematic diagram of the inner and outer hair cells of the organ of Corti. (After J. Wersäll *et al.* 1965. Symp. Quant. Biol. 30:115–132.)

Important cellular components in addition to the hair cells are the various kinds of sustentacular cells (pillar cells, phalangeal cells, Deiter's cells, Hensen's cells, and the cells of Claudius) and the cells of the *stria vascularis.* The latter is believed to be the site of secretion of the endolymph. Reissner's membrane, another important constituent, consists of two layers of cells: squamous epithelial cells of the cochlear duct lining, and squamous mesothelial cells that are continuous with the periosteal (or perichondral) lining of the scala vestibuli.

Correlated with sound perception in nature is sound production. Although not within the province of this book, sound production has had a pronounced bearing on the development of many aspects of the acoustic system in vertebrates. The seas have been shown to be a bedlam of noises created in various ways by both vertebrate and invertebrate animals. Some fishes grind or clack their teeth, others whistle by ejecting air from their swim bladder. Still others use their swim bladders as drums. Toothed whales, notably the porpoises, produce a wide range of noises and doubtless converse with each other, as well as emitting sounds for echolocation. The external and middle ears of cetaceans have been greatly modified for hearing under water. Air-filled sinuses in the skull provide the necessary mechanism for sound interception. Seals also use echolocation in the water. Terrestrial vertebrates that have highly developed adaptations for echolocation are the cave swiftlets of Malaya and Borneo and the chiropterans. The extent of refinement of echolocation in bats is truly astonishing, for not only do they seem to perceive an auditory "picture" of extraordinary refinement, but they also do not lose track of their own ultrasonic emissions in the background of similar emissions from perhaps hundreds or even thousands of their cave mates.

Another aspect of acoustic systems that merits some mention is mimicry. This ability is highly developed in several species of birds, as is well known, and it testifies to the excellence of the avian lagena as an auditory receptor. This is surprising, because the avian lagena is far less complex structurally than the mammalian cochlea, and yet its capabilities for sound discrimination seem to be as good.

THE PHOTORECEPTOR SYSTEM

Photoreceptors of one kind or another occur throughout the living world, but they attain their highest complexity in animals that move swiftly, either to secure food or to elude capture. Hence one finds photoreceptor organs best developed in the cephalopod mollusks, the arthropods, and the vertebrates. The cephalopod eye and the vertebrate eye are both image-forming eyes and they have many anatomical characteristics in common, although they are in no way homologous organs. Just how the vertebrate eye evolved is speculative, but doubtless it appeared very early in vertebrate evolution for it utilizes the second cranial nerve, and it is essentially the same in its basic structure and physiology in all classes of vertebrates.

The eyes begin their development in ontogenesis from bilateral outpocketings of the forebrain almost as soon as, if not before, closure of the neural tube occurs in this region. These outpocketings, the *optic vesicles*, become indented along their lateral and ventral surface while simultaneously constricting where they join the forebrain. Thus from the optic vesicle a sort of two-layered, inverted dipper is formed that is connected to the brain by its slender, two-layered handle (Fig. 13.13). The bowl of the dipper is known as the *optic cup*, and the handle is the *optic stalk*. The edges of the grooved optic stalk soon come together and fuse, making it a tube that later will serve as the sheath for the optic nerve. Fusion of the edges proceeds out over the optic cup, until only a uniformly round opening is left, the *pupil*.

The outer layer of the optic cup differentiates to become the pigmented retinal epithelium, for the most part, whereas the inner layer thickens to form primarily the photosensitive retina. Where the rim of the cup approximates the covering ectoderm of the embryo, it induces the inner layer of that ectoderm to invaginate a vesicle of cells that becomes the *crystalline lens*, which immediately detaches and comes to lie within the pupil. The remaining ectoderm of the skin opposite this point is transparent and differentiates into the corneal epithelium.

Meanwhile mesenchyme has become concentrated into two layers that surround the optic cup. The deeper of the two layers becomes pigmented and is known as the *uvea*. The portion of the uvea that covers the pigmented retinal epithelium becomes richly vascularized and is known as the *choroid coat*. The part of the uvea that lies at the rim of the optic cup differentiates into the ciliary muscles, the ciliary processes, and the outer layer of the iris. The more superficial layer of mesenchyme surrounding the optic cup becomes densely fibrous and differentiates into the *sclera* of the eye—a tough, thin, opaque membrane of very high tensile strength. The part of the sclera that lies opposite the rim and pupil of the optic cup joins with the corneal ectoderm and similarly is transparent. Together they constitute the *cornea*. The boundary between opaque sclera and transparent cornea is very sharp and is marked by a thickening of the former, the *scleral roll*. This appears as a ridge encircling the cornea and it serves as the origin for the ciliary muscles (Fig. 13.14).

Before considering the differentiation and microscopic structure of the eye, some basic ground rules of orientation should be established. As indicated in Fig. 13.14, the optic axis is a straight line that runs through the center of the cornea, the pupil, the lens, and the area centralis of the retina. This axis is also the anterior-posterior axis of the human eye. In some vertebrates the eyes are lateral, in others they are directed anteriorly, and in still others they are directed dorsally, to say nothing of the flounder, in which they are both on the same side of the head. To avoid confusion, the axes of orientation that apply to the human eye will be used in reference to the eyes of all vertebrates. Thus the cornea is considered to be in the anterior part of the eye, the fovea and optic nerve in the posterior part, and the side nearest the nostrils or rostrum is considered to be medial.

The Sclera

The sclera is thin in most vertebrates (about 0.5 mm in human eyes), but it is strong and inelastic. It is composed of flat ribbons of collagenous fibers that run in various directions over the surface of the eye, interlacing with each other to form a tight feltwork. Networks of fine elastic fibers occur between the ribbons, as well as fibroblasts and occasional melanocytes. The sclera is virtually avascular except adjacent to the scleral roll, where a system of sinuses, the *spaces of Fontana* and

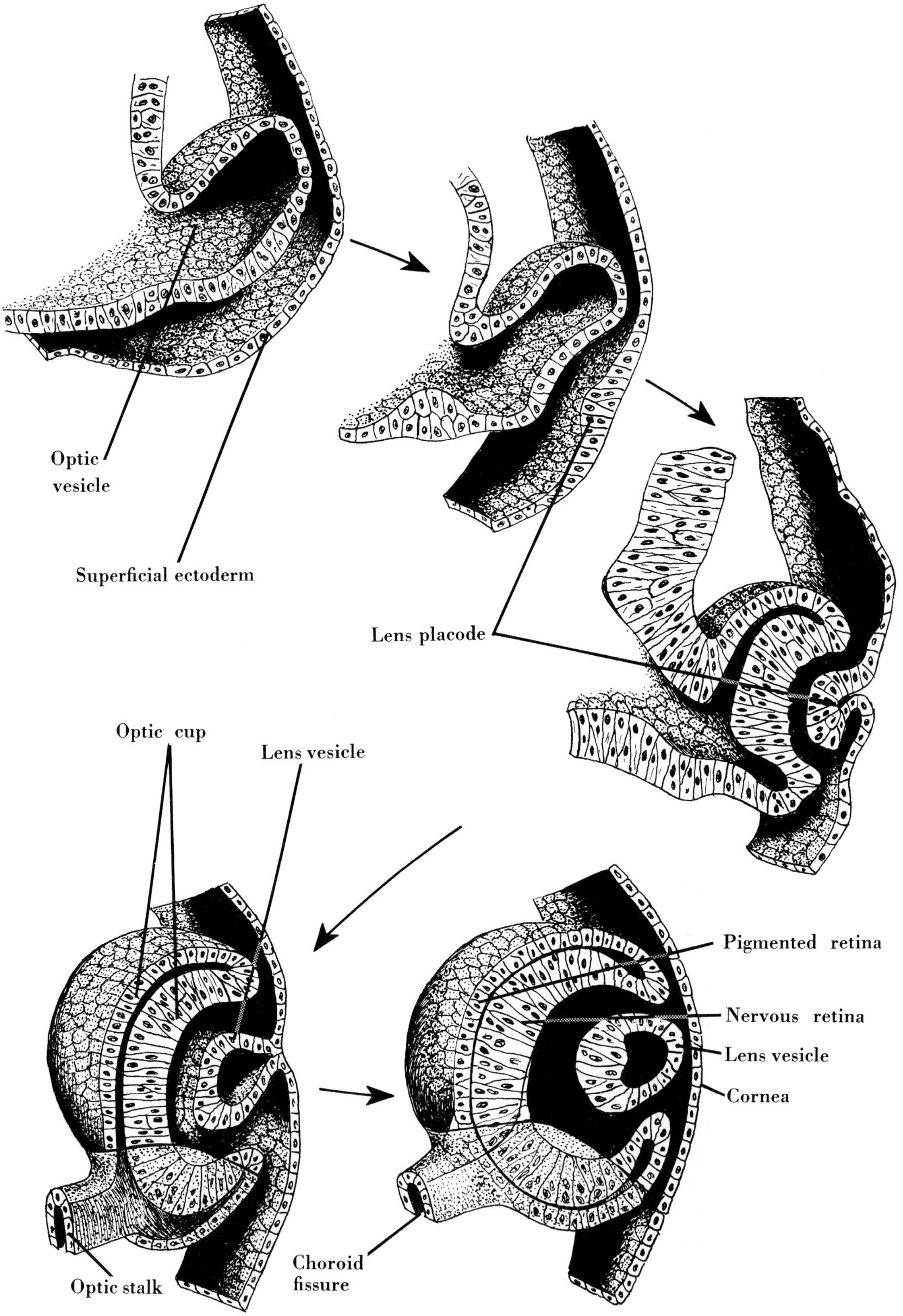

Fig. 13.13 A diagrammatic representation of the ontogony of the vertebrate eye. See text for description.

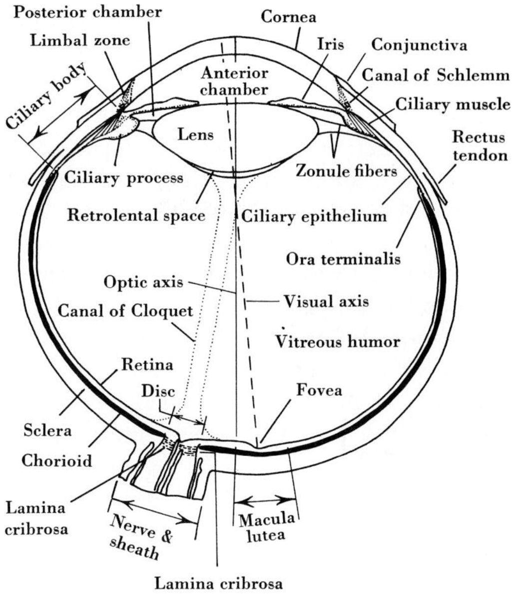

Fig. 13.14 The structure of the human eye in horizontal section. (After G. L. Walls. 1963. The vertebrate eye and its adaptive radiation. Hafner, New York. Copyright Cranbrook Institute of Science.)

the *canals of Schlemm,* are found. These parts serve to transport the aqueous humor of the eye to a small system of scleral veins, the *ciliary veins.* The elasmobranch sclera is quite vascular, however. Cartilage is present in the scleras of many vertebrates, and it is particularly well developed in species that have nonspherical eyes, such as owls and some of the deep-sea fishes. Most elasmobranchs and many of the bony fishes have scleral cartilages. Scleral ossicles, which are small bony plates, occur in all reptiles (except Crocodilia and ophidians) and in birds. They have an important function in accommodation in these forms.

As stated above, the sclera is very thin in most vertebrates, for most vertebrates have spherical eyes. It is only as thick as it need be to serve its two primary functions: to serve as points of insertion of the extrinsic eye muscles, and to withstand the intraoptic pressure or turgor, which in man amounts to about 25 mm Hg. In some vertebrates, however, the sclera is quite thick. It is thick in such giant species as whales, the whale shark (*Rhinodon typicum*), the basking shark (*Selache maxima*), and in elephants. The sclera is believed to be thick in these species because the eyes are so large that the force of gravity alone, at least in respect to elephants, would tend to distort the spherical shape of the eye unless the intraoptic pressure were greatly elevated. In the case of the marine species, the turbulence of the water occasioned by diving would probably be sufficient to distort such large eyes, which are oblate spheroids to begin with. The fact that thick scleras are associated with deep-diving species has misled some authorities to believe that the thick sclera is an adaptation to high pressure, but elephants do not dive deeply enough even to cover their eyeballs if they can help it, and one of the deepest diving fishes known, the elasmobranch *Etmopterus,* has an extremely thin sclera.

The Cornea

The cornea of petromyzonts and of some fishes is solely of mesodermal derivation, for the epithelium covering it is not fused to the cornea. In other species the fusion is complete, so that the cornea is a definitive unit of structure. It is transparent and usually unpigmented. In some bony fishes (*Amia,* perches, wrasses), a light yellowish or greenish color is evident. In Torpedo the upper half of the cornea is heavily pigmented. In the blenny, *Dialommus fuscus,* the cornea is pigmented except for two circular areas, which gives the impression that this fish has double pupils. The cornea is not contractile, except in Petromyzon, whose cornea is equipped with a muscle by which it can be flattened for optic accommodation.

The epithelial layer of the cornea is stratified squamous. The superficial cells are highly irregular in surface view, with many processes that interdigitate with one another. A profuse network of bare nerve endings ramifies between the cells, particularly in the basal and intermediate levels. The epithelium is completely avascular and is nourished in tetrapods by lacrimal secretion, as well as by the aqueous humor of the anterior chamber. The epithelium is characterized by a remarkable capacity for rapid repair in case of injury.

The epithelial layer is separated from the dermal layer in *Petromyzon* and a few other species (conger eels, flounder) by a gelatinous secretion. But in all other vertebrates no such separation exists. Underlying the epithelium is a thick substantia propria that accounts for more than 90 percent of the thickness of the cornea. The junction between the two layers is marked in higher vertebrates by a narrow acellular layer of fine, irregularly arranged collagenous fibers known as Bowman's membrane. The substantia propria itself is mainly collagenous also, but the fibers are arranged in regular layers. All the fibers within a given layer lie flat and run parallel to each other, but the orientation of fibers of adjacent layers is oblique to each other, as in plywood. There is some intermeshing of fibers between adjacent layers, however, which serves to tie the layers together. Between layers there are a glycoprotein cementing substance and flattened, irregular fibroblasts. Like the corneal epithelium, the substantia propria is avascular.

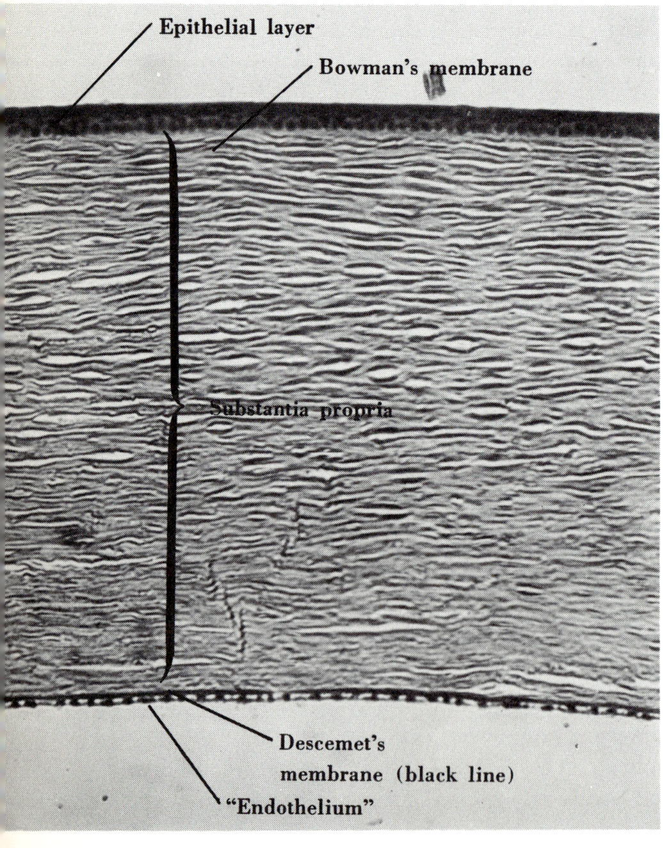

Fig. 13.15 Photomicrograph of a vertical section of the cornea of a monkey. X100.

The innermost or posterior layer of the cornea is an epithelium of mesenchymal origin that is sometimes referred to as an endothelium. It consists of a single layer of low cuboidal cells. It is separated from the substantia propria by a narrow refractile membrane (*Descemet's membrane*) which consists of unusual protein fibers resembling collagen (Fig. 13.15). The filaments of these fibers are arranged in a very fine hexagonal pattern.

The cornea of snakes and some lizards is surmounted by a second transparent membrane or *spectacles;* these represent the skin of transparent eyelids the edges of which have fused over the front of the eye. The outer epithelial layer of the spectacles is shed along with the rest of the outer epithelium of the skin during ecdysis.

The cornea functions as the external "window" of the eye in all species. Its index of refraction in marine and fresh water species is very similar to that of the outside medium, and so it has no refractive function. Because of this situation, fishes' corneas may be ridged, flat, or irregular in thickness without distorting the image. The corneas of terrestrial species are the major refractory component of the eye, however, and their contour and smoothness are of immense importance optically.

The Choroid Layer

The choroid layer, or simply *choroid,* is the part of the uvea that lies between the sclera and the retina and is bounded anteriorly by another modification of the uvea, the ciliary body. The choroid is both richly vascularized and heavily pigmented. At least three layers are distinguishable: an outer *vessel layer,* an intermediate *capillary layer,* and an inner *vitreous layer* (Fig. 13.16). The vessel layer supports the profusely branching blood vessels that have entered through the sclera. The connective tissue stroma between the vessels includes numerous melanocytes, fibroblasts, and macrophages. Smooth muscle cells may occur in the outer part of the vessel layer anterior to the equator. The outer part of the vessel layer is modified in many vertebrates into a light reflecting "carpet" or *tapetum lucidum,* as described below.

The capillary layer is sharply delimited from the choroid body anteriorly at the *ora serrata,* a distinct landmark on the inner surface of the eye. The ora serrata is a scalloped or wavy edge that marks the posterior limit of the scleral insertions of the ciliary muscles and the anterior limit of the photosensitive cells of the retina. The capillaries of the capillary layer lie in a single layer or plane and are among the largest in cross-sectional diameter in the body, appearing nearly as large as sinusoids. In other respects they are typical capillaries. The intercapillary stroma consists mainly of delicate collagenous and elastic networks, fibroblasts, and occasional melanocytes. The microscopic structure of the capillary layer reflects the high metabolic needs of the tissues it nourishes, mainly the photosensitive retina.

The vitreous layer of the choroid is thin (about 4μ), glassy, and microscopically homogeneous. Composed chiefly of elastic fibers that are continuous with those of the capillary layer and of a cuticular matrix that is contributed to by the pigmented retina, it serves as a diffusion membrane between the capillary layer and the retina. The functional role of the vitreous layer is not yet fully understood. This layer is very thick in ostriches and is reflective.

Although primarily a vascular tissue, the choroid coat has assumed an important secondary function in a large number of vertebrates, namely that of reflecting light back through the retina. This phenomenon is observed as the "eyeshine" that is characteristic or so many of the vertebrates, particularly the carnivores and the ungulates. Such reflective areas are known as *tapeta lucida* and they represent an adaptation to low intensities of illumination, for, although reflected light obviously produces some blurring or "halation," it also increases the amount of photic energy

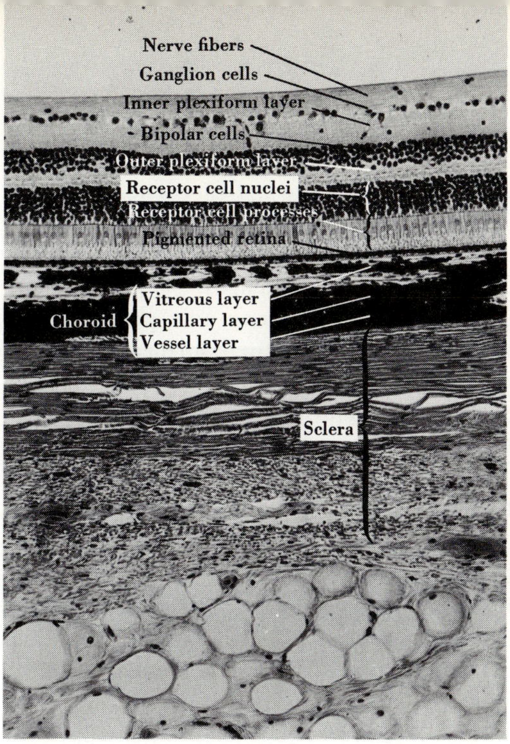

Fig. 13.16 Photomicrograph of the posterior wall of a monkey's eye, showing the layers of the retina and uvea. X100.

passing through the retina sufficiently to make the difference between seeing and not seeing. The area of the choroid that is modified to this function varies from a small triangular area bordering on the area centralis (sometimes including it), such as in dogs, to most of the choroid area, such as in seals and many fishes. Except in seals, the tapetal area is usually dorsal, lying behind the part of the retina that receives light from below the horizon. In seals it is both dorsal and "lateral," thereby intensifying the visual effect from fields both below and ahead. Retinal and iridial tapeta also occur as described below.

Three kinds of choroid tapeta have been described: fibrous, cellular, and guanin-containing tapeta. Fibrous tapeta owe their reflective property to the dense arrangement of the collagenous fibers in the part of the vessel layer next to the capillary layer, and they glisten as fresh tendons do. Fibrous tapeta are perforated for the passage of arterioles and venules between the outer part of the vessel layer and the capillary layer. Cellular tapeta are formed of several layers (for example, 3 to 4 in wolverines, 12 to 15 in cats, and 16 to 35 in seals) of platelike cells that are stacked in register so that the arterioles and venules can pass vertically between them. The cytoplasm of these cells consists of little more than dense bundles of refractile rods that measure 0.1μ in diameter and 3–4μ in length. They are packed in groups with their long axes perpendicular to the plane of the retinal surface. The third type of choroid tapetum owes its reflective property to inclusions of the purine, guanin (see Chapter 2), either as pure crystals or as the calcium salt. The guanin imparts a silvery luster and is a very effective reflector.

Elasmobranchs have an occlusible tapetum lucidum. The cells that contain guanin are arranged as flat plates that are pitched at a sharp angle relative to the plane of the retinal surface and are spaced so that they have the appearance of fixed ventilating louvers (Fig. 13.17). At their bases is a continuous layer of pigment cells the pigment of which is retracted when the light is dim but migrates between the guanin cells when the light is bright. Hence in dim light the guanin cells are exposed and reflect back through the retina, whereas in bright light they are covered by pigment and reflection does not occur.

Fibrous choroid tapeta have been described in a few teleosts, in certain marsupials, cetaceans, proboscidians, night monkeys, and in all ungulates that have a tapetum. Cellular choroid tapeta are characteristic of lemuroids and carnivorans. Guanin-containing choroid tapeta occur in elasmobranchs, bathypelagic teleosts, and some chondrosteans (e.g., *Acipenser fulvescens*).

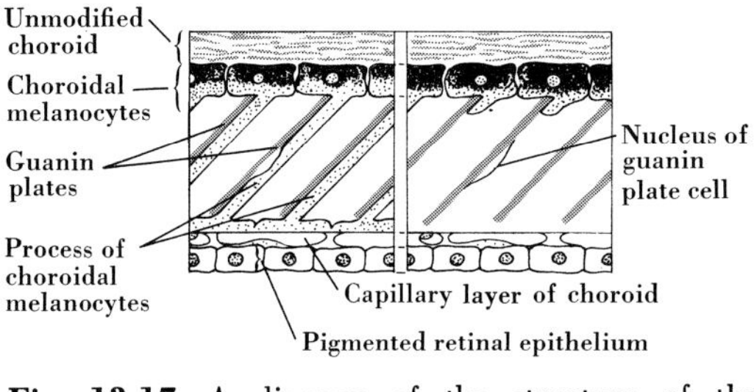

Fig. 13.17 A diagram of the structure of the occlusible tapetum in the smooth dogfish, *Mustelus canis*, when light-adapted (*left*) and dark-adapted (*right*). Processes of the pigmented cells migrate over the lighted surface of the guanin plates during light adaptation, thereby preventing reflection back through the retina. In dark-adaptation, the pigment is withdrawn. (After G. L. Walls. 1963. The vertebrate eye and its adaptive radiation. Hafner, New York. Copyright Cranbrook Institute of Science. After V. Franz. 1931.)

The Ciliary Body

The ciliary body consists of muscular and fibrous tissue derived from the part of the uvea that lies anterior to the ora serrata, and of epithelium derived from the two retinal layers of the optic cup. It is lacking in cyclostomes, and in elasmobranchs it is nonmuscular except for a ventral papilla. Elsewhere in the elasmobranchs the ciliary body is the portion of the uvea that encircles the lens and holds it in position, mainly by its thickened dorsal part and by an intervening gelatinous membrane. It is in teleosts that for the first time a uniformly muscular ciliary body is found, but it is thin, inconspicuous, and of unknown function. The amphibian ciliary body is similar to that of the teleosts.

In reptiles and birds the ciliary body is thick and muscular. It is held firmly about the lens by the externally supporting sclerotic ossicles (except in snakes). Its musculature is striated and originates at the conjunction of the sclera and cornea. When these muscles contract, they draw the ciliary body as a whole anteriorly into a space of smaller diameter. As a consequence the ciliary body squeezes the circumference of the lens, causing it to thicken and thereby shorten its focal length. Thus for the first time the ciliary body assumes the function of *accommodation*, i.e., adjusting the focal point of the eye to the distance between the eye and object being visualized. A secondary effect of contraction of the ciliary muscles in sauropsidans is believed to be reduction of the circumference of the eyeball at the sclerocorneal junction or *sulcus*, thereby lessening the radius of the cornea, which also shortens its focal point.

The ciliary body of mammals is neither so massive nor so complex as that of sauropsidans. As in sauropsidans the mammalian ciliary body is used in accommodation. The first placental mammals were doubtless small, nocturnal animals that had little need to accommodate, and so their ciliary bodies regressed. When larger, diurnal mammals evolved, they had a small ciliary heritage to work with and only a few groups actually put it to really effective use. These were the primates, the carnivorans, and some of the ungulates.

The human ciliary body is the best developed and the most effective accommodator in any of the mammals, but even here its muscle is slow-acting smooth muscle, it does not reach the lens, and it is not so efficient as the sauropsidan ciliary body. The fibers of the human ciliary muscles are arranged in three groups: an outer *meridional* group, a *radiating* group, and an inner *circular* group. The last group is found only in seals, primates, and in some toothed whales and some ungulates. The posterior portion of the ciliary body connects by means of the fibrous *zonule* (Fig. 13.14) to the equatorial periphery of the lens capsule, holding it taut and therefore flattening the lens. When the ciliary muscles contract, they draw the ciliary body anteriorly, thereby relaxing the tension on the zonule and permitting the lens to thicken along the optic axis. In this way the eye accommodates for near objects. Hence it is evident that the way in which the ciliary body accommodates in mammals is opposite to that in sauropsidans.

The ciliary body also has vascular and epithelial components that are important; these appear as folds or as processes that extend from the ciliary body a short distance toward the lens anterior to the zonule. Their function is chiefly secretory, for they are the source of the *aqueous humor*. The aqueous humor nourishes the lens, the iris, and the posterior wall of the cornea, and it is responsible for the turgor of the eyeball. Regulation of intraoptic pressure is effected by (1) a balance between the rate of secretion of the aqueous humor and the rate of its removal by the complex system of sinuses that constitute the canal of Schlemm and the spaces of Fontana at the angle of the iris, and (2) tonic contraction of the extrinsic eye muscles. The canal of Schlemm and the spaces of Fontana are drained by venules into the anterior ciliary veins of the adjacent sclera.

The ciliary body, zonule, and lens partition the lumen of the optic cup. The part of the lumen that is anterior to this partition is filled with the aqueous humor, and that which lies behind it is filled by a firm, gelatinous body, the *vitreous body* (Fig. 13.14). The vitreous body is elaborated by cells that proliferated from the photosensitive layer of the retina during differentiation of the optic cup. Its function in fishes, which lack a well-developed ciliary body or zonule, is to hold the lens in position, as well as to provide the proper refractive properties for projection of light rays from the lens to the retina, as in all vertebrates. The anterior compartment that contains the aqueous humor is divided into *anterior* and *posterior chambers* by the iris. The aqueous humor passes from its origin in the posterior chamber to the anterior chamber either through the pupil, or, in species in which the pupil is fully occluded by the lens, such as fishes and amphibians, via slits or foramina in the basal part of the iris.

The Iris

The vertebrate iris, like the ciliary body, has a dual embryonic origin. It appears as a circular membrane that originates peripherally along the anterior aspect of the ciliary body and terminates centrally at the pupil. The anterior surface and stroma of the iris are mesodermal in derivation and represent the anterior limit of the uvea. The posterior surface and the iridial muscles are ectodermal in origin, being derived from the two retinal layers of the rim of the optic cup.

The human iris, which is fairly representative of vertebrate irises, is organized into five more or less distinct layers (Fig. 13.18). The first and anterior-most is called the *endothelial layer*, and it is an incomplete pavement of thin, delicate cells that are very difficult to discern in vertical section. Underlying this layer, except when it is lacking (over the so-called *pupillary crypts*), is the *anterior border layer*. This is a fairly dense cellular layer that is composed mainly of chromatophores and tissue cells. The chromatophores are irregular or branched cells that contain yellow, brown, or brownish-yellow pigment granules and are

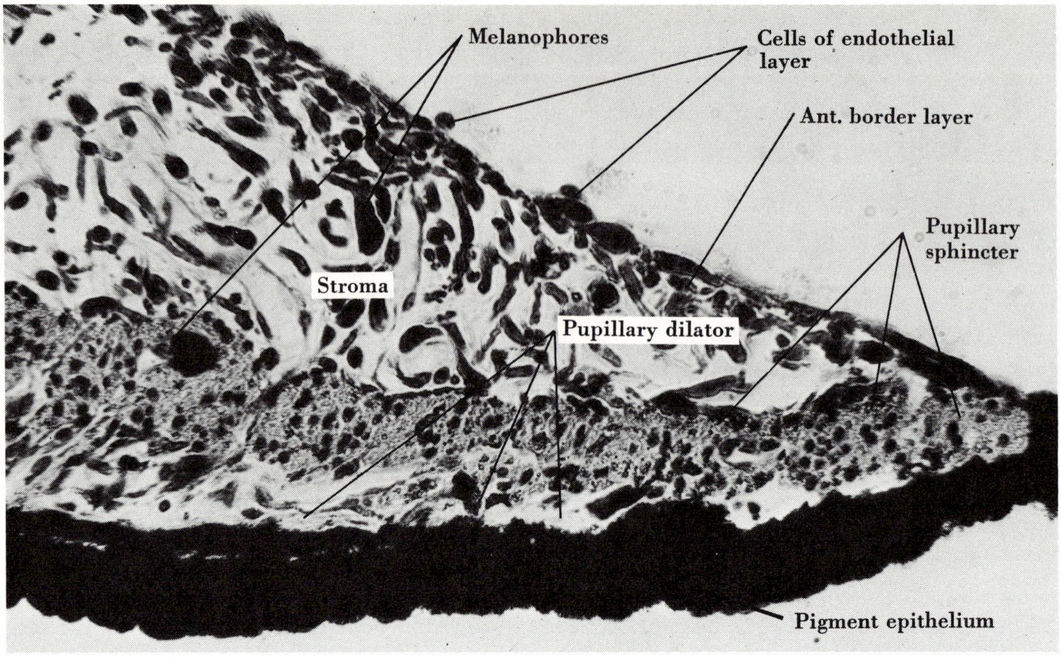

Fig. 13.18 Vertical section of a primate iris, showing the various layers. X450.

responsible for the color of the iris. Guanin-containing cells (guanophores) occupy this layer in most fishes, and in some species are believed to enable the iris to function like a physician's head mirror, casting reflected light along the line of sight. The third iridial layer is the *stroma,* which consists of a rather loose network of collagenous and elastic fibers, chromatophores, fibroblasts, and a large number of blood vessels. Near the pupil the stroma includes smooth muscle fibers of the *pupillary sphincter,* which are derived from cells of the pigmented retina in this area. Contraction of these fibers, which is effected by parasympathetic stimulation, results in pupillary constriction. The pupillary sphincter of the alligator contains a mixture of smooth muscle, striated muscle, and cells resembling myoblasts. The endoplasmic reticulum of the striated muscle is poorly organized, however, lacking the ordered appearance of the sarcoplasmic reticulum of skeletal muscle, and it is thought that this indicates that it is slow in acting, as is smooth muscle. Birds and some species of lizards have striated iridial muscles that are capable of very rapid constriction. There is some indication that the iridial muscles of birds are under voluntary control as well.

The fourth iridial layer is the *pupillary dilator layer* in which is found an unusual kind of primitive smooth muscle in the form of spindle-shaped epithelial cells the tapered end processes of which are contractile. They are derived directly from cells of the pigmented retina and are classed as myoepithelial cells, similar to those (also of ectodermal derivation) associated with sweat glands. These cells often contain pigment, except in the contractile processes. They are innervated by the sympathetic division of the autonomic nervous system and their contraction causes dilatation of the pupil.

The posterior-most and fifth layer of the human iris is the *pigment epithelium.* This layer represents the anterior continuity of the nervous retina between the ciliary body and the pupil. Its cells are not photosensitive, however, but they are heavily pigmented with granules of melanin that obscure all cellular detail. Their pigmentation is not evident when one looks into one's eye, because the color one sees occurs mainly because of the pigment in the relatively thicker anterior border layer and stroma, or, if there is little or no pigment in these layers, because of diffracted light reflected from the surfaces of the cells and fibers of these layers so that the light appears blue. If there is no pigment in any of the iridial layers, as in albinos, the iris appears pink because of the reflection of light from the choroid layer behind the retina.

The primary functions of the iris are to regulate the amount of light that reaches the retina and to restrict the path of incoming light as much to the center of the lens as feasible. Animals whose habits expose them to little variability in light intensity generally have limited iridial excursion—i.e., difference in pupil diameter between fully constricted and fully dilated iris. Maximal excursion is characteristic of animals that are active both under conditions of dim light and in full sunshine, such as the domestic cat, seals, alligators, and penguins, to name a few. In most fishes the lens lies within the pupil so that the iris is incapable of movement. Accordingly, the irises of most fishes are thin and devoid of muscle.

The shape of the vertebrate pupil is basically round, but other shapes appear as special adaptations. The slit pupil of the cat and others permits maximal excursion to meet their 24-hr. requirements. The irises of some fishes, such as the ray *Raja clavata* or the flatfish *Scophthalmus rhombus,* are equipped with a sort of flap on the dorsal edge, which can spread out over the surface of the lens or contract, thereby occluding more or less of the pupil. Some mammals have similar "pupil shades," such as the hyracoidean *Procavia* and camels. Most ungulates and many whales have pupils that are wider horizontally than vertically, and the notched edges of the gecko's slit pupil correspond so that on closure several pinhole apertures result.

The Lens

The vertebrate crystalline lens is induced embryologically from an area of ectoderm of the sense plate, the *lens placode,* by the optic cup. The lens placode thickens, invaginates, then separates away from the rest of the sense plate by pinching off as a hollow ball of cells (Fig. 13.13). The posterior wall of the ball is thicker than the anterior and becomes even thicker, obliterating the lens cavity so that the lens becomes a solid ball of cells. The cells of the anterior wall are flattened and form the lens epithelium (Fig. 13.14). The cells of the posterior wall are very long (up to 10 mm in the human lens), for each one extends from the posterior lens capsule to the lens epithelium. They appear in cross section as six-sided prisms (about 8μ wide by 2μ in thickness) that are fitted together in a honeycomb pattern and held together by an intercellular cementing substance. The lens prisms or "fibers" in the center are enucleate, but those located peripherally are mononucleate and decrease in length in correspondence with the peripheral taper of the lens (Fig. 13.19). At the very periphery the cells are columnar and they multiply slowly so that the lens grows steadily in size throughout life.

The lens is covered by a refractile, elastic capsule of amorphous intercellular substance about 12–18μ in thickness. Zonular fibers, when present, attach to the capsule around the periphery of the lens. The refractive index of the lens is highest in fishes, where it varies parabolically from 1.51 at the center to 1.38 peripherally (compared with 1.42 and 1.36,

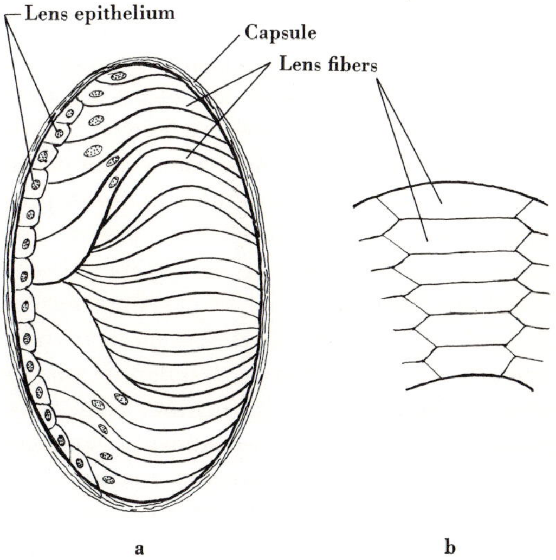

Fig. 13.19 A diagram showing the structure of the vertebrate lens in sagittal section, *a;* and an enlarged view of lens fibers in transverse section, *b.*

respectively, for the human lens), with an effective refractive index of about 1.65 or more in some species. The lens is the only refractive body in the eye of aquatic vertebrates, whereas the cornea assumes the major part of this function in terrestrial forms. This is because the refractive indices of the cornea and aqueous humor are very close to that of water but considerably higher than that of air.

The lenses of most vertebrates are colorless, but in a few species they have a yellow tint, as in lampreys, snakes, geckoes, tree shrews, human beings, and diurnal squirrels. Yellow filters are effective in enhancing color contrast, particularly in the shorter wavelengths of the visible spectrum.

The lenses of aquatic vertebrates are spherical, or nearly so, as in all fishes. It is also nearly spherical in amphibians and in most small, nocturnal, terrestrial animals, such as most rodents and the small species of bats. In other terrestrial forms the lens is a biconvex disk, with the posterior curvature greater than the anterior curvature.

The Pigmented Retinal Epithelium

The cells of the pigmented retinal epithelium are flattened and hexagonal in surface view. The basal part of the cell, which lies closely adherent to the choroid coat, includes the nucleus and a number of round granules of melanin called *fuscin.* The part of the cell next to the photosensitive retina is thrown into a number of processes, some short and some very long. The fuscin granules in these processes are rod- or needle-shaped. The tips of the photoreceptor cells of the retina are enclosed by these processes to varying degrees of depth. In some vertebrates, such as certain fishes and the alligator, crystals of guanin occur in the pigmented epithelium, giving it a silvery luster (as described above in connection with the choroid tapeta). In the cytoplasm of the pigmented epithelial cells of the frog, curious rhomboid lamellae of smooth-surfaced endoplasmic reticulum, the so-called myeloid bodies, are found (Fig. 13.20). They are believed to have something to do with the migration of pigment into and out of the cell processes. The function of the pigmented retinal epithelium is to absorb light after it has passed through the transparent photosensitive retina, so that the light will not scatter and blur the retinal image.

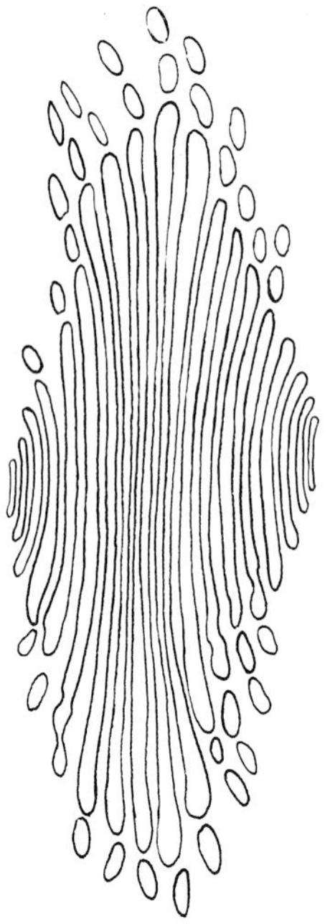

Fig. 13.20 The structure of myeloid bodies in the amphibian pigmented retina.

The Nervous Retina

The nervous retina is in almost every way an exteriorization of the brain, with its neurons in synaptic series with a radial alignment. The order of the neurons, from receptors centrally, is inverted in relation to the direction of the stimulus because of the manner in which the brain develops embryologically from invaginated ectoderm. The photoreceptors (rods and cones) trace their lineage to ependymal cells, which, as described in Chapter 12, trace their ancestry to ciliated epithelial cells on the surface of the body.

Three cell layers are distinguishable in vertebrate retinae (Fig. 13.21). Starting from the pigmented retinal epithelium and working toward the vitreous body in order, the cell layers are the layer of photoreceptors, the layer of bipolar neurons, and the layer of ganglion cells. Separating the cell layers from each other are the outer and inner plexiform layers, which consist mainly of cell processes, as described below.

THE PHOTORECEPTORS

Two kinds of photoreceptor cells occur in vertebrate retinae. They differ in their sensitivity to light, one being characterized by a low threshold (rods) and the other by a high threshold (cones). In human retinae the two types of cells usually show pronounced morphological differences as well. The terminal segment of a typical rod cell is an elongate, isodiametric rod, whereas the terminal segment of a cone cell is somewhat shorter. The inner segment or cell body of a typical rod is also slender and isodiametric, whereas that of a typical cone is rather thicker and broad at the base (Fig. 13.22). The cones show considerable variability in form in human retinae and also in those of other vertebrates. The cones of the fovea of the human retina, for example, resemble rods more than typical cones. Because cone cells are associated with color vision and with high visual acuity, it has been assumed that they are the less primitive of the two kinds of photoreceptors, whereas in fact there is compelling evidence that they are the more primitive.

Rods and cones alike consist of an outer segment, which is the photosensitive part, and an inner segment, which includes the nucleus and the cytoplasmic organelles. The ultrastructure of the outer segment of rods and cones has occasioned considerable speculation on the origin of photoreceptors. These outer segments consist of stacks of flat membranes or disks (Fig. 13.23) that resemble the lamellae of chloroplasts. It is tempting to try to trace the origin of the outer segments back to the photosynthetic organelles of autotrophic protistans because in addition to these lamellae is found a modified cilium or flagellum, remnants of whose nine pairs of circularly arranged tubules are clearly evident (Fig. 13.23). The similarity between the flagellum-eyespot complex of an organism such as *Euglena* and the terminal segment of a rod or cone is most striking.

Associated with the cisternae of the outer segment lamellae occurs the visual pigment, which consists—in all vertebrates in which it has been investigated—of a specific kind of protein, *opsin*, and a particular configuration of an aldehyde of vitamin A, *11-cis-retinaldehyde.* When a sufficient number of photons of energy strike the visual pigment, they isomerize the retinaldehyde from the 11-*cis* to the highly unstable *all-trans* configuration (Fig. 13.24). This transformation triggers a series of biochemical transformations that lead to the generation of an action potential, which is then transmitted ultimately to the visual centers of the brain. The specific visual pigment of vertebrate rods is a reddish-purple compound called *rhodopsin.* Pigment is less abundant in cones and less is known about it for this reason. The visual pigment, *iodopsin,* has

Fig. 13.21 Photomicrograph of a monkey retina. X500.

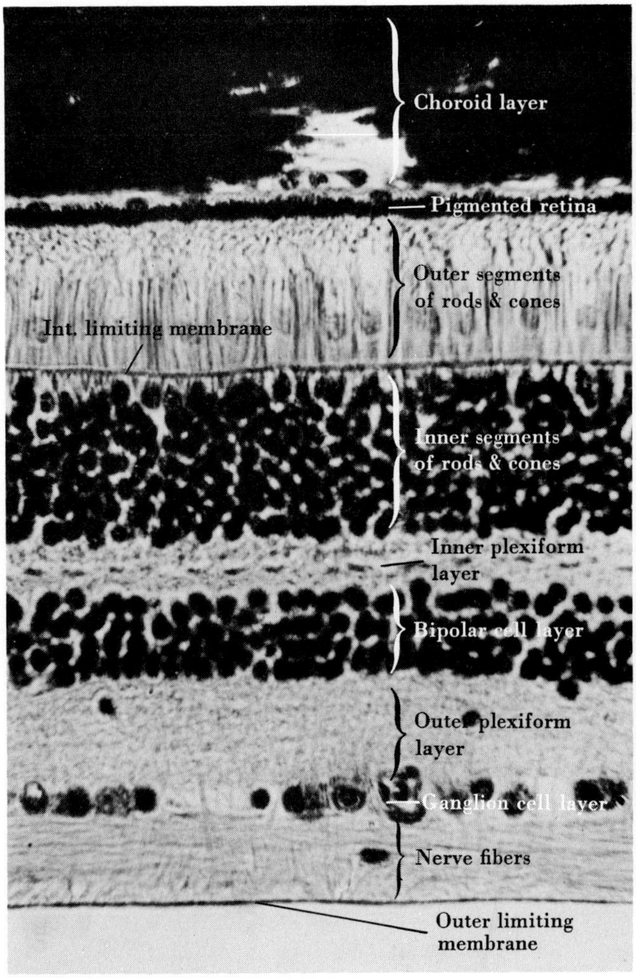

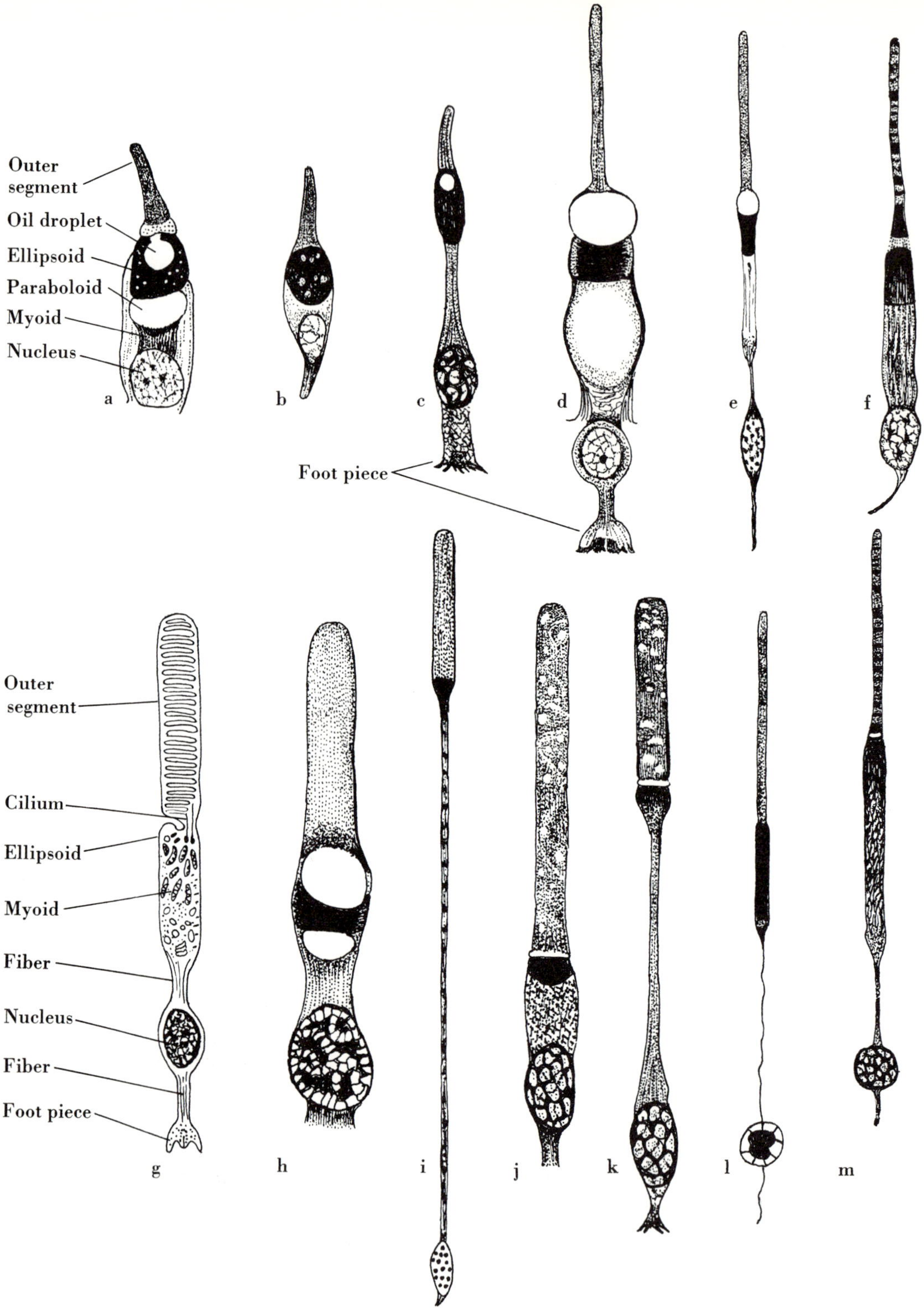

Fig. 13.22 A variety of photoreceptors. Single cones, *a–f: a,* Sturgeon (*Acipenser*) ; *b,* goldfish (*Carassius*) ; *c,* frog (*Rana*) ; *d,* turtle (*Chelydra*) ; *e,* hawk (*Circus*) ; and *f,* man (*Homo*). Rods, *g–n: g,* ultra structure of a generalized rod; *h,* lungfish (*Protopterus*) ; *i,* goldfish; *j,* "red" rod of frog; *k,* "green" rod of frog; *l,* flying squirrel (*Glaucomys*) ; *n,* human. (All figures except *g* after G. L. Walls. 1963. The vertebrate eye and its adaptive radiation. Hafner, New York. Copyright Cranbrook Institute of Science. Part *g* after R. W. Young. 1967. J. Cell Biol. 33:61.)

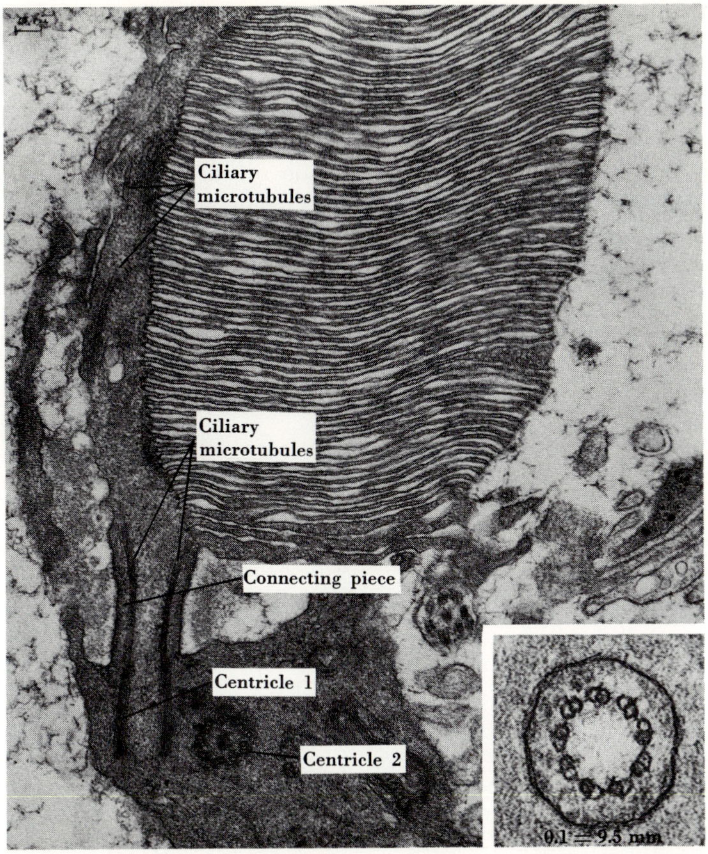

Fig. 13.23 Electron micrograph of a sensory receptor from the median eye of the lizard *Sceloporus occidentalis.* Inset: Cross section of the connecting piece showing the 9 + 0 pattern of microtubules. Note that the microtubules lack the arms typical of ciliary microtubules. (From R. M. Eakin. 1968. *In* Evolutionary biology, vol. 2, Dobzhansky, Hecht, and Steeve, eds. Appleton-Century-Crofts, New York.)

been identified in cones of domestic fowl and pigeons and is suspected of being present in amphibian, reptilian, and feline cones as well.

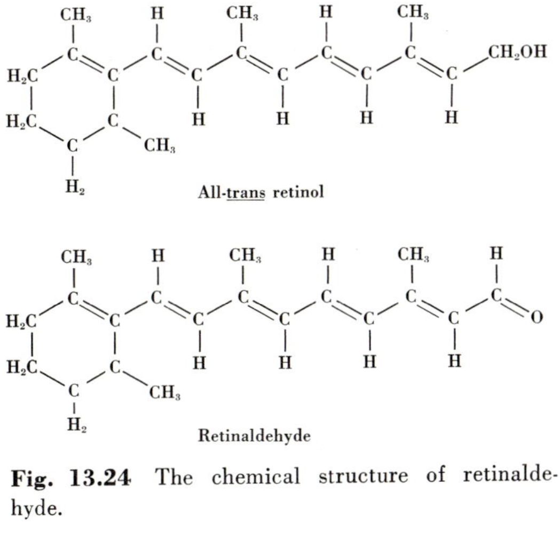

Fig. 13.24 The chemical structure of retinaldehyde.

In any case, the difference in pigment between rods and cones resides in the opsin, for the retinaldehyde is evidently the same and the chemical response to photic stimulation is the same. The rhodopsin of rods does not discriminate specifically between the various wave lengths of light in the visible spectrum, and thus stimulation of them by light fails to elicit any sensation of color. Cone cells in most vertebrates, on the other hand, possess three different kinds of iodopsin pigment which are sensitive (respectively) to red, green, and blue light. By the combination of these three color sensitivities, all colors, hues, and shades are perceived. Although most cone cells possess only one or the other of the three color-sensitive pigments, there is evidence that some of the red-sensitive cones contain a small amount of green-sensitive pigment. Color vision, at least in human beings, is therefore trichromatic.

The threshold sensitivities of rods *vis-à-vis* cones is attributable to the greater abundance of visual pigment in the former. The effect of this difference is enhanced still further by the fact that several rods synapse with a single bipolar cell, so that a summation effect is achieved—although it is at the expense of high visual acuity. Cone cells, on the other hand, show typically a one-to-one relation to bipolar cells; hence there is no summation effect but maximal visual acuity is facilitated (Fig. 13.25).

As described above, the terminal segments of the photosensitive cells of many vertebrates are embedded within or surrounded by processes of the cells of the pigmented epithelium. Interposed between the terminal segments and the cell bodies or *inner segments* is a constricted segment, the *cilium* (Fig. 13.22). The cytoplasm of the cilium contains little more

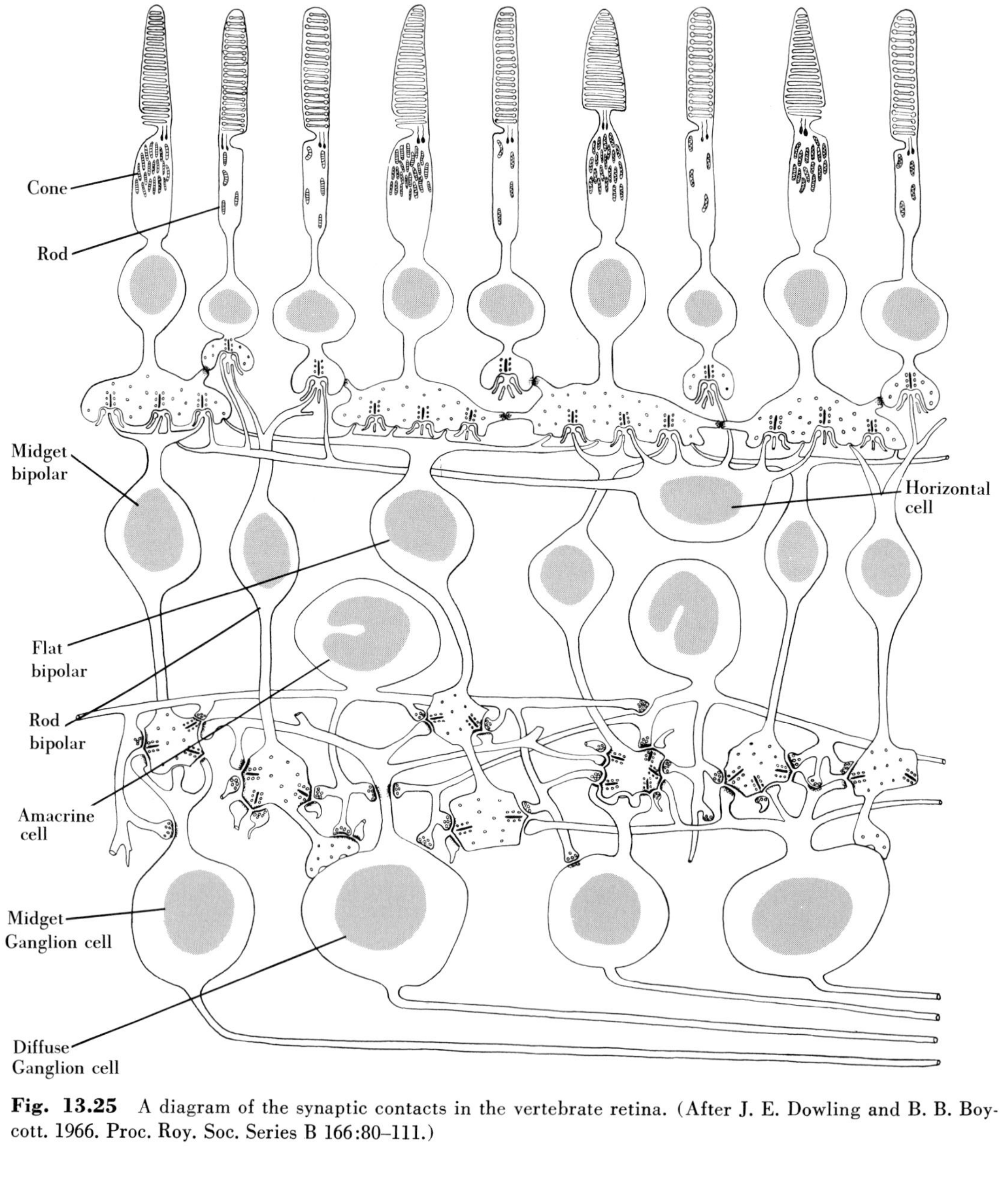

Fig. 13.25 A diagram of the synaptic contacts in the vertebrate retina. (After J. E. Dowling and B. B. Boycott. 1966. Proc. Roy. Soc. Series B 166:80–111.)

than the outer ring of ciliary tubules. However, the inner segment is large and shows regional differences. The cytoplasm adjacent to the cilium is packed with mitochondria and it also contains the ciliary tubules, rough-surfaced endoplasmic reticulum, rosettes of ribosomes, and numerous filaments that measure 150–200A in diameter. These organelles are clumped to form an elliptical mass, hence this region of the inner segment is frequently called the *ellipsoid*. The rest of the inner segment is elongated and has fewer organelles but numerous clear vesicles; however, it is expanded to accommodate the single round nucleus. The nucleus has an eccentric nucleolus and its chromatin is evenly distributed in cones but condensed into zones in the rods.

An interesting adaptation is found in the inner segments of certain cone cells in the guppy, *Lebistes*. As in many teleosts, and many other nonmammalian vertebrates, double cone cells occur. Underlying the plasma membranes where the two members of a double cone contact each other are trilaminar, rectangular membranous sheets. Whether these serve as a layer of insulation or represent a specialized cell-to-cell contact zone is not clear.

The inner segments of the receptor cells are supported and separated from each other by processes of the Müller cells, described below. The rods project a single attenuated process from the inner segment toward the outer plexiform layer, where it expands into a rounded knob. Mitochondria, microvesicles and a unique kind of synaptic organelle, the *synaptic ribbons* (Fig. 13.26) are found in the terminal knobs. The terminal process characteristic of cones is generally larger than that of rods, and it forms a triangular foot. It, too, contains mitochondria, microvesicles, and synaptic ribbons, but in human cones the ribbons are far more numerous than in those of other vertebrates that have been examined, and they are ten times as numerous as those in rod terminal processes. Although the full significance of the synaptic ribbons is not understood, they are believed to be concerned with the synthesis of a neural transmitter substance

Fig. 13.26 Electron micrograph of the outer plexiform layer of the retina showing synaptic ribbons. X31,900. (From G. Raviola and E. Raviola. 1967. Am. J. Anat. 120:403–426.)

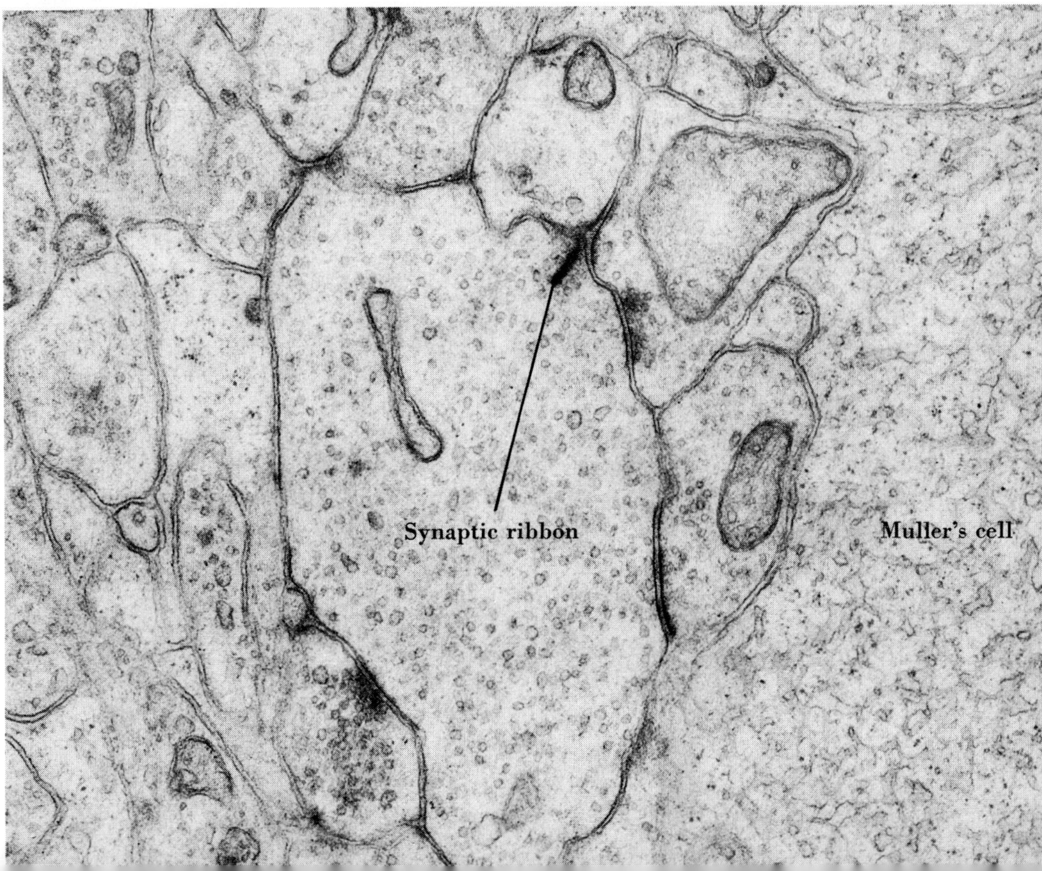

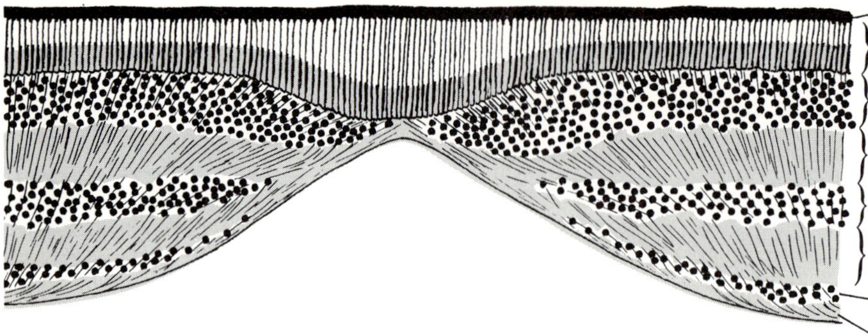

Fig. 13.27 Drawing of the fovea of a primate retina.

and thus are diagnostic of a presynaptic contact.

The distribution of rods and cones in the retina varies considerably among the vertebrates. In human beings, whose retinas are fairly typical of vertebrates, the cones are the only receptors in the area immediately surrounding the optic axis. This area, the *area centralis,* tends to be the thickest part of the retina. The area centralis is lacking in cyclostomes and in all elasmobranchs so far described—except the dogfish shark, *Mustelus canis,* which is the only elasmobranch known to have cones. The centralis is present in most of the higher vertebrates that are diurnal, although it may be imperfect because of the presence of some rods. Whales and seals, however, have only rods.

The very center of the optic axis in diurnal vertebrates characterized by high visual acuity, such as birds, primates, and certain teleosts, is marked by a depression or pit, the *fovea* (Fig. 13.27). By virtue of its geometry and the refractive index of the retina, the fovea magnifies the image, thereby facilitating increased visual acuity. In many birds two foveae are present, one of which is central for lateral vision, the other being temporal for forward vision. The central fovea is the deeper of the two, except in eagles, in which the temporal fovea is the deeper.

Beyond the centralis, if present, the proportion of cones decreases rather rapidly with a compensatory increase in rods, until in peripheral areas only rods are present.

THE BIPOLAR CELL LAYER

In the bipolar cell layer are found the cell bodies of the bipolar cells, the horizontal cells, the amacrine cells, and the Müller cells (Fig. 13.25). The bipolar cells predominate. They are characterized by a large, round nucleus and long processes that connect the cells with both inner and outer plexiform layers (Fig. 13.25). Three types of bipolar cells have been described in primate retinas: rod bipolars, the dendritic processes of which synapse with the foot processes of several rods; flat bipolars, which synapse with several cones; and midget bipolars, which ordinarily synapse with cones in a one-to-one relation (Fig. 13.25).

Along the outer level of the bipolar cell layer are found the cell bodies of the horizontal cells. These are large cells with a round or oval nucleus and with more perinuclear cytoplasm than one finds in bipolar cells. The cell processes are generally thick ($0.6–1.5\mu$) and one or more may be quite long and run horizontally through the outer plexiform layer. These processes synapse with both rod and cone end feet and appear to have an integrative function. Horizontal cells are lacking in the parafoveal parts of the area centralis.

The nuclei and cell bodies of the amacrine cells occur along the outer level of the bipolar cell layer. They are large, round cells (about 12μ in diameter) with a central nucleus that is deeply indented. The cell body has a rich complement of cytoplasmic organelles. The processes of these cells are distributed in the inner plexiform layer, where they synapse with ganglion cell processes, with the axonal processes of rod bipolar cells, and with processes of other amacrine cells. As with the horizontal cells, the amacrine cells are believed to play an integrative role; however, by virtue of their synapsing with each other, their integrative influence is believed to be much more extensive. The exact details of how integration is achieved are not fully known or understood, but it is believed that the retina may approach the cerebral cortex in its integrative complexity.

Müller's cells are the principal non-neuronal cells of the retina. Their cell nuclei lie in the layer of bipolar cells, but their cytoplasm is distributed throughout all layers of the retina between the external and internal limiting membranes. All of the other retinal cells and their processes are embedded in and supported by the folds and processes of the Müller cells, and so these cells serve as the "connective tissue" of the retina. The outer and inner limiting membranes (Fig. 13.21) actually represent the outer and inner limits, respectively, of the cell membranes of the Müller cells that are firmly joined to each other by cell junctions. The nucleus is large, ovoid, and clear, and contains typically a single eccentric nucleolus. The cytoplasm is highly irregular and contains few organelles.

THE GANGLION CELL LAYER

This layer includes the nuclei and cell bodies of the retinal ganglion cells and it also includes a small number of astrocytes, microglia and, rarely, oligondendrocytes. The ganglion cell bodies are also large (15–20μ in diameter) and have a round, eccentric nucleus and abundant cytoplasm with a full complement of organelles. The dendritic synapses of the ganglia cells lie in the inner plexiform layer, although somatic synapses of rod bipolars and amacrine cells frequently occur on the cell bodies of the ganglion cells. The axonal processes of the ganglion cells emerge from the retina, run along its inner surface and converge at a common point, the *head of the optic nerve*, where they join to form the optic tract. Two kinds of ganglion cells occur in mixed rod and cone retinas: *midget ganglia*, which receive impulses from midget bipolars on a one-to-one relation and also synapse with amacrine cells; and *diffuse ganglia*, which make synaptic contact with both rod bipolars and flat bipolars and with amacrine cell processes.

Because a one-to-one relation, or nearly so, is maintained between cones and midget bipolars as well as between midget bipolars and midget ganglion cells, each cone cell has a single fiber representative in the optic nerve. By this arrangement, maximum visual acuity is facilitated, although it is at the expense of stimulus sensitivity. In contrast to this, each rod bipolar receives impulses from several rods, and several rod bipolars relay these impulses to a single diffuse ganglion. This arrangement utilizes the principle of summation, with resultant high stimulus sensitivity, albeit at the expense of resolving power.

Accommodation

Accommodation in most vertebrates is effected either by moving the lens in the optic axis or by changing its shape, thereby changing its focal point. The lamprey eye is myopic when relaxed, i.e., it is focused for near objects. It accommodates for distant objects by contracting an extrinsic corneal muscle, which flattens the cornea and, because the lens is in contact with the cornea, pushes the lens closer to the retina. The feasibility of this mechanism depends on the facts that the lamprey is not normally subjected to rapid changes in external pressure, the cornea is not fused to the corneal epithelium, and most of the time the animal is "interested" only in very near objects.

Accommodation in fishes and amphibians is effected by swinging the lens outward or inward. Elasmobranch and amphibian eyes are set for distant focus, and by contraction of a papilla of smooth muscle in the anterior-ventral part of the ciliary body, the lens is swung outward to focus on near objects. Teleosts, like the lampreys, are myopic, but they accommodate for distant vision by swinging the lens inward and caudad by contraction of a lens retractor muscle (*campanula Halleri*). This action results in the focal point's shifting laterally from the middle of the retina—which is helpful to those species so affected, because their fovea is laterally (or *caudally* from their standpoint) located.

Except for snakes, reptiles and birds have the most efficient method of accommodation.

The lens is padded around its periphery by a gelatinous ring (the *annular pad*) against which the ciliary muscles press. The ciliary muscles are striated and hence act quickly. Moreover, contact between the ciliary body and annular pad is discontinuous, by virtue of ciliary processes, so that the aqueous humor can move freely past the lens in response to changes in lens displacement. The ophidians differ from other reptiles because they evidently evolved from burrowing forms that all but lost their eyes. Consequently they had to rebuild their eyes from what was virtually a vestigial organ. Accommodation is effected by contraction of striated muscle in the root of the iris. This constricts the eye at the corneal-scleral junction and elevates the pressure in the vitreous body. The vitreous body consequently pushes the lens forward and elongates the eyeball in its optic axis. The lens does not change its shape, remaining essentially spherical (except in water snakes). Accommodation powers reach their maximum in birds. Not only do they change the shape of the lens by the contraction of strong, striated muscles, they change the curvature of the cornea as well, except for diving birds, in which corneas are very thick and inflexible. The extent of accommodation achieved in birds may be as great as 40–50 diopters (cormorants), as compared with a maximum of 10 diopters in mammals (human 2-year-olds).

Accommodation mechanisms are unnecessary in many species for a variety of reasons. Animals with very small eyes have little need for accomodation, for the length of their photoreceptor cells is as great as that in large eyes but the distance through which the focal point moves from objects at a distance and nearby is very small, perhaps no more than the length of the photoreceptor cell itself. The retina of fruit bats is papillated, appearing in vertical section as myriads of sharp hills and valleys. This arrangement in effect thickens the retina so that there are photoreceptors at any point where the focal point may fall. To be sure, the visual effect in these animals must be like that of looking through a screen, but this defect is easily compensated for by transverse motion.

In rays and horses, and possibly other ungulates, the retinal surface slants away from the lens from ventral to dorsal. Consequently the retina is in the correct position for focusing on near objects on the ground, the rays from which would reach the dorsal part of the retina, and for more distant objects the rays of which come in from higher and higher levels. Presumably, then, horses need only to tilt the eye up or down to get a sharp image of an object at any distance between its near point and infinity.

Finally, the use of stenopaic or "pinhole" apertures minimizes the need for accommodation, for, within limits, the smaller the aperture the greater the range of distances within which a given object will be in fairly sharp focus. Stenopaic apertures require rather intense illumination, however, as they greatly restrict the total amount of light that reaches the retina. Stenopaic apertures are achieved by extreme pupillary constriction in some sharks and rays, in geckoes, in certain ungulates, such as camels, and in domestic cats.

SELECTED REFERENCES

Amoore, J. E. 1965. Psychophysics of odor. Cold Spring Harbor Symp. Quant. Biol. 30:623–636.

Beidler, L. M. 1965. Comparison of gustatory receptors, olfactory receptors, and free nerve endings. Cold Spring Harbor Symp. Quant. Biol. 30:191–200.

Berger, E. R. 1967. Subsurface membranes in paired cone photoreceptor inner segments of adult and neonatal *Lebistes* retinae. J. Ultrastruct. Res. 17:220–232.

Bernstein, U. H., and D. C. Pease. 1959. Electron microscopy of the tapetum lucidum of the cat. J. Biophys. Biochem. Cytol. 5(1):35–39.

Bloom, G. 1954. Studies on the olfactory epithelium of the frog and toad with the aid of light and electron microscopy. Z. Zellforsch. 44:89.

Cordier, R. 1964. Sensory cells. *In* The cell. J. Brachet and A. E. Mirsky, eds. Academic Press, New York. Vol. VI.
de Lorenzo, A. J. 1958. Electron microscopic observations on the taste buds of the rabbit. J. Biophys. Biochem. Cytol. 4(2):143–149.
Dowling, J. E., and B. B. Boycott. 1966. Organization of the primate retina: electron microscopy. Proc. Roy. Soc. B, 166:80–111.
Eakin, R. M., and J. A. Westfall. 1959. Fine structure of the retina of the reptilian third eye. J. Biophys. Biochem. Cytol. 6:133–134.
Frisch, D. 1967. Ultrastructure of mouse olfactory mucosa. Am. J. Anat. 121:87–120.
Gregory, R. L. 1967. Origin of eyes and brains. Nature 213(5074):369–372.
Iwai, T. 1964. A comparative study of the taste buds in gill rakers and gill arches of teleostean fishes. Bull. Misaki Mar. Biol. Inst. Kyoto Univ. 7:19–34.
Lettvin, J. Y., and R. C. Gesteland. 1965. Speculations on smell. Cold Spring Harb. Symp. Quant. Biol. 30:217–225.
Moulton, D. G., and L. M. Beidler. 1967. Structure and function in the peripheral olfactory system. Phys. Rev. 47:1–52.
Murray, R. W. 1965. Receptor mechanisms in the ampullae of Lorenzini of elasmobranch fishes. Cold Spring Harbor Symp. Quant. Biol. 30:233–243.
Naftalin, L. 1965. Some new proposals regarding acoustic transmission and conduction. Cold Spring Harbor Symp. Quant. Biol. 30:169–180.
Obsche, A., and H. Kirschstein. 1967. Die ultrastruktur der Sinneszellen im Pinealorgan von *Phxinus laevis L.* Z. Zellforsch. 68:151–166.
Reese, T. S. 1965. Olfactory cilia in the frog. J. Cell Biol. 25(2):209–230.
Reger, J. F. 1966. The fine structure of iridial constrictor pupillae muscle of *Alligator mississippiensis*. Anat. Rec. 155:197–216.
Rochon-Duvigneaud, A. 1943. Les yeux et la vision des vertebres. Masson et Cie, Paris.
Smith, C. A., and F. S. Sjöstrand. 1961. Structure of the nerve endings on the external hair cells of the guinea pig cochlea as studied by serial sections. J. Ultrastruct. Res. 5:523–556.
Villegas, G. M. 1964. Ultrastructure of the human retina. J. Anat. 98:501–513.
Wald, G., and D. K. Brown. 1965. Human color vision and color blindness. Cold Spring Harbor Symp. Quant. Biol. 30:345–361.
Walls, G. L. 1963. The vertebrate eye and its adaptive radiation. Hafner Publishing Company, New York.
Wersäll, J., A. Flock, and P. G. Lundquist. 1965. Structural basis for directional sensitivity in cochlear and vestibular sensory receptors. Cold Spring Harbor Symp. Quant. Biol. 30:115–132.

chapter 14
THE ENDOCRINE SYSTEM

THE ENDOCRINE and nervous systems share the job of functionally integrating the various parts of the organism and of coordinating all suborganismal functions so that they serve the needs of the whole. However, there are major differences between the two systems insofar as how and when each operates, and in matters of how long a period of time is involved between stimulus and response.

The organs of the endocrine system are glandular and the secretions, *hormones,* pass directly into the blood capillaries immediately after being produced or following a period of storage within the gland. For this reason these glands are also called *ductless glands,* implying the direct dialization of hormone across the membranes of the secretory cell and capillary endothelium. In contrast to this method of dispersion of secretion is the situation encountered when examining the exocrine, multicellular, ducted glands. The secretory product of these structures leaves the gland via a well-defined duct to reach an internal or external body surface.

The name *hormone* is applied to the secretory products of endocrine structures. To be qualified as a hormone the following characteristics must be manifest. (1) The secretion must be demonstrably liberated directly into the blood and carried throughout the body. (2) Usually only minute amounts are needed to produce a response. (3) Only certain parts of the body will respond to the secretion (the target organ or organs). In some cases, as with the thyroid, parathyroid, or pancreatic hormones, the targets are diffuse, but in most other cases, the target organs are localized as definitive structural entities. (4) The target organ can be relatively far from the site of hormone production because the connecting link is the blood stream and not some other mechanism which depends on physical contact. (5) The chemical agent that is produced must be useful and not primarily an excretory product. All cells produce carbon dioxide as a result of glycolysis and/or the citric acid cycle, and the CO_2 so produced may function positively in the stimulation of certain respiratory reflexes. However, CO_2 is usually excluded in light of its being primarily an excretory product. Moreover, CO_2 fails to qualify because there is too much of it being produced. This particular characteristic also eliminates the liver as a true endocrine gland, although there are a few individuals who believe that the production of glucose by the liver is an endocrine function.

To be considered a hormone, then, a chemical agent must meet the criteria listed above. If the chemical under consideration is a true hormone, the cell that produces it is an endocrine cell. However, such cells are not all part of what is known as the endocrine system. The tissue hormones, for example, are produced by loose cells or very small aggregates of cells which are distributed within other organs. Certain of the duodenal cells produce hormones that affect the secretory functions of the liver and pancreas. These glandular cells do not form a definitive structure or organ, and, hence, are not usually included in a discussion of the anatomy of the endocrine system. Other examples of tissue hormones are substances produced by certain of the blood cells or of the kidney. The bulk of this chapter will be devoted to the so-called glandular hormones and the endocrine organs which produce them.

A number of interesting generalizations can be made with respect to the vertebrate endocrine organs and the hormones they produce. Embryologically, these organs are of mesodermal, ectodermal, or endodermal origin. Interestingly, those of mesodermal origin secrete steroid hormones, such as estrogen and progesterone from the ovary and the cortical hormones from the interrenal glands, whereas the glands which are epithelial modifications secrete simple or conjugated protein or polypeptide hormones. Generally the steroid hormones are chemically either very similar or identical throughout the vertebrates, whereas the polypeptide or protein hormones differ with respect to exact amino acid order. Hence, the insulin of one species may be antigenic when it is administered to another species. Continued chemical evolution has characterized the nonsteroid substances, whereas the steroids were probably selected early in vertebrate evolution and subsequent changes have been concerned more with the site of their action and the nature of the response.

Although details vary, cellular aggregation within endocrine glands can be categorized in one of three ways. In most glands the cells are arranged in branching cords, with blood capillaries and/or sinusoids in the interstices. The thyroid gland differs from this plan by the presence of follicular organization. The third variation is seen in regions of the pituitary, wherein nerves function as secretory agents.

Finally, it must be remembered that endocrine cells exhibit cycles of activity. Although the basic control of secretion may reside in the genes, the variations in secretory activity are influenced by chemical feedback mechanisms or by nervous control.

THE THYROID GLAND

The thyroid gland represents a vertebrate adaptation for the efficient manufacture of an iodine-bound amino acid secretion which has effects so numerous and diverse that its mode of action must somehow be connected to a basic part of cellular metabolism.

The two most common secretions of the thyroid gland are *triiodotyrosine* and *tetraiodotyrosine* (*thyroxine*). The former is the more powerful hormone, but it usually forms only a small fraction of the thyroidal output. The bulk of the hormone of the gland consists of thyroxine, although under conditions of iodine deprivation the relative quantities of these two secretions are changed. Experimental data suggest the role of triiodotyrosine in iodine conservation; when necessary, the organism can make do with a smaller amount of a more potent hormone.

The biosynthesis of iodinated tyrosine is not limited to the vertebrate thyroid gland. Marine algae, especially of the red and brown phyla, manufacture monoiodotyrosine. Quantities of protein-bound monoiodo- and diiodotyrosine are found in the exoskeletons as well as in the pharyngeal region of invertebrates. Presumably the hormone in its protein-bound form finds its way to the digestive tract, where hydrolysis of the "masking" protein can occur. As the chordates evolved, the pharyngeal site gained ascendancy owing, of course, to the progressive loss of an exoskeleton, until, in Ammocetes, the cells concerned with manufacture of iodinated tyrosine became grouped and isolated as an exocrine gland. The transition to the endocrine state can be seen most vividly during the development of the lamprey. During the prolonged larval stage of these organisms, functional thyroid hormone is produced by a tubular exocrine structure, the subpharyngeal gland. During metamorphosis the subpharyngeal gland is converted to an endocrine thyroid gland. In higher vertebrates the thyroid similarly begins its development with a duct to the pharynx, but this duct usually breaks or becomes nonfunctional during the early stages of embryogenesis.

In addition to the presence of similar chemicals in marine algae and invertebrates, iodinated tyrosine occurs in heterotopic vertebrate sites as well as in the pharyngeal thyroid gland itself. Cells of the anterior digestive tract, the chloride cells of the gills, the notochord of the lamprey, the kidney and mammary gland have all been included as sites which synthesize an iodine-bound tyrosine. In cyclostomes and many teleosts, typical thyroid tissue can often be found in the brain, eye, spleen, esophagus, or wall of the ventral aorta and its branches going to the gills. The higher vertebrates usually have a pharyngeally located compact gland consisting of one or two lobes which may be separate from each other, or, as in man, connected to each other via an isthmus.

The secretion of the thyroid gland is under the control of the thyroid stimulating hormone (TSH) of the anterior pituitary. The relationship is one of negative feedback whereby (1) an increase of the trophic hormone results in an increased thyroid output and (2) an increased level of thyroid hormone depresses the rate of secretion of the anterior pituitary with respect to thyroid stimulating hormone.

The effects of thyroid hormone can be studied on many different levels. Many of the gross changes which can be caused or manipulated by thyroid hormone are maturational in nature. Metamorphosis in amphibians can be accelerated with thyroid hormone administration, and conversely, the neotenous state seems to be associated with a decreased or depressed rate of thyroid function. The proper differentiation of skeletal elements, including bone as well as cartilage, depends, at least in part, on thyroid hormone. It is not unusual for cretins (individuals with depressed thyroid function during and beyond early childhood) in their thirties still to have their deciduous teeth in place with the permanent set still deep in the jaw. The hair of mammals, horns of sheep, antlers of deer, fin rays of fishes, and moulting and feather production of birds are all dependent upon proper thyroid stimulation. Thyroid hormone, in conjunction with other substances, plays an important role in the stimulation of most facets of growth, regeneration, and acquisition of sexual maturity, and in many vertebrates seasonal changes relating to mating or hibernation are reflected in the thyroid gland.

The foregoing list is meant neither to be exhaustive nor medically oriented, but rather to indicate the over-all dependence of the body on thyroid secretion. For this reason the mode of action of the hormone is thought to be basic to cell metabolism in general. Because administration of the hormone increases the oxygen consumption of cells and because the size and shape of mitochondria are changed in treated cells, the following hypothesis has been put forward. Thyroid hormone is thought to uncouple oxidative phosphorylation. During normal oxidative phosphorylation, a process which occurs in mitochondria, the energy which is extracted from food is converted in the presence of oxygen to energy of ATP by the addition of a high-energy phosphate group to ADP. If phosphorylation of ADP is uncoupled from oxidatively derived food energy, the cell would become metabolically less efficient. Normally usable energy would be lost as heat, and the cell would require greater supplies of food and oxygen in order to manufacture enough ATP to provide energy for maintenance. This hypothesis would also explain the medical syndrome of hyperthyroidism. Such individuals have overactive thyroid glands and secrete an overabundance of functional hormone. They require a large food intake and have an elevated body temperature and an increased basal metabolic rate—indicating increased oxygen consumption, when compared to that of normal individuals. The converse condition is manifest in hypothyroidism, for in this case the body temperature, oxygen consumption, and necessary food intake are all depressed. Because the lack of thyroid hormone enables the mitochondria to convert food energy to ATP efficiently, the cells in the hypothyroid individual do not have to be so active. In severe cases of hypothyroidism, as in cretins or in people with myxedema, the individual as a whole suffers even though his cells may be performing at greatest efficiency. It seems as if living things must insure themselves a sufficient quantity of wasted energy in order to maintain an optimum temperature for enzyme action, and the evolution of thyroid hormone represents a sure way of generating heat. Maximum efficiency, in terms of complete utilization of food to produce ATP, appears then to be an unattainable and undesirable goal for the organisms or species.

Considering the basic metabolic functions of thyroid hormone, it would be logical and indeed correct to assume that few cytological differences would be evident within the vertebrates. Gross structure may vary with respect to size, shape, and number of lobes, but in all cases the basic functional and structural unit is the *follicle*.

Excluding some of the fishes in which dispersed thyroid tissue is found, the entire gland is surrounded by a connective tissue capsule which invaginates to divide the gland into lobes and the lobes into lobules. Numerous blood vessels course through the septa to the follicles, as do vasomotor nerve fibers and lymphatics. Within each lobule are a number of follicles of uniform or varying size, each surrounded by a meshwork of reticular fibers, a dense capillary network, and a basement membrane. The cells of the follicle itself are arranged in a single epithelial layer surrounding a fluid-filled space (Fig. 14.1). Usually the follicular cells are cuboidal in shape, although variations from low squamous to tall columnar are frequent in follicles of varying activity. In general, the taller the cell, the more active it is presumed to be. In the central or basal portion of the cells a spherical nucleus containing one or more nucleoli can be found. The cells contain the usual array of organelles, with mitochondria especially abundant in the apical region. The supranuclear Golgi body can be seen to increase and decrease in prominence that parallels the secretory cycle of the cell. There are a variety of granules, droplets, vacuoles, and even crystals within the cell, and their number and distribution can also be correlated with secretory activity. Basally, the cell membrane is usually thrown into an extensive array of folds and grooves whereas on the apical surface, abutting the follicular space, microvilli

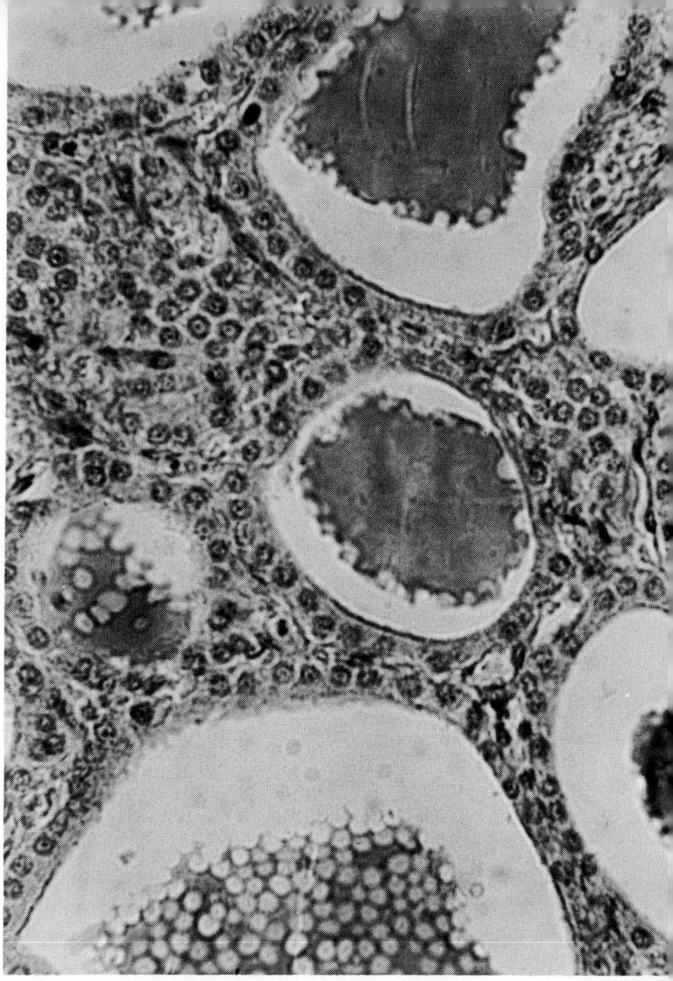

Fig. 14.1 The thyroid of a dog. Note the colloid within the different-sized follicles and the single layer of cuboidal epithelium abutting the lumina. X450.

can be seen, the number of which increases after pituitary stimulation. The higher vertebrates seem to have a greater density of microvilli than do the elasmobranchs and teleosts. In some vertebrates, such as elasmobranchs, and some birds and the dog, the cells are flagellated; in others, such as the lamprey and some follicles in mammals, the epithelium bears functional cilia on the inner surface.

In addition to the follicular cells, there is at least one other active cell type to be considered. Although originally called *parafollicular cells*, cells of this type can lie inside of the basement membrane surrounding a follicle as well as alongside a follicle. In addition, references can be found for *C* cells, clear cells, and light cells, but the consensus seems to be that all of these cells are of the same type and are also equivalent to the parafollicular cells.

Usually abutting the basement membrane but not making contact with the follicle lumen, this second type of cell is often larger and rounder than its neighboring follicle cells. Typically these cells are paler than follicular cells, although with proper staining the cytoplasm can be shown to be full of granules. Some of the granules are similar to those in follicular cells, but others can be demonstrated only in these so-called parafollicular cells (Fig. 14.2).

Although this second cell type has been studied for many years, its functional significance is still not well understood. Recently, however, a second thyroid hormone has been isolated. It is called *calcitonin* or *thyrocalcitonin,* and its action seems to be antagonistic to parathormone because it lowers the level of serum calcium. There have been a number of

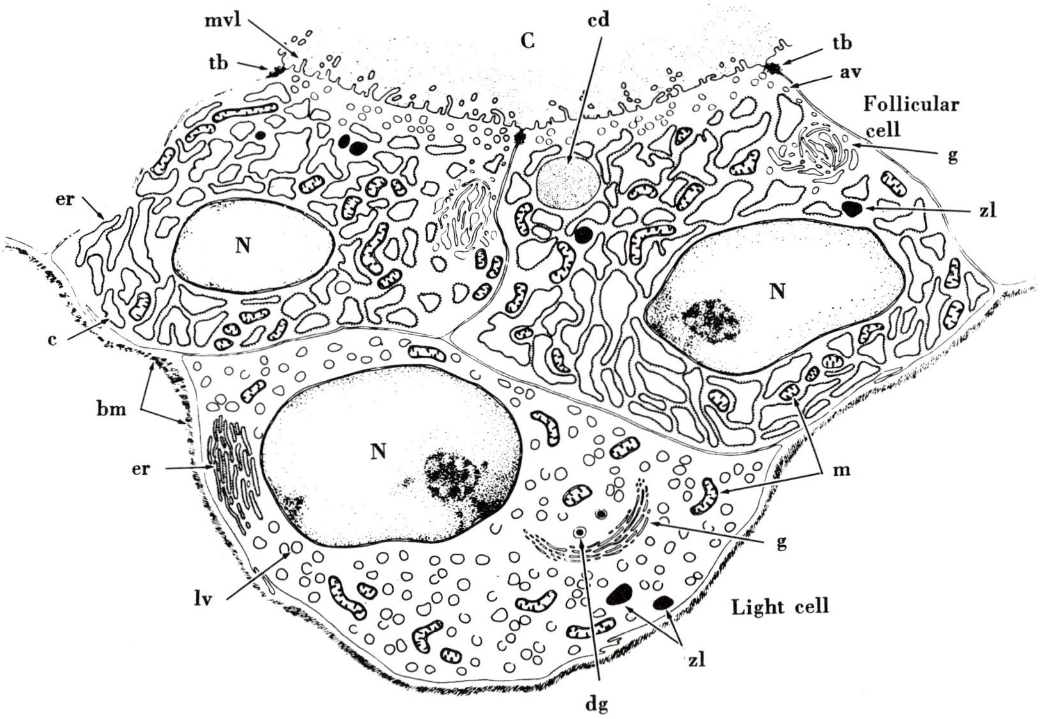

Fig. 14.2 A diagrammatic illustration of the typical shape and appearance of the follicular cell and the light (parafollicular) cell. The basement membrane, *bm*, surrounds both cell types, but only the follicular cells contact the colloid, *c*. A colloid droplet, *cd*, can be seen in one follicular cell. Microvilli, *mvl;* terminal bars, *tb;* distended Golgi vesicles, *g;* and apical vesicles, *av*, are typical of the follicular cells. The endoplasmic reticulum, *er*, in follicular cells appears more extensive and dilated than that in the light cell. Although the light cell appears to be sparsely populated with organelles, dark granules, *dg;* light vesicles, *lv;* and zymogen-like granules, *zl*, can be found. (After B. A. Young and C. P. Leblond. 1963. Endrocrinology 73:669–687.)

attempts to prove that calcitonin is produced by the parafollicular cells but the evidence is not clear-cut. In almost all cases the hormone can be traced to the parafollicular cells, although other studies show involvement of the follicular cells as well.

Moreover, there are some follicles in the avian and mammalian thyroid gland which are of ultimobranchial origin. Their location varies from animal to animal, and in some cases the epithelium of the follicle is stratified and/or ciliated. Again, functional evidence is inconclusive. Some investigators claim that these follicles produce hormone normally, whereas others maintain that they do not metabolize iodine. The possibility still exists that these follicles may play a role in calcitonin synthesis. Although there is logic behind the thought, the only firm evidence that supports the theory has been derived from studies with chickens.

The follicular lumina contain a fluid known as *colloid*. Colloid is thought to represent a site of some facet of hormone synthesis as well as to provide storage of the protein-bound form of the hormone. These facts single out the thyroid gland as the only vertebrate alveolar endocrine gland in which hormone is stored extracellularly within the gland. In general the colloid within a follicle is not uniform, but rather contains spaces which appear to be vacuoles. In inactive follicles the colloid is eosinophilic, changing to basophilic during activity. Colloid has been found to contain globulins as well as a variety of enzymes.

The details of the secretory cycle leading to functional hormone are still clouded by uncertainty, although the main features of thyroid secretion are understood. After stimulation with TSH, secretory droplets which appear to be colloid can be found at the apical ends of the follicular cells. These are released into the lumen. Within the colloid, a number of events are believed to occur. Iodide that has been extracted from the blood by the follicular cells and which is secreted into the colloid is converted to iodine. The iodine is then enzymatically attached to tyrosine in a stepwise fashion: monoiodo-, diiodo-, triiodo-, and finally tetraiodotyrosine being produced (Fig. 14.3). The tyrosine is derived from the protein thyroglobulin, which is also a component of colloid. Indeed, thyroglobulin must remain within the follicle, because an individual can form potent antibodies against his own thyroglobulin. It is one of the classic and few autoimmune substances in the normal body. During hormone production, it is believed that the intermediates are bound to the globulin. After triiodotyrosine and thyroxine have been synthesized, the complex is thought to undergo pinocytosis whereby vacuoles of colloid enter the follicular cells. Within these vacuoles, the hormones are removed from the binding sites on the globulin and are secreted by the basal ends of the cell into the blood stream. The globulin reenters the colloid to await further iodination.

THE PARATHYROID GLANDS

The parathyroid glands most often occur as two pairs of round to ovoid structures lying within or near the thymus or thyroid gland (Fig. 14.4). In certain species, such as some lizards as well as the rat, mouse, seal, pig, and others, only one pair of glands develop to maturity. In other vertebrates, such as man and some frogs, accessory glands may occasionally be found. The glands develop from the endoderm of the corners of the second, third, and/or fourth visceral pouches as the gills degenerate. Therefore, in fish, larval amphibia, and the neotenous amphibia, all of which retain gills, parathyroid glands are lacking. The regulation of calcium levels in the blood of these animals, normally the function of the parathyroid hormone, is thought to be taken over by the ultimobranchial bodies with some assistance rendered by extremely high levels of vitamin D.

Parathormone, the hormone of the parathyroid glands, plays an important role in the maintenance of the balance between calcium ion and phosphate ion in the blood. These ions appear to exist in a reciprocal relationship; as the serum level of one increases, the rate of excretion of the other also increases,

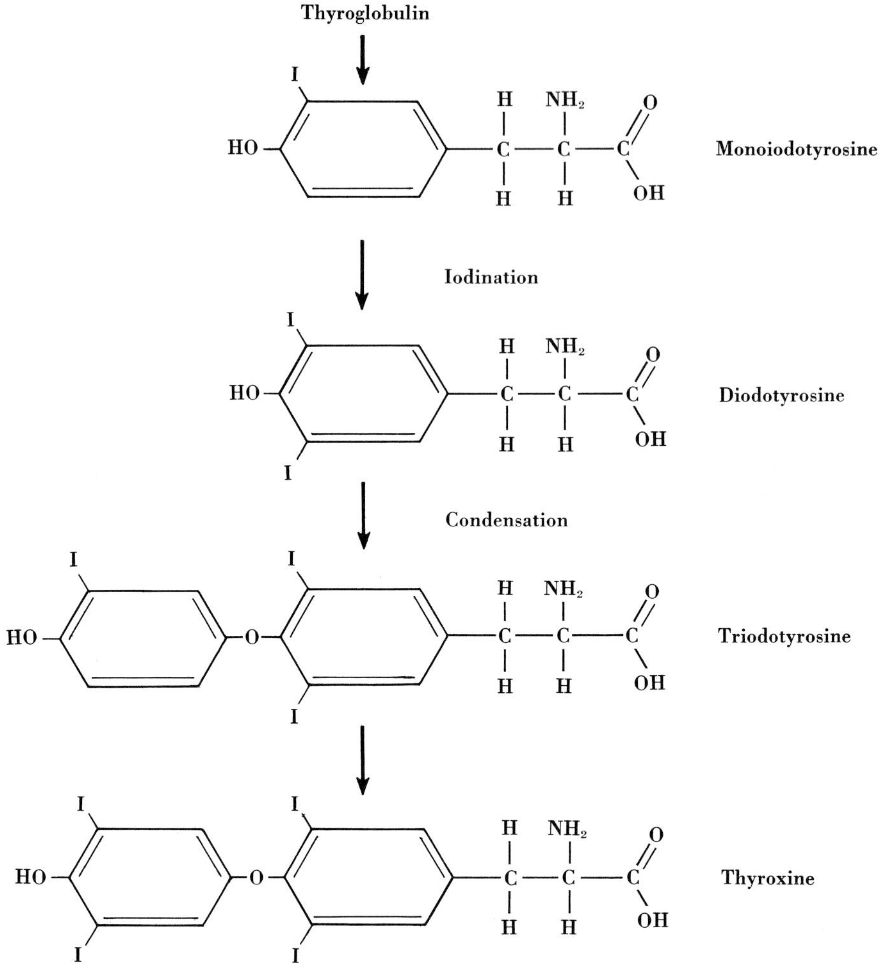

Fig. 14.3 The biosynthesis of thyroxine. Beginning with tyrosine, which is a component of thyroglobulin, the molecule is sequentially iodinated until tetraiodotyrosine (thyroxine) is obtained. This sequence is believed to occur within the folicular colloid.

resulting in a fall in blood concentration of the latter.

Parathormone exerts its effects on this balance by acting directly on the cells of bone and kidney tubules. If the level of serum calcium ion should fall, a negative feedback reaction is initiated which results in either an increased rate of formation of new osteoclasts, or an increased rate of activity of already-existing osteoclasts, or a combination of both. The direct effect is the release of calcium ion into the blood. In addition there is an increase in phosphate excretion by the kidney—caused either by inhibited tubular reabsorption or increased tubular secretion of phosphate. Therefore, it is not difficult to imagine that hypertrophy of the parathyroid, perhaps caused by a tumor, would lead to excessive depletion of bone calcium followed by softening, deformity, and fibrous replacement within the skeleton. In addition, as the level of serum calcium rises, the renal threshold may be exceeded. Calcium will then be present in large amounts in the kidney tubules. Often, instead of or in addition to being passed in the urine, the excess ion will be precipitated and deposited within the kidney. For this reason, tumors of the parathyroid glands can lead to death by kidney failure.

The converse condition, the lack of parathormone, leads to a severe neuromuscular sensitivity known as tetany. One of the roles

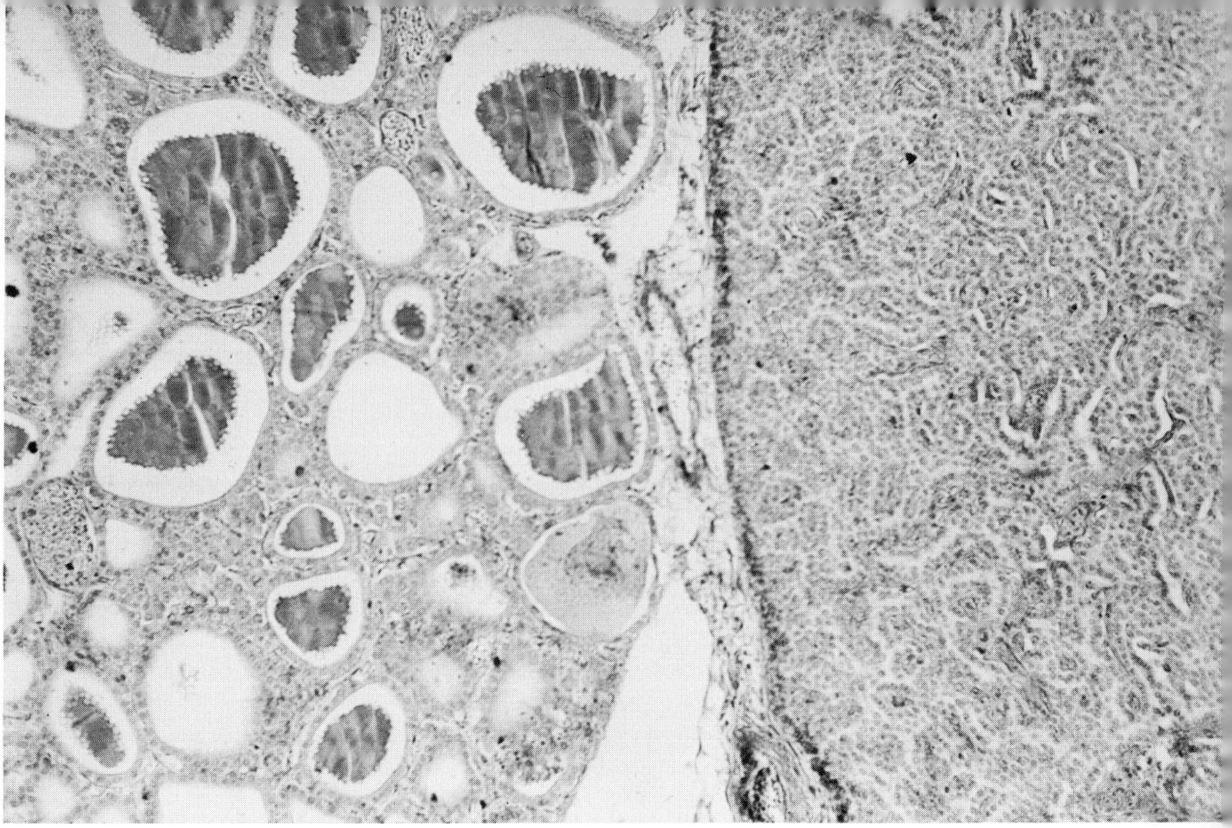

Fig. 14.4 Portions of the thyroid and parathyroid glands of a dog. Note the close association between the two structures and the structural differences in cell arrangement. A thin capsule separates the parathyroid (*above*) from the thyroid (*below*). X100.

of calcium ion is that of a pacifier and stabilizer of the skeletal muscle fibers and nerves. Insufficient parathormone will cause depressed serum calcium levels, and this, in turn, will result in a lowered level of the ion in the tissue fluids bathing the muscles and nerves. Without the proper level of calcium ion, the irritability of these tissues rises precipitously until slight stimulation can cause uncontrolled muscular spasms and convulsions. The degree of severity of such tetanic seizures varies considerably throughout the vertebrates. In man and most other tetrapods death can result if the laryngeal muscles are affected. On the other hand, parathyroidectomy in some anurans will merely result in slight hyperexcitability. The same operation in urodeles will be manifest by no gross change whatsoever. In general, a lack of parathormone affects carnivores most severely. The differences between animals result not from differences in the mode of action of the hormone. In cases in which measurements of blood calcium were taken after parathyroidectomy, depressed serum calcium values were noted. Rather, the differences are thought to be caused either by the effectiveness of ancillary means of calcium and phosphorus control, or by the degree to which normal irritability depends on unbound or free calcium ion rather than in some other bound form.

The entire job of regulation of the level of serum calcium is not thought to be left entirely to the parathyroid glands. In mammals, injection of anterior pituitary extracts results in a dramatic rise in serum calcium as well as an increased frequency of mitotic figures within the parathyroid glands. However, proof of the existence of a parathyroid stimulating hormone produced by the anterior pituitary is lacking. In submammalian vertebrates, estrogen administration results in a rapid rise of serum calcium. By this method, however, the calcium does not appear in ionized form, but rather is bound to a phosphoprotein which is normally produced in the liver. Phosphoprotein capable of binding calcium does not occur in mammals, and this has been postulated to be one reason that mammalian ova are noticeably poor in yolk. The rationale behind this hypothesis is that a high yolk content depends on elevated serum calcium because of the role which is played by calcium in active transport of material across the plasma membrane.

Finally, the thyroid gland is believed to be the site of production of a hormone that is antagonistic to parathormone. This hormone is called either *calcitonin* or *thyrocalcitonin.*

Parathormone has been prepared in pure form only from bovine extracts. The active agent appears to be an unbranched polypeptide having alanine as the terminal amino acid and containing no cysteine. The molecular weight of the pure extract is 8500, although two presumed degradation products of molecular weights 3700 and 6500 have also been isolated. The nature of all of the vertebrate parathormones, however, is still in the realm of speculation. Rabbits can be made to produce anti-bovine parathormone antibody. Injection of bovine extracts into amphibians has yielded limited or negative results. Complicating these experiments, however, is the fact that one major site of parathormone activity is the kidney, and mesonephric kidneys of amphibians might react differently than the metanephric kidneys of higher forms do. In addition, ultimobranchial bodies are present in amphibia, and are presumed to be active at least during the premetamorphic stages when parathyroid glands are lacking. Administration of bovine parathormone to birds results in more typical, if fleeting, changes. The complicating factor in these experiments is the obvious calcium dependency on estrogen. Hence, the suggestion has been made that differences in polypeptide structure exist throughout the vertebrates.

Although the particular parathormone produced may be species- or class-specific, the histologic structure of the cells of the parathyroid is remarkably similar throughout the vertebrates from amphibians to mammals. Actually, there should be no reason for great histologic difference because differences in amino acid arrangement within a polypeptide should be manifest more as a chemical phenomenon at the level of DNA and messenger RNA. More to the point would be the fact that a cell is a secretory one, the product of which is a polypeptide hormone.

The parathyroid glands are covered by a thin, vascular, fibrous connective tissue capsule. Extensions, varying in delicacy, pass inward from the capsule to provide an avenue for blood vessels and vasomotor nerves. The septa are usually of such delicate nature that the gland is not divided into lobes or lobules, but rather remains as a compact, unified entity. In some animals, such as the toad, the interior of the gland contains no blood capillaries; cellular nourishment is provided by prodigious capsular blood vessels and acellular slitlike sinusoids within the parenchyma of the gland.

The secretory cells are compactly arranged in clumps, cords, or whorls that are separated by small amounts of connective tissue fibers, fat, and, in most cases, blood capillaries (Fig. 14.5). Known as the *chief cells,* the secretory cells are more or less cuboidal and possess a rather spherical nucleus that has one or two prominent nucleoli. The chief cells of the parathyroid gland are smaller than most other endocrine secretory cells and this can serve as the diagnostic feature of the gland. Usually two stages in the secretory cycle can be noted. One type of chief cell appears almost clear with only a small degree of cytoplasmic granulation. These are the clear or light cells. Dark, more active chief cells can also be found, and they contain smaller nuclei than the clear chief cells do and have PAS-positive granulation in the cytoplasm. (The term "dark" is used relatively here, and in contrast only to the clear cells.)

Electron micrographs of chief cells exhibit a typical complement of organelles, including a highly porous nuclear membrane and the so-called juxtanuclear body. The latter is composed of dense, lamellar stacks of rough endoplasmic reticulum (Fig. 14.6). The PAS-positive granules are either ovoid or rodlike and are enveloped by a membrane. That these granules represent the secretory material has been deduced by tracing their movement from the chief cells through adjacent capillary endothelium and into the capillaries.

In some vertebrates deviation from the fore-

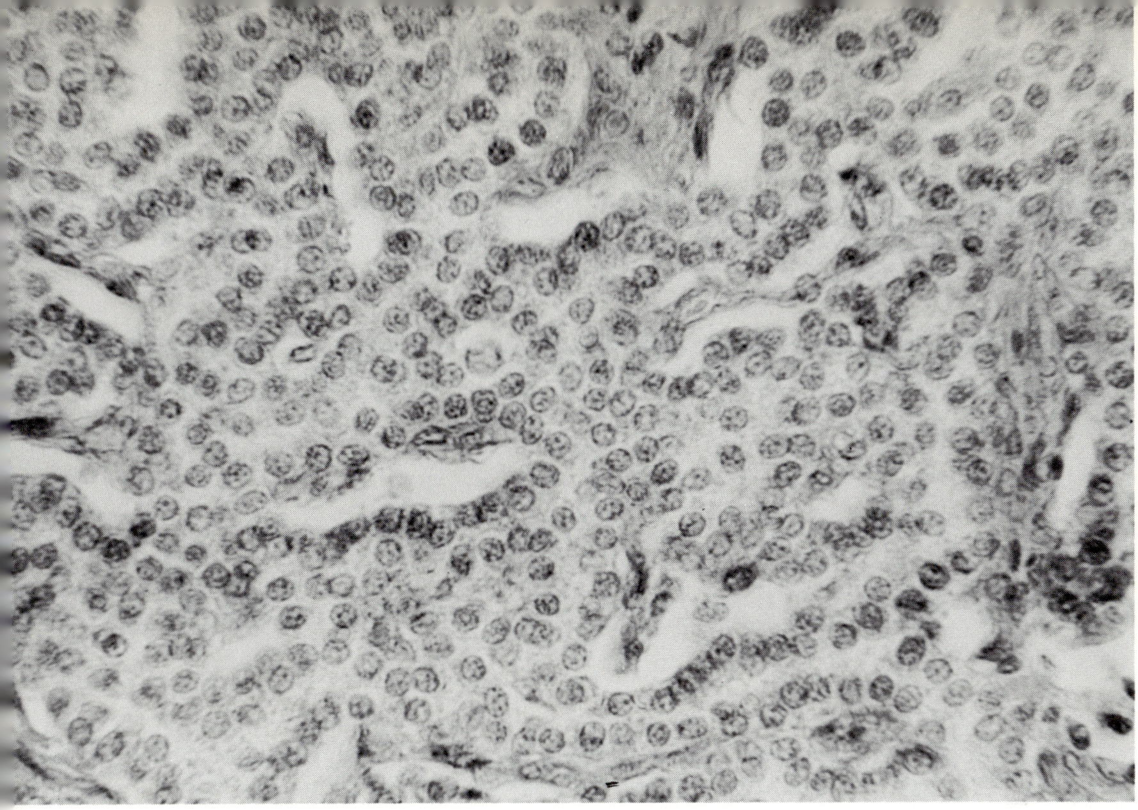

Fig. 14.5 Cords of closely packed cells in the parathyroid gland of the dog. X450.

going pattern exists. Perhaps the most dramatic are the seasonal changes seen in the anuran *Rana catesbeiana*. During the winter the internal cells of the parathyroid of this species seem to dissolve, leaving only a subcapsular ring of cells to provide for regeneration in the spring. Other amphibia show degenerative changes of lesser magnitude. The glands in birds can enlarge up to ten times their normal size; such change is also seasonal and is accompanied by a parallel fluctuation in the estrogen and mineral metabolic cycles during shell deposition and egg laying.

The location of the parathyroid glands in the turtle makes some "detective work" mandatory. The more anterior of the gland pairs is usually found buried deep within the thymus and the left posterior gland becomes engulfed by an extremely large left ultimobranchial body. In addition, there have been reports of reptilian chief cells that are arranged in follicles with intrafollicular storage of secretion. This becomes interesting in light of the absorption of the ultimobranchial body into the thyroid in mammals, the follicular structure of the latter gland, and the origins of the thyroid, parathyroid, thymus and ultimobranchial bodies. In a variety of sub-Eutherian mammals, thymic nodules have been found embedded within and intermingled between the cell cords of the parathyroid gland. In light of the common site of origin of the parathyroid and thymus, questions have been raised about the ability of these two cells types to migrate or perhaps to be interconvertible during embryonic stages.

In man, but not most other mammals, a second cell type can be found after puberty. These cells, the *oxyphiles,* are usually arranged in clumps but are occasionally seen in folliclelike array. They are larger than the chief cells but have smaller and more densely staining nuclei. The cytoplasm of the oxyphiles contains fine glycogen granules, lipid inclusions, and an extremely dense blanket of mitochondria (Fig. 14.7). The Golgi body in these cells is either poorly developed or not present. As further indication of the nonsecretory nature of these cells, few, if any, secretion granules can be found.

The lack of parathyroid glands in the various classes of fishes has elevated the ultimobranchial body to a position of major importance as the possible substitute seat of calcium

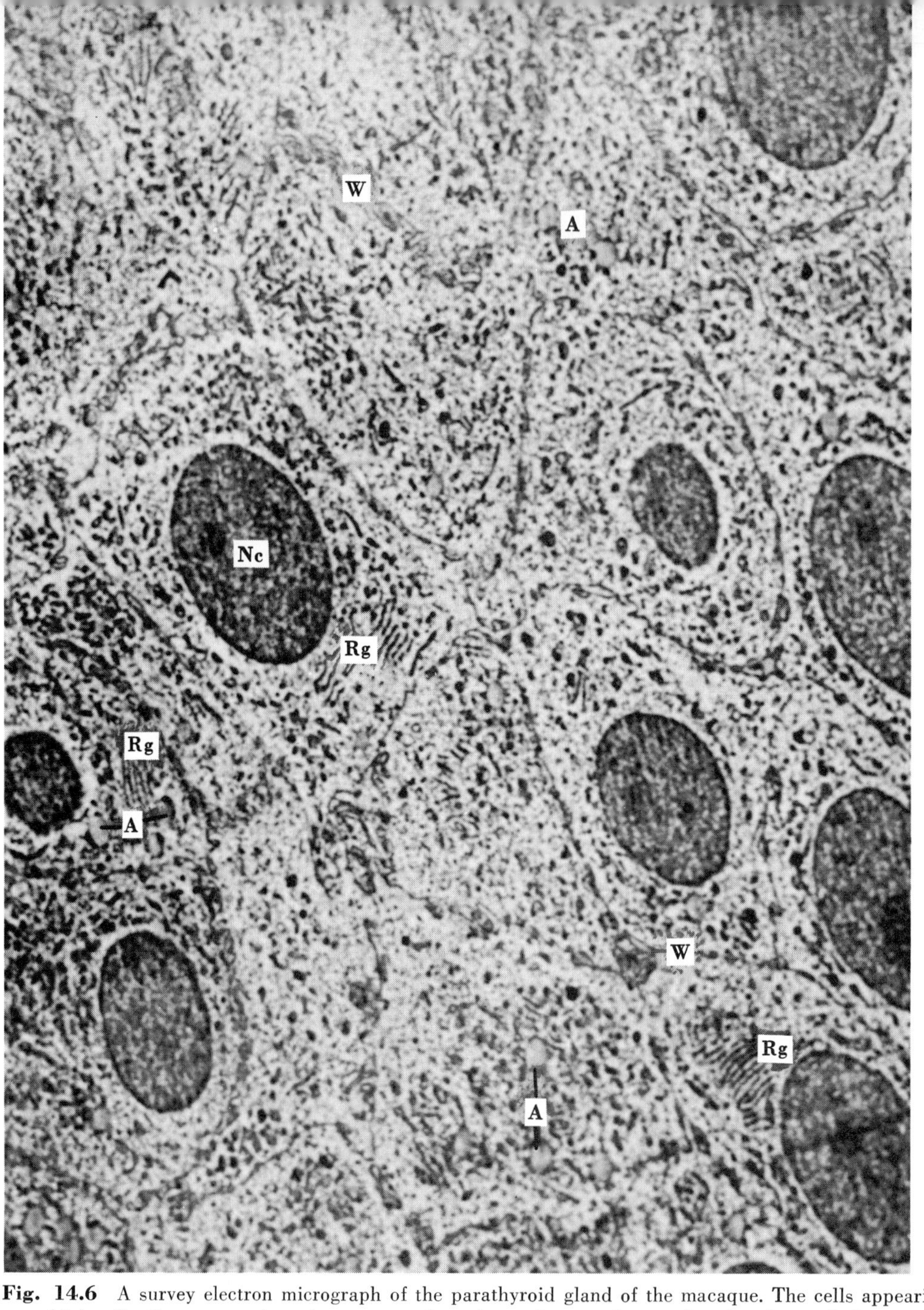

Fig. 14.6 A survey electron micrograph of the parathyroid gland of the macaque. The cells appearing here are chief cells. Large granules, *A*, are seen throughout the cytoplasm, often in close relation to granular endoplasmic reticulum, *Rg*. Nucleus is marked *Nc;* interdigitating cell membranes at cell junctions are marked *W*. X6000. (From J. S. Trier. 1958. J. Biochem. Biophys. Cytol. 4:13–22.)

ion control. Although ultimobodies develop in fishes, amphibians, reptiles, birds, and begin to develop in embryonic mammals before being engulfed and incorporated into the rapidly enlarging thyroid gland, we will consider them only in the fishes as providing a possible mechanism for completing the vertebrate story with regard to parathyroid or parathyroid-type secretions.

The ultimobranchial body is a small mass of epitheloid cells which can usually be found between the ventral wall of the esophagus and the sinus venosus or aorta. In untreated fishes, the ultimobranchial body is in the form of a

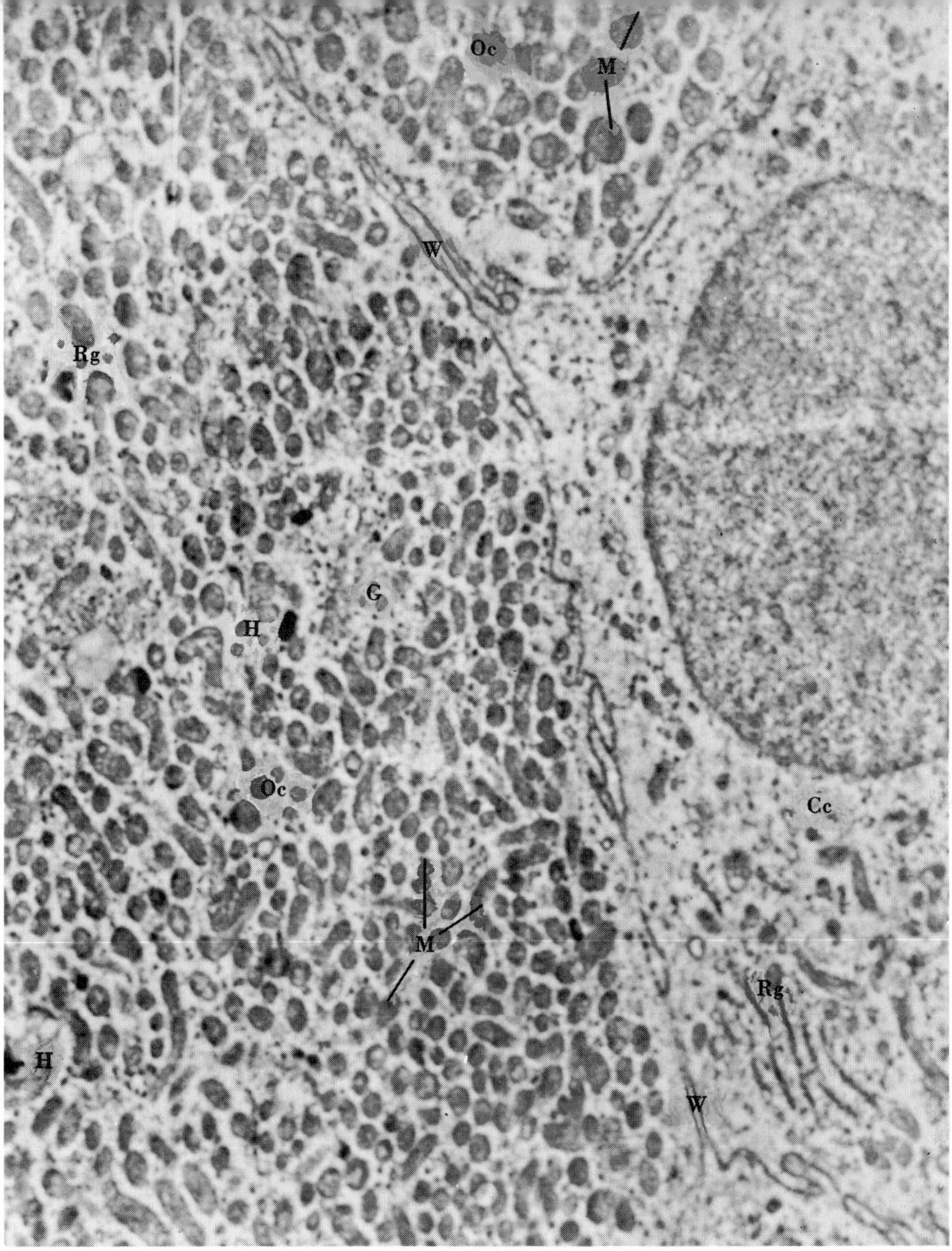

Fig. 14.7 An electron micrograph showing portions of two oxyphil cells, *Oc*, and a chief cell, *Cc*. Numerous large mitochondria, *M*, are found within the oxyphils. Golgi vacuoles, *G;* granular endoplasmic reticulum, *Rg;* and lipid inclusions, *H*, can be seen. The interdigitating cell junctions are indicated at *W*. X15,000. (From J. S. Trier. 1958. J. Biochem. Biophys. Cytol. 4:13–22.)

thin, long crescent which is glandular in appearance. It is covered by a delicate connective tissue capsule, and filled with epithelial cells which approach a columnar shape at the periphery. On occasion the cells appear to be grouped into acini or cords. The lumen of the gland is wide and contains many lymphocytes.

After dark stress, histological changes can be seen which parallel those of any secreting gland. The epithelial cells enlarge and appear more active. Not only do the cells enlarge, but the entire structure can hypertrophy to an amazing degree. Furthermore, skeletal defects, such as curving and decalcification of the spinal column and pathological changes in the kidneys, can be observed. It has been suggested that dark stress which led to a changed functional activity of the pituitary, thyroid, and interrenal glands, eventually resulted in renal insufficiency. This, in turn, led to enlargement of the ultimobranchial body with concomitant secretion of a material similar to parathormone. The secretion, in turn, precipitated the observed skeletal deformities.

Additional evidence for the parathyroidlike function of the ultimobranchial body is derived from mildly positive reactions of fishes that have been treated with extracts of bovine parathormone. However, no extracts of the ultimobranchial body have been purified, and thus direct, positive evidence for admitting the structure as a *bona fide* endocrine organ on a par with the parathyroid is lacking.

THE PANCREAS

The endocrine pancreas consists of small clusters of cords of cells, the *islets of Langerhans,* which originate directly or indirectly from the intestinal epithelium. In the lamprey development of islet tissue proceeds directly from outgrowths of the mucosa, resulting in small aggregations of endocrine cells in the submucosa. These clusters are not directly associated with the diffuse groupings of pancreatic exocrine tissue, so that in the cyclostomes the endocrine and exocrine portions of the pancreas are separated as two distinctly different groups of structures. In teleosts the endocrine pancreas has evolved into a compact structure that is clearly separate from, although near, the exocrine pancreas. In higher vertebrates islet tissue is seen to be intimately associated with exocrine tissue, the only demarcation being a delicate boundary of reticular connective tissue encapsulating the islets (Fig. 14.8). As would be expected of an endocrine structure, capillaries are numerous and in close juxtaposition to the islet cells.

Special staining methods suitable for use with optical microscopy have been used traditionally to differentiate the various cell types within the islet. These methods, developed for use with mammalian tissues, indicated the presence of four different kinds of cells, designated as alpha (α) or *A*, beta (β) or *B*, *C* or clear, and delta (δ) or *D* (Figs. 14.9 and 14.10). The same methods, when carried over for use with other vertebrates, failed to give

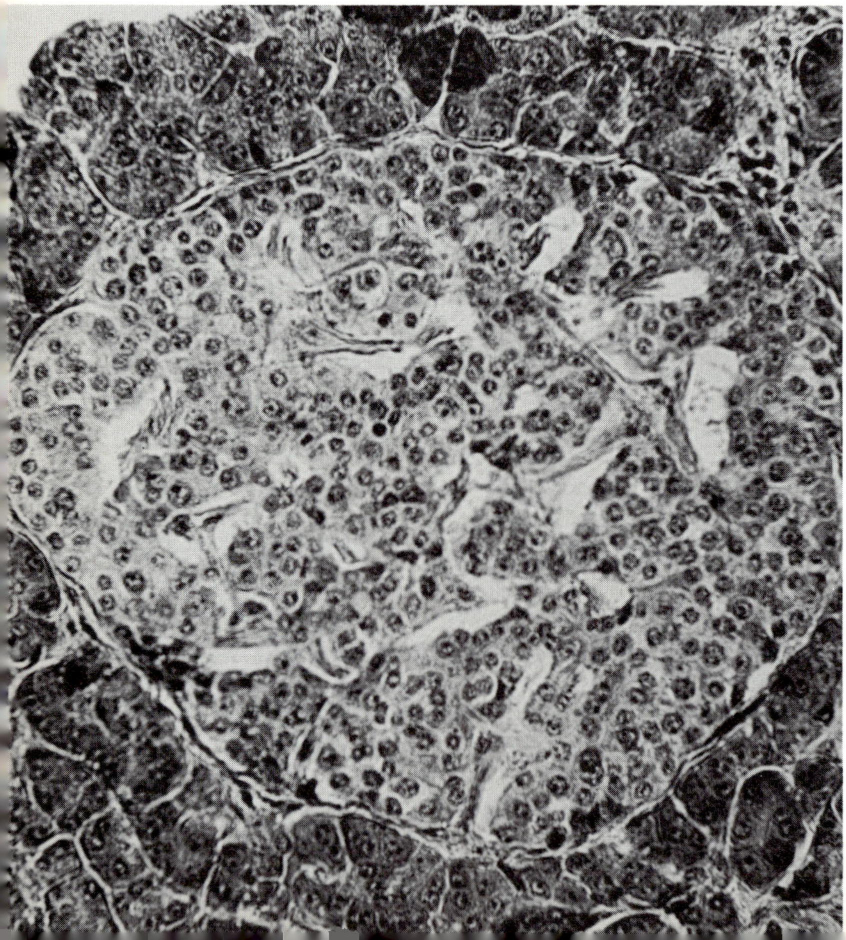

Fig. 14.8 An endocrine islet within the human pancreas. The exocrine acini are easily distinguished from the cell cords of the islet. At this magnification it can be seen that the islet cells are not all identical. X400.

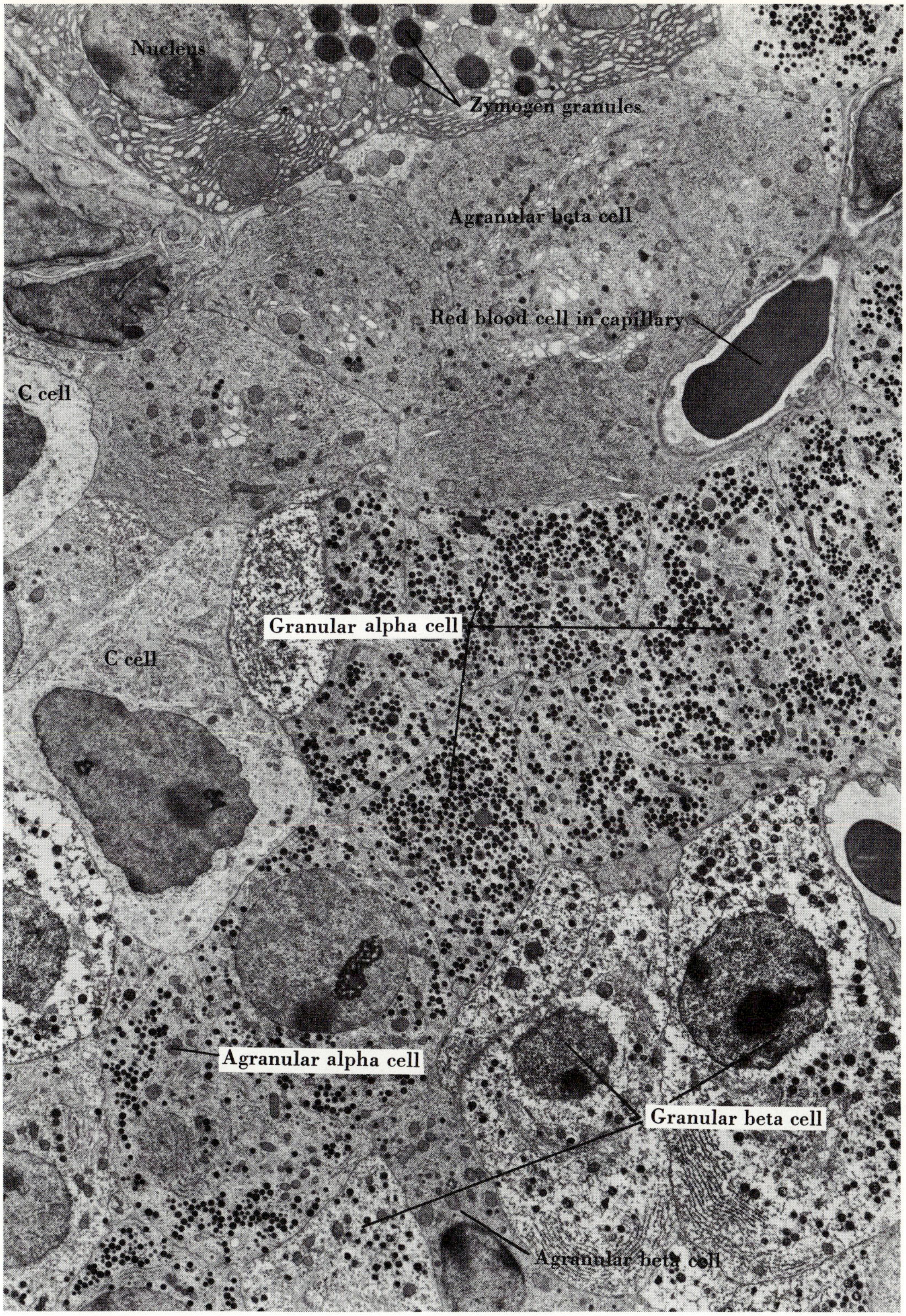

Fig. 14.9 A low-power electron micrograph of the pancreatic islet of a guinea pig. X3780. (From T. Sato, L. Herman, and P. J. Fitzgerald, Dept. of Pathology, State University of New York, Downstate Medical Center, Brooklyn, N.Y. 1966. The comparative ultrastructure of the pancreatic islet of Langerhans, Gen. & Comp. Endocrinology 7:132–157.)

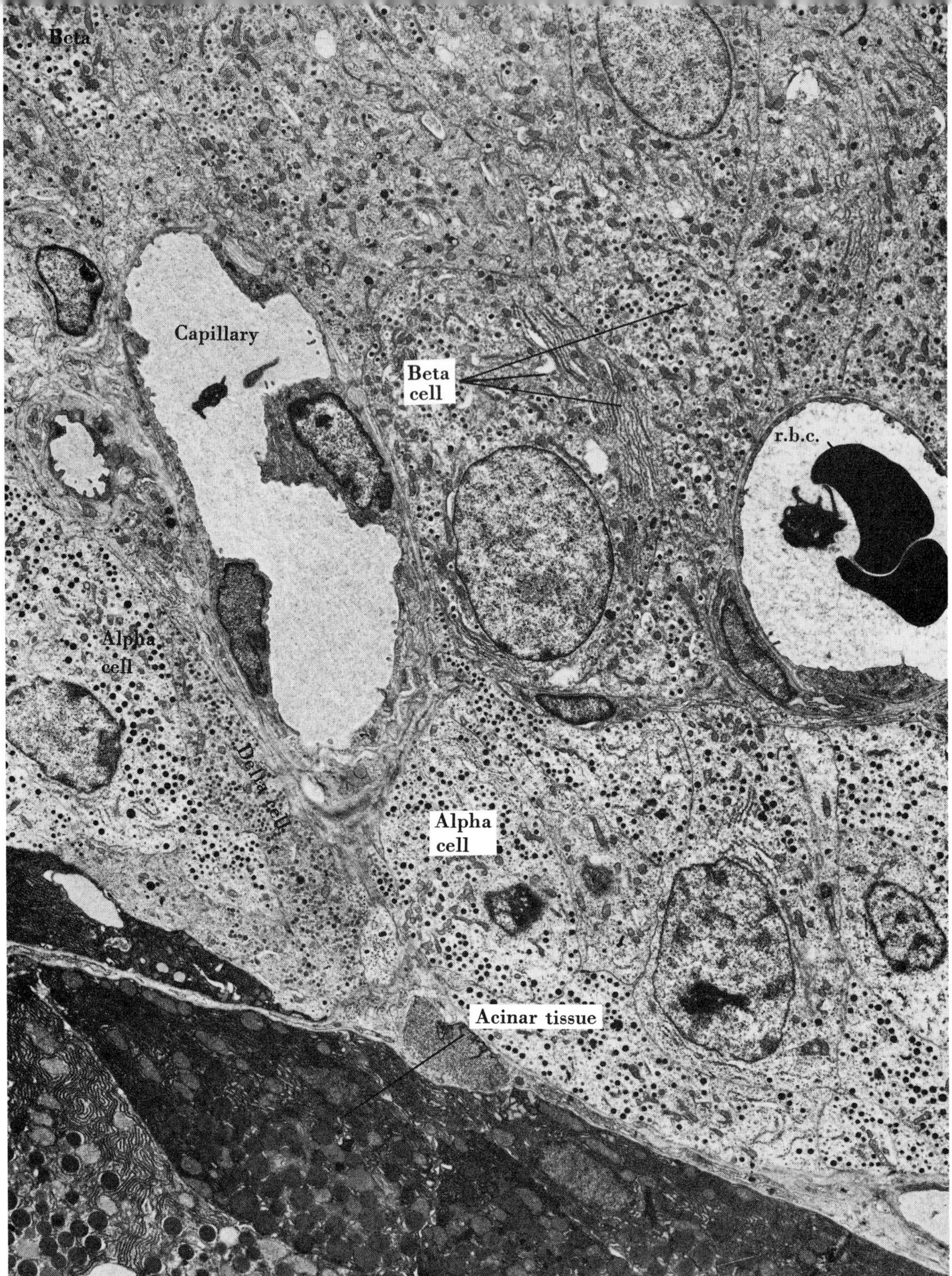

Fig. 14.10 A low-power electron micrograph of the pancreatic islet of a rat. X3780. Compare with Figs. 14.9 and 14.11. (From T. Sato, L. Herman, and P. J. Fitzgerald, Dept. of Pathology, State University of New York, Downstate Medical Center, Brooklyn, N.Y. 1966. The comparative ultrastructure of the pancreatic islet of Langerhans, Gen. & Comp. Endocrinology 7:132–157.)

the same kind of clear distinction. When optical staining methods are used in conjunction with electron microscopy, however, it is seen that—with the possible exception of the cyclostomes—all vertebrates possess at least three if not all four cell types. However, cytologic constancy does not preclude histologic variation (Fig. 14.11). A particular islet in any given

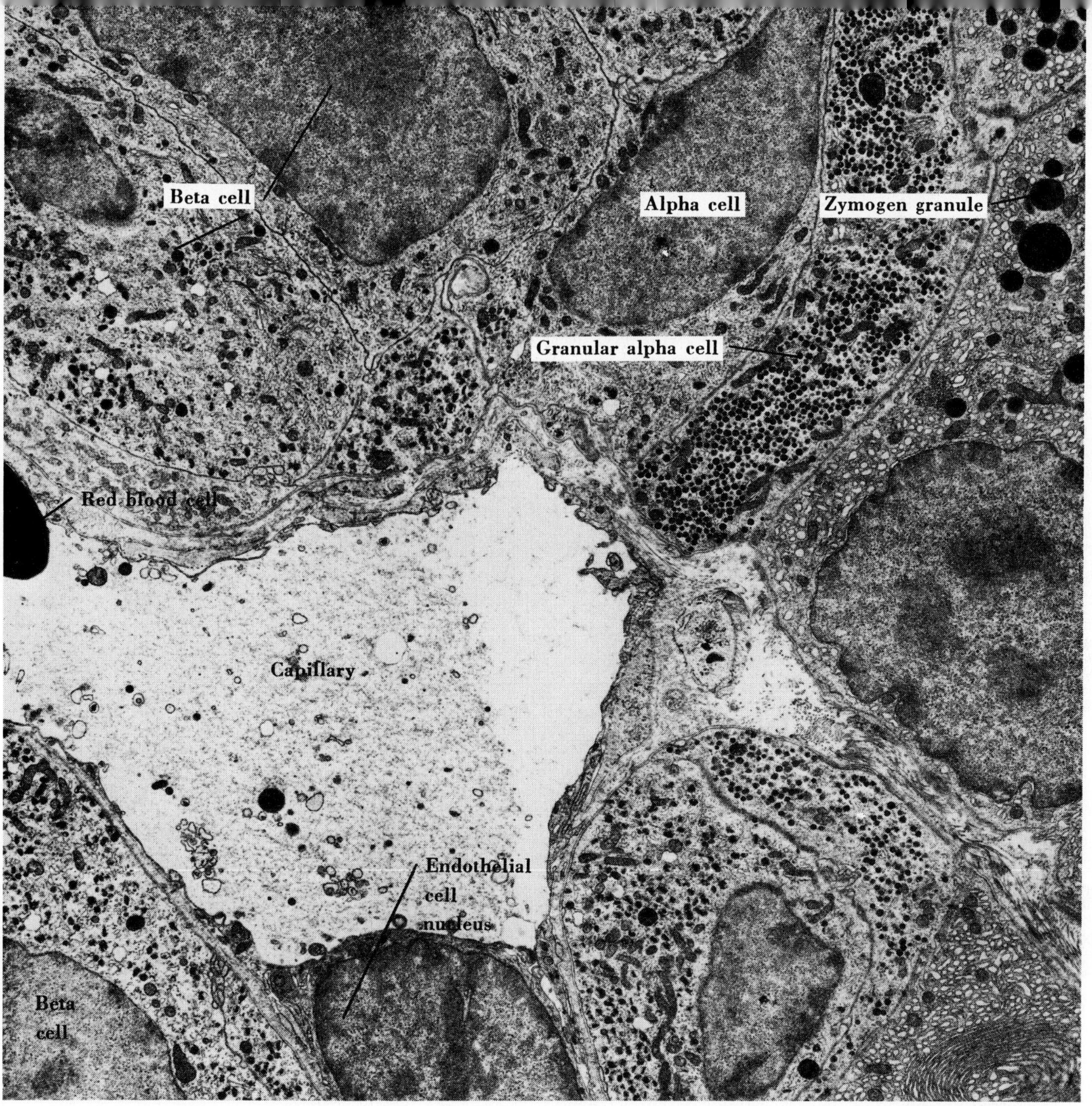

Fig. 14.11 A low-power electron micrograph of the pancreatic islet of a salamander. X3780. (From T. Sato, L. Herman, and P. J. Fitzgerald, Dept. of Pathology, State University of New York, Downstate Medical Center, Brooklyn, N.Y. 1966. The comparative ultrastructure of the pancreatic islet of Langerhans, Gen. & Comp. Endocrinology 7:132–157.)

species may be composed of only *B* cells, with the other cell types segregated into other islets. In other species there may be complete intermingling of the cells within an islet. Finally, there may be intraislet segregation, with *A* cells at the periphery.

A CELLS

The spherical or ellipsoidal *A* cells have traditionally been recognized by virtue of their large population of alcohol-insoluble granules. In man, rodents, and some carnivorans these cells are found at the periphery of the islet whereas in dogs and cats the cells are more centrally located. *A* cells of the bird pancreas usually are clustered in separate islets within the pancreas.

In addition to the granular component, these cells contain rod-shaped mitochondria, an array of rough endoplasmic reticulum, and free ribosomes. A definitive Golgi body is present capping the nucleus. The granules are smooth, spherical, and usually of uniform density—except in birds, in which they posses a rough or serrated edge and may be less homogeneous. The membrane surrounding the granule appears to be tightly adherent in some cases or to be separated from the granule by a fine feltwork of filaments in others. Granules usually appear in greatest concentration on the side of the cell nearest a capillary, although this probably represents a stage in the secretory cycle of the cell. The secretory product is presumed to be synthesized on the ribosomes and, indeed, there have been reports of granules visible within the vesicles of the endoplasmic reticulum.

The function of the *A* cell is the synthesis and secretion of glucagon. Mammalian glucagon has been crystallized and found to be polypeptide, with a molecular weight of about 3480, and containing 29 amino acid residues. Although glucagon is considered to be hormonal in its mode of action, it is short-lived and at present there seems to be no direct and clear-cut way of finding out exactly how its synthesis is stimulated and the exact effect it induces. Direct assay methods for glucagon are lacking at present. It is clear that glucagon must interact with glucocorticoids secreted by the interrenals, and that the final effect is an elevation in the blood sugar as a consequence of increased depolymerization of liver glycogen. Glucagon appears also to prevent the synthesis of glycogen, protein, and fat. Because of this inhibition, existing protein and lipids are metabolically utilized, releasing urea and ketone bodies into the blood. In these ways glucagon acts as an antagonist to insulin. Moreover, glucagon, or the hyperglycemia that results from its action, seems to stimulate the production of insulin. If sufficient glucagon is administered, the insulin-producing abilities of the *B* cells may be depleted, resulting in insulin deficiency and diabetes. Administration of nonphysiological doses of glucagon to liver cells can be correlated with increased size and activity of liver lysosomes, resulting in cellular autophagy.

B CELLS

The *B* cells are the most numerous of the pancreatic endocrine cell types. Similar to the *A* cells in shape, the *B* cells are characterized by the secretory granules they contain. Less homogeneous in density and shape than the *A* granules, the *B* granules do not fill the membranous sac in which they are contained. Hence a clear area, termed the halo, can be used as a diagnostic aid. At high magnification a repeating fibrillar pattern can be discerned within some of the secretory sacs, although the degree and shape of the mass varies with the species. This difference has been postulated to be a reflection of the species-specificity of the secretory product, insulin. The crystalline-like structure which is sometimes observed within the granule is thought also to represent artifactual crystallization of insulin. A typical secretory cycle has been noted with *B* cells. Just prior to secretion the granule concentration is dense, with reduction in the visible quantity of other membranous organelles. During synthesis and replenishment of the secretory product, the reverse situation obtains: a vast array of granular endoplasmic reticulum and small numbers of granules. Insulin has been crystalized and identified as a small protein, with a molecular weight of about 6000, and composed of two polypeptide chains. The α-chain contains 21 amino acid residues and is connected via two disulfide bridges to the β chain of 30 amino acids. Differences in at least seven of the amino acids in the two chains have been described in different species, and it is known that insulin of one species can be antigenic to another.

The activity of insulin is of major importance in maintaining the balance between the utilization of glucose and the production of glucose by hepatic depolymerization of glycogen. The secretion of insulin is believed to be regulated by the quantity of blood sugar in the islet vessels. A stimulatory effect of growth hormone on *B* cells has also been noted. Administration of insulin to an intact animal results in a decrease of blood glucose, probably owing to increased tissue uptake of the sugar. The mode of insulin action is still not clear. One hypothesis ascribes to insulin the function of controlling directly or via coenzymes the enzymatic conversion of glucose to glucose-6-phosphate, a compound which is vital to further glucose metabolism. An alternative hypothesis states that insulin controls membrane permeability and thereby regulates glucose transport across it.

Insulin deficiency leads to a disease known as *diabetes mellitus*. The metabolic abnormalities associated with the disease are widespread and are most severe in carnivorous vertebrates. The usual symptoms include increased glycogen breakdown, leading to an elevated level of blood glucose. If the renal threshold for glucose is exceeded, glucose is found in the urine, which, in turn, is produced in excessive amounts. The body cells, unable to utilize glucose as an energy source, turn instead to fats and protein, resulting in increased nitrogen excretion, protein depletion within the tissues, and accumulation of large amounts of ketone bodies produced during fat degradation.

OTHER CELLS

In some animals such as guinea pigs, a category of cell called the *C* cell is distinguished. Although they are considered to be clear cells, poorly differentiated granules or secretory sacs of low density have been found. With its filamentous mitochondria and extensive Golgi network, the *C* cell is thought either to be similar or identical to the *D* cell, or to be a stage in the differentiation or regression of an *A* or *B* cell.

The delta or *D* cells are the least numerous of the islet cells. They contain low-density granules only partially enclosed by a membrane. The function of these cells is not clear. It is believed that they either secrete a hormone as yet undiscovered, or they are *A* cells in the process of regression.

THE INTERRENAL AND CHROMAFFIN TISSUES

The interrenal and chromaffin tissues are endocrine structures of distinctly different function, structure, and embryonic derivation. However, in many vertebrates they are found within the same organ, either in a random interspersion or organized in a very definite way. The development of such organization reaches its peak in mammals in the form of the *adrenal gland*, in which the interrenal tissue constitutes the *cortex* and the chromaffin substance appears as the central *medulla.*

Interrenal cells are derived from coelomic epithelium. They are, therefore, of mesodermal origin as are all of the steroid-secreting endocrine structures. The chromaffin cells are descended from the neurectoderm of the neural crests, and are homologous to the postganglionic elements of the sympathetic division of the autonomic nervous system (see Chapter 12). Irrespective of these differences, in all classes of vertebrates the two kinds of tissue tend to form structural relationships to each other and to the kidney. This is suggestive of some functional relationship, similar to that linking the neuro- and adenohypophysis. Functional dependence of interrenal on chromaffin tissue is thought to exist, but the dependence may be caused by nonspecific actions of the chromaffin cell hormones.

Clusters of both types of tissues are found in the lamprey. They are located along the major abdominal vein (the postcardinal), although some of the other large blood vessels may also support clusters of the chromaffin cells. The two tissues are usually separate in elasmobranchs and some teleosts. In the latter the interrenal clusters are intimately associated with the kidneys as islands of tissue embedded in the anterior ends of the kidneys surrounding the postcardinal vein and its branches. Chromaffin clusters are encountered in a similar relationship with vascular elements and the kidneys and may even intermingle with the interrenal cells.

In amphibians and reptiles the two tissues either form mixed or contiguous structures near, on, or in the kidney. In some species they are combined in a pair of discrete glands, whereas in others they appear as diffuse clusters.

The tendency for fusion of clusters to form a definitive adrenal organ is seen in birds and mammals. In the former the two tissues are dispersed within the gland whereas in mammals the interrenal cortex completely or partially surrounds the chromaffin medulla.

Many different steroid substances have been

isolated from interrenal tissues of vertebrates. Some of these isolates are hormonally active and others, which are inactive, are thought to be intermediates in cholesterol metabolism. These steroid hormones are necessary for the maintenance of life in many, if not all, vertebrates, and removal of their sources may be manifested in such cellular events as the loss of integrity of subcellular membrane systems. Three generalized functions are attributed to the interrenal hormones, and correspondingly, three categories of interrenal steroid hormones are recognized. The *glucocorticoids* function in the regulation of carbohydrate metabolism. They increase the rate of glucose formation from fats and proteins (gluconeogenesis), facilitate the formation of liver glycogen, and inhibit the use of glycogen by peripheral tissues. Other effects of glucocorticoids are widespread, as would be expected if one considers that the ability to regulate carbohydrate, lipid, and protein metabolism is rather a central phenomenon in the economy of the body—involving all cells and their interaction with many other hormones. The second major category of interrenal secretions is the *mineralocorticoids*. These substances act in the maintenance of salt and water balance

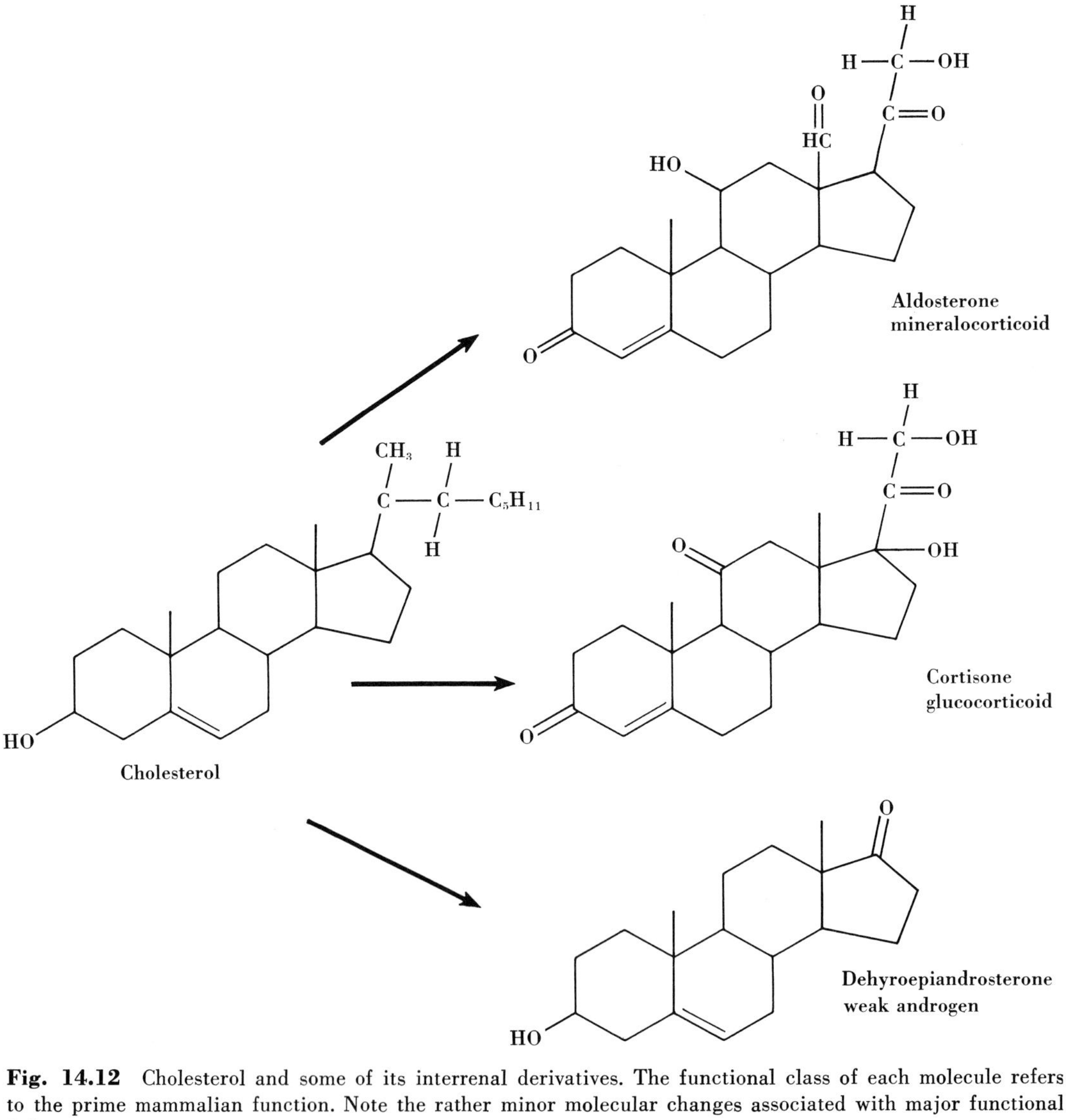

Fig. 14.12 Cholesterol and some of its interrenal derivatives. The functional class of each molecule refers to the prime mammalian function. Note the rather minor molecular changes associated with major functional differences.

within the body. The sites of action are also widespread, including such prime regions as the lung and gill membranes, skin, intestinal lining, salivary glands, and avian salt glands. Finally, there is a category of secretions which complement or inhibit various gonadai hormones. These are the androgenic, estrogenic, and progestational cortical hormones.

Although a complete discussion of the chemistry of these hormones would be out of place in a text such as this, there are a few chemical points which are relevant. All of the interrenal steroids are derivatives of cholesterol (Fig. 14.12). Although functional effects of the various hormones may show wide variation, the chemical differences are reflected as rather simple side-chain substitutions. Hence, the histologic or even cytologic appearance of interrenal cells is characteristic of steroid secretion in general rather than of any specific steroid in particular. Furthermore, it is believed that the potentiality to produce any of the steroids resides in all of the steroid-secreting cells. Secretions which are typical of the interrenals have, on occasion, been isolated from the gonads, and gonadlike hormones are produced in the interrenal cells. Finally, a particular chemical configuration that acts predominantly as a glucocorticoid in one species may exhibit salt regulation as its major activity in another species. The distinction between different chemical configurations and different functions may be made in only the most general way, and some of the corticoids are active in all three kinds of interrenal function.

Control of interrenal secretion is at least partially dependent on the pituitary. Liberation of ACTH from the anterior pituitary causes hypertrophy of the interrenal cells, although the most noticeable hormonal increase is found with glucocorticoids. However, secretion of ACTH is in itself a complex phenomenon that involves releasing factors that are presumably manufactured in the hypothalamus. The interrenal cells also manifest dependent relationships with such other factors as circulating levels of various ions, fluid volume, and angiotensin produced by the kidney.

The chemistry of the secretions of the chromaffin tissue is more circumscribed, although—because it is a functional analogue of the sympathetic nervous system—the effects of these secretions may be almost as widespread as those of the steroid hormones. Chromaffin cells are not limited to the adrenal system, being especially prominent in the paraganglia associated with the sympathetic chain. Many parallels can be drawn between chromaffin cells and other neurosecretory cell aggregations, including nervous control and cytologic appearance. The secretions of the chromaffin cells are the typical catecholamines found in other regions of the nervous system of vertebrates and some of the invertebrates.

Noradrenalin and its methylated partner, adrenalin, are the two products of the chromaffin tissue (Fig. 14.13). Both are found in most vertebrates, although the predominance of one or the other seems to be a class or species characteristic. Both are sympathomimetic, and presumably act at the basic level of phosphorylation of glucose prior to its metabolic breakdown.

Interrenal cells in nonmammalian vertebrates are generally found as interlacing cords or clusters near or around blood vessels (Fig. 14.14). Cells engaged in active synthesis react positively to cytochemical stains for neutral fats and steroid compounds. A spherical nucleus is located in the central or basal portion of the usually prismatic cell and is surrounded by a full orchestration of organelles. A few features of these organelles are outstanding and appear to be characteristic of steroid-

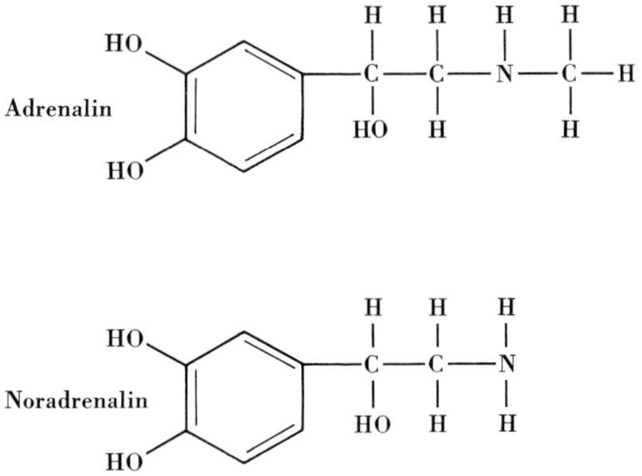

Fig. 14.13 Structural formulae of the catecholamines secreted by the chromaffin tissue.

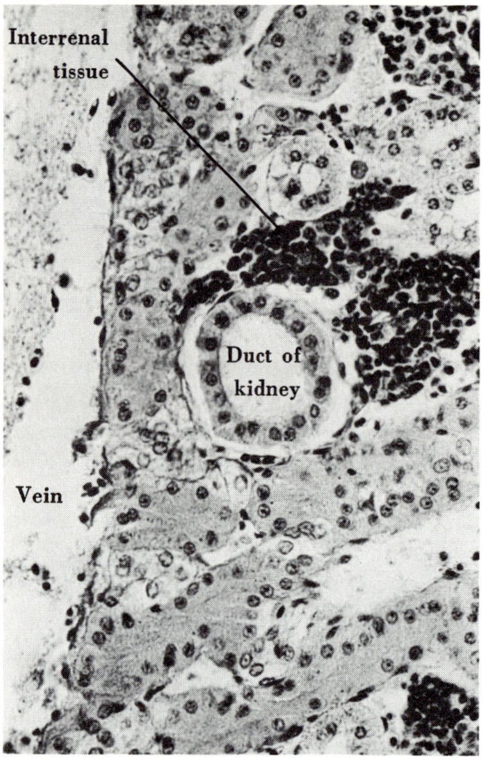

Fig. 14.14 The head kidney of the fish *Bodianus bilunulatus*. The large vein and its branches are seen to be in close association to masses of interrenal cells. The pale cells within the interrenal masses are presumably chromaffin. X370. (From J. Nandi and H. Bern. 1960. Endocrinology 66:295–303.)

secreting cells in general. Endoplasmic reticulum is present in great quantity. Typically, it is smooth-surfaced and vesicular, and because of its abundance and the topography in cells of this sort, it has been postulated to be the site of hormonal production. Also noteworthy are the mitochondria, which appear as round or oval bodies containing tubules and vesicles instead of the usual cristae structure. Various dense bodies can also be observed within the mitochondria and in the rest of the cell (Fig. 14.15).

In the mammalian gland the interrenal cells appear as a cortex underlying a dense connective tissue capsule. Extensions of the capsule enter the gland to serve as avenues for blood vessels and nerves. The vascular system within the entire gland is extensive. Numerous sinusoids exist which extend as subendothelial canaliculi between the cells. The cortex in some species is invaded by hillocks of medullary tissue whereas in other species the margin is straighter and more definitive.

The mammalian cortex differs from other interrenal clusters by its pattern of zonation (Fig. 14.16). The *zona glomerulosa* forms the outermost region. These columnar cells are found in loose whorls and clusters. Microvilli may be found in the portion of the cell that abuts a blood capillary or sinusoid. In some mammals the smooth-surfaced endoplasmic reticulum is vesicular or tubular, and free, unattached polyribosomes can be found. In other species, such as the monkey, a typical granular endoplasmic reticulum predominates, although free polyribosomes are also found; the mitochondria appear more usual, containing platelike extensions of the inner membrane. Lipid accumulations are present, as are small, electron-dense bodies (Fig. 14.17). Because ACTH affects the level of glucocorticoids most strongly, and because ACTH administration leads to only minor changes in the cells of the glomerulosa, it has been postulated that this region primarily secretes the mineralocorticoid hormones. If the adrenal gland is removed, leaving just a capsule and a few adherent glomerulosa cells, these, in turn, can serve as a cell source to repopulate the adrenal cortex. However, under normal conditions of gradual cell replacement, the new cells are thought to arise in each of the layers of the cortex.

The middle and largest cortical layer is the *zona fasciculata*. This zone, a continuation of the zona glomerulosa, contains cells arranged in tall columns. Occasional cilia may be observed. The cells themselves are larger than those in the outer region, and some of them are so highly populated with lipid droplets that they are termed spongiocytes. The cells in this layer most closely resemble the nonmammalian interrenal cells. Mitochondria containing tubular cristae and quantities of vesicular, agranular endoplasmic reticulum have been found in the zona fasciculata of all mammals that have been studied (Fig. 14.18). ACTH exerts its primary effect on the cells of the fasciculata; hence this region is thought to be the site of glucocorticoid secretion.

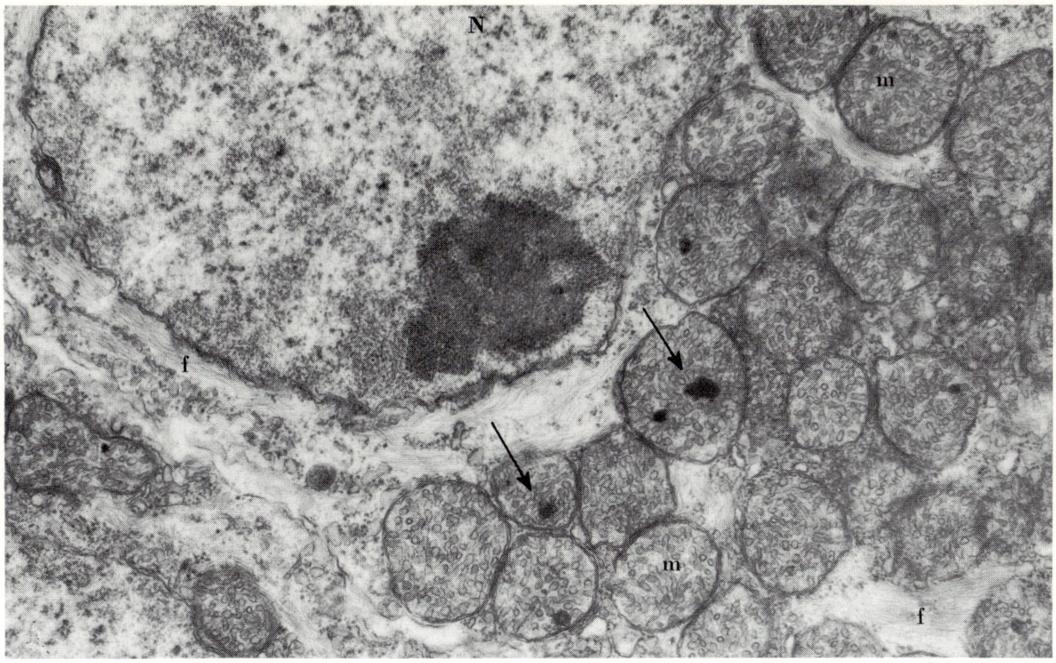

Fig. 14.15 Part of an interrenal cell of a normal goldfish. This electron micrograph clearly shows the vesicular mitochondria, *m*, and cytoplasmic filaments, *f*, found in this type of cell. The arrows indicate intramitochondrial deposits. X15,000. (From M. Ogawa. 1967. Fine structure of the corpuscles of Stannius and the interrenal tissue in goldfish, *Carassius auratus*. Zeitschrift für Zellforschung and mikroskopische Anatomie. Bd. 81, S. 174–189. Berlin-Heidelberg-New York: Springer.)

The cell columns of the fasciculata continue into the next cortical layer, the *zona reticularis,* becoming more loose in their organization as the interior is approached. Within the reticularis the cell cords form a loose network in close proximity to endothelial sinusoids. The cells of this layer are smaller than those in the zona fasciculata and although they contain less lipid material, the configuration of the other organelles is similar (Fig. 14.19). However, in this innermost cortical zone, there is an accumulation of deeply staining material within the cells that gives them a pyknotic appearance. Accordingly, there is some reason to believe that the cortical cells migrate inward to degenerate within the reticularis. On the other hand, in mammals such as the mouse and rabbit, the inner cortex gives evidence of producing gonadlike steroids.

The corpuscles of Stannius, found in or near the kidney of ray-fin fishes, have been postulated to be homologous to the mammalian zona glomerulosa. However, the evidence for the homology is equivocal and no definitive statement of the corpuscles' function can be made at this time.

Chromaffin cells can occur singly or in small clusters or cords along blood vessels (Fig. 14.20). In the mammal, cords of these cells are bound together by reticular fibers to form the adrenal medulla. The vascular supply to chromaffin tissue is extensive, with subendothelial canaliculi appearing between chromaffin cells. Two kinds of cells are found as part of the chromaffin tissue. Both are ovoid or slightly irregular in shape. Both contain a full complement of organelles and large (1800A), dense, membrane-bound secretion granules, which stand out as the most abundant cytoplasmic component (Fig. 14.21). They can be oxidized by potassium dichromate to form the coloration from which the cells derive their name. One kind, the noradrenalin-containing cells, are autofluorescent and hence can be distinguished from the other kind, the adrenalin-containing cells. In addition the granules

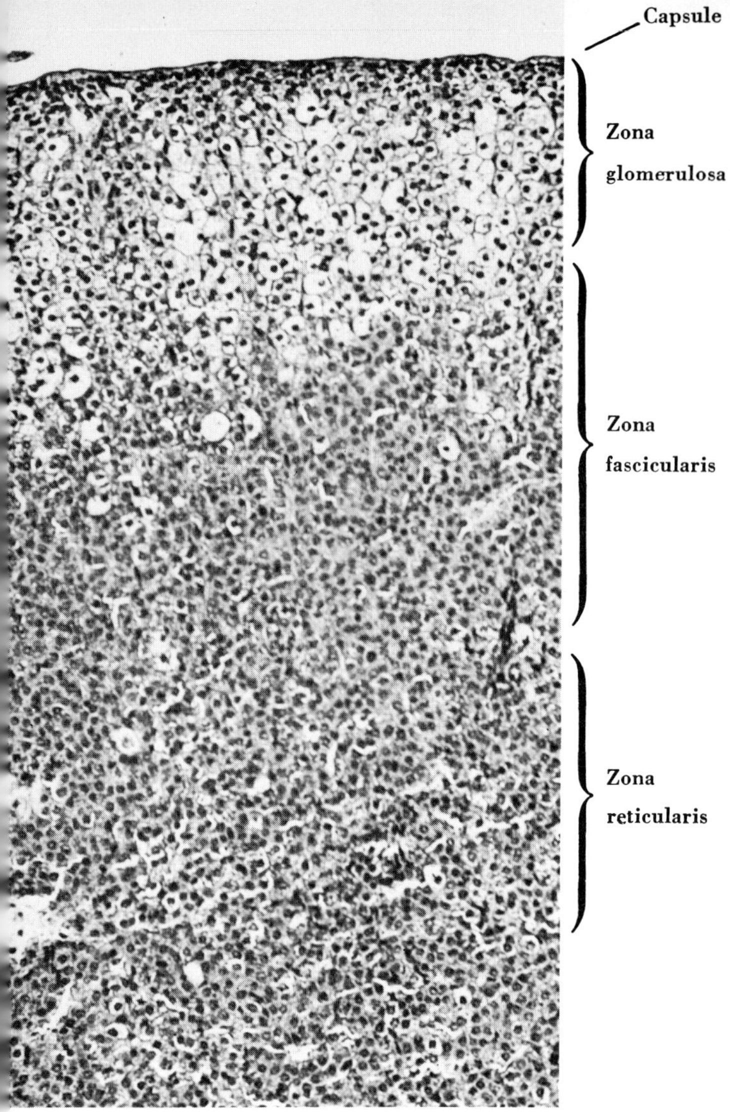

Fig. 14.16 The adrenal gland of the gerbil showing characteristic zonation. See text for an explanation. X100.

of noradrenalin cells are more dense than those of adrenalin cells, and this has provided another rationale for cytochemical distinction. Each cell type is innervated separately and can, therefore, secrete independently of the other.

THE PITUITARY GLAND

The pituitary gland is characteristic of all vertebrates and, indeed, is foreshadowed in some invertebrates and prevertebrate chordates. It is customarily called the "master gland" of the body because of its complex feedback or stimulatory relationships to other endocrine glands, such as the thyroid, interrenals, and gonads, and to such nonendocrine structures as mammary glands, kidney, pigment cells, and the growing regions in bone.

Much of the pre-eminence of the pituitary, or hypophysis, lies in its coordination role between the central nervous system and the rest of the endocrine system. By virtue of feedback control, the internal secretions can be maintained at homeostatic levels, and the internal environment can be stabilized. However, the nervous-endocrine aspects of the pituitary enable the secretory structures within the body to adjust to changes in the external environment as well. The survival value of such a connection cannot be overestimated. Our only knowledge of our surroundings comes via the nervous system, and much of our metabolic adjustment is controlled by blood-borne

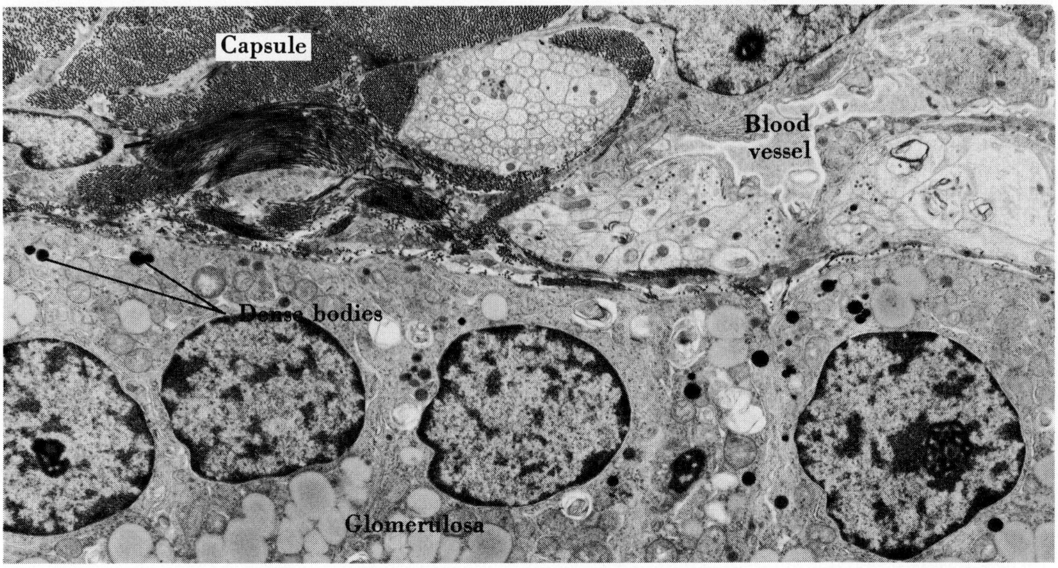

Fig. 14.17 An electron micrograph of the capsule-glomerulosa junction in the Rhesus monkey adrenal cortex. X4680. (From R. M. Brenner. 1966. Am. J. Anat. 119:429–454, Fig. 6.)

hormones. Without a link between the two systems, we would be operating (or shortly not operating!) in an environmental vacuum.

This dual nature of the pituitary derives from its embryology. The pituitary originates as an evagination of oral epithelium (Rathke's pouch), which is met by a downgrowth from the diencephalon (infundibulum). Although the connection between Rathke's pouch and the mouth is usually lost, the connection with the brain is retained by most vertebrates. From these outpocketings are developed the two functional areas of the pituitary. The *adenohypophysis,* which is of epithelial origin (as are most of the other endocrines), contains cells which are truly glandular in the usual sense. The *neurohypophysis* contains neurons which also can manufacture and secrete hormones; this property is spoken of as *neurosecretion.*

In tetrapods and lung fishes the relationship between the functional areas, the adenohypophysis and neurohypophysis, and the gross anatomical areas varies slightly from species to species. Generally, however, there are three types of tetrapod pituitaries (Fig. 14.22).

The first type of pituitary gland is characteristic of amphibians, lungfishes, reptiles, and many mammals. From the adenohypophysis three anatomical areas develop. The first of these is the *pars distalis,* commonly called the anterior lobe, and this is the major glandular portion of the pituitary gland. Its cells are arranged in cords or small aggregates, with an extensive blood-sinus system occupying the interstices (Fig. 14.23). From the pars distalis are secreted the trophic hormones of the pituitary. The second area which comprises the adenohypophysis is the *pars intermedia* or intermediate lobe. This zone of epithelial tissue may be separated from the pars distalis by a cleft, such as exists in the rat, or the cleft may be missing, as it is in the rabbit. In either case the zone is in contact with the neural part of the pituitary. The pars intermedia secretes a hormone which in the lower tetrapods influences melanin coloration or activity. The third region of the adenohypophysis is the *pars tubularis.* Presumably nonsecretory, the pars tubularis is so named because in higher tetrapods it forms a tube through which passes the stalk that connects the pars nervosa to the brain. It is missing or poorly developed in lizards, snakes, ant eaters, and sloths.

The neurohypophysis gives rise to two major areas with a third region included either as part of the pituitary or part of the brain. Connected to the hypothalamus, and really a part of it, is the *median eminence.* Containing a complex capillary pattern, it extends downward and becomes the *infundibu-*

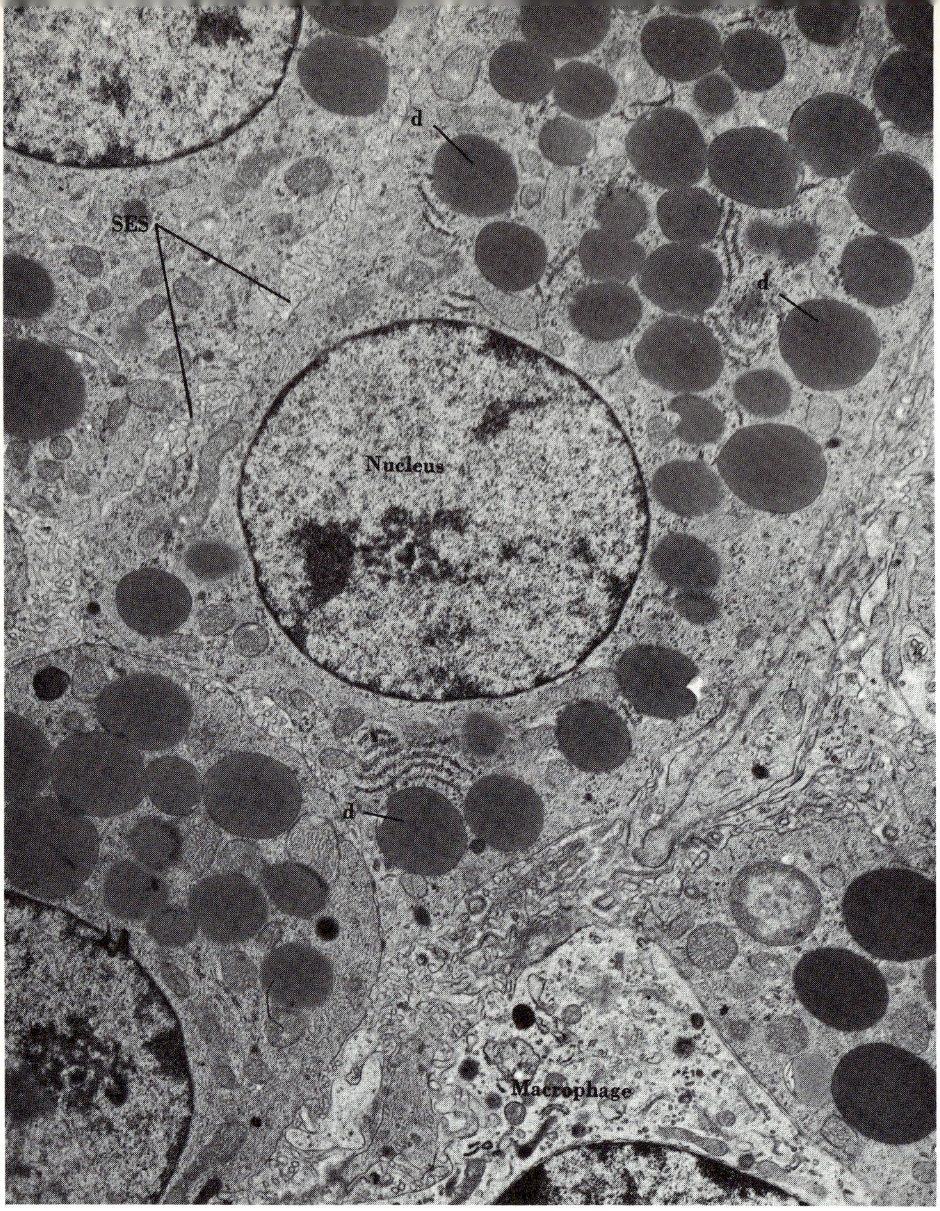

Fig. 14.18 An electron micrograph of an outer fasciculata cell in the adrenal cortex of a male Rhesus monkey. Note the large number of lipid droplets, *d*, and the microvilli protruding into the subendotrelial space, *SES*. X7400. (From R. M. Brenner. 1966. Am. J. Anat. 119:429–454, Fig. 11.)

lar stalk, through which pass nerve fibers from the brain. The infundibular stalk terminates in the *pars nervosa,* also called the posterior lobe. The pars nervosa is the major region of storage and liberation of the neurosecretory hormones.

The second major type of pituitary gland is found in all birds and also in armadillos, elephants, manatees, pangolins, porpoises, and whales. The neurohypophysis develops into the same three areas as described above, but the adenohypophysis forms only two. The pars distalis is present, and it adheres to the pars nervosa via the pars tuberalis. The pars intermedia is lacking.

The third type of pituitary gland is characteristic of the chimpanzee, gorilla, and man. It is similar to the first type of cleftless gland—except that the pars intermedia degenerates, leaving a line of epithelial cysts in its place. Hence, the pars distalis is adherent to the pars nervosa. The secretory cells of the degenerate pars intermedia have infiltrated the pars nervosa, so that the posterior lobe contains cells of adenohypophyseal as well as neurohypophyseal origin.

The pituitary gland of cyclostomes and gnathostomatous fishes (except for the lungfishes) differs from the foregoing in its anatomy. The adenohypophysis is zonated to a greater or less degree to form the *proadenohypophysis* or rostral pars distalis, the *mesoadenohypophysis* or proximal pars distalis, and the *metaadenohypophysis* or pars intermedia (Fig. 14.24). In the hag fish, the adenohypophyseal cells form islets near the infundibulum. Functionally the proadenohypophysis is believed to be analogous to the pars tuberalis of higher vertebrates. The mesoadenohypophysis reacts in a manner comparable to that of the pars distalis proper. The entire adenohypophysis underlies a thin-walled section of the infundibulum called the *neurohypophysis*. This region is thought to be functionally similar to the median eminence. There is no true pars nervosa. However, neurohypothalamic tissue interdigitates to a great extent with the metaadenohypophysis. The degree of penetration by nervous elements is less in the mesoadenohypophysis, and least in the proadenohypophysis. Posterior to the entire pituitary gland another nervous downgrowth can be found. This structure, the *saccus vasculosis,* also receives neurosecretory fibers and may possibly be homologous to the pars nervosa of higher vertebrates (see Chapter 12).

Vascularization of the Pituitary

One of the most interesting aspects of pituitary anatomy is the pattern of its vascularization. It has been stated that the pituitary gland functions to coordinate the external environment with the internal one via the central nervous system; yet, in higher vertebrates, a cellular connection between the neurohypophysis and the pars distalis is lacking. However, control of the secretory activity of the pars distalis is believed to reside in the hypothalamus, the mode of communication between the two being neurosecretory and vascular (Fig. 14.25).

In tetrapods, blood from the internal carotid arteries enters two posterior hypophyseal arteries. These structures lead first to the capsule surrounding the pituitary and then provide branches which terminate in the sinusoids of the pars distalis. Other branches lead directly to the pars nervosa. In addition, however, there is a second supply of arterial blood from the anterior hypophyseal arteries. Blood from these arteries travels through an extensive anastomosing capillary network near the surface of the median eminence. From this plexus the blood is collected into capillaries which either loop into and back to the surface of the median eminence or approach loops of

Fig. 14.19 An electron micrograph of the reticularis in the adrenal cortex of a male Rhesus monkey. X6930. (From R. M. Brenner. 1966. Am. J. Anat. 119:429–454, Fig. 20.)

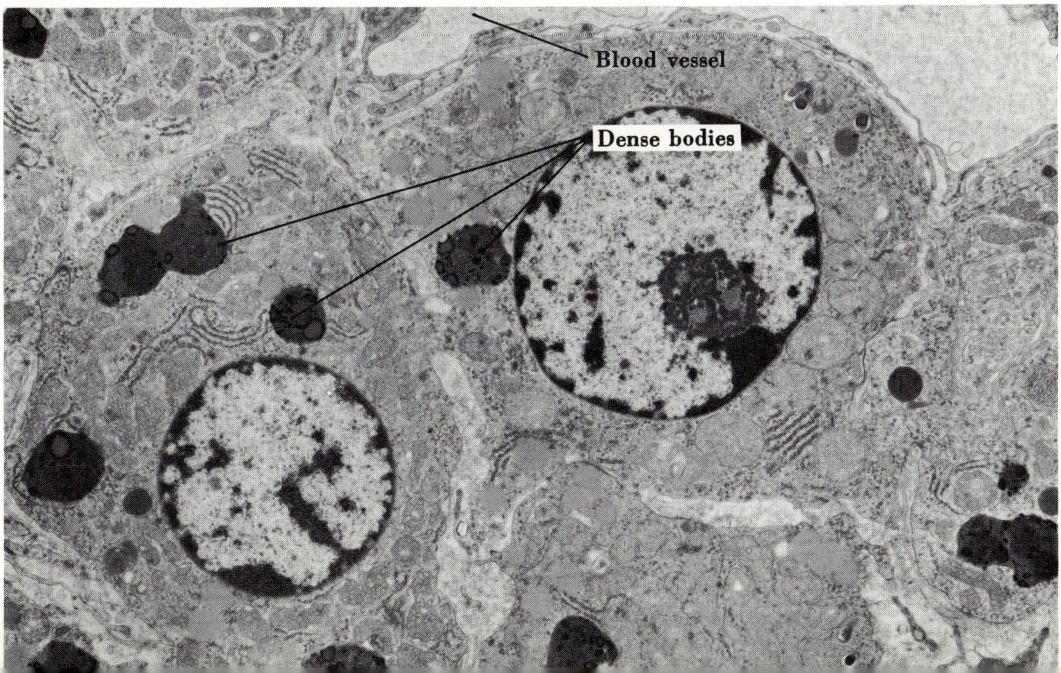

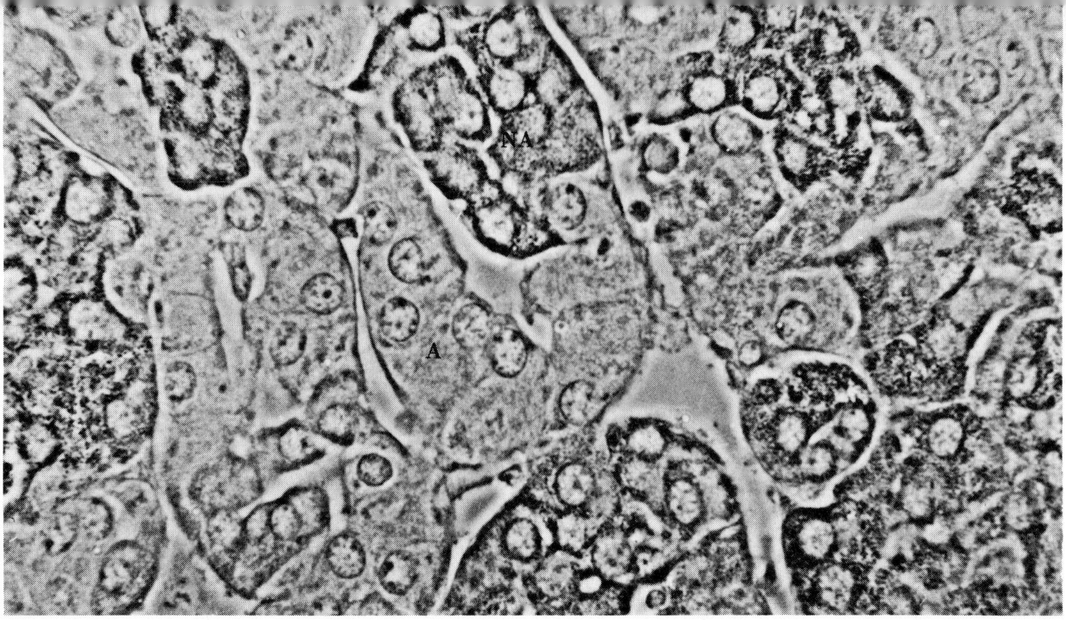

Fig. 14.20 In this phase-contrast photomicrograph the adrenalin-producing cells of the cat adrenomedullary tissue can be distinguished from those that produce noradrenalin. The tissue had been treated with potassium dochromate and sectioned on a freezing microtome. By this technique of tissue preparation, the adrenalin-producing cells, *A*, appear rather homogeneous as opposed to the granular noradrenalin producers, *NA*. X435. (From J. G. Wood, University of Texas Medical School. 1967. Am. J. Anat. 121:671–704, Fig. 1.)

neurosecretory cells in the median eminence. The blood then flows into venules which enter the sinusoids of the pars distalis. Hence, part of the blood supply to the pars distalis has already made its way through neurohypophyseal tissue and has been in contact with neurosecretory cells, the nuclei of which lie in the hypothalamus.

The collection of venules which passes from the first capillary network in the median eminence to the sinusoids of the pars distalis, and which represents the mode of neural control, is termed the *hypophyseal-portal* system.

The Cells of the Pars Distalis

The cordlike arrangement of cells within the pars distalis is generally typical of endocrine glands (Fig. 14.23). However, in some of the lower fishes a tubular or follicular aggregation can be seen, perhaps reflecting the tubular nature of Rathke's pouch. The pars distalis is noteworthy for the varied collection of cells which are found within it. Cell types may be mixed or there may be segregation of like cells to form cephalic and caudal regions, such as in reptiles and birds. The entire region is penetrated by an extensive network of blood sinusoids. These are braced by fibers of collagen which represent ramifications of the collagenous capsule surrounding the gland.

The pars distalis produces six hormones and it is only natural to expect to find six different cell types. This situation has been found to obtain in mammals and those few submammalian vertebrates which have been studied. However, correlative studies assigning a particular hormone to a particular cell have been largely confined to mammals. The reason for this is the difficulty in studying classes in which hormonal effects are not well defined. In order to correlate a cell and a hormone, the feedback arrangement which controls pituitary synthesis must be altered. This alteration must produce the predicted effect in the pituitary cell under study. An example would be the removal of the thyroid gland, resulting in greatly increased production of thyrotrophic hormone by the responsible cell. Conversely, pregnancy ought to be reflected in a decreased secretory level of the cells which manufacture gonadotrophic hormones.

There is probably no area in histologic nomenclature which is filled with greater confusion than the situation which obtains in the pars distalis. The reason for this is understandable but often overlooked. Quite simply, with optical microscopy the cells of the pars

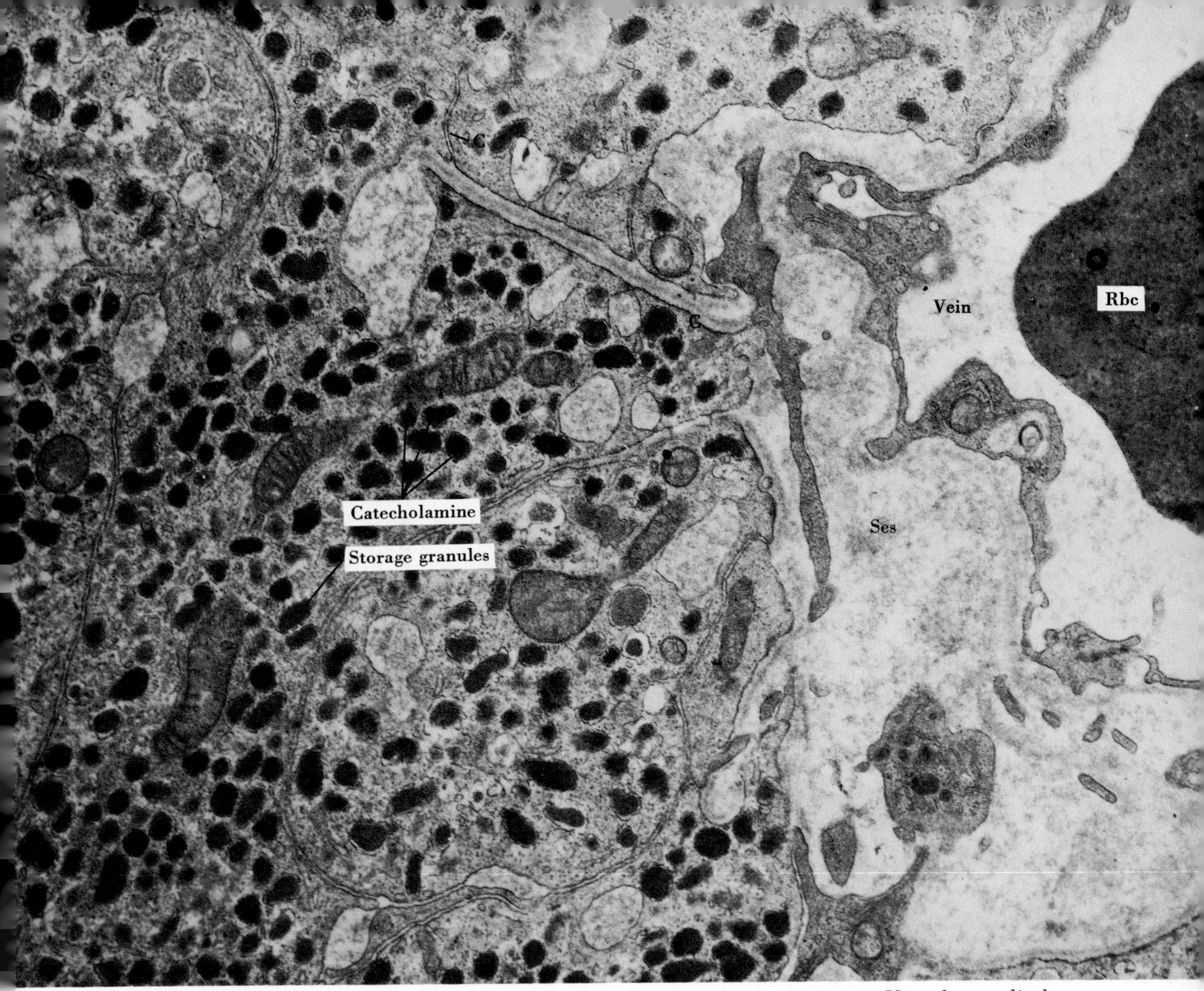

Fig. 14.21 An electron micrograph of cat adrenomedullary tissue adjacent to a vein. Note the canalicular system, *C*, connecting to the subendothelial space, *SES*. X19,500. (From J. G. Wood, University of Texas Medical School. 1967. Am. J. Anat. 121:671–704, Fig. 17.)

distalis are really all similar in appearance. They do differ in function, however, and this becomes evident when different staining techniques are used. Basically the cells can be classified on the staining properties of the granules they contain. *Acidophils* contain granules that stain well with acid dyes. *Basophils* are similarly named, but this is not to be confused with cells containing much basophilic material in the form of ribosomal RNA. Finally, the *chromophobes* are cells in which the granules stain with neither acidic nor basic dyes (Fig. 14.26). The confusion in nomenclature arises because of the multiplicity of acidic and basic stains which have been used by different investigators, hence the cells which in mammals secrete thyrotrophic hormone have been called *B* basophils, aldehyde-fuchsin positive cells, theta cells, cyanophils, blue beta cells, beta 2 cells, and S_2 cells by different investigators. Furthermore, it has not been unequivocally established that the amphibian or avian theta cell secretes the same hormone as the mammalian blue beta cell. However, the confusion is beginning to clear because, with electron microscopy, mammalian cells that secrete thyrotrophic hormone look like their amphibian counterpart and these, in turn, are distinguishable from the other cell types. A new system of nomenclature has been proposed which is based on function rather than on more subjective staining techniques. According to this system, those cells which secrete thyrotrophic hormone would be simply called thyrotrophs. After first distinguishing acidophils, baso-

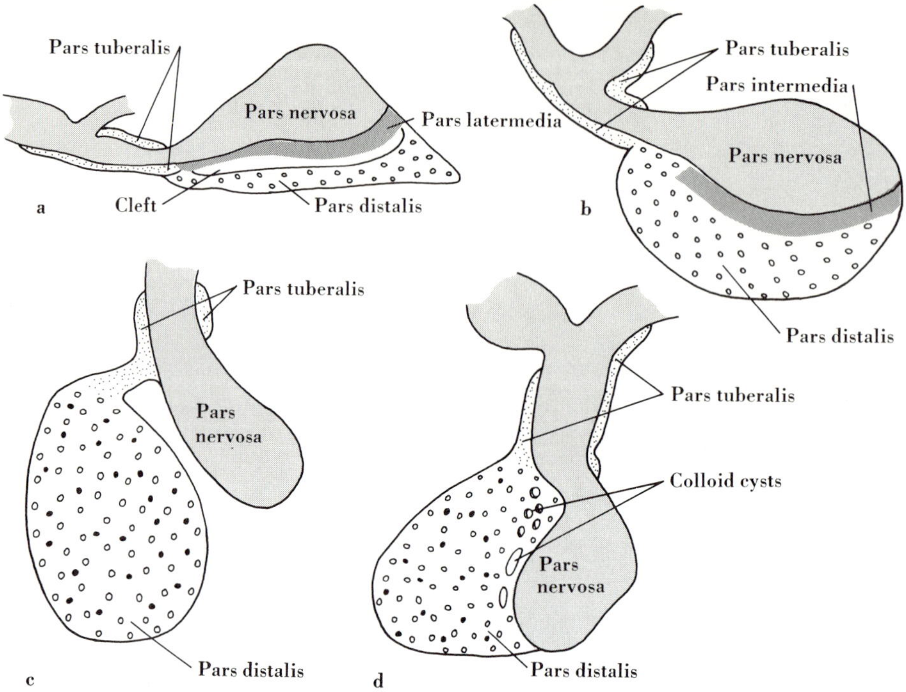

Fig. 14.22 Types of tetrapod pituitary glands. The rat pituitary, *a*, and the rabbit pituitary, *b*, are examples of a Type 1 gland. The whale pituitary is shown in *c* and the human pituitary in *d*. See text for explanation. (After H. D. Purves. 1966. Cytology of the adenohypophysis, p. 147–233. *In* G. W. Harris and B. T. Donovan, eds., The pituitary gland, vol. I, University of California Press, Berkeley.)

Fig. 14.23 Cell cords in the adenohypophysis of a chicken. This tissue had been treated as for electron microscopy but cut in thick sections for optical microscopy. X450. (Courtesy of F. O. Craft.)

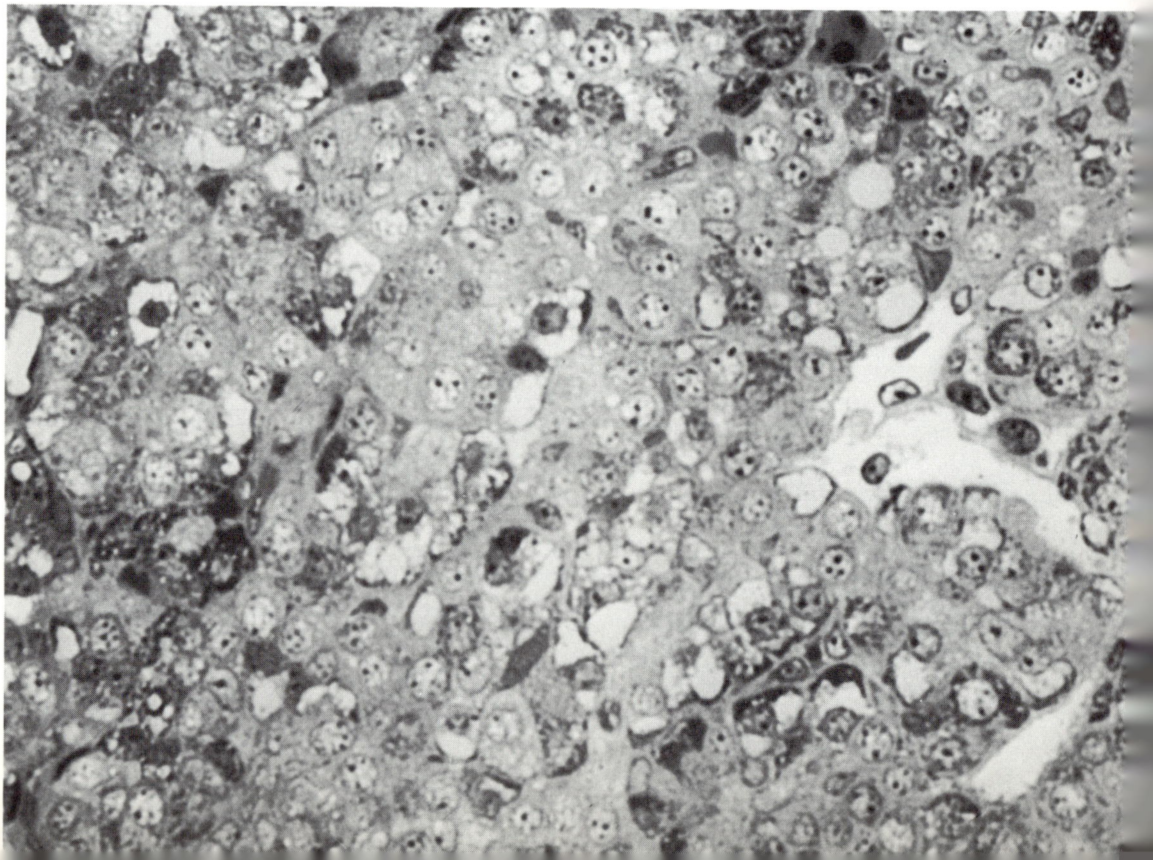

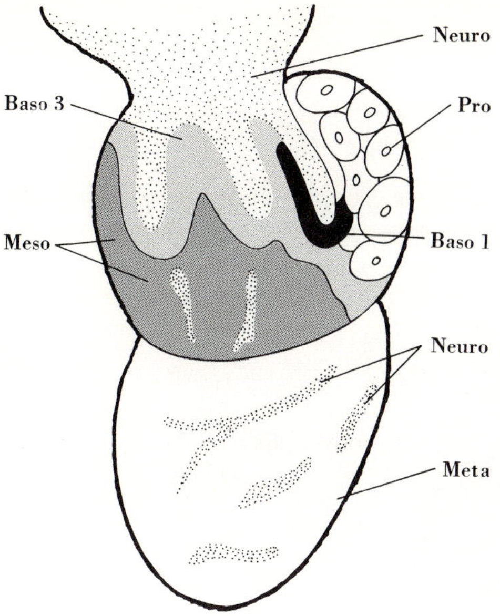

Fig. 14.24 A diagram of the hypophysis of the trout showing general areas and distribution of cell types. Cells of the proadenohypophysis are acidophils which are thought to secrete prolactin. Cells of the mesoadenohypophysis are acidophils and presumed somatotrophs. The *X* cells are believed to be corticotrophs. *Baso 1* cells are basophilic thyrotrophs whereas the cells in the region labelled *Baso 3* are basophilic, and presumed to be gonadotrophs. The metaadenohypophysis contains presumptive melanotrophs. (After H. D. Purves. 1966. Cytology of the adenohypophysis, p. 147–233. *In* G. W. Harris and B. T. Donovan, eds., The pituitary gland, vol. I, University of California Press, Berkeley. After an original drawing by B. I. Baker.)

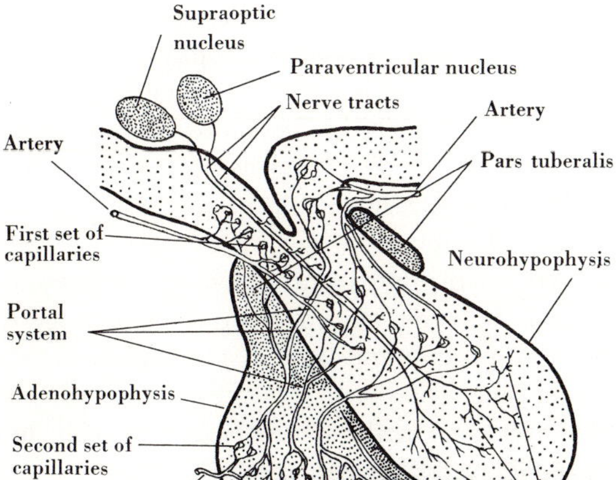

Fig. 14.25 A diagram illustrating the neural and vascular interrelationships within the pituitary. See text for explanation.

phils, and chromophobes, this latter system will be used. It should be remembered that most of the ultrastructural studies thus far have been conducted with mammals. The few investigations that have centered around submammalian species show good agreement with the mammalian material and, on the basis of this agreement, we can tentatively assume consistency. There are, however, species and class differences between similarly functioning polypeptide hormones. These variations may possibly reflect adaptive changes depending on environment.

ACIDOPHILS

Those cells which contain acidophilic granules are generally spherical to ovoid in shape. Containing the usual complement of organelles, these cells stain with eosin and appear red with the Mallory's trichrome technique. The different categories of acidophils can be further distinguished on the basis of granule size and density, cellular stability in relation to the reproductive cycle, and further differential staining. These cells have been found in representatives of all vertebrate classes (Figs. 14.27 and 14.28).

SOMATOTROPHS. The somatotrophs include those acidophils which are presumed to secrete growth or somatotrophic hormone (STH). In mammalian females these cells remain stable with respect to reproductive cycles. Although there are species differences, the very dense, membrane-bound granules of somatotrophs are numerous and in the order of 300–350 mμ in diameter. In mammals these cells are the most numerous of the chromophils and are usually concentrated in the posterior and lateral sections of the pars distalis. In fishes the somatotrophs are found in the mesoadenohypophysis, whereas in reptiles and birds they are most numerous in the caudal pars distalis.

Somatotrophic hormone, of minimum molecular weight 21,000, is a branched protein containing at least 200 amino acid residues. There is great variaton in growth hormones from various sources, although it is believed that a common amino acid core exists in all.

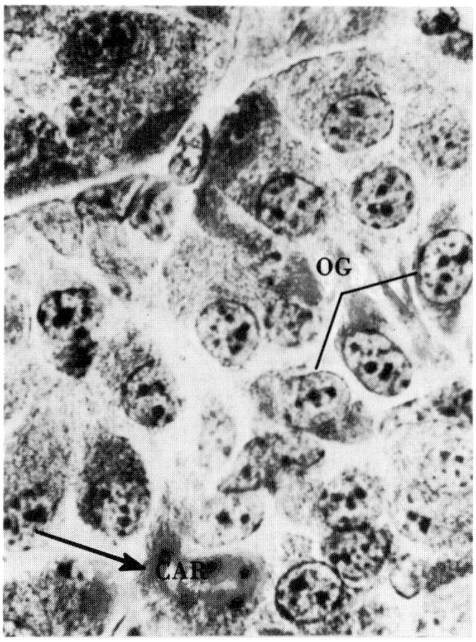

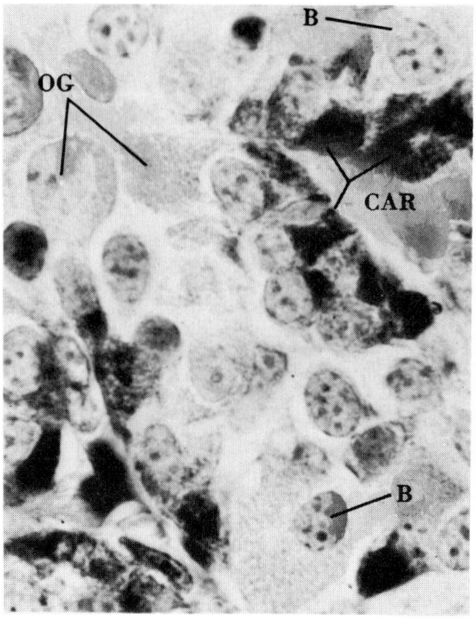

Fig. 14.26 These photomicrographs were taken of sections of frog adenohypophysis. After staining with Mallory-azan, three types of cells can be tinctorally identified: *OG*, orange G acidophils; *CAR*, azocarmine acidophils; and *B*, basophils. It should be remembered that without correlative physiological tests assignment of synthetic function to the various cell types is purely speculative. (From R. Ortman. 1956. Two types of acidophils in the anterior pituitary of the frog and their relation to the periodic acid Schiff reaction. J. Histochem. and Cytochem. 4:471–475. © The Williams and Wilkins Co., Baltimore.)

Interestingly enough, although growth hormone of higher vertebrates is usually active in lower ones, the reverse is not usually so.

STH particularly affects growth of the skeletal and muscular systems. Excessive secretion in children leads to gigantism; undersecretion results in dwarfism. Because STH affects endochrondral bone formation as well as the epiphyseal cartilage formation, oversecretion after closure of the epiphyseal disks results in acromegaly. The action of STH can be enhanced by thyroid hormones, and is antagonized by adrenocortical trophic hormone. STH also appears to have another target organ: the *A* cells in the endocrine pancreas can be stimulated to release glucagon after STH administration.

Fig. 14.27 An electron micrograph of the anterior pituitary of a male mouse. A uniformly granulated β cell and a smaller, peripherally granulated γ cell are present. The α cells and their granules are larger than those of the γ cells. α granules are more electron dense than β granules. X1850. (From B. Barnes. 1962. Endocrinology 71:618–628.)

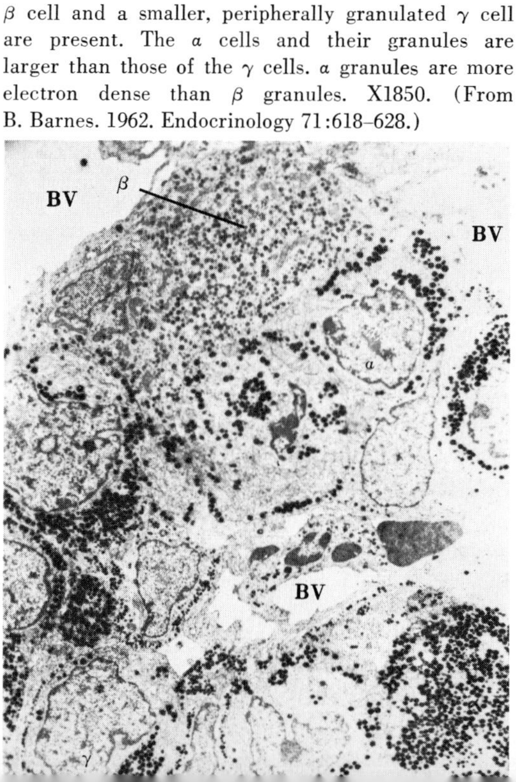

Any hormone which promotes cellular growth will have far-reaching metabolic effects. In order for growth to occur, there must be protein and lipid accumulation in the target cells. Therefore, it is not unexpected that the eventual mode of STH action will be found at such a molecular level.

LACTOTROPHS. Those acidophils which are thought to secrete lactogenic or lactotrophic hormone evidence cyclical changes in mammalian females that can be correlated with the reproductive cycle. The granules contained in

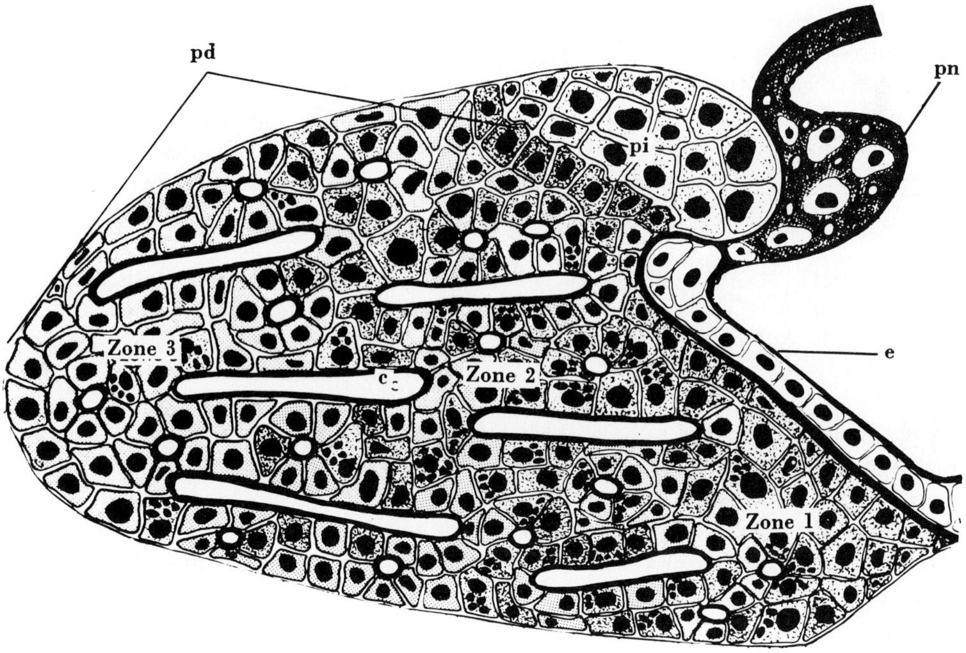

Fig. 14.28 Pituitary morphology in the salamander. The pars distalis, *pd*, is divided into three zones according to cell type. *Zone 1* contains aglobular basophils. *Zone 2* is made up of globular basophils, aglobular basophils, and acidophils. *Zone 3* is mainly composed of acidophils. Capillaries are labeled *c;* pars nervosa, *pn;* ependyma, *e;* pars intermedia, *pi.* (From R. R. Cardell, Jr. 1964. J. Ulstruct. Res. 10:317–334.)

lactotrophs are less numerous than those of somatotrophs, but they are larger (600–900 mμ in diameter) and denser. In birds and reptiles, lactotrophs are located in the cephalic region of the pars distalis and in fishes they are concentrated in the proadenohypophysis.

Lactotrophic hormone (LTH) is a protein with a molecular weight of at least 25,000 and is chemically similar to STH. Its secretion is controlled by the hypothalamus, the nature of this control being inhibitory. LTH exemplifies the notion that the hormones have not been drastically altered during vertebrate evolution, but rather that the hormones have been adapted to do different things. LTH is involved with reproduction. In mammals, LTH functions in milk secretion by mammary glands after birth of the young. In birds its effect is in the crop, resulting in the secretion of so-called crop-milk. In amphibia LTH seems to initiate the behavioral pattern of seeking water, in which mating and egg laying can take place. In all of the vertebrates, LTH seems to act in maintaining the corpus luteum. In mammals, at least, this would be of great value in preventing the arrival of new offspring before sufficient milk was available for their nourishment.

BASOPHILS

The cells containing basophilic granules are generally round to ovoid or angular and stain with hematoxylin and aniline blue. Usually 15–25μ in diameter, they contain granules which are smaller (150 mμ) than those in acidophils (Figs. 14.27 and 14.28). Individually the granules are below the resolution of optical microscopy. However, there is a tendency for flocculation to occur, so that granule aggregates can usually be seen. Basophilic granules are PAS-positive, indicating the presence of glycoprotein. This has been correlated with the glycoprotein nature of thyrotrophic hormone and the two gonadotrophic hormones, follicle stimulating hormone and leuteinizing hormone (in males often called interstitial cell stimulating hormone). All three types of basophils have been found in representatives of each of the vertebrate classes.

THYROTROPHS. The cells that are presumed to secrete thyrotrophic hormone are the irregularly or angularly shaped basophils. They stain well with aldehyde-fuchsin and contain 100-mμ granules, usually concentrated near the boundaries of the cell. In mammals and reptiles thyrotrophs are distributed in all regions of the pars distalis. They are most numerous near the boundary between the pro- and mesoadenohypophysis in fishes. In birds the thyrotrophs tend to be located in the cephalic region of the pars distalis.

Although thyrotrophic hormone (TSH) has not been purified, it is believed to be a glycoprotein with a minimum molecular weight of 10,000. The target organ of TSH is the thyroid gland, and all aspects of production of the thyroid hormones are accelerated by TSH. The control of TSH secretion is a dual one. A fall in the level of blood-borne thyroid hormone will stimulate production of TSH. Additionally, environmental effects, such as decreased temperature, acting on the hypothalamus can stimulate the secretion of TSH.

GONADOTROPHS. Gonadotrophs are generally round to ovoid, slightly larger than the thyrotrophs, and contain an abundance of endoplasmic reticulum. The 150-mμ granules have a tendency to flocculate as either a coarse or fine suspension. The cells containing small aggregations of granules are presumed to be leuteotrophs. Coarser flocculation has been correlated with secretion of follicle stimulating hormone. In the lamprey gonadotrophs are most numerous in proadenohypophysis, whereas in higher fishes they are concentrated in the middle region. In other classes they are found throughout the pars distalis.

Follicle stimulating hormone (FSH) is a glycoprotein of minimum weight 30,000. In vertebrate females its action causes growth of ovarian follicles, whereas in males its effect is on the seminiferous tubules. FSH secretion can be inhibited by the hormones of the ovaries and testes, but, in part, FSH control is hypothalamic and mediated via the pars nervosa. In birds, for example, an increased day length can reflexly stimulate FSH secretion.

Luteinizing hormone (LH) or interstitial cell stimulating hormone (ICSH), its twin in males, is also a glycoprotein of minimum molecular weight 26,000. LH acts on the interstitial cells of the ovaries and testes, causing luteinization of the graffian follicle after ovulation in females and production of androgenic hormones in males. LH itself may act to induce secondary sex characteristics; in birds, LH clearly affects feather pigmentation of the males. Secretion of LH is also under dual control. It is stimulated by estrogen but can also be reflexly influenced by the hypothalamus.

CHROMOPHOBES

Chromophobic cells vary in size, although they are usually smaller than the chromophils, with a granule population which is scanty and neutral or nonexistent. The different kinds of chromophobes are presumed to develop into the various acidophils and basophils, and recent evidence suggests that adrenocortical trophic hormone is secreted by mature chromophobes.

Adrenocorticotrophic hormone (ACTH) is a polypeptide containing 39 amino acids with a molecular weight of about 4500. Although species differences are evident, a common core of 24 amino acids does exist. The target of ACTH is the interrenal gland, which, after ACTH stimulation, undergoes hypertrophy and increased hormonal synthesis. In turn, the hormones of the interrenals regulate the level of ACTH. This negative feedback reaction is mediated by the hypothalamus via a corticotrophic releasing factor.

Within the chain of amino acids comprising ACTH can be found much of the melanotrophic hormone. Therefore, it is not surprising to find that ACTH does have some melanocyte stimulating ability, especially in lower vertebrates.

Cells of the Pars Intermedia

A discussion of the pars intermedia of vertebrate pituitaries can become most confused because not all vertebrate glands possess this area. However, the hormone associated with the intermediate lobe, *intermedin* or melanocyte stimulating hormone (MSH) is produced even in those animals lacking the appropriate region. The cells which produce MSH have been pinpointed in many vertebrates, and they appear to be similar regardless of their location.

Melanotrophs are generally polygonal cells, sometimes binucleate, which react similarly to the basophils of the pars distalis in most vertebrates. The granules are between 200 and 300 mμ in diameter and are PAS-positive. This cytologic test is indicative of glycoprotein, a fact which is interesting if one considers that MSH is a polypeptide of low molecular weight. In order to reconcile these facts it is assumed that MSH is bound to some other material in the granule, and finally released in pure form. Additionally, vacuoles containing lipoprotein material can be found.

In fishes the melanotrophs are located in the metaadenohypophysis. Also present in this region are a few chromophobic cells, and invading neurohypophyseal tissue. In birds that lack an intermediate lobe, MSH is secreted in the cephalic or rostral section of the pars distalis.

In those tetrapods which have a pars intermedia (Type I pituitary), the melanotrophs are organized as an avascular or poorly vascularized epithelium one to several layers in depth and resting on a basement membrane (Fig. 14.28). The cells are usually of uniform cytologic appearance, are occasionally ciliated, and are well innervated. Neurosecretory material often is present.

In man, the gorilla, and the chimpanzee, the cleft between the pars distalis and posterior lobe degenerates during early life to become replaced by colloid-filled cysts of ciliated epithelial cells. Some melanotrophs may be found as part of the cysts, but most of the MSH activity is localized in typical basophilic cells which have theoretically infiltrated the pars nervosa.

Two chemicals with MSH activity have been isolated: α MSH is a basic polypeptide containing 13 amino acid residues; β MSH is acidic and includes 18 amino acid residues. Both of these hormones are structurally identical to a portion of the ACTH molecule.

The primary effect of MSH in cold-blooded vertebrates is the dispersion of pigment granules within chromatophores. Such dispersion results in the positioning of pigment closer to the surface of the skin, so that the skin becomes darker. In the absence of MSH, pigment granules become concentrated near the center of pigment cells, resulting in lightening of the skin. In mammals injections of MSH will also produce skin darkening, but not by the same mechanism. In vertebrates generally, MSH appears to stimulate the synthesis of melanin if such stimulation is of long-term duration. Synthesis of MSH is controlled by nervous mechanisms. The reflex begins in the retina and arrives at the melanotroph via the hypothalamus. The action of MSH is considered unusual when compared with the other vertebrate hormones because its effects are directed toward entirely behavioral or social ends rather than the internal chemistry of the body.

Pars Tuberalis

The pars tuberalis is not considered to have hormonal function. It is well vascularized and consists primarily of large cells which are cuboidal in shape. These cells contain nonsecretory granules and may be arranged in follicular fashion. Groups of squamous cells, undifferentiated cells, and basophilic cells may also be found in the pars tuberalis, but they are believed to have infiltrated from other areas.

Neurohypophysis

The neurohypophysis can essentially be considered a system for the storage and release of hormones manufactured by neurons in the brain. Neurons which are classified as neurosecretory usually do not synapse with other neurons, but rather travel from their site of origin to their endpoint, usually abutting a capillary. Neurosecretory cells are spread throughout the diencephalic region in the lower fishes, but as higher forms evolved these cells have come to be located in a few brain centers. The bulk of the neurosecretory material is derived from neurons the cell bodies of which are located in the paraventricular and supraoptic nuclei of the hypothalamus (Fig. 14.25). The axons of these neurons converge near the infundibular area as the hypothalamohypophyseal tract. Additional fibers may be added from the tuberohypophyseal tract. The bulk of the median eminence of tetrapods and the neurohypophysis of the fishes consists of parallel bundles of these unmyelinated fibers. The inner layer of the median eminence is composed of ependymal cells and the outer regions contain delicate fibers which are branches of the neurosecretory axons and which terminate on the capillaries of the portal system.

The pars nervosa initially develops as a saclike tubular structure. In tetrapods one finds a tremendous increase in complex infoldings of the sac, so that in some forms the neural lobe appears as a solid knob of tissue. Because part of the vascular drainage of the neurohypophysis is completely separate from the portal system and adenohypophysis, and because the pars nervosa develops most fully in those classes which have pronounced systemic effects from the neurosecretory material, the evolution of the pars nervosa has been correlated with the storage and discharge of neurosecretory hormones directly into the systemic circulation. In fishes the neural region of the pituitary interdigitates to a great degree with the meta- and mesoadenohypophysis (Fig. 14.24), and in higher vertebrates nerve endings and neurosecretory material can be found in the median eminence and the intermediate lobe. From these connections, neural control of the adenohypophysis is thought to derive.

The pars nervosa consists of axons of the hypothalamohypophyseal tract, pituicytes, Herring bodies, capillaries, and some few cells which may have invaded from the pars distalis (Fig. 14.29). As mentioned above, the axons are nonmyelinated and contain characteristic swellings along their length. Within the axons much Nissl substance can be found that represents aggregations of rough endoplasmic reticulum (Fig. 14.30). The secretory material is synthesized in the cell body, stored in the vesicles of the Golgi substance, and finally moved the length of the axon to the axon terminal, which terminates as a bulb along pericapillary spaces. However, stainable neurosecretory material, visible within the axon, is not found in the pericapillary space. This is postulated to be caused by the intra-axonal binding of the hormone to a larger

Fig. 14.29 Human pars nervosa. X430.

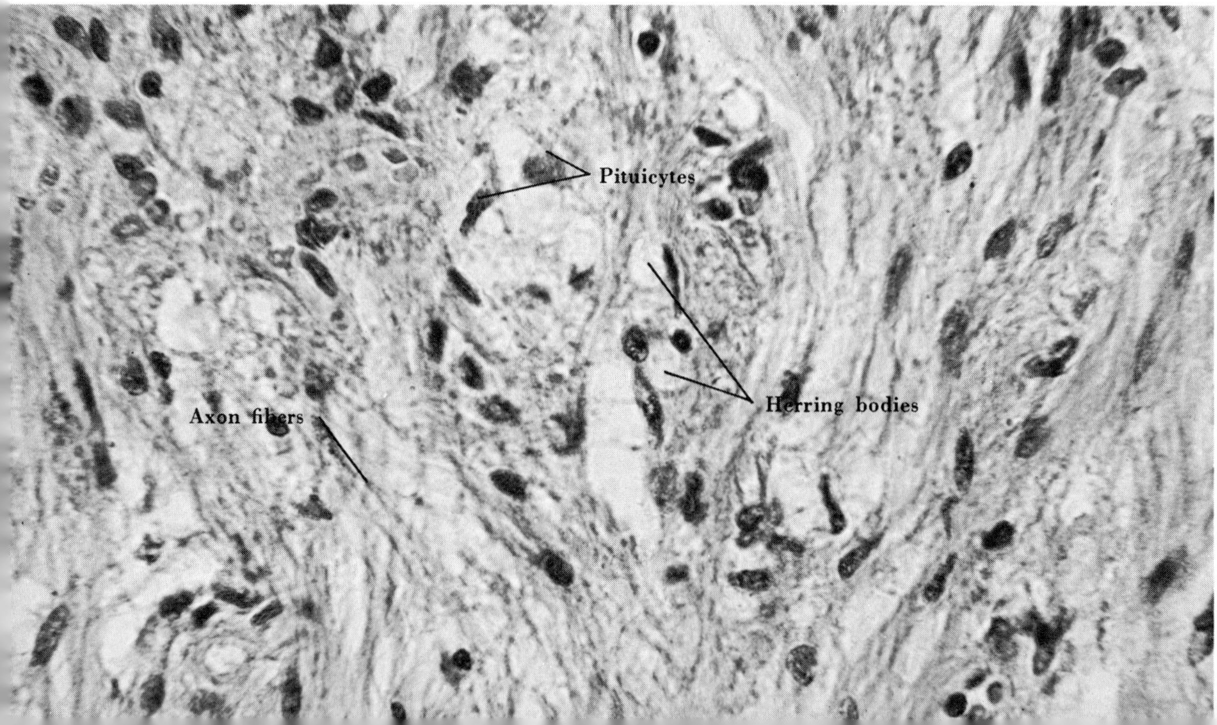

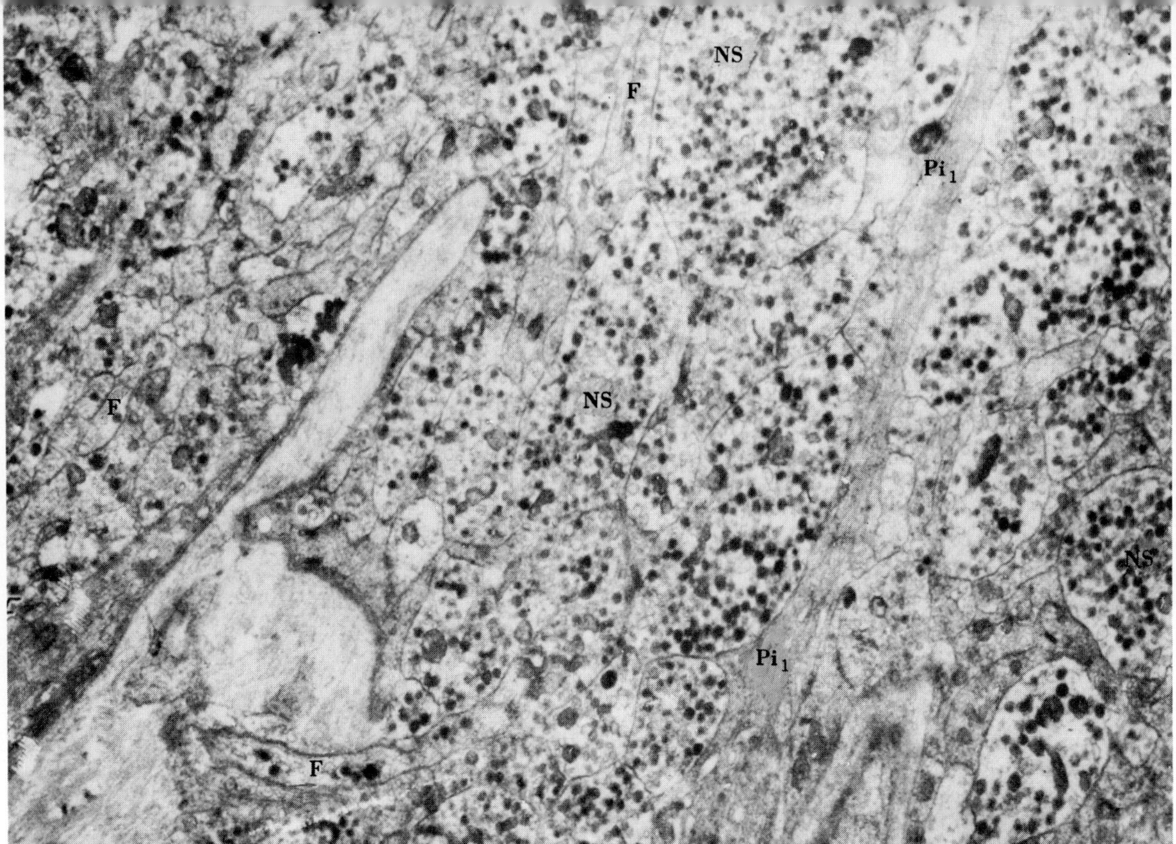

Fig. 14.30 A low-power electron micrograph of the human neural lobe. *NS*, cut nerve fibers; *Pi*, pituicyte; Pi_1, pituicyte about 20 μ in length; *C*, capillary with an erythrocyte, *E*, within it. X7800. (From K. Lederis. 1965. An electron microscopical study of the human neurohypophysis. Zeitschrift für Zellforschung und mikroskopische Anatomie, Bd. 65, S. 847–868. Berlin-Heidelberg-New York: Springer.)

protein, and the severing of this relationship before secretion. The capillary endothelium is unusual in the degree of porosity that can be seen with the electron microscope.

It is generally agreed that the pituicytes are nonsecretory glial cells. Although pituicytes exist in many shapes, two categories can be distinguished. *Fiber pituicytes* contain scanty cytoplasm surrounding an oval nucleus. Their processes are long and delicate. *Protoplasmic pituicytes* contain more cytoplasm in the cell body and this is often granulated.

Herring bodies are spherical masses which are thought to be accumulations of material within the nerve fibers. Present in them are mitochondria, portions or remnants of endoplasmic reticulum, and a variety of electron-dense, membrane-bound particles. Additionally, and most importantly, Herring bodies contain stored quantities of neurosecretory material.

The posterior lobe hormones are, of course, hypothalamic in their site of origin. With the exception of the cyclostomes, all vertebrates have two closely related substances released by the pars nervosa. Because there can be selective release of one hormone at a time, it is believed that there are two different classes of neurons, each of which is responsible for the synthesis of one of the hormones.

Although any one vertebrate secretes two neurosecretory hormones, four different chemicals are represented within the vertebrate group. All of them are cyclical octapeptides, i.e., they contain eight amino acid residues. Those containing phenylalanine in the *A* position are called vasopressins whereas those containing isoleucine as the same position are

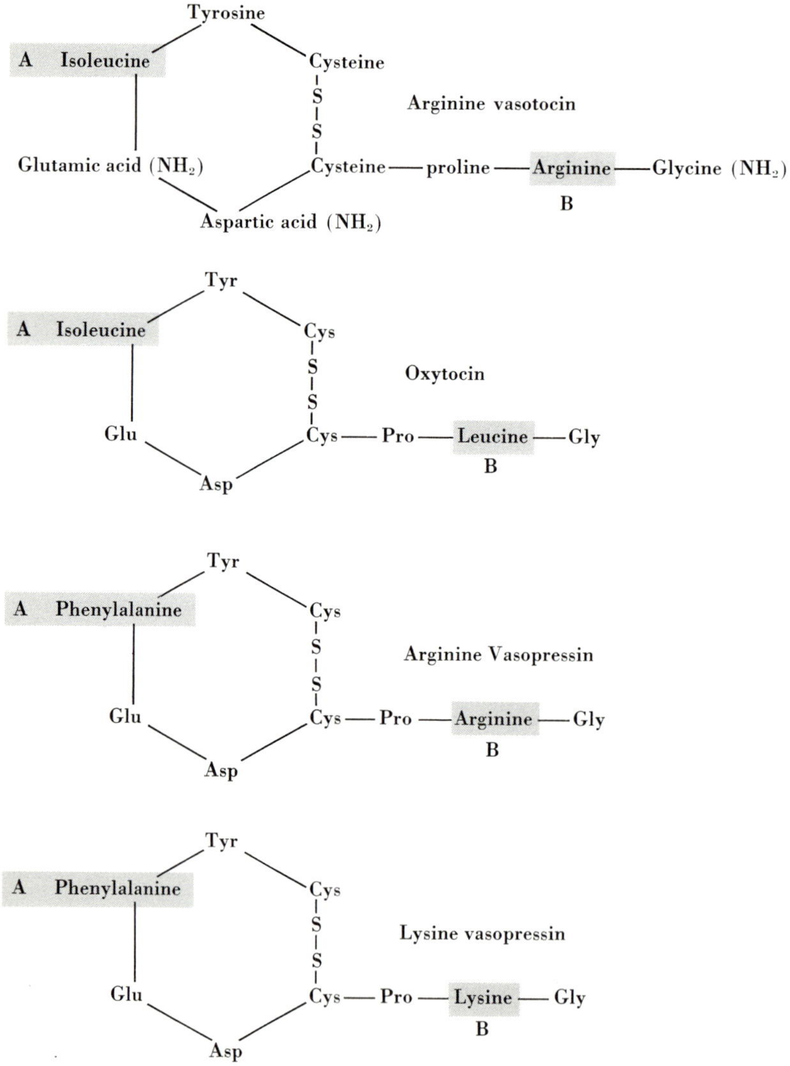

Fig. 14.31 The vertebrate neurosecretory hormones. Except for the amino acids in the *A* and *B* positions (*shaded*), they are identical. See text for an evolutionary explanation.

called oxytocins. One other point, the so-called *B* position, has also undergone evolutionary change. Lysine vasopressin has lysine at the *B* position; arginine vasopressin has its *B* position occupied by arginine. Oxytocin is marked by leucine at *B*, and arginine vasotocin has isoleucine at *A* and arginine at *B* (Fig. 14.31).

The functions of these hormones can be categorized into four separate actions. The evolutionarily oldest action is the local effect of stimulation of the pars distalis. Next in time, as land vertebrates developed, is the neurosecretory control of water retention by the kidney. This is the antidiuretic function or hormone that, in all land animals, inhibits fluid loss in the urine. In frogs ADH even promotes storage of water in the bladder and the transfer of water into the animal through the skin. Finally, these hormones have become involved with contraction of uterine muscle (the oxytocic principle) and the ejection of milk from the previously stimulated mammary glands.

Lampreys secrete arginine vasotocin. Presumably its only effect is anterior lobe stimulation. In the bony fishes, we see the beginnings of a pars nervosa and also the appearance of an antidiuretic principle. In these fishes, as well as in amphibians, reptiles, and birds, arginine vasotocin is retained. The second "new" hormone is oxytocin. With the evolution of the viviparity and lactation in mammals, we see the retention of oxytocin. The phylogenetically old arginine vasotocin disappears and is replaced by arginine vasopressin in most mammals and lysine vasopressin in pigs. Possibly more than with any other system, biochemical evolution can here be correlated with evolution of function in target organs.

SELECTED REFERENCES

Barnes, B. G. 1962. Electron microscope studies on the secretory cytology of the mouse anterior pituitary. Endocrinology 71:618–628.

Barrington, E. J. W. 1963. General and comparative endocrinology. Clarendon Press, Oxford.

Belt, W. D., M. N. Sheridan, R. A. Knouff, and F. A. Hartman. 1965. Fine structural study of a possible mechanism of secretion by the interrenal cells of the brown pelican. Z. Zellforsch. 68:864–873.

Brenner, R. M. 1966. Fine structure of adrenocortical cells in adult male Rhesus monkeys. Am. J. Anat. 119:429–454.

Cardell, J. R., Jr. 1964. Observations on the cell types of the salamander pituitary gland: An electron microscope study. J. Ultrastruct. Res. 10:317–333.

Carvalheira, A. F., and A. G. E. Pearse. 1967. The cytology and cytochemistry of the C cells in the thyroid gland of the pig. J. Roy. Mic. Soc. 86:203–210.

Deter, R., and C. DeDuve. 1967. Influence of glucagon, an inducer of cellular autophagy, on some physical properties of rat liver lysosomes. J. Cell Biol. 33:437–449.

Epple, A. 1966. Islet cytology in urodele amphibians. Gen. & Comp. Endocrin. 7:207–215.

Farquhar, M. G., and J. F. Rinehart. 1954. Cytologic alterations in the anterior pituitary gland following thyroidectomy: An electron microscope study. Endocrinology 55:857.

Fugita, H., and M. Machino. 1965. Electron microscope studies on the thyroid gland of a teleost, *Seriola quinqueradiata*. Anat. Rec. 152:81–98.

Gorbman, A., and H. A. Bern. 1962. A textbook of comparative endocrinology. John Wiley & Sons, New York.

Herman, L. 1960. An electron microscope study of the salamander thyroid during hormonal stimulation. J. Biochem. & Biophys. Cytol. 7:143.

Kerr, T., P. G. W. J. van Oordt. 1966. The pituitary gland of the African lungfish Protopterus sp. Gen. & Comp. Endocrin. 7:549–558.

Lederis, K. 1965. An electron microscopal study of the human neurohypophysis. Z. Zellforsch. 65:847–868.

Munger, B., and S. I. Roth. 1963. The cytology of the normal parathyroid glands of man and Virginia deer. J. Cell Biol. 16:370–400.

Munson, P. L., and P. F. Hirsch. 1966. Thyrocalcitonin: newly recognized thyroid hormone concerned with metabolism of bone. Clin. Orthopedics 49:209–232.

Nandi, J., and H. A. Bern. 1960. Corticosteroid production by interrenal tissue of teleost fishes. Endocrinology 66:295–303.

Ogowa, M. 1967. Fine structure of the corpuscles of Stannius and the interrenal tissue in goldfish, *Carassius auratus*. Z. Zellforsch. 81:174–189.

Ortman, R. 1956. Two types of acidophils in the anterior pituitary of the frog and their relation to the periodic acid Schiff reaction. J. Histochem. Cytochem. 4:471–475.

Pearse, A. G. E. 1966. The cytochemistry of the thyroid C cells and their relationship to calcitonin. Proc. Roy. Soc. Lond. B. 164–478.

Purves, H. D. 1966. Cytology of the adenohypophysis. *In* The pituitary gland. G. W. Harris and B. T. Donovan, eds. University of California Press, Berkeley, Vol. 1.

Sato, T., L. Herman, and P. Fitzgerald. 1966. The comparative ultrastructure of the pancreatic islet of Langerhans. Gen. & Comp. Endocrin. 7:132–158.

Sehe, C. T. 1965. Comparative studies on the ultimobranchial body in reptiles and birds. Gen. & Comp. Endocrin. 5:45–60.

Trier, J. S. 1958. The fine structure of the parathyroid gland. J. Biochem. Biophys. Cytol. 4:13–22.

von Euler, E. S. and H. Heller, eds. Glandular hormones. 1963. *In* Comparative endocrinology. Academic Press, New York. Vol. 1.

Wood, J. G. 1967. The relationship of nucleotidase activity to catecholamine storage sites in adrenomedullary tissue. Am. J. Anat. 121:671–704.

Young, B. W., and C. P. Leblond. 1963. The light cell as compared to the follicular cell in the thyroid gland of the rat. Endocrinology 73:669–686.

INDEX

NOTE: Numbers in italics refer to the illustration on that page.

69 70 71 7 6 5 4 3 2 1